Commercial Flowers

— Volume 3 —

THE EDITORS

Professor T.K. Bose obtained his M.Sc. (Ag.) and Ph.D. degree from Calcutta University, West Bengal, India. Trained at the Royal Botanic Garden, Kew, UK. Former Secretary of the Agri-Horticultural Society of India; Project Coordinator, All India Coordinated Floriculture Improvement Project (ICAR); Professor and Head, Department of Horticulture; Dean, Faculty of Agriculture; Director of Research, Bidhan Chandra Krishi Viswavidyalaya, West Bengal; Member, State Agriculture Commission, West Bengal; Member of Scientific Advisory Committees of ICAR, CSIR (Govt. of India) and Government of West Bengal. Awarded (D.Sc.) Honoris Causa from OUAT, Bhubaneshwar and UBKV, Coochbehar, India. Guided 45 Ph.D. students, published more than 250 research papers and authored 30 books on horticultural crops.

Dr. Chenna Reddy Aswath obtained his B.Sc. (Hort), M.Sc. (Hort.) in Floriculture from University of Agricultural Sciences, Bengaluru and Ph.D. degree in Biotechnology from Guwhati University, Assam. Serving as Principal Scientist in ICAR-IIHR, Bengaluru and has made significant contributions in the field of Floriculture and Biotechnology. Recipient of "Dr BP Pal Best Scientist Award for Agriculture" from NABS, Chennai, became FELLOW of the NABS, Chennai, ISOH, New Delhi and SAB, Dharwad, CHAI, New Delhi. Received the 2012 SAB Award of Excellence in Horticultural Biotechnology. Recipient of Brain pool Post-Doctoral Fellowship from South Korea Science Federation for the outstanding research contributions in Horticultural Biotechnology. His short term fellowship in Japan under INSA has recognized him among the Orchid tissue culture group. Published more than 110 research papers besides several books. Guided more than 10 post-graduate students.

Dr. Reeta Bhatia is working as senior scientist at the Division of Floriculture and Landscaping, ICAR- Indian Agricultural Research Institute, New Delhi. She has more than 13 years of research and teaching experience in the field of floriculture. She has obtained her Master's degree from Punjab Agricultural University, Ludhiana and doctoral degree from Indian Agricultural Research Institute, New Delhi. She has been awarded with the prestigious Endeavor Research Fellowship by the Government of Australia and pursued her postdoctoral research at the University of Western Australia. She has an excellent academic and research career and awarded with several medals. She has published more than 65 research papers in the reputed high impact scientific journals and written more than 30 review/popular/scientific articles. She is also serving as editor for the Journal of Ornamental Horticulture.

Dr. Rajiv Kumar obtained his B.Sc. (Ag and AH) and M.Sc. (Ag) Horticulture from C.S. Azad University of Agriculture and Technology, Kanpur, Uttar Pradesh, India and completed his Ph.D. (Horticulture) from I.A.R.I., New Delhi. Presently working as Principal Scientist at ICAR-IIHR, Bengaluru. Recipient of Young Scientist Associate Award by HTHS, U.P.; Young Scientist Award by SSDAT, U.P.; Gold Medal by HTHS, Meerut, and BR and IC Pvt. Ltd.; Fellow (Horticulture) by HAHS, Muzaffarnagar, U.P., AABS, Karnakata, and ISOH, New Delhi. Published 110 research articles, 25 popular articles, 6 edited books, 20 book chapters, 8 technical bulletins and 6 extension folders. Developed two varieties of China aster, associated in five and three varieties of Gerbera and gladiolus, respectively; registered 17 genetic stocks of China aster, Gerbera, Chrysanthemum and Gladiolus with ICAR-NBPGR, New Delhi. Guided two M.Sc. (Hort.) and two Ph.D. (Hort.) students in Floriculture.

Dr. Koushik Dutta obtained his M.Sc. (Hort.) from CCS Haryana Agricultural University, Hisar, Haryana and Ph.D. from Uttar Banga Krishi Viswavidyalaya, West Bengal. Acting as Assistant Secretary of The Agri-Horticultural Society of India, Kolkata, West Bengal. Recipient of Inspire Fellowship from Department of Science and Technology, New Delhi. Published several scientific papers in national journals on different aspect of horticulture particularly in tissue culture and written book chapters on nursery management and jasmine. Associated with micropropagation of different of horticultural crops during last 7 years.

Dr. Tarak Nath Saha graduated in Agriculture from Banaras Hindu University, Varanasi (Uttar Pradesh), obtained his Masters and Doctoral degree in Horticulture with specialization in Floriculture from India Agricultural Research Institute (IARI), New Delhi. Presently working as Senior Scientist at ICAR-Directorate of Floricultural Research, Pune. He has an experience of more than 12 year in research, teaching and extension, besides coordinating work of All India Coordinated Research Project (AICRP) on Floriculture. He has contributed 25 research papers, 41 technical articles, 5 book chapters, 20 technical bulletins, 6 folders, besides in the editorial team of 30 Annual Reports and 10 AICRP workshop proceedings. Developed one variety and registered one genetic stocks of tuberose with ICAR-NBPGR, New Delhi. He is actively involved in PG teaching; recognized as faculty of MPKV, Rahuri (also at IARI, New Delhi 2009-11) and guided more than 5 students in Floriculture as advisory.

Commercial Flowers

— *Volume 3* —

— *Senior Editor* —

C. Aswath
Division of Flower and Medicinal Crops
ICAR-Indian Institute of Horticultural Research
Bengaluru 560089, Karnataka, India

— *Editors* —

T.K. Bose
Faculty of Horticulture
Bidhan Chandra Krishi Viswavidyalaya
West Bengal 741 252, India

Reeta Bhatia
Division of Floriculture and Landscaping, ICAR-Indian Agricultural Research Institute, New Delhi 110012

Rajiv Kumar
Division of Flower and Medicinal Crops, ICAR-Indian Institute of Horticultural Research, Bengaluru 560089, Karnataka, India

Koushik Dutta
The Agri-Horticultural Society of India
1, Alipore Road, Kolkata 700027
West Bengal, India

Tarak Nath Saha
ICAR-Directorate of Floricultural Research, College of Agriculture Campus, Shivajinagar
Pune 411005, Maharashtra, India

2022

Daya Publishing House®
A Division of
Astral International Pvt. Ltd.
New Delhi – 110002

Third Revised & Illustrated edition 2021

Publisher's Note:

Every possible effort has been made to ensure that the information contained in this book is accurate at the time of going to press, and the publisher and author cannot accept responsibility for any errors or omissions, however caused. No responsibility for loss or damage occasioned to any person acting, or refraining from action, as a result of the material in this publication can be accepted by the editor, the publisher or the author. The Publisher is not associated with any product or vendor mentioned in the book. The contents of this work are intended to further general scientific research, understanding and discussion only. Readers should consult with a specialist where appropriate.

Every effort has been made to trace the owners of copyright material used in this book, if any. The author and the publisher will be grateful for any omission brought to their notice for acknowledgement in the future editions of the book.

Cataloging in Publication Data--DK

Courtesy: D.K. Agencies (P) Ltd. <docinfo@dkagencies.com>

Commercial flowers / senior editor, C. Aswath ; editors, T.K. Bose [and four others]. -- Third revised edition.

Volumes cm

Contributed articles.

Includes index.

ISBN 978-93-90435-37-1 (HB)

1. Floriculture. 2. Flowers--Pictorial works. I. Aswath, C. (Principal scientist at the Indian Institute of Horticulture Research), editor. II. Bose, T. K., editor.

LCC SB405.C66 2021 | DDC 635.9 23

Published by : **Daya Publishing House®**
A Division of
Astral International Pvt. Ltd.
– ISO 9001:2015 Certified Company –
4736/23, Ansari Road, Darya Ganj
New Delhi-110 002
Ph. 011-43549197, 23278134
E-mail: info@astralint.com
Website: www.astralint.com

Digitally Printed at : Replika Press Pvt Ltd

From the Desk of Prof. T.K. Bose

The undersigned had initiated the publications entitled 'Fruits: Tropical and Sub-tropical', 'Vegetable Crops' and 'Commercial Flowers', one volume each in the 1980's in collaboration with co-editors, humbly expresses from the core of his heart that the updated revised editions of the above mentioned publications with colour illustrations in thirteen volumes which was planned in June, 2020, are being printed from January this year, a rare event in Horticultural Science has been possible by the blessing of Almighty God, with the cooperation of dedicated horticulturists as editors and contributors.

Kolkata

T.K. Bose
Editor Emeritus

Preface to the Third Edition

The first edition of the book entitled **'Commercial Flowers'** published in 1989, reprinted in 1993 and 1998 was greatly appreciated and proved valuable scientific review of commercially important flowers, to the students, teachers, researchers and extension specialists in horticulture in general and floriculture in particular, in different parts of the world. The book was revised in 2002 with the same title in two volumes due to its increasing demand, which contained updated scientific information of commercially grown flower crops and considered as valuable contribution on the subject. After about 18 years from the last revised publication in 2002, voluminous information pertaining to crop improvement, production, protection, postharvest management, biotechnology has been generated necessitating further revision of the existing chapters and inclusion of more ornamentals plants of commercial importance in four volumes, with relevant colour illustrations.

Third volume in this edition includes chapters on breeding and improvement of ornamental plants and updated research review on all major aspects on Orchids, Narcissus, Freesia, Tuberous begonia, Cyclamen, Zantedeschia, Water lily, Crocus, Heliconia, Agapanthus, Gloriosa, Crinum, Hymenocallis, Importance and uses of annual ornamental plants, Gomphrena, Cosmos, Lisianthus, Campanula, Astilbe and *Bellis perenis* with relevant colour illustrations, covering more than 700 pages.

Apart from own collection of photographs, editors have also taken some photographs which are in public domain from different websites, as far as possible. Therefore, we acknowledge with enormous thanks to all State Agricultural Universities and Research Institutes under Indian Council of Agricultural Research, New Delhi, India; Public and Private International Agricultural Institutes across the world for using photographs from their websites to make this book a valuable source of knowledge for the readers. We think images can be powerful teaching tools, as illustrations to in-class lectures, or for studying concepts outside of the classroom. We also think that identification of flower crops, their importance and uses, species and cultivars, methods of propagation, cultivation practices including greenhouse cultivation, diseases and pests, harvesting and post harvest technology and achievement of biotechnology, with the help of suitable illustrations in a consolidated manner will make a dent for the students, researchers and teachers of floriculture and ornamental horticulture.

We are grateful to Dr. Arup Chattopadhayay, Professor, Bidhan Chandra Krishi Vishwavidyalaya, West Bengal for his enormous support in preparing the revised illustrated manuscripts. We sincerely thank the contributors for preparing updated revised manuscripts. Our thanks are due to Shri Prateek Mittal, Director, Astral International (P) Ltd., New Delhi, for his keen interest in bringing out this publication.

Kolkata **Editors**

Contents

List of Contributors

Agnihotri, R.

Dr. Rajendra Prasad Central Agricultural University, Pusa, Samastipur 848125, Bihar, India

Aswath, C.

Division of Flower and Medicinal Crops, ICAR-Indian Institute of Horticultural Research, Bengaluru 560089, Karnataka

Bal, S.

Department of Vegetable Science, Faculty of Horticulture, Bidhan Chandra Krishi Viswavidyalaya, Mohanpur 741252, Nadia, West Bengal, India

Bhagavati, P.P.

ANGRAU, Agricultural College, Naira, Srikakulam Dist. 532185, Andhra Pradesh

Bhatia, R.

Division of Floriculture and Landscaping, ICAR-Indian Agricultural Research Institute, New Delhi 110012

Biswas, B.

Department of Horticulture, Bidhan Chandra Krishi Vishwavidyalaya (State Agricultural University), Mohanpur 741 252, West Bengal

Bose, T.K.

Department of Horticulture, Bidhan Chandra Krishi Vishwavidyalaya, Mohanpur 741252, West Bengal

Chandrashekhar, S.Y.

University of Agricultural and Horticultural Sciences, Navile, Shivamogga 577204, Karnataka

Chawla, S.L.

Navsari Agricultural University, Navsari 396450, Gujarat

Chaithra, M.L.

Division of Flower and Medicinal Crops, ICAR-Indian Institute of Horticultural Research, Bengaluru 560089, Karnataka

Choudhury, T.

The Agri-Horticultural Society of India, 1, Alipore Road, Kolkata 700027, West Bengal

Dalawal, B.

University of Agricultural and Horticultural Sciences, Navile, Shivamogga 577204, Karnataka

Dhiman, M.R.

ICAR-Indian Agricultural Research Institute, Regional Station Katrain, Kullu, Himachal Pradesh

Dubey, R.K.

Department of Floriculture and Landscaping, Punjab Agricultural University, Ludhiana 141004, Punjab

Durga, M.L.

ANGRAU, Agricultural College, Naira 532185, Srikakulam Dist, Andhra Pradesh

Dutta, K.

The Agri-Horticultural Society of India, 1, Alipore Road, Kolkata 700027, West Bengal

Ganga, M.

Department of Floriculture and Landscape Architecture, Horticultural College and Research Institute, Tamil Nadu Agricultural University, Coimbatore 641003, Tamil Nadu

Gurung, A.

Division of Floriculture and Medicinal Crops, ICAR-IIHR Bengaluru 560089, Karnataka

Kadam, G.B.

ICAR-Directorate of Floricultural Research, Pune 411005, Maharashtra

Kumar, K.R.

Dr. YSRHU, Horticultural Research Station, Kovvur, W.G. Dist. 534350, Andhra Pradesh

Kumar, P.N.

ICAR-Directorate of Floricultural Research, Pune 411005, Maharashtra

Kumar, R.

Division of Flower and Medicinal Crops, ICAR-Indian Institute of Horticultural Research, Bengaluru 560089, Karnataka

Laishram, N.

Division of Vegetable Science and Floriculture, Sher-e-Kashmir University of Agricultural Science and Technology of Jammu 180009, Jammu and Kashmir

Mitra, M.

Department of Horticulture, Bidhan Chandra Krishi Vishwavidyalaya, Mohanpur 741252, West Bengal

Mukherjee, D.

Department of Vegetable Science, Faculty of Horticulture, Bidhan Chandra Krishi Viswavidyalaya, Mohanpur 741252, Nadia, West Bengal

Madhubala

Deparment of Floriculture and Landscaping, College of Horticulture, Punjab Agricultural University, Ludhiana 141004, Punjab

Naik, B.H.

University of Agricultural and Horticultural Sciences, Navile, Shivamogga 577204, Karnataka

Naveena, N.

Department of Floriculture and Landscape Architecture, Horticultural College and Research Institute, Tamil Nadu Agricultural University, Coimbatore 641003, Tamil Nadu

Namita

Division of Floriculture and Landscaping, ICAR-Indian Agricultural Research Institute, New Delhi 110012

Parthasarathy, V.A.

ICAR-Indian Institute of Spices Research, 32/482 C Narmada Nilayam, Bharathan Bazar, Chelavoor, Calicut 673571, Kerala

Patil, S.

Navsari Agricultural University, Navsari 396450, Gujarat

Pal, P.

Department of Horticulture, Bidhan Chandra Krishi Vishwavidyalaya (State Agricultural University) Mohanpur 741 252, West Bengal

Pawar, R.D.

RCSM College of Agriculture, MPKV, Kolhapur 416013, Maharashtra

Panwar, S.

Division of Floriculture and Landscaping, ICAR-Indian Agricultural Research Institute, New Delhi 110012

Prasad, K.V.

ICAR-Directorate of Floricultural Research, Pune 411005, Maharashtra

Rajamani, K.

Department of Floriculture and Landscape Architecture, Horticultural College and Research Institute, Tamil Nadu Agricultural University, Coimbatore 641003, Tamil Nadu

Saha, S.

ICAR-IARI Regional Station, Pune 411008, Maharashtra

Saha, T.N.

ICAR-Directorate of Floricultural Research, Pune 411005, Maharashtra

Sharma, S.P.

The Agri-Horticultural Society of India, 1, Alipore Road, Kolkata 700027, West Bengal

Singh, A.

Division of Vegetable Science and Floriculture, Sher-e-Kashmir University of Agricultural Science and Technology of Jammu 180009, Jammu and Kashmir

Singh, S.

Department of Floriculture and Landscaping, Punjab Agricultural University, Ludhiana 141004, Punjab

Singh, K.P.

Division of Floriculture and Landscaping, Indian Agricultural Research Institute, New Delhi 110012

Sk. Masudul Islam

Department of Vegetable Science, Faculty of Horticulture, Bidhan Chandra Krishi Viswavidyalaya, Mohanpur 741252, Nadia, West Bengal, India

Tejukumar, B.K.

Division of Floriculture and Landscaping, ICAR-Indian Agricultural Research Institute, New Delhi 110012

Thamaraiselvi, S.P.

Department of Floriculture and Landscape Architecture, Horticultural College and Research Institute, Tamil Nadu Agricultural University, Coimbatore 641003, Tamil Nadu

Vaishali, C.

Division of Floriculture and Landscaping, ICAR-Indian Agricultural Research Institute, New Delhi 110012

Yadav, L.P.

Department of Horticulture, Bidhan Chandra Krishi Vishwavidyalaya, Mohanpur 741 252, West Bengal

1

Breeding and Improvement of Ornamental Plants

Reeta Bhatia, Madhu Bala, Sapna Panwar and Namita

1.0 Introduction

Ornamental plants play a vital role in the life of human kind due to their aesthetic beauty and economic importance. The cultivation of flower crops has been recognized as a lucrative and income-generating venture. Besides fulfilling the aesthetic needs of humans, ornamentals plants are also vital for mitigating pollution and ecological sustenance (Sharma and Messar, 2017). The ornamental plants are broadly classified as cut flowers, loose flowers, pot plants, flowering and foliage plants, ornamental grasses, trees, shrubs, annuals and other plants of ornamental value. Due to rising demands from the consumers, ornamental industry constantly requires new plant varieties with elite traits such as attractive flower colours, novel flower forms, improved longevity, better fragrance, pigments, stress tolerance and disease resistance.

Despite the tremendous economic importance, the development of sophisticated breeding strategies is lagging behind in ornamentals than most of the agricultural crops. This may be attributed to the complex genetic systems and lack of genetic variability in the available gene pools of the major ornamental species (Debener, 2001). Many of the floricultural crops are vegetatively propagated and are highly heterozygous, which causes a complex inheritance of genetic factors. Besides, the polyploidy nature of some of the commercially important ornamentals further complicates their improvement through conventional breeding. This chapter provides the thorough insight into the improvement of ornamentals using classical and modern breeding strategies.

2.0 Breeding Behaviour and Breeding Methods

Before initiating the improvement program in any crop, the thorough knowledge about the breeding behaviour (mating system) and mode of inheritance of the traits of interest is of utmost importance. The breeding/reproductive behaviour determines the genetic constitution of a crop and has impact on gene action and transfer of genes. Based on the breeding behaviour or mating system, most ornamental plant species are cross-pollinated or preferentially cross-pollinated, *i.e.*, outbreeding types. Only a few ornamentals are obligate or preferentially self-pollinated, *i.e.*, inbreeding types. Many species have an intermediate type of breeding system with varying proportions of crossing and selfing. China aster, balsam, sweet pea, lupin, *Gypsophilla*, bells of Ireland are among the inbreeding types of herbaceous annuals. The group of outbreeding ornamentals includes ageratum, carnation, chrysanthemum, gladiolus, gerbera, hollyhock, cornflower, antirrhinum, delphinium, verbena, calendula, cosmos, gazania, poppy, marigold, zinnia, petunia, primula, pansy and rose *etc.* Breeders use methods and techniques that are based on the mode of reproduction of the species *i.e.*, self-pollinating, cross-pollinating, or clonally propagated. The general strategy is to breed a cultivar whose genetic purity and productivity can be sustained by its natural mating system.

The breeding strategy also depends upon the mode of inheritance of the trait of interest. If a trait is quantitatively inherited, heritability, *i.e.*, the coefficient between genetic (or its additive component) and phenotypic variance, is the basis of selection procedures. If major genes control a character, recombination of genes by crossing is the choice for improvement. The breeder should always pay attention to the relative contribution of the components of genetic variance (additive, dominance, epistasis) and environmental variance in choosing the best selection method. Additive genetic variance can be exploited for long-term genetic gains by concentrating desirable genes in the homozygous state in a genotype. The breeder can make rapid progress where heritability is high by using selection methods that are dependent solely on phenotype (*e.g.*, mass selection). However, where heritability is low, the method of selection based on families and progeny testing are more effective and efficient. When overdominance predominates, the breeder can exploit short-term genetic gain by developing hybrid cultivars for the crop (Acquaah, 2015). The breeding methods used for improvement of ornamentals can be broadly classified as conventional (classical or traditional) breeding methods and modern breeding tools. The classical breeding, which is based on the search for exotic plants, mutants and/or crosses within and between related species and selection of the most promising offspring, has played a significant role in the introduction of many traits and the production of a large number of varieties in many ornamentals. However, although classical breeding is still a powerful tool in the breeder's hands, the available gene pool for new traits is limited. Furthermore, the selection of a desired trait in the siblings is performed on the genetic background of their parents, which together with the genetic variability of the offspring, complicates and limits the controlled breeding. Hence, the recent advancements in the plant tissue culture techniques such as embryo rescue, protoplast fusion, polyploidization and molecular cytogenetic have tremendous potentials for the improvement of ornamentals.

2.1.0 Conventional Breeding Methods

Conventional (traditional or classical) plant breeding is based on the development or improvement of cultivars using conservative tools for manipulating plant genome within the natural genetic boundaries of the species. The important conventional breeding techniques used for the improvement of ornamentals includes introduction, selection, hybridization, mutation and polyploidy breeding. Selection is the oldest technique of crop improvement and remains the most widely used technique to date. Besides, breeders also assemble desirable traits from different but usually closely related plants into the new cultivar using the techniques of crossing (hybridization). New varieties are developed through conscious/selective artificial cross-pollination, in various ornamentals. The classical hybrid breeding uses deliberate interbreeding (crossing) of closely or distantly related individuals to produce new varieties with desirable properties. Plants are crossbred to introduce traits/genes from one variety or line into a new genetic background. Using conventional breeding strategy, only those target traits which pre-exist in the genetic potential of the species are improved without introducing new genes. In traditional breeding, the choice of breeding methods is determined by mode of reproduction. The breeding methods that have proved successful with self-pollinated species are: mass selection; pure-line selection; hybridization, with the segregating generations handled by the pedigree method, the bulk method, or by the backcross method;

and development of hybrid varieties. The important methods of breeding in cross-pollinated species includes, mass selection; development of hybrid varieties; and development of synthetic varieties. Since cross-pollinated species are heterozygous for many traits and lose vigour as they become purebred (homozygous), hence the goal of breeding methods is to preserve or restore heterozygosity (Acquaah, 2015). Hybrid cultivar breeding exploits the phenomenon of heterosis, and is applicable to both self- and cross-pollinated species. Mutation breeding may be resorted to when the gene of interest is non-existent in nature and may be induced. Also, sometimes, the desired trait found in wild relatives of the species and may be introduced into cultivated species through pre-breeding. In conventional breeding, breeders assemble desirable traits from different but usually closely related plants into the new cultivar using the techniques of crossing (hybridization) (Acquaah, 2015)

2.2.0 Modern Breeding Tools

The improvement of ornamentals through classical breeding approaches has several limitations including, high degree of heterozygosity, long breeding cycles, sterility, complex and huge genome sizes, limited gene pool, lack of gene of interest in the natural populations, unfavourable linkages of gene of interest with other unfavourable genes. Hence, it is unrealistic to rely upon classical breeding tools like hybridization and selection for improvement of ornamental crops when a particular trait controlled by a specific gene or genes is needed to be modified (Sharma and Messar, 2017). The obstacles of classical breeding can be overcome by the use of modern biotechnological tools such as molecular markers and genetic engineering. Recent development in plant tissue culture techniques such as embryo rescue, protoplast fusion, polyploidization and molecular cytogenetic has introduced the possibility of more precise and faster genetic improvement of plants. In ornamentals, plant tissue culture techniques have been now routinely applied to facilitate interspecific and intergeneric crosses for introgression of genes from wild sources. Recently, the *in vitro* techniques (*in vitro* screening and selection) have proven beneficial in screening against biotic and abiotic stress. The doubled haploid based technique has tremendous potential for speed breeding in ornamentals. The development of new biotechnological tools for the introduction of foreign genes into plants, combined with the growing knowledge and technology related to gene identification and isolation, has revolutionized all aspects of the biological sciences, particularly agriculture. Genetic engineering approaches have created almost unlimited possibilities for the molecular breeding of novel/improved crops. Using modern biotechnological tools, breeders may violate natural biological boundaries in the manipulation of the genetics of plants.

3.0 Introduction

It refers to the process of introducing the plants to a new environment from their natural habitat. Germplasm can be introduced from different countries or within the same country. It is one of the oldest and very effective method of plant breeding. The main function of plant introduction is to enrich the germplasm of a crop so that it can be utilized in plant breeding program. The role of introduction in ornamentals is very important under Indian context as most of the annuals, biennials, herbaceous perennials, bulbous crops, trees and shrubs have been introduced from other countries. The Mughal brought several ornamental plants from Persia and Central Asia. Mughal emperors from Persia and Afghanistan came to rule India during the 16th century have introduced roses in India. It is believed that the damask rose was first introduced in India by the Mughal emperor, Babar. Most of the ornamentals have been introduced from countries like South America, Europe, Africa, China, Japan and USSR. The first introduction of ornamental *Bougainvillea spectabilis* to India from Europe was done in 1860. *B. glabra* 'Splendens' was introduced at Agricultural and Horticultural Society of India at Alipore, Calcutta in 1869 (Roy *et al.*, 2012). The orchid flowered Canna was first introduced in 1897.

In the eighteenth and nineteenth century, large number of trees, shrubs, annuals, bulbous flowers and many other ornamental plants have been introduced in India, mainly through the Agri Horticultural Society of India, Indian Botanical Gardens and Forest Research Institute. The major objective of Indian Botanical Gardens and Forest Research Institute was to introduce plants of economic importance particularly tree species yielding timber. Some of the most important introductions are *Swietenia* spp. *Eucalyptus* spp., *Populus* spp., *Acer* spp. *etc.* Apart from their value as timber for forest plantation, many of them bear beautiful flowers which are being grown in parks, gardens and road sides. Among the three organizations of national importance, the Agri Horticultural Society of India, established in 1820 by Rev. William Carey, introduced a wide range of agricultural and horticultural importance including field crops, fruits, vegetables and large number of species and cultivars of ornamental plants. During the past 150 years, the society has introduced hundreds of beautiful flowering trees, shrubs, climbers, bulbous plants, annuals, foliage plants, orchids, roses, which are very popular as garden plants in the country. The society has been credited with the introduction of many species of trees including *Peltophorum, Tabebuia, Plumeria, Brownea, Cassia, Bombax, Callistemon* spp., *Grevillea* spp., *Cryptomeria japonica, Amherstia nobilis, Delonix regia, Kigelia africana, Pachira* spp., *Podocarpus* spp. and *Chorisia* sp. *etc.* Besides, a large number of ornamental shrubs (species and cultivars) have been introduced in India including, Bougainvillea, Calliandra, Ixora, Gardenia, Rhododendron, *Camellia* spp., *Dombeya* spp., *Ixora* spp., *Hibiscus* spp., *Jasminum* spp., *Mussaenda* spp. and *Magnolia* spp. Many bulbous flowers such as *Narcissus*, Tulip, Lilium, Iris, Freesia and *Alstroemeria etc.* were introduced in India in the 19th century. Foliage ornamentals or indoor plants have been imported from tropical America and southeast Asia. National Bureau of Plant Genetics Resources (NBPGR) is also importing species and cultivars of ornamental plants which are being used for cultivation, breeding and improvement. In India, though efforts are being made to produce hybrids of rose, chrysanthemum, carnation, dahlia, and gladiolus, majority of the commercially grown cultivars have been imported through various sources. Seeds of F_1 hybrids of annual flowering plants are also largely imported for marketing by various private seed firms and nurseries.

Though many orchid genera and species have originated in various parts of India, however most of the breeding works in *Vanda, Dendrobium, Cymbidium, Cattleya, Paphiopedilum* and other have been undertaken in countries like Thailand,

Malaysia, Holland, USA, UK and Japan *etc.* and the hybrids of unusual colour and appearance are being imported in India as cut flower and pot plants. *Anthurium andraeanum*, alstroemeria, carnation and lilium the important cut flower in world trade have been introduced in India and grown in green house for flower production. In recent times, new flowers like iris, liatris, lisianthus and ranunculus *etc.* have been introduced as ornamental germplasms.

The systematic introduction and exchange of plant germplasm of agriculture/ horticultural importance started in 1946 under a scheme initiated at the Division of Botany, Indian Agricultural Research Institute (IARI), New Delhi, which was later replaced by the Plant Introduction and Exploration Organization in 1956 and then by the Division of Plant Introduction in 1961. The National Bureau of Plant Genetic Resources (NBPGR) came into existence in 1976, which is serving as the apex organizations for the plant introductions. The Bureau has acquired more than 1293 new accessions of diverse ornamental crops, the majority of which are introduced from USA (55 per cent) followed by UK (9 per cent), Slovenia (6 per cent) and Netherlands (5 per cent). Among the introduced germplasm, the number of accessions introduced mostly of rose (33 per cent) followed by marigold (17 per cent), and gladiolus (12 per cent). Besides NBPGR, Forest Research Institute, Dehradun, Botanical Survey of India are also actively involved in introduction of plant germplasm. Some of the notable introductions in important ornamentals are listed in Tables 1 and 2.

3.1.0 Type of Introduction

3.1.1 Primary Introduction

When the introduced germplasm is well adapted to the new environmental conditions, it is directly released for commercial cultivation without any alteration in the original genotype. Some of the notable primary introductions in the commercial ornamental crops are as follows:

- **Rose:** First Red, Konfetti, Grand Gala, Noblesse Pink, Golden Gate, Passion, Superstar, Queen Elizabeth
- **Chrysanthemum:** Thai Chen Queen, Snowball, Casa Grande, Mountaineer, Nob Hill
- **Carnation**: Arthur Sim, William Sim, Pink Dona, Tempo, Raggio-de-Sole, Bright Rendezvous, Sunrise
- **Gladiolus**: Hunting Song, Yellow Stone, White Prosperity, Jacksonville Gold, Green Woodpecker

3.1.2 Secondary Introduction

The newly introduced germplasm is subjected to improvement through selection, or hybridization, so that newly evolved improved variety possess some novel trait of interest. The examples of secondary introduction in rose includes, 'Abhisarika: Mutant of Kiss of Fire'; 'Anurag: Sweet Afton x Gulzar'; 'Apsara: Sonia Mielland x Sabine Red'.

Table 1: Some of the Promising Introductions in Ornamental Plants

Name	*Species/Cultivars Introduced*
Rose	**Cultivars:** Red Cushion, Pilgrim, Torchy, China doll, Daydream, Contempo, Yellow Cushion, Touch of Venus, First Prize, Top Secret, Yellow Doll, Gold Coin, Starina *etc.* **Species:** *R. centifolia, R. damascena* var. *Kiese, R. moyesii, R. canina, R. nitida, R. rubiginosa, R. rubrifolia, R. roulatie, R. chinensis* cv. *Brasiba, R. rugosa, R. acicularis, R. amblyotis, R. beggeriana, R. bracteata, R. cinnamomea, R. davidi, R. elliptica, R. ferreri, R. fedstchenkoana, R. filipes, R. helenae, R. horrida, R. longicuspis, R. manettii,R. marginata, R. multiflora, etc.*
Gladiolus	**Cultivars:** Peter Pears, Friendship, White Prosperity, Jacksonville Gold, Silver Moon, Sonata, Yellow Emperor, Snow Princess, Picardy, Autumn Charm, Autumn Rose, Melody, Oscar, Rose Spire, Hunting Song, Camellia, Pink Parassol, Blaze, Julia, Green wood Pecker, *etc.* **Species**: *Gladiolus cardinalis, G. byzantinus, G. communis. G. tristis*
Carnation	William Sim, Red Sim, Dusty Sim, White Sim, Arthur Sim, White Elegance, Master, Scarlet Elegance, Scania, Exquisite *etc.* **Species:** *Dianthus plumarius, D. spiculiformis, D. fragrans, D. pinifolius var. lilacinus, Dianthus balbissii, D. barbatus, D. deltoides, D. prateaus and D. armeria*
Chrysanthemum	**Cultivars:** Casa Grande, Mountaineer, Nob Hill, Snow Ball, Icicles, Orange Fair Lady, Maurice White, Princess Anne, Thai Chen Queen
Gerbera	**Cultivars:** Intense, Stanza, Dalma, Silvester, Imperial, Donna Ellen, Rosalin, Savannah, Cream Clementine, Maron Clemetine
Marigold	**Cultivars:** French Marigold: Harmony, Petite Orange, Petite Yellow, Golden Girl, Crack jack African Marigold: Giant Orange, Giant lemon, Golden Jubilee
Dahlia	**Cultivars:** Maitre Noyer, Spectacular, Chinese Lantern, Elegance, Marrice Porset, Brio, Toppaffir, Deutschland, Golden Autumn, Light Music, Orange Customer *etc.*
Lilium	**Cultivars:** Asiatic: Brunello, Pollyanna, Navona, Eyeliner, Pavia, Prato, Bright Diamond, Celesta, Montego Bay, Celesta, Best Seller, Elite Oriental: Cassandra, Siberia, Stargazer, Sorbonne, Siberia
Tulip	**Cultivars:** Apeldoorn, Banja Luka, Red Impression, Christmas Dream, Negrita, Ali Bi, Golden Oxford, Oxford Wonder, Apeldoorn Elite, Ganders Rhapsody, Lle de France, Montecarlo, American Dream
Amaryllis	**Cultivars:** Candidum, Honeymoon, Snow White, Five Star, Fire Fly, Fairy Queen, Cardinal, Vermilion Record
Alstroemeria	**Cultivars:** Aladdin, Pluto, Rosita, Serena, Pink Panther, Tierra, Serena, Capri, Pink, Cinderella
Narcissus	**Cultivars:** Single Yellow, Scarlet Elegance, Golden Harvest, Trumpet, Trumpet mutant, Texas Semi double, Cheerfulness, Emperor, Wrestler, Geranium, Sicily White
Bougainvillea	**Cultivars:** Scarlet Queen, Bridal Banquet, Mrs. Butt, Mrs. H C Buck, Cream Delight, Salmon Double, Allison Devy, Crimson Glory, James Walker, Lady Hudson, Orange Glory, Princess Elizabeth, 'Rosenka', Santiago, Snow White, Los Banos Beauty, Mahara, Cherry Blossom

Source: Advances in Horticulture (Eds. K.L. Chadha and S. K. Bhattacharjee).

Promising Introduction of Rose in India

First Red

First Prize

Vivaldi

Noblesse

Bianca

Konfeti

Promising Introduction of Chrysanthemum in India

Snowball

Yellow Star

Yellow Misty

Snowdon White

Snowdon Yellow

Thai Chen Queen

Promising Introduction of Tulip in India

Apeldoorn

Ganders Rhapsody

Apeldoorn Elite

Queen of Night

Banja Luka

Promising Introduction of Narcissus in India

Trumpet

Golden Harvest

Trumpet Mutant

Wrestler

Geranium

Cheerfulness

Promising Introduction of Lilium in India

Cilesta

Pavia

Bright Diamond

Brunello

Siberia

Sorbonne

Promising Introduction of Alstroemeria in India

Polyanna

Elite

Yellow Tycoon

Aladdin

Riana

Serena

Source: Dr M. R. Dhiman, IARI, RS, Katrain, Kullu, HP.

Promising Introduction of Gladiolus in India

White Prosperity

Peter Pears

Jacksonville Gold

Hunting Song

Rose Supreme

Yellow Stone

White Friendship

Priscilla

Green Woodpecker

Promising Introduction of Iris in India

Day Light

Mount Everest

Miss Saigun

Promising Introduction of Carnation in India

Master

Kiliyosh

Ambrose

Pink Dona

White Libery

Loris

Promising Introduction of Gerbera in India

Intense

Shimmer

Salvadore

Dana Ellen

Stanza

Silvester

Source: https://agritech.tnau.ac.in/horticulture/horti_flower per cent 20crops_carnation.html; http://agrigoaexpert.res.in/icar/category/horitculture/floriculture_landscaping/gerbera.php

3.2.0 Purpose of Introduction

The main purpose of plant introduction is to improve the plant wealth and enhance genetic variability. Through the process of introduction, an entirely new crop species, which have not been grown earlier at a particular location, can be brought in to cultivation. Most of the ornamentals including, carnation, lilium, gerbera, tulip, lisianthus, bougainvillea, alstroemeria, anthurium *etc.* have been introduced in India from various sources. The newly introduced germplasm can be directly used as commercially variety. Subsequently, it can be utilized for the development of new varieties through selection or hybridization. In ornamentals, most of the commercial cultivated varieties have been introduced from other country for direct cultivation *e.g.*, 'Vivaldi' in rose, 'Snowball' in chrysanthemum, 'Brunello' in lilium *etc.* Introduction is one of the fastest means to create variability and prevent genetic erosion. It is one of the quickest and economical method of crop improvement, particularly when introductions are released as a variety directly or after simple selection. If a particular location is prone to some particular disease or insect attack, plants may be directly introduced to new disease-free areas to protect them from damage. Introductions may be useful for biosystematics and evolutionary studies.

Table 2: List of Few Important Ornamental Trees, Shrubs, Climbers, Flowering Annuals and Herbaceous Plants Introduced in India

Category	*Genera/species introduced*
Ornamental trees	*Acacia pycnantha* (Australia), *Bauhinia esculenta* (USA), *Brownea macrophylla* (USA), *Chaenomeles sinensis, Ceratonia Siliqua* (USA), *Colvillea racemosa* (USA), *Cryptomeria japonica* (USA), Cupressus spp. (Algeria, USA), *Erythrina poeppigiana* (USA), *Eucalyptus* spp., *Paulownia* spp. (China), *Sapium sebiferum* (USA), *Peltophorum* spp., *Tabebuia* spp. (USA), *Plumeria* spp., *Cassia* spp., *Bombax* spp., *Callistemon* spp., *Grevillea* spp., *Amherstia nobilis, Delonix regia, Kigelia africana, Pachira* spp., *Podocarpus* spp, *Chorisia* spp., *Juniperus* spp., *Jacaranda* spp., *Gustavia* spp., *Clitoria arboria, Cassia* spp., *Brownea hybrida, Brachychitom* spp *Acer* spp., *Araucaria* spp., *Agathis* spp., *Adansonia digitata*
Ornamental shrubs	*Bougainvillea* spp (Kenya and New Zealand), *Buddleia* (*B. davidii and B. nivea;* Hungry, *B. japonica;* Australia), *Chaenomeles* (*C. japonica*; Japan, *C. superba*; USSR), *Cotoneaster* (*C. frigidus*; USSR), *Forsythia* (*F. suspensa*; Japan), *Calliandra* (*C. calothyrsus; USA and Australia*), *Hibiscus* (*H. syriacus, H. archeri* and *H. Manihot*; USA, *H. grewinefolius, H. rosa-sinensis, H. schizopetalous*; Indonesia) *Nerium* (*N. oleander*, USA), *Dombeya, Mussuenda, Spirea* (*S. chamaedrifolia*; Holland, *S. japonica; Italy*), *Viburnum* (*V. tinus*; USSR)., *Weigela* (*W. hortensis* and *W. coracensis*; Japan), *Gardenia* spp., *Camellia* spp., *Hydrangea* spp., *Fuchia* spp., *Azalia* spp.
Flowering annuals and herbaceous perennials including ornamental bulbous plants	Aster (*Callistephus sp.*; Australia, France and Canada), *Antirrhinum* (*A. majus*; USA and Australia), *Calendula* (*C. officinalis*; Australia and France), Cock's comb (*Celosia sp.*; Australia), *Eschscholzia* (*E. caespitosa, E. Mexicana*; USA), Pansy (*Viola tricolor* var. *hortensis*; UK, France, Canada and USA), *Petunia* (Japan, USA, USSR, France and Australia), Marigold (*Tagetes erecta, T. patula*; Holland, Australia, USA, USSR and Thailand), *Zinnia* (*Z. angustifolia*; USSR, Thailand, Australia and France), Sweet pea, Dianthus, Salvia, Cosmos, Sunflower, *Amaryllis* (*A. belladonna*; Japan), *Anemone* (*A. sylvestris*; USSR and Australia), *Begonia* (*Begonia sp.* Denmark and Australia) *Cyclamen* (*C. persicum, C. hederifolium*; Germany and UK) *Dieffenbachia* (*D. amoena, D. maculata, D. hoffmanni*), *Gloxinia* (*G. sp.*; Denmark), Tulip, Narcissus, Tuberose, Zantedeschia, Hemerocallis, Pelargonium, Spathiphyllum, Dahlia and others

Source: Advances in Horticulture (Eds. K.L. Chadha and S. K. Bhattacharjee).

3.3.0 Procedure for Introduction of Germplasm in India

3.3.1 Procurement of Germplasm

The germplasm can be obtained as gifts or in exchange of germplasm in consideration of past gifts to the Bureau or in anticipation of future gifts or it can be purchased. All requests for introduction of germplasm from abroad need to be sent through NBPGR along with the specific details of the required material, the source/country as well as address of organisation/scientist through a prescribed application. On receiving the application, the Import Permit is issued and sent to concerned scientist(s) for sending the same to exporting organization or the Bureau be requested to pursue the import of desired material(s) from abroad. The requests can also be sent directly by an individual scientist to any foreign source without an 'Import Permit', however the NBPGR needs to be kept informed of such requests for the issuance of 'Import Permit'.

While introducing the germplasm from abroad, the following points should be taken into consideration by the concerned person/organisation abroad:

- ✰ Only healthy, viable and clean seed materials (free from soil, pests, pathogens and weeds) should be forwarded without any seed treatment so as to facilitate proper quarantine examination. It may, however, be fumigated, if considered necessary.
- ✰ All the introduced material must be accompanied with an 'Import Permit' and 'Phytosanitary Certificate' with additional declaration, if any, based on crop inspection, certifying that the material is free from particular pathogen(s)/insect(s).
- ✰ The seed/planting material should not be sent to any scientist by name directly, it should be preferably addressed to the Director, NBPGR, New Delhi, who has to take delivery of the seed/planting material and conduct quarantine examination.
- ✰ The perishable plant propagules (scion/woods, budwoods, plant rhizomes, suckers, *etc.*) may preferably be sent by air freight through any commercial airlines so as to avoid delay in receipt and clearance. If unavoidable, the material can be sent on charge collect basis. An advance intimation of the dispatch of such perishable materials to the NBPGR will help in prompt receipt and quick clearance of the material soon after its arrival.

3.3.2 Quarantine

The introduced germplasm is kept in isolation to prevent the spread of weeds, insect pests, and diseases. All introduced material is thoroughly inspected for contamination with weeds, insects and diseases. The introduced material that are suspected to be contaminated are fumigated or are grown under isolation after proper treatments and observed for insect pests and disease. This entire process is known as quarantine and the rules which are being followed are known as Quarantine rules. It is essential that all the material being introduced must be accompanied by an authentic phytosanitary certificate. The plant material being introduced must confirm to certain quarantine regulations and quarantine control is exercised by NBPGR at different points of entry.

3.3.3 Cataloguing

All the introduced plant material is assigned an entry number and information regarding agency, place of origin, adaptation *etc.* is well documented. The plant material is classified in three categories *viz.* Exotic collection (EC), Indigenous collection (IC), Indigenous wild collection.

3.3.4 Evaluation

The newly introduced plant material is sent to sub stations of the bureau and evaluated with respect to various characters to assess the potential of new introductions.

3.3.5 Acclimatization

It is the process that leads to the adaptation of a variety to a new environment. The extent of acclimatization is determined by range of genetic variability in original population and duration of life cycle of crop.

3.3.6 Multiplication and Distribution

Plant material which is introduced is multiplied and further tested at various locations. The suitability of cultivation in different regions of the country should be assessed before using it as a commercial variety.

Some rare and uncommon beautiful flowering trees introduced in India from South America and Africa include *Baikea insignis, Tabebuia avellanedae, Monodora grandiflora, Brownea hybrid, Brownea macrophylla, Clitoria arborea, Peltophorum africanum, Cassia moschata etc.* Many species of flowering climbers like *Beaumontia grandiflora, Pyrostegia venusta, Lonicera* sp., *Ipomea* sp., *Wisteria chinensis* were introduced which are now extensively grown in gardens. Many species, hybrids and cultivars of orchids have been introduced in India.

Large number of foliage ornamental plants, commonly known as house plants or greenhouse plants, often planted indoor and shady places in garden have been introduced in India in 20th century from Tropical America, South East Asia *etc.* Species and cultivars of *Aglaonema, Calathea, Dieffenbachia, Dracaena, Peperomia, Cordyline terminalis, Caladium, Philodendron* and *Begonia rex* are now extensively grown as ornamental plants. The Agri-Horticultural Society of India first introduced many species of annual flowering plants like *Antirrhinum, Aster, Phlox, Petunia, Dianthus, Clarkia etc.* in 1828 from UK.

4.0 Selection

Selection is the oldest and the simplest breeding method for improvement and development of new varieties in ornamentals. It is defined as preferential propagation/reproduction of the plants having more desirable characters than others. In simplest form, selection means choosing plants with desirable traits. It is

Introduction of Trees in India

Acacia pycnantha

Ceratonia Siliqua

Plumeria bicolour rubra

Amherstia nobilis

Kigelia Africana

Tabebuia chryotricha

Sesbania punicea

Baikea insignis

Tabebuia avellanedae

Monodora grandiflora

Brownea hybrid

Brownea macrophylla

Introduction of Shrubs

Budliea davidii

Calliandra calothyrsus

Hibiscus syriacus

Hibiscus manihot

Nerium oleander

Dombeya spectabilis

Mussaenda erythrophylla

Viburnum tinus

Weigela hortensis

Introduction of Annuals

Aster (*Callistephus* sp)

Antirrhinum majus

Calendula officinalis

Viola tricolor var. *hortensis*

Marigold

Petunia

Introduction of Climbers

Wisteria sinensis

Ipomea purpurea

Pyrostegia venusta

Beaumontia grandiflora

Lonicera japonica

Petrea volubilis

Introduction of House Plants

Aglaonema commutatum

Dieffenbachia seguine

Calathea makoyana

Monstera deliciosa

Peperomia argyreia

Maranta leuconeura

Begonia rex-cultorum

Dracaena marginata

Cordyline australis

the basis of all crop improvement. The natural hybridization, spontaneous mutation and recombination over generations form the base for the selection. The choice of selection procedure depends upon the mode of reproduction. Self-pollinated crops are naturally homozygous and show no inbreeding depression, so selection can be done to isolate superior genotypes to establish pure lines. The cross-pollinated crops are highly heterozygous and the progenies of these plants are different from parent population due to segregation and recombination and exhibit inbreeding depression. In self-pollinated crops, pure line selection is widely used, however mass selection is used to a limited extent. In cross-pollinated species, the mass selection, recurrent selection or family selection is followed, while clonal selection is the major method of selection in vegetatively propagated crops.

4.1.0 Types of Selection

4.1.1 Natural Selection

This is a natural process that operates in the nature without any human interference. Natural selection follows the Darwin's principle of 'Survival of the fittest', which means plants that survive through the adversities of nature are preferred and the weaker ones are wiped out. Thus, nature itself selects the fittest organisms. Natural selection favors those characters which are essential for survival of a species. The selection pressure ultimately results in the appearance of many differences between species and subspecies. Natural selection has given the cultivated crops and 'ecotypes' in plants.

4.1.2 Artificial Selection

Artificial selection is the deliberate reproduction of plants that have desirable traits. This is done by man, with the purpose to produce new varieties of plants that are higher yielding, resistant to pests and diseases, drought resistant and adapted to different environments and growing conditions. This would ensure that the speed and efficiency of production and quality of crops are maximized. Thus, takes a relatively short amount of time as compared to the natural selection. Artificial selection can add new genetic varieties to species and can help to eliminate diseases. If used with hybridization, artificial selection allows for genetic diversity to continue. The various methods of artificial section include, pure line selection, mass selection, recurrent selection, half-sib selection, full-sib selection and clonal selection.

4.2.0 Methods of Selection

4.2.1 Pure-line Selection

A Pure-line is the progeny of a single, self-pollinated, homozygous plant. Pure-line selection involves selecting and breeding progeny from superior organisms for a number of generations until a pure line of organisms with only the desired characteristics has been established. This is an important method of selection in self-pollinated crops. In pure line selection, a large number of plants are selected from a self-pollinated crop and harvested individually; individual plant progenies are evaluated and the best progeny is released as a pure line variety. All the individuals within a pure line have identical genotype and any variation present within a pure line is solely due to environment. Pure-line selection is used for the improvement of local varieties, selection in introduced varieties, improvement of old pure-line varieties.

4.2.2 Mass Selection

It is the simplest, easiest and oldest method of selection applicable to both self and cross pollinated crops. In mass selection, desirable individual plants are selected on the basis of their phenotypic performance. It involves the selection of a large number of superior plants that are harvested in bulk, the bulked seeds are used to raise the next generation. Mass selection is generally practiced in mixed population of cultivars or land variety or unimproved strains to isolate superior genotypes. It plays an important role in improvement of ornamentals, particularly for those traits that are governed by simple genetic system and are highly heritable. It can even be exploited for complex traits, provided some secondary traits can be identified that have high heritability and are known to contribute to the improvement of the trait. This method exploits additive gene effects plus additive x additive interactions. The efficiency of this method depends greatly on the heritability of the trait under selection. It can be utilized for improving qualitative characters, and, if applied over many generations, it is also capable of improving quantitative characters with low heritability. In ornamentals, mass selection can be deployed to select for photo-period insensitive genes, to permit use of exotic germplasm and to broaden the germplasm base in a breeding programme.

4.2.3 Progeny Selection

A modern refinement of mass selection is to harvest the best plants separately and to grow and compare their progenies. The poorer progenies are eliminated and the seeds of the remainder are harvested. In this method, the selection is based not solely on the appearance of the parent plants but also on the appearance and performance of their progeny. Progeny selection is usually more effective than phenotypic selection when dealing with quantitative characters of low heritability. However, the progeny testing requires an extra generation; hence gain per cycle of selection must be double that of simple phenotypic selection to achieve the same rate of gain per unit time.

4.2.4 Recurrent Selection

Recurrent selection is an important breeding method designed to improve the performance of populations for quantitative traits of economic importance. The recurrent selection gradually increases the frequency of favourable alleles and maintain genetic variability for future selection. It may be undertaken in one (intra-population improvement) or two populations (inter-population improvement) to upgrade the performance as such or combining ability. Intra-population improvement may involve evaluation of individuals (mass selection) or of progenies (full-sibs, half-sibs or selfed progeny selection) that have been developed within a population or test-cross progenies developed by using a tester (related or unrelated, narrow or broad-genetic base). Inter-population improvement involves simultaneous improvement of two heterotic populations with a focus on

Varieties of Rose Developed at IARI through Hybridization and Selection

Hybrids

Pusa Ajay
(Pink Parfait x
Queen Elizabeth)

Pusa Arun
(Queen Elizabeth x
Jantar Mantar)

Pusa Mohit
(Suchitra x
Christian Dior)

Pusa Bahadur
(Cara Mia x
Century Two)

Pusa Priya
(Jantar Mantar x
Queen Elizabeth)

Pusa Gaurav
(Pink Parfait x Arjun)

Selections from Open Pollinated Seedlings

Pusa Mahak
(Selection from
'Century Two')

Rose Sherbet
(Selection from
'Gruss an Teplitz'

Pusa Pitamber
(Jantar Mantar x
Banjaran)

Pusa Virangana
(Selection from cv.
Jantar Mantar)

Pusa Barahmasi
(Selection cv. Sadabahar)

Sadabahar
(Selection from
cv. Frolic)

Varieties of Rose Developed through Hybridization and Selection in India

Himalayan Wonder (Selection from First Red)

Himalayan Glory (Selection from First Red)

Him Basant (Half sib Selection of *R. damascena*)

Arka Pride (Hybrid Seedling of IIHRSG-1)

Arka Swadesh (Hybrid Seedling of IC617922)

Arka Ivory (Hybrid Seedling of IIHRSG-1)

Arka Kinnari

Arka Sinchana 'Five Star'

YCD-1 Rose

Mrinalini

Neelambari

Pusa Komal

Promising Hybrids and Selections in Chrysanthemum

Him Aditya (Yellow Puma x White Star)

Him Ujjwala (Yellow Puma x Shyamal)

Him Pushkar (Poornima x Shyamal)

Him Shringar (Yellow Puma × Shyamal)

Him Shikhar (Yellow Puma × Poornima)

Punjab Shyamli (Kelvin Victory × Flirt)

Yellow Delight (White Bouquet × Gul-e-Sahir)

Punjab Mohini (Yellow Charm x Chidori)

Himanshu

Punjab Shingar 'Seedling Selection of Flirt'

Garden Beauty 'Seedling Selection of Flirt'

MDU 1

Promising Selections of Chrysanthemum

Pusa Guldasta
(Open pollinated seedling selection from Lal Pari)

Pusa Shwet
(Open pollinated seedling selection from Lal Pari)

Pusa Aditya
(Open pollinated seedling selection of Jaya)

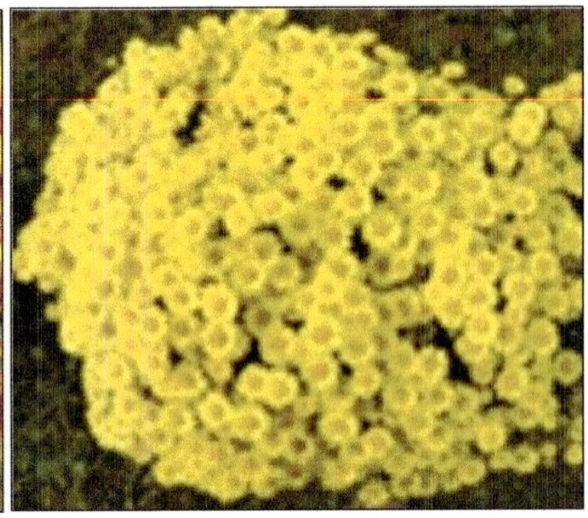

Pusa Sona
(Open pollinated seedling selection of 'Sadbhawana')

Pusa Chitraksha
(OP seedling selection of 'Lal Pari')

Nilima

their combining ability. The inter-population improvement methods, also known as reciprocal recurrent selection, there are two important alternatives, based on the evaluation of half-sib or full-sib progenies developed by using the other heterotic population or inbred line from that as a tester (Vasal *et al.*, 2014). The recurrent selection procedures may involve individuals, families and in some instances, even test-crosses. The methods of recurrent selection depend on the types of gene action, inheritance of the traits under selection, relative efficiency of selection, and the complexity of the trait under selection.

Recurrent selection has proved very effective for improving quantitative traits in gerbera. Harding *et al.* (1996) successfully improving the cut-flower yield in gerbera population using recurrent selection technique for 16-generation. In each generation, 40 gerbera lines were selected and crossed to produce the next generation. Cut-flower yield (per plant) was increased from 15.3 in generation 1 to 28.3 in generation 16.

4.2.4.1 Simple Recurrent Selection

The oldest form of recurrent selection is called simple recurrent selection (SRS) and each cycle of breeding involves only two steps: mass selection of individuals based on their phenotypic appearance; and intermating these selections to produce the offspring for the next generation. The recurrent selection that does not include tester is referred as simple recurrent selection.

4.2.4.2 Half-sib Recurrent Selection

Selections are made based on progeny test performance instead of phenotypic appearance of the parental plants. Seed from selected half-sibs, which have been pollinated by random pollen from the population (meaning that only the female parent is known and selected, hence the term "half-sib") is grown in un-replicated progeny rows for the purpose of selection. A part of the seed is planted to determine the yielding ability, or breeding value, for any character of each plant. The seed from the most productive rows or remnant seed from the outstanding half-sibs is bulked to complete one cycle of selection.

4.2.4.3 Full-sib Recurrent Selection

The method involves developing full-sibs by making plant-to-plant crosses. Direct and reciprocal crosses can be attempted depending on the need for the seed quantity. A number of full-sib families, each produced by making crosses between the two plants from the base population are evaluated in replicated trials. A part of each full-sib family is saved for recombination. Based on evaluation the remnant seed of selected full-sib families is used to recombine the best families.

4.2.4.4 Reciprocal Recurrent Selection

Reciprocal recurrent selection (RRS) is a repeated breeding procedure designed to improve the cross of two populations from different heterotic groups by using both general and specific combining ability. In this procedure, genotypes from two populations are evaluated in reciprocal crosses and the best genotypes of each population are selected and recombined to give rise to improved population crosses. Interpopulation half-sib or full-sib progenies are used as evaluation units and intrapopulation S1 progenies as recombination units. In RRS breeding programs selection intensity is usually high.

4.2.5 Clonal Selection

The procedure of selection used for asexually propagated crops is known as clonal selection, since the selected plants are used to produce new clones. Clone is the progeny of a single plant, produced by asexual reproduction. Clones remain stable until somatic mutations (sports) occur spontaneously. Genetical variation within a clone is due to mutation, mechanical mixture and segregation and recombination due to occasional sexual reproduction. The phenotypic variation within a clone might be due to the environment only and phenotype is due to the effects of genotype (G), the environment (E) and the genotypic x environment interaction over the population mean. Selecting spontaneous mutants within a clone is also termed as clonal selection. The procedure is to select individual plants of a clonal cultivar, and to grow A-clones and usually B-clones in replicated trials to test their performance, as well as their distinctness and superiority compared with the original cultivar.

Most of the commercial ornamentals including chrysanthemum, carnation, gerbera, gladioli, tulips, and daffodils are vegetatively propagated. Due to the high heterozygosity and polyploidy nature, most of the ornamentals do not produce true-to -the-type progeny on seed propagation. In order to improve such species, either crosses followed by one or more cycles of sexual propagation and selection, or clonal selection are required. Clone breeding is often the breeding scheme of ornamentals. In the seedling generation the greatest genetic variability exists and each genotype is represented by a single plant. In later clonal generations, replicated testing is possible. Many cultivars of rose, carnation, chrysanthemum, tulip, bougainvillea, hyacinth, begonia and pot azalea arose as a result of clonal selection.

4.3.0 Improvement of Major Ornamentals through Selection

The selection from seedlings raised from controlled/open pollinated seeds resulted in evolution and release of several outstanding cultivars in several ornamentals (Table 3). Some of the popular varieties of chrysanthemum released as open pollinated seedling selection includes 'Pusa Guldasta', 'Pusa Shwet', 'Pusa Sona', 'Birbal Sahni', 'Hemant Singar', 'Suhag Singar', 'Kundan', 'Rim Jhim', 'Sharada', 'Sharad Bahar', 'Sharad Mala', 'Sharad Shoba', 'Sharad Singar', 'Varsha' and 'Vasantika'. IIHR, Bangalore has developed 'No pinch, No stake' type cultivars namely, 'Sharad Singar', 'Hemant Singar', and 'Guldasta' through selection. At PAU, Ludhiana cvs. 'Punjab Shingar' (Seedling Selection of Flirt), 'Garden Beauty' (Seedling Selection of Flirt)', 'Santi', 'Vasanti', and 'Baggi' were developed.

A large number of rose varieties has been developed at IARI through selection. Some of the prominent selections of rose includes 'Pusa Mahak', 'Rose Sherbet' and 'Banjaran'. CSIR-IHBT, Palampur has developed three varieties namely, 'Himalayan Wonder' (First Red), Himalayan Glory (First Red) and 'Him Basant' (Half-sib selection in *R. damascene*) as a result of selection. Similarly, 'Arka Sinchana' is a seedling selection from half-sib population of variety 'Five Star'. A large number of promising varieties of gladiolus are evolved as a result of open pollinated seedling selection. Some of the notable sections in gladiolus are: 'Pusa Red Valentine', 'Pusa Kiran', 'Pusa Sindhuri', 'Gazel', 'Jwala' *etc.*

Table 3: Promising Cultivars of Ornamentals Developed through Selection

Crop	*Varieties*	*Institute*
Rose	Pusa Mahak, Rose Sherbet, Banjaran	IARI, New Delhi
	Himalayan Wonder (First Red), Himalayan Glory (First Red) and 'Him Basant'	CSIR, IHBT, Palampur
Gladiolus	Pusa Red Valentine, Pusa Kiran, Pusa Suhagin, Pusa Sindhoori, Agnirekha, Bindiya, Mayur	IARI, New Delhi
Chrysanthemum	Sharada, Sharad Bahar, Birbal Sahni, Hemant Singar, Suhag Singar, Jyoti, Kundan, Sharad Mala, Sharad Shoba, Sharad Singar, Varsha and Vasantika	NBRI, Lucknow
	Indira, Rakhee and Red Gold	IIHR, Hessaraghatta
	Santi, Basanti, and Baggi	PAU, Ludhiana
Bougainvillea	Shubhra, Dr. B.P. Pal, Parthasarthy and Surekha	

In gerbera, selection in the progeny from sexual crosses is done at several levels, initially on single plants, followed by un-replicated clonal trials and replicated clonal trials. Single-plant selection is widely attempted in gerbera for improvement of qualitative traits such as flower type and color and those characters with high heritabilities and little influence by environmental factors. In gerbera, two lines *viz.*, 'Arka Aswa' and 'Arka Nesara' have been developed indigenously through half-sib selection from IIHR-3 and IIHR-1, respectively.

In carnation, repeated crossing and selection resulted in significant improvements in the vase life in each generation and particularly in the ethylene production of many lines. The mean vase life in carnation was increased from 7.4 days in the first generation to 15.9 days in the seventh generation with a net increase of 8.5 days. As a result, two cultivars, 'Miracle Rouge' and 'Miracle Symphony', with genetically determined long vase lives of 17.7–20.7 days (3.2–3.6 times that of 'White Sim') were developed. The repeated crossing and selection for seven generation resulted in the development of two lines with ultra-long vase life, namely '532–6' and '806–46b, having vase life of 32.7 days and 27.1 days respectively (Onozaki, 2018).

In bougainvillea, varieties, 'Sholay' and 'Usha' are the half sib selection of cv. 'Red Glory' and 'Lady Hope' respectively, have been developed at IIHR, Bangalore. 'Kamini', 'Poornima', 'Shashank' and 'Violet Cushion' varieties of China Aster have been developed through pedigree selection method.

5.0 Hybridization

It is one of the powerful classical breeding methods for developing new varieties. It refers to the process of crossing two homozygous but genetically dissimilar genotypes. The seeds and the progeny resulting from the hybridization are called F_1 hybrids. The main objective of hybridization is to create genetic variation and to exploit heterosis. It improves a genotype by transferring to it one or a few characters from another genotype of the same or another species or to achieve new character expressions. When two genetically different plants are crossed the genes from both the parents are brought together in F_1. The natural hybridization encourages genetic diversity and better adaptability into the existing or new environments. On the other hand, manual artificial hybridization can only produce a population that is superior in selected traits. These can yield beautiful combinations of colors and characteristics not otherwise produced by nature. Based on the taxonomic relationship of the genotypes involved in hybridization, it can be classified as Intervarietal hybridization or Distant hybridization. In intervarietal hybridization, different strains or varieties within the same species are crossed, hence it is also known as intraspecific hybridization. The distant hybridization involves crossing of different species belonging to same genus (interspecific hybridization) or different genera (Intergeneric hybridization). Intervarietal hybridization has been widely used for the development of new hybrids in several ornamentals (Table 4).

The F_1 hybrids in ornamental were known even before Shull (1911) proposed the concept of heterosis. The first artificial hybrid 'Fairchild Mule' was produced by crossing Sweet William with carnation (*Dianthus barbatus* × *Dianthus caryophyllus*) by Thomas Fairchild in 1717. The development of F_1 hybrids were first initiated only in those crops in which parental lines and cultivars were homozygous, or could be easily selfed and also produce sufficient seeds through hand pollination. The first F_1 hybrid variety 'Prima Donna' in begonia was released by Benary Seed Company (Raghava and Singh, 2006). Though the exploitation of heterosis in ornamentals initiated much early, however the development of F_1 hybrids in these crops is lagging behind than most of the field and vegetable crops. The first commercial F_1 hybrid in petunia was developed in Japan in 1942. The first true single-flowered multiflora F_1 hybrid petunia, 'Silver Medal,' was introduced in 1949 by Charles Weddle. It was followed in 1952 by 'Ballerina,' an F_1 hybrid grandiflora petunia. Hybrids typically display hybrid vigour (heterosis) with superior performance over either the better parent or mid-parent values (Anderson *et al.*, 1993). Traditionally, F_1 hybrid flower breeding has been applied primarily to annual bedding plant species that were mainly seed propagated. In the past few decades, there has been renewed interest in the development of seed-propagated hybrids even in the crops that are vegetative propagated *e.g. Pelargonium* and *Gerbera etc.* In recent years, growing of F_1 hybrids are gaining tremendous popularity in India and other countries as they have plentiful advantages like early and uniform flowering, profuse branching, striking range of colours, large flower size, higher yield, biotic-abiotic tolerance, enhanced blooming season, dwarf and compact growth habit and novelty in flower colour and form (doubleness *etc.*). In several ornamentals such as marigold, petunia, carnation and balsam, the F_1 hybrids exhibits dwarf and compact growth habit, profuse basal branching, earliness and longer blooming season. Doubleness in petunia was achieved only through controlled hybridization. Presently, large number of F_1 hybrids are available in *Ageratum, Antirrhinum, Begonia, Campanula, Gerbera, Primula, Petunia,* Pansies, *Pelargonium,* marigold and lisianthus from private seed companies based in USA, China, Israel, Germany, the Netherland and Japan.

5.1.0 Development of Hybrids

The conventional method of F_1 hybrid production involves selection of parents, development of inbred lines for desired characters, testing their combining ability, production of hybrid seeds and evaluation of F_1 hybrid performance. During hybrid seed production, the male parent must cross-pollinate the female parent, and the maternal parent must be prevented from self-pollinating. The important steps involved in hybridization are:

5.1.1 Selection of Parents

The selection of parents mainly depends upon the objectives of breeding programme. Besides the targeted breeding objective, increased yields are always an objective of the breeder. Therefore, it is essential that at least one of the parents involved in a cross should be a well-adapted and established variety in the area for which the new variety is being developed. The other variety/parent should possess the traits that the breeder intended to transfer into the new variety. It is essential that all the characters which breeders want to improve should be present in one or the other parent. If the desirable character is not present in either of the parents then one can go for use of three or more parents leading to a complex cross. In order to develop uniform F_1 hybrids, homozygous parental lines (Inbreds/Pure lines) are indispensable for commercial hybrid seed. The two-basic prerequisite for hybrid seed production are ease of emasculation of the female parent and effective pollen dispersal from the male parent to ensure a satisfactory seed set in the female parent.

5.1.2 Development of Inbred Lines

An inbred is a nearly homozygous line obtained through continuous inbreeding of a cross pollinating species with selection accompanying inbreeding; it is maintained by close inbreeding, preferably by self-pollination or sib-mating. The development of inbreds through conventional methods of selfing requires 6-7 generations of selfing to attain near homozygosity. Hence, these days breeders are emphasizing the use of doubled haploids (DHs) based techniques for the development of homozygous lines in different crops. Using this approach, the homozygosity can be obtained within one to two years' time. In cross-pollinated crops, the continuous inbreeding/selfing may leads to loss of vigour, reduction in plant height, plants may become susceptible to lodging, insects and pests and many other undesirable effects. This phenomenon is known as inbreeding depression. Hence, after each selfing desirable plants are selected and self-pollinated or sib pollinated to avoid inbreeding depression. In cultivated allohexaploid chrysanthemums, *D. grandiflora*, inbreeding depression significantly decreased the percentage germination, yield potential, the proportion of individuals reaching reproductive maturity (percentage survivorship), seed set and pollen stainability (Datta, 2018). The purpose of inbreeding is to fix the desirable characters in homozygous condition in order to maintain them without any genetic change. These inbred lines can be used in genetic research *e.g.*, mapping genes and quantitative trait loci, allele discovery, and directly as cultivars in selfing species or as parents of hybrids and synthetic cultivars.

5.1.3 Evaluation of Inbred Lines

After the development of inbred line, it is crossed with other inbreds and its performance in single and double cross combination is evaluated. The ability of an inbred to transmit desirable performance to its hybrid progenies is referred as its combining ability. The average performance of an inbred line in a series of crosses with other inbred lines is known as general combining ability (GCA). While, the excessive performance of a cross over and above the excepted performance based on GCA of the parents is known as specific combining ability (SCA). Combining ability analysis helps to identify superior parents to be used in breeding programs or to identify promising cross combinations for cultivar development. The GCA is the characteristic of parents and SCA is characteristic of crosses or hybrids. General combining ability is directly related to the breeding value of a parent and is associated with additive genetic effects, while specific combining ability is the relative performance of a cross that is associated with non-additive gene action, predominantly contributed by dominance, epistasis, or genotype × environment interaction effects. The inbreds are evaluated in following way:

5.1.3.1 Phenotypic Evaluation

It is based on phenotypic performance of inbreds themselves. It is effective for characters, which are highly heritable *i.e.*, high GCA. Poorly performing inbreds are rejected. The performance of inbreds is tested in replicated yield trials and the inbreds showing poor performance are discarded.

5.1.3.2 Top Cross Test

The inbreds, which are selected on phenotypic evaluation, are crossed to a tester with wide genetic base *e.g.*, an open pollinated variety, a synthetic variety or a double cross.

5.1.3.3 Single Cross Evaluation

Outstanding single cross combinations can be identified only by testing the performance of single cross. The remaining inbred lines after top cross test are generally crossed in diallel or line x tester mating design to test for SCA.

5.2.0 Pollination Control Mechanism

Emasculation is essential to prevent self-fertilization in the flowers of the female parent. It involves the removal of stamens or anthers of a flower without affecting the female reproductive organs. In ornamental crops, emasculation by hand followed by bagging of plants or individual flowers, to exclude pollen-carrying insects, and hand-pollination are often used for control crossing. In most of the commercial ornamentals including, Antirrhinum, Gloxinia, Gerbera, Petunia, Impatiens, lisianthus *etc.* hybrids are produced by hand emasculation and hand pollination, hence hybrid seed production suffers from limitations of high cost. The availability of suitable pollination control mechanism is prerequisite for large scale hybrid seed production. The mechanism of male sterility and self-incompatibility have been exploited for hybrid development in few ornamentals including marigold, zinnia and petunia. When a male sterile or self-incompatible plant is used as a female parent, the emasculation procedure can be omitted and the cost of production of hybrid seed can be decreased substantially.

The separation of male and female flowers in monoecious *Begonia* species, where they are on different parts of the same inflorescence, also permits control of cross-pollination. In species carrying hermaphrodite (including both sexes) flowers, obligate cross pollination can be achieved by protandry or protogyny, as well as through self-incompatibility. Male sterility and self-incompatibility offer an alternative to hand-emasculation. Male sterility (Genic or Cytoplasmic) has become an important tool for hybrid seed production in many economically important crops. In petunia, though both SI and CMS system is present, breeders prefer manual operations because of other advantages offered in that species, such as easy flower handling and many seeds per pollinated flower. In pansy, herkogamy (position of male and female organs that prevents self-pollination) eliminates the need for hand-emasculation

5.2.1 Self-incompatibility

It is the inability of a functional pollen grain to fertilize the same flower or other flowers of the same plant. The term self-incompatibility (SI) refers to the partial or complete incapacity of a plant, producing functional gametes, to set viable seed on self-pollination. Alternatively, it is the failure of the genetically identical gametes to unite during fertilization. SI systems show a characteristic lack of pollen germination or stigmatic penetration (sporophytic SI) or the slow growth of pollen tubes in the style relative to compatible crosses (gametophytic SI). Thus, SI is a pre-fertilization reproductive barrier controlled by a single locus, S, with multiple alleles that encodes the specificities. In diploids, SI is controlled by a single gene, S; while polyploids may have additional genes. The hexaploid chrysanthemums possess three genes controlling SI.

Based on the floral morphology, the SI is broadly classified as homomorphic SI or heteromorphic SI. In homomorphic SI system, all plants in a population have the same floral morphology; *i.e.*, no visible difference in anther and style lengths between genotypes. In heteromorphic SI system, either two (distyly) or three (tristyly) flower forms are found in same plants. In distyly, two different anther and style lengths, short (thrum) and long (pin), are found. The tristylous plants possess three style lengths, shorts, mids, and long; flowers of each genotype have the same style (one) and anther morphs (two). Ornamental crops with heteromorphic SI systems include *Narcissus*, purple loosestrife (*Lathyrus salicaria*), *Primula* spp. and oxalis.

The homomorphic SI system can be classified as: sporophytic self-incompatibility (SSI) and gametophytic self-incompatibility (GSI). In case of GSI system, the SI reaction of a pollen is determined by its own genotype not by the genotype of the plant on which the pollen is produced. The gametophytic SI systems has been reported in petunia, rose, antirrhinum, lily (all *Lilium* species with the exception of *Lilium formosanum*), primrose and heliconia. In Sporophytic SI system, the SI reaction of pollen is governed by the genotype of the plant on which the pollen is produced and not by the genotype of the pollen. The SSI systems has been reported in ageratum, chrysanthemum, zinnia, marigold, gerbera, dahlia, ipomoea, and sweet alyssum. The major difference between GSI and SSI is as under:

Gametophytic SI (GSI)	*Sporophytic SI (SSI)*
The pollen-pistil interaction is governed by haploid genome of each male gametes and diploid genome of pistil tissue (Haplo-Diplo)	The pollen-pistil interaction is governed by genome of the plant on which the male and female gamete produced (Diplo-Diplo)
The stigma is smooth and wet	The stigma is papilate (hairy) and dry
Pollen tube inhibition take place in style	Pollen tube inhibition take place on the stigmatic surface itself
Homozygotes not obtained	Homozygotes can be obtained

There are three known major types of SI systems, the Papaveraceae, the Brassicaceae and the Solanaceae systems, which have been extensively studied systems. In all of these, the SI responses are all controlled by a single polymorphic locus, termed the S locus. In the Papaveraceae, a pistil-specific protein encoded by the S locus mediates pollen-tube growth inhibition through a Ca^{2+} signaling cascade involving programmed cell death. In the Brassicaceae, the two S-locus proteins known as SRK (S-locus receptor-like kinase), which is expressed in the stigma, and SCR (S-locus cysteine-rich)/SP11(S-locus pollen 11), which is expressed in pollen, have been shown to control pollen recognition and rejection through a ligand– receptor binding mechanism (Yang *et al.*, 2007). The gametophytic self-incompatibility (GSI) system found in three plant families, Rosaceae, Solanaceae and Scrophulariaceae, the S-locus product in the pistil is a glycoprotein with ribonuclease activity called S-RNase, whereas the S-locus pollen product is one of a class of F-box proteins called SLF (S-locus F-box) or SFB (S-haplotype-specific F-box). Molecular studies have demonstrated that the S-RNase and SLF/SFB genes encoded by the single polymorphic S locus, which control the pollen and pistil functions of SI in three distantly related families, the Solanaceae, Scrophulariaceae and Rosaceae, are organized in a haplotype-specific manner (Yang *et al.*, 2007).

The GSI occurs in the majority of self-incompatible roses, and is more prevalent in diploid species. In tetraploid and hexaploid species of rose, the break-down of self-incompatibility reaction is observed and it is thought that such self-compatibility arises through competitive interaction of S-alleles in diploid and triploid pollen. This means that all compatible or incompatible pollen starts to germinate, but growth of incompatible pollen tubes stops in the style while compatible tubes go on to fertilize the egg in the ovule. The SI response is typically manifested within the upper third portion of the style where incompatible pollen tubes exhibit reduced rates of elongation, loss of membrane integrity, disrupted organelles, and wall thickening, all of which can lead to swelling and bursting of the tube tip.

Gametophytic Self incompatibility (GSI) has been reported in petunia (Sims *et al.*, 2009). The GSI in Petunia is governed by a single, multiallelic S-locus, and recognition and rejection of self-pollen is controlled gametophytically by alleles expressed in pollen. The GSI prevents growth of a pollen tube through the style, provides a means of preventing self-pollination. In gerbera sporophytic self-incompatibility operates in dominance relationship between two SI alleles and influence seed set.

In chrysanthemum, sporophytic type of self-incompatibility (Drewlow *et al.*, 1973) involving more than one locus (Petty *et al.*, 2003) have been reported. Type of SI and its mechanism facilitates the development and production of F_1 hybrids in chrysanthemum for use in cut flower, bedding and pot plant industries. High temperature treatment (40°C for 60 min. or 50°C for 30 min.) to stigma and pollen have been found effective in overcoming varietal-cross incompatibility in chrysanthemum as reflected by high seed set in treated cultivars than control (Myung *et al.*, 2006a). Besides, treatments of stigmas with amino acids (alanine, arginine and glycine), IAA and nicotinic acid have also proved effective in overcome the incompatibility in chrysanthemums. Microscopic study revealed that low callosic deposit, no secretion on the stigma and collapsed papillae were observed in compatible crosses (Myung *et al.*, 2006b). Basic steps in the utilizing the SI for hybrid seed production

- Identification of self-incompatible plants in diverse population/genotypes.
- Development of homozygous self-incompatible lines.
- Identification of S-alleles in the homozygous self-incompatible lines.
- Establishment of inter-allelic relationships among the S-alleles.
- Identifying the best combining lines.
- Maintenance of parental self-incompatible lines.
- Commercial hybrid seed production.

Identification of self-incompatible plants is carried out by making different types of pollination *viz.*, self-pollination in freshly opened flowers (OP), compatible pollination with unrelated S-allele pollen (CP) and bud pollination to ascertain whether the plant in question is male and female fertile and also to get selfed seeds to raise its progeny. For quicker (within 24-48 hours of pollination) identification of degree of SI, the flower styles are observed under fluorescence microscopy for the penetration of pollen tube few hours after pollination. The self-incompatible lines to be used for hybrid seed production must be stable in its SI reaction and produce good seed set on selfing,

The SI lines are maintained via bud pollination/sib mating as self-incompatibility reaction is weak at early stage of flower development. Treatment with CO_2 gas (CO_2 enrichment), use of sodium chloride sprays, exposure to high temperature and double pollination techniques can also be utilized for overcoming self-incompatibility. Removal of stigmatic surface or whole stigma, useful in sporophytic system. Recently, many private seed companies have started exploiting tissue culture techniques for *in vitro* maintenance of SI/male sterile lines.

The major problems associated with use of SI system in hybrid seed production are weakening of S-allele lines by continuous inbreeding, pseudo-compatibility, the effect of environmental factors on the level of self-incompatibility and higher proportion of selfs/sibs in hybrid seed due to lack in proper synchronization of flowering. These could be managed by resorting to vegetative propagation, using the S-allele lines which behave stable under diverse environments and selecting the parental lines which have perfect male gametophytes.

Commercial hybrid seed production using SI

Line A — Line B

S_{11} — S_{22}

Maintained through bud pollination

S_{11} X S_{22}

S_{12}
(Commercial Hybrid)

5.2.2 Male Sterility

Male sterility refers to a condition in which pollen is either absent or non-functional in flowering plants, while female gametes function normally. There are three basic types of male sterility. Genic/nuclear male sterility (GMS), Cytoplasmic male sterility (CMS) and Cytoplasmic genetic male sterility (CGMS). Besides, two other types, *viz.*, chemically induced male sterility and transgenic male sterility have also been used in field crop.

5.2.2.1 Genic Male Sterility (GMS)

This type of male sterility is governed by the nuclear genes. The phenomenon of GMS is present in marigold, zinnia, delphinium antirrhinum, calceolaria, salvia and impatiens. In most cases, GMS is controlled by a pair of recessive alleles "*msms*", though few dominant genes have also been reported. Closely linked markers are rare to identify the male fertile plants, hence identifying the male fertile plants is difficult and time consuming. Maintenance of the genetic stock is difficult as there will be continuous segregation of the fertile and sterile individuals in 1:1 proportion. Male sterile line is advantageous and desired in flower crops not just as a parent for production of hybrid seeds; but as a variety itself provided the flowers are attractive. Male sterility as a mechanism to enhance flower longevity can be a viable approach in flower crops.

While using GMS system, the hybrid seed is produced either under greenhouses or in the open field by maintain proper isolation distance. The female and male lines are planted alternatively in the ratio of 2:1. The female line produces both the male fertile (*Msms*) and male sterile (*msms*) plants in the ratio of 1:1. The male fertile plants should be rouged/removed from the female line well before cross pollination for hybrid seed production. The hybrid seed should always be harvested from male sterile lines.

Tejaswini *et al.* (2016) have identified three distinct male sterile systems in marigold and classified them as apetalloid, petaloid and gynomonoecious types based on their floral morphology. The Apetaloid type of male sterility (flowers without petals) in marigold is reported to be genetic male sterile system controlled by single recessive gene in the nucleus. He *et al.* (2009) reported that the male sterility trait was controlled by the recessive gene, *Tems*. They utilized inter simple sequence repeat (ISSR) and sequence-related amplified polymorphism (SRAP) techniques combined with bulked segregant analysis to develop markers linked to the male sterility trait. One SRAP marker that was closely linked to the target trait was identified and successfully converted into sequence characterized amplified region (SCAR) marker that was located within 2.4 cM from *Tems* locus. The development of SCAR marker therefore would permit the efficient marker-assisted selection of male sterile individuals in breeding programmes of marigold and will greatly facilitate the breeding of F_1 cultivars.

5.2.2.2 Cytoplasmic Male Sterility (CMS)

This type of male sterility is determined by the cytoplasm. Since the cytoplasm of a zygote comes primarily from egg cell, the progeny of such male sterile plants would always be male sterile. The expression of male sterility in CMS plants is the result of incompatibility between recessive nuclear gene (called maintainer gene; *rf*) and male sterile specific cytoplasmic genome. Now it is well documented that specific open reading frame (ORFs) in mitochondrial genome (mt-genome) are responsible for the expression of male sterile trait. Once dominant restorer (*Rf*) gene (located in nuclear genome) responsible for pollen fertility of a cytoplasmic male sterile line is identified, it is commonly known as cytoplasmic- genic male sterility (CMS). Therefore, those cytoplasmic male sterile lines for which *Rf* gene(s) have been identified are widely known as genic-cytoplasmic male sterility (G-CMS) and more often treated as a separate class of male sterility system. However, both CMS and G-CMS can be described under common head (*i.e.*, CMS) because of the fact that in both these systems, expression of male sterility is due to the defect in the cytoplasm (mt-genome).

CMS can be transferred easily to a given strain by using that strain as a pollinator (recurrent parent) in the successive generations of backcross programme. After 6-7 backcrosses, the nuclear genotype of the male sterile line would be almost identical to the recurrent pollinator strain. Cytoplasmic male sterility can be maintained by crossing a male sterile line (A line) with the pollinator strain (maintainer line) used as a recurrent parent in the back cross programme since the nuclear genotype of the pollinator is identical with that of the new male sterile line. Such a male fertile line is called the maintainer line or B line as it is used to maintain the male sterile line.

In petunia, cytoplasmic male sterility has been observed but the use of this system is not practical because of breakdown of male sterility in the maternal parents or malformation of flowers in F_1 plants. Similarly, in Sunflower, cytoplasmic male sterility has been commercially exploited to produce commercial hybrids like, 'Sunrich Orange' and 'Orit'.

5.2.3 Double Flower Condition

The double flower condition results from transformation of anthers into petals and hence this character is regarded as one type of male sterility. This condition arises as a result of mutations in the homeotic box genes. They are usually recessive, although the double flower mutation in carnations exhibits incomplete dominance. The double flower condition has been commonly observed among the members of Asteraceae family including ageratum, aster, chrysanthemum, cornflower, dahlia, daisy, *Gaillardia*, marigold, *Rudbeckia* and sunflower. For effective use in hybrid seed production, the doubleness should be inherited as dominant trait to be manifested in F_1 hybrids. Some other type of floral abnormalities like *'Cindrella'* character in Begonia and *'Femina' in Tagetes erecta* and *Zinnia elegans* are also considered as form of male sterility. The double-flowered plants, such as *Aquilegia vulgaris, Begonia, Calendula, Chamelaucium ucinatum,* have been characterized anatomically, but not genetically. None of the genes responsible for double-flowered mutants in *Camellia, Narcissus, Primula, Rosa,* and many other horticulturally important plants, have yet been identified at the molecular level. The double and semi-double forms of *Dianthus caryophyllus* (carnation) have been shown to be caused by a dominant mutation at a single locus that shows meristic effects and homeotic conversion of stamens to petals. By contrast, a single recessive allele causes stamens to become petaloid in a *Nicotiana alata* mutant and three recessive alleles are known to cause double-flowered phenotypes in *Papaver somniferum*.

5.2.4 Triploidy

The advantage of growing triploid varieties is that of continuous blooming period as there is little or no seed set. Among the flower species, in *Tagetes*, commercial triploid hybrid (2n = 36) results from a cross between diploid female *Tagetes erecta* (2n = 24) and tetraploid male *Tagetes patula* (2n = 48). The triploid hybrid 'Nugget' is sterile and has the unique ability of holding the flowers on the plants for a longer period. However, in *Begonia semperflorens*, the loss of uniformity in triploids has been observed.

5.3.0 Types of Crosses

Commercial hybrid seed production is done by way of developing single cross, three-way cross or double cross or triple cross. In single cross, two genetically diverse inbreds are crossed to develop F_1 hybrids. For the development of single cross hybrids, a planting ratio of 2:1 is maintained between female and male line so that hybrid seeds can be harvested from two-third of the plants in the field. The single cross hybrids are more uniform and maximum hybrid vigour is manifested in this type of crossing. This method is widely used for production of F_1 hybrids in gladiolus, chrysanthemum, lisianthus and paeony *etc.* The three-way cross hybrids result from the cross between an F_1 hybrid and an inbred line. The double cross hybrids are produced when two different single-cross hybrids are cross-pollinated. The development of a double-cross hybrid requires two stages of crossing involving two pairs of inbreds. Double cross method has been used in Orchids and *Viola tricolor*. The top cross hybrids are developed by crossing an open pollinated variety and inbred line. It is employed in *Ageratum houstonianum, Bellis perennis, Gerbera jamesonii.*

While using SI system for developing single cross hybrids, two self-incompatible but cross-compatible best combiners are planted in alternate rows in isolated plots. The hybrid seed is harvested on both the lines. In three way cross, one single cross and a self-incompatible line are planted in alternate rows. Similarly, in double cross, two single crosses are used. For top cross, for every 2 or 3 rows of a self-incompatible line, one row of a good open pollinated (OP) cultivar as a pollen parent is provided. However, the hybrid seed is harvested from the self-incompatible plants only.

Types of Crosses

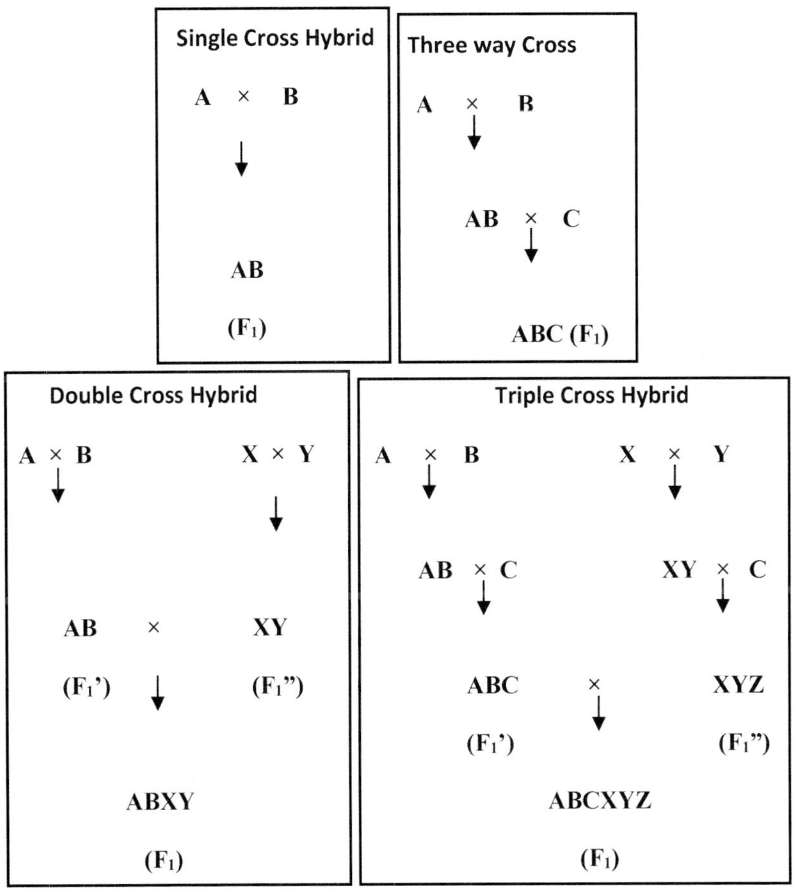

In most of the vegetatively propagated ornamentals, the process of hybridization is often followed by selection of superior clone which is termed as clonal hybridization. Large number of cultivars of perennial ornamental plants *e.g., Bougainvillea, Chrysanthemum, Dahlia, Hibiscus, Gladiolus, Rosa etc.* have been obtained by this technique.

5.4.0 Methods of making a Cross and Hybrid Seed Production

1. Hand emasculation and hand pollination: This technique is widely used for production of F_1 hybrids seeds in those ornamentals which are relatively easy to emasculate and produce large number of seeds on pollination *e.g.*, *Pelargonium, Eustoma, Antirrihnum majus, Petunia, Dianthus caryophyllus, Gladiolus, Rose etc.*
2. Hand emasculation and natural pollination.
3. Elimination of male plants and pollination.
4. Genetic male sterility as in marigold and zinnia, ageratum and calceolaria.
5. Cytoplasmic male sterility as in petunia.
6. Self-incompatibility as in petunia, pansy, stocks and ornamental kale.
7. Chemical emasculation- selective elimination of pollen production, that is, use of gameticides.
8. Use of marker genes to identify the selfs so that they can be eliminated as seedlings.

5.5.0 Selection Procedures with Hybridization

In most of the vegetatively propagated ornamentals, hybridization is often followed by selection in order to isolate the desirable genotypes from the segregating progeny. Two methods of selection namely, pedigree method and bulk method are commonly used for this purpose.

5.5.1 The Pedigree Method

This is widely followed by the plant breeders now, who maintain a detailed record of relationships between the selected plants and their progenies. It consists of the selection of individuals plants with the desired combination of characters in the F_2 generation and reselection of the progenies of each selected F_2 plant in succeeding generations until genetic purity is reached.

5.5.2 The Bulk Method

This method differs from the pedigree method in that the hybrids are grown in bulk till the F_5 or F_6 generation. Desirable individual plants are selected only in the F_5 or F_6 generation and these are then carried forward as families, which are evaluated in the same way as in the case of pedigree method.

5.6.0 Progress of Hybridization in Important Flower Crops

The systematic hybrid breeding efforts at Indian Agricultural Research Institute (IARI), New Delhi, Indian Institute of Horticultural Research (IIHR), Bengaluru, National Botanical Research Institute, Lucknow *etc.* have led to the development of several promising hybrids in commercial ornamentals (Table 4)

5.6.1 Rose

Modern day roses are highly heterozygous and polyploidy in nature (diploid, triploid, tetraploid, aneuploid, *etc.*) which make it practically impossible to forecast the result of any specific cross. Besides, female sterility, chromosomal disorders, poor seed setting and seed germination, are some of the major hurdles for hybrid breeding in roses (Datta, 2018). At IARI, New Delhi, Dr. B.P. Pal has taken up rose breeding and developed some hybrid seedlings and released his first rose variety 'Rose Sherbet' in 1962. Other hybrids released by him are; 'Delhi Princess' (Floribunda) 'Dr. Homi Bhaba' (Hybrid Teas) 'Kanakangi', 'Poornima', 'Hasina', 'Lalima', 'Nayika', 'Rat Ki Rani', 'Raja of Nalagarh', 'Ranjana', and 'Surkhab'. As a result of intensive hybridization, a series of new rose hybrids have been developed at IARI, New Delhi, including 'Anurag', 'Apsara', 'Pusa Ajay', 'Pusa Komal', 'Pusa Muskan', 'Pusa Priya', 'Pusa Pitamber', 'Pusa Shatabdi', 'Pusa Bahadur' *etc.*

5.6.2 Orchids

Orchids are fascinating, exquisitely beautiful, hardy ornamentals of enormous diversity, universally accepted as cut flowers, pot plants, also grown in garden in suitable location. Among the ornamentals cut blooms of orchids have the longest vase life and the flowers remain fresh on the plant for a long duration. Apart from large number of genera and species, orchids have numerous cultivars from spontaneous mutation and natural selection, inter specific and inter generic hybrids. Intensive investigations and efforts of production of new cultivars are being made in Thailand, Singapore, Malaysia, Japan, Holland, UK and Brazil and in many other countries, by nurseryman and institutions.

It has been possible to combine characters up to six genera (hexageneric) and the hybrid is name *Brilliandera* Gary. Many bigeneric and trigeneric hybrids of *Cattleya* and *Vanda* as one of the parents are valuable and colourful orchids in world market. Inter specific and intervarietal hybrids of *Dendrobium* are widely accepted ornamental plants across the world. Species and hybrids of *Cymbidium, Phalaeonopsis, Paphilopedilum, Miltonia, Odontoglossum* and *Oncidium* also comprise beautiful flowering orchids. Some of the importance hybrids of orchids used for cut flower and pot plants are given below.

Ascocenda- 'Dechjum Yip Sum Wah', 'Rakpaibulsombat, 'Meda Arnold'; ***Cattleya-*** 'George Angus', 'Rain Gold' 'Yue', 'Esmeralda Heritage', Greenwich 'Killarney', 'Glorious Sunset', 'Chinese Beauty'; ***Cymbidium-*** 'Sensation Chianti', 'Sungold', La costa 'Pink Lady', 'Pearlbel Pink', 'Beautiful Spirit', 'Cherry Red'; ***Dendrobium-*** 'Tomi', 'Spellbound', 'Ceaser Alba', 'Jaquelyn', 'Thomas', 'Kanjana Snow', 'Ekapol No. 1', 'Lovely Virgin Angel', 'Stardust Chiyomi', 'Deep Blush', 'Kasem White'; ***Miltonia-*** 'Playgirl 'Cha-Cha', Saint Heiler 'Red Jewel', Zorro 'Yellow Delight'; ***Mokara*** × 'Khaw Phaik Suan', 'Mak Chin On'; ***Odontoglossum-*** 'Elegans', 'Crispania', 'Kalkarstern', 'Geyser Gold'; ***Oncidium-*** 'William Thurston', 'Noel Schoenrock', 'Florida Gold', 'Saulsee', 'Red Velvet', ***Phalaenopsis-*** 'Happy Valentine', 'Taipei Gold', 'Blue Ridge Bubblegum', 'Dragon's Gem'; ***Paphiopedilum-*** 'Harrisianum', 'Charles Richman', 'Winnifred', 'Alma's Moth', 'Amarone Dream', 'Alpine Pleasure'; ***Vanda-*** 'Thananchai', 'Kultana Gold', 'April Rose Motes', Kalyana, 'Motes Sparkle', 'William Motes'. (see chapter 2).

Improved Cultivars of Cattleya

Evergreen
(Blc. Envy × *C. bicolor*)

Fireball Fame
(*C.* Netrasiri Fireball × *C.* Thospol Spot)

Angelwalker 'Easter'
(*C.* Little Angel × *C. walkeriana*)

Yellow Doll
(*C. luteola* × *Sophrolaelia* (Sl.)

Beautiful Lady Pat
(*Blc. Tampico* × *C. percivaliana*)

Victoria-Reginae
(*C. labitata* × *C. leopoldii*)

Improved Cultivars of Cymbidium

Alexanderi
(*C. eburneolowianum* x *C. insigne* 'Westonbirt')

Jean Brummitt
(*C. devonianum* x *C. eburneum*)

Happy Alex
(*C.* Happy Promenade x *C. Alexandra* Rose),

Improved Cultivars of Dendrobium

Bangkok Fancy

Red Bon

Thongchai Gold

Improved Cultivars of Miltonia

Sunset

Edmonds

Pink Lady

Improved Cultivars of Odontoglossum

Alispum

Chayenne

Black Diamond

Improved Cultivars of Oncidium

Golden Afternoon

Red Twinkle

Fragrance Fantasy

Improved Cultivars of Paphiopedilum

Cocoa Green

Silent Knight

Mountain Lass

Improved Cultivars of Phalaenopsis

Pink Marble

Red Rose

Leopard Prince

Improved Cultivars of Vanda

Jairak Glory

Pachara Dellight

Boonyarit Gold

5.6.3 Gerbera

Controlled hybridization has been as one of the breeding approaches for development of essentially all the cut-flower gerbera cultivars, including garden-type and some of the pot-type cultivars. Cultivars developed by this approach are then asexually propagated, using tissue culture techniques. Hybrid breeding has enriched gerbera flower types and colors, generated new combinations of qualitative traits in new cultivars, and improved important quantitative traits such as cut-flower quality, flower yield, and vase life in cut-flower gerberas. Recently, IIHR, Bangalore has developed two gerbera hybrids 'Arka Ashwa' and 'Arka Nesara', through hybridization using half-sib method of breeding where superior lines, IIHR-3 and IIHR-1, were crossed with pollen mixed from different varieties. Both of these hybrids have novel flower colour with double type of flowers. The Institute of Himalayan Bioresource Technology (CSIR-IHBT), Palampur have successfully developed five cultivars of gerbera namely, 'Him Saumya', 'Him Gaurav', 'Him Aabha', 'Him Apoorva', 'Him Keerti' through hybridization and selection.

5.6.4 Chrysanthemum

Chrysanthemum is a highly heterozygous plant that shows inbreeding depression and self-incompatibility; as a result, conventional crossbreeding is a powerful method for developing modern chrysanthemum cultivars. Many commercial cultivars with desirable traits, such as the Mammoth series garden chrysanthemum 'Lavender Daisy', have been developed via crossbreeding. Hybridization is simple and effective, and the F_1 progeny derived from two parents with contrasting target traits generally exhibit a wide phenotypic variation in chrysanthemum. However, there are several factors that must be taken into consideration when crossbreeding chrysanthemum, such as the fertility of a certain cross combination, qualitative analysis of the target traits, and superior hybrid progeny selection. Selection through hybridization resulted in development of more than 95 new promising varieties at CSIR-NBRI, Lucknow, some promising varieties are mentioned in Table 3. Cultivars of chrysanthemum 'Him Aditya' (CSIR-IHBT-CH-14-1), 'Him Ujjwala' (CSIR-IHBT-CH-14- 2), 'Him Shringar' (CSIR-IHBT-CH-14-4), 'Him Pushkar' (CSIR-IHBT-CH-14-8) and 'Him Shikhar' (CSIRIHBT-CH-14-18)' were developed by CSIR-Institute of Himalayan Bioresource Technology, Palampur, Himachal Pradesh through hybridization and selection programme.

5.6.5 Gladiolus

Hybridization is an important breeding technique for the improvement of gladiolus and several commercial cultivars have been developed as a result of controlled crossing. Systematic hybridization involving gladiolus cv. 'Frendship' (2n = 60) with *G. tristis* (2n = 30) resulted in the development of tripolids (2n = 45) namely, 'Manmohan', 'Monohar,' 'Manhar', 'Mukta', 'Manisha', 'Mohini', 'Triloki' and 'Sanyukta'. Two aneuploid cvs. 'Archana' and 'Arun' were also evolved. Several commercial hybrids of gladiolus including 'Pusa Srijana', 'Pusa Unnati', 'Pusa Vidushi', 'Pusa Manmohak' 'Pusa Shanti' have been developed at IARI, New

Delhi. Two promising hybrid selections namely 'Arka Ranjini' and 'Arka Pratham' have been released by IIHR, Bangalore.

Table 4: Promising Hybrids of Important Ornamental Crops Developed in India

Crop	*Cross with Parentage*	*Organization*
Gladiolus	Pusa Srijana: Berlew × Heady Wine Pusa Unnati: Berlew × Heady Wine Pusa Vidushi: Melody × Berlew Pusa Manmohak: Mayur × Hunting Song Pusa Shanti: Yellow stone × melody Pusa Chandni: Green woodpecker × White friendship	IARI, New Delhi
	Arka Pratham: Junior prom × Arka Poonum Arka Ranjini: Arka Poonum x Gold Medal Meera: G.P. 1 × Friendship Nazrana: Black Jack × Friendship Poonam: Geliber Herald × R.N. 121 Sapna: Green Woodpecker × Friendship Aarti: Shirley × Melody Apsara: Black Jack × Friendship	IIHR, Bengaluru
	Archana: Sylvia × Friendship Arun: Sylvia × Fancy Hans: Friendship × G. *tristis* Indrani: Friendship × *G. tristis* Kohra: Sylvia × King Lear Manhar: Friendship × *G. tristis* Manisha: Friendship × *G. tristis* Manmohan: Friendship × *G. tristis* Manohar: Friendship × *G. tristis* Mohini: Friendship × *G. tristis* Mridula: Friendship × *G. tristis* Mukta: Friendship × *G. tristis* Pitamber: Friendship × *G. tristis* Roshni: Friendship Pink × Red Beauty Sanyukta: Friendship × *G. tristis* Triloki: Friendship × *G. tristis* Neelima: Snow Princess × Tropic Sea JC Bose: Friendship Pink × Red Beauty Urvashi: Snow Princess × Eurovision SP Lancaster: Friendship Pink × Priscilla Suvarna: Fidelio × Hall Mark	NBRI, Lucknow

Crop	*Cross with Parentage*	*Organization*
	Heerak: Yellow Stone × My Love Amethyst: Lavender Puff × Tropic Sea Lalima: Jester × Fidelio	
	Chaubattia Ankur: Oscar × Friendship Chaubattia Arunima: Oscar × Motherfisher Chaubattia Shobhit : Meria Goretti × Tropic Seas Chaubattia Tripti: Sunny Boy × Oscar	Horticultural Experiment and Training Centre, Chaubattia, Almora (U.P.)
Rose	Anurag: Sweet Afton × Gulzar Apsara: Sonia Mielland × Sabine Red Arjun: Blithe Spirit × Montezuma Pusa Ajay: Pink Parfait × Queen Elizabeth Pusa Komal: Pink Parfait × Suchitra Pusa Manhar: Jantar Mantar ×Lahar Pusa Ranjana: Pink Parfait × Iceberg Pusa Muskan: Pink Parfait × Alinka Pusa Priya: Jantar Mantar × Queen Elizabeth) Pusa Pitamber: Jantar Mantar × Banjaran, Pusa Bahadur: Cara Mia × Century Two Pusa Shatabdi: Jadis x Century Two	IARI, New Delhi
Chrysanthemum	Punjab Mohini: Yellow Charm × Chidori Punjab Shyamli: Kelvin Victory × Flirt Anmol: Puncho × Rage Royal Purple: Bindiya × Rage Yellow Delight: White Bouquet × Gul-e- Sahir Winter Queen: White Bouquet × Flirt Punjab Gold: Flirt ×Gul-e-Sahir Autumn Joy: White Bouquet x Flirt	PAU, Ludhiana
	Him Aditya: Yellow Puma × White Star Him Ujjwala: Yellow Puma × Shyamal Him Pushkar: Poornima × Shyamal Him Shringar: Yellow Puma × Shyamal Him Shikhar: Yellow Puma × Poornima	CSIR-IHBT, Palampur
Carnation	Arka Tejas: *Dianthus carophyllus* × *Dianthus chinensis*	IIHR, Bengaluru
Gerbera	Him Saumya: CSIR-IHBT-Gr-1× CSIR-IHBT-Gr-3 Him Keerti: CSIR-IHBT-Gr-1 ×CSIR-IHBT-Gr-3 Him Apoorva: CSIR-IHBT-Gr-1× CSIR-IHBT-Gr-4 Him Abha: CSIR-IHBT-Gr-4× CSIR-IHBT-Gr-3 Him Gaurav: CSIR-IHBT-Gr-2× CSIR-IHBT-Gr-6	CSIR-IHBT, Palampur

Crop	Cross with Parentage	Organization
Hibiscus	Arunodaya: IIHR- H.2 × Rachaiah Ashirwad: H.S.21 × Hombe Gowda Basant: IIHR-1 × Rachaiah Chitralekha: Debby Ann × H. S. 203 Nartaki: H. S. (red) × H. S. 123 Nazneen: H. S. 203 × Rashtrapati Phulkari: H. S. 139 × H. S. 181 Ratna: H. S. 127× Ruffle Red Saturn: H.S. 182 × Red Double Smt. Indira Gandhi: H.S. 182 × Ruffle	IIHR, Bengaluru
	Aurobindo: Leberacea × White Pink Centre Mother: Honi Honi × Comet	Lalbagh, Bengaluru

Table 5: Promising Hybrids of Bougainvillea Developed through Intervarietal and Interspecific Hybridization

Sl.No.	Crosses	Cultivars	Hybridization
	Intraspecific hybridization		
	Bougainvillea glabra		
1	Trinidad × Formosa	Dr. H. B. Singh	Artificial
2.	Formosa × Trinidad	Purple Wonder	Artificial
	Bougainvillea spectabilis		
3.	Lalbagh × Red Glory	Chitravati	Artificial
	Bougainvillea peruviana		
4.	Dr. B. P. Pal × Princess Margaret Rose	Mary Palmer Special	Artificial
	Interspecific hybridization		
1.	*B. peruviana* × *B. glabra*	Begum Sikander, Mrs Butt	Artificial
2.	*B. peruviana* × *B. spectabilis*	Wajid Ali Shah	Artificial
3.	*B. buttiana* × *B. peruviana*	Chitra	Artificial
4.	*B. spectabilis* × *B. buttiana*	Dr. R. R. Pal, Summer Time, Spring Festival	Natural
5.	*B. spectabilis* × *B. glabra*	Maharaja of Mysore	Natural
6.	*B. glabra* × *B. spectabilis*	Pink Beauty, Pixie, Rose Queen	Natural

6.0 Mutation

Mutation refers to sudden heritable change in the characteristic of an individual. The genetic improvement of crop plants for various economic characters through the use of induced mutations is termed as Mutation breeding (Miri, 2018). It is an innovative breeding tool and ornamental crops are ideal candidates for improvement

Promising Hybrids and Varieties of Gladiolus Developed at IARI, New Delhi

Pusa Srijana
(Berlew × Heady Wine)

Pusa Unnati
(Berlew × Heady Wine)

Pusa Vidushi
(Melody x Berlew)

Pusa Manmohak
(Mayur ×
Hunting Song)

Pusa Shanti
(Yellow stone ×
Melody)

Pusa Chandni
(Green woodpecker x
White Butterfly

Selection from Open Pollinated Seedlings

Pusa Red Valentine
(Selection from OP
seedlings of Regency)

Pusa Kiran
(Selection from OP
seedlings of Ave)

Pusa Sindhuri
(Selection from OP
seedlings of Little fawn)

Promising Hybrids of Gladiolus

Roshni (Friendship Pink × Red Beauty)

Neelima (Snow Princess × Tropic Sea)

J C Bose (Friendship Pink × Red Beauty)

Urvashi (Snow Princess × Eurovision)

SP Lancaster (Friendship Pink × Priscilla)

Suvarna (Fidelio × Hall Mark)

Arka Ranjini (Arka Poonum × Gold Medal)

Arka Pratham (Junior prom × Arka Poonum)

Lalima (Jester × Fidelio)

Source: Singh *et al.*, 2016.

Promising Selections in Marigold

Pusa Deep

Pusa Bahar

Pusa Arpita

Punjab Gainda No. 1

Arka Bangara

Arka Bangara 2

Promising Varieties of Bougainvillea Developed through Hybridization and Selection

Wajid Ali Shah
(A hybrid seedling of
Bougainvillea peruviana
'Dr. B.P. Pal' × *B. spectabilis*
'Mrs. Chico')

Chitra
(A hybrid seedling of
Bougainvillea xbuttiana
'Tetra Mrs. McClean' × *B. peruviana* 'Dr. B.P. Pal')

Dr R R Pal
(*Bougainvillea glabra*
'Thomasii' ×
Bougainvillea × *buttiana*
'Louise Wathen')

H B Singh
(A hybrid seedling of
B. glabra 'Trinidad' ×
B. glabra 'Formosa')

Mary Palmer Special
(Hybrid seedling of
B. peruviana 'Dr. B.P. Pal'
× *B. peruviana* 'Princess
Marget Rose')

Begum Sikander
(A hybrid seedling of
Bougainvillea peruviana
'Dr. B.P. Pal' × *B. glabra*
'Jennifer')

Summer Time
(Hybrid seedling of *B. spectabilis* var. 'Thomasaii'
and *B.* × *buttiana* var.
'Louise Wathen')

Spring Festival (Hybrid
seedling of *B. spectabilis*
var. 'Thomasaii' and
B.×*buttiana* var. 'Louise
Wathen')

Pixie
(Hybrid seedling of
B. glabra × *B. spectabilis*)

through mutation breeding as novelty is highly desirable in these crops. The Induced mutations lead to alteration in one or a few characters of an otherwise outstanding cultivar without altering the remaining and often unique traits of the genotype. Any changes in the phenotypic characteristics such as colour, shape or size of flower and variegation in leaves can be easily detected in ornamentals. Further, due to the heterozygous nature of many ornamentals, they have high mutation frequency (Datta and da Silva, 2006). The majority of economically important ornamentals such as rose, carnation, chrysanthemum *etc.* are vegetatively propagated and thus any desirable change in the phenotype of ornamentals can be easily fixed through vegetative means of propagation. Most of the ornamental species are highly heterozygous and often propagated vegetatively; this allows the detection, selection and conservation of mutants in the M1 generation. Natural mutations have played a significant role in the evolution of modern day ornamental species. However, the systematic work on induction of mutation in ornamentals was initiated only during 1930s. The first mutant variety of tulip 'Faraday' was developed by De Mol in 1936 through X-ray irradiation of bulbs of cv. 'Fantasy'. Another mutant cv. 'White Sim no. 1' of carnation was introduced in 1962 in the USA by Mehlquist. It was developed through irradiation of rooted cuttings with c-rays. Mutation breeding has been widely utilized in ornamental crops like Alstroemeria, Bougainvillea, chrysanthemum, rose, carnation, canna, tuberose *etc.*

At molecular level, mutation causes permanent and relatively rare change in the number or sequence of nucleotides. The mutations often cause alteration in the nuclear DNA due to change in genes or chromosome structure as a result of addition, deletion, transition and transversion in nucleolus of cell, or in cytoplasmic DNA (*i.e.,* male sterility). Mutation that causes changes in chromosomal structure are called as chromosomal mutations. While, the mutation that alter the base sequences of genes are termed gene or point mutations. Among the various kinds of mutational changes at the molecular, base substitutions, that involve substitution of one base for another play an important role in induction of mutation. Base substitution can occur due to mis-pairing of the base analogue in the treated DNA during replication, leading to mutation through transitions (when exchanges occur either between purines; A-G or between pyrimidines; T-C) and transversions (when purines are exchanged for pyrimidines or vice versa). Transitions and transversions are considered to be the simplest kinds of base pair changes, but they may result in phenotypically visible mutations. Further, addition or deletion of a nucleotide base pair (when one of the bases manages to pair with two bases or fails to pair at all) may result in an alteration in the reading frame of the gene's DNA, such changes are known as frame-shift mutations. The frame shift mutations are more drastic in their effect as they may completely change the message of the gene starting with the point of deletion/ addition. Besides, the gene mutations may also arise as a result of rearrangement of bases in the DNA in which a small or large sequence of bases may be inverted as a result of chromosome breakage, and reunion of the broken ends may involve different DNA molecules in a reciprocal rearrangement or in loss of a fragment. The mechanism of duplication of a DNA sequence may also leads to change in the structure of a gene leading to gene mutation (Oladosu *et al.,* 2016).

Induced mutations are highly effective to enhance natural genetic resources and have successfully assisted in developing improved and new cultivars among both seed and vegetatively propagated crops. So far, there are more than 2300 officially released mutant varieties worldwide, of these 566 represent ornamental plants (http://www-mvd.iaea.org). Mutation breeding is considered as a pillar of modern plant breeding. In crops that do not produce seeds, mutation induction may be the only reliable way of increasing variability for developing new cultivars (Table 6).

Table 6: Mutant Ornamental Varieties Registered in the Joint Food and Agriculture Organization of the United Nations/ International Atomic Energy Agency database (FFO/IAEA, 2017)

Genus	*Varieties Obtained*	*Varieties Obtained with Gamma Rays*
Chrysanthemum	283	150
Rosa	67	52
Dahlia	35	18
Alstroemeria	35	11
Streptocarpus	30	0
Dianthus	28	8
Begonia	25	17
Portulaca	17	17
Bougainvillea	16	14
Rhododendron	15	9
Hibiscus	10	10
Cytisus	9	9
Tulipa	9	0
Achimenes	8	0
Canna	8	8
Lilium	6	4
Limonium	6	6
Antirrhinum	5	0
Iris	5	5
Tibouchina	5	0
Abelmoschus	4	4
Gladiolus	4	2
Hoya	4	4
Juncus	4	3
Kalanchoe	4	1

Source: Selene *et al.* (2019)

6.1.0 Types of Mutation

Mutations can be broadly divided into intragenic or point mutations (occurring within a gene in the DNA sequence); intergenic or structural mutations within chromosomes (inversions, translocations, duplications and deletions) and mutations leading to changes in the chromosome number (polyploidy, aneuploidy and haploidy).

6.1.1 Spontaneous Mutations

They are also called as natural mutations or 'bud sports. 'Sport' is a natural mechanism by which a plant of an existing variety puts forth a shoot in which genetic change has occurred. This genetic mechanism plays an important role in increasing the range of variation and giving rise to a new variety (Datta, 2018). Bud mutations arise from the mutation of a meristematic cell at a certain stage during its development. Spontaneous mutations occur in natural population (without any artificial treatment of physical and chemical agents) at a very low frequency. The frequency of spontaneous mutations is generally very less *i.e.*, 10^{-6}. Several kinds of radiations and cosmic rays from the sun are responsible for spontaneous occurrence of natural mutations in plants. Spontaneous mutants have been reported in several genera of ornamental *e.g.*, *Oenothera, Godetia, Paeonia, Bougainvillea, Dahlia, Rosa etc.*

Natural mutations are frequently noticed in many species of foliage plants which are popular as house plants grown indoors or in partially shady location in gardens. Variation in foliage colour are common in the species belonging to genera *Aglaonema, Begonia, Calatheas Dieffenbachia, Dracaena, Hedera, Monstera, Philodendron, Peperomia, Neoregelia etc.* Apart from hybrids produced as a result of hybridization among species and cultivars, natural mutants frequently occur in many bulbous plants as in *Hippeastrum, Amaryllis, Hemerocallis, Dahlia, Canna indica, Alstroemeria etc.* Among the succulent or ornamentals wide range of flower colour, single and double flower forms is noticed in *Adenium obesum,* which is widely grown in pots and landscape garden. In cacti and succulent, which comprise of many genera and species show variation in colour of plants, the most spectacular colour is noticed in *Gymnocalycium mihanovichii.* In herbaceous perennial flowering plants like dahlia, chrysanthemum, pelargonium, selection of natural variant has given rise to many attractive cultivars which are maintained by vegetative propagation. The important spontaneous mutants of different ornamentals that has been utilized as commercial cultivars are listed in Table 7.

In roses, about 18 per cent of the varieties of the Hybrid Tea group have originated as sports. Some of the promising sports of Hybrid Tea roses includes 'Mme Butterfly', 'Lady Sylvia', 'Rapture', 'Better Times', 'Jewel', 'Royal Beauty', *etc.* The most important bud sport is the climbing habit in Hybrid Teas. Some important climbing sports are 'Crimson Glory', 'Climbing Blue Moon', 'Climbing Cinderella', 'Climbing Fragrant Cloud', 'Climbing Ladies', 'Climbing Peace', 'Climbing Queen Elizabeth', Climbing Sterling Silver', 'Climbing Yellow Doll', 'Climbing Zambra', *etc.* Similarly, in Dwarf polyanthas, around 54 per cent of varieties including, 'Cameo', 'Ideal', 'Juliana Rose', 'Miss Edith Cavell', 'Coral Cluster' and 'Locarno', have been developed through bud sports. A number of striped roses such 'Careless Love' from 'Red Radiance'; 'Candy Stripe' from 'Pink Peace' have been developed as sports (Datta, 2018).

Table 7: Important Mutants in Ornamental Plants Derived through Spontaneous Mutations

Mutant	*Parent*	*Organization*
Rose		
Chitra	Janina	ICAR–IARI, New Delhi
Pink Montezuma	Montezuma	ICAR-IARI, New Delhi
Pusa Mansij	Raja Surendra Singh of Nalagarh	ICAR-IARI, New Delhi
Pusa Abhishek	Jantar Mantar	ICAR–IARI, New Delhi
Pusa Urmil	Jantar Mantar	ICAR-IARI, New Delhi
Nav Sadabahar	Sadabahar	ICAR-IARI, New Delhi
Chrysanthemum		
Kasturba Gandhi	Mahatma Gandhi	NBRI, Lucknow
Sonar Bangla	Snow Ball	NBRI, Lucknow
White Cloud	Pink Cloud	NBRI, Lucknow
Sharda	Sharad Shobha	NBRI, Lucknow
Bougainvillea		
Jawahar Lal Nehru	Lalbaugh	IIHR, Bengaluru
Archana	Rosevilles Delight	NBRI, Lucknow
Mary Palmer	Mrs. H. C. Buck	Kolkata
Alick Lancaster	*B.* × *buttiana*	
Vishakha	Mrs. H. C. Buck	ICAR–IARI, New Delhi
Shubhra	Mary Palmer	NBRI, Lucknow
Thimma	Mary Palmer	Lalbagh Botanic Gardens, Bengaluru
Partha	Scarlet Queen	NBRI, Lucknow
Mahara	*B.* × *buttiana* 'Mrs. Butt'	Philippine
Cherry Blossom	*B.* × *buttiana* 'Mrs. Butt'	Philippine
Los Banos Beauty	*B.* × *buttiana* 'Mrs. Butt'	Philippine
Roseville's Delight	*B.* × *buttiana* 'Mrs. Butt'	Philippine
Louis Wathen	Scarlet Queen 'Mrs. Butt'	Madras
Mrs. McClean	*B.* × *buttiana* 'Mrs. Butt'	NBR, Lucknow
A.P.J. Abdul Kalam	Fantasy	NBR, Lucknow

In carnation, bud mutation is widely utilized to establish novel cultivars because carnation mutates spontaneously at a high frequency. The well-known examples of bud mutation in carnation is the 'Sim types', which originate from 'William Sim', include more than 300 bud mutants with a range of petal colors. Somatic mutations have played very important role in the evolution of many varieties throughout the world. According to Datta and Janakiram (2016), approximately 30 per cent of the chrysanthemum cultivars originated as sports. Some cultivars have given rise to a great number of sports such as 'Sweetheart', 'The Favourite' and 'Indianapolis'. Some of notable chrysanthemum cultivars that have originated in India through spontaneous mutations have been listed in Table 7.

In bougainvillea, several spontaneous mutants have been isolated and established as new varieties. Some of the important bud sports includes 'Shweta' (Trinidad); 'Shubhra' (Mary Palmer); 'Partha' (Scarlet Queen) and 'Surekha' (Scarlet Queen).

Among the plants of different groups of horticultural crops, ornamental plants are the largest both in number of species, cultivars and diversity. Thousands of ornamental plants have developed as natural mutants which are being grown in garden and as indoor plants. Among the trees grown in tropical garden, variation in leaf colour in *Ficus elastica,* usually a large green and glossy leaves, several attractive varieties are popular and elegant house plants are grown in pots in most parts of the world. *Ficus benjamina* another popular tree in the genus, grown in road side in garden in tropics, has variant with attractive deep yellow and variegated foliage which are also grown as tall hedge, pot plants and are grown as colourful tree. Natural mutations are frequently noticed in many species of foliage plants which are popular house plants grown indoor and partially shady location in the gardens, variation in foliage colours is common in the species belonging to genera *Aglaonema, Begonia, Calathea, Dieffenbachia, Dracaena, Monstera, Peperomia, Cordyline, Caladium, etc.*

Apart from hybrids produced as a result of hybridization among the species and cultivars, natural mutations occur in many bulbous plants like *Amaryllis and Hippeastrum, Dahlia, Canna, Hymenocallis, Tuberose etc.* Among the succulent ornamental plants wide range of flower colour, single and double are frequently notice in *Adenium obesum* and *Euphorbia milii* which are widely grown in pots and rock gardens. In *Gymnocalycium mihanovichii,* a species of cactus has spectacular display in colour in different mutant.

Among the shrubs, spontaneous mutants with attractive foliage colours are observed in *Acalypha, Wilkinsiana, Duranta macrophylla, Lemonia spectabilis, Hydrangea macrophylla* and *Codieum varigatum* a popular shrub extensively grown in tropical and mild sub-tropical garden in pots and in ground.

Several species of plumeria, trees with large beautiful flowers introduced from South America produce large number of mutants when grown from seeds, particularly in *Plumeria bicolour* and *Plumeria bicolour 'Rubra'* when grown from cuttings and seeds (Bose and Mukherjee, 1967).

6.1.2 Induced Mutations

This type of mutation is artificially induced by using various physical or chemical agents. Induced mutagenesis is widely used for the development of new varieties in ornamental crops (Table 8). NBRI, Lucknow and IARI new Delhi have developed and released several novel mutants in commercial ornamental including rose, chrysanthemum, bougainvillea and gladiolus. The induced mutations can be categorized into two groups.

6.1.2.1 In vivo Mutagenesis

This method involves the treatment of seeds or vegetative plant parts with physical or chemical mutagens and then the treated material is further raised under field conditions for isolation of mutants. The major disadvantage of *in vivo* mutagenesis in higher plants is the chimera formation which does not allow the isolation of solid mutants. The other drawbacks of this techniques are limited population size and requirement of large space for screening. The *in vivo* mutagenesis has been successfully utilized for creation of variability and cultivar development in Chrysanthemum (Kapadiya *et al.*, 2016), Rose (Singh and Bala, 2019; Gupta and Shukla, 1970 and 1971) and Gladiolus (Yadav *et al.* (2016).

Table 8: Important Mutants in Ornamental Crops Derived through Induced Mutations

Mutant	*Parent*	*Mutagen*	*Organization*
Rose			
Pusa Christiana	Christian Dior	Gamma Rays	ICAR-IARI, New Delhi
Abhisarika	Kiss of Fire	Gamma Rays	ICAR-IARI, New Delhi
Madhosh	Gulzar	EMS	ICAR-IARI, New Delhi
Angara	Montezuma	Gamma Rays	NBRI, Lucknow
Sharada	Queen Elizabeth	Gamma Rays	NBRI, Lucknow
Sukumari	America's Junior Miss	Gamma Rays	NBRI, Lucknow
Yellow Contempo	Contempo	Gamma Rays	NBRI, Lucknow
Pink Contempo	Contempo	Gamma Rays	NBRI, Lucknow
Tangerine Contempo	Contempo	Gamma Rays	NBRI, Lucknow
Curio	Imperator	Gamma Rays	NBRI, Lucknow
Twinkle	Imperator	Gamma Rays	NBRI, Lucknow
Light Pink Prize	First Prize	Gamma Rays	NBRI, Lucknow
Chrysanthemum			
Pusa Centenary	Thai Chen Queen	Gamma Rays	ICAR-IARI, New Delhi
Pusa Anmol	Ajay	Gamma Rays	ICAR-IARI, New Delhi
Pusa Kesari	Thai Chen Queen	Gamma Rays	ICAR-IARI, New Delhi
Pusa Arunodaya	Thai Chen Queen	Gamma Rays	ICAR-IARI, New Delhi
Shabnam	D-5	Gamma Rays	NBRI, Lucknow
Sharad Har	Sharad Mala	Gamma Rays	NBRI, Lucknow
Sheela	Himani	Gamma Rays	NBRI, Lucknow
Pitamber	Otome Zakura	Gamma Rays	NBRI, Lucknow
Taruni	Kingsford Smith	Gamma Rays	NBRI, Lucknow
Tulika	M-24	Gamma Rays	NBRI, Lucknow
Aruna	Ashankit	Gamma Rays	NBRI, Lucknow
Asha	Hope		
Vijaykiran	Vijay		
Tuberose			
Rajat Rekha		Gamma Rays	NBRI, Lucknow
Swarna Rekha		Gamma Rays	NBRI, Lucknow
Gladiolus			
Shobha	Wild Rose	Gamma Rays	IIHR, Bengaluru

Mutant	Parent	Mutagen	Organization
Carnation			
Arka Flame	——	Gamma Rays	IIHR, Bengaluru
Hibiscus			
Anjali	Alipore Beauty	Gamma ray	NBRI, Lucknow
Bougainvillea			
Los Banos Variegata	Los Banos Beauty	Gamma ray	NBRI, Lucknow
Los Banos Variegata -Jayanthi	Los Banos Beauty	EMS	NBRI, Lucknow
Los Banos Variegata Silver Margin	Los Banos Beauty	Gamma ray	NBRI, Lucknow
Mahara Variegata	Mahara	Gamma ray	NBRI, Lucknow
Pallavi	Roseville's Delight	Gamma ray	NBRI, Lucknow
Arjuna	Partha	Gamma ray	NBRI, Lucknow
Pixie Variegata	Pixie	EMS	NBRI, Lucknow

6.1.2.2 *In vitro* Mutagenesis

This method involves the induction of mutation in combination with the use of *in vitro* tissue culture techniques. The *in vitro* mutagenesis involves, treatment of any explant or callus with mutagens and then the treated explants are allowed to regenerate under *in vitro* conditions. The *in vitro* mutagenesis allows for the uniformly treatment of cells/explants with physical and chemical mutagens and thus improves the frequency of recovery of solid mutants. This technique offers several advantages over the conventional methods in terms of mutation induction and selection of desired traits. The chance of getting a solid mutant is higher during *in vitro* mutagenesis. The main advantage of this technique is to overcome chimera formation. The other benefits of *in vitro* mutagenesis include, wider choice of explants, large population size, uniform treatment of physical and chemical mutagens, year-round activity *etc.* (Datta and da Silva, 2006). *In vitro* mutagenesis has opened new possibilities for inducing increased number of mutants and solid mutants in ornamentals. The *in vitro* mutagenesis has been successfully employed for induction of novel traits in number of ornamental crops including, Rose (Bala and Singh, 2015; Namita *et al.*, 2014), Chrysanthemum (Hossain *et al.*, 2006; Verma *et al.*, 2012), Carnation: (Singh *et al.*, 2000, Tejaswini *et al.*, 2006).

6.2.0 Techniques of Mutagenesis

6.2.1 Plant Material for Induction of Mutations

Any types of planting materials, *e.g.*, whole plants, seeds, seedlings, vegetative cuttings, scions, bulbs, corms rhizomes, tubers or *in vitro* cultured tissues like leaf and stem explants, anthers, calli, cell cultures, microspores, ovules, protoplasts and *in vitro* cultured cells can be used as starting materials for the induction of mutations. The seeds are most preferred as starting material, however multiple forms of plant propagules, such as cuttings, scions, bulbs, tubers, corms and rhizomes have been frequently utilized in vegetatively propagated ornamentals for efficient induction of novel mutants in ornamentals. Over the last three decades, researchers have started exploiting the totipotency using single cells and other forms of *in vitro* cultured plant tissues such as leaf, stem explants, anthers, calli, cell cultures, microspores, ovules, protoplasts for mutagenesis.

6.2.2 Mutagens

Physical or chemical agents which are used to enhance the mutation frequency are called mutagens. The major benefit of using induced mutagenesis is the high frequency of induction of mutations as compared to spontaneous mutation which occurs at the rate of 10^{-6}. Based on type of agents used for induction of mutations it is categorized into two categories *i.e.*, physical mutagenesis and chemical mutagenesis.

6.2.2.1 Physical Mutagenesis

The physical mutagenesis involves the use of ionizing or non-ionizing radiations. Ionizing radiations cause mutations by breaking chemical bonds in the DNA molecule, deleting a nucleotide, or substituting it with a new one. Among the various physical mutagens, gamma rays have been widely used for development of new varieties in ornamental crops. Gamma rays have a shorter wave length and thus possess more energy than protons and X-rays, which gives them ability to penetrate deeper into the tissue. Exposure of the planting material to gamma radiation causes breakage in double-stranded DNA and thus leads to mutation. The gamma induced mutants have great potential for improving the plant architecture leading to better crop improvement (Oladosu *et al.*, 2016). It can be used for irradiating whole plants and delicate materials, such as pollen grains. The gamma irradiation has been widely used for induced mutagenesis in several ornamental crops including rose, carnation, chrysanthemum, alstroemeria, bougainvillea *etc.* (Table 8). During the past two decades, ion beams irradiation technique has become popular as a new type of physical mutagen and has been widely used for mutagenesis in flower crops (Table 9). These radiations cause enormous structural changes at the chromosome level or in DNA and thus leads to high mutation frequency as compared to gamma rays and x- rays. The ion beam radiation has been successfully utilized in Chrysanthemum (Imakiire *et al.*, 2006; Ueno *et al.*, 2013; Nagatomi *et al.*, 2003; Yamaguchi *et al.*, 2009), Cyclamen (Ishizaka *et al.*, 2018), Carnation (Okamura *et al.*, 2003), African daisy (Iizuka *et al.*, 2011; Okada *et al.*, 2012).

Physical mutagens are being preferred over the chemical mutagens as the handling is very easy, there are no problems associated with mutagenic agents reaching the target cells, results are reproducible and there is no associated problem of toxic waste (Schum, 2003). The physical mutagens have high degree of accuracy and sufficient reproducibility, particularly for gamma rays, which have a uniform penetrating power in the tissue. The mutagenic frequency of the physical mutants is determined by mutagen dose (a factor that depends on radiation intensity and duration of exposure). In most of the breeding programs, doses that lead to 50 per cent lethality (LD50) is considered as optimum dose (Oladosu *et al.*, 2016).

Table 9: Important Mutant Cultivars of Ornamental Plants Derived through Ion Beams Technology

Name	*Mutant*	*Parent*	*Major Feature Altered*	*Reference*
Madagascar periwinkle	Setofuku CAP	Setofuku CAMR	Flower colour and petal shape	-
	Setofuku GPN	Kamihon BPN	Flower colour	-
Chrysanthemum	Aladdin	Jimba	Few lateral buds	Imakiire *et al.* (2006)
	Imajin	Jimba	Few lateral buds	-
	Aladdin 2	Aladdin	Early flowering at low temperature	Ueno *et al.* (2013)
	Ion-no-Kouki Ion-no-Mahou Ion-no Koumyou Ion-no-Seikou Ion-no-Hatsune	Taihei	Flower colour variation	Nagatomi *et al.* (2003)
Cyclamen	Tennyo-no-Mai	Uruwashi-no-Kaori	Flower colour variation	Ishizaka *et al.* (2018)
Carnation	Misty Pink Vital Ion Dark Pink Vital Ion RedVital Ion	Vital	Flower colour variation	Okamura *et al.* (2003)
African daisy	Viento Flamingo	Mother Symphony	Flower colour variation	Iizuka *et al.* (2011)
	Viento Labios	Viento Flamingo	Flower colour variation	Okada *et al.* (2012)
Salvia	TL585	Lady in Red	Plant height	-

Source: Yamaguchi (2018).

6.2.2.2 Chemical Mutagens

The chemical mutagenesis involves the use of chemical mutagens for the induction of mutation. It generates stable and inheritable changes as it generates variations in simple nucleotides which create new allele series. The chemical mutagens including, alkylating agents: Ethyl Methane Sulphonate (EMS), Methyl Methane Sulphonate (MMS), Ethylene Imines (EI) Sulphur mustard, Mustard gas, Epoxides), base analogues (5-bromo uracil, 2-amino purine), acridine dyes (acriflavine, proflavine, acridine orange, acridine yellow, ethidium bromide) and colchicine are commonly used for mutagenesis. It has been reported that over 80 per cent of the mutants registered with the International Atomic Energy Association (IAEA) database have mainly utilized three alkylating agents namely, EMS, methylnitrosourea (MNU) and ethylnitrosourea (ENU) (Pacher and Puchta, 2017). The alkyl group of alkylating agents reacts with DNA which may change the nucleotide sequence and cause a point mutation. (Broertijes and Harten, 1988).

EMS alkylates with guanine bases and leads to mispairing, alkylated-G pairs with T instead of C, resulting in primarily G/C to A/T transitions (Bhat *et al.*, 2007). Another category of chemical mutagens, base analogues are closely related to the DNA bases and can be wrongly incorporated during replication thus induce gene mutations. The use of colchicine for the induction of flower colour mutants in rose has been reported by Gupta and Datta (1983) and Datta and Gupta (1985). The work on mutagenesis using chemical mutagens has been reported by many researchers in various crops including, Chrysanthemum (Datta, 1987; Hossain, 2006; Lertsutthichawan *et al.*, 2017), Rose (Kaicker and Swarup, 1972; Nonomura *et al.*, 2001; Senapati and Rout, 2008, 2011; Toyoda *et al.*, 2015), Carnation (Hentrich and Glawe, 1982; Singh *et al.*, 2000; Tejaswini, 2006).

The chemical mutagens are relatively easy to use and can be applied without any complicated equipment or facilities. The ratio of mutational to undesirable modifications is generally higher for chemical mutagens than for physical mutagens. The mutagenesis efficiency of a chemical agents is affected by concentration of the mutagen, the length of treatment and the temperature at which the experiment is carried out. However, chemical mutagens are generally carcinogenic and therefore, extra care must be taken for health protection during the procedure.

The choice of a mutagen often depends upon the safety of usage, ease of use, availability of the mutagens, effectiveness in inducing certain genetic alterations, type of tissue, cost and available infrastructure. The chemical mutagens are primarily used to induce point mutations, while physical mutagens induce gross lesions, such as chromosomal abbreviation or rearrangements. The frequency and types of mutations are direct results of the dosage and rate of exposure or administration of the mutagen rather than its type (Oladosu *et al.*, 2016).

6.3.0 Doses of Mutagen

The appropriate doses for mutation induction may varies with the sensitivity of the given plant species and material to a particular mutagen. It is important to determine the optimal doses of desired mutagens in advance before actually starting large scale mutagenic experiments. The adequate dose of mutagen is usually determined based on the extent of physiological damages which manifest themselves in a delay or inhibition of cell division activity. Based on the differences in the dose rate, gamma-ray treatment is divided into two categories—acute and chronic irradiations. In acute irradiation, plant materials are irradiated at high doses for periods ranging from several minutes to a day, whereas in chronic irradiation, plant materials are exposed to gamma rays at a low dose rate over several weeks or even longer. It is hypothesized that chronic irradiation is useful for mutation breeding of vegetatively propagated crops because the mutants obtained would be less radiation damaged and could be directly used as new varieties. The frequency of mutational events increases with increasing doses of mutagens; however, it causes severe reduction in the vitality and regeneration capacity of the treated plant material. The application of higher doses increases the risk of inducing more than one mutational event per cell and leads to induction of negative characters which is highly undesired. In many cases the LD 50 which causes a 50 per cent lethality, is considered as the optimum dose for mutation induction. The standardized doses of mutagens for important ornamentals have been mentioned in Table 10.

Table 10: Doses of Mutagens and Type of Explant Used for Induction of Mutations in different Ornamentals

Crop	*Mutagen*	*Dose*	*Plant Part Exposed*
Rose (*Rosa* spp.)	EMS	2.5 per cent	Budwood
Carnation	EMS	2.5 per cent	Rooted cuttings
Amaryllis	Gamma-rays	250 rad – 5 Krad	Bulb
Bougainvillea	Gamma-rays	250 – 1250 rad	Stem cuttings
Gerbera	Gamma-rays	1 and 2 Krad	Rooted plantlet
Rosa spp.	Gamma-rays	2 – 6 Krad	Stem with budding eyes
Tagetes erecta	Gamma-rays	500 rad – 2 Krad	Rooted cuttings
Chrysanthemum sp.	Gamma-rays	1.5 to 2.5 krad	Rooted cuttings
Polianthus tuberosa	Gamma-rays	250 rad – 8 Krad	Bulb
Perennial portulaca	Gamma-rays	250 – 1250 rad	Stem cuttings

Source: Datta, 2009.

Genetic status of the plant material, its heterozygosity and ploidy level, is a key factor in determining the success of mutagenesis. Mutation breeding is more likely to be successful in diploid genotypes with a high degree of heterozygosity. Acute irradiation, in which a total dose is administered in a single exposure, often permits the detection and conservation of mutants within the M1-generation. However, chances for obtaining valuable mutants can be enhanced by application of split doses, which allows to increase mutation efficiency by ensuring sufficient regeneration at the same time (Schum, 2003).

6.4.0 Recurrent Irradiation Treatments

The recurrent irradiation *i.e.,* irradiating the plant material that has already been irradiated for one or more subsequent generations has proved an effective method in accumulating and expanding genetic variability in ornamental crops. The application of fractionated doses of mutagen and recurrent irradiation combined with *in vitro* culture techniques have; proved to be cost-effective and have wide potential in ornamental plants. Since various physical and chemical agents are known to cause different types of mutations; hence recurrent treatments using various chemical mutagens such as ethyl methanesulphonate (EMS) or hydroxylamine (HA) separately in a combination or alternatively associated with physical mutagen such as X-rays and also the combination of UV light and X-rays have proved advantageous in ornamentals.

6.5.0 Detection of Mutation

The final step in a mutation breeding program is the identification and selection of desired mutants. Visual selection is recognised as the most effective method of identification and detection of somatic mutations. It can be used effectively for identifying common traits and characters such as color changes, plant morphology, earliness, resistance to pests and diseases, *etc.* Somatic mutations in vegetatively propagated plants are mostly detected in M_1V_1. However, in some cases, mutations can even be detected in M_1V_2, M_1V_3 and later vegetative generations from normal-looking irradiated plants in M_1V_1, in which a mutated cell expresses its mutant character if it gets the chance. It has been observed that the chance of getting solid mutants is higher in M_1V_2 and later generations. The mutated cells of the lower auxiliary buds remain in the dormant stage and express their mutant character when included during vegetative propagation in M_1V_2 (Datta, 2001).

Selection usually starts from M_1V_2 and continues for confirmation in M_1V_3 or M_1V_4 generation. Since the probability of identifying desired mutants in a homogenous M_1V_2 population is quite low, selection is necessary in the M_1V_3 generation or in some cases in M_1V_4 generation. In the case of ornamental plants such as tulip, selection is continued for at least 4 years to allow mutated cells to grow out and to produce either a solidly mutated bulb or at least a larger visibly mutated sector that will be expressed in further generations.

6.6.0 Major Milestones of Mutation Breeding in Ornamental plants

In ornamentals, mutagenesis has proved to be highly effective in inducing novel mutants and these mutants have been successfully utilized as commercial cultivars (Tables 7, 8, 9). Induced mutations have been successfully utilized to alter the flower characters (colour, size, morphology, fragrance); leaf characters (form, size, pigmentation); growth habit (compact, climbing, branching); and physiological traits such as changes in photoperiodic response, early flowering, free-flowering, flower keeping quality, and tolerance to biotic and abiotic stresses (Schum and Preil, 1998).

6.6.1 Mutation for Altered Flower Morphology

Schum and Preil, (1998) have reported that 55 per cent of the recorded mutations in ornamentals have altered flower colour and only 15 per cent have altered flower morphology. In ornamentals, induced mutations have successfully altered the flower colour, size, inflorescence and flower shape and petal number. In, chrysanthemums, mutagenic treatment of pink flowering genotypes have induced wide spectrum of flower colours including bronze, brown, yellow, white, red and orange. Alteration in flower colour has been successfully obtained through mutagenesis in chrysanthemum, carnation, bougainvillea, hibiscus, cyclamen, rose, gladiolus and tulip (Table 11). In Hibiscus, mutants with altered flower colour and form have been successfully induced in 'Alipur Beauty' and 'Cruenthus' cultivars using semi-acute gamma-rays (Das *et al.*, 1977). Five petalled single flower type mutant has been induced and isolated from double flower type 'Alipur Beauty' treating stem cuttings with 4 Krad c-rays (Banerjee and Datta 1986). The Double flower mutants have been successfully induced in several ornamentals including *Begonia elatior*, Gladiolus and Hyacinthus. While, single flowered mutant has been obtained from a double-flowered in Hibiscus and Portulaca. In Roses, both an increase as well as reduction in petal number have been obtained through mutagenesis. Mutants with changed corollas (broad, dentate/dissected margin) were detected in *Petunia nyctaginiflora*, while, increased in peduncle length was obtained in Dahlia after mutagenesis.

The morphology of flowers and inflorescences have been successfully modified in several ornamental species through induction of mutation mutations. An increase as well as unfavourable reduction in flower size have been observed in large number of ornamentals. In many species, mutation induction has led to changes in petal shape, which sometimes can be ornamental value. In Compositae family, an increase in whorls of ligulate florets as well as a conversion from ligulate into tubular florets was obtained on mutagenesis. Some of promising mutants with altered flower morphology has been listed in Table 11.

Table 11: Traits Improved through Mutation Breeding in Ornamental Plants

Traits Improved	*Plants*
Floral Traits Improved	
Increased inflorescence size	*Streptocarpus, Crossandra, Begonia, Dahlia, Dendranthema, Dianthus, Kalanchoe*
Decreased inflorescence size	*Alstroemeria, Brachycome, Callistephus, Dendranthema, Eustoma, Ornithogalum, Pelargonium, Rhododendron, Rose,*
Altered petal shape	*Begonia, Brachycome, Dahlia, Dendranthema, Gerbera, Kalanchoe, Pelargonium, Petunia, Portulaca, Rhododendron, Rudbeckia, Streptocarpus, Tulipa*
Increase/decrease in number of petals	*Begonia, Brachycome, Dianthus, Gladiolus, Hibiscus, Hyacinthus, Portulaca, Rose*
Colour of bracts and spathes	*Anthurium, Bougainvillea, Euphorbia*
Leaf traits improved	
Leaf size	*Dendranthema, Anthurium, Begonia, Brachycome, Callistephus, Euphorbia, Eustoma, Forsythia, Gerbera, Kalanchoe, Lonicera, Petunia, Rosa*
Variegation	*Begonia, Bougainvillea, Canna, Coleus, Ficus, Hedera, Hibiscus, Lantana, Lonicera, Petunia, Polyanthes, Populus, Saintpaulia, Tulipa*
Pigmentation	*Acalypha, Begonia, Dahlia, Lilium, Lonicera, Passiflora, Pelargonium, Saintpaulia, Zinnia*
Number of leaflets	*Rosa, Zinnia*
Growth Characteristics Improved	
Compactness	*Begonia, Callistephus, Dendranthema, Euphorbia, Ficus, Kalanchoe, Lonicera,* Rose
Semi-dwarf or dwarf	*Anthurium, Begonia, Brachycome, Canna, Coleus, Dahlia, Dianthus, Euphorbia, Impatiens, Jasminum, Kalanchoe, Petunia, Rhododendron*
Branching habit	*Kalanchoe, Lantana, Zinnia*
Physiological Traits Improved	
Early flowering	*Alstroemeria, Brachycome, Euphorbia, Gerbera, Kalanchoe, Streptocarpus*
Extended flowering duration	Alstroemeria, Bougainvillea, Brachycome, Gladiolus
Altered response to daylength	*Begonia, Dendranthema*

Traits Improved	*Plants*
Tolerance to low temperature	*Dendranthema, Euphorbia, Iris, Thuja*
Tolerance to salt stress	*Dendranthema*
Tolerance to low light conditions	*Alstroemeria, Dendranthema, Kohleria, Lilium, Streptocarpus*
Increased resistance against pathogens	*Begonia, Dianthus, Ficus, Iris, Jasminum, Rosa, Tulipa*

Source: Scum (2013).

6.6.2 Mutation in Leaf Morphology

Mutation breeding has been widely used to alter the leaf shape and to induce leaf variegations in the commercial cultivars. Induced mutagenesis often causes alterations in leaf characteristics, improve the balance between leaves and inflorescences or alter the leaf. Leaf variegations provide additional beauty to plants at the time of blooming and even when there is no flower. Chlorophyll variegations have been successfully induced in *Abelia, Begonia,* Canna, *Ficus, Polianthes, Tulipa, Weigela,* Bougainvillea, Lantana and Hibiscus. Some of the promising mutants with chlorophyll variegation in bougainvillea include 'Arjuna', 'Pallavi', 'Mahara Variegata', and 'Los Banos Variegata'. Similarly, the prominent tuberose cultivars 'Swarna Rekha' and 'Rajat Rekha' are the mutant varieties exhibiting leaf variegation.

6.6.3 Alteration in Growth Habit

Development of compact plant types with good branching is an important breeding goal in flowering and foliage pot plants and in woody ornamentals. Compact and/or dwarf types of putative mutants have been successfully induced after mutagenic treatment in several ornamental including begonia, aster, coleus, canna, chrysanthemum, dahlia, kalanchoe, rose and *Euphorbia* (Table 11). Alteration in leaf size, leaf shape and form of leaf margins, and growth habit has been successfully obtained through induced mutagenesis techniques. Plants with slender leaves and elegant growth habit has been successfully induced in the spray-type Eustoma, following mutation treatment (Nagatomi *et al.*, 1996). In *Zinnia,* a trifoliate leaf mutant has obtained via induced mutagenesis (Venkatachalam and Jayabalan, 1992). In a few cases induced changes in branching habit have been observed. In Zinnia, a conversion to dichotomous branching was found in yellow and red flowered colour mutants (Venkatachalam and Jayabalan, 1991). In Kalanchoe, a compact branching mutant has been isolated and marketed as new cultivar 'Flores'.

6.6.4 Induce Mutagenesis for Biotic and Abiotic Stress Resistance

Induced mutagenesis has been considered as an innovative approach for the improvement of physiological traits including, earliness, perpetual and rich flowering habit, changes in photoperiodic response, tolerance against abiotic and biotic stresses. However, the genetic basis of most of these physiological traits is poorly understood. Most of these traits exhibit complex inheritance and their expression is governed by more than one gene. Hence, selection of physiological mutants requires more sophisticated methods and is by far more difficult as

Natural Mutants: Bud Sports in Important Ornamental Plants

Pusa Mansij
(Natural bud sport of variety
Raja Surendra Singh
of Nalagarh)

Pusa Abhishek
(Bud sport of
Jantar Mantar)

Chitra
(Bud sport of
var. Janina)

Pusa Sundari
Bud sport of 'Himanshu'

Pusa Sundari
'Bud sport of 'Himanshu'

Sonar Bangala

Mary Palmer
Bud sport of
B. peruviana var.
'Mrs. H.C. Buck'

Vishakha
Bud sport of
B. peruviana var.
'Mrs. H.C. Buck'

Cherry Blossom
Bud sport of
B.x buttiana 'Mrs. Butt'

Los Banos Beauty
Bud sport of
B. x buttiana 'Mrs. Butt'

Thimma
Bud Sport of
'Mary Palmer'

Roseville's Delight:
Bud sport of
B. x buttiana 'Mrs. Butt'

Spontaneous Mutation in Ornamental Plants

Ficus elastica 'Tricolour'

Ficus elastica 'Tineke'

Ficus elastica 'Shivereana'

Ficus elastica 'Belize'

Ficus benjamina 'Starlight'

Ficus benjamina 'Variegata'

Codiaeum variegatum 'Rainbow'

Codiaeum variegatum 'Golden Dust'

Codiaeum variegatum 'Mammy'

Aglaonema commutatum maculatum

Aglaonema commutatum 'Treubii'

Aglaonema crispum 'Silver Queen

Spontaneous Mutation in Ornamental Plants

Aglaonema crispum
'Deborah'

Begonia rex
'Merry Christmas'

Begonia rex
'Escargot'

Caladium hortulanum
'White Wing'

Caladium hortulanum
'Pink Beauty'

Caladium hortulanum
'Candidum'

Calathea insignis
'Lancifolia'

Dracaena deremensis
'Lemon Surprise'

Maranta arundinacea
'Variegata'

Cordyline terminalis
'Rainbow'

Cordyline terminalis
'Tricolor'

Cordyline fruticosa
'Bolero Tricolor'

Spontaneous Mutation in Ornamental Plants

Monstera deliciosa
'Compacta'

Monstera deliciosa
'Albo variegata'

Monstera deliciosa
'Marmorata'

Peperomia caperata
'Pink Lady'

Peperomia caperata
'Silver'

Peperomia caperata
'Rosso'

Pelargonium
'Madame Salleron'

Pelargonium hortorum
'Mrs. Pollock'

Spontaneous Mutation in Ornamental Plants

Amaryllis belladonna
'Pink'

Hippeastrum
'Santiago'

Hippeastrum
'Elvas Double'

Canna indica
'Coltivazione'

Canna indica
'Tropical Yellow'

Hymenocallis
'Sulphur Queen'

Colour Variations in ***Adenium obesum*** due to Mutation

Euphorbia milii
'Monaco'

Euphorbia milii
'Sunny Border'

Euphorbia milii
'Endless Love'

Spontaneous Mutation in Ornamental Plants

Gymnocalycium mihanovichii

Acalypha godseffiana
'Compacta'

Acalypha wilkesiana
'Tahiti'

Acalypha wilkesiana
'Copperleaf'

Acalypha wilkesiana
'Obovata'

Duranta repens
'Variegata'

Duranta repens
'Golden Edge'

Hydrangea macrophylla
'Greatful Red'

Hydrangea macrophylla
'Green Lips'

Hydrangea macrophylla
'Magical Blue Cloud'

Induced Mutants in Ornamental Plants

Pusa Anmol
(A Gamma ray
induced mutant of
'Ajay'

Pusa Centenary
(A Gamma ray
induced mutant of
'Thai Chen Queen')

Pusa Kesari
(A gamma ray
induced mutant of
Tai Chen Queen')

Shekhar
Induced Mutant of 'Su-Neel'

Abhisarika

Arka Flame

Los Banos Variegata
'Silver Margin'-
mutant of Bougainvillea
× buttiana 'Los Banos
Beauty'

Pallavi-A mutant
of *Bougainvillea ×
buttiana*
'Roseville's Delight'

Los Banos Variegata -
mutant of *Bougainvillea
× buttiana* 'Los Banos
Beauty

Promising Mutants in important Ornamental Plants

Double-Flowered Mutant: *Impatiens walleriana*

Dahlia Mutant var. Babylon

Geranium (Natural Mutant)

Double-Flowered Mutant: Rose (Var.Bridal Pink)

Double-Flowered lilac Mutant: French

Double-Flowered Mutant: Camellia

Double-Flowered Mutant: Carnations

Double-Flowered Mutant: Petunia

Double-Flowered Mutant: Peony

Mutant Sunflower var. Teddy Bear

Mutant Black Dahlia Var. Karma Choc

Peloric Foxglove (*Digitalis purpurea*)

Promising Mutants in important Ornamental Plants

African Violet (*Saintpaulia lonantha*)
Mutant Var. Fantasy

Mutant *Chrysanthemum morifolium*
cv. "DelistarWhite"

Spray Rose Mutant:
Julieta (Coral colored)

Mutant Flower: White Camelia

Mutant: Galaxy Petunia

Conjoined Mutation
Gerbera var.Siamese twin

compared to selection of morphological traits (Schum, 2003). The mutants for different physiological traits have been enlisted in (Table 11)

Jung-Heiliger and Horn (1980), reported the selection of first cold-tolerant variants in chrysanthemum following the irradiation of unrooted cuttings. Further, genetically stable winter-flowering mutants were successfully induced by gamma-irradiation of chrysanthemum cuttings (Hu *et al.*, 1989), the selected mutants have been reported to initiate flowering even below 9°C. Preil *et al.* (1983) exploited the in-vitro cell suspension culture and low temperature screening techniques for the induction of LTT mutant chrysanthemum. Induced mutagenesis has been successfully exploited to induce low temperature tolerance in chrysanthemum, poinsettia, *Saintpaulia*, *Fuchsia* and *Iris* (Schum, 2003). In Northern countries, during winter months, light intensity becomes a major limiting factor for greenhouse plant production. The low light intensities often suppress the plant growth, inhibit flower initiation, cause flower bud abortion, and thus leads to overall reduction in plant quality. The mutants with low light tolerance have been isolated in chrysanthemum, lilium, alstroemeria and *Streptocarpus*.

In carnation, two mutants namely, 'Maiella-lonchabi' and 'Galatee-lonvego' have been isolated through gamma-irradiations, they have shown high degree of resistance against *Fusarium oxysporum* f.sp. *dianthi* (Mitteau and Silvy, 1982). Another mutant 'Loncerda' have been proved to be 'very resistant' to *Fusarium* (Mitteau and Silvy, 1983). In *Begonia elatior*, resistance to mildew has been improved in 'Schwabenland Pink', a mutant (cv. Heirloom) following induced mutation by fast neutrons. In rose, the mutant cultivar 'Pink Hat', have shown complete resistance against powdery mildew (James, 1983). Similarly, disease-resistant mutants have been successfully obtained in Iris, tulips, *Ficus benjamina* exotica and *Jasminum grandifiorum* (Schum, 2003).

7.0 Polyploidy Breeding

Polyploidy has the utmost importance in ornamentals for the development of novel genotypes with desirable morphological traits such as plant size and vigour, leaf thickness, larger flowers with thicker petals, intense colour of leaves and flowers, long lasting flowers, compactness, dwarfness and restored fertility. Although most plant species are diploid, several species are natural polyploids with three or more sets of chromosomes. Polyploidization is widely recognized as a major mechanism of adaptation and speciation in plants and it has been estimated that approximately 2–4 per cent of speciation events in angiosperms are associated with polyploidy (De Storme and Mason, 2014). Polyploidy has been reported to lead changes in gene dosage that causes chromosomal rearrangements, epigenetic remodeling and reunion of divergent gene regulatory hierarchies. These changes may bring out novel variations or may result in speciation and reproductive isolation. Polyploidy is often associated with enhanced adaptive skills, providing significant benefits for polyploids in harsh, selective environments and allowing them to easily colonize new environments (Eeckhaut *et al.*, 2018). Polyploids can be classified as autopolyploids and allopolyploids according to their origin. Autopolyploids originate from the doubling of the chromosome number of a diploid species, which results in two (or more) pairs of homologous chromosomes. While, allopolyploids are obtained after hybridization between diverse species and subsequent polyploidization, it thus combines two distinct genomes (Thomas, 1993).

Induced polyploidization has wide potential in improvement of ornamentals through creating the new genetic resources (Kazi, 2012). Any change in the basic chromosome (x) number of the plant genome often results in altered plant morphology and cytology in plant species. The morphological effects of induced polyploidization includes, increased leaf thickness, increased leaf width/length ratio, dark green leaves, sturdy and stronger stems, compact growth habit, larger flowers, more number of petals, bright coloured flowers, and delay of flowering time (Dhooghe *et al.*, 2011). In general, polyploid plants, with their larger flowers and bigger leaves, are more robust and more vigorous than their corresponding diploids. The induction of polyploidy may also be employed to overcome sterility, as frequently found in the F1-progeny of interspecific crosses, or to generate genotypes with equal ploidy level when interploidy crosses are problematic.

7.1.0 Induction of Polyploidy

The polyploidy can occur in nature (Spontaneous) and can be induced (Artificial). Spontaneous chromosome doubling is due to endomitosis and endoreduplication. In nature, polyploids occur spontaneously through the formation of unreduced (2n) gametes. These spontaneous polyploids can also be exploited as seed or pollen parents in a breeding program for the evolution of triploid and/or tetraploids. Naturally produced sexual polyploids, particularly triploids produced through formation of 2n gametes, have been studied as important features in the evolution of flowering plants. Polyploids can also be induced artificially through the heat and cold treatments and use of mitosis inhibitors, or other chemicals.

Three mechanisms are generally used for induction of polyploids namely, mitotic polyploidization, which is based on the doubling of somatic tissues, meiotic polyploidization, which generates 2n gametes (Ramsey and Schemske, 1998). Since 2n gametes can be used immediately in crossing experiments, meiotic polyploidization can reduce the breeding process by one generation. Additionally, 2n gametes enhance the genetic variation in the progeny. The third mechanism includes haploidization *i.e.*, induction of haploids through regeneration of either male or female gamete or distant hybridization techniques.

7.1.1 Mitotic Polyploidization

Most of the polyploidization protocols are still based on mitotic polyploidization, which involves the treatment of somatic cells with mitosis suppressing chemicals that interfere with tubulin polymerization required for the formation of the spindle fibres, thus prevent the segregation of chromatids during cell division. The chromosome duplication in somatic cells form additional copies of chromosome and existing genes which leads to variations in plant phenotypes. This type of chromosome duplication is widely used to induce tetraploids through anti-mitotic chemicals.

7.2.0 Antimitotic Agents

The different antimitotic agents that affect the plant's cell cycle are colchicine, colcemid, oryzalin, trifluralin, benfluralin, ethalfluralin, pendimethalin, butralin, dinitramin, phosphoroamidates (amiprophos-methyl, butamiphos), pyridines (dithiopyr, thiazopyr), benzamides (propyzamide), benzoic acid (chlorthaldimethyl) and carbamates (chlorpropham, propham, carbetamide). Colchicine is a widely used anti mitotic agent for the induction of polyploidy in plants. It prevents the spindle fiber formation thus inhibits chromosome segregation during cell division. It is an alkaloid extracted from bulbs and seeds of the wild meadow saffron (*Colchicum autumnale*) (Eigsti and Dustin, 1955) that was used earlier for the treatment of various inflammatory diseases. Since the 1930s, it has been widely used to increase plant ploidy levels in agricultural cops (Blakeslee and Avery, 1937). As microtubules function in chromosome segregation, colchicine induces polyploidy by preventing the segregation of chromosomes during meiosis that results into half of the gametes (sex cells) containing double the chromosome number than usual. The major advantage of colchicine is that it does not reduce its polyploidizing capacities even after being autoclaved (Zhang *et al.*, 2007). However, in many plant species, it causes side effects such as sterility, abnormal growth, chromosome losses or rearrangements and gene mutation (Luckett, 1989). Colchicine binds only poorly to plant tubulins; thus, it must be used in relatively high concentrations. Two novel chemicals with significantly increased affinity for plant tubulin have also been developed *i.e.*, 3,5-dichloro-N-benzamide and trans-4-nitro-5-cyclohexene (Young *et al.*, 2001; Young and Lewandowski, 2000).

7.3.0 Type of Explants

Different plant organs/tissues such as roots, seeds, flowers, axillary and apical meristems can be exploited for polyploidy induction. It is effective to induce polyploidy in those organs which have highly morphogenetic responses such as embryo, apical meristems and young inflorescence. The efficiency of polyploidization depends on the type of explant used (Kermani *et al.*, 2003, Petersen *et al.*, 2003), however, within an explant type, success of ployploidization depends on the permeability of the tissue and its capability to transport the antimitotic agent to the meristem (Allum *et al.*, 2007). The success rate is also genotype dependent (Chauvin *et al.*, 2005, Khosravi *et al.*, 2008). Each crop species has different responses towards polyploidization, which depends on their genomic structure, reproduction patterns, ancestral ploidy level and the purpose for which the crop has been grown.

Recently, the *in vitro* induction of polyploidization is gaining popularity in ornamentals as tissue culture protocols have already been standardized in large number of ornamentals. Various types of explants including, shoot tips, shoots, plantlets, buds or callus, somatic or zygotic embryos, seeds, seedlings, nodal and tuber/bulb segments can be used for *in vitro* polyploidization (Tables 12 and 13). This technique allows for uniform treatment of the explants with anti-mitotic agent and screening of large population under controlled and standardized environment.

7.4.0 Method of Application

The methods of application of anti-mitotic chemicals depend upon the plant type. One of the simplest, easiest and effective methods is to use a large number of seedlings that have small and actively growing meristematic tissues. Seedlings can be dipped/soaked or apical meristems can be submerged in anti-mitotic agent solution of different concentrations at different exposure times or frequencies. Shoots of older plants can also be used, however this method may lead to the formation of large number of cytochimeras. Alternatively, treatment of sub-axillary and small axillary meristematic tissues is considered to be effective. The anti-mitotic agents are generally applied on the growing buds with the help of cotton, lanolin and agar or by dipping branch tips in solutions of chemicals. Addition of wetting agents and surfactants can enhance the penetration of a chemical. The most successful method for tetraploid induction is through seed treatment. Colchicine treatment of pre-germinated seeds having emerging roots is more effective as compared to dry or germinated seeds because large numbers of tetraploid plants are produced through it.

Sometimes, a pre-incubation is given under *in vitro* condition before the antimitotic agent is applied, to synchronize the mitotic divisions and thus ameliorate the effect of the antimitotic agent (Chauvin *et al.*, 2005). Besides, the treatment of buds or shoots with mitotic inhibitors in the tissue culture medium (liquid or solid) during one propagation cycle, polyploidization is often induced during a short application in liquid medium (on a shaker) followed by culture of new polyploid shoots on a regeneration medium. The *in vitro* chromosome doubling protocols also varies for different antimitotic agents, type of explant, exposure time, concentration, method of application and confirmation technique (Dhooghe *et al.*, 2011).

7.5.0 Doses and Exposure Time

The success rate for induction of polyploidy may depend upon the concentration and duration of exposure. Sometimes very low doses are not useful, while excessively high doses are lethal. Too high concentrations or longer exposure periods sometimes result in redoubling, which leads to induction of higher ploidy levels than required (Allum *et al.*, 2007). The solvent used to dissolve antimitotic agent is also important as it contributes to the effect and toxicity of the treatment. DMSO (dimethyl sulfoxide) has been widely used in many studies to dissolve antimitotic agent (Allum *et al.*, 2007). DMSO increases cell permeability and allows increased absorption of chemicals, colchicine dissolved in 2 or 4 per cent DMSO decreased the survival rate compared to colchicine dissolved in water and also resulted in increased number of tetraploids (Hamill *et al.*, 1992). The doses of mutants, exposure time and methods of the application of mutants for different ornamentals is given in the Tables 12 and 13.

7.6.0 Determination of Ploidy Levels

For crop breeding, it is important to identify the correct ploidy level at various growth stages through quick, simple and easy methods. The Flow cytometry and chromosome counting, with an observation of morphological and physiological traits are routine procedures for the determination of ploidy levels.

Table 12: Explant Types, Method of Application and Doses and Exposure Time for *In vitro* Polyploidization in different Ornamentals

Plants	*Explant and Antimitotic Agent*	*Concentration and Exposure Time*	*Reference*
Dieffenbachia	Shoot clumps, liquid medium with colchicine	Colchicine 1.25 mM, 24 h	Henny *et al.* (2009)
Lagerstroemia indica	Nodal buds, medium with colchicine	Colchicine 750 mM, 10 d	Zhang *et al.* (2010)
Rhododendron	Seedlings, seeds or shoots, solid medium with oryzalin, trifluralin or colchicine	Trifluralin 0.3 mM, 3 d (seedlings)	Eeckhaut *et al.* (2002)
Rosa	Shoot tips and thin nodal sections, liquid or semisolid medium with oryzalin	Oryzalin 5 mM, 1 d (thin nodal sections)	Kermani *et al.* (2003)
Rosa hybrida and *Rosa persica*	Nodal sections, liquid medium with trifluralin, oryzalin or APM	Trifluralin 24 mM, 48 h APM 24 mM, 48 h	Khosravi *et al.* (2008)
Rosa rugosa	2 or 10 mm nodes, semi-solid medium with oryzalin	Oryzalin 2.5 mM, 48 h (2 mm nodes)	Allum *et al.* (2007)
Syringa	Nodal sections, liquid medium with colchicine	Colchicine 0.05 to 0.25 mM, 1 or 2 days	Rose *et al.* (2000)
Alstroemeria	Plantlets, colchicine solution	Colchicine 5 to 15 mM, 6 to 24 h	Lu and Bridgen (1997)
Cattleya	PLB, liquid medium with colchicine Clone dependent	Colchicine 1.25 mM, 8 days	De Mello e Silva *et al.* (2000)
Cyclamen	Tuber segments, solid medium with colchicine	Colchicine 0.25 mM, 4 days	Takamura and Miyajima (1996)
Dianthus	Nodal segments, liquid medium with colchicine or APM	APM 32.9 mM, 24 h	Nimura *et al.* (2006)
Gladiolus spp.	Callus, treatment with colchicine	Colchicine 1.25 mM, 72 h	Suzuki *et al.* (2005)
Lilium longiflorum	Scales, solution with oryzalin or Surflan	Surflan (0.1 mM oryzalin), 3 h	Takamura *et al.* (2002)
Phlox subulata	Shoot tips, solid medium with colchicine	Colchicine 125 mM, 20 days	Zhang *et al.* (2008)
Ranunculus asiaticus	Plantlets, medium with oryzalin or trifluralin or colchicine	Oryzalin 0.5 mM, 10 week, Trifluralin 2.0 mM, 10 weeks	Dhooghe *et al.* (2009)
Spathiphyllum wallisii	Secondary somatic embryos, medium with colchicine, oryzalin or trifluralin	Trifluralin 10 mM, 6 weeks	Eeckhaut *et al.* (2004)
Tulipa gesneriana	Flower stem discs, liquid medium with oryzalin, Cultivar dependent	Oryzalin 2.88 mM, 2 weeks	Chauvin *et al.* (2005)
Zantedeschia	Shoot cultures, medium with colchicine	Colchicine 1.25 mM, 1, 2 or 4 days	Cohen and Yao (1996)

Table 13: Enhancement of Quality Traits in Ornamental Plants through Colchicine Treatment

Crop	*Method of Application and Plant Part Used*	*Colchicine Concentration and Duration*
Orchid (*Dendrobium nobile*)	Immersing whole plant in solution	0.1 per cent, 96 h
Phlox (*Phlox drummondi*)	Seed soaking method	0.5 per cent, 36 h
Marigold (*Tagetes erecta*)	Pre-germinated seeds	0.1 per cent, 3-6 h: 0.2 per cent, 3 h
Lily (Lilium)	Treatment of bulb scales with colchicine solution	1 per cent, 24 h
Calendula (*Calendula officinalis*)	Soaking of seeds in colchicine solution	0.08 per cent, 4 h
Orchid (*Dendrobium nobile*)	Plant of height 5cm was dipped in colchicine solution	0.1 per cent, 96 h
Rose (*Rosa multiflora* var *inermis, Rosa roxburgii* var *normalis*)	Germinated seeds soaked in solution Seedlings at cotyledon stage	0.2 per cent, 12 h
White orchid tree (*Bauhinia acuminata*)	Treatment of seeds	0.1 per cent, 12 h (3 consecutive days
Gladiolus (*Gladiolus grandiflora*)	Corm soaked in solution	0.1-0.3 per cent, 24 h
Garden balsam (*Impatiens balsamina*)	Colchicine application to seedlings using cotton wool	0.4 per cent, 3 days
Petunia (*Petunia hybrida*)	Shoot apices of seedlings treated with colchicine using cotton wool	0.2 per cent, 48 h
African marigold (*Tagates erecta*)	Whole plant with roots kept in a colchicine solution and treatment of apical buds with colchicine-soaked cotton	0.0005 per cent, 6 h
Chrysanthemum (*Dendranthema indicum var. aromaticum*)	Grin seeds soaked in solution, Pre-germinated seeds dipped in colchicine aqueous solution, Cotton balls placed on shoot tips; colchicine poured on cotton balls through micropipette	0.1 per cent, 24 h (grin seeds); 0.1 per cent, 7 d (shoot tips)
Chrysanthemum (*Chrysanthemum carinatum*)	Apical buds treated with colchicine-soaked cotton swab	0.2 per cent, 3 days with 6 h duration per day
Moth orchid (*Phalaenopsis pulcherrima*)	Treatment of seeds with colchicine through dropping method	Colchicine, 0.5 per cent
Madagascar periwinkle (*Catharanthus roseus*)	Seedling shoot apices treatment	0.4 per cent, 7 d
Bougainvillea (*Bougainvillea* spp cvs Lalbagh, Mahara)	Colchicine-soaked cotton placed on dormant nodal buds	0.4 per cent for 72 h (cv. Lalbagh); 0.3 per cent for 72 h (cv. Mahara)

Source: Manzoor *et al.* (2019)

7.7.0 Meiotic Polyploidization

Meiotic polyploidization is the polyploidization of plants through unreduced (2n) gametes in which the chromosome number is equal to the chromosome number of the plant that produces them. The most common chances of such anomalies are omission of meiosis I or II, abnormal spindle orientation, or defective synapsis. Gametes can be of two types on the basis of their chromosomal makeup, *i.e.*, first and second division restitution. In first division restitution (FDR) gametes, non-sister chromatids and centromeres are co-localized, making the gametes relatively heterozygous. Whereas, in Second division restitution (SDR), meiosis II is omitted along with sister chromatids together that resulted reasonably homozygous gametes. Gamete formation after abnormal spindle formation in meiosis II (such as parallel, tripolar, or fused spindles which result in dyads or triads) can also occur. This is also classified as FDR (Bretagnolle and Thompson, 1995) although the cytological anomaly occurs during meiosis II and is different from the anomaly that generates "classical" FDR gametes. The parental heterozygosity at the centromeres is however retained (Zamariola *et al.*, 2014). It has also been reported that environmental factors can also induce 2n gametes.

The most potential approach for meiotic polyploidization is to use the mitosis inhibitors, *viz.* colchicine, oryzalin, trifluralin, or related compounds that interfere with the formation of the spindle fibre. The most ideal stage to induce 2n gametes is the pachytene stage (Younis *et al.*, 2014; Zhao *et al.*, 2017). However, mitosis inhibitors like colchicine also cause toxic effects that are very common in bulbous plants (Van Tuyl *et al.*, 1992). Treatment of plant tissues with N_2O at high pressure (± 6 bar) is an effective alternative tool for 2n gamete induction. Nitrous oxide gas has a similar mode of action, but because of the high pressure, the gas can penetrate tissues far more efficiently. The ideal stage for treatment is metaphase I (Dewitte *et al.*, 2010). Alternative chemical 'caffeine' has been reported in Lilium to induce 2n gametes. The exact mode of action is unclear, although three possible mechanisms have been postulated, including prevention of homologous recombination, cytokinesis inhibition, and post meiotic restitution (Younis *et al.*, 2014). Heat or cold stress can significantly increase unreduced gamete formation (Zhou *et al.*, 2015). Interference of exogenously applied gibberellic acid with unreduced pollen formation has also been demonstrated (Liu *et al.*, 2017). Jaap Van Tuyl's research group has created hybrids between different lilium hybrid groups (Longiflorum, Asiatic, Oriental, and Trumpet) which were then used as a source for 2n gametes, leading to triploid interspecific hybrids. The obtained cultivars cover more than 45 per cent of the lily production acreage. Other modern polyploids such as *Narcissus*, *Tulipa*, and *Begonia* cultivars find their origin at least partly in unreduced gametes (Marasek-Ciolakowska *et al.*, 2016).

7.8.0 Role of Polyploidy in Ornamental Plants

Polyploidy has been exploited in the evolution of species and cultivar development in various ornamental crops like tulip, dahlia, anthurium, bougainvillea, lily, cacti, primula, dahlia, narcissus, rose *etc.* Induced tetraploids are reported in marigolds, petunia, snapdragon, portulaca, chrysanthemum, calendula and lily.

7.8.1 Improvement in Morphological Traits

One of the immediate effects of polyploidization is an increase in cell size due to increase in the nuclear content, which causes reduction in the cell division during their growth and development. Induced polyploidization have resulted in an increase in number of leaves, number of branches, plant height and stem length in large number of ornamentals including Salvia, Lilium, calendula, chrysanthemum and orchids. Colchi-tetraploids of chrysanthemum had larger flowers with thicker petals helping them to improve their vase life. In gladiolus, novel variation in flower morphology like serrated margins with ruffled edges along with pointed outgrowth has been observed in flower petals. The induced polyploids also produced larger flowers with more vase life. While, polyploidization in African marigold induced early flowering, higher number of flowers with increased diameter and weight of flowers.

The variation in colour and shape of leaves has been observed in tetraploid plants of pelargonium (*Pelargonium graveolens*). Induced polyploidy resulted in altered flower colour in African violets, increased number of petals in garden balsam and increased flower diameter in *Limonium* and dense flowering with erect, compacted and short inflorescence in gymnostachyum. Similarly, in vegetatively propagated crops like Lilium, induced chromosome doubling produced wider bulb scales.

7.8.2 Improvement in Physiological Traits

Polyploidy can also have significant impact on various plant physiological processes such as water relations. Large-sized stomata with lower density per unit area have been observed in polyploid plants of African marigold, chrysanthemum salvia, phlox, petunia and celosia. The lower stomatal density has been associated with reduction in the transpiration rate (overall gaseous exchange rate) of polyploid plants. Further, the increased vacuole size and thicker leaves of polyploids increases the water content retention capacity, which can be further utilized during drought period conditions. Therefore, these polyploid plants could be grown in water scarcity conditions and can be utilized as valuable germplasm for drought stress breeding.

7.8.3 Resorting Fertility in Wide Hybrids

Distant hybrids are usually sterile due to the failure of chromosome pairing during meiosis. Mitotic polyploidization has been exploited to restore fertility in interspecific hybrids, where synopsis between different chromosome sets is hampered and prevents regular meiosis. This technique has been used to restore fertility in lavandin (*Lavandula × intermedia*), which is grown for both oil and ornamental purpose. Colchicine (Asano, 1982) and oryzalin have been utilized in sterile F_1 lily hybrids to restore fertility (Van Tuyl, 1990)

7.8.4 Overcoming Hybridization Barrier

In distant hybridization, desirable crosses are sometimes difficult to obtain due to different ploidy level of parents. In order to break these barriers in hybridization, ploidy levels of one or both parents are manipulated to match before the hybridization. Polyploidization has been frequently attempted to equalize ploidy levels between crossing partners or to induce sterile triploids. For example, *Hibiscus syriacus* 'DVPAzurri' Azurri®, obtained after a cross between *H syriacus* 'Oiseau Bleu' and it's polyploidized counterpart, combines better growth vigour with longer flowering period and sterility. Similarly, in Chrysanthemum, Scented chrysanthemum was developed by crossing *D. indicum* var. *Aromaticum* (2n=2x=18) and *D. grandiflora* (2n=6x=54), chromosome doubling has been induced in *D. indicum* var. *Aromaticum*, in order to increase the chance of cross between both the parents.

7.8.5 Improvement in Biotic and Abiotic Stress Tolerance

Polyploid plants could have a better nutrient uptake, cold tolerance and improved resistance to insect/pest and pathogens. Improved tolerance of polyploids to various biotic and abiotic stresses can be attributed to increase in the nuclear content, that ultimately increases the gene expression causing increase in secondary metabolites production. Induced triploids and tetraploids plants of *Hibiscus moscheutos* showed resistance to aerial *Phytophthora* disease, while the induced tetraploids of garden impatiens exhibited improved resistance to downy mildew as compared to their diploid counterparts. In ornamental plant breeding, increased abiotic stress tolerance has been associated with *in vitro* induced polyploidy, *e.g.*, against drought, cold, and salinity in *Dendranthema* (Liu *et al.*, 2011) and against drought in Spathiphyllum (Eeckhaut *et al.*, 2004; Van Laere *et al.*, 2011). In roses, drought stress tolerances of 'Yesterday' x *Rosa wichuriana* diploids and their tetraploidized counterparts were compared. Some polyploidized hybrids had an improved tolerance to drought stress, whereas others were less tolerant and in some cases the drought tolerance was not affected by the ploidy level.

7.9.0 Limitation of Induced Polyploidization

Though induced polyploidization is considered as a valuable tool for improvement of ornamentals, there are still several limitations associated with this method. It is expected that the doubling of the cell genome will double the nucleus volume which can severely disrupt the balance between the chromosome and nuclear components. This imbalance can cause various abnormalities during mitosis and meiosis. It can also affect cell physiology (metabolism), regulation of gene expression and gene stability (Manzoor *et al.*, 2019).

Further, high levels of polyploidy often result in stunted and malformed phenotypes due to somatic instability and extreme gene redundancy, which leads to production of chimeric tissues. Polyploidization has been associated with several negative effects including floral abnormalities, delayed flowering and sterility. Induced polyploidy has led to scabbing and thickening of leaf tissue, leaf cupping, deformities, stunting and death of growing points in Zinnia. In Pelargonium, hexaploids and octoploids exhibited failure in blooming, and mixploid showed colour loss of leaves. Induced tetraploid plants did not produce any pollens in Hibiscus, whereas in triploid plants, non-viable pollen grains were produced but the fruit were aborted after pollination, thus leading to infertility of triploids (Manzoor *et al.*, 2019).

8.0 Haploids and Doubled Haploids

Haploids and doubled haploids are valuable tools for speeding up the breeding programs, improving selection efficiency, detecting linkage and gene interactions, QTL mapping, substitutions and chromosome addition lines (Basu *et al.*, 2011). The plants with gametophytic chromosome number in their sporophyte (whether diploid or polyploid) are referred as haploids (Palmer and Keller, 2005). The Haploid plants induced from a true diploid (2n=2x) are termed as monoploids as they contain only one set of chromosomes. In contrast, haploids from polyploid species have more than one set of chromosomes and are polyhaploids; for example, dihaploids (2n=2x) from tetraploid French marigold (*Tagetes patula L*, 2n=4x), trihaploids (2n=3x) from allohexaploid (*Chrysanthemum* × *morifolium*, 2n=6x) *etc.* The haploids contain only one set of chromosomes in the sporophytic phase (2n=x); hence they are much weaker and less vigorous than the normal plants. They are often sterile due to the inability of their chromosomes to pair during meiosis and difficult to maintain. In order to propagate them through seed and to include them in breeding programs, their fertility has to be restored (Dunwell, 2010). In order to produce Double Haploids (DHs), the chromosome number of haploids are doubled either spontaneously or through induced chromosome doubling agents. The obtained DHs are fully fertile, homozygous at all loci and can represent a new variety in self-pollinated crops or parental inbred line for the production of hybrids in cross-pollinated crops (Murovec and Bohanec, 2012). Plant breeders and geneticists had long known the benefits of pure lines, and it became apparent that haploid plants would be beneficial in generating these pure lines, as complete homozygosity can be accomplished by doubling the chromosomes of a haploid plant (Ferrie *et al.*, 2011).

Haploid cells occur naturally in the gametophytic phases of higher plants in their ovules and pollen. Thus, the haploids can be induced from pollen or egg cells. By manipulating the gametic cells, it is possible to produce embryos rather than mature pollen grains or ovules. The major advantage of DH technology lies in instantaneous development of homozygous lines instead of 6–10 generations of inbreeding by selfing or sib-crossing (Prigge *et al.*, 2012), which is a major breakthrough to speed up cultivar development (Dunwell, 2010). The first report of the haploid plant was published by Blakeslee *et al.* (1922) in *Datura stramonium*. Subsequently, haploids were reported in many other species. Naturally occurring haploids have been observed in over 70 species (Ferrie, *et al.*, 2011). Guha and Maheswari (1964) developed anther culture technique for the production of haploids in the laboratory. Haploid production by wide crossing was reported in barley (Kasha and Kao, 1970) and tobacco (Burk *et al.*, 1979). At present, doubled haploid methodologies have been applied to over 250 species (Maluszynski *et al.*, 2003). An extensive review of haploidization in ornamental species have enlisted 46 ornamental genera in which haploidization had been accomplished through one or multiple techniques, and ovule/ovary culture was reported in only six genera. Some of the naturally occurring haploids has been directly used as cultivars including the horticultural variety of Pelargonium 'Kleine Liebling' (Daker, 1966).

8.1.0 Production of Haploids

Haploids can be produced *in vivo* or *in vitro*. Haploid embryos are produced *in vivo* by parthenogenesis, pseudogamy, or chromosome elimination after wide crossing. The haploid embryo is rescued, cultured, and chromosome-doubling produces doubled haploids. The *in vitro* methods include androgenesis (anther and microspore culture) and gynogenesis (unfertilized ovary or ovule culture).

8.2.0 Induction of Maternal Haploids

Inter-specific and inter-generic pollinations exploit haploidy from the female gametic line. *In situ* induction of maternal haploids can be initiated by pollination with pollen of the same species (*e.g.*, maize), pollination with irradiated pollen (Petunia, Rose, Iris and Carnation), pollination with pollen of a wild relative (*e.g.*, Petunia) or unrelated species (*e.g.*, chrysanthemum). Pollination can be followed by fertilization of the egg cell and development of a hybrid embryo, in which paternal chromosome elimination occurs in early embryogenesis or fertilization of the egg cell does not occur, and the development of the haploid embryo is triggered by pollination of polar nuclei and the development of endosperm.

8.2.1 Wide Hybridization

Wide crossing between species has proved to be a very effective method for haploid induction and has been used successfully in several cultivated species. It exploits haploidy from the female gametic line and involves both inter-specific and inter-generic pollinations. The fertilization of polar nuclei and production of functional endosperm can trigger the parthenogenetic development of haploid embryos, which mature normally and are propagated through seeds (*e.g.*, potato). In other cases, fertilization of ovules is followed by paternal chromosome elimination in hybrid embryos. The endosperms are absent or poorly developed, so embryo rescue and further *in vitro* culture of embryos are needed (*e.g.*, barley). Wang *et al.* (2014) used this approach to induce haploids in chrysanthemum cultivar "Zhongshanzigui" (2n = 6x = 54) by utilizing diploid *Argyranthemum frutescens* (2n = 2x = 18) as a pollen parent.

8.2.2 Induced Parthenogenesis through Irradiated Pollen Technique

Pollination with irradiated pollen is another possibility for inducing the formation of maternal haploids using intra-specific pollination. Embryo development is stimulated by pollen germination on the stigma and growth of the pollen tube within the style, although irradiated pollen is unable to fertilize the egg cell. Pollen irradiation represents a valuable method for inducing haploid plants especially when combined with embryo rescue. The induction of haploids through this technique is influenced by different factors such as the irradiation dose, the developmental stage of the embryo during *in vitro* culture, the culture media composition and the culture conditions (Germana, 2011). Ionizing and non-ionizing radiation has been used for pollen irradiation to induce *in situ* haploid plants. Radiation destroys the generative function of pollen without affecting its capacity in egg cell stimulation, thus allowing development of parthenogenic embryo. Irradiation dose is the key factor controlling *in situ* haploid production. If irradiation dose is low, the generative nucleus in pollen maintains its ability to fertilize the egg cell. Therefore, obtained embryos will be hybrid and mutant phenotype. An increase in the dose of irradiation will causes a decline in the number of embryos developed but these embryos will be frequently of haploid origin. For most plant species, *in vitro* embryo rescue is necessary to recover haploid plants. This approach has been successfully utilised for induction of maternal haploids in petunia (Raquin *et al.* (1989), rose (Meynet *et al.*, 1994; Lim *et al.*, 2005), carnation (Sato *et al.*, 2000), sunflower (Todorova *et al.*, 1997) and iris (Grouh *et al.*, 2015).

Using the irradiated pollen technique, Raquin *et al.* (1989) have successfully induced the haploids in *Petunia hybrida* or *Petunia parodii* by *in vitro* culture of ovaries. The gamma radiation doses ranging from 200-1000 Gy lead to the successful development of haploids in petunia. The dihaploid plants of roses (*Rosa* x *hybrida*, cv 'Sonia') were developed through parthenogenesis by using irradiated pollen and *in vitro* culture of immature seeds (Meyneta *et al.*, 1994). They reported that a 500-Gy minimum dose was sufficient to inactivate pollen and induce *in situ* parthenogenesis. DHs were obtained from the sunflower hybrids 'Albena' and 'Viki', through the pollination of irradiated pollen and *in vitro* culture of immature embryos. Todorova *et al.* (1994); Todorova *et al.* (1997) found that the efficiency of this method was affected by the interaction of pollen source genotype, γ radiation strength and the genotype of the initial forms for parthenogenic induction.

Similarly, Sato *et al.* (2000) utilised the pseudo-fertilized ovule culture technique for induction of DHs in carnation. They used the X ray (100 kR or 200 kR) irradiated inactivated pollen to pollinate the emasculated flower buds and subsequently recovered the haploids following *in vitro* culture of pseudo-fertilized ovaries 2 to 3 weeks after pollination. Among several methods of haploid induction, pollination with gamma-irradiated pollen proved to be the efficient in Iris. Grouh *et al.* (2015) successfully induced the haploid lines in *Iris pseudacorus* via *in situ* parthenogenesis followed by embryo rescue. They reported irradiation of pollen with 300 or 400 Gy of X-ray as the best dose for haploid production in *I. pseudacorus*.

8.2.3 *In vitro* Induction of Maternal Haploids-Gynogenesis

The culture of unfertilized ovaries or ovules for *in vitro* induction of maternal haploids from egg cell or other haploid cells of the embryo sac is termed as gynogenesis. It can be achieved with the *in vitro* culture of various un-pollinated flower parts, such as ovules, placenta attached ovules, ovaries or whole flower buds. Generally, culture of whole flower, ovary, and ovules attached to placenta respond better, but in gerbera and sunflower isolated ovules show better response The haploid plants generally originate from egg cells in most species (*in vitro* parthenogenesis), but in some cases they may arise from synergids; and even from antipodal cells. Although gynogenetic regenerants show higher genetic stability and a lower rate of albino plants compared to androgenetic ones, it is used mainly in those plant species for which androgenesis is not effective to produce haploids (Murovec and Bohanec, 2012). Further, *in vitro* induction of the maternal haploids through unfertilized ovule culture technique (gynogenesis) is the only feasible method for the production of completely homozygous lines from fully double, petalous type of male sterile hybrids/lines (Ravindra *et al.,* 2020). The female gametophyte is usually immature at inoculation and, in contrast to androgenesis, its development continues during *in vitro* culture, leading to a mature embryo sac. Mature embryo sacs contain several haploid cells theoretically capable of forming haploid embryos, such as the egg cell, synergids, antipodal cells and non-fused polar nuclei. However, under optimal conditions, the egg cells in most gynogenetic responsive species undergoes sporophytic development (haploid parthenogenesis; Bohanec, 2009). They can develop into haploid plants directly, avoiding the risk of gametoclonal variation, or through an intermediate callus phase. Gynogenic regeneration has been obtained though *in vitro* culture of unfertilized ovules, ovaries and whole flowers in several economically important ornamentals like Gerbera (Li *et al.,* 2020; Miyoshi *et al.,* 1996; Tosca *et al.,* 1999), Marigold (Ravindra *et al.,* 2020), chrysanthemum (Watanabe, 1977; Wang *et al.,* 2014; Miler and Muszczyk, 2015) and *Spathiphyllum wallisii* (Eeckhaut *et al.,* 2001). Watanabe (1977) used pseudo-fertilized ovary culture in chrysanthemum and succeeded in the production of haploids and F_1 hybrids in Japanese *Chrysanthemum* species. whereas, Wang *et al.* (2014) developed the *in vitro* haploids and doubled haploid in *chrysanthemum morifolium* plants via unfertilized ovule culture.

Many factors are responsible for production of successful doubled haploid plants *viz.,* genotype, physiological condition of donor plant, developmental stage of the male and female gametophytes, pre-treatment, composition of the culture medium, physical factors like light and temperature during tissue culture (Afza *et al.,* 2000; Zhang *et al.,* 2009 and Qi *et al.,* 2011).The plant genotype plays a predominant role in determining the gynogenic response and even the clones derived from the same variety may exhibit varied response. Multiple studies have demonstrated that genotype is a fundamental factor influencing the successful induction of haploid through *in vitro* gynogenesis (Ravindra *et al.,* 2020; Li *et al.,* 2020)

Donor plant conditions also influence the embryogenic response. Temperature, photoperiod, light quality, fertilizer regimes, time of year, and where the plants are grown (*i.e.,* environmental growth unit, greenhouse, or field) can influence the number of haploid embryos in culture. Temperature is very crucial factor because sudden change in the temperature regime mainly after floral differentiation could greatly diminish the embryogenic response, as stress alters the level of growth regulators, amino acids, carbohydrate or lipid in the plant. For better gynogenic response, it is recommended to grow the donor plants in a controlled environment where the stress can be minimized. In case of marigold, for obtaining effective gynogenic response, the day temperature is maintained between 20 to 25 °C and the night temperature between 15 to 18°C during the active plant growth and flowering periods (Ravindra *et al.,* 2020).

In addition to genotype, other factors that critically affect unpollinated ovule culture include season and temperature. The seasonal effects on ovule culture have been observed in gerbera and marigold (Cappadocia *et al.,* 1988; Tosca *et al.* (1999); Li *et al.,*2020; Ravindra *et al.,* 2020). Cappadocia *et al.* (1988) observed seasonal effect on *G. jamesonii* callus induction via unfertilized ovules and showed that callus produced in the spring had a higher morphogenetic capacity.

The developmental stage of female gametophyte also plays a crucial role in gynogenic haploid induction and reprogramming of pathway from gametophytic to sporophytic (Chen *et al.*, 2011). Various developmental stages of ovules are responsive to gynogenesis, and later stages are more responsive. It has been reported that a gynogenic response from ovules occurs at a broad range of developmental stages, in contrast to androgenesis (Doi *et al.*, 2011; Bohanec, 2009). In ovule culture of *Gerbera jamesonii*, developmental stages of ovules can be classified as young ovules occupying the half of the ovary cavity and ovules near maturity and it is reported that only ovule in second stage is responsive (Sitbon, 1981). In most cases the optimum stage for ovary culture is the nearly mature embryo sac, but in case of rice ovaries at free nuclear embryo sac stage are most responsive. Musial *et al.* (2005) reported that in Allium, large (containing mature embryo sac) and medium (containing mature embryo sac and 2–4 nucleate embryo sac) flowers were more responsive than small flowers (containing megaspore mother cell), and gynogenic embryos developed in mature embryo sac. The mature embryo sac with eight nuclei is usually cultured and any of the haploid cells can potentially develop into haploid embryos.

Stress treatments such as low/high temperatures, dark period, starvation medium *etc.* applied to ovary or ovule may have a strong influence on embryo induction. The stress pre-treatments are highly crucial for diverting the normal gametophytic pathway into a sporophytic development which would therefore promote the formation of haploid embryo (Lakshmi Sita, 1996). In gerbera, 4°C cold pre-treatment of flower buds for 7 days, promoted adventitious bud induction (Li *et al.*, 2020). Similarly, in gentians (*Gentiana* spp), cold pretreatment (4°C) of flower buds for 1–2 weeks also promoted effective induction of embryo-like structures from unfertilized ovule culture (Doi *et al.*, 2013).

The composition of the culture media can greatly influence embryogenesis. There have been countless studies evaluating all aspects of the culture medium and it's role in embryogenesis. This includes carbon source and concentration, nitrogen source and concentration, growth regulators, physical factors, and additives (*e.g.*, polyamines, brassinosteriods, colchicine). The pH of the medium can influence the embryogenic response. Several researchers have studied the effect of growth regulators especially, auxins and cytokinins on parthenogenic embryo induction from un-pollinated female gametophytes and reported that the optimum growth regulator combination may vary considerably from species to species (Ravindra *et a.l*, 2020 in marigold; Sibton, (1981) in gerbera).

Ovary culture has mainly two limitations (i) it is successful only in few species and (ii) the frequency of responding ovaries and the number of plantlets/ ovaries is quite low. Therefore, ovary culture is preferred over anther culture only where anther culture fails and in cases of male sterile lines, ovary culture assumes significance.

8.3.0 *In vitro* Induction of Paternal Haploids-Androgenesis

Haploid plants may be induced from male gametophytes through *in vitro* culture of anthers or isolated pollen/microspores on a suitable medium. The process of regeneration of haploids from male gametic cells is termed as androgenesis. The haploid induction through this approach mainly relies on the ability of microspores or immature pollen grains to switch their normal developmental pathway from gametophytic (leading to mature pollen grain) to sporophytic, resulting in cell division at a haploid level followed by formation of calluses or embryos. Due to its high effectiveness and applicability in numerous plant species, androgenesis has tremendous potential for plant breeding and commercial exploitation of DH (Germana, 2011; Murovec and Bohanec, 2012). In ornamentals, in *vitro* induction of androgenic haploids through anther culture have been achieved in anthurium (Winarto *et al.*, 2010) cone flower (Zhao *et al.*, 2006), Chrysanthemum (Yang, 2005; Gao *et al.*, 2011), Gentians (Doi *et al.*, 2011), lisianthus (Zhou, 2014), lupinus (Kozak *et al.*, 2012), primula (Jia *et al.*, 2014), Phlox (Razdan *et al.*, 2008), marigold (Yingchun *et al.*, 2011; Ravindra *et al.*, 2019). The anther culture is more laborious technique and requires manual excision of anther and there is time lag before anther can be seen successful. Hence, isolated microspore culture (IMC) is advantageous over anther culture for *in vitro* induction of haploids. The culture of microspore does not involve the risk of regeneration from somatic tissues and embryo yield from microspore culture is higher than anther culture. In ornamentals, the isolated microspore culture techniques have been standardized in Anemone (Paladine *et al.*, 2012) camellia (Pedroso and Pais, 1994), cow cockle, (Kernan and Ferrie, 2006), ornamental kale (Wang *et al.*, 2011), Rose (Ardakani *et al.*, 2015), Zantedeschia (Wang *et al.*, 2011) and sunflower (Gurel *et al.*, 1991).

A number of factors influence the androgenic response of cultured anther and isolated microspores. The plant genotype plays a predominant role in determining the androgenic response; hence culture procedure should be optimized for each genotype. Genotypic difference has been observed within and in-between the genotypes of African and French marigold groups (Ravindra *et al.*, 2019). The conditions under which donor plant are grown are an important consideration for successful haploid production. Temperature is very crucial factor because sudden change in the temperature regime mainly after floral differentiation could greatly diminish the embryogenic response as stress alters the level of growth regulators, amino acids, carbohydrate or lipid in the plant. Seasonal variations in *in vitro* androgenesis have been observed in marigold.

The developmental stage of male gametes at the time of anther or microspore isolation, in combination with suitable stress treatments, are the main factors determining the androgenetic response. Microspores at a specific stage in their development can undergo embryogenesis, whereas those cells at other developmental stages do not. It can be triggered within a relatively wide developmental window around the first pollen mitosis, when uninucleate microspores divide asymmetrically resulting in a generative cell embedded in a vegetative cytoplasm (Touraev *et al.*, 1997; Maraschin *et al.*, 2005). The presence of younger and older cells can be toxic to embryogenic cells. The most frequently used stage for anther and isolated microspore culture is between the mid-uninucleate to early binucleate stages. In mid uninucleate stage microspores are round and developed a thick exine with the nucleus in the centre. At late uninucleate stage, the nucleus migrates to the cell wall and exine become more yellow. In certain species, bud sizes can be correlated with the developmental stage of the pollen. In French marigold genotypes, the bud length of 1.2 -1.4 cm (unopened) has high frequency of disc florets (3.5-4.0 mm) with optimum stage of microspore development (mid-uninucleate to early binucleate). Whereas, in case African marigold genotypes, the predominance of this stage was observed in half opened flower buds (2.5 cm length). In Gentian, anthers collected from buds 9–12 mm in length induced the highest frequency of androgenesis.

Abiotic stresses play an important role in androgenic induction. The application of suitable physiochemical factors promotes a stress response, which arrests the microspores or young pollen grains in their gametophytic pathway. Their development is triggered through embryogenesis by promoting cell divisions and the formation of multicellular structures contained by the exine wall. Finally, the embryo-like structures are released from the exine wall (Maraschin *et al.*, 2005). The most widely used triggering factors are temperature pre-treatment, sucrose and nitrogen starvation and osmotic stress. Depending on the plant species and genotype, temperature stress can be applied by subjecting excised flower buds, whole inflorescences or excised anthers to low or high temperature (Murovec and Bohanec, 2012). Exposure of excised flower buds to a low temperature for some time, *e.g.*, 3-5°C for 2 days or at 7-8°C for 12 days, prior to excision of anthers can markedly enhance the recovery of haploid plants. Cold treatment of detached anthers is less effective than that of detached flowers. Low temperature kills weak or non-viable pollen and microspores; it also arrests many of pollen grains in the first mitosis and blocks starch accumulation. Microspore and tapetal developments then become asynchronous, influencing the switch from gametophytic to sporophytic phase. In chrysanthemum, callus induction was enhanced when cultured anthers were pre-treated at 4 °C for 48h (Khandakar *et al.*, 2014). In *Gentiana triflora* too significant improvement in plant regeneration from cultured anthers was noticed when flower buds were treated at 4°C for 48 h prior to culture (Pathirana, 2011). The exposure of excised flower buds to a low temperature at 4°C for 10 days, prior to removal of anthers significantly enhanced the recovery of haploid plants in marigold (Ravindra *et al.*, 2019). In ornamental kale (*Brassica oleracea* L. var. *acephala*), cold treatment (4°C for 2 days or 5 days) significantly improved plant regeneration from microspore-derived embryos (Wang *et al.*, 2011). The heat shock pre-treatment (30-32°C for 3-5d) has been found beneficial, both in anther and microspore culture, in several plant species. Heat stress for 3-7 days (33°C) stimulated production of microspore-derived embryos in petunia (Chaar *et al.*, 2016). In addition, treatments like centrifugation, irradiation with X-ray and gamma rays and reduced atmospheric pressure are reported to promote androgenesis.

The composition of the culture media is a major factor influencing embryogenesis. This includes carbon source and concentration, nitrogen source and concentration, growth regulators, physical factors, and additives (*e.g.*, polyamines, brassinosteriods, colchicine). The pH of the medium can also influence the embryogenic response (Ferrie *et al.*, 2012). Sucrose plays a key role in induction of embryogenesis, while other medium constituents are needed for post induction development of embryo. High sucrose level may play an osmoregulatory role during induction, but it is not necessary or even detrimental during embryo development. The addition of ovaries to anther or microspore cultures has also been observed to induce microspore embryogenesis.

Gao *et al.* (2011) obtained high callus induction in anther culture of chrysanthemum with 9 per cent sucrose. In gentian (*Gentiana triflora*), higher androgenic response was obtained when cultured anthers were subjected to heat shock treatment and cultured on medium containing 13 per cent sucrose (Doi *et al.* (2011). For microspore culture of ornamental kale, a relatively higher concentration of sucrose between 13 per cent or 17 per cent is used. The presence of an auxin may determine the mode of subsequent development or androgenic cell mass. Qi *et al.* (2011) studied the effect of growth regulator on callus formation and shoot differentiation in anther culture of French marigold. They proposed that 1.82 µM NAA and 6.6 µM BA or 2.7 µM NAA and 2.2 µM BA were the best combinations for both callus induction and shoot regeneration. Murthy *et al.* (2000) reported direct embryo induction from the niger anthers when the medium was supplemented with higher concentrations of sucrose (4-8 per cent) along with different combinations of 2,4-D and BA. They noticed optimum embryo induction in the medium supplemented with 2 µM 2, 4-D and 1 µM BA with 6 per cent sucrose.

8.4.0 CENH3 Mediated Haploid Induction

Ravi and Chan, (2010) discovered that haploids could be obtained in *Arabidopsis* through the manipulation of the centromere-specific histone 3 variant, CENH3. The CENH3, held promise of being translated to crop species, and has been successfully employed in maize (Kelliher *et al.*, 2016). Refinements of this technology have since been developed which indicate that non-transgenic modifications to CENH3 will also induce haploids. The complementation of a *cenh3* null by CENH3 from closely related plant species can result in plants that are fertile but haploid-inducing on crossing by *CENH3 wt* plants- suggesting that introgression of alien CENH3 may produce non-transgenic haploid inducers. Similarly, a remarkably wide variety of point mutations in CENH3, induced by chemical agents, have recently been shown to result in haploid induction on crossing by wild-type CENH3 plants. These CENH3-variant plants grow normally, are fully fertile on self-pollination, and may be present in existing mutagenized collections. The alternative approaches to CENH3-mediated HI may also be employed. Non-GMO point mutants can be obtained by EMS mutagenesis, and may already be present in existing populations. The recent findings of complementation and HI by CENH3 from related species (Maheshwari *et al.*, 2015) offers an alternative non-GMO approach for species in which EMS mutagenesis is inefficient. This approach has been used in chrysanthemum.

8.5.0 Identification of Haploids

A fast and reliable haploid identification method is needed for large scale production of DHs. Morphological markers expressed at the embryo, seed or early seedling stages are preferentially used. Several direct and indirect approaches are available for determining the ploidy level of regenerated plants. Indirect approaches are based on comparisons between regenerated and donor plants in terms of plant morphology (plant height, leaf dimensions and flower morphology), plant vigor and fertility. Chloroplast counting in a pair of stomata guard cells, the ratio of length and width of stomata, stomatal size and frequency, size of leaves and pollen grain size have been utilized as indirect methods for ploidy analysis, but these alternative methods are also time-consuming process (Winarto *et al.*, 2010; Ravindra *et al.*, 2019). They are fairly unreliable and subjected to environmental effects but do not require costly equipment. Direct methods for ploidy determination are more robust and reliable and include conventional cytological techniques, such as counting the chromosome number in root tip cells and measurement of DNA content using flow cytometry. Chromosome counting is conventional method based on counting mitotic cells of root tip and it has been used in many plants including ornamentals. When few plants need to be assessed, chromosome counting is commonly used and successfully employed method. However, it is costly, time consuming and laborious method to evaluate big amount of plant material.

At present, flow cytometry method, based on the detection of the fluorescence emitted by compounds that specifically bind to DNA have been used in many ornamental plants such as carnation (Agullo-Anton, 2013), cyclamen (Ishizaka, 2009), Gentian (Pathirana, 2011), phlox (Zale and Jourdan, 2015), marigold (Ravindra *et al.*, 2019) and gerbera (Li *et al.*, 2020). This provides a rapid and simple option for large-scale ploidy determination as early as in the *in vitro* culturing phase. It also enables detection of mixoploid regenerants (having cells with different ploidy) and the determination of their proportion. Additionally, ploidy level can be determined utilizing multiallelic molecular markers (microsatellites). Both of the methods (flow cytometry and molecular markers) can help to determine ploidy levels rapidly in closely related taxa in the absence of meristematic tissue. Several markers can be used for assessing the origin of diploids, depending on their availability for a particular plant species. In the past, evaluation of regenerants mainly relied on phenotypic markers, progeny testing after self-pollination and isozyme analysis. Nowadays, DNA molecular markers, such as AFLP (Amplified Fragment Length Polymorphism), RAPD (Random Amplified Polymorphic DNA), SCAR (Sequence Characterized Amplified Regions) or SST (Simple Sequence Repeat), are commonly used for homozygosity testing and assessment of plant origin. There is a considerable difference in interpretation of dominant or co-dominant electrophoretic profiles. Co-dominant molecular markers, as well as isozyme markers, have the advantage that a single locus, when heterozygous in donor plants, might be used for homozygosity determination with dominant markers. In such a case, bands missing from the donor profile indirectly indicate homozygosity

8.6.0 Diploidization of Haploids Plants

Haploid plants obtained either from anther or ovule culture may grow normally under *in vitro* conditions up to the flowering stage but viable gametes are not formed due to the absence of one set of homologous chromosomes and, consequently, there is no seed set. The only mechanism for perpetuating the haploids is by duplicating the chromosome complement in order to obtain homozygous diploids. In pollen derived plants duplication of chromosomes may occur spontaneously in cultures but due to the small percentage of such double-haploids it is necessary to diploidize the haploids by chemical means. A simple procedure designed to achieve diploidization involves immersion of very young haploids in a filter sterilized solution of colchicine (0.4 per cent) for 2-4 days, followed by their transfer to the culture medium for further growth. In this procedure chromosome or gene instabilities are minimal compared to other methods of colchicine or chemical treatment. Colchicine causes an arrest of mitosis due to a failure of the mitotic spindle to form and function in the normal manner. As colchicine is highly toxic, alternatively, three microtubule depolymerizing herbicides trifluralin, oryzalin, and amiprophosmethyl (APM) are being employed for chromosome doubling of haploid plants. The haploid genome doubling rate of trifluralin treatment in *B. napus* was 85.7 per cent, for colchicine 74.1 per cent and for oryzalin 66.5 per cent, compared to only 42.3 per cent without any treatment (Klima *et al.*, 2008). APM combined in an optimum dosage with pronamide has similar rates of genome doubling as colchicine in maize (Melchinger *et al.*, 2016). Kato and Geiger, (2002) developed an effective genome doubling procedure in maize using a nitrous oxide (N_2O) gas treatment at the six-leaf stage, with about 44 per cent of the haploids producing seed after selfing.

8.7.0 Applications of DHs in Breeding

Haploids and doubled haploids have wide applications in basic research (genomics, biochemical, and physiological studies) and for practical application (varietal development, mutagenesis, transformation). The DH populations are highly valuable for construction of genetic maps, locate genes of agronomic and economic importance, identify markers for trait selection, accelerate crop improvement programs and increase plant breeding efficiency (Forster and Thomas, 2005; Ferrie and Caswell, 2011).

8.7.1 Development of Homozygous Inbred Lines and Accelerating Breeding Cycle

Homozygous parental lines are indispensable for commercial hybrid seed production in any crop. Particularly, in F_1 hybrid breeding, development of homozygous parental through conventional techniques of selfing lines is cumbersome, time-consuming and highly laborious procedure. In a classical plant breeding program, a total of about eight inbreeding (selfing) generation is required to get an almost complete homozygous plant (99.2 per cent) with traits of interest. Through double haploid based breeding, 100 per cent homozygosity for all the traits can be achieved in only one generation (Rahman and de Jimenez, 2016). Using conventional techniques such as backcrossing and selfing, it generally takes about 12-15 years to develop a new cultivar. Using doubled haploid based technology in crop improvement program, the breeding cycle can be shortened by three to four years (Ulrich *et al.*, 1984). There are various ways in which DHs can be used in cultivar production. The DH lines themselves can be released as cultivars; they may be used as parents in hybrid cultivar production or more indirectly in the creation of breeder lines and in germplasm conservation.

8.7.2 Haploid Mutant Production

The use of haploid system for mutant induction and selection is one of the most important applications of haploid technologies. Doubled haploidy provides the fastest route to homozygosity with the greatest fidelity. The ability to fix mutations via doubled haploidy is a key factor, especially as induced mutations are predominantly recessive and cannot normally be detected until the M2 generation at the earliest. The DH systems themselves provide an opportunity to target haploid as well as doubled haploid cells for mutation treatment and capture the mutation in a homozygous, pure line. Further, there is possibility of applying *in vitro* selection methods at the haploid or double haploid level which can facilitate the generation of genetic variation and its identification. Using the DH technology, there is possibility to screen both recessive and dominant mutants in the first generation after mutagenic treatment. The DH technology allows for easier identification of recessive traits as there is no masking by dominant genes. Some of the benefits of using DH systems include immediate fixation of mutated genotypes, which saves time in production of pure mutant lines, lack of chimerism and being able to work with smaller population sizes. Using this technique, the selection efficiency of desired mutants can be greatly improved due to the gametic versus zygotic segregation ratios (1:1 vs. 3:1 respectively).

8.7.3 Gene Stacking from Biparental Crosses

Nowadays, introgression of one or a limited number of genes into elite inbreds by marker-assisted backcrossing is routine in plant breeding. At the end of backcross programmes, a heterozygous plant is selfed to produce a fixed line. For single gene introgression, the expected probability of individuals with desired homozygous genotype is 1/4. The frequency of expected genotypes decreases exponentially following the formula $1/4^n$, where n is the number of independently segregating genes. In contrast, haploid target genotypes are generated with a frequency of $1/2^n$. For example, for five loci, the frequency of the desired homozygous genotype is 1/1024 in selfed diploid progeny and 1/32 in haploid progeny. Application of doubled haploids thus significantly reduces the population size required to find desirable genotypes (Ren *et al.*, 2017).

8.7.4 Double Haploid in Genetic Studies and Molecular Marker Development

Double haploid lines are used with high efficiency in genetic study for traits of interest. In traditional bi-parental crossing population a simple trait controlled by a single dominant gene segregates in the F_2 with a 1:2:1 ratio for homozygous dominant, heterozygous, and homozygous recessive, respectively. However, the same trait will segregate in the double haploid population with a 1:1 ratio for homozygous dominant and homozygous recessive, respectively. In the case of multiple genes, double haploid technology could fix the genotypic recombinations for the homozygosity for many loci in a single generation. Thus, this method offers an advantage to use a smaller population for genetic study of quantitative traits.

DH populations are ideal material for constructing genetic linkage maps and QTL analyses. For construction of genetic maps, a segregating population, such as F_2 or back cross or near isogenic lines or double haploid is required. The major advantage of DHs over other commonly used populations such as F_2 and backcross is that they produce completely homozygous lines which can be easily multiplied without any genetic alteration occurring and thus facilitates conducting replicated trials across different locations and years for various agronomic traits (Collard *et al.*, 2005). In addition, DH populations allow for assignment of components of environmental variation, which can be valuable for analysis of physiological effects (Pink *et al.*, 2008). The DH populations are also ideal for bulked segregation analysis (BSA).

8.7.5 Doubled Haploids in Marker Assisted Breeding (MAS) and Genomic Selection (GS)

The availability of cheap and abundant molecular markers allows breeders to apply MAS and genomic selection (GS) in crop improvement. MAS depends on the identification of markers significantly associated with the trait. MAS allows breeder to discard a large number of plants with undesired gene combination, pyramid beneficial genes in subsequent generations, minimize field testing and reduce the number of generations (Collard and Mackill, 2008; Dwivedi *et al.*, 2015). The combination of MAS and DHs offers new opportunities for increasing genetic gain and shortens the time required to cultivar breeding. MAS and DH have been successfully used to accelerate resistance breeding in many crops. MAS was used for the selection of resistance genes and DH technology was used for the generation of homozygous lines. Hence, integration of MAS and DH technology into conventional breeding processes can increase the speed of cultivar development.

In contrast to MAS, GS employs genome wide markers to predict genomic estimated breeding values (GEBVs). GS requires a training population to estimate GEBVs based on phenotypic and genotypic data. In a breeding cycle, the estimated marker effects in the training population can then be used for GEBVs prediction without phenotyping. As an individual's GEBV can be predicted before or without phenotypic characterization, this enables breeders to make early selection decisions which could increase genetic gains and shorten breeding cycles (Daetwyler *et al.*, 2015). Mayor and Bernardo, (2009) studied GS and MAS in DH versus F2 populations and found that GS was superior to MAS and DH populations are superior to F2-derived population using GS.

8.7.6 Haploids and Doubled Haploids for Genetic Transformation

Microspores contain gametic number (n) of chromosomes. A highly synchronized microspore or haploid cell culture where embryogenesis occurs at high frequency is an ideal target for transformation, as transgenics can be recovered rapidly and the inserted trait get fixed in the homozygous state by chromosome doubling. Haploid microspores can be targeted in genetic engineering studies for transforming a gene prior to chromosome diploidization process, that allow a stable fixation of homozygosity of the integrated gene. The haploid plants can also be used for gene transformation followed by the chromosome diplodization process. Drought tolerant gene (HVA1) was successfully transformed in haploid bread wheat followed by chromosome doubling to fix the homozygosity of the integrated gene. The highly efficient microspore embryogenesis system in *Brassica napus* has been used successfully to obtain various mutants after microspore mutagenesis, and to achieve gene transfer mediated by *Agrobacterium tumefaciens*.

8.7.7 Development of Reversible Male Sterility

Male sterility is pre-requisite for hybrid seed production but, most crops lack a naturally occurring male sterility system. The achievement of complete male sterility in the female parent and the restored fertility in F_1 hybrids is the major bottleneck in the commercial hybrid seed production (Shariatpanahi and Ahmadi, 2016). Recent progresses in the area of metabolic engineering in pathways of microgametogenesis, biotechnology has added new possibilities to obtain reversible male-sterile plants (Ribarits *et al.*, 2007; Toppino *et al.*, 2011). It has been indicated that glutamine plays an essential role in pollen development, as isolated and *in vitro* cultured microspores were unable to develop into functional pollen grains in glutamine lacking medium (Ribarits *et al.*, 2007). Transformation of *Nicotiana tabacum* L. with mutated tobacco glutamine synthetase genes, which synthesize glutamine from glutamate, fused to the tapetum- and microspore-specific promoters caused pollen abortion closed to the first pollen mitosis and thus resulted in male sterility. Complete homozygous male-sterile lines were rapidly produced by microspore embryogenesis technique. Furthermore, foliage sprays with glutamine and *in vitro* maturation restored the fertility of male-sterile lines (Ribarits *et al.*, 2007). Using this technique, F_1 hybrid seed production technology can be applied in many commercially important crops (Shariatpanahi and Ahmadi, 2016).

Lots of advancements has been made in use of doubled haploid based technology in agronomically important crops, however DH based research in ornamental crops is still lagging behind than majority of other crop species. There are many ornamentals that still need to be evaluated for microspore culture response. Further research is required to develop and optimize protocols for production of doubled haploids in many ornamentals (Thaneshwari *et al.*, 2018).

Table 14: Progress in Development of Doubled Haploids in different Ornamentals

Crop	Technique Used for Haploid Induction	References
Anemone (*Anemone coronaria*)	Anther Culture	Laura *et al.* (2006)
	Microspore culture	Paladine *et al.* (2012)
African violet (*Saintpaulia ionantha*)	Anther Culture	Weatherhead *et al.* (1982)
		Smith *et al.* (1981)
Anthurium (*Anthurium andreanum)*	Anther Culture	Winarto (2010)
Calla Lily (*Zantedeschia aethiopica*)	Anther Culture	Ko *et al.* (1996); Zhang *et al.* (2011)
	Microsore culture	Wang *et al.* (2011a)
Carnation (*Dianthus caryophyllus* L.)	Pseudofertilized ovule culture	Sato *et al.* (2000)
		Dolcet-Sanjuan *et al.* (2001)
Camellia (*Camellia japonica*)	Microspore culture	Pedroso and Pais (1994)
Cyclamen (*C. persicum* and *C. purpurascens*)	Anther culture	Ishizaka (1998)
Chrysanthemum (*Dendranthema grandiflorum*	Anther culture	Yang *et al.* (2005); Gao *et al.* (2011); Khandakar *et al.* (2014
	Wide crossing	Wang *et al.* (2014)

Crop	*Technique Used for Haploid Induction*	*References*
Cone Flower (*Echinacea purpurea*)	Anther Culture	Zhao *et al.* (2006)
Cow Cockle (*Saponaria vaccaria*)	Microspore culture	Kernan and Ferrie (2006)
Ornamental pepper (*Capsicum annuum* L.)	Shed microspore culture	Ari *et al.* (2016)
Gentian (*Gentiana* spp.) (*Gentiana triflora, G. scabra*)	Unfertilized ovule culture	Doi *et al.* (2011); Doi *et al.* (2013)
Gentian (*Gentiana triflora*)	Ovule and anther culture	Pathirana *et al.* (2011)
Gerbera (*Gerbera jamesonii*)	Unfertilized ovule culture	Sitbon (1981); Cappadocia *et al.* (1988); Miyoshi, and Asakura (1996), Li *et al.* (2020)
Iris (*Iris pseudacorus*)	Induced parthenogenesis via pollen irradiation	Grouh *et al.* (2015).
Lisianthus	Anther culture	Zhou (2014)
Lupinus	Anther culture	Kozak *et al.*, 2012
Petunia	Induced parthenogenesis using irradiated pollen	Raquin (1985)
	Pollen culture	Sangwan and Norreel (1975)
	Ovule culture	De Verna and Collins (1984)
	Microspore Culture	Chaar *et al.* (2016)
Primula (*Primula forbesii*)	Anther culture	Jia *et al.* (2014)
Ornamental Kale (*Brassica oleracea* var. *acephala*)	Microspore Culture	Wang *et al.* (2011b)
Phlox (*Phlox drummondii*)	Anther culture	Razdan *et al.* (2008)
Peace lily (*Spathiphyllum wallisii*)	Ovary culture	Eeckhaut (2001)
Roselle (*Hibiscus sabdariffa*).	Microspore culture	Maarup *et al.* (2012)
Rose (*Rosa* x *hybrida*, cv 'Sonia')	Parthenogenesis induced using irradiated pollen	Meyneta *et al.* (1994);Lim *et al.* (2005)
	Microspore culture	Ardakani *et al.*, 2015
Geranium (*Pelargonium hortorum*)	Anther culture	Abo El-Nil (1973); Becker-Zens (1983)
Marigold	Anther culture	Yingchun *et al.* (2011); Ravindra *et al.* (2019)
	Unfertilized ovary culture	Ravindra *et al.* (2020)
Tradescantia paludosa	Anther Culture	Dietert *et al.* (1982)
Tulip (*Tulipa gesneriana*)	Isolated Microspore culture	Bulk *et al.* (1994)
Easter Lily (*Lilium longiflorum*)	Anther culture	Arzate-Fernandez (1997)
Asiatic Lily Hybrids	Anther culture	Han *et al.* (1999)
Sunflower (*Helianthus annuus*)	Induce parthenogenesis using irradiated pollen	Todorova *et al.* (1997)
	Microspore culture	Gurel *et al.* (1991).

9.0 Wide Hybridization for Gene Introgression

The economic importance of the ornamental plant industry requires constant development of novel and high-quality varieties. Traits attractive for production of new ornamental plants may not be available within the commercial cultivars, but broad genetic variation is present within the plant genera. Wide hybridization is one of the oldest breeding techniques that contributed enormously to the development of modern plant cultivars. Hybridization among different species and genera is one of the most important breeding approaches used for the improvement of ornamental plants. The main aim of this strategy is to merge distant gene pools, hence broadening the genetic variability. Interspecific and intergeneric crosses are very effective in terms of broadening genetic variability and introgression of desirable traits from wild species. In breeding of ornamental plants, interspecific hybridization and polyploidization have successfully been used to produce novel cultivars with blended traits of both parents and introgress useful traits of one species to another (Van Tuyl and Lim, 2003). Recent advancements in the breeding strategies, plant tissue culture techniques such as embryo rescue, protoplast fusion, polyploidization, and molecular cytogenetic methods have contributed enormously in the development and characterization of interspecific hybrids in various ornamentals (Mii, 2009). Production of interspecific hybrids is useful not only for development novel cultivars with intermediate characters between the parental species, but also to introgress the useful genes possessed by one species to another species through alien chromosome addition line. Thus, interspecific or intergeneric hybrids have the enormous potential to extend not only their qualitative but also quantitative traits such as the type of flower, plant phenotypes, other single dominant traits from parent species with an environmental adaptation or improve the flower vase life (Van Tuyl and Lim, 2003).

9.1.0 Barriers to Interspecific Hybridization

Successful sexual hybridization involves a series of events including pollen germination, pollen tube growth, fertilization, embryo and endosperm development, and seed maturation. However, there are several barriers to wide hybridization that impede gene flow between species/genera. These barriers can be broadly divided into two major classes *viz*. pre-zygotic (pre-fertilization) and post-zygotic (post-fertilization) barriers. Pre-fertilization barriers prevent mating or fertilization, while post-fertilization barriers affect and reduce viability of hybrid zygote and later development of hybrid progeny (Rieseberg and Carney, 1998).

9.1.1 Pre-fertilization Barriers

Pollination with other species is restricted by their natural habitat (geographic isolation), flower structure (*i.e.*, shape or color), flowering time or pollen mediator, which are involved in pre-zygotic barriers (Lowry *et al.*, 2008). Pollen-pistil incompatibilities and incompatibilities of gametes further prevent crossing among unrelated parents during wide hybridization. In the modern theory of speciation, several types of pre-zygotic barriers are recognized, including geographical isolation, habitat isolation, temporal isolation (phenological differences *i.e.*, two species may flower in different seasons), pollinator isolation, and mechanical isolation. The most obvious form of reproductive isolation is spatial isolation. Spatial isolation occurs between populations that are separated by great distances, but it can also take place between populations that inhabit different parts of the same area. If members of two populations never encounter each other, they will never mate and no gene flow will occur. Staggered planting and planting under climatic controlled glasshouses can be employed to overcome the barriers arising as results of spatial/temporal or ecological isolation,

Gametic isolation is another form of prezygotic barrier that acts after mating but before fertilization. In plants, gametic isolation prevents interspecific hybridization from occurring through mechanisms such as pollen–pistil incompatibility and competition between conspecific and hetero-specific pollen for fertilization. The mechanism of pollination can be broadly divided into four characteristic stages that include: (a) pollen hydration and germination; (b) growth of the pollen tube through the stigma, style and ovary; (c) pollen tube guidance to the ovule micropyle; and (d) delivery of the sperm to the embryo sac (Wilhelmi and Preuss, 1997). The time required to accomplish this series of events varies considerably depending on the species, from a few hours to several month. Pollen tube guidance in angiosperms is a multi-stage mechanism, which leads the pollen tube to the ovule from the stigma via the pistil by recognition of its own pollen. The first pre-zygotic barrier is pollen rejection; the inhibition of pollen hydration, germination, or pollen tube growth. Hence, cross incompatibility during wide crossing may be due to lack of pollen germination, insufficient growth of pollen tube to reach ovule and inability of male gamete to unite with the egg cell. In course of distant hybridization, pollen hydration and germination are inhibited on the stigma and pollen tube elongation through stigma and style is prevented (Tonosaki *et al.*, 2016).

Interspecific incompatibility (II) acts as an active selection mechanism and prevents production of hybrid progeny. This process can include recognition and rejection of unsuitable pollen and it is considered to be related to the self-incompatibility (SI) system (Kuligowska *et al.*, 2016a). In wide hybridization, the existence of SI systems is important as asymmetric pollen rejection has been observed when self-compatible (SC) and SI plants were crossed. This phenomenon is called as unilateral incompatibility (UI) and implicates the role of the S-locus in interspecific pollen rejection. UI has been observed in several plant genera including Brassica, Nicotiana and Solanum. In both, Nicotiana and Solanum rejection of hetero-specific pollen in SI and SC crosses is possibly due to the activity of S-RNase that acts as cytotoxin specifically directed against incompatible pollen. The pistil of the flowers also plays major role in discriminating pollen from plants that are too closely or too distantly related (Hancock *et al.*, 2003). Inhibition on the stigma surface causes arrest of pollen germination or prevent the entry of pollen tube into the stigma. This is mainly due to lack of effective adhesion, lack of full hydration and absence of pollen germination factors on the stigma. The role of stigmatic lipids and proteins in specific pollen adhesion in distant crosses has been partly related to SI mechanisms. In ornamentals, the existence of SI system as pre-fertilization barrier has been suggested in *Dianthus* (Nimura *et al.*, 2003), *Aechmea, Tillandsia* (Vervaeke *et al.*, 2001) and Rosa (Dugo *et al.*, 2005).

Interactions between pollen and ovules play an important role in the regulation of compatibility among distant species. Synergid cells play crucial role in pollen tube guidance (Higashiyama *et al.*, 2006). In *Torenia fournieri*, LUREs, which are secreted cysteine-rich polypeptides derived from synergid cells, have been identified as female-derived chemo-attractants. LUREs of *T. fournieri* do not attract the pollen tube of the closely related species such as *Lindernia micrantha*, indicating pollen tube guidance by LUREs acts in a species-preferential manner (Okuda *et al.*, 2009). The pollen tube bursts after entering into the micropyle in the last steps of pollen tube growth, and discharges two sperm cells into the female gametophyte for double fertilization. Another protein, a receptor-like kinase encoded by FERONIA/SIRENE (FER/SRN) gene has been associated with the control of pollen–ovule interaction. This protein is believed to be involved in a reception of pollen tubes and localized predominantly at the surface of synergid cells. Species-specificity of the FER/SRN proteins was suggested based on the variability of their extracellular domains (Huck *et al.*, 2003; Rotman *et al.*, 2003). This suggests the contribution of FER/SRN to interspecific cross-compatibility via strong regulation of pollen tube growth in the step of ovular guidance.

9.1.2 Overcoming Pre-fertilization Barriers

Several techniques have been utilized to overcome prezygotic barrier arising as a result of inter-specific incompatibility. These incompatibility barriers are weaker at the early stage of floral development; hence, the method of bud-pollination (pollinating stigmas of buds two to three days before anthesis) has often been used in Brassicaceae and Solanaceae species (Tonosaki *et al.*, 2016). Several techniques such as reciprocal crosses, mixed pollen (*i.e.*, mixture of compatible and incongruous pollen), mentor pollen, bridge crosses, use of pollen mixtures, pistil manipulations, use of growth regulators, heat treatment of pistil and pollen, *in vitro* pollination and fertilization have been used to circumvent the pre-zygotic barriers. For obtaining fragrant evergreen azaleas, reciprocal crosses have been found effective in overcoming the pre- and post-fertilization barriers in interspecific hybridization between evergreen azalea species and *Rhododendron uwaense* (Okamoto and Ureshino, 2015). The use of mixed pollen (*i.e.*, mixture of compatible and incongruous pollen) and mentor pollen have been suggested to overcome the pre-fertilization barriers occurring at style (Van Tuyl, 1982). The mentor pollen may be defined as the compatible pollen which may be genetically inactivated (by irradiation) to inhibit its fertilization ability. The mentor pollen itself may or may not germinate on stigmas but still retain its power to stimulate incompatible pollen to accomplish fertilization during wide hybridization.

9.1.2.1 Stylar or Pistil Manipulation

Inter-specific incompatibility occurring at the stigma or stylar region can be successfully overcome by using cut-style pollination method (CSM). In this technique pollen grains are directly applied on to the top of the ovary after removing the style and have been well established in Liliaceae. When parents with different pistil lengths are crossed, pollen tube growth and development is often impeded in (too) short or long pistils (Williams and Rouse, 1990). The cut style pollination technique allows pollen to circumvent stylar barriers which normally inhibit pollen tube growth. The CSM has been successfully utilized during wide hybridization in different ornamental genera, like *Tulipa* (Van Creij *et al.*, 1997), *Fritillaria* (Wietsma *et al.*, 1994) *Lilium* (Van Tuyl *et al.*, 1988; Van Tuyl *et al.*, 1991; Janson 1993) and Vriesea (Vervaecke, 2001). The style can even be completely removed in order to pollinate the ovules directly (placental pollination). Style grafting technique, where a compatible style is grafted on an incompatible ovary, have been used in lilium to remove the reproductive barriers (Van Tuyl *et al.*, 1991). This technique is referred to as 'ovary grafting' when next to the donor style (a part of) the donor ovary is grafted on the ovary of the seed parent. In lilies, hormone treatment of the style was more efficient than ovary grafting, placental pollination and mentor pollination to circumvent the inhibition of ovule penetration (Van Creij *et al.*, 1997).

The use of high temperature to overcome incongruity has been applied in breeding programs of lily (Van Tuyl *et al.*, 1982; Okazaki and Murakani, 1992) and Rhododendron (Preil and Ebbinghaus 1985; Okonkwo and Campbell, 1990; Rouse, 1993). The high temperature may lead to the inactivation of heat sensitive inhibitors of pollen tube growth and hence influence pollen tube growth. The organic solvents, such as hexane and ethyl acetate, have been applied on stigmas to overcome interspecific inhibition (Kuligowsk *et al.*, 2016). Application of plant hormones, especially gibberellins act as a promotor of pollen tubes growth in interspecific crosses. The phyto hormones influence the signaling pathways responsible for the regulation of pollen tube growth in interspecific crosses and thus overcome pre fertilization barrier. To speed up pollen tube growth, application of growth regulators *e.g.*, IAA, 2,4-D or GA_3 or use of radiation or an immunosuppressant like alpha amino caproic acid have been recommended. Application of plant hormones, especially gibberellins promoted growth of pollen tubes in interspecific crosses of Chrysanthemum (Sun *et al.*, 2011).

9.1.2.2 In vitro Pollination and In vitro Fertilization

Various *in vitro* methods have been developed to overcome incongruity barriers in a number of plant species. The techniques of *in vitro* pollination and *in vitro* fertilization have been utilized to bypass the pre fertilization barriers resulting from the hindrance of the growth of pollen tube on the stigma or style in number of ornamentals including lilium (Van Tuyl *et al.*, 1991). The techniques of *in vitro* pollination including intra ovarian pollination, *in vitro* placental pollination and *in vitro* ovular pollination *etc.* have been established in *Papaveraceae, Bromeliacea* and *Liliacea* families (Vervaeke *et al.*, 2003; Van Tuyl *et al.*, 1991). *In vitro* cut-style pollination and *in vitro* grafted style technique have been developed and applied in various interspecific crosses using *Lilium longiflorum, L. dauricum, L. henryi,* and both Asiatic and Oriental hybrids as the parents. The possibility of introducing pollen grains directly into the ovary (intra-ovarian pollination) is important approach to bypass the pre-fertilization barriers. Here, either the style or part of it is excised and pollen grains are either placed on the cut surface of the ovary or transferred through a hole in the wall of the ovary. Direct pollination of cultured ovules (*in vitro* ovular pollination) or excised ovule with placenta (*in vitro* placental pollination) have been successfully employed in Lilium to circumvent the barrier to pollen tube growth (Chi,2000). In bromeliads, placental pollination results in a low fertilization rate as pollen tubes were not guided to the ovule micropyle. However, when a style with active growing pollen tubes was grafted to the ovules (placental grafted style pollination) fertilization rate increases (Vervaeke *et al.*, 2003). The intra ovarian pollination opens a gateway for subsequent development of *in vitro* fertilization. In this technique ovules and pollen grains are excised and cultured in the same medium for the successful fertilization of pollen and ovules under *in vitro* conditions. This application of plant tissue culture successfully overcomes the incompatibility barriers existing at the sexual level.

9.1.2.3 Somatic Hybridization

The gametic in-compatibilities during wide hybridization can be overcome by somatic hybridization. Development of hybrid plants through the fusion of somatic protoplasts of two different plant species/genera is called somatic hybridization. It is a non-conventional technique used for production of symmetric hybrids between species, which cannot be hybridized sexually. During sexual crossings, the plasmid and mitochondrial genomes are inherited maternally; however, the protoplast fusion technique combines the nucleus and cytoplasm of both the parents in the hybrid cell. Further, cybrids (cytoplasmic hybrids) and organelle recombinants, not possible through conventional methods, can also be developed. Cybrids in contrast to conventional hybrids possesses the nucleus genome from only one parent but cytoplasmic genes of both the parents. Recombinant organellar genomes, especially of mitochondria, are generated in somatic hybrids and cybrids. In comparison with sexual hybridization, somatic cell fusion is a much more powerful method for increasing cytoplasmic diversity in a population as it is possible to transfer chloroplasts and mitochondria in a single step, even between incongruent species, and to obtain novel combinations of nucleus and cytoplasm organelles. In cybridization, heterozygosity of extra-chromosomal material can be obtained, which has direct application in plant breeding. The production of cytoplasmic hybrids (cybrids) is particularly attractive when one is interested in the transfer of cytoplasmic-encoded traits into a particular genetic background. In most of the crop species, any two protoplasts can be fused by chemical or physical means under *in vitro* conditions. However, production of unique somatic hybrid plants is limited by the ability to regenerate the fused product and sterility in the interspecific hybrids rather than the production of protoplasts. The technique of protoplast fusion is a valuable tool for the induction of genetic variability and introgression of traits which do not pre-exist in crop species.

Somatic hybridization can be a precious tool especially for the improvement of flower crops. As flower crops are mainly cultivated for their ornamental value, hence novel somatic hybrids with certain ornamental value can directly be used as variety without any further additional breeding. In addition, the somatic hybridization between distantly related species often results in the sterile hybrids, the problem of sterility can be easily overcome in ornamentals by clonal propagation and use of tissue culture techniques.

In ornamental plants, somatic hybrids have been produced since the first successful result obtained from the fusion between *Petunia hybrida* and *P. parodii* (Power *et al.*, 1976). However, only limited successful results of the somatic hybridization have been reported so far because of the difficulty in establishing efficient plant regeneration systems from protoplasts in the target plant species and genotypes. Nakano and Mii, (1993) successfully used the somatic hybridization approach for development of intergeneric hybrids between *Dianthus caryophyllus* and *Gypsophila paniculata*. The somatic hybrid plants have been successfully obtained in *D. chinensis* + *D. barbatus* (Nakano and Mii, 1993a), *D. chinensis* + *D. caryophyllus* (Nakano and Mii, (1993b) and *D. acicularis* and *D. caryophyllus*. In chrysanthemum, somatic hybrids having resistance against white rust has been successfully created through protoplast fusion of *C.×morifolium* and *Artemisia sieversiana* (Furuta *et al.*, 2004).

9.1.3 Post Fertilization Barriers

Even if pollen germination and fertilization are successful, embryo abortion or abnormal growth of endosperm (over- and under-proliferation of endosperm) may impede normal seed development during distant hybridization (Haig and Westoby, 1991). Among the post-fertilization barriers, the occurrence of hybrid incompatibilities such as unviability, chlorosis, necrosis, stunted growth and albinism were the main drawback in production of hybrids (Kuligowska *et al.*, 2016a). The post-fertilization barriers often affect development of hybrid embryo and further growth of hybrid plants. Hybrid embryos may fail to develop to maturity due to abnormalities in endosperm development, the nourishing tissue within the seed. The endosperm abnormality is governed by cross combination of species, direction of hybridization, and ploidy levels of parental species. Consequently, the failure of endosperm growth often causes embryo starvation and hybrid inviabilty (in viability of the hybrid zygote or embryo). In some wide crosses, zygote formation occurs, but further development of the zygote is arrested, while, in other cases, after the completion of the initial stages of development, the embryo gets aborted. This is mainly because of unfavourable interactions between the chromosomes of the two species or unfavourable interaction of the endosperm with the embryo.

Hybrid in-viability is caused due to disharmony between cytoplasm and nuclear genes. The ploidy differences between parent plants might cause malformation of endosperm and the inhibition of germination. The studies have proved that when the ploidy level of the female parent is higher (maternal excess), under-proliferation of endosperm and the promotion of seed storage synthesis are observed. However, over-proliferation of the endosperm and alteration of seed storage synthesis are observed when the ploidy level of the male parent is higher (Scott *et al.*, 1998). Endosperm absence or retarded development is the main cause of spontaneous abortion of hybrid embryos during wide crossing in lilium (Van Tuyl *et al.*, 1991). The genome balance between male and female parents also plays significant role in hybrid endosperm development. Reciprocal crosses, application of growth hormones and embryo rescue techniques have been widely used to overcome the problem of hybrid inviabilty in ornamentals. The development of hybrids can also be hampered as a result of improper interactions between genes or alleles coming from different parents that are brought together in hybrid progeny. Deleterious attributes of hybrid plants caused by these interactions are called hybrid incompatibilities (HIs) and they are a consequence of genetic divergence (Bomblies, 2010; Johnson, 2010). Other examples of HIs are the dysfunctional interactions between cytoplasmic genomes and nuclear genomes from two different species. They are called cytonuclear incompatibilities (CNI). Recently, it has been suggested that epigenetic remodelling leading to transposon activation, gene expression changes, and loss of gene imprinting after interspecific hybridization can contribute to HIs.

Another important post fertilization barrier frequently encountered during wide hybridization is hybrid weakness and albinism of hybrid seedlings. Interspecific hybrids show several developmental defects compared to the parental lines, which is termed hybrid necrosis or hybrid weakness. The hybrid progeny often exhibits reduced growth rate and dwarfism, wilting, tissue necrosis and chlorosis, that often lead to hybrid plant death (Bomblies, 2010). Incompatibility between the plastome and nuclear genome often inhibits chloroplast development and chlorophyll formation. The studies conducted on hybrid *Zantedeschia* albinos have revealed that albinism occurs due to a block in chloroplast biogenesis resulting in failure to properly develop chlorophyll-containing photosystems in the thylakoids (Yao and Cohen, 2000). The problem of albinism occurs frequently in interspecific hybrids of Campanula (Roper *et al.*, 2015a), Hibiscus (Kuligowska *et al.*, 2016b) Pelargonium (Metzlaf *et al.*, 1982) and Rhododendron ((Michishita *et al.*, 2002). The typical strategy to avoid albino plants in wide crosses includes selection of suitable genotypes of parental species. Though, interspecific hybridization of Rhododendron the interploidy crosses were effective in overcoming albinism in the progeny (Sakai *et al.*, 2008).

Hybrid sterility and hybrid breakdown are common feature in wide hybridization. Hybrid sterility refers to the inability of a hybrid to produce viable offspring. This is more prominent in the case of intergeneric crosses. The lack of structural homology between the chromosomes of the two species often result in hybrid sterility. Lack of homology between chromosomes results in incomplete pairing of chromosomes and often lead to several meiotic abnormalities like chromosome scattering, chromosome extension, lagging of chromosome in the anaphase, formation of anaphase bridge, development of chromosome rings and chains, irregular and unequal anaphase separations. These irregularities may also cause aberrations in chromosome structure. Although sterility of the hybrids sometimes may not cause serious problems in some vegetatively propagated ornamental species because they can be cultivated as clonal cultivars. In most ornamental species it is essentially important to restore the fertility for utilizing them for further breeding. When F_1 hybrid plants of an interspecific cross are vigorous and fertile but their F_2 progeny is weak and sterile it is known as hybrid breakdown. Hybrid breakdown hinders the progress of interspecific gene transfer. This may be due to the structural difference of chromosomes or problems in gene combinations. To overcome the problem of hybrid breakdown growing of larger F_2 population is recommended.

9.1.4 Overcoming Post Fertilization Barriers

9.1.4.1 Embryo Rescue

Embryo rescue through embryo or ovule culture is a valuable *in vitro* tool for circumventing the postzygotic barrier and to incorporate valuable genes across the species/genera in various ornamental crops, *e.g.*, Gypsophila (Kishi *et al.*, 1993), Lilium (Van Tuyl *et al.*, 1991), Primula (Amano *et al.*, 2006) and Sandersonia (Morgan *et al.*, 2001). Embryo culture has been used to rescue hybrid plants from wide crosses, which often fail to produce mature viable seeds. The *in vitro* culture of immature embryos on artificial media can replace the endosperm and allow maturation of the hybrid embryo and thus overcome the lack of endosperm. Embryo culture enables the breeder to successfully make wide crosses with a greater number of related species of wild plants and have access to a much wider range of genes that can be used for genetic improvement of crop plants. The success of embryo rescue in wide crossing is influenced by the age of the embryo at the time of its culture. Based on the nutritional independence, the developmental stages of the embryo are categorised as heterotrophic and autotrophic. The heterotrophic stage of growth is the period during which the embryo depends on the endosperm for its nutrition; it extends from the fertilization to approximately the heart stage. Hybrid embryos which start to degenerate at the heterotrophic phase, after isolation and culture in artificial media, die soon after transfer (Zenkteler, 1990). The autotrophic phase starts from the late heart stage. This stage is significant for the *in vitro* culture as the embryo becomes sufficiently independent of the endosperm for the subsequent growth. Despite considerable progress in the field of embryo culture, embryo rescuing seems to be difficult where embryo abortion occurs at a very early stage of development. Embryo rescue can be performed in one step (embryo culture) or two steps (*in ovule* culture + embryo culture). The most applied culture methods are ovary slice culture, ovule with placenta culture, single ovule culture, and embryo sac culture (Van Tuyl and De Jeu, 1997; Chi 2002). To culture very young embryos successfully, the embryo of a particular species is implanted in the endosperm from another seed of the same species. Embryo rescue through embryo or ovule culture is a valuable *in vitro* tool for circumventing the postzygotic barrier and to incorporate valuable genes across the species/genera in several ornamental crops, *e.g.*, *Alstroemeria* (De Jeu and Jacobsen, 1995; Burchi *et al.*, 1998), Gypsophila (Kishi *et al.*, 1993), Hydrangea (Reed, 2000; Reed *et al.*, 2001), Lilium (Van Tuyl *et al.*, 1991), Primula (Amano *et al.*, 2006), *Tulipa* (Custers *et al.*, 1995; Van Creij *et al.*, 1997), Sandersonia (Morgan *et al.*, 2001) and Zantedeschia (Yao and Cohen, 1996).

9.1.4.2 Mitotic Polyploidization

Mitotic polyploidization (somatic chromosome doubling) is one of the approaches to restore fertility of sterile interspecific hybrids. When hybrid sterility is a result of disturbed chromosome division during meiosis, fertility can be restored by chromosome doubling that adds homologous chromosome set and allows their normal pairing. The metaphase inhibitors like colchicine, oryzalin or trifluralin are frequently used for mitotic polyploidization. In some species, alloploid plants can be regenerated under *in vitro* cultures from cells with spontaneously duplicated chromosomes. This phenomenon called endoreduplication and often results from replication of nuclear genome in the absence of cell division. Spontaneous chromosome duplication has been reported in plants possessing small genomes such as *Phalaenopsis equestris* (Lin *et al.*, 2001; Chen *et al.*, 2009) and several Kalanchoe species (Aida and Shibata, 2002; Kuligowska, 2015). Colchicine (Asano, 1982) and oryzalin have been utilized in sterile F_1 lily hybrids to restore fertility (Van Tuyl, 1990). Though fertility is restored, there is no scope for recombination between the parental genomes due to autosyndetic pairing. Hence, this method offers little possibilities to create genetic variation and can be defined as a 'dead end' for further breeding programs. (Lim *et al.*, 2000; Wendel, 2000; Van Tuyl *et al.*, 2002).

9.1.4.3 Meiotic Polyploidization

Meiotic polyploidization offers new opportunities for the interspecific hybridization in ornamentals. Homologous recombination is crucial for introgression of desired traits, however, mitotic polyploidization often hamper segregation in the progeny. Hence, the utilization of unreduced '2n' gametes (meiotic polyploidization) is an alternative to mitotic polyploidization. Through this approach, heterozygosity is not fixed; hence, recombination between the parental genomes is possible in the sexual polyploids (Bretagnolle and Thompson, 1995; Soltis and Soltis, 2000). Additionally, during formation of 2n gametes homologous recombination can take place leading to trait segregation. The '2n' gametes originate due to deviating meiosis in plants. The 2n gametes contribute with two chromosome sets to the progeny, most of the times generating triploid and tetraploid progenies, depending if 'n' or '2n' gametes are contributed by one or both parents. The spontaneous occurrence of 2n gametes and its use to produce progeny is reported in a number of LA and OA interspecific lily hybrids (Lim *et al.*, 2001a, 2001b; Barba-Gonzalez *et al.*, 2004). The major attribute of 2n gametes is that introgression of chromosome segments of the parental species is achieved and even if backcrosses are performed, the introgressed chromosome segments remain in the progeny (Lim *et al.*, 2001a, 2001b)

Research in *Begonia* (Dewitte *et al.*, 2009) and *Rosa* (Crespel *et al.*, 2015) has demonstrated that environmental factors such as low or high temperature and high irradiance can contribute to the formation of 2n gametes. Unreduced gametes can also be induced by chemical treatments as documented in Lilium (Kitamura *et al.*, 2009) and Begonia (Dewitte *et al.*, 2010).

9.2 Progress of Wide Hybridization in Ornamentals

9.2.1 Alstroemeria

Interspecific hybrids have been successfully developed in Alstroemeria using embryo rescue techniques and the problem of hybrid sterility was overcome by restoring the fertility of the interspecific hybrids using antimitotic agents (Lu and Bridgen, 1997). The interspecific hybrids possessing novel flower colors and shapes, winter hardiness, compact growth habit for potted plant growth, and flower fragrance have been developed by crossing Brazilian and Chilean species of Alstroemeria (Aros *et al.*, 2012, 2015). The first fragrant cultivar of Alstroemeria 'Sweet Laura' was developed as a result of crossing the Chilean species, *Alstroemeria aurea*, with the Brazilian species, *A. caryophyllaea* (Bridgen *et al.*, 1997). Two other interspecific hybrids of *Alstroemeria* namely 'Mauve Majesty' and 'Tangerine Tango' have been recently patented for their winter hardiness, everblooming growth habit, and long flower stems.

9.2.2 Rose

In roses, the major problems associated with inter-specific breeding include low seed set and lack or low per cent of seed germination. Low seed set is usually due to non-amenable parents, which may have different ploidy level or other divergences that result in embryo abortion at early stages of development. Interploidy and interspecific hybridization is found in the genetic background of modern roses. High percentages of successful embryo germination under *in vitro* conditions were observed when interspecific hybridization was attempted at the tetraploid level between *R. damascena* ($2n = 4x$) and tetraploid cultivar roses resulted in (Abdolmohammadi *et al.*, 2014). Interspecific crosses between diploid rose species and tetraploid cultivars often yield triploids. Using such triploids as pollen parents and tetraploid plants as seed parents have resulted in relatively high seed set (Van Huylenbroeck, 2005). A high priority in rose (Rosa sp.) breeding research is the transfer of disease resistance, especially to black spot (*Diplocarpon rosae* Lib.) from wild diploid Rosa species to modern rose cultivars. However, breeding roses for disease and pest resistance lagged behind than most other ornamentals. One of the major reasons behind this could be the tetraploid ($2n=4x=28$) or triploid($2n=3x=21$) nature of the cultivated roses, while the wild species possessing resistance are diploid in nature. The triploid hybrids produced between diploid and tetraploid Rosa species are generally infertile and have little significance in introgression of disease resistance trait through backcross breeding. Hence, amphidiploids ($2n = 4x = 28$) have been induced in interspecific diploid ($2n = 2x = 14$) hybrids involving the black spot resistant diploid species *R. wichuraiana* Crep, *R. roxburghi* Thratt, *R. banksiae* Ait., *R. rugosa rubra* Hort., and *R. setigera* Michaux (Ma *et al.*, 1996).

9.2.3 Chrysanthemum

The important traits from some of the wild Chrysanthemum species have been successfully incorporated into the gene pool of cultivated chrysanthemums through interspecific and intergeneric hybridization approach (Deng *et al.*, 2010). Majority of chrysanthemum cultivars are susceptible to a range of diseases, pests and abiotic stresses. Deng *et al.* (2009) have reported the successful transfer of aphid resistance from *Artemisia vulgaris* L. into *C. grandiflorum* using intergeneric hybridization approach. The somatic hybridization techniques have been successfully used in chrysanthemum to introduce rust resistance from wormwood (*Artemisia sieversiana* J.) by Furuta *et al.* (2004). Interspecific hybridization followed by backcross breeding has tremendous potential for cultivar improvement in chrysanthemum. This approach has been successfully utilized to create novel germplasm and introduce aphid resistance and heat tolerance from *C. makinoi* to commercial cultivar of chrysanthemum, (Cheng *et al.*, 2011). With the aim to introgress cold hardiness, Cheng *et al.* (2010) employed ovary rescue to create six interspecific hybrids from the cross between *Dendranthema morifolium* (Ramat.) Kitamura 'rm20-12' (2n = 54) and its wild diploid relative *D. nankingense* (Nakai) Tzvel. (2n = 18). The cold tolerance of five hybrids was significantly superior to that of their *D. morifolium* parent.The drought tolerance have been successfully improved in chrysanthemum through interspecific hybridization approach involving *C. indicum* as the donor parent for drought tolerance Sun *et al.* (2010).

One of the most promising approaches for development of multi-resistant genetic resources is the exploitation of multigeneric hybridization in breeding, involving species from different genera. Multigeneric hybrids (incorporating germplasm from three or more genera) may enable the transfer of different alien genes to a cultivated crop, and help to establish evolutionary relationships among different genomes included in the same genetic background. The trigeneric hybridization between different bigeneric hybrids havc proved a promising method for combining the multiple stress-resistance traits for improvement of chrysanthemum (Deng *et al.*, 2012). The trigeneric hybrids were successfully generated in florist's chrysanthemum by crossing two intergeneric F_1 hybrids *Chrysanthemum grandiflorum* × *Artemisia vulgaris* and *Chrysanthemum crassum* × *Crossostephium chinense*. The enhanced aphid resistance of the hybrid line reflected the inheritance of chromosomes from *A. vulgaris*, which carries genes that encode bioactive components. The enhanced salt tolerance of the trigeneric hybrid was attributed to inheritance of genetic materials from *Chrysanthemum crassum* and *Crossostephium chinense*, which act to maintain the compartmentation of Na^+ and K^+ ions and their selective transportation among different organs to avert deleterious effects and protect the photosynthetic apparatus.

9.2.4 Carnation

Several desirable traits such as early flowering, resistance to bacterial wilt disease, and heat tolerance has been introgressed through interspecific hybridization in carnation (Sparnaaij and Koehorst-van Putten, 1990; Nimura *et al.*, 2008a). Onozaki *et al.* (2002) succeeded in introducing resistance to bacteria wilt (*Burkholderia caryophylli*) from *D. capitatus* ssp. *andrzejowskianus* into cultivated carnations. Using the interspecific hybridization approach, the world's first bacterial wilt resistant carnation cultivar, 'Karen Rouge' has been developed. It is a deep yellowish-red standard-type cultivar selected from a cross between carnation cultivar 'Miracle Rouge' and bacterial wilt resistant line 4AZ31-5 (*D. capitatus* ssp. *Andrzejowskianus*-wild species resistant to bacterial wilt). The cultivar 'Karen Rouge' has a high degree of resistance to bacterial wilt, and its productivity and other traits desirable for cut flowers are similar to those of modern commercial cultivars. Nimura *et al.* (2006) attempted reciprocal interspecific crosses between seven lines of *D. caryophyllus* and one line of *D. isensis* in order to introduce the characters of extremely early flowering and unique flower shape from *Dianthus isensis* into *Dianthus caryophyllus* L. Interspecific hybridization between carnation lines with long vase life and *D. superbus* var. *longicalycinus* was carried out to combine early flowering, high productivity, and long vase life (Onozaki *et al.*, 2011). Breeding using *D. superbus* var. *longicalycinus* was effective for selecting early-flowering progeny.

9.2.5 Tuberose

Interspecific hybridization has been exploited for the development of novel flower colors in tuberose. More than 50 coloured hybrids have been bred with shades of pink, reddish-purple, purple, orange and yellow flower colors by *Polianthes tuberosa* with *P. howardii* (Huang, 2002). A double flowered lilac pink hybrid 'Chia Nong Pink Sapphire' has been patented by Huang Kuang-Liang in 1997 through interspecific crosses between *Polianthes tuberosa* × *Polianthes howardii.*

9.2.6 Lily

In the genus *Lilium*, interspecific hybridization is the most important breeding program that had been conducted to produce novel cultivars. Asiatic hybrid lilies are derived from interspecific hybridization among *L. dauricum*, *L. lacifolium* and *L. maculatum* and so on. While, *L. auratum*, *L. speciosum*, *L. japonicum*, *L. rubellum*, *L. alexandrae* and *L. nobilissimum* have contributed to the development of modern-day Oriental hybrid lilies. However, genetic cross combinations in *Lilium* are limited by incompatibility and incongruity due to sexual barriers, which can be separated into pre- and postfertilization barriers. *Lilium longiflorum* (2n = 2x = 24) is an important species for flower production and acts as a major parent in interspecific hybridization programs. *L. lophophorum* var. *linearifolium* is as important germplasm resource of *Lilium* sp. due to its vivid color, low temperature and disease resistance. Wang *et al.* (2009) successfully attempted interspecific hybridization between *Lilium longiflorum* and *L. lophophorum* var. *linearifolium* using cut style pollination and *in vitro* embryo rescue techniques. The ovules excised at 10 DAP developed into seedling. Obata *et al.* (2000) obtained 9 and 3.6 per cent seedlings respectively, from the cross-combination of *L. nobilissimum* x *L. regale* and *L. regale* x *Lilium nobilissimum* via ovules-with-placental-tissue culture at 30 or 40 DAP. *Lilium Brownie* var. *colchesteri* (BRO) possess some unique ornamental traits *viz.*, graceful harmony between flower and anther color (brown color), elegant fragrance and petal color changing from yellowish cream to white during anthesis. In order to introduce these traits to cultivated lilium, interspecific crosses of *L.×formolongi*, *L. formosanum*, *L. longiflorum* (as seed parents) and BRO (as a pollen parent) were carried out (Hai *et al.*, 2012). Hybrid plantlets were successfully obtained from crosses of *L.* × *formolongi* × BRO and *L. formosanum* × BRO, but not from the cross of *L. longiflorum* × BRO. The hybrids showed changes in flower colour from yellowish cream to white during anthesis as does the pollen parent.

9.2.7 Tulip

Wild *Tulipa* species constitute a potential genetic resource for the improvement in tulip cultivars through interspecific hybridization because they harbor several favorable genes, such as tolerance to drought and cold, resistance to diseases and viruses. Hence, there is tremendous scope for improvement of tulip cultivars by integrating desirable traits from wild *Tulipa* species using interspecific hybridization approach. The arrest of pollen tube growth at different positions within the pistils has been observed as the main pre-fertilization barrier for interspecific hybridization in tulips. Van Eijk *et al.* (1991) successfully obtained interspecific hybrids from crosses between *T. gesneriana* cultivars and 10 species belonging to the Eichleres

Promising New Interspecific Hybrids of Ornamental Plants

Karen Rouge: A carnation cultivar resistant to bacterial wilt developed through interspecific hybridization

(a) *D. capitatus* ssp. *andrzejowskianus:* A highly bacterial wilt resistant species.
(b) Carnation Nou No. 1'- a breeding line resistant to bacterial wilt developed through interspecific hybridization (*D. carophyllus* 'Super Gold' × *D. capitatus* ssp. *andrzejowskianus* (*Source*: Onozaki, 2018)

Chia Nong Pink Sapphire: A colored hybrid of tuberose developed by crossing *Polianthes tuberosa* × *P. howardii*

Interspecific Hybridization of Alstroemeria

Sweet Laura
(*A. aurea x A. caryophyllaea*)

Mauve Majesty

Tangerine Tango

Source: Bridgen *et al.* (2009).

section, except for the crosses with *T. micheliana, T. kaufmanniana and T. praestans*. In these interspecific crosses, the cross of *T. gesneriana* L. with *T. fosteriana* was most important, because it has resulted in Darwin Hybrids tulips. Custers *et al.* (1992, 1995), successfully obtained intraspecific crosses of *T. gesneriana* × *T. gesneriana* using non-abortive embryos collected from flowers, 3 to 9 weeks after pollination. Custers *et al.* (1995) used the direct ovule culture technique to obtain interspecific cross where *T. gesneria*na Christmas Marvel (CM) and Gander (GA) were used as female parents, and *T. kaufmanniana* as the male parent. Okazaki, (2005) applied embryo culture to rescue abortive embryos in the cross of *T. gesneriana* with several species of section Eichleres. As a result, hybrids were effectively produced with relatively high bulblets formation rates per cultured embryo.

9.2.8 Hibiscus

One of the major breeding goals in hibiscus is to develop interspecific hybrids between winter-hardy hibiscus species and the tropical Chinese hibiscus to create 'tropical'-looking hibiscus hybrids that are adapted to temperate environments. The herbaceous winter-hardy species belonging to the section Muenchhusia and native to North America (*H. coccineus, H. moscheutos, Hibiscus militaris, Hibiscus dasycalyx*, and *Hibiscus lasiocarpos*) are mostly diploid (2x = 2n = 38; Hinsley, 2012), while the chromosome numbers in the tropical *H. rosa-sinensis* group may vary (2n = 36, to 144), depending on the ecotype and degree of hybridization. Numerous pre- and postfertilization barriers prevent interspecific hybridization and/or production of fertile F_1 plants in Hibiscus species belonging to different sections. One of the first successful attempts to create interspecific winter-hardy hibiscus hybrids belonging to the section Muenchhusia resulted in an amphidiploid hybrid named "Ai-Fuyo" (*H. muta-moscheutos* Kuwada), where *H. mutabilis* (2x = 2n = 92) was used as maternal parent and *H. moscheutos* (2x = 2n = 38) was the pollen donor (Kuwada, 1964). Recently, Pounders and Sakhanokho, (2016) successfully hybridized *H. mutabilis* (seed parent) with *H. moscheutos* (pollen donor) and selected three interspecific cultivars namely, 'Hapa White', 'Hapa Red', and 'Hapa Pink'. Cold hardiness of these cultivars was inherited from the maternal component, the confederate rose. Kuligowska *et al.* (2016 b) have observed incompatibility reactions and post-fertilization barriers among the interspecific crosses between of *H. rosa-sinensis* and cold hardy Hibiscus species (*H. syriacus* and *H. moscheutos*). *H. syriacus* was successfully crossed with *H. paramutabilis* and *H. sinosyriacus* (Van Laere *et al.*, 2007). The interspecific hybridization between *H. sinosyriacus* and *H. syriacus* lead to the development of the cultivar 'Tohagol Red', with unique flower shape and color, uniform growth habit and hardiness (Ha *et al.*, 2014). A number of multiple hybrids namely "Angelique", "Cherub", "Pink Comet" have been developed by crossing *H. moscheutos* and *H. rosa-sinensis*.

9.2.9 Hydrangea

Distant hybridization techniques have been used successfully utilized to broaden the genetic base of commercial cultivars of *Hydrangea macrophylla* (Kudo *et al.*, 2008). With the aim to combine the evergreen foliage and winter flowering characteristics of *H. scandens ssp. chinensis* with ornamental characteristics of *H.*

macrophylla, interspecific hybrids have been successfully recovered between *H. macrophylla* and *H. scandens* ssp. *chinensis* using ovule culture techniques (Kudo *et al.*, 2008).

9.2.10 Primula

Primula × *polyantha*, which is one of the most popular primulas, have its origin from fertile interspecific hybrids between *P. vulgaris* and *P. veris* (Richard, 1993) and numerous seed propagation cultivars have been bred in this hybrid species without any problem. Therefore, genomic difference between these two parental species is considered to be small or none. Inter-section hybrids were successfully obtained by rescuing the immature embryos produced in the cross between *P. sieboldii* of Sect. Cortusoides and *P. obconica* of Sect. Obconicolisteri (Kato *et al.*, 2001).

9.2.11 Campanula

Wide hybridization between Campanula species could be used to increase phenotypic variability in this genus. However, pre- and post-fertilization barriers often prevent successful fertilization between Campanula species. Embryo rescue techniques such as embryo-, ovary- and ovule culture are often applied to overcome post-fertilization barriers, as in *Dendranthema* and *Cyclamen* (Cheng *et al.*, 2010). One of the first steps in the establishment of an efficient protocol for ovule culture is the determination of the optimal ovule developmental stage for *in vitro* culture. A positive effect of ovule isolation time on germination was observed for most of the interspecific hybrids of *C. medium* and *C. formanekiana*. Longer *in vivo* maturation of ovules results in higher ovule germination rates in *C. medium* and *C. formanekiana interspecific* crosses (Roper *et al.*, 2012).

Roper *et al.* (2015) employed ovule culture technique for the generation of interspecific hybrids in *Campunala*. They reported that interspecific hybrids between *C. medium* and *C. formanekiana* exhibited an increased number of viable ovules with later isolation time, but with different ovule germination rates. A parent-of-origin effect on both the number of collectible ovules and ovule germination was observed for *C. medium* x *C. formanekiana*. The occurrence of mis-coloured seedlings in interspecific crosses indicated incompatibilities between the parental lines that could not be prevented by reciprocal crossing. The low number of collectible ovules and germination rates might be inhibited due to fertilization barriers.

9.2.12 Cyclamen

The interspecific hybridization approach has successfully utilized in cyclamen to introgress the scent from *C. purpurascens* into *C. persicum* (Ishizaka and Uematsu, 1995). They used ovule culture techniques to rescue the abortive hybrid embryos and to recover the interspecific hybrids.

Kuwayama *et al.* (2005) successfully circumvented the cross-compatibility barriers among the Colchicaceous ornamentals using ovule culture technique and produced several interspecific and intergeneric hybrid plants among *Gloriosa* spp., *Littonia modesta* and *Sandersonia aurantiaca*. Similarly, wide hybridization in the family Colchicaceae: *S. aurantiaca* x *L. modesta* and *L. modesta* x *S. aurantiaca* have been attempted (Morgan *et al.*, 2001), and *S. aurantiaca* x *G. rothschildiana* (Nakamura *et al.*, 2005). In Gypsophila, new hybrid with big and pale pink flowers and thick stems have been developed by crossing *G. paniculata* L. 'Red Sea' with double-pink flowers with G. *manginii* having thick stems (Kishi *et al.*, 1994).

Table 15: Successful Examples of Interspecific Hybridization in Ornamental Plants

	Interspecific Cross	*Method*	*Trait*	*Reference*
Campanula	*C. portenschlagiana* × *C. poscharskyana*	Ovule culture	Phenotypic variability	Roper *et al.* (2015)
	C.medium × *C. formanekiana*	Ovule culture	Phenotypic variability	Roper *et al.* (2015)
Cyclamen	*C. purpurascens* × *C. persicum*	Ovule Culture	Scent	Ishizaka and Uematsu (1995)
Hydrangea	*H. scandens* ssp. *Chinensis* × *H. macrophylla*	Ovule culture	Large, double type, pink color flowers	Kudo *et al.* (2008)
Chrysanthe-mum	*Chrysanthemum grandiflorum* 'Yuhuaxingchen' × *C. indicum*	Repetitive Backcrossing	Drought tolerance	Sun *et al.* (2010)
	Chrysanthemum grandiflorum (Ramat.) Tzvel. × *C. makinoi*		Aphid resistance and heat tolerance	Cheng *et al.* (2011)
	Dendranthema morifolium (Ramat.) Kitamura 'rm20-12' × *D. nankingense*	Ovary culture	Cold tolerance	Cheng *et al.* (2010)
	C. grandiflorum × *Artemisia vulgaris* L.	Embryo rescue	Aphid resistance	Deng *et al.* (2009)
	C. grandiflorum +*Artemisia sieversiana*	Somatic hybridization	Rust resistance	Furuta *et al.* (2004)
	[*Chrysanthemum grandiflorum* × *Artemisia vulgaris*] × [*Chrysanthemum crassum* × *Crossostephium chinense*]	Trigeneric hybrids	Aphid resistance, Salt tolerance	Deng *et al.* (2012)
Lily	*L. Brownie* var. *colchesteri* × *L. × formolongi* and *L. formosanum*	Ovary Slice culture	Flower color	Hai *et al.* (2012)
	Lilium longiflorum × *L. lophophorum* var. *linearifolium*	Ovule culture	Low temperature and disease resistance	Wang *et al.* (2009)
Alstroemeria	*A. pelegrina* × *A. caryophyllaea* segregant	Embryo rescue	Flower scent	Aros *et al.* (2019)
Alstroemeria	*Alstroemeria aurea* × *A. caryophyllaea*	Embryo rescue	Fragrance, winter hardiness, and long flower stems	Bridgen *et al.* (2009)
Iris	*Iris ensata* + *Iris germanica*	Somatic hybridization	Genetic diversity	Shimizu *et al.* (1999)
Carnation	*Dianthus chinensis* + *D. barbatus*	Somatic hybridization	Flower color	Nakano and Miii (1992)
	D. caryophyllus f. *spray* × *D. chinensis, D. carthusianorum*	Interspecific hybridization	Fusarium wilt resistance	Slavov *et al.* (2001)

	Interspecific Cross	Method	Trait	Reference
Marigold	*Tagetes erecta* × *T. patula*	Interspecific hybridization	Long flower duration	Namita *et al.* (2011)
Eustoma	*E. grandiflorum* × *E. exaltatum*	Interspecific hybridization	Heat tolerance; Genetic diversity	Barba-Gonzalez *et al.* (2017)
Tuberose	*Polianthes tuberosa* × *P. howardii.*	Interspecific hybridization	Colour	Huang (2002)

10.0 Marker Assisted Breeding

To strengthen the conventional breeding approaches, the advent of molecular-DNA marker has been proved instrumental to facilitate the marker assisted selection (MAS) and marker assisted backcrossing (MAB) for the genetic improvement of ornamental crops (Su *et al.*, 2019). The marker assisted approaches have been utilized to improve the genetic gain and disease resistance in commercial flower crops. In particular, marker-assisted selection (MAS) and marker-assisted backcrossing have been widely adopted to improve resistance to diseases and other relatively simple traits. Genomic selection (GS), genetic mapping, identification of quantitative trait loci (QTLs), and marker-assisted selection have been used to facilitate the introgression of desirable alleles from wild species to elite cultivars. Genotyping by sequencing (GBS) has been widely used to construct high-coverage linkage maps and conduct QTL analyses in multiple important ornamental crops. Plants have inbuilt immunity to endure various biotic and abiotic stresses (Younis *et al.*, 2020). The PCR (polymerase chain reaction)-based markers have improved the ability of crop plants to defend themselves against various stresses. Currently, the sequence-characterized amplified regions (SCAR), single nucleotide polymorphism (SNPs) and simple sequence repeats (SSRs), diversity array technology (DArT) and cleaved amplified polymorphic sequences (CAPS) are most commercially exploited PCR-based markers for the genetic improvement of crop plants (Younis *et al.*, 2020).

Another category of markers, functional markers (FM), which are derived directly from the genomic region of the gene controlling the target trait of interest are effectively utilized for precision plant breeding (Salgotra and Jr Stewart, 2020). The FM is also called as precision or diagnostic markers. These molecular markers are efficient in screening of large germplasm for allelic diversity analysis with high accuracy and fixing useful alleles in breeding populations. The generation of functional markers requires the pre-existing information on allele sequences of functionally characterized genes from which polymorphic, functional motifs affecting plant phenotype can be identified (Andersen and Lubberstedt, 2003).

With the advancements in next-generation sequencing (NGS) technologies many QTLs have been identified and characterized with the help of mapping populations like doubled haploids (DHs), Near Isogenic Lines (NILs) and Recombinant inbred lines RILs. The NGS tools have enhanced the applications of Restriction enzyme-assisted sequencing methods, including restriction-site-associated DNA sequencing (RAD-seq), genotyping by sequencing (GBS) and specific-locus amplified fragment sequencing (SLAF-seq) for the crop improvement (Song *et al.*, 2020). The possibility of fine mapping and ultrahigh-density genetic linkage map construction has been enhanced in ornamental crops using SLAF-seq. For instance, exploiting SLAF-seq technology in hybrid *Ginkgo biloba* plants a total of 204,361 SNP markers were developed and a high-density genetic map with 12,263 markers was constructed (Song *et al.*, 2020). With the help of SLAF-seq approach in *Paeonia suffruticosa* (Cai *et al.*, 2015) the high-density genetic linkage map was constructed. In case of ornamental crop, the flower type is an important quality trait for ornamental breeders. Mining tightly linked markers to the target traits based on QTL analysis is a crucial approach to improve the quality of crop plants. The advent of molecular markers has enabled the construction of high-density genetic linkage map in crop plants including ornamental flower crops for mapping complex quantitative traits. Exploiting the SLAF-seq technology, Song *et al.* (2020) constructed the high-density genetic linkage map of chrysanthemum with a genetic distance of 0.76 cM and identified three major QTLs governing corolla tube merged degree trait and four QTLs for relative numbers of ray florets. The study rendered the development of 6452 high-quality SNP primers for chrysanthemum. Yagi *et al.* (2013) generated reference genetic linkage map in carnation for QTL mapping, map-based cloning and identification of markers.

10.1.0 Molecular Breeding and Gene Functions in Ornamental Plants

10.1.1 Flower Traits

The QTLs/genes and linked markers for different quantitative and qualitative traits in ornamental crops have been demonstrated in Table 16. For instance, flower colour is an important qualitative trait in ornamental crops. The conventional breeding, interspecific hybridization and mutation breeding has been successfully utilized for developing novel colours in flower crops. In the rose cultivars the petal colours exhibit a great diversity. Numerous studies in genetics of flower colour in roses have demonstrated the role of QTLs for red, pink or yellow colour petals (Byrne, 2009). Recently, Henz *et al.* (2015) reported two major stable QTLs for anthocyanin content in roses on LG2 and LG6. More recently, within the QTL region of LG3 a rose *APETALA2/TOE* homologue was reported as candidate gene controlling the petal number and knockdown of which reduced the petal numbers (Hibrand Saint-Oyant *et al.*, 2018). Ray petal colors of ancestral chrysanthemums are limited to pink, yellow, and white, which derived from anthocyanins and/or carotenoids, and absence of these pigments, respectively (Kishimoto *et al.*, 2004). In chrysanthemum, array of flower colours like purplish red, orange, red, and dark red, have been developed by using different breeding approaches. Double flower type is another important breeding trait in flower crops. In this context, Wang *et al.* (2020) constructed a linkage map of carnation exploiting double-digest restriction site-associated DNA sequencing (ddRAD-seq) with 2353 single nucleotide polymorphisms (SNPs). Based on the integrative analysis of QTL mapping and BSR-seq analysis they suggested that an SNP mutation in the *miR172* cleavage site of the A-class flower organ identity gene *APETALA2* (*DcAP2L*) is responsible for double flower formation in *Dianthus*. Flowering time is a potent trait in chrysanthemum. Zhang *et al.* (2013) reported the QTLs (*qDS-Y5, qDS-Y15.1* and *qDS-Y15.2*) for flowering time in chrysanthemum using mapping population of cross between the autumn-flowering cultivar 'Yuhualuoying' and the summer-flowering 'Aoyunhanxiao'.

Table 16: Molecular Mapping of Genes/QTLs in Ornamental Plants

Crop	*Gene/QTLs/TFs*	*Trait*	*Marker Type/ Methodology*	*Reference*
Lily	*LFCc*	Flower colour	AFLP	Shahin *et al.* (2011)
	lfs	Flower spot	DArT	
	LSC	Stem colour	NBS	
	lal	Antherless phenotype		
	lfd	Flower direction		
Carnation	*Cbw1, Cbw2, Cbw4, Cap1, Cap2*	Bacterial wilt resistance	RAD, SNP	Yagi *et al.* (2013)
Gerbera	RBQWI1, RBQWI3, RBQB5	Botrytis resistance	SNP	Fu *et al.* (2017)
Carnation	Cbw1, Cbw2, Cbw3	Carnation bacterial wilt	STS markers (STS-WG44)	Yagi *et al.* (2013)
Rosa chinensis	RhCyc2, AB201048, AB038247	Early to late flower development	Microarray (Rosa1_ Affyarray) tool)	Dubois *et al.* (2011)
Chrysanthemum	*CmHSFA4*	Salt tolerance	Sequence analysis and expression profiling	Li *et al.* (2015)
	CmDREB6	Heat tolerance		Du *et al.* (2018)
Chrysanthemum	*CmCYC2 genes*	Increase flower number and length of ray floret	Sequence analysis and expression profiling	Huang *et al.* (2016)
	Cm-BBX24	Salt and drought tolerance	RNAi mediated transgenic approach	
Rose	*Rdr1*	Leaf spot resistance	Sequencing and de novo assembly	Terefe-A *et al.* (2011)
Tulip	*QTL1 to QTL6*	Fusarium resistance	SNP genotyping and SSRs	Tang *et al.* (2015)
Chrysanthemum	*CtA32 52 GT, encoding UDP-glucose : anthocyanin 32,52 -O-glucosyltransferase*	Blue color flower	Transgenic	Noda *et al.* (2017)

10.1.2 Disease Resistance

A linkage map provides a basis for the identification of genomic regions related to traits of interest and the necessary infrastructure for MAS breeding. Quantitative trait locus (QTL) mapping is a highly effective approach for studying genetically complex forms of plant disease resistance. Yagi *et al.* (2012) reported one major QTL (*Cbw1*) and two minor QTLs (*Cbw2* and *Cbw3*) for carnation bacterial wilt (CBW) resistance derived from *D. capitatus* ssp. *Andrezejowskianus.* Another QTL (*Cbw4*), derived from carnation line 85–11, based on SSR based genetic linkage map. Gerbera is an important cut flower crops and its production is limited by several biotic and abiotic stresses. While its transportation the prevalence of high humidity leads to incidence of grey mould (Fu *et al.*, 2017). Based on the genetic linkage map analysis and SNP markers, Fu *et al.* (2017) identified major QTLs (*RBQWI2, RBQWI4 and RBQWI6*) for resistance to *Botrytis cinerea* in Gerbera. Likewise, in rose using virus induced gene silencing Li *et al.* (2020) identified AP2/ERF family gene *'RcERF099'* as an important regulator for *B. cinerea* resistance in the rose petal. Members of the large WRKY transcription factor (TF) family are responsible for the regulation of plant growth, development and the stress response. Li *et al.* (2015) isolated five WRKY TF genes from chrysanthemum and reported that the overexpression of *CmWRKY48* is associated with aphid resistance. Chrysanthemum white rust disease (*Puccinia horiana* Henn) severely reduces the ornamental quality and yield chrysanthemum. In this context, Bi *et al.* (2021) recently reported that over expression of *CmWRKY15-1* gene induce resistance to *Puccinia* fungus. The genes/major QTLs identified in important ornamentals have been listed in Table 16.

10.1.3 Abiotic Stress Resistance

Abiotic stress such as cold, high salinity, metal ions, drought and other abiotic stress will severely inhibit plant growth progress, cause morphological and physiological changes, and even death. In plant stress resistance system, transcription factors regulate the expression of functional genes, which are essential for plant stress response (Liang *et al.*, 2017). Many transcription factors related to plant stress tolerance have been identified, such as DREB, bZIP, NAC and WRKY. In line with this, Liang *et al.* (2017) reported that expression of *WRKY* gene *DgWRKY5* in transgenic chrysanthemum induces tolerance to salt stress. Unlikely, CmWRKY17 was found to negatively regulate the salt stress in chrysanthemum (Li *et al.*, 2015). In recent years, the threat of global warming and the wide-reaching implications of the adverse effects on plant growth have encouraged the study of heat stress. High temperatures typically stimulate the expression of heat shock protein genes (HSPs), and these can have key roles in enabling the survival of an individual in the face of potentially fatal high temperatures. Wan *et al.* (2015) identified heat responsive genes in carnation via RNA-seq analysis. Similarly, Lee *et al.* (2020) transferred the superoxide dismutase 2 (*SOD2*) (from *Escherichia coli*) gene to *Rosa hybrid* and obtained drought tolerance transgenic rose lines. They suggested that the over-expression of *SOD2* enhances drought tolerance in rose plants. The enzyme *RNAPII CTD* phosphatase-like 1 is known as a transcriptional regulator of the plant response to various abiotic stresses. Qi *et al.* (2018) isolated *CmCPL1*; a chrysanthemum (*Chrysanthemum morifolium*) gene encoding this enzyme and reported that overexpression of this gene enhances heat tolerance in chrysanthemum. Similarly, in another study Du *et al.* (2018) reported that overexpression of *CmDREB6* induces tolerance to heat stress in chrysanthemum.

11.0 Genetic Transformation in Ornamentals

Genetic engineering involves the transfer of a single or a few genes between unrelated species by making use of the recombinant DNA technology. It is also occasionally referred to as transgenic biotechnology. Transgenic plants referred to those plants in which one or more genes have been introduced from unrelated organisms through the process genetic engineering. In the modern era of plant breeding, the techniques of genetic manipulation can greatly enhance the pace of improvement of ornamentals. The technique of genetic engineering has tremendous potential to incorporate those characters into a crop species which are either difficult or impossible to introgress/improve through the conventional breeding approaches. In ornamentals, this approach can be widely employed as a valuable tool to modify floral attributes like color, forms, fragrance, vase life and appearance as well as for increasing resistance against biotic and abiotic stresses. Transgenic strategies possess immense potential to produce novel flower phenotypes that are not found in nature (Noman *et al.*, 2017).

11.1.0 Techniques of Genetic Transformation

Genetic transformation refers to the transfer of desired genes, their stable integration and expression of foreign DNA in a particular organism. There are several methods of genetic transformation, namely, Agrobacterium mediated, particle bombardment, protoplast transformation, polyethylene mediated, electroporation, microinjection, physical abrasives, laser beams, in planta transformation, and vacuum infiltration. However, only two approaches *i.e.*, Agrobacterium and particle bombardment-mediated gene transfer are widely employed for the development of transgenic plants.

11.1.1 *Agrobacterium tumefaciens*-Mediated Gene Transfer

This is one of the widely employed approaches to develop transgenics in plants. The extraordinary capacity of *Agrobacterium* to transfer its genetic material to host cell have evolved this bacterium from phytopathogen to a powerful transgenic vector. Both *A. tumefaciens* and *A. rhizogenes* are considered as natural genetic engineers due to their ability to transfer and integrate DNA into plant genomes through a unique intergeneric gene transfer mechanism. Both of these bacteria are reported to be phytopathogenic, *A. tumefaciens* causes the crown gall disease, while *A. rhizogenes* is responsible for hairy root disease. The large plasmid of *A. tumefaciens* induces tumor in the infected plant cells, and hence called tumor-inducing (Ti) plasmid. It has been revealed that only a particular gene segment called transfer DNA (T-DNA) from the Ti plasmid of Agrobacterium is transferred and stably integrated into the chromosomal DNA in the nucleus of the host plant. The expression of T-DNA genes of Ti-plasmid in plant cells causes the formation of tumors at the infection site. The process of Agrobacterium-mediated transformation involves the stable integration of a desired DNA sequence (T-DNA) from Ti plasmid of *A. tumefaciens* into the plant genome. The major advantages of this gene transfer methods include; wide host range of plants, integration of low copy number of T-DNA into host genome, stable gene expression and transformation of large size DNA segments. This approach has been widely employed for genetic transformation of large number of flower crops including Rose (Li *et al.*, 2002a, 2002b; Pourhosseini *et al.*, 2012; Condliffe *et al.*, 2003), Chrysanthemum (Courtney-Gutterson *et al.*, 1994; Shinoyama *et al.*, 1998), Anthurium (Kuehnle and Sugii, 1991), Carnation (Lu *et al.*, 1991) and Gerbera (Elomaa *et al.*,1993).

11.1.2 Particle Gun Method

This method is also termed as direct method of gene transfer or Biolistic Particle Delivery method. The gene gun propels tiny gold or tungsten particles (1mm in diameter) coated with DNA containing the genes of interest into the plant tissue. Those cells which incorporate the new DNA into their genome are selected and allowed to grow into full transgenic plants. Particle bombardment method is widely used in the transformation of organelles such as chloroplasts, which enables engineering of organelle-encoded herbicide or pesticide resistances in crop plants and to study photosynthetic processes. Transformation using this method has been successfully done in Gladiolus (Kamo *et al.*, 1995), Rose (Marchant *et al.*, 1998b), Orchid (Kuehnle and Sugii, 1992), Lily (Penafiel and Kamo, 1994). The major disadvantages of this method include, frequent integration of multiple copies of the transgene at a single insertion site, rearrangement of the inserted genes, and incorporation of the transgene at multiple insertion sites.

11.2.0 Traits Improved through Genetic Transformation

The major traits that have been improved through genetic engineering techniques are broadly classified as:

- ✰ **Abiotic stress**: Herbicide resistance, Salt tolerance, Drought tolerance, Heat tolerance, cold tolerance
- ✰ **Biotic stress resistance**: Insect resistance, Disease resistance
- ✰ **Quality improvement**: Plant architecture, Flower colour, Fragrance Enhanced shelf life, Blooming rooting characteristics *etc.*

11.2.1 Abiotic Stress Tolerance

Abiotic stresses such as salinity, water scarcity, high and low temperatures, and drought *etc.* can adversely affect the yield and quality of ornamentals. Ornamental plants exhibit different responses upon exposure to abiotic stresses. The technique of genetic engineering has been widely used for developing abiotic stress tolerant varieties. The cold tolerance in *Rosa chinensis* Jacq. was improved by using transgenic approach by integrating the Medicago gene *MtDREB1C* (Chen *et al.*, 2010). The frost tolerance in petunia has been enhanced by transforming it with *AtCBF3* gene (Warner, 2011). Low temperature resistant chrysanthemum lines have been successfully developed by integrating the AtDREB1A gene (Hong *et al.*, 2006). The constitutive expression of *CcSOS1* encoding for plasma Na^+/H^+ antiporter has improved the salt stress tolerance in chrysanthemum (An *et al.*, 2014; Song *et al.*, 2014). Two constructs *i.e.*, 35S:DREB1A and rd29A:DREB1A were integrated into chrysanthemum genome for enhancing tolerance to salinity and water deficiency. Some of the examples of abiotic stress tolerance in ornamental crops through transgenic approach is presented in Tables 17 and 18.

11.2.2 Biotic Stress Resistance

Plant growth and productivity are severely affected by abiotic and biotic stresses. Due to inadequate number of resistance genes, breeding for stress resistance/tolerance in ornamental plants is difficult (Azadi *et al.*, 2016).

Table 17: Important Transgenic Ornamental Plants for Abiotic Stress Tolerance

Plants	*Gene*	*Resistant to*	*Reference*
Chrysanthemum	*CmWRKY1* transcription factor (derived from *C. morifolium*	Drought stress tolerance	Fan *et al.* (2016)
	CgHSP70 gene (from *C. morifolium*)	Heat tolerance	Song *et al.* (2014)
	CmWRKY17 transcription factor;	Salinity stress	Li *et al.* (2015)
China rose (*Rosa chinensis*)	*RcXET* and *MtDREBIC* genes	Freezing and drought tolerance	Chen *et al.* (2016)
	AtDREB2A-CA gene	Salinity stress	Josine *et al.* (2015)

11.2.2.1 Insect Pest Resistance

Insect pest management is an important concern in floriculture industry as insect damage reduces the ornamental as well economical value of flowers. Though chemical control measures are available but they are hazardous and are not eco-friendly. GE technology has tremendous potentials in ornamentals for improving the resistance against the insect-pests. The insect resistance has been conferred in various ornamental by utilizing the *Bt* genes from *Bacillus thuringensis* (Christou *et al.*, 2006). However, this gene has proved less effective against several pests of ornamental crops. Alternatively, the insecticidal genes such as lectins, protease inhibitors, *etc.* can also be utilized for integration of insect resistance in ornamentals.

The introduction of *Bacillus thuringiensis* (*Bt*)-based resistance gene (*Cry1ab*) has successfully improved resistance to caterpillars, *e.g.*, common cutworm (*Spodoptera litura*) and tobacco budworm (van Wordragen *et al.*, 1993; Shinoyama *et al.*, 2003, 2008, 2012). Aphid resistance was obtained by transferring the *Lycoris longituba* agglutinin (*LLA*) gene into chrysanthemum (He *et al.*, 2009). Combined resistance toward insects (Caterpillars) and pathogens (white rust) has been successfully obtained in chrysanthemum through the introduction of modified *Cry1Ab* gene of *Bacillus thuringiensis* and modified *sarcotoxin I1* gene of *Sarcophaga peregrine* (Shinoyama *et al.*, 2015). Some of the examples of insect pest resistance in ornamental crops through transgenic approach is presented in Table 18.

11.2.2.2 Virus Resistance

Flower crops are reported to be attacked by wide range of viruses. Approaches like coat protein genes, mutated replicase (RP) genes RNAi technology is found effective in imparting resistance against various viral diseases. In chrysanthemum, the resistance to Cucumber mosaic virus (CMV) has been successfully conferred by the introduction of coat protein (cp) gene of CMV (Kumar *et al.*, 2012). The introduction nucleocapsid genes was found effective against Tomato spotted wilt virus (TSWV) in chrysanthemum (Sherman *et al.*, 1998). In gerbera, resistance against TSWV have been obtained by introduction of the NP-gene (nucleoprotein gene) (Korbin, 2006). Resistance to CMV was introduced in lily by use of a defective CMV replicase gene (Azadi *et al.*, 2011). It was revealed that CMV-GDD Replicase gene was an effective construct to confer protection against CMV (Cucumber mosaic virus) in *Lilium*. In *Poinsettia*, virus-derived hairpin RNA constructs conferred resistance to Poinsettia mosaic virus (PnMV) (Clarke *et al.*, 2008). In gladiolus, transgenic approach was used to introduce resistance against Bean yellow mosaic virus (BYMV) and CMV. Gladiolus plants transformed with the bean yellow mosaic virus coat-protein gene in either sense or antisense orientation exhibited variable degree of resistance to BYMV and was not stable (Kamo *et al.*, 2005). Resistance to Cucumber mosaic virus (CMV) was incorporated in gladiolus using CMV coat protein and CMV replicase genes, or using genes encoding for single chain variable fragment antibodies (scFv) to CMV (Kamo *et al.*, 2012). Some of the examples of virus resistance in ornamental crops through transgenic approach is presented in Table 18.

Table 18: List of Transgenes for Biotic and Abiotic Stress Tolerance in Ornamental Plants

Trait	*Gene/Genes Responsible for the Trait*	*Source of Gene*	*Transgenic Plants Developed*
Biotic Stress			
Septoria leaf spot disease resistance	*Rice chitinase* (*chi II*)	*Oryza sativa*	*Chrysanthemum*
Tolerance to *Botrytis cinerea*	*Rice Chitinase 10* (RCH10)	*Oryza sativa*	*Lilium oriental 'Star Gazer'*
Tolerance to *Powdery mildew*	*Rice-chitinase* gene	*Oryza sativa*	*Rosa*
Burkholderia disease resistance	*Oat thionin*	*Avena sativa*	*Dianthus caryophyllus*
Gray mold (*Botrytis cinerea*) disease tolerance	*RCC2*	*Oryza sativa*	*Chrysanthemum*
Alternaria leaf spot disease resistance	*hpaGxoo* gene	*Xanthomonas oryzae* cv. *oryzae*	*Chrysanthemum morifolium*
Alternaria leaf spot disease resistance	*PGIP*	*Prunus mumei*	*Chrysanthemum*
Insect resistance	*cry1Ab*	*B. thuringiensis*	*Chrysanthamum*
Abiotic Stress			
Salt (Iron)	*Fro2*	*Arabidopsis thaliana*	Rhododendron
Drought	*P5CS D1* (D1-pyrroline-5-carboxylate Synthetase) (proline)	*Arabidopsis thaliana*	Petunia
Heat	*AtDREB1A*	Arabidopsis	Chrysanthemum
Heat and Water stresses	*DvDREB2A*	*Dendranthema grandiflora*	*Chrysanthemum*
Freezing stress tolerance enhancement	*MtDREB1C*	*Medicago truncatula*	*Rosa chinensis*
Cold	*DREB1C*	*Medicago truncatula*	China Rose
Cold	LTP(lipid transfer protein)	Rice	Orchid (*Phalaenopsis amabilis*)
Cold	*DREB1C*	*Medicago truncatula*	China Rose
Drought and Salinity	*RhNAC2* and *RhEXPA4*	Rose	Rose

Trait	*Gene/Genes Responsible for the Trait*	*Source of Gene*	*Transgenic Plants Developed*
Salt, cold and drought	*ICE1*	*Chrysanthemum dichrum*	*Chrysanthemum grandiflorum*
Salt	*NnCIPK* (calcineurin B-like protein kinase)	Lotus	Lotus
Salt, drought and temperature	P5CS1 and P5CS2	*Chrysanthemum lavandulifolium*	*Chrysanthemum × morifolium*
Salinity and drought	*ClCBF1* gene	*Chrysanthemum lavandulifolium*	*Chrysanthemum morifolium* 'White Snow'

11.2.2.3 Fungal and Bacterial Disease Resistance

Ornamental plants are attacked by different kinds of fungal and bacterial pathogens. Various genes like chitinase, glucanase, osmotin, defensin, *etc.* plays an important role for imparting resistance against bacterial and fungal diseases. In chrysanthemum, resistance to *Alternaria tenuissima* was successfully achieved through the overexpression of *hpaGXoo* gene (Xu *et al.*, 2011) and the *Prunus mumei PGIP* (polygalacturonase-inhibiting protein) gene (Miao *et al.*, 2010). The integration of the rice chitinase gene (*Chi-II*) has improved against *Septoria* leaf spot disease in chrysanthemum (Sen *et al.*, 2013). Further, the rice chitinase gene (Takatsua *et al.*, 1999) and N-methyltransferase genes; *CaXMT1*, *CaMXMT1*, and *CaDXMT1* (Kim *et al.*, 2011) have been reported to confer resistance *Botrytis cinerea* in chrysanthemum. Partial black spot and powdery mildew resistance in roses has been induced through the expression of antifungal proteins (Dohm *et al.*, 2002; Marchant *et al.*, 1998; Li *et al.*, 2003). The integration of antimicrobial protein AceAMP1(Li *et al.*, 2003) and the rice chitinase gene (Pourhosseini *et al.*, 2013) have reduced the susceptibility of transgenic roses to powdery mildew in roses. In lily, transgenic plants have been developed with improved resistance against *Botrytis cinerea* by overexpression of the Rice Chitinase 10 (*RCH10*) gene (De Caceres Gonzalez *et al.*, 2015). Some of the examples of fungal and bacterial disease resistance in ornamental crops through transgenic approach is presented in Table 18.

11.2.3 Flower Colour

It is one of the most important traits in ornamentals where consumer always desires for novelty. Genetic engineering is an innovative approach for modification of flower colour and production of novel types in ornamentals. The flower colour is governed by three major pigments namely, flavonoids, carotenoids, and betalains. Anthocyanins are recognized as a colored class of flavonoids that accumulate in vacuoles. The majority of flower color pigments in ornamentals are primarily based upon six anthocyanidins types *i.e.*, cyanidin, pelargonidin delphinidin, peonidin (3′ O-methyl cyanidin), petunidin (3′ O-methy delphinidin), malvidin (3′, 5′ O-dimethy delphinidin. The presence of high level of delphinidin and its derivatives often produces blue flower colour while, pelargonidin often results in intense red flower color is (Tanaka *et al.*, 2009).

The recent progress in biochemical and molecular biological technologies have led to the identification of key genes responsible for the enzymes involved in anthocyanin synthesis (Tanaka and Brugliera, 2013). Genes encoding for flavonoid pathway enzymes have been characterized and cloned. Hence, it is possible to obtain novel flower colour in ornamentals through modification of flavonoid biosynthetic pathway. The modification of colour is possible *via* over expression of structural genes; use of sense or antisense enzyme construct; inhibiting the production of key biosynthetic enzyme and addition of an enzyme of a particular biosynthetic step. Petunia, tobacco and torenia are considered as model plant species for flower colour modification by means of genetic engineering.

The first successful modification of flower colour through manipulation of anthocyanin metabolism using genetic engineering technique has been reported in petunia (Meyer *et al.*, 1987). The production of pelargonidin based anthocyanins was successfully achieved in petunia by the expression of a maize *DFR* gene encoding for Dihydro-flavonol 4-reductase enzyme in white petunia variety that was deficient in *F3′5′H* and *F3′H* activities. The Introduction of the *F3′ 5′ H* gene encoding flavonoid 3′,5′ -hydroxylase enzyme has led to the production of delphinidin in various flowers such as roses and carnations. Similarly, it is possible to obtain white flowers from anthocyanin producing flowers by down-regulating the expression of one of many structural or regulatory genes of the flavonoid biosynthetic pathway. The *CHS* gene is the most common target for down-regulation of anthocyanin biosynthesis. Successful reduction of anthocyanin biosynthesis has been reported in petunia, gerbera, chrysanthemum, rose, carnation lisianthus and torenia (Tanaka *et al.*, 2005). The transformation of blue gentian (*Gentian atriflora*) using an antisense gentian *CHS* gene led to the development of transgenic plants whose flower colour varied from white to pale blue (Nishihara *et al.*, 2003). Dark pink flowers have been generated in a normal blue coloured *Torenia* by following a two-step transformation approach. Initially the *F3′5′H* gene was down regulated to produce a pink flowered torenia. It was further transformed with a *CaMV35S* promoter driving a torenia *F3′H* gene to generate dark pink torenia flower (Ueyama *et al.*, 2002)

11.2.3.1 Development of Moon® Series Carnation

Due to deficiency of the *F3′ 5′ H* gene encoding Flavonoid 3′,5′- hydroxylase enzyme, carnations, roses, and chrysanthemums do not produce delphinidin-based anthocyanins and thus lack blue/violet flower color. The successful attempt to generate carnation flowers with blue hues was made by Australian company called Florigene, which isolated the gene responsible for blue color in petunia and then transferred into carnation. It was done by selecting white carnation cultivars which lacks *DFR* gene. The expression of petunia *F3′ 5′ H* (under the control of a promoter region from the snapdragon *CHS* gene) and petunia *DFR* (under the control of a constitutive promoter) genes in one such DFR mutant lead to exclusive accumulation of delphinidin pigments, generating flowers with significant colour change toward blue colour thereby generated cultivar FLORIGENE®Moondust™. It was the first genetically engineered flower crop marketed in the world. The expression of a pansy *F3′ 5′ H* gene (under the control of a promoter region from the snapdragon *CHS* gene) and a petunia *DFR-A* gene (under the control of its own promoter and terminator regions) resulted in exclusive accumulation of higher concentration of delphinidin pigments which produces dark violet colour flower cultivar FLORIGENE®Moonshadow™. Further same gene combinations were introduced into a white standard-type carnation (also a mutant in DFR) which led to production of 4 cultivars *i.e.*, FLORIGENE®Moonvista™, FLORIGENE®Moonshade™, FLORIGENE®Moonlite™ and FLORIGENE®Moonaqua™ (Tanaka *et al.*, 2009).

11.2.3.2 Orange Petunia

Petunia *DFR* is unable to reduce dihydrokaempferol and so petunia flowers rarely contain pelargonidin-type anthocyanins and therefore do not produce orange to brick-red colours. Transgenic petunias with brick-red colour, accumulating pelargonidin- type anthocyanins have been developed by modifying the expression of *DFR* genes from different species *viz.*, maize, gerbera and rose in a mutant petunia line that accumulated dihydrokaempferol (deficient in *F3'5'H*, *F3'H* and *FLS*). Orange petunia produced from red petunias producing cyanidin by downregulation of the F3' H gene and expression of rose *DFR* gene (Mizutani *et al.*,2003).

11.2.3.3 Development of Blue Rose

Katsumoto *et al.*, 2007 revealed that rose is deficient in *F3'5'H* gene encoding for Flavonoid 3', 5'- hydroxylase, a key enzyme for delphinidin synthesis. Besides, other factors such as co-pigments and vacuolar pH also affect the flower colour development in rose. Even if delphinidin-type anthocyanin accumulates, rose petals tend to express redness due to the low vacuolar pH of its petal epidermal cells (3.69 to 5.78). In order to express blue color, it is important to either introduce a gene that can slightly improve the pH to about 5.6 to 6.2, or to use strains or cultivars that naturally exhibit a relatively higher pH. The research group of Suntory and Florigene developed the world's first blue roses 'Applause' through genetic transformation approach. Katsumoto *et al.* (2007) developed the first true blue roses, by down regulation of endogenous dihydroflavanol reductase gene (*DFR*) via RNAi-mediated silencing and overexpressing the *Iris hollandica DFR* gene, viola *F3' 5' H* gene in the rose cultivars having higher vacuolar pH, large amount of flavonols (co-pigments) with weak or no *F3'H* activity. Roses that have petals with a high flavonol content and relatively high pH were selected for gene introduction. Among various *F3' 5' H* genes, the pansy *F3' 5' H* gene was found to be effective for producing delphinidin-based anthocyanins in roses.

11.2.3.3 Development of Blue Chrysanthemum

After the successful development of blue carnation and rose, lots of research has been under taken to modify the color in flower crops. The Florigene and Suntory group have successfully developed purple and violet-colored chrysanthemums by suppressing the endogenous *F3' H* gene and expressing pansy *F3' 5' H* under the control of the rose *CHS* promoter (Brugliera *et al.*, 2013). Recently, world's first blue chrysanthemum has been developed by introducing the butterfly pea (*Clitoria ternatea*) UDP (uridine diphosphate)–glucose: anthocyanin 3',5' -O-glucosyltransferase (*A3' 5' GT*) gene encoding anthocyanin 3',5' -O-glucosyltransferase, in addition to the expression of the Campanula *F3' 5' H*, into the host plant (Noda *et al.*, 2017). Similarly, "blue dahlia" was created by introducing the *F3′ 5′ H* gene derived from dayflower, *Commelina communis* L. The expression of *F3′ 5′ H* gene in the transgenic dahlia successfully produced delphinidin derivatives and violet color (Nakano *et al.*, 2016). The transient expression of Phalaenopsis *F3′ 5′ H* gene in a pink flowered lily has led to the accumulation of blue colour in petals (Qi *et al.*, 2013). In addition, purple lilies have been developed by expressing the Campanula *F3′ 5′ H* gene under the control of the *CaMV 35S* promoter (Nakano *et al.*, 2016; Tanaka *et al.*, 2012). Some of the examples of flower colour modification in ornamental crops through transgenic approach is presented in Tables 19 and 20.

Table 19: Important Transgenic Ornamental Crops for Flower Colour Modification

Crop	*Gene*	*Trait modified*	*Reference*
Carnation	*F3'5' H* gene from *Petunia hybrida*	Transgenic violet carnations	Holton *et al.* (1993)
Red rose	*DFR* gene in RNAi vector, delphinidin gene from blue pansy and a *DFR* gene from iris	Transgenic roses with flowers having blue hues	Katsumoto *et al.* (2007)
Chrysanthemum	RNAi suppression of *CmF3'H* gene and over-expression of *Senecio cruentus* F3'5'H gene	Transgenic chrysanthemum developed bright red flowers	He *et al.* (2013)

Source: Parmar *et al.* (2017).

Table 20: Flower Colour Modifications by Regulating Flavonoid Biosynthesis in Ornamentals

Plant Species	*Original Colours*	*Gene Sources*	*Methods*	*Produced Flower Colors*	*References*
Cyclamen persicum	Purple	Endogenous *F3'5'H*	Antisense	Red to pink	Boase *et al.* (2010)
Gentiana sp.	Blue	Endogenous *CHS*	Antisense	White	Nishihara *et al.* (2011)
	Blue	Endogenous *F3'5'H*	RNAi	Lilac to purple blue	Nakatsuka *et al.* (2011)
Nierembergia sp.	Violet	Endogenous *F3'5'H*	Antisense	Pale blue	Pavasant *et al.* (2005)
Ostespermum hybrida	Magenta	Endogenous *F3'5'H*	RNAi	reddish	Seitz *et al.* (2006)
Petunia hybrida	Blue	*Mazus psonilum CHS*	Dominant negative	Pale blue	Hanumappa *et al.* (2007)
Rosa hybrida	Red to pale cyanic	*Viola* sp. *F3'5'H*	Over expression	Bluish	Katsumoto *et al.* (2007)
Torenia hybrida	Blue	Endogenous *ANS*	RNAi	Whitish to pale blue	Nakamura *et al.* (2006)
	Blue	Endogenous *ANS*	Antisense	Pale blue	Nakamura *et al.* (2010)
	Blue	*Antirrhinum majus AS*	Over expression	yellow	Ono *et al.* (2006)

Transgenic Varieties in Commercial Flowers

Applause: The First Blue Rose Variety

Blue Chrysanthemum

Florigene® Moondust: The First True Blue Carnation

Florigene® Moonvista

Florigene® Moonshade

Florigene® Moonlite™

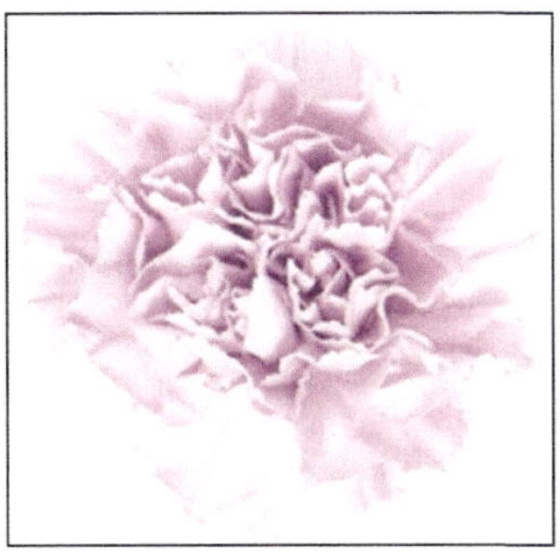

Florigene® Moonaqua™

Florigene® Moontea™

Florigene® Moonpearl™

Source: https://www.florigene.com/product/; https://www.suntoryapplause.com

11.2.4 Morphological Modification

The novel floral form and plant architecture play an important role in improving the aesthetic values of ornamentals. The GE techniques can be utilized as valuable tool for modification of plant architecture and floral morphology. The manipulation of homeotic genes regulating flower development can be utilized for altering the floral morphology. Several transgenics with altered morphology has been generated in ornamentals using GE approach. The over expression of tobacco phytochrome b1 gene in chrysanthemum led to production of small sized plants but with larger branch angles. The transformation of chrysanthemum with Arabidopsis GA insensitive (*gai*) gene resulted in significant reduction in plant height. The abundant branching in chrysanthemum was achieved by the introduction *DgLsL* gene with vector *pCAMBIA1301*-sense and antisense *DgLsL*. Aida *et al.* (2008) demonstrated that suppression of CAG gene through antisense insertion has led to the alteration gynoecium and androecium to corolla-like tissues in *C. morifolium*. Xie *et al.* (2015) reported that collective silencing of *DmCPD* and *DmGA20ox* genes is a feasible strategy to create dwarf chrysanthemum varieties (Table 21). The over-expression of *GSQUA2* in transgenic Gerbera plants exhibited several vegetative modifications like elongation of vegetative axis. The transgenic plants *Gypsophila paniculata* transformed with the *rol* C gene exhibited increased lateral branch and abundant bud formation when (Zvi *et al.*, 2008). Similarly, transformation of *Kalanchoe* spp. with a GA biosynthesis inhibition gene (Lutken *et al.*, 2010) or *rol* genes (Christensen *et al.*, 2008) produced compact plants. The transgenic approach has been widely utilized for alteration of several morphological traits of plants like plant height, branching, rooting traits *etc.* Some of the examples of morphological modification in ornamental crops through transgenic approach is presented in Table 22.

Table 21: The Genetic Modification for Plant Height in Important Ornamentals

Plants	*Gene*	*Resistant to*	*Reference*
Rose	*rol A, B* and *C* genes	Improved rooting characteristics	Van der Salm *et al.* (1997).
Chrysanthemum	*ipt* gene	Reduced plant height with a greater number of flowers, but with smaller size and late flowering period	Khodakovskaya *et al.* (2009)
	gai gene from *Arabidopsis thaliana*	Decrease in plant height	Petty *et al.* (2003)
	Tobacco *phytochrome B1* (*PHYB1*) gene	Reduction in plant height	Zheng *et al.* (2001)
Petunia	*gai* gene from *Arabidopsis thaliana*	Decrease in plant height	Tanaka *et al.* (2005)
	rolC	Reduced plant height, leaf and flower size and increased branching	Winefield *et al.* (1999)

Source: Parmar *et al.* (2017).

Table 22: Modification of Morphological and Quality Traits of Important Ornamentals using Genetic Engineering Approach

Trait Improved	*Gene Introduced*	*Source of Gene*	*Transgenic Crop Developed*
Shorter plants with larger branch angles	*Tobacco phytochromeb 1 (PHYB1)*	*Nicotiana tabacum*	*Chrysanthemum, Iridon'*
Reduction in plant height	*A. thaliana GA insensitive gene (GAI)*	*Arabidopsis thaliana*	*Chrysanthemum morifolium*
Profuse lateral branching	*DgLsL*	*Dendranthema grandiflorum*	*Chrysanthemum morifolium*
Reduced fertility for both male and female	*CmETR1/H69A*	*Cucumis melo*	*Chrysanthemum*
Dwarfism and vegetative axis elongation	*Overexpression of GSQUA2*	*Gerbera hybrida*	*Gerbera hybrida*
Fragrance	*Clarkia breweri* gene (*BEAT*)	*Clarkia breweri*	*Eustoma grandiflorum*
Alteration in flowering time	*OMADS1*	*Orchid*	*Eustoma grandiflorum*
Fragrance production	*linalool synthase* (*LIS*)	*Clarkia breweri*	*Dianthus caryophyllus*
Production of fringed leaves	*TCP3* chimeric repressor	*Arabidopsis*	*Chrysanthemum*
Production of purple colored petal cells	*DifF*	*Petunia*	*Lilium*
Dwarfism	*RolC*	*Agrobacterium rhizogenes*	Pelargonium, Petunia
Delayed yellowing of leaves in Chrysanthemum	Mutated gene *ERS1*	*Chrysanthemum*	*Chrysanthemum*
Increased vase life in carnation	*ACO/ACS*-coding genes	*Carnation/apple*	Carnation
Reduces ethylene sensitivity	*mDG-ERS1* (*etr1-4*)	*Dendranthema grandiflora*	*Chrysanthemum morifolium* 'Sei-Marine'

11.2.5 Prolong Shelf-Life

Shelf life is one of the complex traits, difficult to improve through classical breeding approaches. One of the effective ways for enhancement of shelf life is down regulation of internal ethylene biosynthesis genes such as *ACO* (*ACC* oxidase) or *ACS* (1-aminocyclopropane-1- carboxylic acid synthase). Insertion of *ACS* or *ACO* genes in antisense orientation have been reported to improve the shelf life of ornamentals. Transgenic carnation plants exhibiting delayed petal senescence have been developed by silencing the *ACO* gene that down regulate ethylene synthesis (Savin *et al.*, 1995). Bovy *et al.* (1999) developed transgenic carnations harboring Arabidopsis *etr1-1* gene. The expression of Arabidopsis *etr1-1* gene reduced ethylene production and enhanced flower longevity. Inokuma *et al.* (2008) observed reduced ethylene production and longer vase life in the transgenic lines of 'Lillipot' carnation harbouring an *sACO* transgene.

Further, extended shelf life can be achieved by maintaining resistance to ethylene or by the inhibition of ethylene biosynthesis genes. The internal ethylene production can be inhibited via RNA interference (RNAi) and by mutated or modified ethylene receptor gene (*etr1*). The vase life of *Oncidium* and *Odontoglossum* have been improved by mutating ethylene receptor gene (Raffeiner *et al.*, 2009). The transformation of petunia with *etr1-1* gene resulted in delayed senescence in petunia. The upregulation of *ipt* gene in miniature rose transformed with *PSAG12*-ipt gene resulted in better post-harvest life and delayed leaf senescence and ethylene production (Zakizadeh *et al.*, 2013). Sumitomo *et al.* (2008) revealed that the transgenic chrysanthemums lines harboring mutated ethylene receptor genes (*DG-ERS1*) exhibited reduced ethylene sensitivity and delayed leaf senescence. Some of the examples of prolonged shelf life in ornamental crops through transgenic approach is presented in Table 22.

11.2.6 Floral Fragrance

The floral scent is an important trait which can greatly enhance the aesthetic characteristics of ornamental plants. Most of the floral scent volatiles fall into the category of terpenoid, phenylpropanoid/benzenoid, and aromatic amino acid. A large number of scent genes encoding enzymes that directly catalyze the formation of volatile compounds have been identified and characterized. However, production of specialized metabolites for scent biosynthesis is not solely dependent upon enzymatic actions and greatly relies upon several transcriptional factors (TFs). The manipulation of these scent genes through genetic engineering approach can be employed for the modification of scent. Zuker *et al.* (2002) revealed that down regulation of carnation *F3'H* gene led to the increase in methylbenzoate levels in the transgenic plants and thus improved the fragrance. Lavy *et al.* (2002) revealed that the ectopic expression of *Clarkia breweri lis* (Linalool synthase) gene in the transgenic carnations have led to the production of linalool and its derivatives' cis and trans-linalool oxide. In order to induce fragrance in Lisianthus petals, *Clarkia. breweri BEAT* gene (benzyl alcohol acetyl transferase for benzyl acetate production) was introduced using transgenic technique (Aranovich *et al.*, 2007). The production of volatile compounds including benzyl acetate was noticed in the transgenic leaves and flowers.

12.0 Genome Sequencing and Transcriptome Analysis in Ornamentals

Genome information is useful for functional analysis of genes, comparative genomic analysis, breeding of new varieties by marker-assisted selection, and map-based gene isolation. Genome-related research in ornamentals plants has been relatively slow to develop because of their heterozygosity or polyploidy (Yagi, 2018). The advent of revolution in next generation sequencing (NGS) techniques and computational tools has enabled the investigation of nearly every nucleotide in the genome (Koboldt *et al.*, 2013). The elucidation of sequence composition of genetic material in the cell of an organism using computational processing and myriad short DNA sequence reads of target DNA is referred to as genome assembly. The comprehensive whole transcriptome sequences had been obtained for most of the major ornamentals, but the whole genome sequences are available for only few of the ornamental crops.

The whole genome sequences in ornamentals have been reported for carnation (Yagi *et al.*, 2014), *Phalaenopsis equestris* (Cai *et al.*, 2015), *Dendrobium officinale* (Yan *et al.*, 2015), *Primula veris* (Nowak *et al.*, 2015), *Dendrobium catenatum* (Zhang *et al.*, 2016), *Phalaenopsis* (Huang *et al.*, 2016), *Petunia* (Bombarley *et al.*, 2016), *Ipomoea nil* (Hoshino *et al.*, 2016), *Hibiscus syriacus* (Kim *et al.*, 2017) and *Helianthus annuus* (Badouin *et al.*, 2017). The remarkable genome size variation is present in the ornamental crops (Table 23). Recently, Lv *et al.* (2020) published draft genome sequence of tree Peony ornamental plant (*Paeonia suffruticosa*). They reported the draft genome of a wide grown *Paeonia suffruticosa* cultivar "Luo shen xiao chun," representing the largest sequenced genome in dicots to date with a genome size of 13.79 Gb. The identification and expression analysis of MADS box genes based on the genome assembly and de novo transcriptome assembly of *P. suffruticosa* revealed that the function of C class genes was restricted in flower development, which might be responsible for the stamen petaloidy in tree peony cultivars. Rose is the world's most important ornamental plant, with economic, cultural and symbolic value. Hibrand Saint-Oyant *et al.* (2018) reported a reference genome for *Rosa chinensis* cultivar Old Bush with a genome size of 512 Mb.

Table 23: Estimated Genome Size of Few Important Ornamental Plants

Crop Species	*Genome Size*	*Target Cultivar*	*Reference (s)*
Carnation	622 Mb	Francesco	Yagi *et al.* (2014)
Phalaenopsis equestris	1.2 Gb	Inbred line	Cai *et al.* (2015)
Dendrobium officinale	1.3 Gb	Green	Yan *et al.* (2015)
Primula veris	479 Mb	Unknown cultivated strain	Nowak *et al.* (2015)
Dendrobium catenatum	1.1 Gb	Wild plant	Zhang *et al.* (2016)
Phalaenopsis	3.5 Gb	KHM 190	Huang *et al.* (2016)
Petunia axillaris	1.4 Gb	N	Bombarely *et al.* (2016)
Ipomoea nil	750 Mb	Tokyo Kokei Standard	Hoshino *et al.* (2016)
Hibiscus syriacus	1.9 Gb	Serial number 520	Kim *et al.* (2017)
Paeonia suffruticosa	13.79 Gb	Luo shen xiao chun	Lv *et al.* (2020)
Rosa chinensis	512 Mb	Old Bush	Hibrand Saint-Oyant *et al.* (2018)

12.1.0 Transcriptome Analysis and Gene Expression Studies in Ornamentals

In the post-genome era, people have started studying the phylogenetics and dynamic expression mechanisms systematically based on transcription of the whole genome. The transcriptome represents the complete set of transcripts in a cell, and their quantity, for a specific developmental stage or physiological condition. The transcriptome, which represents all RNA transcripts in cell or tissue, reflects the expressed genes at various life stages, tissue types, physiological states, and environmental conditions (Capobianco, 2014). Understanding the transcriptome is essential for interpreting the functional elements of the genome and revealing the molecular constituents of cells and tissues, and also development and disease. The identification of candidate genes influencing any important trait can be approached through an analysis of complementary DNA (cDNA), copies of messenger RNA (mRNA). Devoid of intronic and intergenic sequences, whose biological significance is still obscure, these mRNAs represent only a small percentage of the total genome (about 1-3 per cent in eukaryotes). However, they do contain valuable information on gene activity since they correspond to the proteins expressed in a specific tissue and responsible for the identity of that tissue.

Various technologies have been developed to deduce and quantify the transcriptome, including hybridization or sequence-based approaches. Hybridization-based approaches typically involve incubating fluorescent-labelled cDNA with custom-made microarrays or commercial high-density oligo microarrays. Specialized microarrays have also been designed; for example, arrays with probes spanning exon junctions can be used to detect and quantify distinct spliced isoforms. Genomic tiling microarrays that represent the genome at high density have been constructed and allow the mapping of transcribed regions to a very high resolution, from several base pairs to ~100 bp. Hybridization-based approaches are high throughput and relatively inexpensive, except for high resolution tiling arrays that interrogate large genomes. However, these methods have several limitations, which include: dependence upon existing knowledge about genome sequence; high background levels owing to cross-hybridization; and a limited dynamic range of detection owing to both background and saturation of signals. Moreover, comparing expression levels across different experiments is often difficult and can require complicated normalization methods. Transcriptome RNA sequencing (RNA-Seq) is a recently developed approach to transcriptome profiling that uses deep-sequencing technologies. RNA-seq technology enables the analysis of whole transcriptomes and allows for generation of large-scale gene expression datasets that can guide future research. As RNA-Seq is quantitative, it helps to determine RNA expression levels more accurately than microarrays.

RNA-seq provides significant advantages for ornamental crops where genomic resources are still scarce and high levels of heterozygosity are expected. As most of the ornamentals have relatively large genome size, sequencing transcripts as a genome complexity reduction not only reduces cost and time significantly, but also contributes to establishment of resources by the focus on genes. De novo transcriptome analysis has become a useful tool in the discovery of genes involved in various metabolic pathways, including the determination of sequence and expression patterns without the need for a reference genome. Transcription expression profiling, data assembly, and analysis can provide broad and deep insights into gene regulatory networks and biological pathways because they can reveal the genes downstream of key transcription factors in related pathways. NGS-sequencing are valuable resources for genetic research and accepted as markers for MAS in ornamentals. RNA- Seq can generate numerous transcripts with sufficient read- depth to guarantee high quality SNP identification (Kim *et al.*, 2014). Using this technique, it is feasible to develop the SNP markers in highly heterozygous ornamentals, which can be further exploited for construction of a genetic map for QTL mapping. The transcriptome is an important intermediate link between genome and phenotype, playing a critical role in genome evolution. In recent years, researchers have commenced studying cross-species evolutionary by comparison of their transcriptomes

Gene expression and regulation are very important in plant growth, development, cell proliferation, differentiation and apoptosis, substance metabolism and tumor formation of organisms. Key genes which play an important role in activating various responses can be searched through the transcriptome data, and their transcriptional regulation mechanism can be investigated. The differences in expressed genes between two transcriptomes and the cluster analysis of gene expression data are important tools for discovering gene functions. The transcriptome datasets are also helpful in discovering genetic markers. Compared to the other horticultural crops, genome-related research on ornamentals has been delayed mainly because of the polyploid nature and highly heterozygous genetic background of most of the important ornamentals. Transcriptome have proved a valuable tool in the breeding of ornamentals. In ornamentals, the transcriptome analysis has provided insight into various traits including, floral development, floral transition, vase life, floral scent, secondary metabolite biosynthesis, photoperiodic response and resistance to biotic and abiotic stresses and has been utilized for better understanding of the mechanisms of these traits. With the advent of next-generation RNA sequencing and genome sequencing, comparative transcriptome studies have been successfully applied to investigate the signaling pathways and gene networks of Heat Stress Resistance (HSR) in non-model plant species. A series of heat stress-responsive genes and several signal regulation pathways have been identified in *Chrysanthemumn nankingense* (Sun *et al.*, 2014), *Dianthus caryophyllus* (Wan *et al.*, 2015) and *Rhododendron hainanense* (Zhao *et al.*, 2018), which provides many interesting gene candidates for genetic engineering to reduce negative effects on the growth of plants. Using transcriptome sequencing, Pei *et al.* (2013) identified ethylene-responsive transcripts in modern rose petals. Recently, the genome sequence of *R. chinensis* 'Old Blush' was released (Raymond *et al.*, 2018), providing a valuable database for further functional and genomic research in rose. The progress of transcriptome based research in important ornamentals has been enlisted in Table 24.

12.1.1 Rose

Kim *et al.* (2012) collected transcripts from three rose cultivars of *R. rugosa* Thunb., and Yan *et al.* (2014) acquired transcript data from another *R. chinensis* cultivar. Pei *et al.* (2013) created a rose floral transcriptome using 454 technology and exploited the transcriptome to identify potential key regulators of ethylene-influenced cell expansion. The floral transition is a crucial developmental event, but little is known about the underlying regulatory. Hence, Guo *et al.* (2017) compared the genetic basis of flowering in two rose species, *R. chinensis* 'Old Blush', which flowers continuously, and *R. odorata* var. *gigantea*, which blooms in early spring. The differentially expressed genes (DEGs) involved in vernalization, photoperiod, gibberellin (GA), and starch metabolism pathways, *e.g.*, LFY, AGL24, SOC1, CAL, and COLs, regulated the floral transition in *R. odorata* var. *gigantea*, while DEGs related to photoperiod, sugar metabolism, and GA pathways, including COL16, LFY, AGL11, 6PGDH, GASA4, and BAM, modulated the floral transition in *R. chinensis* 'Old Blush'. Han *et al.* (2017) provided insight into the processes underlying rose petal development at the transcriptional level and identified several potential candidate genes involved in petal development during the transcriptome analysis of *R. chinensis* 'Old Blush' petals. Huang *et al.* (2019) performed Hybrid-Transcriptome Sequencing and associated metabolite analysis to reveal putative genes involved in flower color difference in rose mutants. They observed significant differences in expression of genes involved in the flavonoid biosynthesis pathway *e.g.*, *CHS, FLS, DFR, LDOX* and indicated the possible involvement of these genes in the flower color difference in the rose mutants. The transcriptome profiling has also provided insights into the complex gene regulatory networks mediating the rose petal response to Grey mold rot (*Botrytis cinerea*; Liu *et al.*, 2018). The Comparative RNA-Seq analysis revealed the critical role for brassinosteroids in petal defense against *Botrytis cinerea* infection. Further, the comparative transcriptome analysis of *R. chinensis* 'Slater's crimson China' provided insights into the crucial factors and signaling pathways in heat stress response (Li *et al.*, 2019). The transcriptome analysis using Illumina and single-molecule real-time (SMRT) sequencing platforms revealed the potential candidate lignin-related genes and transcription factors in *Rosa roxburghii* during fruit ripening (Lu *et al.*, 2020).

12.1.2 Carnation

Transcripts have been identified for almost every gene involved in flower chlorophyll and carotenoid metabolism and anthocyanin biosynthesis. Transcripts have also been identified for every step in the ethylene biosynthesis pathway. Ohmiya *et al.* (2013) prepared a custom oligonucleotide array based on the carnation EST database generated by Tanase *et al.* (2012) and compared carotenogenic gene expression levels in pale-green and white petals. Ohmiya *et al.* (2013) suggested that the low levels of carotenoids in carnation petals is not due to enzymatic degradation but rather by low rates of carotenoid biosynthesis caused by the suppression of the genes encoding phytoene synthase and lycopene ε-cyclase. The transcriptomic studies have shed light on the complexity of gene expression and regulatory networks at various developmental stages, including flower opening and senescence, as well as response to biotic/abiotic stress.

The comparative transcriptome studies on vase life and carnation type in *Dianthus caryophyllus* L has been conducted by Boxriker *et al.*, 2016. Transcriptome sequencing of carnations was conducted by Tanase *et al.* (2012) and Wan *et al.* (2015) using RNA-Seq (Wang *et al.*, 2009b). Wan *et al.* (2015) performed transcriptome analysis using the RNA-seq approach to generate expression profiles of heat stress (HS)-responsive genes in carnation. Their study revealed induction of numerous genes in response to HS, few of the candidate genes involved in thermo-tolerance responses were characterized.

Table 24: Transcriptome Analysis for Gene Expression Studies in Ornamentals

Species	*Trait*	*Technique*	*Reference*
Chrysanthemum	Low temperature stress	cDNA Library construction, Illumina sequencing technology	Wang *et al.* (2018)
Chrysanthemum	White rust	cDNA Library construction, Illumina HiSeq™ 2000	Lu *et al.* (2018)
Chrysanthemum morifolium	Flavonoid biosynthesis	cDNA Library construction, RNA-Seq technology, qRT-PCR	Yue *et al.* (2018)
Chrysanthemum	Flower color change	single-molecule real-time sequencing and RNA-sequencing	Dong *et al.* (2019)
Chrysanthemum	Anther Dehiscence	cDNA Library construction, Illumina HiSeq platform	Li *et al.* (2019)
Chrysanthemum	Flower bud differentiation	Illumina sequencing, de novo assembly	Liu *et al.* (2015).
Gerbera	Disease resistance	cDNA Library Construction, Illumina paired-end sequencing, in silico characterization	Fu *et al.* (2016)
Gerbera	Stem bending	cDNA Library Construction, Illumina sequencing technology	Ge *et al.* (2019)
Lilium	Plant Height Heterosis	RNA–sequencing (RNA–Seq)	Howlader *et al.* (2020)
Marigold	Secondary metabolite biosynthesis during flower development	cDNA Library construction Illumina HiSeq™ 2000	Feng *et al.* (2018)
Marigold	Flower Organ Development	Digital Gene Expression Profiling	Ai *et al.* (2016)
Petunia	Corolla senescence	Virus-Induced Gene Silencing (VIGS) analysis, Illumina sequencing	Wang *et al.* (2018)
Rose	Petal development	RNA-seq technology	Han *et al.* (2017)
Rose	Petal defense against *Botrytis cinerea* infection	RNA-Seq analysis	Liu *et al.* (2018)
	Heat stress response	cDNA library construction, and illumina sequencing	Li *et al.* (2018)
Rose	Floral transition	Comparative transcriptome analysis, qRT PCR	Guo *et al.* (2016)
Rose	Patterns of natural selection	Illumina sequencing	Li *et al.* (2018)
Rose	Floral Scent	Illumina HiSeq, De Novo Transcriptomics Analysis	He *et al.* (2020)
Rose	Flower Color Difference	Hybrid RNA-sequencing	Huang *et al.* (2019)
Rose	Fruit Ripening	Short-read Illumina and long-read SMRT sequencing	Lu *et al.* (2020)
Rose	Flower blooming	Illumina sequencing technology	Yan *et al.* (2014)

12.1.3 Chrysanthemum

An extensive transcriptome assembly has been published by Xu *et al.* (2013), based on an Illumina RNA-sequencing experiments to investigate dehydration stress, and by Sasaki *et al.* (2017) based on 454 RNA-sequencing of a wide range of tissues. In *Chrysanthemum nankingense*, next-generation sequencing technology produced over 53 million sequencing reads and a large number of putative EST-SSRs were identified. Among them 81 markers had amplicons and 20 were polymorphic for genotypes analysis in Chrysanthemum (Wang *et al.*, 2013). Using this approach, researchers have succeeded in detecting the genes encoding florigens and antiflorigens, the key regulators of photoperiodic flowering in chrysanthemum, using the wild diploid chrysanthemum, *C. seticuspe* (Higuchi *et al.*, 2013; Oda *et al.*, 2012). Li *et al.* (2014) acquired transcription profiles of chrysanthemum leaf following inoculation with *Alternaria tenuissima* and revealed that most of the differentially transcribed genes were up-regulated by the presence of the pathogen, while those involved in photosynthesis were down-regulated. The comparative transcriptomes of chrysanthemum using Illumina paired-end sequencing technology have revealed significant differences in gene expression and signaling pathway components of vegetative buds, floral buds and buds (Liu *et al.*, 2015). This study has indicated that the phase transition from vegetative to reproductive growth is a complex biological process regulated by a wide range of factors, during which many physiological processes, such as sugar metabolism and phytohormone regulation, become activated.

In Chrysanthemum flowering is often promoted by artificial short-day treatment. However, the flower colour of Chrysanthemum induced under artificial short-day treatment are lighter than those induced by the natural photoperiod. Dong *et al.* (2020) performed transcriptome analysis to explore the intrinsic molecular mechanism of fading of flower color under short-day photoperiods in chrysanthemum. It was revealed that the anthocyanin synthesis is tightly regulated by the photoperiod in chrysanthemum and the major regulators of the light signalling ELONGATED HYPOCOTYL5 genes were markedly upregulated under natural daylight conditions. The transcriptome data of *Dendranthema grandiflorum* var. 'Jinba' have provided the deeper insight into the mechanism of cold tolerance. This data can be further exploited to explore the new candidate genes for chilling-tolerance and freezing-tolerance and molecular breeding in chrysanthemum.

12.1.4 Gerbera

The main focus of transcriptome researches in gerbera is the identification of gene families related to organ growth and development including flower organ differentiation, petal organogenesis, petal growth and disease resistance. Transcriptome analysis of the inflorescence of *Gerbera hybrida* have uncovered novel regulators of flower type (Laitinen *et al.*, 2005). The transcriptome of the gerbera ray floret sequenced by NGS sequencing have predicted the genes involved in gibberellin metabolism and signal transduction (Kuang *et al.*, 2013). The gerbera EST database, generated through next-generation sequencing has been exploited by Fu *et al.* (2016) to identify a series of SNP markers for linkage mapping and also identify transcripts that might be involved in Botrytis. High-throughput transcriptome sequencing of gerbera during stem bending using the Illumina sequencing technology revealed that stem bending is caused by water stress and regulation of abscisic acid (Ge *et al.*, 2019).

12.1.5 Tuberose

The understanding about the molecular mechanism responsible for the floral scent in *Polianthes tuberosa* is limited. Recently, the transcriptome sequencing and de novo assembly, identified 13 and 17 candidate genes associated with terpene and benzenoid biosynthesis, respectively in tuberose (Fanm *et al.*, 2018). The possible role of PtTPS1, PtDAHPSs, PtPAL1, and PtBCMT2 in regulating the formation of floral volatiles has been revealed by this study. The transcriptome analysis of *P. tuberosa* also laid the foundation for the elucidation of the molecular mechanism underlying the regulation of floral scents in monocots

12.1.6 Lilium

Transcriptome analysis using RNA–Seq technique has been utilized to identify heterosis related DEGs involved in plant height and to interpret underlying genetic reasons that control heterosis in intra-specific hybrids in *L. longiflorum* (Howlader *et al.*, 2020).

12.1.7 Narcissus

Lack of transcriptomic and genomic information hinders understanding of the molecular mechanisms underlying the biosynthesis of *Narcissus* floral scent volatiles. He *et al.*, 2019 used the RNA-sequencing (RNA-Seq) on the Illumina HiSeq system and de novo transcriptome assembly, to study the gene expression in narcissus corona and petal tissues at the early flowering (day 1) and full-bloom (day 7) stages. Significant differences were observed in the expression profiles of 14 fragrance-related genes, that were further analyzed by qRT-PCR. Additionally, 46 and 71 significantly differentially expressed genes encoded enzymes and transcription factors, respectively, associated with floral volatiles biosynthesis pathways, were analyzed in depth.

12.1.8 Petunia

Wang *et al.* (2018) performed a transcriptome analysis in petunia corolla at four developmental stages, including corolla fully opening without anther dehiscence (D_0), corolla expansion, 2 days after anthesis (D_2), corolla with initial signs of senescence (D_4), and wilting corolla (D_7). They identified large numbers of DEGs, ranging from 4626 between the transition from D_0 and D_2, 1116 between D_2 and D_4, a transition to the onset of flower senescence, and 327 between D_4 and D_7, a developmental stage representing flower senescence. The KEGG analysis showed that the auxin- and ethylene-related hormone biosynthesis and signaling transduction pathways were significantly activated during the flower development and highly upregulated at onset of flower senescence. Further, it was revealed that hormonal interaction between auxin and ethylene may play critical regulatory roles in the onset of natural corolla senescence in petunia.

12.1.9 Marigold

Several studies on functional characterization of *T. erecta* genes have been performed. Five MADS-Box genes, *TePI, TeAP3-1, TeAP3-2, TeTM6-1* and *TeTM6-2* have been cloned and their putative functions were investigated. Very little is known for the genomic and transcriptomic sequences of *T. erecta*. With the advancement of sequencing technology and decrease of sequencing cost, there is a good chance to reveal its transcriptomic sequences and gene profiling during different development stages. Transcriptomic analysis of differentially expressed genes during flower organ development in genetic male sterile and male fertile *Tagetes erecta* revealed the possible involvement of 12 MADS-box genes in male sterility (Ai *et al.*, 2017). Feng *et al.* (2018) carried out the RNA-seq-based comprehensive transcriptional analyses on leaves, immature flowers and mature flowers of *T. erecta* to reveal the molecular mechanisms underlying the secondary metabolite biosynthesis and flower development. They revealed that the genes involved in photosynthesis and secondary metabolite biosynthesis were differentially expressed between immature flowers and mature flowers. The transcriptome analysis of *Tagetes erecta* provides novel insights into secondary metabolite biosynthesis during flower development.

Recently, the New breeding techniques (NBTs) have created high expectations, both for research and applied breeding. The use of NBT to improve biotic and abiotic stress tolerance has attracted increasing interests. These techniques include zinc finger endonucleases, transcription activator-like effector nucleases (TALENs), and CRISPR-Cas9 technology. Genome editing techniques like CRISPR/Cas9 are currently attracting the most attention. The mode of action of CRISPR/Cas9 is the inactivation, change, or insertion of a gene, which leads to far superior precision than other techniques.

13.0 References

Abdolmohammadi, M., Kermani, M.J. and Zakizadeh, H. (2014) *Euphytica,* **198**: 2-7.

Abo EI-Nil, M.M. and Hildebrandt, A.C. (1973) *Can. J. Bot.,* **51** (1973): 2107-2109.

Acquaah G. (2015) *In*: Al-Khayri J., Jain S., Johnson D. (eds) *Advances in Plant Breeding Strategies: Breeding, Biotechnology and Molecular Tools*. Springer, Cham. pp. 115-158

Ai, Y., Zhang, Q. and Wang, W. (2016) *PLoS ONE,* **11**(3): e0150892.

Aida, R. and Shibata, M. (2002) *Plant Biotechnol. J.,* **19**: 329-334.

Aida, R., Komano, M. and Saito, M. (2008) *Plant Biotechnol.,* **25**: 55–59.

Allum, J.F., Bringloe, D.H. and Roberts, A.V. (2007) *Plant Cell Rep.,* **26**: 1977–1984.

Amano, J., Kato, J., Nakano, M. and Mii, M. (2006) *Sci Hortic.,* **110**: 223-227.

An, J., Song, A. and Guan, Z. (2014) *Mol. Biol. Rep.* **41**: 4155–4162.

Andersen, J.R. and Lubberstdt, T. (2003) *Trend Plant Sci.,* **8**: 554-560.

Aranovich, D., Lewinsohn, E., and Zaccai, M. (2007) *Postharvest Biol.* Biotechnol., **43**, 255–260.

Ardakani, M. D., Shariatpanahi, and M.E., Kafi, M. (2015) *EC Agriculture,* **2**(4): 413-425

Aros, D., Gonzalez, V., Allemann, R.K. (2012) *J. Exp. Bot.,* **63**(7): 2739–2752.

Aros, D., Spadafora, N., Venturi, M. (2015) *Sci Hortic.,* **185**: 183–192

Aros, D., Suazo, M., Rivas, C. (2019) *Euphytica,* **215**(5).

Asano, Y. (1982) *Euphytica,* **31**: 121-128.

Azadi, P., Bhageri, H., Nalousi, A.M. (2016) *Biotechnol. Adv.* **34**(6).

Azadi, P., Otang, N.V., Supaporn, H. (2011) *Biotechnol. Lett.,* **33**: 1249–1255.

Badouin, H., Gouzy, C.J., Grassa, F. (2017) *Nature,* **546**: 148–152.

Bala, M. and Singh, K.P. (2015) *Indian J. Biotechnol.,* **14**: 525-531.

Banerjee, B.K. and Datta, S.K. (1986) *J. Nucl. Agric. Biol.,* **17**: 113-114.

Banyal, N., Kanwar, P. S., Bharadwaj, C. (2011) *Indian J. Agric. Sci.* **81**.

Barba-Gonzalez, R., Lokker, B., Lim, K. (2004) *Theor. Appl. Genet.,* **109**: 1125-1132.

Barba-Gonzalez, R., Tapia-Campos, E., Lara-Bannuelos, T.Y. (2017) *Acta Hortic,* **1167**: 197–204.

Basu SK., *et al.* (2011) *Aust. J. Crop Sci,* **5**(9): 1087-1093.

Becker-Zens, R. (1983) *Acta Hortic.,* **131**: 209-214

Bhat, R.S., Upadhyaya, N.M., Chaudhury, A. (2007) *In Rice functional genomics, Springer* 148-180.

Bi, M., Li, X., Yan, X. (2021) *Hortic. Res.,* **8**: 6.

Blakeslee, A. and Avery, A. (1937) *J Hered.,* 28: 393–411.

Blakeslee, A.F., Belling, J., Farnham, M.E. (1922) *Science,* **55**: 646-647.

Boase, M.R., Lewis, D.H., Davies, K.M. (2010) *BMC Plant Biol.,* **10**: 107.

Bose, T.K. and Mukherjee, D. (1967) *Bulletin Botanical Society of India,* **20**, pp. 162-168.

Bohanec, B (2009) Springer, Berlin, pp 35–46.

Bombarely, A., Moser, M., Amrad, A. (2016) *Nat. Plants,* **2**: 16074.

Bomblies, K. (2010) *Annu Rev Plant Biol,* **61**: 109–124.

Bovy, A.G., Angenent, G.C., Dons, H.J., and van Altvorst, A.C. (1999) *Mol. Breed.,* **5**: 301–308.

Boxriker, M., Boehm, R., Krezdorn, N. (2007) *Sci. Hort.,* **217**: 61-72.

Bretagnolle, F. and Thompson, J. (1995) *New Phytol,* **129**: 1-22.

Bridgen, M.P., Kollman, E.E. and Lu, C. (2009) *Acta Hortic,* **836**: 73–78.

Bridgen, M.P., Lu, C. and Neuroth, M. (1997) U.S. Patent 10,030.

Brugliera, F., Tao, G.Q., Tems, U. (2013) *Plant Cell Physiol.,* **54**: 1696–1710.

Burchi, G., Mercuri, A., Bianchini, C. (1998) *Acta Hortic,* **508**: 233-235

Byrne, D.H. (2009) *In*: Folta, K.M. Gardiner, S.E. (eds) *Genetics and genomics of rosacea, Plantgenetics and genomics: crops and models* 6. Springer Science and Business Media, New York, pp. 353–379.

Cai, C., Cheng, F.Y., Wu, J. (2015) *PLos ONE*, **10**: e0128584.

Cappadocia, M., Chrétien, L. and Laublin, G. (1988) *Can. J. Bot.*, **66** (6): 1107-1110.

Chaar M., Pinker, I. and Bohme, M. (2016) *Acta Hortic.*, **1127**: 163-170.

Chan, Y.L., Lin, K.H., Sanjaya, (2005) *Transgenic Res.*, **14**: 279–288.

Chauvin, J.E., Label, A. and Kermarrec, M.P. (2005) *J Hortic Sci Biotech.*, **80**: 693–698.

Chen, J.F., Cui, L., Malik, A.A. (2011) *Plant Cell Tiss. Org.*, **104**(3): 311-319

Chen, J.R., Chen, Y.B., Ziemian´ska (2016) *Plant Growth Regul.*, **35**: 586–599.

Chen, J.R., Lu, J.J., Liu, R. (2010) *Plant Growth Regul.*, **60**: 199-211.

Cheng, X., Chen, S. and Deng, Y. (2011) *Euphytica*, **177**(1): 45-53.

Cheng, X., Chen, S., Chen, F. (2010) *Euphytica*, **172**: 101–108.

Chi (2000) *Bot. Bull. Acad. Sin.*, **41**: 143-149.

Chi, H. (2002) *Bot. Bull. Acad. Sin.*, **43**: 139-146.

Christensen, B., Sriskandarajah, S., Serek, M. and Muller, R. (2008) *Plant Cell Rep.*, **27**, 1485–1495.

Christou, P., Capell, T., Kohli, A. (2006) *Trends Plant Sci.*, **11**: 302–308.

Clarke, J.L., Spetz, C., Haugslien, S. (2008) *Plant Cell Rep.*, **27**: 1027–1038.

Cohen, D. and Yao, J.L. (1996) *Plant Cell Tissue Organ Cult.*, **47**: 43–49.

Collard, B.C.Y., Jahufer, M.Z.Z., Brouwer, J.B. (2005) *Euphytica*, **142**: 169–196.

Condliffe, P.C., Davey, M. R., Power, J. B. (2003) *Acta Hortic.*, **612**: 115-120.

Courtney-Gutterson, N., Napoli, C., Lemieux, C. (1994) *Biotechnol.*, **12**: 268-271.

Crespel, L., Le Bras, C., Relion, D. (2015) *Plant Breed.*, **134**(3): 356–364.

Custers, J., Eikelboom, W., Bergervoet, J. (1995) *Euphytica*, **82**: 253-261.

Daker, M. (1996) *Nat.* **211**, 549–550.

Das, P.K., Dube, S., Ghosh, P. (1977) *Indian J. Hortic.*, **34**: 169-174.

Datta, S. K. (2018) *Curr. Sci.*, **114**(6), 1194-1206.

Datta, S. K. (2020) *Int. J. Life Sci.*, **9**(2), 115-133.

Datta, S. K. and Gupta, M. N. (1985) *The Indian Rose Annual*, **4**: 80-83.

Datta, S. K. and Janakiram, T. (2015) *Indian J. Agric. Sci.*, **85**(10), 1379-1395.

Datta, S.K. (1987) *The chrysanthemum*, **43**(1): 40.

Datta, S.K. (2001) *Scientific Horticulture*, **7**: 159-199.

Datta, S.K. (2009) *In*: Shu, Q.Y. (editor) Vienna: *Joint FAO/IAEA Division of Nuclear techniques in Food and Agriculture, International Atomic Energy Agency*; pp. 300–2.

Datta, S.K. and Teixeira da Silva, J.A. (2006) *Floriculture, ornamental and plant biotechnology: advances and topical*, **1**: 640-645.

De Cáceres González, F.F.N., Davey, M.R. (2015) *Plant Cell Rep.*, **34**: 1201–1209.

De Jeu, M. and Jacobsen, E. (1995) *Euphytica,* **86**: 15-23.

De Mello, E., Silva, M., Callegari, J. and Zanettini, B. (2000) *Ciencia Rural.*, **30**: 105–111.

De Storme, N. and Mason, A. (2014) *Curr Plant Biol.*, 1: 10–33.

Debener, T. (2001) *Acta Hortic.,* **522**

Deng, Y., Chen, S., Aimin, L. (2009) *Planta,* **231**(3): 693-703.

Deng, Y., Jiang, J., Chen, S. (2012) *PLoS ONE,* **7**(8): 1-15.

Deng, Y., Teng, N., Chen, S. (2010) *Euphytica,* **174**(1): 41-50.

Dewitte, A., Eeckhaut, T. and Van Huylenbroeck, J. (2010) *Euphytica.,* **171**: 283–293.

Dewitte, A., Eeckhaut, T., Van Huylenbroeck, J. and Van Bockstaele, E. (2009) *Euphytica.*, 168: 81–94.

Dewitte, A., Eeckhaut, T., Van Huylenbroeck, J. (2009) *Euphytica,* **168**(1): 81–94.

Dewitte, A., Eeckhaut, T., Van Huylenbroeck, J. (2010) *J. Hered.*, **104**(2): 215–223.

Dhooghe, E., Denis, S., Eeckhaut, T., Reheul, D. and Van Labeke, M.C. (2009) *Euphytica.*, **168**: 33–40.

Dhooghe, E., Van Laere, K., Eeckhaut, T., Leus, L. and Van Huylenbroeck, J. (2011) *Plant Cell Tissue Organ Cult.*, **104**: 359–373.

Dohm, A., Ludwig, C., Schilling, D., and Debener, T. (2002) *Acta Hortic.*, **572**: 105–111.

Doi, H. (2013) *Breed. Sci.* 63: 400–406.

Doi, H., Yokoi, S., Hikage, T. (2011) *Plant Cell Rep.,* **30**(6): 1099-1106.

Dolcet-Sanjuan, R., Clavería, E., Llauradó, M. (2001) *Acta Hortic.,* **560**: 141-144.

Dong, L., Huang, Z., Liu, D. (2018) *Sci Hort* 233: 421-430.

Drewlow, L.W., Ascher, P.D. and Widmer, R.E. (1973) *Theor. Appl. Genet.,* **43**: 1-5.

Du, X., Li, W., Sheng, L. (2018) *BMC Plant Biology,* **18**: 178.

Dubois, A., Remay, A., Raymond, O. (2011) *PLoS ONE,* **6**(12): e28455.

Dugo, M.D.L.L., Satovic, Z., Millan, T. (2005) *Theor Appl Genet,* **111**(3): 511–520.

Dunwell JM. (2010) *Plant Biotech. J.,* **8**: 377–424.

Eeckhaut, T, Jeroen Van der, V., Dhooghe, E. (2018) In J. Van Huylenbroeck (ed.), Ornamental Crops, **11**: 145-173.

Eeckhaut, T., Samijn, G and Van Bockstaele, E. (2002) *Acta Hort.*, 572: 43–49.

Eeckhaut, T., Werbrouck, S., Dendauw, J. (2001) *Plant Cell Tiss. Organ.,* **67**: 181-189.

Eeckhaut, T., Werbrouck, S., Leus, L., Van Bockstaele, E. and Debergh, P. (2004) *Plant Cell Tissue Organ Cult.*, **78**: 241–246.

Eigsti, O. and Dustin, P. (1955) *Annu Rev Ecol Syst.,* **29**: 467–501.

Elomaa, P., Honkanen, J., Puska, R. Seppanen, P., Helariutta, Y., Mehto, M., Kotilainen, M., Nevalainen, L. and Teeri, T.H. (1993) *Biotechnol.*, **11**: 508–511.

Fan, Q., Song, A., Jiang, J., Zhang, T., Sun, H., Wang, Y., Chen, S. and Chen, F. (2016) *PLoS One*, **11**: e0150572.

Fanm, R., Chen, Y., Ye, X., Wu, J., Lin, B. and Zhong, H. (2018) *PLoS One* 13(9): e0199261.

Ferrie, A. M. and Caswell, K. L. (2011) *Propag. Ornam. plants* **11**, 63–77.

Forster, B.P., Heberle-Bors, E., Kasha, K.J. (2007) *Trends Plant Sci.*, **12**(8): 368-375.

Fu, Y., Esselink, G.D., Visser, R.G.F. (2016) *Front. Plant Sci.*, **7**: 24.

Fu, Y., van Silfhout, A., Shahin, A. (2017) *Mol. Breed.*, **37**: 13.

Furuta, H., Shinoyama, H., Nomura, Y. (2004) *Plant Sci*, **166**(3): 695-702.

Gao, Y., Chen, B. and Zhang, J. (2011) *Acta Hortic.*, **923**: 103-110.

Ge, Y., Lai, Q., Luo, P. *et al.* (2009) *BMC Genom.*, **20**: 600.

Grouh, M. S. H., Sousaraei, N., Akbari, M. (2015) *Turk. J. Agric. For.*, **39**: 596-600.

Guha, S. and Maheshwari, S.C. (1964) *Nat.*, **204** (4957): 497.

Guo, L., Wang, Y., da Silva, J.A.T. (2019) *Plant Physiol. Biochem.*, **138**: 130-139.

Guo, X., Chao, Yu., Le, L. (2017) *Scientific reports*, **7**(1): 1-12.

Guo, X., Yu, C., Luo, L. (2017) *BMC Genom.*, **18**: 199.

Guodong, F., Shengxiong, H., Ying, L. (2018) *Gene*, **16**: 18-27.

Gupta, M.N. and Shukla, R. (1970) *Recent Adv. Hort.*, 81-86.

Gupta, M.N. and Shukla, R. (1971) *Jap. J. Breed.*, **21**: 129-136.

Gurel, A, Kontowski S., Nichterlein, I. (1991) *Helu*,**149**(14): 123-128.

Ha, Y. M., Shim, K. K., Kang, H. C., and Lim, K. B. (2014) *Flower Res. J*, **22**: 278-282.

Hai, N.T.L., Masuda, J.I. Miyajima, I. (2012) *J. Japan. Soc. Hort. Sci.*, **81**: 366–373.

Haig, D., and Westoby, M. (1991) *Philos. Trans. R. Soc. Lond. B Biol. Sci.*, **333**: 1–13.

Hamill, S., Smith, M. and Dodd, W. (1992) *Aust J Bot.*, **40**: 887–896.

Han, Yu., Huihua, W., Tangren, C. (2017) *Scientific reports*, **7**(1): 1-14.

Hancock, P., Lesch, M., and Simmons, L. (2003) *Accid. Anal. Prev.*, **35**(4): 501-14.

Hanumappa, M., Choi, G., Ryu, S. (2007) *J. Exp. Bot.*, **58**(10): 2471–2478.

Harding J., Huang, H. and Byrne, T. (1996) *Euphytica*, **88**(1): 55–60.

He, H., Ke, H., Keting, H. (2013) *PLoS One*, **8**(11): e74395.

He, J.P., Chen, F.D., Chen, S.M. (2009) *Acta Bot. Boreal. Occident. Sin.*, **29**: 2318–2325.

He, Y., Min, Xu., and Xiaojing, C. (2020) *Tropical Plant Biology*, 1-17.

Henny, R.J., Holm, J.R., Chen, J.J. and Scheiber, M. (2009) *Hort Sci.*, **44**(3): 646–650.

Hentrich, W. and Glawe, M. (1982) *Archiv-fur- Zuchtungsforschung.*, **12**: 197-207.

Henz, A., Debener, T. and Linde, M. (2015) *Mol. Breed.*, **35**: 190.

Hibrand Saint-Oyant, L. (2018) *Nat Plants*, **4**: 473–484.

Holton, T.A., Brugliera, F. and Tanaka, Y. (1993) *Plant J.*, **4**: 1003–1010.

Hong, B., Tong, Z., Ma, C. (2006) *PLoS ONE*, **49**: 436.

Hoshino, A., Jayakumar, V., Nitasaka, E. (2016) *Nat. Commun.*, **7**: 13295.

Hossain, Z., Mandal, A.K.A., Datta, S.K. (2006) *Plant Biol.*, **8**(4): 450-61.

Hu, Y., D.C. Yin and Q.C. Yu. (1989) *Jiangsu Agric Sci.*, **6**: 31-32.

Hua, L., Ming, S., Dongliang, Du. (2015) *PLoS ONE*, **10**(5): e0128009.

Huang, C.H., Zhang, C., Liu, M. (2016) *Mol. Biol. Evol.*, **33**: 2820-2835.

Huang, D., Li, X., Sun, M. (2016) *Front. Plant Sci.*, **7**: 1633.

Huang, G., Han, M., Yao, W. (2015) *Peer J.*, **5**: e3382.

Huang, J.Z., Lin, C.P., Cheng, T.C. (2016) *Peer J.*, **4**: e2017.

Huang, K. L., Miyajima, I., Okubo, H. (2000) *Acta Hortic.*, **570**: 367-371

Huang, P., Lin, F., Li, B. (2019) *Plants* (*Basel*), **8**(8): 267.

Huck, N., Moore, J.M., Federer, M. (2003) *Development*, **130**(10): 2149–2159.

Iizuka, M., Kimura, Y., Okada, T. (2011) *Agric. Tech. Cent.*, **8**: 89–92.

Ikeda, N., Niimi, Y. and Han, D.S. (2003) *Plant Cell Tissue Organ Cult.*, **73**: 159-166.

Imakiire, S., Nagayoshi, S., Koriyama, K. *et al.* (2006) *Agric. Res. Cent.*, **34**: 15–19.

Inokuma, T., Kinouchi, T. and Satoh, S. (2008) *J. Appl. Hortic.*, **10**(1): 3-7.

Ishizaka, H. (1998) *Plant Cell Tiss. Org.*, **54**: 21–28, 1998.

Ishizaka, H., Oku T., Mishiba, K. I., *et al.* (2009) *Caryologia*, **74** (4), pp. 457-465.

Ishizaka,H., Kondo, E. and Kameari, N. (2018) *Breed. Sci.* **68**.

James, J. (1983) *Am Rose Ann. 99-101.*

Janson, J. (1993) *Plant Sci.*, **90**: 105-115.

Jewel, H., Arif, H.R., Natarajan, S., *et al.* (2020) *Scientific reports*, **10**: 9082.

Jia, Y., Zhanga, Q. X., Pana, *et al.* (2014) *Sci. Hortic.*, **176**: 273–281.

Josine, T., Ji, J., Wang, G., *et al.* (2015) *Int. J. Agri. Crop Sci.*, **8**: 463–47.

Jung-Heiliger, H. and Horn, W. (1980) Z *Pjlanzenzüchtg.* **85**: 185-199.

Kaicker, U.S. and Swarup, V. (1972) *Indian J. Genet.*, **32**: 257-265.

Kamo, K., Aebig, J.A., Gauaranga, M., *et al.* (2012) *Plant Cell Tisue Organ Cult.*, **110**(1).

Kamo, K., Blowers, A., Smith, F., *et al.* (1995) *Plant Sci.*, **110**(1): 105-111.

Kamo, K., Gera, A., Cohen, J., *et al.* (2005) *Plant Cell Rep.*, **23**(9): 654-663.

Kapadiya, D.B., Chawla, S.L., Patel, A.I. *et al.* (2016) *Indian J. Hortic.*, **73**: 141-.144.

Kasha, K.J. and Maluszynski, M. (2003) In Doubled haploid production in crop plants (pp. 1-4) Springer, Dordrecht.

Kato, J., Ishikawa, R. and Mii, M. (2001) *Theor. Appl. Genet.* **102,** 1129–1135.

Katsumoto, Y., Fukuchi-Mizutani, M., Fukui, Y., *et al.* (2007) *Plant Cell Physiol.*, **48**: 1589–1600.

Kazi, N.A. (2012) Effect of colchicine on growth and flowering of China aster (*Callistephus chinensis* Nees.).Post-Graduation Thesis.

Ke, W., Zhen-yu, B., Qian-yu, L., *et al.* (2018) *BMC Genom.*, **19**: 319.

Kermani, M.J., Sarasan, V., Roberts, A.V., *et al.* (2003) *Theor Appl Genet.*, **107**: 1195–1200.

Kernan, Z. and Ferrie, A.M.R. (2006) *Plant Cell Rep.*, **25**: 274–280.

Khandakar, R.K.M.D., Yu, J., Min, S.K., *et al.* (2014) *Not. Bot. Horti. Agrobot. Cluj. Napoca.*, **42**(2): 482-487.

Khodakovskaya, M., Vankova, R., Malbeck, J., *et al.* (2009) *Plant Cell Rep.*, **28**: 1351–1362.

Khosravi, P., Kermani, M.J., Nematzadeh, G.A., Bihamta, M.R. and Yokoya, K. (2008) *Euphytica.*, 160: 267–275.

Kim, J., Park, J.H., Chan, J.L., *et al* (2012) *BMC Genom.* 13: 657

Kim, Y.M., Kim, S., Koo, N., *et al.* (2017) *DNA Res.*, **24**: 71–80.

Kim, Y.S., Lim, S., Yoda, H., *et al.* (2011) *Plant Signal. Behav.*, **6**: 409–412.

Kishi, F., Kagami, Y., Shinohara, M. *et al.*(1993) *Euphytica* 74: 85–90.

Kishimoto, S., Maoka, T., Nakayama, M., *et al.* (2004) *Phytochemistry*, **65**: 2781–7.

Kitamura, S., Akutsu, M. and Okazaki, K. (2009) *Sex. Plant Reprod.* **22**(1): 9–14.

Ko, J.A., Kim, Y.S. and Eun, J.S. (1996) *Korean Society of Horticultural Science*, **37**: 468–474.

Koboldt, D.C., Steinberg, K.M., Larson, D.E., *et al.* (2013) *Cell*, **155**: 27-38.

Kozak, K., Galek, R., Waheed, M.T. *et al.* (2012) *Plant Growth Regul.*, **66**(2): 145-153.

Kuang, Q., Li, L., Peng, J., *et al.* (2013) *PLoS ONE*, **8**: e57715.

Kudo, N., Matsui, T. and Okada, T. (2008) *Plant Biotechnol.*, **25**(6): 529-533.

Kuehnle, A.R. and Sugii, N. (1991) *Hort. Sci.*, **26**: 1325–1328.

Kuehnle, A.R. and Sugii, N. (1992) *Plant Cell Rpt.*, **11**: 484–488.

Kuligowska K., Lutken, H., Christensen, B., *et al.* (2015) *BMC Plant Biol.*, **15**(1): 1-15.

Kuligowska, K., Lutken, H. and Muller, R. (2016a) *Planta*, **244**: 1–17.

Kuligowska, K., Lutken, H., Christensen, B., and Müller, R. (2016b) *Breed. Sci.*, **66**(2), 300-308.

Kumar, S., Raj, S.K., Sharma, A.K., and Varma, H.N. (2012) *Sci. Hortic.*, **134**: 40–45.

Kuwayama, S., Nakamura, T., Mizuta, Y., *et al.* (2005) *Acta Hort.* **673**: 421–427.

Laitinen, RA., Pöllänen, E., Teeri, TH., *et al.* (2007) *Planta.*, **226**: 347–60.

Laura, M., Safaverdi, G. and Allavena, A. (2006) *Plant Breed.*, **125**: 629-634.

Lavy, M., Zuker, A., Lewinsohn, E., *et al.* (2002) *Mol. Breed.*, **9**: 103–111.

Lear, B.G.A., Marchbank, A., Kent, N.A., *et al.* (2019) *Acta Hortic.*, **1263**: 351-358.

Lee, S.Y., Cheon, K.S., Kim, S.Y., *et al.* (2020) *Hortic. Environ. Biotechnol.*, **61**: 569–576.

Lertsutthichawan, A., Ruamrungsri, S., Duangkongsan, W., *et al.* (2017) *Int. J. Agric.*, **13**(7.3): 2325-2332.

Li, D., Liu, X., Shu, L. *et al.* (2020) *BMC Plant Biol.*, **20**: 533.

Li, F., Cheng, Y., Zhao, X. *et al.* (2020) *Sci Rep.*, **10**: 1702

Li, F., Zhang, H., Zhao, H., *et al.* (2018) *Plant Biotechnol. J.*, **16**: 1311-1321.

Li, L., Zhang, W., Zhang, L., *et al.* (2015) *Front. Plant Sci.*, **6**: 168.

Li, P., Song, A., Gao, C., *et al.* (2015) *Plant Cell Rep.*, **34**(8): 1365-78.

Li, P., Song, A., Gao, C., *et al.* (2015) *Plant Physiol. Biochem.*, **95**: 26-34.

Li, S., Micai, Z., Xue, D., *et al.* (2018) *BMC plant biology*, **18**(1): 1-12.

Li, X., Gasic, K., Cammue, B., *et al.* (2003) *Planta*, **218**: 226–232.

Li, X., Krasnyanski, S. and Korban, S.S. (2002) *J. Plant Physiol.*, **159**: 313–319.

Li, Z.Q., Wen, X., Ping, L., *et al.* (2019) *Plant Physiol. Biochem.*, **142**: 312-331.

Liang, Qy., Wu, Yh., Wang, K. *et al.* (2017) *Sci. Rep.*, **7**: 4799.

Liao, L.J., Pan, I.C., Chan, Y.L., *et al.* (2004) *Mol. Breed.*, **13**: 229–242.

Lim, K., Chung, J., Van Kronenburg, B., *et al.* (2000) *Chromosome Res*, **8**: 119-125.

Lim, K., Ramanna, M. and Van Tuyl, J. (2001b) *Acta Hortic.*, **552**: 65-72

Lim, K., Ramanna, M., De Jong, J., *et al.* (2001a) *Theor. Appl. Genet.*, **103**: 219-230.

Liu, B., De Storme, N. and Geelen, D. (2017) *Plant Physiol.*, **173**: 338–353.

Liu, H., Sun, M., Du, D., *et al.* (2015) *PLoS ONE*, **10**(5): e0128009.

Liu, S., Chen, S., Chen, Y., *et al.* (2011) *Sci Hortic.*, **127**: 411–419.

Liu, X., Cao, X., Shi, S., *et al.* (2018) *BMC Genet.*, **19**(1): 1-10.

Lowry, D.B., Modliszewski, J.L., Wright, K.M., *et al.* (2008) *Philos. Trans. R. Soc. Lond. B. Biol. Sci.*, **363**(1506): 3009–3021.

Lu, C. and Bridgen, M. (1997) *Euphytica*, **94**: 75-81.

Lu, C. and Bridgen, M.P. (1997) *Euphytica.*, **94**: 75–81.

Lu, C.Y., Nugent, G., Wardley-Richardson, T., *et al.* (1991) *Biotechnol.*, **9**: 864–868.

Lu, M., Wen-Tao, M., Yu-Qian, L., *et al.* (2020) *Plant Molecular Biology Reporter*, 1-12.

Luckett, D. (1989) *Euphytica.*, **42**: 177–182.

Lutken, H., Jensen, L.S., Topp, S.H *et al.* (2010) *Plant Biotechnol. J.*, **8**: 211–222.

Lv, S., Cheng, S., Wang, Z., *et al.* (2020) *Ecol. Evol.*, **10**: 4518-4530.

Ma, Y., Byrne, D. H., and Chen, J. (1997) *Hort. Science*, **32**(2): 292-295.

Maarup, R., Aziz, M. B. and Osman, M. (2012) *Sci. Hortic.*, **145**: 52–61.

Maluszynski, M., Kasha, K.J. and Szarejko, I. (2003) *In*: Maluszynski, M., Kasha, K.J., Forster, B.P. and Szarejko, I. (eds.), *Doubled Haploid Production in Crop Plants: A Manual*, Kluwer Academic Publishers, Dordrecht. pp. 309–335.

Manzoor, A., Ahmad, T., Bashir, M. A., *et al.* (2019) *Plants*, **8**(7), 194.

Marasek-Ciolakowska, A., Arens, P. and Van Tuyl, J. (2016) In: Mason A (ed.),Taylor and Francis Group, Boca raton. Chapter 7, pp 159–181.

Marchant, R., Power, J.B., Lucas, J.A. and Davey, M.R. (1998b) *Ann. Bot.*, **81**: 109–114.

Metzlaf, M., Pohlheim, F., Borner, T., *et al.* (1982) *Curr Genet*, **5**(3): 245–249.

Meyer, P., Heidmann, I., Forkmann, G. and Saedler, H. (1987) *Nature*, **330**: 677–678.

Meynet, J., Barrade, R., Duclos, A., *et al.* (1994) *Agronomie*, **14** (3): 169-175.

Miao, H., Jiang, B., Chen, S., *et al.* (2010) *Plant Breed.* **129**: 707–714.

Michishita, A., Ureshino, K. and Miyajima, I. (2002) *J. Japan. Soc. Hort. Sci.*, **71**: 375–381.

Mii, M. (2009) *Acta Hortic.*, **836**: 63-72.

Miri, S.M. (2018) *2nd International and 3rd National Congress on Flower and Ornamental Plants*, Mahallat, Iran.

Mitteau, Y. and A. Silvy. (1982) *Mutat. Breed Newsl.*, **39**: 1992: 23.

Miyoshi, K. and Asakura, N. (1996) *Plant. Cell Rep.* **16**, 1–5.

Mizutani, M., Tsuda, S., Suzuki, K., *et al.* (2003) *Plant Cell Physiol.*, **44**: s122.

Morgan, E.R., Burge, G.K., Seelye, J.F., *et al.* (2001) *Euphytica*, **121**: 343–348

Murovec, J. and Bohanec, B. (2012) *In Tech. Dr. Ibrokhim.*

Musial, K., Bohanec, B., Jakse, M. *et al.* (2005) *In Vitro Cell. Dev. Biol. Plant*, **41**(4): 446-452.

Myung, S.S., Hak, K.S. and Ki, S.K. (2006a) *Hort. Environ. Biotechnol.*, **47**: 376-381.

Myung, S.S., Hak, K.S. and Ki, S.K. (2006b) *Hort. Environ. Biotechnol.*, **47**: 288-293.

Nagatomi, S., Watanabe, H., Tanaka, A., *et al.* (2003) *Institute of Radiation Breeding*, Technical News 65.

Nakamura T, Kuwayama S, Tanaka S, *et al.* (2005) *Euphytica* **142**: 283–289

Nakamura, N., Fukuchi-Mizutani, M., Fukui, Y., *et al.* (2010) *Plant Biotechnol.*, **27**(5): 375–383.

Nakamura, N., Fukuchi-Mizutani, M., Suzuki, K., *et al.* (2006) *Plant Biotechnol.*, **23**: 13–18.

Nakano, M. and Mii, M. (1992) *Plant Cell Rep.*, **11**: 225-228.

Nakano, M. and Mii, M. (1993a) *Theor. Appl. Genet.*, **86**: 1-5.

Nakano, M. and Mii, M. (1993b) *Plant Sci.*, **88**: 203-208.

Nakano, M., Mii, M., Kobayashi, H., *et al.* (2016) *Breeding Res.,* **18**(1): 34-40.

Nakatsuka, T., Saito, M., Sato-Ushiku, Y., *et al.* (2011) *Euphytica,* **184**: 335–344.

Nakatsuka,T., Sato, Takahashi, K., *et al.* (2008) *J. Exp. Bot.,* **59**: 1241–1252.

Namita, B., Raju, D.V.S., Prasad, K.V. *et al.* (2014) *Proceedings of the International Conference on Radiation Biology, IAEA,* **314**: 115.

Nimura, M., Kato, J., Horaguchi, H., *et al.* (2006) *Breed Sci,* **56**(3): 303–310.Nimura, M., Kato, J., Mii, M., *et al.* (2003) *Theor Appl Genet,* **106**(7): 1164–1170.

Nimura, M., Kato, J., Mii, M., *et al.* (2008) *Sci Hortic,* **115**(2): 183–189.

Nishihara, M., Hikage, T., Yamada, E., *et al.* (2011) *Mol. Genet. Genom.,* **286**: 371-382.

Nishihara, M., Nakatsuka, T., Hosokawa, K., *et al.* (2006) *Plant Biotech.,* **23**: 25–31.

Nishihara, M., Nakatsuka, T., Mishiba, K. *et al.* (2003) *Plant Cell Physiol.,* **44**: s159.

Noda, N., Aida, R., Kishimoto, S., *et al.* (2013) *Plant Cell Physiol.,* **54**(10): 1684–1695.

Noda, N., *et al.* (2017) *Sci. Adv.,* **3**: e160278.

Noda, N., Yoshioka, S., Kishimoto, S., *et al.* (2017) *Sci. Adv.,* **3**: 7,e1602785.

Noman, A., Fahad, S., Aqbeel, M., *et al.* (2017) *Biotechnol Lett.* doi: 10.1007/s10529-017-2302-9.

Nonomura, T., Ikegami, Y., Morikawa, Y., *et al.* (2001) *Plant Biotechnol.,* **18**(3): 233-236.

Nowak, M.D., Russo, G., Schlapbach, R., *et al.* (2015) *Genome Biol.,* **16**: 12.

Obata, Y., Niimi, Y., Nakano, M., *et al.* (2000) *Sci Hortic,* **84**: 191-204.

Okada, T., Iizuka, M., Hase, Y., *et al.* (2012) *Hort. Res.,* **11**(1): 428.

Okamoto, A., and Ureshino, K. *(2015) Hort. J.,* **84**.

Okamura, M., Yasuno, N., Ohtsuka, M., *et al.* (2003) *Nucl. Instrum. Methods Phys. Res. B.,* **206**: 574–578.

Okazaki, K. and Murakami, K. (1992) *J. JPN. SOC. HORTIC. SCI.,* **61**: 405-411.

Okazaki, K., Kurimoto, K., Miyajima, I. *et al.* (2005) *Euphytica,* **143**: 101–114.

Okonkwo, C. and Campbell, R. (1990) *Genetics and Breeding,* **23**: 9-15.

Okuda, S., Tsutsui, H., Shiina, K., *et al.* (2009) *Nature.* **458**: 357-61.

Oladosu, Y., Rafii, M. Y., Abdullaha, N. *et al.* (2016) *Biotechnology and Biotechnological Equipment,* **30**(1): 1-16.

Ono, E., Fukuchi-Mizutani, M., Nakamura, N., *et al.* (2006) *Proc. Natl. Acad. Sci.,* **103**: 11075–11080.

Onozaki T (2008) *JARQ,* **42**: 137–144.

Onozaki, T., Ikeda, H., Yamaguchi, T., *et al.* (2002) *Hortic. Res.,* **1**: 13–16.

Onozaki, T., Yagi, M., Fujita, Y., *et al.* (2011) *Hortic. Res.,* **10**: 161–172.

Onozaki, T., Yoshinari, T., Yoshimura T, *et al.* (2006) *Hortic. Res.,* **5**: 363–367.

Pacher, M. and Puchta, H. (2017) *Plant J.,* **90**(4): 819- 833.

Paladines, R., Jurado, D., Riksen-Bruinsma, *et al.* (2012) *Biotechnol.*, **6**(2): 140-145.

Parmar, N., Singh, K.H., Sharma, D., *et al.* (2017) *Biotech.*, **7**: 239.

Pathirana, R., Frew, T., Hedderley, D., *et al.* (2011) *Plant Cell Rep.*, **30**: 1055-1065.

Pavasant, P., Apiratikul, R., Sungkhum, V., *et al.* (2006) *Bioresour. Technol.*, **97**(18): 2321-9.

Pedroso, M.C. and Pais, M.S. (1994) *Plant Cell Tiss. Org.*, **37**: 129-136.

Pei, H., Ma, N., Chen, J. (2013) *PLoS ONE*, **8**: e64290.

Peñafiel, J.M. and Kamo, K. (1994) *In Vitro*, **30**A: 67.

Petersen, K.K., Hagberg, P. and Kristiansen, K. (2003) *Plant Cell Tissue Organ Cult.*, **73**: 137–146.

Petty, L.M., Harberd, N.P., Carre, I.A. *et al.* (2003) *Plant Sci.*, **164**: 175–182.

Pounders, C. T., and Sakhanokho, H. F. (2016) *Hort. Science*, **51**(12): 1616-1617.

Pourhosseini, L., Habashi, A.A., Kermani, M.J., *et al.* (2012) *Ann. Biol. Res.*, **3**(6): 2843-2850.

Pourhosseini, L., Kermani, M.J., Habashi, A.A., and Khalighi, A. (2013) *Plant Cell Tissue Organ Cult.*, **112**: 101–108.

Power, J., Frearson, E., Hayward, C. *et al.* (1976) *Nature*, **263**: 500–502.

Preil, W. and Ebbinghaus, R. (1985) Rhododendron und immergrüne Laubgehölze, Jahrbuch, pp: 85-92.

Preil, W., Engelhardt, M., and Walther, F. (1983) *Acta Hortic.*, **131**: 345-351.

Qi, W., Chen, X., Fang, P., *et al.* (2018) *BMC Plant Biol.*, **18**: 119.

Qi, Y., Liu, Y., Zhang, Z. *et al.* (2018) *Hortic. Res.*, **5**: 37.

Qi, Y., Lou, Q., Quan, Y., *et al.* (2013) *Plant Cell Tissue Organ Cult.*, **115**: 263–273.

Qi, Y., Ye, Y. and Manzhu, B. (2011) *Afr. J. Biotechnol.*, **10**(75): 17332-17338.

Qian, Li., Ze, Wu., Huijun, Wu., *et al.* (2019) *Int. J. Mol. Sci.*, **2**: 5865-5887.

Raffeiner, B., Serek, M., and Winkelmann, T. (2009) *Plant Cell Tissue Organ Cult.*, **98**: 125–134.

Ramanna, M. and Jacobsen, E. (2003) *Euphytica*, **133**: 3-18.

Ramsey, J. and Schemske, D.W. (1998) *Annu Rev Ecol Syst.*, **29**: 467–501.

Raquin, C. (1985) *Zeitschrift Fur Pflanzenzuchtung*, **94**: 166-169.

Raquin, C., Cornu, A., Farcy, E., *et al.* (1989) *Theor. Appl. Genet.*, **78**: 337-341.

Raymond, O., Gouzy, J., Just, J., *et al.* (2018) *Nature genetics*, **50**(6) : 772-777.

Razdan, A., Razdan, M. K., Rajam, M. V. *et al.* (2008) *Plant Cell Tiss. Org.*, **95**: 245–250.

Reed, S. (2000) *J. Environ. Hortic*, **18**: 34-39.

Reed, S., Riedel, G. and Pooler, M. (2001) *J. Environ. Hortic.*, **19**: 85-88.

Ren, J., Wu, P., Trampe, B., *et al.* (2017) *Plant Biotechnol. J.*, **15**: 1361–1370.

Ribarits, A., Mamun, A.N.K., Li S., *et al.* (2007) *Plant Biotechnol. J.*, **5**: 483–494.

Richards, J. (1993) *Primula*. Timber Press, Portland, Oregon.

Rieseberg, L.H. and Carney, S.E. (1998) New Phytol, **140**: 599–624.

Roper, A. C., Lutken, H., Christensen, B. *et al.* (2015a) *Euphytica*, **203**(3), 643-657.

Roper, A., Lütken, H., Hegelund, J., *et al.* (2012) *Acta Hortic.*, **953**: 161-166.

Roper, A., Orabi, J., Lütken, Henrik., *et al.* (2015b) *PloS one*, **10**.e0137537.

Rose, J.B., Kubba, J. and Tobutt, K.R. (2000) *Plant Cell Tissue Organ Cult.*, **63**: 127– 132.

Rotman, N., Rozier, F., Boavida, L., *et al.* (2003) *Curr. Biol.*, **13**(5): 432–436.

Rouse, J. (1993) *JARS*, **47**: 23-45.

Sakai, K., Ozaki, Y., Ureshino, K., *et al.* (2008) *Sci. Hortic.*, **115**(3): 268–274.

Salgotra, R.K. and Stewart, C.N. (2020) *Int. J. Mol. Sci.*, **21**: 4792.

Sato, S., Katoh, N., Yoshida, H., *et al.* (2000) *Sci. Hortic.*, **83**: 301- 310.

Savin, K.W., Baudinette, S.C., Graham, M.W., *et al.* (1995) *Hortic Sci.*, **30**: 970- 72.

Schum, A. and Preil, W. (1998) In *Somaclonal variation and induced mutations in crop improvement* (pp. 333-366) Springer, Dordrecht.

Schum, A. (2003) *Acta Hortic.*, **612**: 47-60.

Scott, R.J., Spielman, M., Bailey, J., *et al.* (1998) *Development*, **125**(17): 3329–3341.

Seitz, C., Eder, C., Deiml, B., *et al.* (2006) *Plant Mol. Biol.*, **61**: 365–381.

Seitz, C., Vitten, M., Steinbach, P., *et al.* (2007) *Phytochemistry* **68**(6): 824-833.

Sen, S., Kumar, S., Ghani, M. and Thakur, M. (2013) *Plant Pathol. J.*, **12**: 1-10.

Senapati, S.K. and Rout, G.R. (2008) *Indian J. Hort.*, **65**(4): 452-460.

Senapati, S.K. and Rout, G.R. (2011) *Afr. J. Biotechnol.*, **10**(30): 5705-5712.

Shahin, A., Arens, P., Van Heusden, A.W., *et al.* (2011) *Plant Breed*, **130**(3): 372–382.

Shahin, A., Arens, P., Van Heusden, A.W., *et al.* (2011) *Plant Breed.*, **130**: 372–382.

Shariatpanahi, M. E. and Ahmadi, B. (2016) *Plant Tissue Culture: Propagation, Conservation and Crop Improvement*, Springer, Berlin, pp. 487-507

Sharma, R. and Messar, Y. (2017) *Curr. Sci*, **113**(1): 43-52.

Sherman, J.M., Moyer, J.W. and Daub, M.E. (1998) *Plant Dis.*,**82**(4).

Shimada, A. and Kimura, Y. (2006) *J. Biosci.*, **61**: 578-82.

Shimizu, K., Miyabe, Y., Nagaike, H. and Yabuya, T. (1999) *Euphytica*, **107**: 105-113.

Shimizu, K., Yabuya, T. and Adachi, T. (1996) *Euphytica*, 89.

Shinoyama, H., Komano, M., Nomura. Y. *et al.* (1998) *Bull Fukui Agric. Expt. Stn.*, **35**: 13- 21.

Shinoyama, H., Mitsuhara, I., Ichikawa, H. *et al.* (2015) *Acta Hortic.*, **1087**: 485–498.

Shinoyama, H., Mochizuki, A., Komano, M., *et al.* (2003) *Breed. Sci.*, **53**: 359-367.

Shinoyama, H., Mochizuki, A., Nomura, Y., and Kamada, H. (2008) *Plant Biotechnol.*, **25**: 17–29.

Shinoyama, H., Sano, T., Saito, M., *et al.* (2012) *Mol. Breed.*, **29**: 285–295.

Sims T.L., Robbins T.P. (2009) *In*: Gerats T., Strommer, J. (eds) *Petunia*. Springer, New York, NY. *pp*, 85-106.

Singh, K.P., Singh, B., Raghava, S.P.S., *et al.* (2000) *Indian J. Genet.*, **60**(4): 535-539.

Singh, M. and Bala, M. (2019) *Indian J. Expt. Biol.*, **57**: 50-54.

Sitbon, M. (1981) *Agronomie*, **1**: 807–812.

Slavov, S., Batchvarova, R., Atanassova, B. and Zaprjanova, N. (2001) *Acta Hortic.*, **552**: 113-118.

Soltis, P. and Soltis, D. (2000) *PNAS*, **97**: 7051-7057.

Song, A., Zhu, X., Chen, F., *ety al* (2014) *Int. J. Mol. Sci.*, **15**: 5063–5078.

Song, X., Xu, Y., Gao, K., *et al.* (2020) *Hortic. Res.*, **7**: 108.

Sparnaaij, L. and Putten, H. (1990) *Euphytica*, **50**: 211-220.

Sumitomo, K., Narumi, T., Satoh, S. *et al.* (2008) *J. Exp. Bot.*, **59**(15): 4075-82.

Sun, C.Q., Chen, F.D., Teng, N.J., *et al.* (2010) *Euphytica*, **174**(4): 51-60.

Sun, C.Q., Huang, Z.Z., Wang, Y.L., *et al.* (2011) *Euphytica*, **178**(2): 195–202.

Sun, J., Ren, L.P., Cheng, Y., *et al.* (2014) *Gene*, **552**: 59–66.

Suzuki, K., Takatsu, Y., Gonai T., Kasumi, M. (2005) *Acta Hortic.*, **673**: 175–181.

Takamura, T., Lim, K.B., van Tuyl, J.M. (2002) *Acta Hort.*, **572**: 37–40.

Takamura, T., Miyajima, I. (1996) *Sci. Hortic.*, **65**: 305–312.

Takatsu, Y., Nishizawa, Y., Hibi, T., *et al.* (1999) *Sci. Hortic.*, **82**: 113–123.

Tanaka, Y., and Brugliera, F. (2013) *Philos. Trans. R. Soc.*, **368**: 20120432.

Tanaka, Y., Brugliera, F. and Chandler, S. (2009) *Int J Mol Sci.*, **10**: 5350-5369.

Tanaka, Y., Katsumoto, Y., Brugliera, F. and Mason, J. (2005) *Plant Cell Tiss. Org. Cult.*, **80**: 1–24.

Tang, N., van der Lee, T., Shahin, A. *et al.* (2015) *Mol. Breed.*, **35**: 122.

Tang, N., van der Lee, T., Shahin, A., *et al.* (2015) *Mol. Breeding*, **35**: 122.

Tejasvini, Sane, A., Gadre, A. and Ghatke, M. (2016) *Indian J. Agric. Sci.*, **86**(10): 1271–5.

Tejaswini, Chowdury, S., Paramesh, H. *et al.* (2006) *Indian J. Genet.*, **66**(1): 71-72.

Terefe-Ayana, D., Yasmin, A., Le, T.L., *et al.* (2011) *Front. Plant Sci.*, **2**: 35.

Thomas H (1993) In: Hayward MD, Bosemark NO, Romagosa I (eds) *Plant breeding: principles and prospects.* Chapmann and Hall, London, pp. 79–92

Todorova, M., Ivanov, P., Shindrova, P., *et al.* (1997) *Euphytica*, **97**: 249-254.

Tonosaki, K., Osabe, K., Kawanabe, T. and Fujimoto, Ryo. (2016) *Breed. Sci.*, **66**(3).

Toppino, L., Kooiker, M., Lindner, M., *et al.* (2011) *Plant Biotechnol. J.*, **9**: 684–692.

Tosca, A., Arcara, L. and Frangi, P. (1999) *Plant Cell Tiss. Org.*, **59**: 77-80.

Tosca, A., Lombardi, M., Marinoni, L., *et al.* (1990) *Acta Hortic.*, **280**: 337-340.

Toyoda, H., Nonomura, T. and Matsuda, Y. (2015) *Acta Hort.*, **1083**: 487-490.

Ueno, K., Nagayoshi, S., Imakiire, S., *et al.* (2013) *Hort. Res.*, **12**: 245–254.

Ueyama, Y., Suzuki, K., Mizutani, M.F., *et al.* (2002) *Plant Sci.*, **163**: 253-263.

Van Creij, M., Kerckhoffs, D., Van Tuyl, J. (1997) *Acta Hortic.*, **430**: 547-557.

Van der Salm, T.P.M., van der Toorn, C.J.G., Bouwer, R., *et al.* (1997) *Mol.Breed.*, **3**: 39-47.

Van Eijk, J.P., Van Raamsdonk, L.W.D., Eikelboom, W. and Bino, R.J. (1991) *Sex. Plant Reprod.*, **4**: 1-5.

Van Huylenbroeck, J., Leus, L. and Van Brockstaele, E. (2005) *Acta Hortic.*, **690**: 109–112.

Van Laere, K., Dewitte, A., Van Huylenbroeck, J., *et al.* (2009) *J. Hortic. Sci. Biotechnol.*, **84**(2): 240-247.

Van Laere, K., França, S., Vansteenkiste, H., *et al.* (2011) *Acta Physiologia Plantarum*, **33**: 1149–1156.

Van Tuyl, J., and Kim, B. (2003) *Acta Hort.*, **612**: 13-22.

Van Tuyl, J., Marcucci, M., Visser, T. (1982) *Euphytica* **31**: 613-619.

Van Tuyl, J., Meijer, B. and Van Dien, M. (1992) *Acta Hort.*, **352**: 625–630.

Van Tuyl, J., Van De Sande, K., Van Dein, R., *et al.* (1990) *Acta Hortic.*, **266**(266): 317-322.

Van Tuyl, J., Van Diën, M., Van Creij, M., *et al.* (1991) *Plant Science*,**74**: 115-116.

Van Tuyl, J.M. (1990) *Herbertia.*, **45**: 97–103.

Van Tuyl, J.M. and De Jeu, M.J. (1997) *CUP*, 273–292.

Van Tuyl, J.M., Maas, I.W.G.M. and Lim, K.B. (2002) *Acta Hortic.* **570**, 213-218.

Van Wordragen, M.F., Honee, G. and Dons, H.J.M. (1993) *Transgenic Res.*, **2**: 170-180.

Vasal, S.K., Singh, N.N., Dhillon, B.S. *et al.* (2004) *In*: Jain H.K., Kharkwal M.C. (eds) *Plant Breeding*. Springer, Dordrecht. pp. 391-406.

Venkatachalam, P. and Jayabalan, N. (1994) *J. Swamy Bot Cl*, **11**: 58.

Verma, A.K., Prasad, K.V., Singh, S.K. *et al.* (2012) *Indian J. Hortic.*, **69**(4): 562–567.

Vervaeke, I., Deroose, R. and De Proft, M. (2003) *Acta Hortic.*, **624**: 77-83.

Vervaeke, I., Parton, E., Maene, L. *et al.* (2001) *Euphytica.*, **118**: 91-97.

Wan, X.L., Zhou, Q., Wang, Y.Y., *et al.* (2015) *Front. Plant Sci.*, **6**: 519.

Wang, H., Dong, B., Jiang, J., *et al.* (2014) *Front. Plant Sci.*, **5**: 738.

Wang, h., XiaoXiao, C., Jing, Lin., *et al.* (2018) *Horticulture research,* **5**(1): 1-13.

Wang, J., Huang, L., Bao, M. Z. *et al.* (2009) *Euphytica,* **167**(1): 45-55.

Wang, J.M., Yang, Y., Liu, X.H., *et al.* (2014) *BMC Genom.,* **15**(1): 1-23.

Wang, Q., Zhang, X., Lin, S., *et al.* (2020) *J. Exp. Bot.,* **71**: 1915-1927.

Wang, S., Li, X., Yang, L., *et al.* (2011) *Afr. J. Biotechnol.,* **10**(50): 10287-10292.

Wang, Y., Tong, Y., Li, Y., *et al.* (2011) *Sci. Hortic.,***130**: 296–302.

Warner, R. (2011) *Floricult. Int.* June, 15–16.

Watanabe, K. (1977) *J. Hered.,* **68**: 317-320.

Weather head MA., *et al.* (1982) *Sci. Hortic.,* 17(2): 137-144.

Wei, D., Mangmang, Li., Zhongai, Li., *et al.* (2019) *Plant Physiol. Biochem.* **146**: 315-328.

Wendel, J. (2000) *Plant Mol. Biol.,* **42**: 225-249.

Wietsma, W., De Jong, K., Van Tuyl, J. (1994) *Plant Cell Incompatibility Newsletter,* **26**: 89-92.

Wilhelmi, L, and Preuss, D. (1997) *Plant Physiol.,* **113**: 307-312.

Williams, E. and Rouse, J. (1990) *Sex. Plant Rep.,* **3**: 7-17.

Winarto, B. (2010) *J Hortic,* **20(**4): 342–351.

Winarto, B., Mattjik, N. A., Teixeira da Silva, *et al.* (2010) Hortic., **127**(1): 86-90.

Winefield, C., Lewis, D., Arathoon, S. and Deroles, S. (1999) *Mol. Breed.,* **5**: 543–551

Xie, Q., Chen, G., Liu, Q., *et al.* (2015) *Mol. Breed.,* **35**: 1–13.

Xing, W., Wang, Z., Wang, X., *et al.* (2014) *Sci. Hortic.,* **172**: 68–72.

Xu, G., Liu, Y., Chen, S., and Chen, F. (2011) *Russ. J. Plant Physiol.,* **58**: 491–497.

Xu, Y., Gao, S., Yang, Y., *et al.* (2013) *BMC Genom.,* **14**: 662.

Yadav, M., Rajput, V., Saharan, R.S. *et al.* (2016) *TAJH,* **11**(1): 114-118.

Yafei, Ge., Qixian, L., Ping, L., *et al.* (2019) *BMC Genom.,* **20**: 600.

Yagi, M. (2013) *JARQ* **47**: 29–35.

Yagi, M. (2018) *Breed. Sci.,* **68**: 62-70.

Yagi, M., Kimura, T., Yamamoto, T., *et al.* (2012) *Mol. Breed.,* **30**: 495-509.

Yagi, M., Kosugi, S., Hirakawa, H., *et al.* (2014a) *DNA Res.,* **21**: 231–241.

Yagi, M., Yamamoto, T., Isobe, S., *et al.* (2013) *BMC Genomics,* **14**: 734.

Yamaguchi H. (2018) *Breeding Science Preview.*1-8.

Yamaguchi, H., Shimizu, A., Haso, Y., *et al.* (2009) *Euphytica,* **165**: 97-100.

Yan, H., Hao, Z., Min, Chen., *et al.* (2014) *Gene,* **540**(1): 96-103.

Yan, L., Wang, X., Liu, H., *et al.* (2015) *Mol. Plant,* **8**: 922–934.

Yang, J. (2005) *Engei Gakkai Zasshi.,* **74**(1): 78-86.

Yang, M., Zhu, L., Xu, L., *et al.* (2014) *Ann. Appl. Biol.*, **165**: 73–95.

Yang, Q., Zhang, D., Li, Q., Cheng, Z. and Xue, Y. (2007) *Plant J.*, **51**: 140–151.

Yao, J. and Cohen, D. (1996) *N. Z. J. Crop Hortic. Sci.*, **24**: 297-301.

Yao, J. and Cohen, D. (2000) *Theor. Appl. Genet.*, **101**: 400-406.

Yingchun, Q., Yaomei, Y. and Manzhu, B. (2011) *Afr. J. Biotechnol.*, **10**(75): 17332-17338.

Young, D. and Lewandowski, V. (2000) *Plant Physiol.*, **124**: 115–124.

Young, D., Tice, C., Michelotti, E., Roemelle, R., *et al.* (2001) *Bioorg. Med. Chem. Lett.*, **11**: 1393–1396.

Younis, A., Hwang, Y. and Lim, K. (2014) *Plant Cell Rep.*, **33**: 215–233.

Younis, A., Ramzan, F., Ramzan, Y., *et al.* (2020) *Plants*, **9**: 1374.

Yue, J., Zhu, C., Zhou, Y., *et al.* (2018) *Sci. Rep.*, **8**: 13414-13428.

Zakizadeh, H., Lutken, H., Sriskandarajah, S., *et al.* (2013) *Plant Cell Rep.*, **32**: 195–205.

Zamariola, L., Tiang, C., De Storme, N., *et al.* (2014) *Front. Plant Sci.*, **5**: 279.

Zenkteler, M. (1990) *CRC Crit. Rev. Plant Sci.*, **9**: 267-279.

Zhang, F., Chen, S., Jiang, J., *et al.* (2013) *PLoS ONE*, **8**(12): e83023.

Zhang, G.Q., Xu, Q., Bian, C., *et al.* (2016) *Sci. Rep.*, **12**: 19029.

Zhang, J., Zhang, M.and Deng, X. (2007) *Plant Cell Tissue Organ Cult.*, **89**: 211–216.

Zhang, Q.Y., Luo, F.X., Liu, L. and Guo, F.C. (2010) *Plant Cell Tissue Organ Cult.*, **101**: 41–47.

Zhang, X., Wu, Q., Li, X., *et al.* (2011) *Sci. Hortic.*,**129**: 335–342.

Zhang, XQ., Yang, JS., Zheng ZX (2009) *Jiang Su Agri. Sci.*,**108**: 57–58.

Zhang, Z., Dai, H., Xiao, M. and Liu, X. (2008) *Euphytica.*, **159**: 59–65.

Zhao, C., Tian, M., Li, Y. and Zhang, P. (2017) *Euphytica.*, **213**: 94.

Zhao, F. C., Nilanthi, D., Yang, Y. S. *et al.* (2006) *Plant Cell Tiss. Org.*, **86**: 55–62.

Zheng, Z.L., Yang, Z., Jang, J.C. and Metzger, J.D. (2001) *J. Am. Soc. Hortic. Sci.*, **126**: 19–26

Zhou, X., Mo, X., Gui, M., *et al.*(2015) *Plant Physiol Biochem.*, **97**: 255–263.

Zhou, X., Mo, X., Qu, S. *et al.* (2014) *Hortic. Environ. Biotechnol.*, **55**: 125–128.

Zuker, A., Tzfira, T., Ben-Meir, H., *et a.* (2002) *Mol. Breed.*, **9**: 33-41.

Zvi, M.M.B., Zuker, A., Ovadis, M. *et al.* (2008) *Mol. Breed.* **22**, 543–553.

2

Orchids

Debmala Mukherjee, L.P. Yadav,
T.K. Bose, T. Choudhury and K. Dutta

1.0 Introduction

Orchids are considered as the most beautiful plants for the exquisite beauty of the flowers, variety of fragrance, brilliance in colour, unusual shapes, variation in form and attractive growth habits. There are about 30,000 species and 32,000 hybrids of orchid. Among 300,000 species in angiosperms, Orchidaceae, containing 30,000 species, is one of the largest families. Almost every habitats on earth have orchid plants successfully colonized, and it indicates that orchids are among the plants with significant ecological and evolutionary importance (Tsai *et al.,* 2017). It has also taken a significant position in the cut flower industry due to its attractiveness, long shelf life, high productivity, right season of bloom, ease of packing and transportation. Orchids account for a large share of global floriculture trade both as cut flowers and as potted plants and are estimated to comprise around 10 per cent of the international fresh cut flower trade (De *et al.,* 2019). Although orchids have been documented as potential ornamental plant; they are widely known for their medicinal value also. The miraculous curative property of orchid has been exhibited from time immemorial. Incredible diversity, high alkaloids and glycosides content, makes orchid to become potential both in phytochemical and pharmacological point of view (Nongdam, 2014).

India is known for its rich biodiversity, and one of its dominant plant families is Orchidaceae consisting of about 1350 species in 185 genera occurring in diverse phytogeographical conditions ranging from tropical, subtropical and temperate conditions with varying microclimates (Hegde, 2020). However, despite the rich natural occurrence and ideal agroclimatic conditions in India, orchid-based floriculture has not been systematically developed, and the people are yet to reap the benefit out of this natural resource having the least production area and minimum contribution in the overall turnover of floriculture products. In fact, India has lagged other countries in Orchid trade (Hegde, 2020).

2.0 Production and Market

The orchid growing countries supplying major bulk of cut flowers for export are Thailand, the Netherlands and Singapore. Countries like France, Germany, U.K., Australia, New Zealand, Japan, USA, Malaysia, Sri Lanka, Taiwan South Korea, South Africa and several South American countries also grow orchids for pot plants and cut flowers. In Asia, orchids are commercially grown in Thailand, Singapore, Malaysia, Sri Lanka and Indonesia, Thailand being the largest producer and exporter. *Dendrobium* is the most popular orchid in cut flower trade in Asia (80 per cent) followed by *Arachnis, Aranda, Renanthera, Mokara, Oncidium* and *Vanda*. In Europe, the Netherlands is the largest producer. The Dutch growers mostly cultivate *Cymbidium* for cut flowers and *Phalaenopsis* for cut flowers and pot plants. According to Hinsley *et al.,* 2018 the top ten orchid taxa traded are given in Table.1.

Table 1: Top Ten Reported Orchid Taxa Commercially Traded as Artificially Propagated Live Plants in the 10-year Periods 1996–2005 and 2006–2015 (as reported by importers), Including Trade Reported at Genus and Family Levels*

Rank	1996-2005 (Total no live plants: 459857389)		2006-2015 (Total no live plants: 659817913)	
	Reported Taxa	No. Live Plants (Per cent total)	Reported Taxa	No. Live Plants (Per cent total)
1	Orchidaceae species	165962470 (36.1)	Orchidaceae hybrids	189447122(28.7)
2	Orchidaceae hybrids	123939767 (27.0)	*Cymbidium* spp.	177536225 (26.9)
3	*Dendrobium* spp.	92482163 (20.1)	Orchidaceae species	124907316 (18.9)
4	*Phalaenopsis* spp.	31572618 (6.9)	*Phalaenopsis* hybrids	66683709 (10.1)
5	Species of *Cymbidium* Sw.	24672878 (5.4)	*Phalaenopsis* spp.	28954444 (4.4)
6	Species of *Oncidium* Sw.	7077873 (1.5)	*Dendrobium* spp.	22692242 (3.4)
7	*Phalaenopsis amabilis* (L.) Blume	4769951 (1.0)	*Cymbidium* hybrids	21813621 (3.3)
8	*Cattley spp.*	2375391 (0.5)	*Dendrobium* hybrids	15134974 (2.3)
9	*Cymbidium kanran* Makino	1478658 (0.3)	*Cattley spp.*	2276462 (0.4)
10	Species of *Vanda* R. Br.	1130662 (0.3)	*Oncidium* spp.	1489956 (0.2)

* Hinsley *et al.* (2018).

Orchids are consistently ranked among the best sellers in the global potted plant trade and also comprise 10 per cent of all fresh cut flowers traded internationally. Thailand is the largest exporter of orchid cut flowers to India devoting 80.67 per cent of total import followed by Netherlands 15.54 per cent, New Zealand 2.29 per cent and China 1.5 per cent, respectively. Highest import of orchids was recorded in 2013-2014 (Rs. 3425.76 Lacs) followed by 2015-2016 (Rs. 2985.19 Lacs) and 2018-2019 (Rs. 2321.84 Lacs) (Table 2). Maximum export of orchids was found in 2016-2017 (Rs. 5.23 Lacs) followed by 2017-2018 (Rs. 4.89 Lacs) (De, 2020).

Table 2: Export of Orchids from India (Rs. in lacs and quantity in MT)*

2013-2014		2014-15		2015-16	
Quantity	Value	Quantity	Value	Quantity	Value
0.72	2.72	2.09	3.24	0.99	2.11

* De (2020).

2.1.0 Asia

Thailand is a natural habitat for about 1,129 species and 178 genera of orchids. For about 100 years, orchid growing was considered a hobby for the rich, but the introduction of *Dendrobium* Pompadour proved to be a landmark that also brought popularity of orchid cultivation in Thailand. In 1966, only a small amount of orchid cut flowers were exported from Thailand to some European countries, but with a decade, the country attained the status of the world's leading producer and exporter of orchids where it continues to hold the primary orchid exporter in the world. In 2012, the total Thai growing areas of production was 7,420 acres with yields of 2,403 kilograms per acre and cut-flower exports of 2.1 billion 63.6 billion US$ to 148 countries and plant export to 160 countries (17.8 million US$) where. Also in 2012, it is estimated that 46 per cent of the orchid production were consumed locally and 54 per cent were exported. Newly developed technology to increase the production and reduce costs, such as greenhouses, planting materials and containers, pest control, production, plant breeding, tissue culture, postharvest and packaging technology, replaced traditional growing, which imitated natural growing. These improved practices have resulted in higher yields and quality of flowers and plants, as well as reductions in production costs. Orchids will continue to dominate other ornamental crops in Thailand due to better technology know-how, suitable climatic conditions, experienced and skillful growers and exporters, as well as their nationwide popularity and the country's pride (Thammasiri, 2015).

Major markets for orchids in Asia are occupied by Japan and Singapore. The total imports of orchids by Japan accounted for US$ 57.4 mn in 2008 making it the largest importer of orchids in the world. The main sources for these imports include Thailand, Taiwan, New Zealand and Malayasia which together account for as much as 96.5 per cent of the total orchids imports by Japan in 2008. Imports by Singapore of fresh orchids amounted to US$ 6.5 mn in 2007 with Malayasia, Thailand and Taiwan being the main sources of imports for the country. In contrast, imports of fresh orchids by Singapore from India were only US$ 1379.3 representing a share of 0.02 per cent of the country's total imports of the product in 2007. This clearly indicates that there are vast possibilities for increasing India's exports to Singapore particularly considering the proximity of the country and India's East Policy (De, 2020).

2.2.0 Europe

The Netherlands is the top orchid exporting country in the world (39.67 per cent of world orchid market) followed by Thailand (28.41 per cent), Taiwan (10 per cent), Singapore (10 per cent) and New Zealand (6 per cent). Importing countries are mainly Japan (30 per cent), UK (12 per cent), Italy (10 per cent), France (7 per cent) and the USA (6 per cent). The total orchid cut flower trade of the world mostly consists of 85 per cent Dendrobium species and 15 per cent Phalaenopsis and Cymbidium species, and Asia is the main source of orchid to enter the world market (Cheamuangphan *et al.*, 2013 cited by De, 2020).

2.3.0 United States

U.S. domestic production of orchids is widespread, and local orchids (mostly Cattleyas) can be found at all the major wholesale flower markets. Production of dendrobium orchids is concentrated in Hawaii, where approximately 2.6 million square feet are devoted to dendrobium spray cultivation. In 1993, 3.5 million stems, valued at US$2.7 million, were shipped out of state. This number represented a decrease from 1992, when 4.2 million stems, worth US$3.0 million, were exported.

U.S. imports of fresh-cut orchids have increased by 51.2 per cent in volume and 21.8 per cent in value since 1990, from 15.3 million stems (US$10.85 million) to 23.2 million stems (US$7.13 million). Almost all of this increase is accounted for by a rise in imports of dendrobium orchids, from 9.1 million stems (US$3.3 million) in 1990 to 16.8 million stems (US$4.4 million) in 1994 (Anon.,1995).

Thailand, the world's largest producer of dendrobium orchids, is also the biggest supplier to the United States and accounted for 97.0 per cent of total imports in 1994. U.S. imports from Thailand of dendrobium product have nearly doubled in the last five years, from 8.6 million stems in 1990 to 16.4 million stems in 1994. Singapore was the second-largest supplier in 1994, with 289,000 stems supplied, followed by Jamaica (44,611 stems), New Zealand (37,142 stems) and Costa Rica (23,660 stems).

An analysis of the prices from the Chicago Wholesale Flower Market showed that orchid prices there were quite stable during 1992-1994. The local Cattleyas were the most expensive, selling for US$4.75 per stem year round. Prices for other varieties that year were US$2.50 per stem for California cymbidiums, US$1.75-US$1.88 per stem for dendrobiums and US$2.50-US$3.38 per stem for the Thai phalaenopsis variety.

Although India is blessed with a wealth of orchid flora and about 1300 species are estimated to occur in this country (Rao, 1979; Arora, 1983), cultivation of orchids by amateur and professional growers is not a popular hobby or enterprise. In the recent years tropical orchids like *Dendrobium, Vanda, Mokara* and *Oncidium* are being grown commercially for cut flowers in the southern part of the country. In Sikkim and North Bengal though orchids are grown for many decades, mostly because of the favourable climates, commercial production of plants and cut flowers is meagre. Many orchids native to this country have already proved to be important parent plants and contributed to the production of several outstanding hybrids in the world. Hybrids of certain Indian orchids like *Vanda coerulea* and species of *Cymbidium, Dendrobium* and *Paphiopedilum* are considered the monarchs in the orchid world. Due to the diversity of environmental condition in India, it is possible to grow all types of orchids in suitable places in ordinary greenhouses without the control of environment.

Prices of *Cymbidium Cattleya* and *Phalaenopsis* flowers are much more than that of *Dendrobium* and *Oncidium, etc.*

3.0 Importance and Uses

3.1.0 Importance as Ornamental Plants

Orchids with their bewildering range of flowers and beautiful colour combinations provide a source of profound aesthetic pleasure to both owners and visitors. The striking resemblances of their flowers to various forms of animal life behold the attention of everyone who looks at them. A *Brassia* spike suggests a small collection of large and colourful spiders. The flowers of *Phalaenopsis* species producing white or pink blossoms look like moths while those of *Oncidium papilio* and *O. kramerianum* resemble gorgeous tropical butterflies. There are also bee

Orchid Show Around the World

23rd World Orchid Conference at Taichung, Taiwan (2020)

Orchid show at Royal Horticulture Society, London (2015)

Singapore Garden Festival

Orchid Show at Singapore Botanic Garden

Orchid Show at Kennett Square, USA (2020)

Singapore Garden Festival

Orchid Show at Bangalore (2019)

Orchid Show at Japan

Distribution of Orchid throughout the World

Orchid Show Around the World

Orchid Valley in Malaysia

Orchid Valley in Taiwan

Orchid Valley in New Zealand

Orchid Valley in Santa Barbara, California

Orchid Valley in Holland

Orchid Valley in Thiland

orchid, scorpion orchid, lizard orchid, frog orchid, tiger orchid and even man orchid. Growing orchid is a fascinating hobby throughout the world. There is joy in watching new growths and roots waiting for seedling to become a mature plant and seeing the new hybrids in flower. Orchids are excellent items for garden and can be grown in beds, pots, baskets, split hollows of bamboo pieces, *etc.* They are the most beautiful items for indoor decoration. A flowering plant, kept in a living room, remains fresh for many days and attracts the attention of visitors.

Orchids, having flowers of wonderful beauty and very good keeping qualities, are of the highest value as cut flowers. Some orchid flowers last for one to three months if remained attached to the plant, and as cut flowers they remain fresh for one to four weeks. *Cattleya* flowers stay fresh and perfect for two to six weeks and *Phalaenopsis* can last for two to five weeks (Northern, 1970). One cut flower in a simple little vase is a thing of beauty for table or mantel. Because of their long lasting qualities, orchids make attractive corsages. Ladies also use orchid flowers for the adornment of their hair. Commercial cultivation of orchids both for plant sale as well as cut flower production has developed into vast industries in many countries and the sale of flowers runs into millions of dollars.

3.2.0 Food and Therapeutic Uses

Several orchids have been used as food in different parts of the world. *Anoectochilus* leaves are used in Indonesia and Malaysia as vegetable. Dried leaves of *Dendrobium salaccense* cooked with rice add delicate and exotic flavour. On the Island of Cyprus, the villagers make a milk-custard drink with dried ground tubers of *Orchis anatolica* (Withner, 1959). Pseudobulbs of *Cymbidium madidum* and *Dendrobium speciosum* are used as food. In Western Australia, bulbs of *Thelymitra antennifera* are used as yam while tubers of *Microtis uniflora, Caladenia carnea* and *Eriochilus cucullatus* are eaten by the native inhabitants of South Australia (Bose and Bhattacharjee, 1980).

The tubers and pseudobulbs of several orchids like *Orchis latifolia, O. mascular, Cymbidium aloifolium, Zeuxine strateumatica*, some species of *Dendrobium, Eulophia* and *Habenaria* are used for preparing salep which is valued as a restorative and is used in the treatment of various diseases (Puri, 1970). In the Khasi hills, juice from crushed leaves of *Cymbidium giganteum* is said to be utilized for clotting of blood in wound while powdered flowers of *Vanda spathulata* is used in Malabar for hysteria. In North Queensland, fruits and pseudobulbs of *Cymbidium canaliculatum* and *C. madidum* were used in cases of dysentery. The seeds of *Cymbidium aloifolium* are also used for healing wounds. The powdered roots of *Vanda tessellata* is considered to be an antidote for poisoning. It is also used in rheumatic pains and abdominal complaints. *Dendrobium fimbriatum* has been used for liver upsets and nervous debility, while *D. teretifolium* for headache and pain in other parts of the body. Other orchid genera like *Oberonia, Eria, Bulbophyllum, Eulophia, Geodorum, Grammatophyllum* and *Hetaeria* are also reported to have been used as medicine in different parts of the world to cure various diseases (Hawkes, 1943, 1944; Withner, 1959).

Salep is made from the polysaccharide-rich tubers of wild orchids that are traded predominantly in Turkey as part of a seasonal trade recorded from as far back as 1850 (Hinsley *et al.*, 2018). After collection, the orchid tubers are boiled in water, milk or ayran (a yoghurt-based drink) to render the enzymes in them inactive and prevent tubers from regrowing (Tamer *et al.*, 2006). The tubers of terrestrial orchids are used in several African countries in the production of chikanda, a large cake with a meat-like structure, made of ground orchids and peanuts baked with ashes or baking soda (Kaputo, 1996; Bingham, 2009). Chikanda is a dish that was traditionally eaten by the Bemba tribe in northern Zambia (Hinsley *et al.*, 2018).

Indigenous system of medicine, particularly the *Ayurveda* has attracted modern scientists in finding out solutions for many challenging diseases. Orchids are being used in this system of medicine for over 300 years and some of the common drugs prepared from these plants are *Riddhi, Vriddhi, Jeevak* and *Rishabhak, Jeewanti, Munjatak, Amarkand* and *Rasna* (Agrawal and Dey, 1969; Kirtikar and Basu, 1975; Kurup, 1977). They also form useful drugs in combination with drugs obtained from plants of other families. Some of the compound drugs from orchids like *Astavarg, Churn, Jeevaniyo Dashko Mahakshay* and *Chyawan Prash* linctus are well known in the *Ayurveda* (Anon., 1949; Raghunathan, 1976; Kaushik, 1983).

The most prominently cited orchids in Traditional Chinese Medicine are various *Dendrobium* spp. used to make the drug *shi-hu* [particularly *D. catenatum* Lindl. (including *D. officinale* Kimura and Migo), *D. loddigesii* Rolfe, *D. moniliforme* (L.) Sw. and *D. nobile* Lindl.)](Teoh, 2016; Leon and Lin, 2017 cited by Hinsley *et al.*, 2018)

Several orchids are glycosidal plants. Loroglossin glycosides, having fragrance of hay, is found in the genus *Loroglossum*. Coumarine glycosides are derived from *Angraecum fragrans. Paphiopedilum javanicum* has saponine glycosides which possess strong foaming properties. The vanilla glycosides and related aromatic compounds are found in several species of *Vanilla*.

Many of the orchid species contain alkaloids which are important secondary meta-bolites (Luning, 1964, 1967; Lawler and Slaytor, 1969, 1970). Many alkaloids from Orchidaceae have now been isolated, but from chemical considerations, the most interesting ones are dendrobine and nobilonine from *Dendrobium nobile* (Suzuki *et al.*, 1932, 1934; Yamamura and Hirata, 1964), laburine from *Liparis bicallosa* (Nishikawa and Hirata, 1968), malaxine from *Malaxis congesta* (Leander and Luning, 1967) and phalaenopsine from *Phalaenopsis mannii* (Brandange and Luning, 1969).

A few species of orchids have been considered to induce sterility among women. The native inhabitants in some areas use seeds of *Cymbidium madidum* and pseudobulbs of *Dendrobium tokai* as oral contraceptives. Some species of orchids have religious influences over the native inhabitants in some parts of the world. In Malaysia, *Cymbidium finlaysonianum* was used as a talisman to keep the evil spirit away from the village. In Borneo, *Coelogyne asperata* is kept inside home with the belief that it helps to overcome many difficulties in human life. *Dendrobium hookerianum, D. nobile, D. gibsoni* and *Cymbidium grandiflorum*, grown in the eastern corners of the places of worship, symbolize sanctity. In Moluccan Islands, it is believed that the seeds of *Grammatophyllum scriptum* excite love (Bose and Bhattacharjee, 1980). Similarly, few orchids are associated with festivals.

Uses of Orchid

In Hanging Basket

In Pot

Beautification of Wall

Landscaping with Orchid

Orchid Bed in Garden

Home Decoration

Orchid as National Flower of Countries

Wild Banana Orchid (*Schomburgkia thomsoniana*) of Cayman Islands

Cockle Shell Orchid/ Clamshell Orchid/Black Orchid (*Prosthechea cochleata*) of Belize

Bulbophyllum auricomum, another National Flower of Myanmar

Christmas Orchid (*Cattleya trianae*) of Columbia

Orchid (*Rhyncholaelia digbyana*) of Honduras

Cattleya Orchid (*Cattleya labiata*) of Brazil

Orchids in Floral Arrangements

3.3.0 Other Uses

Orchid flowers are also used as national symbols, including the national flower of Myanmar, *Bulbophyllum auricomum* Lindl., and similar species,such as *B. sukhakulii* Seidenf., which are often used to adorn women's hair (Goh, 2013 cited by Hinsley *et al.*, 2018). Besides, several orchis species have been recognized as national flower of diffrent countries *i.e.* Wild Banana Orchid (*Schomburgkia thomsoniana*) of Cayman Islands; Cockle shell orchid or clamshell orchid or **Black orchid (***Prosthechea* cochleata) **of Belize; Christmas Orchid (***Cattleya trianae*) of **Columbia;** Orchid (*Rhyncholaelia digbyana*) of Honduras; Cattleya Orchid (*Cattleya labiata*) of Brazil. Hardy orchids with beautiful flowers are now being used in landscaping and species and cultivars are selected in accordance with the natural environment. In temperate climate, Bailes (1994) suggested rhizomatous genera (including *Cypripedium*, *Bletilla*, *Calanthe* and *Epipactis*) and tuberous genera (including *Dactylorhiza* and *Orchis*) from Europe, *Platanthera* from North America, and *Pterostylis* and *Corybas* from Australasia for woodland habitat and open situation.

Some orchids are useful in small-scale village industries. A few species of *Cyrtopodium*, *Geodorum nutans* and *Bletilla* are utilized for making glues employed in the manufacture of musical instruments and enamel objects. The stems of some *Dendrobium* species are used in making baskets in the Philippines, Indonesia and New Guinea. Split fresh bulbs of *Coelogyne asperata* are reported to have been used in central Sumatra as blackboard erasers (Withner, 1959). Tribal people in New Guinea prepare attractive bracelets from the yellow pseudostems of *Dendrobium utile* and use them as ornaments (Cooray, 1940).

3.4.0 Importance of Orchid in India

According to Hegde (2016), commercial potentials of orchids in India or its strength in orchid development and trade lies in: i) rich orchid diversity/ germplasm; ii) varying an ideal agro climate from tropical to temperate regions to grow varieties of orchids; iii) technical knowhow of orchid growing, propagation technique, biotechnological backing, green house technology; iv) cheap labour; and v) ever growing high end consumer market. Our weaknesses include: i) lack of quality planting materials in adequate quantity; ii) lack of market driven approach in the production of plants and flowers adopting modern technologies; iii) lack of production of our own hybrids that can compete in world market; iv) lack of consistent Research and Devlopment (R and D) back up with new hybrid varieties and technical innovations; v) lack of production of planting material; vi) lack of quality and quantity of cut flowers to feed the market; vii) lack of training and extension programs; and viii) lack of the involvement of communities both at rural and urban levels in suitable agro climatic zones and develop hub of activities leading to market places.

4.0 Origin and History

Majority of the orchids known to cultivation are native of tropical countries and occur in greatest diversity in humid tropical forests of South and Central America, Mexico, India, Sri Lanka, Burma, South China, Thailand, Malaysia, Philippines, New

Guinea, Australia, *etc*. Brazilian Cattleyas, Mexican Laelias and Indian *Dendrobiums*, *Cymbidiums* and *Vandas* have played a major role in the development of modern orchid industry.

Orchids have been very closely associated with human being since ancient times and there is reason to believe that in the Orient they were cultivated as early as 500 BC. The term 'lan', as the orchid is called in Chinese, appeared in the oldest book on Chinese Philosophy, *Eki-Kyo*, of which Confucius (551-479 BC) was a co-author. He is said to have remarked that acquaintance with good men was like entering a room full of 'lan', or fragrant orchids. The earliest reference to orchids in the western hemisphere is recorded in the Badianus Manuscript, an Aztec herbal of 1552 (Emmart, 1940). It depicted *Vanilla* which has been used since ancient times to perfume and flavour cocoa. Though not illustrated there is also record of *tzaconhxo-chitl*, or glue flower, which has been associated with *Bletilla campanulata* and *Epidendrum pastoris*, or possibly *Catasetum maculatum*.

The ancient Greeks were the first to take note of these strange plants. The word orchid derived from the term *orchis*, meaning testis, may be traced back to Theophrastus (370-285 BC), a pupil of Plato and Aristotle. Strangely enough, the Greeks used it not with reference to the flower but with reference to the paired underground bulbs of the Mediterranean orchid plants. It was believed in those days that certain plant parts with some resemblances to the human organs were medicinal and efficacious for the disease of such organs. The twin tubers of orchids were considered valuable in connection with fertility and virility. Later, the name was adopted by Discorides (1st century AD) in his *Materia Medica* where description of two orchids was given along with several species of medicinal plants.

Indian *Vedic* scriptures also mention about these plants under the name *Vanda* which has been adopted as a generic name for one of the most beautiful monopodial groups of orchid now commercially grown in Malaysia, Singapore and Thailand.

The earliest book on orchids was edited by Joan Matsuoka in Japanese, then rewritten in Chinese and republished in 1772 with block illustrations (Withner, 1959; Nagano, 1960).

When plant exploration in various lands gained momentum, orchids, with their beautiful flowers, were also collected. Here again, the first exotic orchid introduced to Europe was not for its flower but for the essence it yielded, the vanilla. Vanilla was introduced into Spain in 1510 while in France it was popular by 1609. Much later, the first ornamental orchid, *Cypripedium reginae* from North America and grown in Leiden (the Netherlands) appeared in its Botanical Garden Catalogue.

In 1778, Dr. John Fothergill brought plants of *Phaius grandi-folius* and *Cymbidium ensifolium* from China. In 1787, *Epidendrum cochleatum* flowered for the first time in England and *E. fragrans* in the following year. By 1794, fifteen species of epiphytic orchids were reported to be grown in the Royal Botanic Gardens at Kew, most of which were brought from the West Indies. They included *Orninthidium coccineum, Oncidium altissimum, Oncidium carthagenense, Lycaste barringtoniae, Epidendrum ciliare, Isochilus linearis, etc*. Due to the establishment of the Royal Horticultural Society in England, there was tremendous impetus to horticulture during the year 1809 and orchids were taken as a subject of serious study. Messrs Loddiges and Sons of Hackney, London, a famous nursery of plants, began to cultivate orchids on a commercial scale about the year 1812. Epiphytic orchids were difficult to culture, but Fairbairn succeeded to obtain flowering in *Aerides odoratum* in 1813. John Lindley suggessted cultural schedule for orchids in 1830, but there were defects of excessive humidity and temperature. In 1835, Paxton made considerable improvement which involved the use of orchid houses. Such attempts resulted in the great interest for the culture of orchids in Europe, and several wild types were acclimatized and used to produce hybrids. John Dominy was the first to produce a man-made hybrid which flowered in 1856 at the Veitch Nursery, England (Withner, 1959).

4.1.0 Evolution of Orchid Cultivation

The evolution of orchid culture from hobbyist to commercial grower was slow as most of the orchids collected from different regions did not adapt to local conditions and died. The methods of propagation were not known and most of the orchids which flowered under utmost care were found to be self-sterile. Further, it was difficult to germinate seeds. The seeds sown in the nursery beds required long period for germination and any disturbance to the soil or physical environment destroyed the whole population (Northen, 1970).

A revolution in orchid growing was brought about by Prof. Lewis Knudson of Cornell University in 1922. He found that orchid seed germinated readily in a glass flask containing nutrients and agar. It meant that instead of fifteen or twenty seedlings, thousands could be grown from one pod. Further researches carried out by Wimber (1963), Morel (1964a) and Hamilton (1965) perfected a method by which an indefinite number of plantlets could be obtained from a single shoot tip.

At about this time sufficient knowledge had accumulated regarding the requirements of orchids, and techniques were standardized for their culture. Different factors like media, water and nutritional requirements, temperature, relative humidity, *etc.*, influencing the growth and flowering were thoroughly studied and suitable procedures were suggested. Today, orchids are grown in extensive glasshouses with controlled environment. The commercial cultivation of orchids to supply cut flower has increased amazingly in some countries. *Vanda, Arachnis* and other genera have been pressed into service with excellent results. A large number of hybrids have been raised invol-ving even three, four and five-way crosses. In California, New York and in some parts of Australia, Cymbidiums are cultivated on a large scale out of doors or under lath roofing and without the aid of artificial heat.

A few orchids figure prominently as national flowers of many states and countries. *Cypripedium reginae* has been adopted by Minnesota, *Lycaste virginalis* or *Monja blanca* by Guatemala, and *Cattleya skinneri* by Costa Rica. The family crests of the former royal house of Manchukuo having *Cymbidium* flower appeared on stamps (Withner, 1959).

5.0 Survey Identification and Distribution

5.1.0 Survey

Extensive survey of orchid flora in different parts of the world which was started many decades ago is being continued and new species are collected, identified and reported adding to the enormous wealth in the family Orchidaceae. The orchids found in various agroclimatic regions are given in Table 3.

Table 3: Selected Orchid Genera for Various Agro-climatic Zones*

Agro-climatic Zones	*Temperature Range*	*Suggested Ornamental Orchid Genera (Both species and hybrids)*	*Suggested Centers/ States for Orchid Development*
Tropical	Warm temp. 21°C-29 °C day to 18°C -21°C nights	Species: *Aerides, Arachnis, Armodorum, Ascocentrum, Calanthe, Dendrobium, Hygrochilus, Paphs, Papilionanthe, Phaius, Renanthera, Spathoglottis, Vanda.*Hybrids: *Aranda, Ascocenda, Cattleyas, Dendrobium clones, Mokora, Oncidium, Phalaenopsis, Paphs.*	Assam, Goa, Karnataka, Kerala, Maharashtra, Tamil Nadu, Tripura, West Bengal
Sub-tropical	Intermediate with 18°C-24°C day to 16°C-18°C nights	As above and other ornamental native species like *Aerides, Anoectochilus, Bulbophyllum,Calanthe, Coelogyne, Dendrobium, Eria, Goodyera etc.*	Foothills of Himalayas in N-E region, western ghats and Eastern India
Temperate	Cool with 16°C-21°C day to 13°C night	*Coelogyne, Cybidium, Cypripedium, Paphiopedilum, Pleione etc.*	Hills of N-E states, Himachal Pradesh, Uttarakhand, West Bengal

* Hegde (2016).

The orchid herbarium at Kew now has over 2,50,000 specimens including some 30,000 types and has an influx of about 1500 specimens per annum (Wood, 1993). Apart from collection, morphology, identification, distribution, ecology, growth behaviour of the new orchid species, their importance as ornamental plants is also being reported. It may be relevant to review some of the major reports on the survey of orchid flora in the last decade from South America (Carnevali and Romero, 1991; Dodson and Atwood, 1992; Mora Retana and Garcia Castro, 1990), Australia (Webb and Webb, 1990, 1992, 1993), Europe (Stewart, 1992), USA (Garay *et al.*, 1995), China (Zi, 1991), India (Pradhan, 1975; Rao, 1996; Bose *et al.*, 1999), Malaysia (Dourado, 1992), the Philippines (Cootes and Banks, 1997), Borneo (Wood, 1992), Africa (Bruyns, 1989; Croix, 1994). Studies on the occurrence, morphology and taxonomy of many Asiatic orchids including *Dendrobium, Ascocentrum, Cadetia, Cordiglottis, Thrixspermum, Tuberolobium, Taprobanea, Trudelia, Habenaria, Vanda, etc.* have been made (Christenson, 1992).

Six previously undescribed species of *Lepanthes* were added to the previously known 18 species in Cuba; *L. comadresina, L. diaziae, L. llamachoi, L. minima-mundana, L. palpebralis* and *L. woodfredensis* (Luer, 1998).

A new species of *Disa*, the largest genus of terrestrial orchids in southern Africa was described by Linder *et al.* (1998) from the Western Cape. *Disa virginalis*, was previously included in *D. maculata*, but is white rather than blue, with shorter and broader lateral sepals, and a less developed spur. In addition, the galea is narrowed in front. Similarities to *D. schizodioides* are also discussed; these indicate that the current sectional classification of subgenus *Disa* may not be correct.

Many natural hybrids in orchids identified through survey have proved to be of scientific and commercial interest. Natural hybrids of *Catasetum* from Venezuela (Romero and Carnevali, 1992), *Orphys* in Europe (Rossi *et al.*, 1990) and many others have been reported.

5.2.0 Identification and Distribution

Orchids occur almost throughout the world, except, perhaps, in the Antarctica. Some species are found even in the frozen areas of Alaska, the snow covered areas of the Himalayas and the sandy deserts of Africa and Australia (Mukherjee, 1983). Seven species have been reported to occur in the Arctics (Stearn, 1960). However, orchids are abundant in tropical countries in South-East Asia, South and Central America and South Africa. They are also distributed in China, Japan, the Philippines, Europe and Australia (Rao, 1979). India is very rich in orchid flora and the major areas are the eastern Himalayas, the western and South Indian hills. The distribution of orchids in different climatic zones and geographical areas in India has been described by Bose *et al.* (1999).

According to genera, the distribution of some important orchids in different parts of the world is given below :

Aerides : Myanmar, Indonesia, Malaysia, southern and northern India.

Anoectochilus : Extending from the Himalayas of India to New Caledonia, with a large number of species in Indonesia.

Arachnis : The Himalayas to New Guinea and the Solomon Islands, with large numbers in Indonesia and Malaysia.

Arundina : Extending from southern China and the Himalayas, throughout Malaysia, Indonesia and the Pacific Islands.

Ascocentrum : Extends from southern China, Formosa and the Philippines to Java and Borneo.

Bulbophyllum : Tropical Asia, Africa, New Guinea, Brazil, South Florida, Cuba, Mexico.

Calanthe : South Africa, Japan, China, Vietnam to Australia, South India.

Cattleya : South America.

Coelogyne : Nepal, Bhutan, Sikkim, China, Upper Myanmar, South-East Asia, the Philippines and Fiji.

Cymbidium : Extends from Korea, Japan, China, throughout the Himalayas, South-East Asia to New Guinea and Australia.

Dendrobium	: Japan, Korea, throughout the Indo-Malayan region, Indonesia to Australia, New Zealand and Polynesia.
Epidendrum	: From North Carolina and Mexico throughout the West Indies and Central America to the Galapagos Islands and Argentina.
Habenaria	: All parts of the globe, but abundant in Brazil, Tropical Africa and north temperate zone.
Laelia	: From Cuba and Mexico to Argentina; the largest number of species found in Brazil.
Lycaste	: American tropics, ranging from Cuba and Mexico to Peru and Brazil.
Masdevallia	: Columbia, Mexico, Brazil, Guiana, Peruvian Andes.
Miltonia	: From Costa Rica and Panama through the Andes to Brazil.
Odontoglossum:	From Mexico to Brazil and Bolivia with the majority of the species occurring in the Andean regions of Columbia, Ecuador, Peru and Venezuela.
Oncidium	: Ecuador, southern Mexico to Columbia, Peru.
Paphiopedilum	: From China to the Himalayas throughout South-East Asia and Indonesia to New Guinea.
Phalaenopsis	: From the Himalayas through Malaysia and Indonesia to Formosa, the Philippines, Australia and New Guinea.
Vanda	: China, the Himalayas, Sri Lanka, the Philippines, South-East Asia, Indonesia, New Guinea and northern Australia.

The orchid flora may also be divided as those distributed in the Old World, those of the New World and those found in both Old and New Worlds. The Old World orchids include *Cymbidium, Dendrobium, Eulophia, Phalaenopsis, Vanda, etc.*, and all monopodial types. Genera like *Brassavola, Cattleya, Epidendrum, Laelia, Maxillaria, Oncidium, Sobralia, Zygopetalum, etc.* occur in the New World, whereas *Calanthe, Cephalanthera, Epipactis, Erythrodes, Habenaria, Goodyera, Galeola, Liparis, Malaxis, Orchis, Pogonia, Spiranthes, Vanilla, etc.* are found in both the regions (Abraham and Vatsala, 1981). According to Dillon (1971), the distinction between Old and New World species is more evident among the epiphytic forms. The terrestrial ones, particularly those belonging to the Neottieae and Orchideae tribes are found in both the hemispheres. Advanced forms like *Calanthe, Liparis, Malaxis, Polystachya, etc.* are worldwide in distribution.

There are some genera which are confined to particular regions, for example, *Angraecum* to Africa, *Anguloa* to Columbia, *Calopogon* to South-East United States, *Cypripedium* to northern temperate regions of Asia, Europe and America, *Paphiopedilum* to India and South Asia and *Phragmipedium* and *Selenipedium* to South-Central America.

Out of 184 genera and about 1200 species reported from India. One hundred and fifteen genera and about 496 species are represented in Sikkim. Their names, phenological data, distribution and phytogeographical affinities are listed (Shukla *et al.*, 1998). Diversity of orchids in different altitudinal zones is discussed. *Bulbophyllum, Dendrobium, Eria, Liparis* and *Oberonia* were the dominant genera in Sikkim; 88 genera were represented by a single species. A total of 37 species were endemic to Sikkim. Over-exploitation of a number of species for medicinal and ornamental purposes, and destruction of their natural habitats, has contributed to the depletion of biodiversity of many species; several species of orchids are threatened with extinction.

6.0 Morphology of Orchids

Orchids are mostly long-lived, evergreen or deciduous herbs.Some individual plants, such as those of Cypripedium calceolus, can live 30 to 100 years (Kull, 1999). The pre-flowering vegetative period for most species usually lasts for four to seven years, but can be even longer (Kull, 1999; Wang *et al.*, 2010 cited by Zhang *et al.*, 2018). This longevity may be attributed to their inherently slow growth and reduced photosynthetic capacity (Schmidt and Zotz, 2002; Shefferson, 2006).

The structure of the orchid plants as well as of flowers is very interesting. These extraordinary plants are perennial herbs of varying habit which is generally influen-ced by different environmental factors. The orchids found in nature grow wild; when grown on trees or shrubs, they are termed epiphytes and when on rocks, they are correctly termed lithophytes. The species growing on ground are referred to as terrestrials and certain seldom-seen orchids which derive their nourishment from dead or decayed organic matter are called saprophytes. A few orchids such as Australian Cryptanthemis and Rhizanthella grow in water as semi-aquatics under special conditions (Hawkes, 1978). They are not parasites, prepare their own food by phtosynthesis and do not easily die even if left uncared for several months.

6.1.0 Vegetative Structure

The vegetative structures of an orchid plant are modified according to the diversified habit of growth which has close relation with the climate. The majority of orchids occurring in temperate climates are terrestrials while those in tropical regions are found frequently on trees or rocks (Salisbury, 1953).

Vegetatively, the orchids are divided into sympodial, monopodial and pseudomonopo-dial types. The sympodial types, for example *Cattleya* and *Epidendrum*, have stems of determinate growth. The growth of the stem sooner or later ceases, usually at the end of one season's growth, and lateral shoots are produced in the following season. This new growth generally begins with the development of leaf-like scales between which the truc leaves arise later. Plants of this group produce pseudobulbs which are swollen stems used for storage of water and food. These pseudobulbs often grow between the first pair of true leaves, and one, two or more leaves are produced from the apex of the pseudobulbs. The flower spikes may be either lateral or terminal. A vast majority of orchids are characterized by the sympodial mode of growth which includes all terrestrial orchids, both tropical and temperate and species possessing pseudobulbs whose stems mature in one season.

The monopodial orchids such as *Vanda* and *Angraecum* have indeterminate terminal growth. They do not make new vegetative growth each season. The word 'Monopodial' is derived from Greek 'mono'- means one and 'podial'mans 'foot', in reference to the fact that monopodial plants have a single trunk or stem. There is one main upright stem which continues to grow and elongate season after season, bearing new leaves at the apex and producing flowers laterally arising from the leaf axils of the older parts of the plant. They have no rhizome and form no pseudobulb. The food materials, when necessary, are stored in leaves. The plants produce aerial roots, which help them to anchor as they extend upwards. *Aerides, Angraecum, Arachnis, Phalaenopsis, Rhynchostylis, Saccolabium, Vanda, etc.* are the examples of monopodial orchids.

The small group of pseudomonopodials is considered to be somewhat intermediate between sympodial and monopodial orchids. They have stems which elongate through seasons and also extend in a lateral direction. The flowers are borne mostly laterally, often near the apices of the stems. Pseudomonopodials belong to the subtribes Dichaeinae, Pachyphyllinae, and possibly Pterostemmatinae (Hawkes, 1978).

Some orchids like *Vanilla* and *Arachnis* are climbing in nature, growing rapidly in length and climb a considerable distance. *Vanilla,* a vine orchid, may reach over 16 metres in length bearing one leaf and a root at each node. For the most part, orchids are evergreen, but some are deciduous, notably the Calanthes and Pleiones. Others are semideciduous which shed leaves from the older stems when the new growths have partly or wholly developed.

6.1.1 Roots

Orchids depending upon their habit have different kinds of roots. However, in general, the roots of most orchids are cylindrical, often thread-like, branched and frequently elongated. In terrestrial orchids, in addition to the simple absorbing roots provided with root hairs, sometimes large root tubers, as in the case of *Habenaria,* are also found. Epiphytic and most lithophytic orchids have clinging roots for attachment to the tree or rock on which they grow, absorbing roots which penetrate the humus collec-ted on the bark and aerial roots which hang free in the air and aid in the absorption of moisture and to a lesser extent, food materials (Hawkes, 1978). In some species of *Chiloschista* and *Taeniophyllum,* flattened green roots capable of carrying photosynthesis are met with (Hegde, 1984). The aerial roots are remarkable structures being covered with a loose spongy tissue called velamen. Dycus and Knudson (1957) reported that the velamens absorb moisture from the atmosphere but they are inca-pable of transmitting it to the interior of the root. According to Shushan (1959) its main function is to protect underlying tissues.

Sakharam (1953) found no histological differences between the aerial roots of the epiphytic orchid *Vanda tessellata* and the ground roots of two terrestrial orchids *Spathoglottis plicata* and *Phaius wallichii.* The velamen was present in both cases. In the epiphytic roots, root hairs, though normally absent, were present under some conditions. The roots of the genera *Phalaenopsis* become considerably flat and assist the plant to creep over the surface while those of *Aerides* and *Vanda* help the plants to climb on the tall trees.

6.1.2 Rhizomes

These underground swollen structures are actually primary stems from which arise secondary stems, and are found only in the sympodial type of orchids. A few pseudo-monopodials also have reduced rhizomes. Rhizomes in orchids show considerable variation in shape and size. They may be short with tufted habit as in many species of *Dendrobium*, elongated with staggering habit as in *Bulbophyllum* and *Coelogyne*, prominently thickened and very strong as in *Cattleya* or secunded as in *Oncidium*.

6.1.3 Pseudobulbs

Pseudobulbs are thickened secondary stems with one or more internodes and found in a large proportion of the epiphytic orchids whose growth habit is sympodial. Their shape varies tremendously, even within the same genus. Some of the *Cattleya* species possess bulky and spindle-shaped pseudobulbs but others have elongated cylindrical stems. Pseudobulbs may be round and globular as observed in several species of *Bulbophyllum*, angular and fusiform as in some species of *Coelogyne*, oblong and thumb-shaped as in *Eria mysorensis* or elongated and club-shaped in *Dendrobium heterocarpum*. Striking variations are also found in their size from minute pin-head-like pseudobulbs to those attaining more than 7.5 m as in *Grammatophyllum speciosum* (Hawkes, 1978). A plant of *G. elegans* which attains a height of more than 3 metres may weigh several hundred pounds (Marv and Ragan, 1973).

6.1.4 Leaves

Orchid leaves, conforming to the monocotyledonous pattern, are plicate strap-shaped, conduplicate or terete with sheathing bases and of variable thickness, occurring in all possible dimensions. They are generally green in colour but sometimes have silvery or golden veins providing beautiful ornamentation as in the species of *Anoectochilus*.

6.2.0 Floral Structure

6.2.1 Inflorescence

Orchids bear flowers on an inflorescence which is either a spike as in *Oncidium* and *Phalaenopsis*, a raceme as in *Dendrobium moschatum* or a panicle as in *Cleisostoma racemiferum*. Sometimes flowers are produced singly as in *Paphtiopedilum*. The origin of this inflorescences varies depending upon genera and species. In terrestrial orchids, the inflorescence is either terminal as in the tuberous perennial forms or lateral as in advanced members like *Phaius, Calanthe, Eulophia, etc.* while in major of epiphytic forms they are lateral. However, in a few epiphytic orchids, such as in *Polystachya*, the various *Eria* species with discoid pseudobulbs, in *Porpax, etc.* inflorescences are terminal. In the case of *Cymbidium* it arises at the base of the stem. Generally, inflorescences are produced from a leafy shoot, but in a few cases like

Bulbophyllum fimbriatum, Dendrobium heterocarpum, D. ovatum, etc. leaves are shed before flowering. In saprophytes, only leafless flowering shoots appear.

6.2.2 Flower

Like vegetative habits, orchid flowers exhibit a wide range of variations. The largest flower known is that of *Sobralia macrantha*, 15 to 30 cm across, and the tiniest one is that of the *Bulbophyllum minutissimum*, the flower being about the size of a pin head (Rao, 1979). The predominant shades are white, yellow, green and purple, occurring in pure form or in every possible combinations. The beautiful azure-blue in species like *Vanda coerulea* is valued much among orchid breeders. Similarly, the bright orange-red colour in species like *Ascocentrum ampullaceum* is very popular among orchid lovers. One of the most important characteristics of all orchid flowers is that they are zygomorphic. The flowers are bisexual or rarely unisexual as in most catasetums.

The orchid flower consists of three sepals, three petals and the column or gynostemium having the reproductive parts. The three uppermost floral parts or sepals are more or less alike in appearance. But in a few species of *Bulbophyllum, Renanthera*, and many species of *Oncidium*, the upper or dorsal sepal is of different size from the laterals. Sepals are narrower than the petals but coloured like them and are known as petaloid sepals. They may be free from one another or somewhat united as in species of *Cryptochilus* and *Masdevallia*, sometimes forming a definite tube which is called the sepaline tube. In some species of *Bulbophyllum* and *Oncidium*, the two lateral sepals are combined together into a single structure. In *Cryptophoranthus*, the fused lateral sepals also unite with the upper or dorsal sepals at the base and at the tip, leaving the middle portion free so that the flower never opens.

The inner floral segments, called the petals, are three in number. Two of these, the lateral ones, are alike and the other one, called the lip or the labellum, is highly modified and enlarged. This is the most prominent and distinctive part of the flower. The petals may be similar to sepals in size, but in a few cases, as in some *Habenaria* and *Coelogyne* species, they are filiform, or fimbriate as in *Bulbophyllum fimbriatum*. The lip is attached to the base of the column and the mode of attachment may be rigid or loose. The colour, size and shape of the lip vary in different genera. They may be simple or lobed, flattened, sac-like, or tubular, and apically and marginally entire, fringed, toothed or notched. Lips of *Cattleya, Laelia* and *Phaius* are tubular and in *Oncidium, Odontoglossum* the lips are extended into a broad area but in *Masdevallia* it is undistinguishable. The lip of *Paphiopedilum* is of quite distinct shape and resembles a slipper.

The column or gynostemium, situated in the centre of the flower, is the unique structure distinguishing the orchids from all other kinds of plants. This is formed by the fusion of male and female organs and is the reproductive part of orchid flower. It adds charm to the flower and is the most delightful part of all. It is attractively shaped and often is decorated with wings, or a cap or a fringed bonnet. The column varies in size, shape and posture in the various genera and species, and bears at the tip or on the sides near the tip, one to rarely three movable or rigidly attached anthers. In Monandrae, where only one anther is fertile, it is borne on the top of

Morphology of Orchid

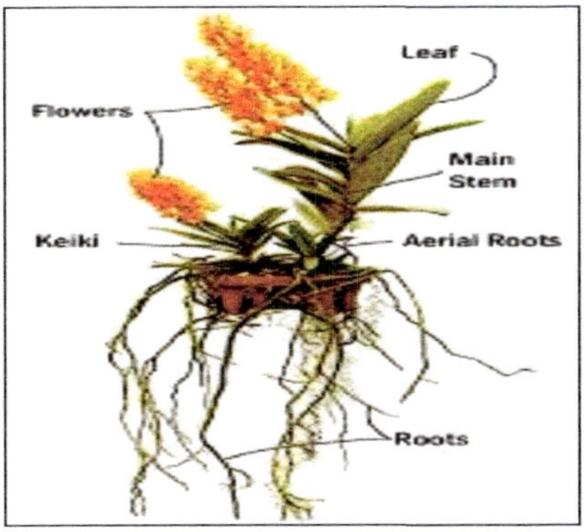

Whole Plant of Orchid

Leaves

Sympodial Orchid

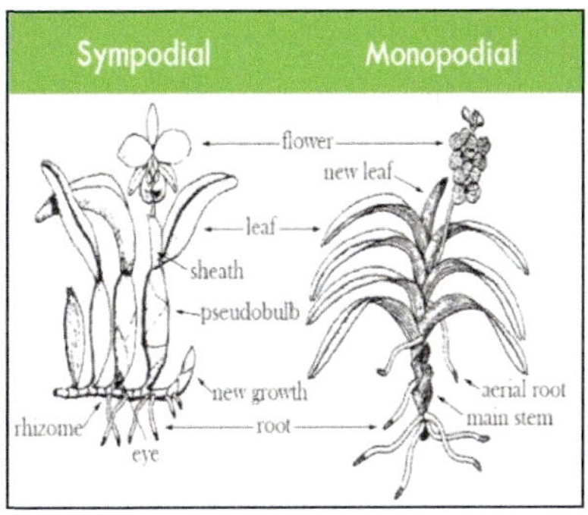

Growth Habit of Orchid

Monopodial Orchid

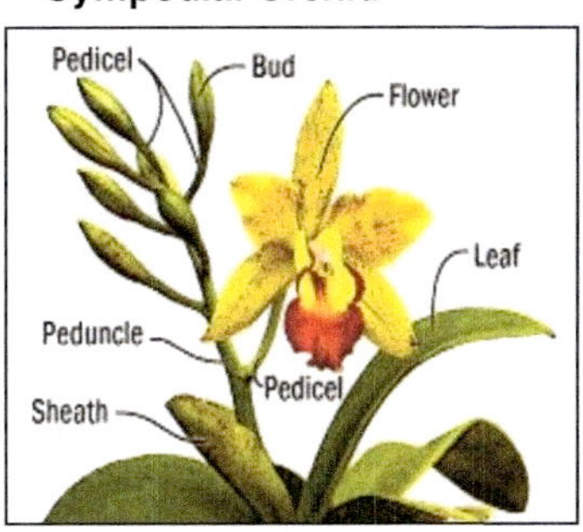

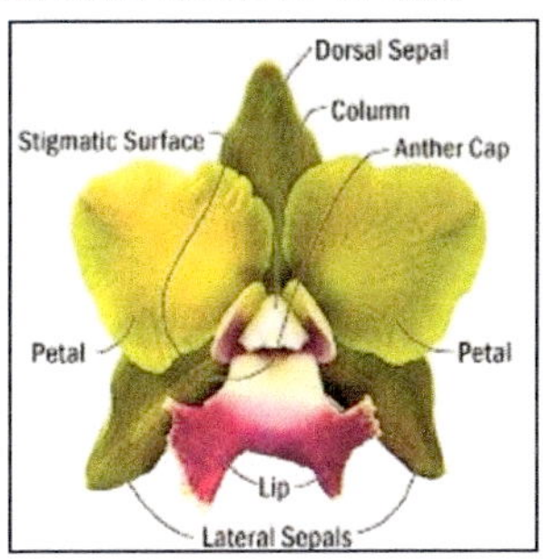

Floral Morphology

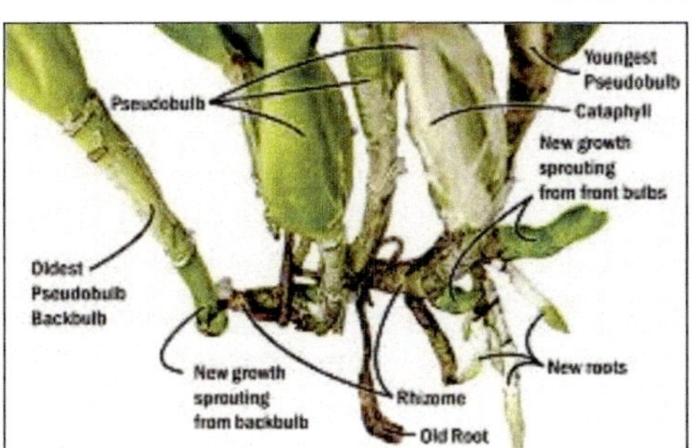

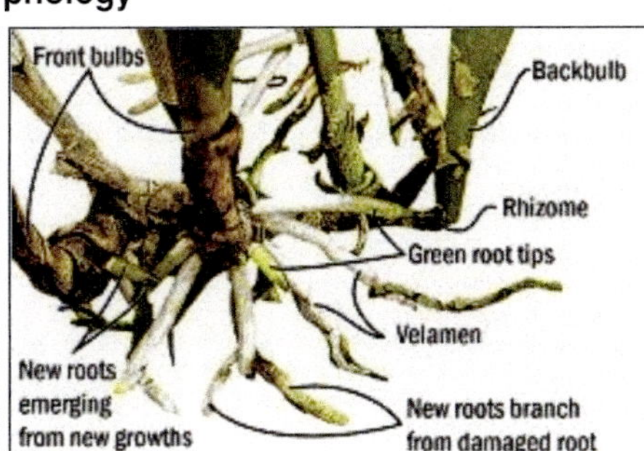

Morphology of Root

Rooting Habit of Orchid

Terrestrial Orchid

Root of Terrestrial Orchid

Epiphytic Orchid (on trees)

Epiphytic Orchid (Roots on air)

Lithophytic Orchid

Roots of Lithophytic Orchid

Adventitious Root (Keiki)

Pseudobulb of Orchid

Backbulb of Orchids

Flower Parts

Seed Capsule

Harvesting of seeds from Capsule

the column whereas in case of Pleonandrae, where two or more stamens are fertile, they are borne on the sides of the column.

The anther contains a mass of male cells pollinium. It may or may not have appendages and the number of pollinia may vary from 2, 4 and 6 to 8. Just below the anther, on the ventral surface of the column, is a hollow cavity of sticky and viscid mass known as stigma or stigmatic surface. Actually, it is the two fertile stigmas of the flower fused together. The partition wall between the stamen and stigma is termed as rostellum. The ovary being located at the base of all floral organs is termed inferior and is another main characteristic of Orchidaceae. It may be stalked or not and is generally tricar-pellary, one-celled, with three parietal placentations and consists of numerous minute ovules.

The orchid fruit is a capsule which takes a long time to ripe. It varies in shape from rather egg-shaped to almost globular or cylindrical. Each mature capsule may contain more than a million seeds and the entire content looks like yellowish or brownish dust (Treub, 1879). The orchid seeds are the smallest amongst the flowering plants. They are extremely light each weighing between 0.3 and 6.3 g. The tiny seed consists of only the embryo with a few surrounding cells. It has no nutritive endosperm to nourish the embryo and the seedlings (Ames, 1948; Arditti, 1967a, 1981).

7.0 Classification

The family Orchidaceae has by far the largest assemblage of flowering plants known to science, and it is, therefore, essential to evolve a system with which plants are correctly identified and named. For this purpose, a large number of orchid species have been studied by the orchidologists or botanists for the last two centuries all over the world and various systems of classification have been suggested, based on the morphological and other characters. The first outstanding figure in systematic botany is Linnaeus (1707-1778) who arranged all the plants known at that time into several groups according to the number and position of the stamens and pistils. All the orchids were placed in class XX-Gynandria (stamens adnate to the pistils), Diandria (two stamens). He listed 8 genera and 69 species. Swartz (1800) described 25 genera under two divisions, *i.e.*, orchids with one anther (comprising 24 genera) and orchids with two anthers (embracing one genus, *Cypripedium*). He was apparently the first man to recognize the two great divisions of the orchid family which are known today as Monandrae and Diandrae.

A significant contribution was made by Brown (1773-1858) who clarified the complicated morphology of the orchid flower and defined the systematic position of the family Orchidaceae. Richard (1818) introduced a key to the genera of European orchids and described species of each genus including their synonyms. Similarly, the key of the orchids of Java was given by Blume (1825) who divided orchids into three groups according to the granular, powdery or waxy nature of the pollen grains.

Orchids in general, like *Cattleya trianae,* have been characterized mostly by taxonomic and visual studies. However, colour is not used for classification. Here, a new method for identifying and classifying orchids of different varieties of *Cattleya trianae* is introduced (Forero *et al.*, 2020). This method is not subjective and uses the colour information obtained from the central axis of the *Cattleya trianae* lip. To this end, a new acquisition protocol was established, which uses a new device for image acquisition of the labellum's central axis from the hippocampus to the epicentre. The colour patterns found between samples of the same variety were adjusted and it was verified by using correlation they can be employed to identify each variety. Finally, a support vector machine was used to classify and identify four *Cattleya trianae* varieties, finding that a linear kernel was enough to classify them with an accuracy of 100 per cent (Forero *et al.*, 2020).

Van Den Berg (2014) proposed a new system of infrageneric taxonomy for *Cattleya* based previous phylogenetic studies as below.

Identification key for infrageneric taxa within genus *Cattleya* (Van den Berg, 2014).

1. Pollinia 8..2
   - Pollinia 4..6
2. Leaves 2, terete..series *Microlaelia*
   - Leaves mostly 1, or 2 not terete..3
3. Spathaceous bract subtending the inflorescence absent..4
   - Spathaceous bract subtending the inflorescence present..5
4. Flowers red, orange or yellow, or pink (in this case pseudobulbs <2 cm long and leaves <4 cm long, more or less flat)..series *Sophronitis*
   - Flowers pink, purple or mauve, plants with conduplicate leaves >4 cm long..series *Hadrolaelia*
5. Plants epiphytic with more or less clavate pseudobulbs and flowers generally larger than 6 cm..series *Cattleyodes*
   - Plants mostly rupicolous or terrestrial with cylindrical or subcylindrical pseudobulbs or epiphytic plants with long, slender pseudobulbs and flowers generally less that 6 cm..series *Parviflorae*
6. Leaves 2 or 3..subgenus *Intermediae*
   - Leaf solitary..7
7. Column tip with unciform projections around the sides of the anther..subgenus *Cattleyella*
   - Column tip without unciform projections..8
8. Pseudobulbs subcylindrical or slender and cylindrical; flowers pink to purple, Peru and Ecuador..subgenus *Maximae*
   - Pseudobulbs more or less clavate or fusiform, flowers not pink to purple if from Peru and Ecuador..9
9. Plants from mountain regions (>600m) [Peru, Ecuador, Colombia and Venezuela, or eastern Brazil..section *Cattleya*

- Plants from low elevation in the Amazon (<400) [northern Brazil and Venezuela], or from lowland, dry habitats [Lara, Yaracuy, Carabobo, Vargas, Aragua States, and Distrito Federal, Venezuela] .... subgenus *Lawrenceanae*

7.1.0 Lindley's Classification

It was John Lindley (1799-1865) who marked the beginning of modern orchidology. He proposed a taxonomic classification of the family dividing it into 7 tribes after giving due consideration to the study of orchidaceous species distributed throughout the world. The various systems of classification of the Orchidaceae starting from that proposed by Lindley (1840) are given below :

I. Anther One Only

A. Pollen Masses Waxy

(a) No caudicula or separable stigmatic gland.
Tribe I. Malaxeae (or Malaxideae)

(b) A distinct caudicula, but no separable stigmatic gland.
Tribe II. Epidendreae

(c) A distinct caudicula, united to a deciduous stigmatic gland.
Tribe III. Vandeae

B. Pollen Powdery, Granular or Sectile

(a) Anther terminal, erect or incumbent, adnate to the top of the column.
Tribe IV. Ophreae (or Ophrydeae)

(b) Anther terminal, opercular (lying over the rostellum).
Tribe V. Arethuseae

(c) Anther dorsal (behind the rostellum).
Tribe VI. Neotteae

II. Anthers Two

Tribe VII. Cypripedeae

The three important modern systems of classification of orchids were put forward during the late 19th and early 20th century. The first of these, an improvement over Lindleyan system, was published by Bentham (1881). Later, he published this system in a much amplified form under co-authorship with Hooker (1883).

7.2.0 Bentham and Hooker's Classification

They recognized 5 tribes in the family, chiefly those proposed by Lindley, placing Malaxeae as a subtribe of Epidendreae and Arethuseae as a subtribe of Neottieae. The tribes (except Cypripedeae which has no subtribe) were divided into 27 subtribes. The outline of Bentham and Hooker's classification is given below :

Tribe I. Epidendreae : Anther one, terminal, posterior, opercular and deciduous. Pollinia waxy, 1-4 in number, held together at the base by a viscid appendage, very rarely or occasionally somewhat attached to the rostellum.

The nine subtribes contain, among many others, the genera *Pleurothallis, Masdevallia, Malaxis, Liparis, Dendrobium, Bulbophyllum, Eria, Bletia,Coelogyne, Calanthe, Epidendrum, Cattleya* and *Laelia.*

Tribe II. Vandeae : Anther one, terminal, posterior, opercular, incumbent, lying upon the horizontal rostellum or applied to the erect rostellum, deciduous after opening. Pollinia waxy, usually 2 or 4, applied to each other in pairs,attached to a gland or part of the rostellum which is carried away when pollinia is removed.

The 8 subtribes contain the genera *Eulophia, Cymbidium, Grammatophyllum, Cyrtopodium, Zygopetallum, Bifrenaria, Lycaste, Stanhopea, Maxillaria, Trichopilia, Odontoglossum, Oncidium, Miltonia, Brassia, Arachnis, Phalaenopsis, Aerides, Vanda, Angraecum* and many others.

Tribe III. Neottieae : Anther one, terminal, posterior, opercular or erect, deciduous or persistent, with 2 distinct parallel cells. Pollinia granular, powdery or sectile.

The 6 subtribes include the genera *Vanilla, Sobralia, Corymborchis, Cranichis, Gomphichis, Spiranthes* and its allies, *Erythrodes, Goodyera, Prasophyllum, Epipactis, Pterostylis, Calopogon, Pogonia, Chloraea, Cephalanthera* and many others.

Tribe IV. Ophrydeae : Anther one, posterior, erect, reclining or reflexed. Pollinia solitary, granular, produced into caudicles attached to a gland or to the rostellum. Terrestrial.

The four subtribes include, among others, the genera *Orchis, Serapias, Ophrys, Habenaria, Satyrium, Disa, Brachycorythis, etc.*

Tribe V. Cypripedeae : Anthers 2, at the sides of the rostellum, sessile or stipitate, with parallel contiguous cells. Pollen granulose. Posterior anther transformed into a sterile polymorphic male organ. Lip inflated, shoe-shaped.

7.3.0 Pfitzer's Classification

The second of the three major systems was proposed by Pfitzer (1889) who emphasized the importance of vegetative characters (stem, leaf and inflorescence) instead of floral structure. He also incorporated into his system several new genera making up the total number to 410, distributed in 32 tribes and several subtribes.

Orchidaceae

Diandrae	*Monandrae*
Two fertile anthers. Lip shoe-shaped. One subfamily, the Cypripedilinae, containing the genera *Cypripedilum* (=*Cypripedium*), *Selenipedilum* (=*Selenipedium*), *Paphiopedilum* and *Phargmopedilum* (=*Phragmipedium*).	One fertile anther. Lip very rarely shoe-shaped.

Basitonae	*Acrotonae*
Anther always persistent. Pollinia developing caudicles towards the base of the anther. One large subfamily, Ophrydinae, containing the genera *Ophrys, Orchis, Serapias, Habenaria, Brachycorythis, Satyrium, Disa, etc.*	Anther deciduous. Pollinia without appendage or developing these towards the tip of anther.

Acranthae	*Pleuranthae*
Inflorescences terminal, at the summit of each shoot.	Inflorescence on special individual lateral shoots.

Convolutae	*Articulatae*	*Duplicatae*
Leaves convolute, Leaf blade not joined to the leafsheath. Anthers withering. Pollinia soft and granular. One large and diverse subfamily, Neottinae, containing the genera *Prasophyllum, Pterostylis, Caladenia, Chloraea, Pogonia, Vanilla, Cephalanthera, Epipactis, Spiranthes, Erythrodes, Goodyera, Macodes, etc.*	Leaves convolute, leaf blades jointed to the leafsheath and eventually falling off. Anther deciduous. Pollinia mostly waxy. Containing three sub-families, the most important one being the Coelogyninae, includes the genera *Coelogyne, Pleione, Pholidota, etc.*	Leaves conduplicate containing seven sub-families of which the Liparidinae includes the genera *Malaxis, Liparis, Calypso, etc.* The Pleurothallidinae includes *Masdevallia, Pleurothallis, Octomeria, etc.* The Laelinae includes the genera *Epidendrum, Cattleya, Laelia, Brassavola, etc.* The Sobraliinae includes the genera *Elleanthus, Sobralia, etc.*

Convolutae	*Duplicatae*
Leaves convolute	Leaves conduplicate

Homoblastae	*Heteroblastae*
Stem-members slender or equal swollen. Containing three sub-families of which the Phajinae includes the genera *Phaius, Calanthe, Bletia, Spathoglottis, etc.* The Cyrtopodinae includes the genera *Lissochilus, Eulophia, Cyrtopodium, etc.* The Catasetinae includes the genera *Mormodes, Catasetum and Cycnoches.*	A single stem-member developed into a pseudobulb. Containing three subfamilies of which the Lycastinae includes the genera *Anguloa, Lycaste, Bifrenaria, etc.* The Gongorinae includes the genera *Acineta, Stanhopea, Houlletia, Gongora, etc.* The Zygopetalinae includes the genus *Zygopetalum, etc.*

Sympodiales	*Monopodiales*
Plants with limited growing point. Containing nine subfamilies of which the Dendrobiinae includes the genera *Dendrobium, Eria, etc.* The Bulbophyllinae includes *Bulbophyllum, etc.* The Cymbidiinae includes the genera *Grammatophyllum, Cyperorchis, Cymbidium, etc.* The Maxilariinae includes *Maxillaria, Scuticaria, Trigonidium etc.* The Oncidiinae includes *Ionopsis, Comparettia, Ada, Trichopilia, Cochlioda, Odontoglossum, Brassia, Miltonia, Oncidium, etc.*	Plants with unlimited growing point. Containing two subfamilies, of which the Sarcanthinae includes *Renanthera, Phalaenopsis, Sarcanthus, Saccolabium, Vanda, Angraecum, Aerides, Rhynchostylis, etc.*

7.4.0 King and Pantling's Classification

King and Pantling (1898) restored Lindley's tribe Malaxideae and split Neottieae into two tribes, *i.e.,* Listereae and Goodyereae. They recognized 7 tribes.

Tribe I. Malaxideae

Tribe II. Epidendreae

Tribe III. Vandeae

Tribe IV. Listereae (anther terminal)

Tribe V. Goodyereae (anther posticous)

Tribe VI. Ophrydeae

Tribe VII. Cypripedieae

7.5.0 Schlechter's Classification

The third system which came to be accepted as the best of the three systems was proposed by Schlechter (1926). In this system there is balance between the emphasis given on floral characters by Bentham and Hooker and on vegetative characters by Pfitzer. The outline of the system is given below.

Orchidaceae

Subfamily 1. Diandrae	*Subfamily 2. Monandrae*
Fertile anthers two. Lip always slipper-shaped. Tribe 1. Cypripedioloideae single, subtribe Cypripedileae.	Fertile anther one. Lip very rarely slipper-shaped.
Division 1. Basitonae	*Division 2. Acrotonae*
Caudicles and viscid disc arising from the base of the anther. Anther persistent, erect or resupinate. Pollinia granular. Tribe 2. Ophrydoideae	Caudicles and viscid disc arising from the top of the anther. Anther erect or incumbent, attached by a short, thin filament.Commonly deciduous. Rarely persistent.

Pollinia granular. Anther persistent. Inflorescence terminal. Tribe 3. Polychondreae. 26 subtribes, Vanilleae, Nervilleae, Spirantheae, Physureae, Tripidieae, *etc.*	Pollinia waxy. Anther deciduous. Inflorescence terminal or lateral. Tribe 4. Kerosphaereae.
Series a. Acranthae. Inflorescence usully terminal or by the abortion of the terminal inflorescence, axillary in the uppermost leaves. 14 subtribes, Liparideae, Colla- bieae, Adrorhizeae, Coelogyneae, Thunieae Dendrobieae, Podochileae, Polystachyeae, *etc.*	Series b. Pleuranthae. Inflorescence lateral, arising near the base of the pseudobulbs or in the axils of the lower leaves of sheaths.
Subseries a. Sympodiales. Growth sympodial. 32 subtribes, Phajeae, Bulbophylleae, Cyrtopodieae, Cymbidieae, Thelasieae, *etc.*	Subseries b. Monopodiales. Growth monopodial. 4 subtribes, Sarcantheae, *etc.*

7.6.0 Mansfeld's Classification

Monandrae

Tribal group Thrauosphaereae. Pollen grains mostly in free or loosely bound tetrads or in masses, very seldom waxy. Anther rarely deciduous. Leaves unjointed, convolute in the bud. Pseudobulbs none, plants often with tubers on the rhizome.	Tribal group Kerosphaereae. Pollen grains mostly in waxy or cartilaginous pollinia. Anther deciduous. Leaves jointed, convolute or conduplicate in the bud. Often with pseudobulbs or tubers on the rhizome.
Tribe Neottieae. Anthers free, rarely adnate. Pollinia not sectile.	Tribe Ophrydeae. Anthers always persistent, mostly adnate to the column at the base. Pollinia sectile.
Tribe Epidendreae. Sometimes with a viscid disc, never a stipe. Growth sympodial. Leaves jointed, rarely unjointed.	Tribe Vandeae.Viscid disc always present. Stipe always present, except in one monopodial group. Growth sympodial or monopodial. Leaves rarely unjointed.

Rendle (1930) divided Orchidaceae into two subfamilies, *i.e.*, Diandreae and Monandreae and in all he recognized 22 tribes within the family. Mitra (1958) also divided Orchidaceae into two subfamilies and in all 5 tribes.

Subfamily I. Diandreae with one tribe Cypripedileae.

Subfamily II. Monandreae with four tribes as follows :

Tribe I. Ophrydeae (3 subtribes)

Tribe II. Polychondreae (10 subtribes)

Tribe III. Karosphaereae (16 subtribes)

Tribe IV. Sarcantheae.

7.7.0 Dressler and Dodson's Classification

Dressler and Dodson (1960) have proposed a phylogenetic classification of the Orchidaceae which has been largely accepted in the recent years. They divided Orchidaceae into two subfamilies and in all 5 tribes.

Orchidaceae

Fertile anthers two or three. Filaments more or less united to the style but arising below the level of stigma. Stigma and terminal portion of the style free. No rostellum. Subfamily Cypripedioideae.	Fertile anther one. Anther and stigma connected by a terminal rostellum. Subfamily Orchidioideae.
Perianth regular. Lip never deeply saccate. Fertile anthers two or three, elongate. Style slender. Tribe 1. Apostasieae. Includes genera *Apostasia*.	Perianth irregular. Lip deeply saccate. anthers two. A flattened medium stami-node. Style thick. Tribe 2. Cypripedieae. Includes genera *Paphiopedilum, etc.*

Anther more or less erect, dorsal. Pollinia granular, sectile or powdery. Stems without corms or other thickenings. Tribe 3. Neottieae. Subtribes – Spiran- thinae, Limodoriinae, *etc.*	Anther erect or reclinate, persistent, broadly jointed to the column. Pollinia in soft masses. Caudicles arising from the base of the pollinia. Mostly terrestrial tuberous plants. Tribe 4. Orchideae. Subtribes – Disinae, Epipoginae, Orchidinae.	Anther terminal and operculate or rarely erect. Usually more or less versatile. Stems often with corms and corms-like thickenings. Tribe 5. Epiden- dreae. Subtribes – Vanillinae, Pogoninae, Collabiinae, Adrorhizinae, Bletiinae, Epidendrinae, Thuninae, Dendrobinae, Cymbidiane, Sarcanthinae, *etc.*

Garay (1960, 1972) proposed the division of Orchidaceae into five subfamilies known as Apostasioideae, Cypripedioideae, Orchidoideae, Neottioideae and Epidendroideae. Holttum (1964) outlined the following system of classification to suit the species available locally in Malaysia. This system appears very simple and easy to follow.

Orchidaceae

Subfamily Pleonandrae. Anther more than one. Apostasia tribe. The slipper orchids.	Subfamily Monandrae. Anther one.

Basitonae	*Acrotonae*
Anther attached by its base *Habenaria* tribe.	Anther attached by the top.
Pollinia granular. Saprophytes and their allies. Tribes - *Goodyera, Corymborchis, etc.*	Pollinia waxy

Sympodial Orchids	*Monopodial orchids. Tribes - Vanda, Arachnis.*	
A. Mainly terrestrial. Pollinia without disc. Tribes – *Phaius, Nephelaphyllum, Arundina, Sobralia, Liparis.*	B. Epiphytes. Pollinia with no disc. Tribes – *Coelogyne, Dendrobium, Bulbophyllum, etc.*	C. Epiphytes and terrestrial. Pollinia with disc, sometimes with a stipe. Tribes – *Appendicula, Thelasis, Eulophia, Bromheadia, Cymbidium, etc.*

Vermeulen (1966) gave to the family the rank of an order and divided Orchidales into three families, *viz.*, the Apostasiaceae, Cypripediaceae and Orchidaceae. In the Orchidaceae he recognized two subfamilies– Orchidoideae and Epidendroideae. Dahlgren (1980) also followed this classification and placed Orchidales under super order Liliiflorae. Dressler (1974) favours the 'Kew' classification of the orchid family in which only three subfamilies– Apostatioideae, Cypripedioideae and Orchidoideae are recognized. The later includes all monandrous orchids which were divided into 17 tribes arranged into three groups, the Neottioid, Epidendroid and Vandoid. Neottioid group includes tribes Neottieae, Diurideae, Cranichideae, Epipogeae and Orchideae. The Epidendroid comprises the tribes Castrodieae, Arethuseae, Epidendreae, Dendro-bieae, Malaxideae. The Vandoid group includes Maxillarieae, Vandeae, Polystachyeae, Cymbidieae, Pachyphylleae, Gongoreae, Oncideae.

In spite of repeated efforts of several taxonomists, the classification of orchids still leaves much to be desired. The position allotted to the three Indo-Malayan genera – *Apostasia, Adachylus* and *Neuweidia* has been one of the major points of contro-versy. Lindley (1840), Bentham (1881), Holttum (1964) and Dressler and Dodson (1960) considered these genera as the most primitive of orchids while Pfitzer and Schlechter did not include them in the family. Summerhayes (1968) treated the *Apos-tasia* group, the *Cypripedium* group and all the rest of the orchid genera as three separate subfamilies under the family Orchidaceae. There are also controversial opinions regarding the order in which the tribes are arranged in the monandrous group. Bentham (1881) arranged them in the order as Epidendreae, Vandeae, Neottieae, Ophrydeae and Cypripedieae. Pfitzer (1889) reversed this order. Schlechter (1926) also followed this arrangement. But Mansfeld (1937) and Dressler and Dodson (1960) considered Neottieae as more primitive than Ophrydeae. There is also controversy regarding the delineation of tribes in the monandrous section.

8.0 Genera, Species and Hybrids of Horticultural Importance

The number of different kinds of orchids which grow wild throughout the world has never been determined with any accuracy, and estimates today range from a low of 7,500 species to a high of more than 35,000 (Schultes and Pease, 1963; Hawkes, 1978; Arditti, 1979; Bose and Bhattacharjee, 1980). According to various assessments, however, a figure of approximately 25,000 apparently valid species in some 700-800 genera will be a safe and reasonable estimate. In addition to the large number of wild orchids, there are about 32,000 or so hybrid forms and more than 1000 new hybrids are being added every year. The genera, species and hybrids of horticultural importance are briefly described below.

Details of Dendrobium, Cymbidium and Paphiopedilum are being dealt with in a separate chapter.

8.1.0 Aerides Lour

This popular genus comprises more than sixty species of monopodial epiphytes or lithophytes. These are widely grown by amateurs and because of unique formaion of the inflorescences in many of the species, are often known as 'Fox-Tail Orchids'.

Plants are attractive, generally erect, closely allied to *Vanda*. Stem long, cylindrical, stout with persistent leaf bases. Leaves sheathing at the base, alternate, flat or channelled, with strong aerial-roots growing from their stalks by which they absorb mois-ture from atmosphere. Inflorescence drooping, racemes consist of few to many fragrant, waxy, attractive flowers. Floral colour is variable, ranging primarily through shades of green, magenta and white in various combinations. Sepals and petals similar, spreading; lateral sepals slightly decurrent on column foot. Lip trilobed, lateral lobes slightly smaller, spurred at the base. Column is short with long foot. Anther is 2-celled. Pollinia are 2, globose, sulcate. These beautiful orchids are natives of the tropical parts of Asia, ranging from China and Japan to New Guniea with many species occurring in the Himalayas, Myanmar and Indonesia.

Important species: *Aerides crispum, A. falcatum, A. fieldingii, A. japonicum, A. lawrenceae, A. maculosum, A. mitratum, A. multiflorum, A. odoratum, A. radicosum, A. vandarum, etc.*

Hybrids: *Aerides* is freely inter-fertile with many of the allied genera like *Arachnis, Ascocentrum, Neofinetia, Phalaenopsis, Renanthera, Rhynchostylis* and *Vanda*. The names of commercially important bigeneric and trigeneric hybrids are given below:

Aeridachnis (*Aerides* × *Arachnis*)

Aeridocentrum (*Aerides* × *Ascocentrum*)

Aeridofinetia (*Aerides* × *Neofinetia*)

Aeridopsis (*Aerides* × *Phalaenopsis*)

Aeridostylis (*Aerides* × *Rhynchostylis*)

Aeridovanda (*Aerides* × *Vanda*)

Renades (*Aerides* × *Renanthera*)

Tanakara (*Aerides* × *Phalaenopsis* × *Vanda*)

Vandopsides (*Aerides* × *Vandopsis*)

Sidran (1984) reported the recent development on the intergeneric hybridization of vandaceous orchid. *Aerides crassifolia*, a dwarf plant with thick leaves from Thailand, has widely been used as a parent, and one cross *Aeridovanda* (*Aerdv.*) Pink Glory × *V. sanderana*, Takakura, has two awarded clones. The clone 'Crestwood' has 25 light pink flowers. *Aer. flabellata* from Thailand, a dwarf plant with tan flowers and is one of the favou-rite *Aerides*. It was used as a hybrid × *Ascocentrum curvifolium* = *Aeridocentrum* (*Aerctm.*) Luke Nok. This latter hybrid × *V.* Josephine Van Brero created the much awarded *Christieara* (*Chtra.*) Malibu Gold. *Chtra.* Manoa 'Orchid Paradise' × *Ascda.* Yip Sum Wah has 45 pink with pink spots round flowers on 1 inflorescence (Ozaki, 1982). Sidran crossed *Christiera* Jeanine Sidran × *Ascda.* Blue Boy to produce blue Ascdas that smelled like sweet peas. Some of the progenies have been a blue-purple, and others a rich deep pink-purple. Another hybrid of *Aer. flabellata*, Sidran used in hybridization in *Chtra.* Jiad, *Aerdv.* Vieng Ping × *Ascda.* Mad. Kenny. Jiad has a lovely yellow colour overlaid with a light dotting of red.

In 1980, Richard Kosel registered *Vandopsides* Cherry Blossom, *Aer. multiflora* var. *godefroyae* × *Vandopsis parishii* var. *mariottiana*. The small plants carry long branched racemes of dark pink flowers.

In 1963, Miller registered the cross, *Aerdv.* Frank Johnston, *Aer. Lawrenceae* × *V. tessellata*. The clone, 'Hawaii' has many fragrant white flowers suffused with lavender on the tips of the segments. The *Aerdv.* Tsuruko Iwasaki (registered in 1951) × *V. sanderana*, *Aerdv.* Blue Spur × *V. coerulea*, *Aerdv.* Henry Furimizo × *V.* Stuart Furimizo (*V.* Green Gold × *V.* Ellen Noa) are also important hybrids of the group (Sidran, 1984). Julia Hamner (*Aer. lawrenoeae* × V. Keeree) (Anon., 2000).

8.2.0 Anoectochilus Bl.

This genus contains about twenty species of terrestrial orchids popularly known as 'Jewel Orchids'. They are allied to genera *Zeuxine, Hetaeria, Vrydagzynea, Dossinia, Macodes, Haemaria* and are among the rarest and desirable cultivated orchids. Unlike the vast majority of orchids, these plants are not cultivated for their flowers, which are rather small and insignificant, but for the magnificent, variegated, velvety foliage. Plants are with soft, creeping stem and beautifully coloured leaves. The flower stalk is upright and about 15 cm high while the flowers are small and white. Lip is adnate to the column. Column is short.

Important species: *Anoectochilus geniculatus, A. regalis, A. roxburghii, A. setaceus, A. sikkimensis, etc.*

Hybrids: *Anoectochilus roxburghii* has been crossed with *Haemaria discolor* to produce the rare and beautiful *Anoectomaria* Dominyi.

8.3.0 Arachnis Bl. (*Arachnanthe* Bl.)

The genus consists of about seventeen species and is closely allied to *Vanda, Renan-thera* and *Dimorphorchis*, differing in shape of sepals and petals. These beautiful orchids are very popular with amateurs and commercial growers. They are found as epiphytes, lithophytes or as terrestrials often in boggy places. Stems elongate and vary in length and thickness. Leaves leathery, abundant, persistent. Flowers many, borne on simple racemes or branched panicles, large, showy. Colour varies, though green, yellow, brown and purple predominate. Sepals and petals generally blotched or otherwise marked with different hue from that of the lip, lip spurless; pollinia 4. The genus is widespread in areas from the Himalayas to New Guinea and the Solomon Island and large concentration is found in Indonesia and Malaysia.

Important species: *Arachnis annamensis, A. cathcartii, A. clarket, A. flosaeris, etc.*

Hybrids: The Malayan *Arachnis* Maingayi is known to be a natural hybrid. Artificially induced hybridization in this genus has been extensive in recent years and even multigeneric hybrids have been raised. Some of these combinations are listed below:

Aeridachnis (*Arachnis* × *Aerides*)

Arachnopsis (*Arachnis* × *Phalaenopsis*)

Aranda (*Arachnis* × *Vanda*)

Aranthera (*Arachnis* × *Renanthera*)

Armodachnis (*Arachnis* × *Armodorum*)

Limara (*Arachnis* × *Renanthera* × *Vandopsis*)

Ridleyara (*Arachnis* × *Trichloglottis* × *Vanda*)

Trevorara (*Arachnis* × *Paraphalaenopsis* × *Vanda*).

An interesting intergeneric hybrid *Ascoarachnis* Mem. Lucille Keywood (*A. cathcartii* × *Ascocentrum curvifolium*) was registered by J. Rives in 1976. The tan flowers have open shape and are dotted with a dark brown colour. It is cold tolerant and blooms frequently throughout the year (Sidran, 1984).

8.4.0 Ascocentrum Schltr.

Once included in Saccolabium, the genus consists of about nine handsome species. These are dwarf, compact monopodial epiphytes (rarely lithophytes) allied to *Ascoglossum* and *Vanda*. Plants with stout stem and leaves are linear, fleshy, toothed at the apex. Inflorescences stiffy, erect, cylindrical racemes. Flowers are bright coloured, long lasting. Lip is with large hanging spur. Pollinia are 2.

Important species: *Ascocentrum ampullaceum, A. curvifolium, A. miniatum, etc.*

Hybrids: Registered hybrids in which *Ascocentrum* figure include *Ascocenda* (× *Vanda*), *Asconopsis* (× *Phalaenopsis*) and *Ascofinetia* (× *Neofinetia*) which are charming groups. *Ascda*. Princess Mikami (*Ascda*. Royal Saphire × *V. coerulea*), *Ascda*. Summer Gold (*V*. Chaiya × *Ascda*. Theptong), *Ascda*. Yellow Pacific (*Ascda*. Muttaparb Rama × *V*. Thananchai), *etc.* are some of the popular *Ascocenda* hybrids. Bebe warren (*Ascda*. Crownfox Golden Dawn × *Ascda*. Fuchs Baby Doll), Chameleon (*Ascda*. Pander × *V*. roeblinglana), Clyde Mackay (*V*. Princess Blue × *Ascda*. Yip Sum Wah), Gall Noyes (*Ascda*. Fuchs Harvest Moon × *V. denisoniana*), Le-Anne Gan (*V*. Josephine van Brero × *Ascda*. Yee Peng), Sandy Hatos (*Ascda*. Memoria Choo Laikeun × *V*. Memoria Lyle Swanson) are also beauliful (Anon., 2000, 2001).

In 1949, a hybrid from Hawaii combining a species called *Ascocentrum curvifolium* with *Vanda lamellata* created a new genus and was known as *Ascocenda*. Since that time many colourful *Ascocenda* hybrids have been made and some of them have won numerous awards from American Orchid Society. Some of the important *Ascocenda* hybrids by using *Ascocentrum curvifolium* and *Vanda* are given below:

Ascocentrum curvifolium × *Vanda Rothschildiana* = *Ascd*. Meda Arnold

A. curvifolium × *V*. Pukele = *Ascda*. Yip Sum Wah

A. curvifolium × *V*. Bill Sutton = *Ascda*. Ophelia

A. curvifolium × *V*. Betsy Summer = *Ascda*. Tropicana

A. curvifolium × *V*. Sanderana = *Ascda*. Erika Reuter

A. curvifolium × *V*.B.B. Fukumina = *Ascda*. Peggy Foo.

About one third hybrids registered with *Ascocentrum miniatum* have been with *Phalaenopsis* or *Doritaenopsis*. *A. miniatum* has been difficult to breed with *Vanda* than *A. curvifolium* and seems to accept *Phalaenopsis* pollen more readily (Sidran, 1984).

Asconopsis Irene Dobkin, a hybrid registered by Thornton in 1968, combined *Phal*. Doris with *A. miniatum*. This hybrid received many prizes for its outstanding quality. *Asconopsis* Irene Dobkin has been further crossed with *Phalaenopsis* to create other *Asconopsis*. Beardara Henry Wallbrunn (*Ascps*. Irene Dobkin × *Dor. pulcherrima*) was registered by Wallbrunn in 1978. It has branched upright spike of red purple flowers (Sidran, 1984).

From 1971 through 1981, Sanders listed 35 Kagawaras (*Kgwra*.), a combination of *Ascocenda* × *Renanthera*, although 13 have *R. storiei* as one parent. Furmizo registered a cross of R. Brookie Chandler × *Ascda*. Meda Arnold, *Kgwra*. Chie Furmizo. The flowers are a rich red colour with excellent shape. The Kagawaras also tend to bloom throughout the year, because of the *Ascda*. parent. *Kgwra*. Lumpur Beauty 'Kasha' (*Ascda*. Peggy Foo × *R. storiei*) has attractive blue-purple flowers (Sidran, 1984).

8.5.0 Brassavola R. Br.

These are very popular, showy, mostly epiphytic orchids, closely allied to *Laelia*. The genus comprises about fifteen species whose original home was tropical America. Plants have cylindrical, stem-like pseudobulbs which are topped by a solitary, terete, fleshy leaf. The flowers, numbering one to several per raceme are produced from the top of the pseudobulbs, and are green, yellowish or white in colour.

Important species: *Brassavola acaulis, B. cordata, B. cucullata, B. flagellaris, B. nodosa, B. perrini, B. tuberculata, etc.*

Hybrids: With *Brassavola digbyana* and *B. glauca* removed from the genus, it seems evident that proportionately little breeding has been done, almost all of it centring round the very popular and handsome *B. nodosa* (Hawkes, 1978). By considering the two first-named species under *Brassavola*, these have been very extensively utilized in the production of artificial hybrids. Some of the intergeneric hybrids are *Brasso-tonia* (× *Broughtonia* × *Brassocattleya* (× *Cattleya*), *Brassodiacrum* (× *Diacrium*), *Brassoepidendrum* (× *Epidendrum*), *Brassolaelia* (× *Laelia*), *Schombavola* × *Schomburgkia*), *Lowiara* (× *Laelia* × *Sophronitis*), *Brassophronitis* (× *Sophronitis*), *Dekensara* (× *Cattleya* × *Schomburgkia*), *Beaumontara* (*Cattleya* × *Laelia* × *Schomburgkia*). Some important hybrids of *Brassocattleya* are- Bride's Rouge (*Bc*. Bride's Blush × *Bc*. Speciosa), Dal's Cherry (*Bc*. Island Charm × *C*. Cherry Chip), Memoria Ed Beth (*Bc*. Mount Anderson × *B. digbyana*), Nina Consentida (*C*. Horace × *B. digbyana*), Tripp's Grandma (*C*. Tripp Johnston × *B. nodosa*) (Anon., 2000).

8.6.0 Broughtonia R. Br.

This is a small genus containing a single species, *B. sanguinea* which is well worthy of cultivation. It is one of the most delightful of the species of the subtribe Laeliinae and is sometimes confused with members of the allied genera, *Cattleyopsis and Laeliopsis*. Plants of this species have roundish pseudobulbs and pale green leaves. Inflorescences slender, wiry, rarely branched and bear 8-15 flowers that are usually rich purplish crimson, about 2.5 cm across and with an orange blotch at the base of the lip.

Species: *Broughtonia sanguinea.*

Hybrids: The cultivars, Aquinii and Red Stripe have colourful flowers. The inter-genetic hybrids developed with *Broughtonia* are *Brassotonia* (× *Brassavola*), *Cattleytonia* (× *Cattleya*), *Diabroughtonia* (× *Diacrium*), *Shipmanara* (× *Diacrium* × *Schomburgkia*), *Epitonia* (× *Epidendrum*), *Lioponia* (× *Laeliopsis*) and *Schombonia* (× *Schomburgkia*). *Broughtonias* breed more readily with bifoliate Cattleyas and a lot of seeds are produced. A good example is *Cattleytonia* Keith Roth (*B. sanguinea* × *C. bicolor*) registered in 1966. This hybrid crossed back to *B. sanguinea* produced *Ctna*. Jamaica Red registered in 1972. *Ctna*. Jamaica Red is more vigorous and floriferous than *Ctna*. Keith Roth. *Ctna*. Rosy Jewel (*C. bowringiana* × *B. sanguinea*), *Ctna*. Alina Froelich (*Ctna*. Rosy Jewel × *B. sanguinea*), *Ctna*. Why Not (*C. aurantiaca* × *B. sanguinea*), *Ctna*. Penny Bassin (*C. chocolate* Drop × *B. sanguinea*), *Ctna*. Sugar Plum 'Yagi's' (*C. guttata alba* × *B. sanguinea alba*), *Ctna*. Breathless 'Alpha' (*C. intermedia* × *B. sanguinea*), *Laeliocatonia* Roy Fields (*Lc*. Mattie Shave × *Ctna*. Keith Roth), *Otaara* Maili's Surprise (*Blc*. Waikiki Gold × *Ctna*. Keith Roth) and *Hawkinsara* Alice 'Susan' (*Slc*. Little Beamche × *Ctna*. Keith Roth) are some other beautiful hybrids. *Stellamizutaara* Kelly (*B. nodosa* × *Ctna*. Keith Roth) produced star shaped flowers, sepals and petals yellow with a tinge of lavender (Chang, 1984).

8.7.0 Bulbophyllum Thou.

This is the largest orchid genus containing about two thousand species distributed in tropical and subtropical belts of the world. The plants are epiphytic having conical or globose pseudobulbs borne on a creeping rhizome bearing a single leathery leaf. Inflorescences arise from the rhizome, at or near the base of pseudobulbs and produce single or many flowers. Flowers small to large and their lateral sepals are more or less joined at the base. The lip is generally very small and inconspicuous, occasionally large and hinged lightly, fleshy and warty.

Pollinia are 4, in pairs with or without gland. Garay *et al.* (1994) reported that the genus *Cirrhopetalum* is reinstated among the *Bulbophyllum* alliance on the basis of the character of its type species.

Important species: *Bulbophyllum barbigerum, B. biflorum, B. bracteolatum, B. campanulatum, B. chinense, B. comosum, B. crassipes, B. falcatum, B. fascinator, B. leopardinum, B. lobii, B. longissimum, B. macranthum, B. maximum, B. medusae, B. ornatissimum, B. reticulatum, B. robustum, B. umbellatum, etc.*

Hybrids: Interspecies crosses have produced some good hybrids like *B.* Fascination (*B. fascinator* × *B. longissimum*), *B.* Louis Sander (*B. longissimum* × *B. ornatissimum*) and *B.* David Sander (*B. lobbii* × *B. virescens*), Ed Gilliland (*B. bicolor* × *B. carunculatum*) and Tsiku Torero (*B. leysianum* × *B. carunculatum*).

8.8.0 Calanthe R. Br.

A very popular genus comprising about 150 species of primarily terrestrial or lithophy-tic orchids, closely allied to Phaius. The genus is divided into two groups, one with evergreen leaves and obscure pseudobulbs, the other with deciduous foliage and very prominent pseudobulbs. Orchids of both groups have folded leaves upto 30 cm long and a large number of blossoms with irregular petals and sepals. These blossoms occur in almost every imaginable colour combination, though white, pink and yellow are predominant. The lip is fused to the column, three-lobed, the middle one large and further bilobed. Column is short and stumpy. Pollinia 8, bunched all together or in two groups of 4.

Important species: *Calanthe biloba, C. brevicornu, C. discolor, C. furcata, C. herbacea, C. masuca, C. veratrifolia, C. vestita, etc.* An unusual *Calanthe* (*C. delavayi*) was recorded from Yunan (Robbins, 1991).

Hybrids: Several hybrids have been developed within the genus *Calanthe* and many of them are grown commercially for cut-flower production in various parts of the world. Some of the interspecific hybrids are C. Norman (*C. vestita* var. *gigantea* × *C. vestita* var. *rubro-oculata*), C. Sedenii (*C. veitchii* × *C. vestita* var. *rubro - oculata*) and C. Veitchii (*C. rosea* × *C. vestita*), Tsiku Flaningo (C. Rozel × C. Kyoto) and Tsiku Honolulu (C. Saint Aubin × C. Rozel).

In *Calanthe*, the seedlings raised from crosses between *Calanthe* Sedenii 'Harrisii' and Baron Schroeder and Diana Broughton mostly died when taken out of the flask. Chromosome counts showed that C. Sedenii 'Harrisii' was triploid, whereas other cultivars were diploid and the seedling proved to be aneuploid. One of the survivors was a tetraploid and received an award of merit and has considerable breeding potential (Moon, 1989).

According to flower colour and the shape of petals, sepals and lips, *Calanthe* cultivars from Cheju Island, Korea Republic are named as follows: Arang, Cheongsu, Sara, Aeyeol, Darigot, Saebaengi, Andeok, Sethyeol, Kyorae, Jasujeong, Backrok, Seokwang, Deokcheon, Seonheul, Saechimi, Jokduri, Baedal, Noksu. Noeul and Dansim for 20 cultivars of *C. discolor*; Hwanghee, Waheul, Illchool, Hwangryong, Dabo, Hwangjinee, Daeha, Chwikwang, Seokyang, Eunpa, Chwiseon, Sinbaram, Ara, Dano, Pakae and Sojeob for 16 cultivars of *C. sieboldii*; and Dongkwang,

Myeongdeok, Wangdolseon, Myeongrycok, Kwibuin, Semi, Cheonwangsa, Mandeok, Seolmundae, Kyeongok, Darangsi, Byeolbang, Honghak, Dongbaeksan, Chejudo, Samda, Kwandeok, Youngjusan, Tamra, Koyeongdi, Yongduam and Jeolbuam for 22 cultivars of *C.bicolor* (Hyun *et al.*, 1998).

A few intergenetic hybrids have also been made, notably with *Phaius* (*Phaiocalanthe*) and *Gastorchis* (*Gastocalanthe*).

8.9.0 Cattleya Ldl.

This magnificent genus, containing about 65 species, innumerable cultivars and several thousands of hybrids, is without doubt the most popular and widely grown in the entire Orchidaceae. Cattleyas are epiphytic orchids, closely allied to *Epidendrum*, *Laelia*, *Sophronitis* and *Brassavola*. The species are wonderfully striking for their extreme richness of colour, their large size and fragrance of bloom. Vegetatively, the genus is divided into two groups. These are unifoliates and the bifoliates. The unifoliate section consists of plants with club-shaped pseudobulbs joined by a creeping rhizome; single large leaf of a heavy texture produced at the top of the pseudobulb, and producing one to a very few large and spectacular flowers. These pseudobulbs are water-storage organs and are large, fleshy roots with a spongy covering useful for capturing moisture. The bifoliates have taller and more cylindrical pseudobulbs which terminate in a pair of shorter leaves and carry more, although smaller flowers of heavier texture. The floral structure of this genus is basically simplified; the sepals are spreading, wide and frilled or crisped petals away from a large different coloured lip. The basal part of the lip in most cases is tubular.

Cattleyas are widespread in the American tropics particularly in the Andean region of northern and western and western South America and Brazil.

Important species: *Cattleya aurantiaca, C. bicolor, C. bowringiana, C. citrina, C. dowiana, C. intermedia, C. labiata, C. leopoldii, C. luteola, C. maxima, C. schilleriana, C. skinneri, C. violacea, C. warscewiezii, etc. Cattleya bicolor*, an orchid for the Rainbow Goddess described by Lima (1993) has three subspecies *bicolor, minasgeraisensis, brasilienses*.

Hybrids: A large number of hybrids of *Cattleya* have arisen as a result of both natural and artificial crosses. ***Cattleya* Hardyana (*C. dowiana aurea* × *C. gigas*)**, C. Guatemalensis (*C. skinneri* × *C. aurantiaca*), C. Intricata (*C. intermedia* × *C. leopoldii*) and C. Victoria-Reginae (*C. labitata* × *C. leopoldii*) are few natural hybrids while C. Ballantineana (*C. trianaei* × *C. warscewiczii*), C. Clotho (*C. enid* × *C. trianaei*), C. Fabia (*C. dowiana* × *C. labiata*), C. Iris (*C. bicolor* × *C. dowiana*), C. Omar (*C. enid* × *C. leda*) and C. Whitei (*C. warneri* × *C. schilleriana*) are some of the artificial crosses. Few hybrid of recent originare are: Dibreball Fame (C. Netrasiri Fireball × C. Thospol Spot), Fireball Beauty (C. Netrasiri Fireball × C. Netrasiri Beauty), Hannyahime (C. Horace × C. Mary Lynn Mckenzie), Rosetta Stone (C. Myrtie × C. Marjorie Hausemann), Tsiku Cannas (C. Condy Tuft × C. loddigesii), Tsikum Eglantine (*C. dolosa* × *C. intermedia*) and Rsikum Violet (C. Gravesiana × C. Sea Breeze) (Anon., 2000).

Menezes (1988) reported a natural hybrid *Cattleya labiata* Orbatchev and *Brassocattleya* × *miranda*. He also found natural hybrid between *Cattleya warneri* and *Brassavola tuberculata* in a forest in the Rio Doce Valley, Minas Gerais State, Brazil together with flowering plants of both parent species.

Many of the cultivated species of this genus and their hybrids have also been crossed with allied genera like *Brassavola*, *Laelia*, *etc.* and thousands of hybrids have been registered. *Brassocattleya* (*Brassavola* × *Cattleya*), *Brassolaeliocattleya* (*Brassavola* × *Cattleya* × *Laelia*), *Brownara* (*Cattleya* × *Broughtonia* × *Diacrium*), *Cattleytonia* (*Cattleya* × *Broughtonia*), *Epicattleya* (*Cattleya* × *Epidendrum*), *Epilaeliocattleya* (***Cattleya* × *Epidendrum* × *Laelia***), *Hasegawaara* (*Cattleya* × *Brassavola* × *Broughtonia* × *Laelia* × *Sophronitis*), *Hawkinsara* (*Cattleya* × *Broughtonia* × *Laelia* × *Sophronitis*), *Laeliocatonia* (*Cattleya* × *Broughtonia* × *Laelia*), *Laeliocattleya* (*Cattleya* × *Laelia*), *Potinara* (*Cattleya* × *Brassavola* × *Laelia* × *Sophronitis*), *Rolfeara* (*Cattleya* × *Brassavola* × *Sophronitis*), *Sophrocattleya* (*Cattleya* × *Sophronitis*), *Sophrolaeliocattleya* (*Cattleya* × *Laelia* × *Sophronitis*), *Yamadara* (*Cattleya* × *Brassavola* × *Epidendrum* × *Laelia*) are the examples of intergeneric crosses.

Some of the beautiful and popular hybrids are *C.* Angelwalker 'Easter' (*C.* Little Angel × *C. walkeriana*), *C.* Brabantiae (*C. aclandiae* × *C. loddigesii*), *C.* Small World (*C. aclandiae* × *C. luteola*), *Brassocattleya* Harriet Moseley (*Bc.* Deese × *C. bicolor*), *Brassolaeliocattleya* (*Blc.*) Dawson City 'Mendenhall' (Laeliocattleya (*Lc.*) Memoria Albert Heinecke x *Blc.* Acapana), *Blc.* Constant Spring 'Evergreen' (*Blc.* Envy × *C. bicolor*), *Blc.* Pamela Hetherington 'Coronation' (*Lc.* Paradisio × *Bc.* Mount Anderson), *Blc.* Verdant Venture (*Blc.* Greenheart × *Blc.* Emerald Meadow), *Blc.* Green Fantasy (*Blc.* Greenheart × *Blc.* Olympic Meadows), *Blc.* Don De Michaels (*Blc.* Fleur de Lys × *Lc.* Oakland), *Blc.* Greenwich (*Blc.* Lester Mc. Donald Greenbrier × *Lc.* Ann Follis), *Lc.* Wrigleyi 'Blue Hawaii' (*L. anceps* × *C. bowringiana*), *Lc.* Mini Purple (*L. pumila* × *C. walkeriana*), *Lc.* Ann Follis (*C. granulosa* Aspirata × *Lc.* Ethel Mermann Pinnacle), *Sophrolaeliocattleya* (*Slc.*), Yellow Doll (*C. luteola* × *Sophrolaelia* (*Sl.*) Psyche), *Slc.* Hazel Boyd 'Royal Scarlet' (*Slc.* California Apricot × *Slc.* Jewel Box), *Potinara* (*Pot.*) Canyon Sands 'Anaconda' (*Slc.* Fort Caroline × *Blc.* Fortune), *Pot.* Rising Moon 'Apricot Glow' (*Sc.* Dorea × Mcllow Vista) *etc.* Two of the more interesting and awarded white cattleyas are *C.* Ruth Gee 'Goldkist' notable for its almost solid golden lip, and *Blc.* Nickie Holguin 'White Heat' which almost entirely lacks any colour in the lip. Pink is perhaps the colour currently most demanded. Two popular hybrids have been *Bc.* Cynthia and *Bc.* Heart's Desire. Among semi-alba hybrids, *Lc.* Garden Delight 'Memorial Day', *Lc.* Starting Point 'Unique' and *Lc.* Persepolis 'Splendor' are outstanding ones. Truly red *Cattleya* hybrids have always been uncommon. *Slc.* Redzac is an older hybrid. Balinese Beauty and Red Celestic are possibly the best of the red *Sophrolaeliocattleya* (Nash, 1984).

Green cattleyas are often produced by crossing a complicated tetraploid parent such as *Blc.* Xanthette 'Chartreuse' with a cattleya-alliance species known to have green or a greenish colour. One of the better greens derived from *C. bicolor* is *Lc.* Erin 'Bill Teel' (*Lc.* Kencolor × *Lc.* Elastead Gem). Hybrids from *C. bicolor* sometimes have yellow or almost white lips. A good example is *Blc.* Constant Spring 'Evergreen' (*Blc.* Envy × *C. bicolor*). This cultivar holds its beautiful green colour for the full life of the flower. Another achievement in green cattleya breeding occurred with the flowering of a hybrid of *C. granulosa* 'Aspirata' × *Lc.* Ethel Merman 'Pinacle'. The cultivar 'Green' derives its intense purple lip from *Laelia purpurata*. A very spectacular new green cattleya is *Blc.* Waikiki Gold × *Blc.* Mem. Helen Brown 'Williette' (Petroski, 1984).

Blc. hybrids produced included Aussle Gold T'nt (*Blc.* Blumen Insel × *Blc.* Makaha Gold), Beautiful Lady Pat (*Blc.* Tampico × *C. percivaliana*), Cracklin Rosie (*Blc.* Memoria Crispin Rosales × *B. nodosa*), Carol Young (*Blc.* Norman's Bay × *Lc.* Irene Finney), Grezaffi-Hill (*Blc.* Joe × *Blc.* Oconee), Jane Brecht (*Blc.* Llewellyn × *Blc.* Pink Suprise), Lawless Aneas (*Blc.* Honolulu Sunset × *Blc.* Jeremy Island), Lawless-Fortune (*Blc.* Fortune × *Lc.* Mamo Gold), Lawless-Freischutz (*Blc.* American Heritage × *Blc.* Hawaiian Lightning), Mishima Honor (*Lc.* Hobon × *Blc.* Pamela Hetherington) and Mishima Orange (*L.* Orange Gem × *Blc.* Memoria Max Goodless) (Anon., 2000).

The *Cattleya* genus stands out for its importance to the global floriculture agribusiness by presenting many varieties of hybrids with lush flowers, high commercial and ornamental value with great morphological and chromosomal variation and high genetic diversity (Cruz *et al.*, 2003 cited by da Silva *et al.*, 2014). *Cattleya* has a natural recombination capacity which make hybridization mechanism possible and produce fertile descendants (Bernacci *et al.*, 2005), and usually has a chromosomal variation with register from $2n$ =12 to $2n$ = 240, more often n =19 or 20 (Felix and Guerra, 2010). Da Silva *et al.* (2017) study the cytogenetic characterization of *Cattleya nobilior* species and *C. bowringiana* and hybrids of orchids *C. violacea* xtype, × *C. granulosa, C. violacea* type × *C. Harrisoniae* xperola × and *C. violacea* xtype × *C. nobilior* xamaliae.× *Cattleya nobilior* has $2n = 2x = 30$ metacentric chromosomes,with size ranging from 0.54 to 1.58 μm and an active NOR in the first chromosome pair, and *C. bowringiana* has $2n = 2x = 34$ metacentric chromosomes, from 0.38 to 1.46 μm of size and an active NOR in the first chromosome pair. As for the hybrids, *C. violacea* xtypex *C. harrisoniae* xpérolax has $2n = 2x = 32$ metacentric chromosomes, with size from 0.47 to 1.73 μm and an active NOR in the second chromosome pair, *C. violacea* xtypex*C. nobilior* xamaliaex has $2n = 2x = 66$ metacentric chromosomes, with size between 0.22 to 1.48 μm and active NORs in chromosome pairs 1 and 3, and *C. violacea* xtypex *C. granulosa* with $2n = 2x = 86$ metacentric chromosomes, with size from 0.26 to 1.56 μm and an active NOR in chromosome pairs 3 and 23 (da Silva *et al.*, 2017).

8.10.0 Cochlioda Ldl.

This genus comprises five species of mostly epiphytic orchids native in the Andean regions of Peru, Ecuador and Columbia. Plants are similar to *Odontoglossum* in vegetative habit but produce flowers of distinct structure and typically of brilliant scarlet colour. Flowers are large and borne in dense racemes.

Species: *Cochlioda noetzliana, C. rosea* and *C. vulcanica* are the important species.

Hybrids: Cochliodas are best known for their contributions to hybridization where they have been used extensively in making crosses with *Odontoglossum, Oncidium* and *Miltonia*. Some of the intergeneric hybrids are *Adioda* (× *Ada*), *Miltonioda* (× *Miltonia*), *Oncidioda* (× *Oncidium*), *Charleswortheara* (× *Miltonia* × *Oncidium*),

Burrageara (× *Miltonia* × *Odontoglossum* × *Oncidium*), *Vuylstekeara* (× *Miltonia* × *Odontoglossum*), *Odontioda* (× *Odontoglossum*), *Symphodontioda* (× *Odontoglossum* × *Symphyglossum*), *Sanderara* (× *Brassia* × *Odontoglossum*), *Wilsonara* (× *Odontoglossum* × *Oncidium*). Arriaga and Rounez Point are two attractive *Odontioda* hybrids.

8.11.0 Coelogyne Ldl.

The genus comprises about 125 species of mostly epiphytic orchids, though some of them grow as terrestrials and still others are found as lithophytes. Plants are variable, but usually with slender pseudobulbs on a creeping rhizome bearing one or a pair of leaves. Many Coelogynes have folded leaves which alongwith angular pseudobulbs give to the plants a most characteristic appearance. The inflorescences one or more erect, arching or pendulous, few to many flowered and arises either at the base of the pseudobulbs or at the top in between leaves. Small to relatively large, fragrant flowers vary in colours from white to brown, yellow or green, often with blotchings of yellow, brown, orange or almost black on keeled and callused lip. Sepals subequal. Petals are narrow and slender. Lip entire or prominently 3-lobed, sometimes sac-shaped basally and occasionally joined to the base of the column, sessile, usually concave. Column long, arched or erect, free or adnate to the lip-base, winged at the top. Pollinia are 4.

Species: *Coelogyne barbata, C. corymbosa, C. cristata, C. dayana, C. elata, C. fimbriata, C. fragrans, C. flaccida, C. nitida, C. ochracea, C. odoratissima, C. speciosa, C. tomentosa* and others.

Hybrids: Only few hybrids are known in *Coelogyne*, the finest of all being C. Mem. William Micholitz (*C. lawrenceana* × *C. mooreana*), registered in 1950. The commonly grown hybrid in this group is C. Stanny (*C. asperata* ×*C. pandurata*).

8.12.0 Cymbidium Sw.

Cymbidium is the loveliest and most popular orchid grown on a commercial scale in many parts of the world. It consists of about 70 species of terrestrial lithophytic or epiphytic habit. Evergreen plants possess abundant fleshy roots and short and stout pseudobulbs enclosed by linear leaves joined to the base. Inflorescences erect or pendulous bearing large, well-spread, long-lasting flowers, white or yellow to dark purple-maroon colour. Sepals and petals alike. Lip 3-lobed. Column boat-shaped with a short foot. Pollinia 2, with short caudicle, jointed to a broad disc. The long spikes bearing showy flowers are excellent for decoration.

Important species: *Cymbidium aloifolium, C. devonianum, C. eburneum, C. ensifolium, C. insigne, C. giganteum, C. grandiflorum, C. longifolium, C. lowianum, C. madidum, C. mastersii, C. pendulum, C. simonsianum, C. tigrinum, C. tracyanum*,etc. Wong and Chan (1993) reported *Cymbidium kinabaleunse* from Borneo which bears 21/23 flowers, yellowish green in colour with brown spots. Five new species of *Cymbidium* (*C. prompovenium, C. longipes, C. induratifolium, C. yongfuense* and *C. nigrovenium*) from Asia were described by Liu and Zhang (1998).

Hybrids: Several thousand hybrids have been produced through interspecific breeding within *Cymbidium*. *Cymbidium* Alexanderi 'Westonbirt' (*C. eburneolowianum* x *C. insigne* 'Westonbirt') is the finest of early hybrids. This hybrid gave rise to several others like C. Rosanna 'Pinkie' and C. Balkis which have been widely used for further improvement. Dark reds like C. Carisbrook and C. Red Beauty, greens like C. Baltic and C. King Arthur and yellows like C. Marcia have also been produced.

Miniature hybrids include C. Pumander (C. Louis Sander x *C. pumilum*), C. Jean Brummitt (*C. devonianum* x *C. eburneum*), C. Flirtation (*C. pumilum* x C. Zebra), *etc. Cymbidium* Putana 'Showpiece' is considered a true miniature cymbidium hybrid. The crossing of *C. pumilum* with C. Rutana has produced C. Putana. Its almost pendulous flower spikes of small blooms and compact plant growth combine to make the typical miniature cymbidium. C. Touchstone 'Mahogany' is another true miniature and it has come from the combination of *C. devonianum* with C. Mission Bay (Monkhouse, 1984). The genus has also been crossed successfully with *Phaius* (*Phaio-cymbidium*), *Grammato phyllum* (*Grammato cymbidium*) and with *Cyperorchis* (*Cypero cymbidium*).

Few hybrids of recent origin are Blanche Ames (C. One free Hill x C. Kusuda Shining), Dick Swain (C. Bagdemagus x *C. devonianum*), Enzan Floss (*C. pumilum* x C. Candy Floss), Enzan Sarah (*C. pumilum* x C. Sleeping Dawn), Emest Hetherington (C. Mighty Sunset x C. Chironla), Globe Trader (C. Globetrotter x C. Richard Tauber), Gustav Mehlquist (C. Kabuki Legend x C. Thanksgiving), Happy Alex (C. Happy Promenade x C. Alexandra Rose), Hebert Geonge Alexander (*C. parishii* x *C. tracyanum*), Honey Pearl (C. Sleeping Beauty ´ C. Supreme Destiny), Jungle King (C. Tender Love x C. Dream Girl), Laragold (C. Moongold x *C. wallara*) and Last Supper (C. Beaujolais x C. Sylvania) (Anon., 2000).

C. Scott's Sunrise 'Aurora' is a magnificent intermediate cymbidium. It is an attractive clone with perfect spike habit and a rich orange shade of flowers. C. Moriah 'Hindu' is a deep brown intermediate from the combination of C. Sweetheart with C. Little Redhead. It has proved useful as the parent of famous hybrid C. Arunta (C. Moriah 'Hindu' x C. Wyanga 'Elanora'). The Arunta grex has produced many outstanding clones, mostly in the brown to smoky shades, all of perfect shape and with flowering period extending from early winter to late spring. A selection of some of the top Aruntas includes : C. Arunta 'Tara', C. Arunta 'Claypan', C. Arunta 'Lovely Lips', C. Arunta 'Ebatajinga', C. Arunta 'Princess' and C. Arunta 'Amber'. Yellows in intermediate cymbidiums are rare and one reasonable clone that flowered was C. Gloria 'Lella', a combination of two intermediates. C. Marry Prinches 'Sunbeam' with C. Showgirl 'Julie Christie' (Monkhouse, 1984).

8.13.0 Cyprepedium L.

Cyprepediums, the true Lady's slippers, are widespread in NorthTemperate Zone with few species extending in warm areas such as Mexico and parts of tropical Himalayas. They are commonly terrestrial, rarely lithophytic or epiphytic. Stem erect, leaves folded, strongly ribbed. Flower solitary or in raceme borne on a scape on leafy stem, usually attractive and colourful. The sepals are spreading, either all free or with lateral petals united to form a synsepal behind the lip; the petals spreading are smaller than sepals. The stalkless lip is inflated, sac or pouch shaped. The short column bears fertile stamens on the sides (Hawkes, 1978).

Important species: *Cyprepedium calceolus, C. candidum, C. parviflorum, C. pubescens,* and *C. reginae.*

Sheviak (1992) reported that *Cyprepedium parviflorum* var. *pubescens,* western American plants similar to *C. parviflorum* var. *parviflorum,* presently indeterminable western forms of *C. parviflorum* and natural hybrids between *C. montanum* and *C. parviflorum.* The binomial *Cyprepedium* × *columbianum* is assigned for those plants intermediate between the 2 species. *Cyprepedium japonicum* was found in Korea Republic in mixed deciduous forest and on uprooted dead trees (Park and Kim, 1995).

Hybrids of *C. candidum* with *C. parviflorum* and *C. pubescens* have been named C. Andrewsii Fuller and C. Favillianum Curtis, respectively (Atwood, 1984). *Cypripedium ventricosum* was derived from the cross *C. calceolus* × C. macranthon (Knyazev *et al.,* 2000).

8.14.0 Dendrobium Sw.

Dendrobium is the second largest genus of *Orchidaceae* consisting of over 1200 species, mainly originate from tropical and subtropical Asia and northern Australia (Wang *et al.,* 2009). Genus *Dendrobium* is the second largest genus of *Orchidaceae* family in India. It consists of 103 species occurring in Eastern Himalayas, Western Himalayas, Peninsula, Andaman and Nicobar Island (Singh *et al.,* 2001). The most incredible species in this genus occur in Eastern Himalayas. The Eastern Himalayas harbour about 83 of 103 species of Dendrobium found in India. The orchids of this marvellous genus are very popular with amateurs for a very long period because of their large showy flowers. Plants are epiphytic with a great range of variation in vegetative as well as floral parts. Stems elongate or pseudobulbous, fleshly, angled, cylindrical, erect or variously shaped. Leaves one to several. Inflorescences terminal or lateral with one to many flowers in racemes. The colour ranges from white, through many shades of yellow to rose, purple and crimson. Sepals and petals usually equal in size. The lip, with various shapes, sizes and colours, is the most attractive part of the flower.

Important species : *D. aggregatum, D. amplum, D. bigibbum, D. bellatulum, D. bensoniae, D. canaliculatum, D. chrysanthum, D. chrysotoxum, D. densiflorum, D. devonianum, D. draconis, D. falconeri, D. farmeri, D. fimbriatum, D. gibsoni, D. gouldii, D. grantii, D. moschatum, D. nobile, D. parishii, D. phalaenopsis, D. sulcatum, D. thyrsiflorum, D. undulatum, etc.*

A new species, *Dendrobium prianganense* with white flowers, delicately scented was reported from Java and *Dendrobium moorei* and *D. agrostophyllum* from the natives of Australia. Dauncey and Cribb (1993) suggested that the name *Dendrobium purpureum* has been applied to distinct taxa. *Dendrobium shiraishii,* a new species from Irian Java, of vigorous growth and strikingly coloured flower was reported by Yukawa and Nishida (1992).

Hybrids and cultivars: There were 14,673 hybrids of Dendrobium registered with International Orchid Registration Authority, London till 2019. The majority of cut flower varieties have been developed in Thailand, Hawaii, Singapore, Indonesia

Genera of Important Orchids

Arachnis **sp.**

Ascocentrum **sp.**

Laelia gouldiana

Epidendrum **sp.**

Phalaenopsis amabilis

Paphiopedilum **sp.**

Aerides **sp.**

Bulbophyllum echinolabium

Oncidium sphacelatum

Ophrys apifera

Brassavola nodosa

Cymbidium iridioides

Genera of Important Orchids

Cattleya sp.

Dendrobium nobile

Calanthe triplicata

Vanda coerulea

Renanthera matutina

Broughtonia

Rhynchostylis retusa

Prosthechea cochleata

Phaius tankervilleae

Lycaste sp.

Anguloa sp.

Spathoglottis plicata

Genera of Important Orchids

***Masdevallia* sp.**

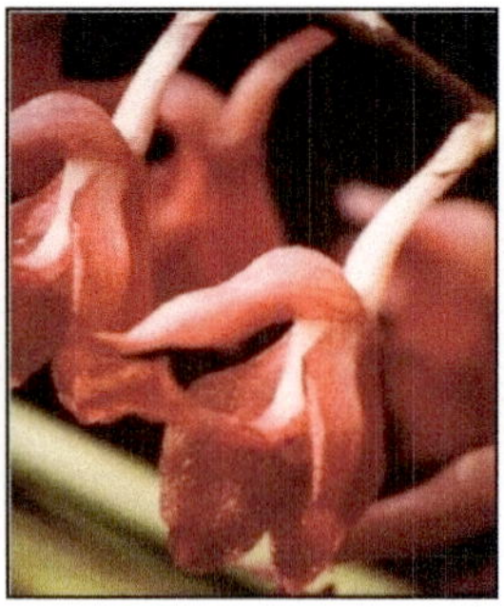

***Rodriguezia* sp.**

***Thunia* sp.**

and Malaysia. In Thailand, Sonia cultivars accounts for 70 per cent of total cut flower production, but In Hawaii, Hawaiian varieties developed by University of Hawaii account for 70 per cent of total cut flower production. Dendrobium cultivars grown in India include *Den* Bankok 'Blue', *Den* Big White 'Jumbo', *Den* Big White '4N', *Den,* Burana Stripe, *Den* Dang Saard, *Den* Earsakul, *Den* Emma White, *Den* Erika, *Den* Julie, *Den* Kating Dang, *Den* Lervia, *Den* Madame Pink, *Den* Thongchai Gold, *Den* Tripple Pink. *Den* Sonia-17, *Den* Sonia Red, *Den* Sirsophol, *Den* Somark White, *Den* Mona Red, *Den* Charming White, *Den* Nopporn Pink. In India, the highest acerage is under Sonia cultivars.

Hybrids : Several hundred hybrids are on record within the genus *Dendrobium*. However, many of the species used as parents in these hybrids have been misnamed and the state of *Dendrobium* hybrids in general is confused. *D.* Wiganiae (*D. nobile* x *D. signatum*), *D.* Ainsworthii (*D. aureum* x *D. nobile*) and *D.* Cybele Oakwood (*D. nobile* 'Burfordiense' x *D. findlayanum*) are few interspecific hybrids. *Dendrobium* Candy Stripe (*D.* Dianne Mc.Farlane x *D.* Margaret Joan Fall), *D.* Kultana (*D.* American Beauty x *D.* Gold Flush), *D.* Royal Velvet (*D.* Chao Phya River x *D.* Bangsaen Beauty), *D.* Tomie (*D.* Theodore Takiguchi x *D.* Jaquelyn Thomas), *D.* Youppadeewan (*D.* Jaq-Hawaii x *D.* Theodore), *D.* Golden Blossom (*D.* Golden Eagle x *D.* Dream), *D.* Honey Leen, (*D.* Miyakodori x *D.* Super Star), *D.* Oriental Beauty (*D.* Golden Blossom x *D.* Golden Eagle), and *D.* Sailor Boy (*D.* Fortune x *D.* Beautiful Egg) are some other attractive hybrids.

Orchid breeders from many countries are producing new hybrids every year and elite cultivars are released for commercial exploitation. Some of the recently produced hybrids include Cobber (*D.* Aussle Treat x *D.* Candice), Donovan (*D.* Elegant Heart x *D. kingianum*), Eddaz Cool (*D.* Edda x *D.* Peter), Florafest Indigo (*D.* Anne's Rainbow Suprise x *D. kingianum*), Green Elf (*D.* alexandrae x *D. convolutum*), Jeannie Woolf (*D.* Anne's Rainbow Surprise x *D.* Red River), Mary Stripes (*D.* Mary Mak x *D.* Candy Stripe), Maxy's Gold (*D.* Brolga x *D.* Colonial Bullion), Regal Gillieston (*D.* Peewee x *D.* Regal Affair), Rod's Eagie (*D.* Red Toro x *D. canaliculatum*), Sally Fiesta (*D.* Hawaiian velvet x *D.* Thailand), Sonobe Memorial (*D.* Pink Beauty x *D.* Yukidaruma), Sonobe Millenium (*D.* Pittero Gold x *D.*

Yukidaruma), Tsiku Pearl (*D.* Pearloysler x *D. lamellatum*) and Tsiku Romp (*D.* Banana Royal x *D. canaliculatum*).

8.15.0 Diacrium Bth.

This is a small genus of about six species native in the area from Guatemala to Trinidad and Venezuela. It is allied to *Epidendrum*. The plants are usually epiphytic, possess hollow pseudobulbs which bear leathery leaves at the upper end. The terminal inflorescences are often long and bear a few to many flowers, white or creamy white.

Species: *Diacrium bicornutum*, the magnificent 'Virgin Orchid' is one of the finest American orchids. *D. bilamellatum* has colourful flowers.

Hybrids: *Diacrium* has been crossed with several genera of the subtribe Laeliine. *Diacattleya* (× *Cattleya*), *Brassodiacrium* (× *Brassavola*), *Diabroughtonia* (× *Broughtonia*), *Dialaelia* (× *Laelia*), *Schombodiacrium* (× *Schomburgkia*), *Epidiacrium* (× *Epidendrum*), *Dialaeliocattleya* (× *Laelia* × *Cattleya*), *Shipmanara* (× *Broughtonia* × *Schomburgkia*), *Iwanagara* (× *Brassavola* × *Cattleya* × *Laelia*).

8.16.0 Epidendrum L.

The *Epidendrum* is among the largest genera of the Orchidaceae, the total number of species being estimated to be approximately one thousand. Plants of this genus are epiphytic, rarely terrestrials and characterized by highly varied forms. Some have stems, while others have a leafy segmented stalk bearing elongated, leathery leaves. With a few exceptions, the inflorescences are terminal. They may be simple or profusely branched, erect to somewhat pendulous, and bear from one to many flowers which are usually of medium to relatively large size. The perianth parts are spreading and the petals narrower than sepals. Lip is either simple or three-lobed and is smooth or furnished with varying types of callosities. The labellum is short to elongate and bears four equal pollinia in a terminal anther.

Important species: *Epidendrum alatum, E. anceps, E. arachnoglossum, E. ciliare, E. cochleatum, E. difforme, E. endresii, E. fragrans, E. ibaguense, E. parkinsonianum, E. polybulbon, E. prismatocarpum, E. radicans, E. radiatum, E. vitellinum, E. wallisii, E. xanthinum, etc.*

Hybrids: *Epidendrum* Endresio-Wallisii (*E. endresii* × *E. wallisii*), *E.* Burtoni (*E. ibaguense* × *E.* O'brienianum), *E.* Charlesworthii (*E.* Frederico-Guilielmi × *E. radicans*), E. Dellense (*E. radicans* × *E. xanthinum*), *etc.* are some of the attractive hybrids. *Epitonia* (× *Broughtonia*), *Epidiacrium* (× *Diacrium*), *Epigoa* (× *Domingoa*), *Epilaelia* (× *Laelia*), *Epilaeliopsis* (× *Laeliopsis*), *Epicattleya* (× *Cattleya*), *Epiphronitis* (× *Sophronitis*), *Epiphronitella* (× *Sophronitella*), *Schomboepidendrum* (× *Schomburgkia*), *Kirchara* (× *Cattleya* × *Laelia* × *Sophronitis*), *etc.* are the intergeneric hybrids.

Aster McMillan (*E. pseudepidendrum* × *C. luteola*) and Gail's True Love (*Lc.* Clayton Waglay × *E. alatum*) are *Epicattleya* and *Epilaelio cattleya* hybrids respectively.

8.17.0 Laelia Ldl.

This genus comprises epiphytic orchids related to those of the genus *Cattleya*.

Plants have long, fleshy stems bearing one or sometimes two leaves at their tip. The inflorescences of several of the laelias may attain a height of 2 metres while others bear solitary flower on very short stalk. The flowers of the genus are diverse, many species bear large colourful blossoms. The blooms are large with conspicuous, cornet shaped, usually simple labellum.

Important species : *Laelia albida, L. anceps, L. autumnalis, L. cinnabarina, L. crispa, L. flava, L. furfuracea, L. grandis, L. harpophylla, L. lobata, L. pumila, L. purpurata, L. rubescens, L. speciosa, L. superbiens, L. tomentosa, L. undulata, L. xanthina, etc.*

Hybrids: Apart from the hybrids among the species and cutivars of *Laelia* a great deal of hybridization has been done with the members of this genus, the most notable resultant group being *Laeliocattleya* (*Laelia* × *Cattleya*) of which several hundred crosses are on record. Button Top (*Cattleya* Angelwalker × *Laelia* Lundii), Cornelia (*C. labiata* × *L. pumila*), Love Knot (*L. sincorana* × *C. walkeranal*), Wrigleyi (*L. anceps* × *C. bowringiana*), Cosmo-Puppinney (*Lc.* Puppy Love × *Lc.* Irene Finney), Heidi Kirkpatrick (*Lc.* Mini Purple × *Lc.* Mariner), Highland Treat (*Lc.* Trick or Treat × *L. albida*), Irma Hottinger (*Lc.* Mermoz × *C. percivaliana*), Magic Road (*Lc.* Tokyo Magic × *C. loddigesii*), Memoria Mildred Byrd (*Lc.* Melody Fair × *C. walkeriana*), Samba Crown (*C. warneri* × *L. sincorana*), Tropical Fashion (*Lc.* Colorama × *C. aurantiaca*) and Tsikum Akakallis (*Lc.* Angel Love × *Lc.* Jungle Elf) *etc.* are few examples of *Laeliocattleya*.

The laelias freely interbreed among themselves and *L.* Canariensis 'Golden Glow' (*L. anceps ´ L. harpophylla*), *L.* Star Sky (*L. pumila ´ L. rupestris*), *L.* Zip 'Fireball', Mishima Bouquet (*L.* Mishima Purple × *L. dayana*), Mishima Purple (*L. purpurata* × *L. rubescens*), *etc.* are few hybrids of such crosses. They have also been successfully bred with many other genera such as *Brassavola, Broughtonia, Diacrium, Epidendrum, Schomburgkia* and *Sophronitis*.

8.18.0 Lycaste Ldl.

The genus *Lycaste* comprises about thirtyfive species of mostly epiphytic, (sometimes lithophytic or even semi-terrestrial) very handsome orchids. The stems are subovate, flattened and bear one or more leaves. Flowers grow singly from the base of the stalks and are fairly large with bent petals and sepals while the labellum is small. The lovely and highly valued *Lycaste virginalis* (usually grown as *L. skinneri*) is the National Flower of Guatemala.

Species: *Lycaste aromatica, L. candida, L. ciliata, L. cochleata, L. costata, L. cruenta, L. deppei, L. gigantea, L. lasioglossa, L. locusta, L macrobulbon, L. macrophylla, L. skinneri, L. tricolor, L. virginalis* are some important species. *Lycaste leucantha, L. lawrenciana, L. luminosa, L. dowiana, L. schillariana* and *L. xytriophora* were reported by Oakeley (1991a). *Lycaste aromatica*, its allies and their hybrids were described by Oakeley (1991b). He (1991c) also reported 16 species of *Lycaste* in 3 flower colour groups (dark green, greenish yellow to tawny orange and bicolours) and described many hybrids. Several yellow-flowered species and hybrids of *Lycaste* were also reported (Oakeley, 1991d).

Hybrids: *Lycaste* Balliae (*L. macrophylla* × *L. skinneri*), *L.* Koolena 'Excellency' (*L.* Auburn × *L. skinneri*), *L.* Brugense (*L. gigantea* × *L. skinneri*), *L.* Gratrixiae (*L. candida* × *L. macrophylla*), *L.* Auburn 'Meadow Mist', *L.* Auburn 'Black Knight', Sun Leo (*L.* Leo x *L.* Sunrise), Sunset Jaguar (L. ipala × L. Macama), Venus (*L.* Aphrodite × *L.* Jackle-Bear), White Christmass (*L.* Memoria Bill Congleton × *L. skinneri*) *etc.* are some of the beautiful *Lycaste* hybrids.

Three intergeneric crosses are on record, *Angulocaste* (× *Anguloa*), *Lycasteria* (× *Bifrenaria*) and *Zygocaste* (× *Zygopetalum*). Forty-seven hybrids of *Angulocaste* intergeneric hybrids between *Anguloa* and *Lycaste* were described by Oakeley (1991e).

8.19.0 Masdevallia R. and P.

Masdevallia is a unique genus comprising approximately 300 species. These orchids produce some of the most unusual and extraordinary flowers in the entire family. Plants vary in habit from epiphytic or lithophytic to terrestrial. They bear solitary, glossy-green narrow leaves on very short secondary stems. The inflorescences grow from the axils of the leaves and usually bear a single flower varying in size from rather small to gigantic. The sepals are relatively large, wide at the base and elongated at the tip with narrow appendages. The petals are always smaller and narrow and the labellum small and strap or mussel shaped.

Important species: *Masdevallia amabilis, M. bella, M. carderi, M. caudata, M. coccinea, M. cucullata, M. davisii, M. elephanticeps, M. erythrochaete, M. floribunda, M. ignea, M. leontoglossa, M. macrura, M. melanopus, M. muscosa, M. nycterina, M. pachyantha, M. peristeria, M. polysticta, M. racemosa, M. radiosa, M. rosea, M. simula, M. tovarensis, M. veitchiana, M. velifera, etc.* Leur (1991) described 8 new species of *Masdevallia* from South America. They are *M. indecora, M. dudleyi, M. eucharis, M. rubeola, M. sertula, M. zapate, M. anceps, M. ximense.* In 1995, three new *Masdevallia* species *M. papillosa, M. renzii and M. zongoensis* were also recorded by Leur (1995). Description and cultural recommendations of miniature masdevallias *M. citrinella, M. andreetana, M. rubifinosa, M. rodolfi, M. pteroglossa* and *M. persicina* were made by Webb (1988).

Hybrids: Several improved hybrids have been developed utilizing *M. caudata, M. coccinea, M. tovarensis, M. veitchiana* and other as parents *M.* Bocking Hybrid (*M. cucullata* × *M. veitchiana*), *M.* Chelsonii (*M. veitchiana* × *M. amabilis*), *M.* Courtauldiana (*M. rosea* × *M. caudata*), *M.* Gairiana (*M. veitchiana* × *M. davisii*), *M.* Hinksiana (*M. tovarensis* × *M. ignea*), *M.* Measuresiana (*M. amabilis* × *M. tovarensis*), *etc.* are some the popular hybrids. Some recent hybrids are Tenepahuay, Angel Frost, Calogras, Marguerite 'J and L'.

8.20.0 Miltonia Ldl.

The genus *Miltonia*, closely allied to *Odontoglossum, Oncidium* and *Brassia* consists of about 20 species and several natural hybrids. These epiphytic orchids bear short pseudobulbs each ending in two or more narrow, flexible leaves. The blossoms are borne singly or in clusters from the base of formed pseudobulbs. The most common colours found in this genus are rich dark red, soft pale pink and white, with a few pale yellow and every shade in between. Baker and Baker (1995) studied the growth habit, climate and cultural requirement of *Miltonia kayasimae, M. regnelli* and *M. russelliana*.

Important species: *Miltonia anceps, M. candida, M. spectabilis, M. stenoglossa, M. vexillaria, M. warscewiczii, etc.*

Hybrids: A large number of very complex hybrids have been raised in the genus *Miltonia*, mostly among the species inhabiting high elevations in the Andes of South America. Some of the outstanding hybrids are *Miltonia* Lycaena (*M.* Lord Lambourne × *M.* Princess Margaret), *M.* Solfatari, (*M.* St. Andre × *M.* Sanderiana), *M.* Leo Holguin (M. Hamburg × M. Franz Wichmann), M. Ned Nash (M. Lycaena × M. Franz Wichmann), *M.* Alderwood (*M.* Solfatari × *M.* Wych Cross), *M.* Golden Butterfly (*M.* Butterfly × *M.* Lixos), *M.* Edmonds (*M.* Solfatari × *M.* Woodlands), *M.* Evergreen Joy (*M.* Etendard × *M.* Edmonds), *M.* Cindy Kane (*M.* Violet × *M.* Woodlands), *M.* Mary Stewart (*M.* Violet × *M.* Hamburg), *M.* Bert Field (*M.* Mulatto Queen × *M.* Woodlands), Belle Hougue Point (*M.* Mount Bingham × *M.* Bel Royal), Burney Falls (*M.* Emotion × *M.* Jolene Carison), La Ponterrin (*M.* Maufant × *M.* Akagi), Maui Sunset (*M.* Mary Anderson Phillips × *M.* Linda Marie Sellon) and Max Cat Suki (*M.* Kay Rinaman × *M.* Goodnews Bay).

Chromosomal studies and parentage of 91 hybrid clones of *Miltonia* revealed that 28 per cent were diploids, 57 per cent were triploids and 15 per cent were tetraploid. The high proportion of triploids explains the difficulty with infertile clones which has been experienced by many breeders (Baker, 1995).

Multigeneric crosses are also very successful and many beautiful and floriferous hybrids have been produced. The common intergeneric hybrids are *Miltonidium* (× *Oncidium*), *Odontonia* (× *Odontoglossum*), *Miltonioda* (× *Cochlioda*), *Miltassia* (× *Brassia*) and *Milpilia* (× *Trichopilia*), *Symphyglossonia* (× *Symphyglossum*), *Charleswortheara* (× *Cochlioda* × *Oncidium*), *Colmanara* (x *Odontoglossum* × *Oncidium*), *Burrageara* (× *Cochlioda* × *Odontoglossum* × *Oncidium*).

8.21.0 Odontoglossum HBK

Odontoglossums, among the loveliest of orchids, are very popular and cultivated widely in many parts of the world. A polymorphic genus containing about three hundred species, is allied to *Oncidium, Brassia, etc.* Plants are epiphytic, lithophytic or rarely terrestrial having smooth, bright green, oval to round pseudobulbs compressed at the sides. From the top of these pseudobulbs are carried, one to three dark green leaves with two shorter leaves embracing the base of the pseudobulb. Erect or arching inflorescences (few species with pendulous spike) appear between the bulb and one of the basal leaves. Flowers are generally large, showy, with spreading sepals and petals having complex lip variable in colour and markings. White, yellow and greenish are the most frequent hues. Lip base parallel to the column with its outer lobe spreading downward, thus differing from *Oncidium*. In most species, complex crests, lamellae, teeth, *etc.* on the disc of lip are present. Column long, slender.

Important species: *Odontoglossum bictoniense, O. cervantesii, O. cirrhosum, O. citrosmum, O. cordatum, O. crispum, O. grande, O. harryanum, O. insleayi, O. lindleyanum, O. luteo-purpureum, O. maculatum, O. nobile, O. rossii, O. stellatum, O. triumphans, etc.*

Hybrids: Several thousand hybrids in this genus are known and the number continues to increase. A few important natural hybrids are *Odontoglossum* Adrianae (*O. crispum* × *O. hunnewellianum*), *O.* × Aspersum (*O. maculatum* × *O. rossii*), *O.* Brandtii (*O. luteo-purpureum* × *O. nobile*), *O.* Elgans (*O. cirrhosum* × *O. hallii*), *O.* Humeanum (*O. rossii* × *O. cordatum*) and *O.* Wilckeanum (*O. crispum* × *O. luteo-purpureum*). Carpenter (1988) discussed hybrids of *Odontoglossum bictoniense*.

White hybrids: *Odontoglossum* Alispum (*O.*Alorcus × *O. crispum*), *O.* Crispelus (*O.* Crispum × *O.* Belus), *O.* Princess Elizabeth (*O.* Faustina x *O. crispum*).

Reddish hybrids: *Odontoglossum* Eudora (*O.* Dictune x *O.* Serapis), *O.* Maronius (*O.* Clemius x *O.* Marcella), *O.* Himachal (*O.* Cloamel × *O.* Eudora).

Yellow hybrids: *Odontoglossum* Crispania (*O. crispum* × *O.* Ascania), *O.* Elise (*O. triumphans* × *O.* Ascania).

Rose or purple hybrids: *Odontoglossum* Ophoon (*O.* Ophelia × *O.* Neron), *O.* Nisus (*O.* Regium × *O.* Toreador), *O.* Perryanum (*O.* Camilla × *O.* Toreador).

Anneliese Rothenberger, Elle's Triumph 'Dahlenburg' are two outstanding hybrids of *Odontoglossum*. *Odontoglossum* Groganiae, a hybrid between *O. edwardii* and *O. uro-skinneri*, registered in 1908 is potentially useful in current breeding programme (Easton, 1992).

Most of the species comprising the genus have also been found to be readily crossable with many other members of the subtribe Oncidiinae to give rise to such artificial bigeneric hybrids as *Odontioda* (× *Cochlioda*), *Odontocidium* (× *Oncidium*), *Odon-tonia* (× *Miltonia*), *Odontobrassia* (× *Brassia*), *Adaglossum* (× *Ada*), *Symphodontoglossum* (× *Symphyglossum*). Among the tri- and quadri-generic hybrids are *Beallara* (*Odontoglossum* × *Brassia* × *Cochlioda* × *Miltonia*), *Burrageara* (*Cochlioda* x *Miltonia* x *Odontoglossum* x *Oncidium*), *Sanderara* (*Brassia* × *Cochlioda* × *Odontoglossum*), *Wilsonara* (*Cochlioda* × *Odontoglossum* × *Oncidium*), *Colmanara* (*Miltonia* × *Odontoglossum* × *Oncidium*), *Lagerara* (*Odontoglossum* × *Aspasia* × *Cochlioda*) and *Vuylstekeara* (*Cochlioda* × *Miltonia* × *Odontoglossum*).

8.22.0 Oncidium Sw.

A large genus contains about seven hundred fifty species, most of which grow as epiphytes, a few are terrestrial or lithophytes. *Oncidiums* are Western Hemisphere orchids found in habitats ranging from sea level in the tropics to high elevations in the Andes Mountains. They are commonly known as "dancing ladies," due to the resemblance of the broad, flat lips of their flowers to ruffled skirts. The genus is closely allied to *Brassia, Odontoglossum, Miltonia* and many other members of the subtribe Oncidiinae. Vegetative and flowering habit is extremely variable. Erect or sharply pendulous plants with conspicuous, compressed pseudobulbs, though some species are without pseudobulbs. Leaves are fleshy, oval and pointed with single prominent mid-vein, or thinner, and many veined, or terete. Inflorescences are long, sometimes drooping, and often erect. In some species, these inflorescences continue to grow and produce flowers for several years. These flowers are most commonly yellow and brown, while some species are pink, red, magenta, green or white, many with exotic markings. Sepals and petals are usually subequal. The lip in most oncidium is 3-lobed to a greater or lesser degree, often fiddle-shaped, clawed or sessile. The disc on the mid-lobe bears conspicuous crests or tubercles, which add much to their charm. Column is winged and anther bears a pair of waxy pollinia.

Important species: *Oncidium ampliatum, O. auriferum, O. cavendishiaum, O. cheirophorum, O. crispum, O. divaricatum, O. forbesii, O. lanceanum, O. macranthum, O. marshallianum, O. papilio, O. sarcodes, O. varicosum, O. variegatum, etc. Oncidium dowiana* and *O. kramerianum*, two endangered species from Costa Rica have been described by Synder (1992).

Hybrids: A large number of spectacular hybrids have been registered in *Oncidium*. Some of the natural hybrids are *Oncidium* Mantinii (*O. forbesii* × *O. marshallianum*), *O.* Punctatum (*O. forbesii* × *O. gardneri*), *O.* Haematochilum (*O. lanceanum* × *O. luridum*) and *O.* Wheatleyanum (*O. crispum* × *O. dasytyle*). Interspecific artificial hybrids include *Oncidium* Burzeffianum (*O. marshallianum* × *O. varicosum* var. *rogersii*), *O.* Kaiulani (*O. flexuosum* × *O. ornithorynchum*), *O.* Vizer (*O. ampliatum* × *O. papilio*), *etc.*

Some hybrids of *Oncidium* registered in 2000 are Tsiku Vanessa (*O.* Cathenne Wilson × *O.* Barbie), Tsikum Verbena (*O.* Hamana Elfin × *O. flexuosum*), Volcano Gold (*O.* Gower Ramsey × *O.* Killer Bees), Candy Clown (*O.* Cotton Candy × *O. triquetrum*), Crimson Vell (*O.* Snow Shower × *O.* Regina), Golden Sunshine (*O.* Little Sunshine × *O.* Golden Sunset) and Growers Choice (*O.* Gower Ramsey × *O.* gravesianum).

Oncidium has also been hybridized with a number of allied genera to produce hybrids like *Trichocidium* (× *Trichocentrum*), *Rodricidium* (× *Rodriguezia*), *Brassidium* (× *Brassia*), *Miltonidium* (× *Miltonia*), *Oncidioda* (× *Cochlioda*), *Charlesworthеara* (× *Cochlioda* × *Miltonia*).

Folksy 'Misty', Mid as 'Willow Pond', Esther, Irina Muray 'Orchidglade', Golden Sunset 'Robsan II', Potpourii 'Scarlet' are some of the spectacular hybrids.

Charansari and Kamemoto (1978) made interspecific cross pollinations between 11 *Oncidium* species in two geographical sections and observed good compatibility and homology of chromosomes, which indicated a close cytogenetic relationship between the two sections despite their disparity in leaf structure and variation in chromosome number.

8.23.0 Paphiopedilum Pfitz.

The genus *Paphiopedilum* consists of 69 species. These are grouped in three subgenera, namely *Parviflorum, Brachypetalum* and *Paphiopedilum*. The subgenus *Paphiopedilum* consists of the largest number, 60 species, followed by section *Pravisepalum* and *Bracypetalum*, 5 and 5 species, respectively. The section *Paphiopedilum* further divided in five sections, namely *Corypetalum, Pardalopetalum, Paphiopedilum,* and *Barbatulum*. The section *Barbata* has the largest number, 27 species followed by section *Paphiopedilum, Coryopedilum, Brachypetalum* and *Paradopetalum*, 13, 11, 5 and 4 species, respectively (Cribb, 1998). The majority of species are terrestrial or lithophytic, and only a few species are epiphytic. Eight species are reported to occur within political boundaries of India. Of them one species *P. druryi* is endemic to India. All the species found in India bear showy flowers and are suitable for cultivation as a pot plant. The Indian species has also been extensively utilized in the breeding of new cultivars. *Paphiopedilum fairrieanum* is an ornamental plant in high demand and is extensively collected for commercial use for horticulture, domestic and international trade. This species was in very high demand after it was discovered in 1904 it became so rare that a reward of 1,000 pounds was announced for anyone who could find the plant (Cribb, 1998). There are a few reports on ethno-botanical utilization of *P. insigne* as a medicine in curing stomach ailments (Friesen and Friesen, 2012). *Paphiopedilum venustum* is an ornamental plant in high demand and is extensively collected for commercial use for horticulture, domestic and international trade. Because of the beautiful attractive flowers, Indian government released a postal stamp on this name. *Paphiopedilum villosum* is an ornamental plant in high demand and is extensively collected for commercial use for horticulture, domestic and international trade. Because of the gorgeous flowers, Indian government released a postal stamp on this name. The major characterstics of these species with their contribution in evolution of new cutivars are described below:

Important species: *Paphiopedilum bellatulum, P. callosum, O. concolor, P. fairrieanum, P. hirsutissimum, P. insigne, P. lawrenceanum, P. niveum, P. parishii, P. spicerianum, P. venustum, P. villosum, etc. Paphiopedilum barbigerum* var. *aureum*, a yellow- flowered form discovered in Malipo, southern Yunnan, China, was reported by Hua (1999).

An uncommon species, *Paphiopedilum exul*, endemic to the islands on the West Coast of Thailand was reported by Cribb and Robbins (1993). *Paphiopedilum serpenti* and *P. labukensis* were added later (Christenson, 1995). The important species for producing white flowers in paphiopedilums are *P. niveum, P. bellatulum, P. godfroyae* and *P. insigne* var. *sanderae* (Rogerson, 1991).

Paphiopedilum druryi (Bedd.) Pfitzer: It is an endemic to southern India in Kerala and Tamil Nadu (Cribb, 1998; Chowdhery, 2004). It grows in xerophytic habitat on southeastern and southwestern facing slopes in thin sandy loam (pH 4.6 – 5.5) in montane grassland with scattered shrubs and small trees, on weathered rocks. It can tolerate full sunshine during March and April in flowering season. This species confined to very local and scarce with a restricted distribution at 1,400-1,550 m asl in the Agastamalaya Hills. This species is always found growing in association with the endemic grass *Zenkeria sebastineii.* The medium-sized, cool to warm growing and single-flowered species has high demand in cultivation for a pot plant. The flower is borne on an erect, 25-30 cm long pubescent, purple, terete inflorescence. Sepal and petal are greenish-yellow to yellow with purple medial stripe and pale yellow lip. This species has also successfully utilized in the breeding of Paphiopedilum hybrids. So far, this species has been used to produce 84 first-generation hybrids (either of the parents is species) including 47 primary hybrids (both the parents are species). In the development of first-generation hybrids, 46 times the species has been used as a female parent and 37 times as a male parent.

***P. fairrieanum* (Lindley) Stein:** It is commonly called as Fairrie's *Paphiopedilum* and occurs in the eastern Himalayas to Assam. The species found in India (Sikkim - Tinkitam, Arunachal Pradesh - Kameng) and Bhutan (White, 1909; Pradhan, 1996; Cribb, 1998; Pearce and Cribb 2002; Lucksom 2007; Misra 2007). *P. fairrieanum* is dwarf, cool to cold growing terrestrial or lithophytic species that grow on limestone outcrops. They bear single flower on a pubescent, 25-35 cm long, purple pubescent inflorescence. The flowers are with broad, white dorsal sepal stripped and suffused with purple, petals deflexed wavy at margins, and lip tan dotted with rusty dots. Overall the flowers are charming and flower during late to spring. So far 380 first-generation hybrids including 66 primary hybrids have been evolved using this species. The important primary hybrids developed using the *P. fairrieanum* are Vexillarium (*P. fairrieanum* x *P. insigne*), Niobe (*P. fairrieanum* x *P. spicerianum*), and Arthurianum (*P. barbatum* x *P. fairrieanum*).

***Paphiopedilum hirsutissimum* (Lindl. ex Hook. f.) Stein:** Medium-sized, warm to cool growing terrestrial or lithophytic species distributed in India (Scattered distribution in Nagaland, Manipur, Mizoram and Meghalaya), China, Myanmar, Thailand, Laos and Vietnam. This species is a facultative epiphyte. It occurs in hilly forest tracts and humus-rich soil, at altitudes ranging from 200 to 1800 m. As it has beautiful flowers, it is clandestinely collected in bulk for its commercial trade. Loss of its natural habitat has affected its natural population (Jain and Sastry, 1984; Nayar and Sastry, 1990). One hundred thirty-nine first-generation hybrids including 59 primary hybrids have been produced using this species. So far 52 times used as the female parent and 87 times as a male parent. Its primary hybrids Ceres (*P. hirsutissimum* x *P. spicerianum*), Godseffianum (*P. villosum* var Boxallii x *P. hirsutissimum*) and Germinyanum (*P. hirsutissimum* x *P. villosum*) have been used developing 9, 8, and 7 hybrids, respectively.

***Paphiopedilum insigne* (Lindley) Pfitz :** Small to medium-sized, cool growing species habited in small pockets of Northeast India, China, Myanmar, Thailand; Laos and Vietnam (Chowdhery, 2004; Tandon and Kumaria, 2011; Diengdoh *et al.*, 2017). Once this species was plentiful in Meghalaya, but due to over-collection and loss of habitat, it has become rare (Jain and Sastry, 1984). In nature, it grows on limestone outcroppings at an elevation of 1000 to 2000 meters Strapped leaved plant bears single, rarely two on flowers dark green-brown, purple-pubescent inflorescence. The species extensively used in the breeding of Paphiopedilum hybrids. So far, 289 first-generation hybrids including 84 primary hybrids have been developed using this species. A primary hybrid, Leeanum (*P. insigne* x *P. spicerianum*) has produced 218 first-generation hybrids. The four progenies of Leeanum, Alcibiades, Actaeus, Christopher and Hara contributed significantly in the development of modern Paphiopedilum hybrids. Another primary hybrid, Niten, (*P. insigne* x *P. villosum*) has produced 125 first-generation hybrids. The worthy progenies of this cross are Aureum, Lady Dilton, Earl of Tankerville and Redstart.

***Paphiopedilum spicerianum* (Rchb.f)Pfitz. :** Small-sized, cool growing species, terrestrial species distributed in India, Bhutan, Myanmar and south China. Although this orchid is assessed as Endangered (EN) globally on the IUCN Red List (Rankou and Kumar, 2015), and has been heavily harvested, because of its horticultural value. The beauty of this species lies in curled white dorsal sepal. The species is well suited for growing as a pot plant. This species has considerably used in hybridization programme. Two hundred and thirty-three first-generation hybrids including 111 primary hybrids have been evolved using this species. This species has contributed to the genomes of 18,717 paphiopedilum hybrids. Its primary hybrid Leeanum (*P. insigne* x *P spicerianum*) has already been discussed under P. insigne. The important primary hybrids of this species are Lathamianum (*P. spicerianum* x *P. villosum*) is involved in the development of 66 first-generation hybrids. It hybrid Cardinal Mercier' has produced many hybrids. Primary hybrid Niobe (P. fairrieanum x P. spicerianum) also has produced 35 first-generation hybrids.

***Paphiopedilum venustum* (Wall. ex Sims) Pfitz. :** It is commonly called as *Charming Paphiopedilum*. This species distributed in India (Assam, Meghalaya, Sikkim), Nepal, Bangladesh, Bhutan and China (Pradhan, 1975; Pearce and Cribb 2002; Chowdhery, 2004; Lucksom, 2007; Misra 2007; Raskoti and Ale, 2011). It grows in the wettest place and found in a variety of habitats, including on cliffs above streams, shaded humus-rich gullies or steep loamy cliffs overgrown with grasses and deep humus at the foot of trees (Cribb 1998; Liu, *et al.*, 2009). Using *P. venustum* 131 first-generation hybrids including 64 primary hybrids have been evolved. The primary hybrid Double Deception (*P. venustum* x *P. sukhkulii*) gave 22 first-generation hybrids. The other primary hybrids of this species Auroreum (*P. lawrenceanum* x *P. venustum*) has produced 23 first-generation hybrids.

***Paphiopedilum villosum* (Lindl.) Pfitzer:** Large-sized, cool to cold growing, epiphytic rarely lithophyte occurs in northeast India (Mizoram, Meghalaya, Assam), China and Indochina Cambodia, Laos, Myanmar, Vietnam and Thailand. The species is found between 1,300 - 2,200 masl (Cribb 1998; Chowdhery 2004; Lucksom, 2007; Misra, 2007). Many varieties were described under this species, some of which are accepted varieties. *Paphiopedilum villosum* var. villosum has been recorded from India, China, Laos, Myanmar, Thailand, and Vietnam. *P. villosum* var. annamensis has been recorded from China Laos, and Vietnam, *P. villosum* var. boxallii found in China, Myanmar and Vietnam and found growing in epiphytic conditions on trees in humid evergreen forests mainly on silicate rock forests (Averyanov, 2008; Liu *et al.*, 2009). *P. villosum* var. densissimum has been recorded from China and grows in forests or thickets or on cliffs along dry and hot valleys at an elevation of *c.* 800-900 masl (Liu *et al.*, 2009). There are 281 first-generation hybrids including 167 primary hybrids have been evolved from this species. The primary hybrid Harrisianum (*P. insigne* x P. *villosum*), Niten (*P. barbatulum* x *P. villosum*), and Lathamianum (*P. spicerianum* x *P. villosum*) have contributed significantly in the breeding of Paphiopedilum orchids.

***Paphiopedilum wardii* Summerhayes:** Small-sized, mottled leaved, cool growing terrestrial species found in India- Arunachal Pradesh (Lohit District), SW China, and Northern Myanmar. It grows on rock surfaces covered by mosses, leaf litter, and earth banks more or less exposed to the sun and at altitudes between 1200 and 1500 m. It is highly exploited for its showy flowers. Nothing is known about the pollinator, population size and dynamics of this species. There are 88 first generation hybrids including 37 primary hybrids. This species has been used 41 times as female parent and 47 times as male parent.

Hybrids : Thousands of hybrids have been raised in the genus *Paphiopedilum*, using the species *P. villosum, P. insigne, P. spicerianum, P. niveum, P. bellatulum, P. hirsutissimum* and others. These hybrids are, however, almost universally known under the incorrect and very confusing name of *Cypripedium*. Some of the hybrids are *Paphiopedilum* Leeanum (*P. insigne* x *P. spicerianum*), *P.* Arthurianum (*P. insigne* x *P. fairrieanum*), *P.* Harrisianum (*P. villosum* x *P. barbatum*), *P.* Aylingii (*P. niveum* x *P. ciliolare*), *P.* Demura (*P. bellatulum* x *P.* Blendia), *P.* Charles Richman (*P. bellatulum* x *P. barbatum*), *P.* Lathamianum (*P. villosum* x *P. spicerianum*), *P.* Winnifred Hollington (*P. niveum* x *P. callosum*), *etc.*

Two bigeneric hybrids under the name of *Selenocypripedium* have also been registered. These are actually hybrids between *Paphiopedilum* and *Phragmipedium* and should be known under the name of *Phragmipaphiopedilum* (Hawkes, 1978).

The maudiae-type slipper orchids have retained an important place among the orchid growers. The Maudiae types are a group of slipper orchids having large dorsal sepals, narrow petals and a ventral synsepalum which is usually insignificant. Shape differs from standard hybrids considerably, especially with regard to petal width and stance.

Exceptionally dark forms of the species, *P. callosum*, have been evolved and they led to a new race of novelty paphiopedilums _ the vinicolours. Vinicolours have deep intensity of colour in the dorsal sepal and an exceptionally dark coloured pouch, as well. The colouration of the pouch is uniformly blackish and does not fade appreciably at the tip or in the rear. The one common feature of vinicolours seen to date is that they have a dark *P. callosum* as one of their parents. Several different types of vinicoloured callosums are known. There is the strain bred from *P. callosum* 'Sparkling Burgundy' and that from *P. callosum* 'JAC'. The 'Sparkling Burgundy' clone was selfed and several of its offsprings have been used for breeding dark coloured hybrids (Koopowitz and Hasegawa, 1984).

Perhaps the most desirable of the first generation vinicoloured crosses produced were *Paphiopedilum* Maudiae and *P.* Goultenianum. These were made with *P. callosum* onto *P. lawrenceanum* and *P. curtisii*, respectively. They tend to have large flowers with good uniform dark colouring in good clones. *P.* Vinter's Treasure, a cross between *P.* Kowloon (which is *P.* New Era x *P. purpuratum*) and *P. callosum* was one of the more productive crosses but is the most variable both in shape and size. *P.* Gloriosum, a cross made between *P.* Holdenii, (*P.* Maudiae x *P. callosum*) and *P. callosum* 'Bion', has produced among its progeny pale colouratums as well as flowers that are darkly vinicoloured. Several other vinicolours were produced in the original group of hybrids and included *P.* Tesselatum (*P. barbatum* x *P. concolor*) x *P. callosum*; *P.* Le Douxiae (*P.* Harrisonianum x *P. callosum*); *P.* Wottonii (*P. bellatulum* x *P. callosum*) (Koopowitz and Hasegawa, 1984).

Some of the bybrids and their parents registered by orchid breeders are Bob Nagel (*P.* Mystic Jewel x *P.* Gowerianum), Bridget Ring Lawless (*P. concolor* x *P.*

Procal), Compact Sioux (*P*. Compton x *P*. Original Sioux), Dazzland (*P*. Dazzler x *P*. Troyland), Dragon Church (*P*. Tree of Dragon x *P*. Winston Churchill), Dragon Quest (*P*. Gibbert Blythe x *P*. Double Deception), Ed Lawless (V. Macabre x *P*. Iantha Stage), Love Yokohama (*P*. Transvaal x *P*. Leo Organo), Lucinda Lachelin (*P*. Faire-Maud x *P. lawrenceanum*), Magiclight (*P*. Magic Mood x *P*. Nulight) and Magic Oak (*P*. Magic Mood x *P*. Green Oak) (Anon., 2000, 2001).

About twenty-eight thousand hybrids have been registered with International Orchid Registration Authority, RHS, London. They are primary hybrids (hybrids developed through interspecific crossing), secondary hybrids (where either of parents is species) and complex hybrids (crossing between the cultivars). All the hybrids registered so far are not under cultivation. Many of them are lost, but new hybrids are being developed and registered every year. In the last five years, they have been registered on an average rate of 550 hybrids per year. Proper care must be taken while selecting the species or hybrids for the successful cultivation. There are thousands of beautiful commercial orchids belonging to different climatic zones, and all of them cannot be grown in a particular climatic condition. So, the choice of cultivar or species depends upon prevailing climatic conditions. Healthy and free from pests and disease plants should be selected. Evaluation of species and hybrids for their performance in a particular area and selection of promising ones are very important for the exploitable commercial orchids. Paphiopedilum hybrids are available in a range of colours including white, yellow, green, sunset tones, red and spotted. Some of hybrids are described below:

***Paph.* Silent Knight:** This is a cross of Paph Greenvale and Paph White Knight. It was originated by Orchid Zone Ltd, USA and registered by Paphanatics Ltd, the USA in 2003. The grex has several superior clones. The clones 'Snow Snow' has large pure white flowers on a strong stem. It also has potential as a good parent.

***Paph.* Cocoa Green:** Cocoa Green was evolved by crossing *Paph.* Hamana and *Paph.* Yi-Ting Sring Mist. This cross was originated and registered by Cocoa Paphiopedilum Farms, Taiwan in 2011. This hybrid produces solid green, long-lasting flowers through the fall and winter months. It is easy to grow and bloom.

***Paph.* Mountain Lass:** Mountain Lass was originated and registered by Lehua Orchids, Hawaii by crossing *Paph.* Mountain Meadow and *Paph. helene* in 2019. It is a beautiful compact growing complex Paphiopedilum hybrid. Due to the influence of *Paph. helene*, the flower colour ranges from bright green to bronze tones.

***Paph.* In-Charm Topaz:** It is a small and compact hybrid. It is cross of *Paphiopedilum helene* and *Paph.* Pacific Shamrock. It was originated and registered by In-Charm Nursery in 2006. There are several superior clones of this cross. It produced beautiful, glossy yellow flowers in the fall and winter months.

***Paph.* Olympic Moon:** Paph. Olympic Moon is cross of Paph Warm Valley and Paph. Olympic Halo. It was originated and registered by Theresa J. Jill, the USA in 2018.

***Paph.* Yerba Ice:** Yerba Ice is cross *Paph.* Yerba Buena and *Paph.* Ice Age. It was originated and registered by David Sorokowsky, the USA in 2018. It is a vigorous and colourful, complex hybrid. The flowers have ranged from spots to pale pinks, greens and yellows.

***Paph.* Snow Tiger:** It is a cross of *Paph.* Frienship and *Paph.* Memoria Toshio Miyata. The cross was originated by Fukashi Tsuchiya, Japan and registered by Seji Matsuzawa, Japan in 1987. The clone 'Legacy' is good for breeding for pink colour breeding.

***Paph.* Tomiko Moon:** Tomiko Moon was evolved by crossing *Paph.* Moon Stone and *Paph.* Tomatillo. It originated and registered by Lehua Orchids, the USA in 2018. It is a beautiful complex Paphiopedilum hybrid with a little sequential influence. It is late blooming and colour range from white through yellow-green.

***Paph.* Ruby Cabochon:** This hybrid was evolved by crossing *Paph.* Bloodline and *Paph.* John Hainesworth. It was originated and registred by Theresa J Hill, the USA in 2018. It is a complex hybrid and blooms during winters.

***Paph.* Tommy Egger:** Paph. Tommy Egger is cross of *Paph.* Cruella and Paph. Don Egger. It was originated and registered by Theresa J Hills, USA in 2018. It is a complex hybrid and blooms in the winter.

***Paph.* Buddy Fun Mark:** *Paph.* Buddy Fun Mark was evolved by crossing *Paph.* Cruella and *Paph.* Spotglen by Theresa J Hills, the USA in 2018. M Buddy Mark, USA registered the cross.

***Paph.* Payakka Kodkod:** Payakka Kodkod is an interspecific hybrid between *P. concolor* and *P. thaianum.* It is a true miniature Paphiopedilum. The flowers that are white and fragrant with no yellow background colour.

***Paph.* Clair de Lune :** It is a hybrid between Paph. Emerald and Paph Alma Gevaert. It has three species *P. lawrenceanum* (50 per cent), *P. curtisii* (25 per cent) and *P. callosum* (25 per cent) in its background. The original Clair de Lune registered in 1927 by Sander's of England.

8.24.0 Paraphalaenopsis A.D. Hawkes

This genus has been erected by A.D. Hawkes in 1963 to accommodate three rare and attractive orchids which have been considered aberrant species of *Phalaenopsis*. These epiphytes are endemics in western Borneo (Indonesia). Leaves are terete. Flowers are of heavy-textured, variable in size and colour, lip 3-lobed.

Species: *Paraphalaenopsis denevei* and *P. laycockii* are the two important species.

Hybrids: When crossed with *Vanda, Paraphalaenopsis* gives *Paravanda*, with *Arachnis, Pararachnis*, and with *Renanthera, Pararenanthera*.

8.25.0 Phaius Lour.

This genus consists of approximately 30 species of large and showy terrestrial orchids. Pseudobulbs are elongated and leaves large, folded and deciduous. The inflorescence is lateral on the pseudobulbs and typically erect. Many flowers usually open successively and the plants remain in bloom for a long period of time. Sepals and petals are large, subequal, spreading or suberect. Lip is slightly joined to the column, erect, tri-lobed and spurred at the base; the tubular lip is rather characteristic of the blossoms. Column is long, stout or slender without any foot. Pollinia are 8 in two groups of 4.

Species: *Phaius amboinensis, P. callosus, P. flavus* (*P. maculatus*), *P. longipes, P. pauciflorus, P. tankervilliae* (*P. grandifolius, P. wallichii, P. blumei*), *P. tuberculosus.*

Hybrids: Phaius has widely been utilized by the breeders and the intergeneric hybrids resistered are *Phaiocalanthe* (× *Calanthe*), *Phaiocymbidium* (× *Cymbidium*), *Gastophaius* (× *Gastorchis*) and *Spathophaius* (× *Spathoglottis*). Colourful hybrids within the genus are also recorded, most of them having been produced many years back. Some of these are *P.* Cooksonii (*P. wallichii* × *P. tuberculosus*), *P.* Marthae (*P. blumei* × *P. tuberculosus*), *P.* Norman (*P. blumei* var. *sanderianus* × *P. tuberculosus*), *P.* Phoebe (*P. blumei* var. *sanderianus* × *P. humblotti*).

8.26.0 Phalaenopsis Bl.

The fabulous 'Moth orchids'; commonly called as members of *Phalaenopsis* are very popular to the amateur and commercial growers. *Phalaenopsis* originate in tropical Asia and the South Pacific. They are the most common type of orchid sold today and the easiest to grow in the home. They have large, broad leaves and long flower stems with blossoms of white, yellow, green, pink, and red. They require constant watering, as they have no major water-storage organ other than their leaves. They are grown in large quantities for cut-flower trade. These fine orchids are short-stemmed epiphytes with a few leaves at the base. Inflorescences are short or elongate, simple or branched, drooping to erect and appear from the axil of the leaves. Flowers are one to numerous, long lasting and with similar sepals and petals. Their colour varies but in the most frequently cultivated species is either white (with some yellow markings on the lip) or different shades of pink. The lip is joined to the column foot, spurless and more or less three-lobed. Pollinia are 2, somewhat round and divided.

Important species: *Phalaenopsis amabilis, P. fuscata, P. mannii, P. parishii, P. schilleriana, P. speciosa, P. stuartiana, P. violacea.*

Hybrids: Many hybrids have been raised within the genus *Phalaenopsis* and some of the beautiful and popular ones are *P.* Riea (*P. amabilis* × *P. formosana*), *P.* Ann Marie (*P. amabilis* × *P. formosana*), *P.* Rothschildiana (*P. schileriana* × *P. amabilis*), *P.* Reve Rose (*P. algers* × *P. schilleriana*). Several *Phalaenopsis* species have also been successfully crossed with allied genera giving rise to such groups as *Vandaenopsis* (× *Vanda*), *Doritaenopsis* (× *Doritis*), *Renanthopsis* (× *Renanthera*), *Arachnopsis* (× *Arachnis*), *Rhynchonopsis* (× *Rhynchostylis*), *Aeridopsis* (× *Aerides*), *Phalandopsis* (× *Vandopsis*), *Rhynodoropsis* (× *Doritis* × *Rhynchostylis*), *Tanakara* (× *Aerides* × *Vanda*).

Phalaenopsis gigantea has a very good form and very interesting patterning. A hybrid produced from *P. gigantea* is *P.* Tiger Butter which has a very beautiful yellow colour and interesting markings. *P.* Tiger Butter 'Larkin Valley' has fine form, good colour and good markings. This clone has been used in crosses with larger *Phalaenopsis* for some really exciting yellows. *P.* Natasha made by Irene Dobking, a cross of *P. gigantea* and *P. samba* is a rather difficult breeder. However, *P. gigantea* and *P.* Star of Rio which was made by Irene Dobkin and named by her as *P.* Eye Dee, has been used in many crosses. *P.* Eye Dee crossed with *P. sumatrana* produced red patterning. *P.* Eye Dee with Golden Pride has also some very beautiful patterning. *Doritaenopsis* (*Dtps.*) George Molar 'Golden Shower' was made by Charles Beard in Florida and was perhaps one of his famous hybrids. *Dtps.* Jerry Sue King with *P. violacae* is a cross made by John Miller. A cross of *P.* Encantadora and *P.* Spotted Moon is sort of rusty yellow. *Dtps.* George Moler with *P.* Career Girl has yellow sepals and petals from the *P.* George Moler and the bright red lip from. *P.* Career Girl. *P.* Golden Liz is a cross of *P.* Artienne and *P.* Golden Pride. This was a very uniform cross with many yellows and yellow-greens (Ewing, 1984). Brother Lady Jane (*Dtps.* Judy Valentine x Dtps. Taisuco Candystripe), Brother Lucky Strike (*Phal.* Brother Little Venus × *Dtps.* Minho Stripes), Janine Vcelka Kennedy (*Dtps.* Pretty Nice × *Phal.* Sandra Tullis Galaxy), John Vcelka (*Phal.* Chickasaw × *Dtps.* Taisuco Candystripe), Joie (*Phal.* Festive Mood × *Dtps.* Magenta Spray), Kaitly's Rose (*Phal.* Spring Slik × *Dtps.* King Shlang's Rose), Lavender Gem (*Dtps.* Inverness Firebrand × *Phal.* Charity Lace), Lavender Haze (*Dtps.* Foderaro × *Phal.* Shahid Ekbal), Moutain Sweetheart (*Dtps.* Ala Mode x *Phal.* Orglade's Hot Spot), Neapolitan Belle (*Dtps.* Silver Treasure × *Phal.* Richfest), Orchidom Bulls-Eye (*Dtps.* Memoria James McPherson × *Dtps.* Hamps hire Bulls-Eye), Orchidom Red Ups (*Phal.* Su's Red Lip × *Dtps.* Hampshire Bulls-Eye), Tai-Zhou Roseherz (*Dtps.* Lochia Beauty × *Dtps.* Luchla Hoseherz), Taida Valbeauty (*Dtps.* Talda Valentine × *Dtps.* Tinny Beauty), Talda Valrose (*Dtps.* Talda Valentine × *Dtps.* Happy Rose) and Talsuco Confidence (*Dtps.* New Candy × *Dtps.* Taisuco Newbeauty) are few reported *Dtps* cultivars (Anon., 2000, 2001).

Norton *et al.* (1995) have reported that the first 10 hybrids of *Phalaenopsis venosa* were registered in 1984 and since then 130 first generation and about 50 second generation and third generation hybrids have appeared using both green and yellow forms. In order to extend the flowering season, expand the range of flower colours and shapes and increase floriferousness of *Phalaenopsis*-type dendrobiums, much inter-group hybridizing has been conducted between members of this section and those in section Spatulata, the so called antelope dendrobium which have twisted petals (Davidson, 1994). The breeding of semi-alba *Phalaenopsis i.e.,* plants with red-lipped flowers began with the crossing of *P. equestris* which has a lavender pink lip, and the solid-white *P. approdite* in 1996. Since then hybridization with *P. equestris* has progressively led to the striking 10 cm semi-albas available today, with their full, flat flowers and white sepals and petals that contrast with the deep red lips (Moses, 1994).

Chen *et al.* (2000b) developed white Taisuco Phalaenopsis varieties with large, round flower shape, solid texture, multiple flowers and uniformity.

A new hybrid, Phalaenopsis 'Pink Marble' was developed in 2005 (Kim *et al.,* 2011). This hybrid was selected from self-crossed progenies of P. '21-1' (collected number) in 1999. In 2001, one line was selected based on the aspects such as flower color, leaf shape, flower stalk, and vigorous growth. Trials were conducted from 2003 to 2005 for evaluation and selection of this cultivar. 'Pink Marble' had a medium flowering habit and a dark pink spot (RHS, RPN74B) on white petal and sepal when fully opened. The number of flowers on each peduncle was 7.5, and flower diameter was 52.3 cm. The general impression of petals and sepals is a plate shape. The thick sepal could extend the long flowering time. The average length of leaf and peduncle were 16.5 cm and 6.8 cm, respectively. It had a half-erect leaf form, and was a fast-growing cultivar. This hybrid is relatively easy to clone (Kim *et al.*, 2011).

8.27.0 Renanthera Lour.

This genus has about 12 valid species which are mostly epiphytic, occasionally they are found to adopt terrestrial or lithophytic habit. Botanically, *Renanthera* is closest to *Vandopsis* and *Arachnis* from both of which it differs in superficial structure of the usually scarlet, orange or crimson flowers which are produced in large numbers in huge, horizontally spreading, usually branched sprays. These tall growing, free flowering orchids are very attractive.

Important species: *Renanthera coccinea, R. imschootiana, R. storiei, R. elongata and R. pulchella.*

Hybrids: The genus is freely compatible with most of the allied genera and many handsome hybrids have been developed by crossing with *Aerides* (*Renades*), *Arachnis* (*Aranthera*), *Vanda* (*Renantanda*), *Vandopsis* (*Renanopsis*), *Phalaenopsis* (*Renanthopsis*), *Ascocentrum* (*Renancentrum*), *Rhynchostylis* (*Renantylis*), *Renanthoglossum* (*Ascoglossum*), *Sarcochilus* (*Sarcothera*), *Paraphalaenopsis* (*Pararenanthera*). The trigenetic hybrids of *Renanthera* are *Moirara* (× *Paraphalaenopsis* × *Vanda*), *Hawaiiara* (× *Vanda* × *Vandopsis*), *Limara* (× *Arachnis* × *Vandopsis*), *Holttumara* (× *Arachnis* × *Vanda*), *etc.* Besides hybridization with allied genera interspecific crossing have been suggested. These hybrids utilizing *Renanthera* in their parentage are often almost ever-blooming. A hybrid between Renanthera and *Renantherella* has been registered and Hawkes (1978) has suggested to name as *Ellanthera* for this intergeneric hybrid.

Of all the interspecific crosses utilizing the *Renanthera*, the most famous is *R*. Brookie Chandler (*R. monachica* × *R. storiei*) registered in 1950 by J. P. Russell. Next is the spectacular flowering of the clone of *R*. Brookie Chandler 'Higgins'. An interspecific hybrid combining *R*. Brookie Chandler and *R. imschootiana* is *R*. Poipu. These hybrids start flowering when small. The clone 'Mauricette' shows the intense yellow orange colour overlaid with heavy red spotting. Sidran (1984) made intergeneric hybrids with *R*. Poipu which include the Kagawaras × *Ascda*. Sauvanee, 'Amy Sidran' in a attempt for blue purple colours and with *Ascda*. yip Sum Wah for intense reds.

Renacentrum combines *Renanthera* × *Ascocentrum* and was first registered by Iwanaga (1962); 14 crosses have been registered, and some of them are very attrac-tive. Little Red (*Asctm. curvifolium* × *R. imschootiana*), by Wilkins (1970) has a branched spray with orange-red flowers resembling the *Renanthera* parent. *Rnctm*. Stella 'Haruko' × *R. storiei*, by Mizuta in 1968 was awarded on cut inflorescence of good red colour and shape. *Rnctm*. Yap Sin Yee × *R*. Brookie Chandler, by Kirsch in 1965 has received 2 awards, the clone 'Kensington' in 1979 (Sidran, 1984).

Renaglottis, combining *Renanthera* × *Trichoglottis* is represented by the cross *R. monachica* x *Trichoglottis fascita* = Clarence Russell. Seven *Rngl.* hybrids have been registered and *Rngl.* Clarence Russell 'McQuerry' has 6 star-shaped yellow flowers barred in red (Sidran, 1984).

The intergeneric hybrid of *Phalaenopsis* × *Renanthera* created *Renanthopsis* (*Rnthps.*) with 115 crosses registered. The majority of these hybrids were initially made using a large white *Phalaenopsis* × *Renanthera.*, hybridizers have used *Phal. stuartiana* and other more star-shaped species. *Rnthps.* Elaine Noa, 'Doris' × *R. storiei* and the clone, 'The Chief' shows a good arrangement of rose coloured flowers. *Rnthps.* Star Fire 'Patricia', *R. storiei* × R. H. Montogomery, is apricot yellow suffused with rose. *Rnthps.* Kamehameha 'Orchidglade' with *R. storiei* resulted in an orange-red hybrid. Whereas, with *R. monachica,* Mildred Jameson, 'Dee Dee' a yellow colour with heavy dotting in the lateral sepals resulted. *Rnthps.* Zuma Canyon × R. Brookie Chandler, 'Trenton' and Freckles' have an orange to orange-red colour with an overlay of dotting to all segments (Sidran, 1984).

Among the *Renanstylis* hybrids, Rnst. Queen Emma (*R. gigantea* × *R. storei*) registered by Kirsch, in 1961 is the best. This hybrid continues to be awarded as in 1980.

Renantandas, combining *Renanthera* × *Vanda,* was first registered in 1947 by Warne with *Rntda.* Gold Nugget (*R. storiei* × *V. spathulata*). Since then 152 *Rntda.* have been registered, 72 of these using *R. storiei* as the Renanthera parent. *Rntda.* Redwings Hemo × *V.* Honolulu shows a pink-red clone with reflexed petals and wide lateral sepals. *Rntda.* Suchen Chen 'Ruth' has orange-yellow flower with red dots. *Rntda.* Lily Aow 'Dream City' is outstanding as a cut flower.

Two unusual *Renanthera* hybrids combine 3 genera in the grex, the first *Gotterara* Mem. Abeey Gotterer (*Rnps. Embers* × *Asctm.curvifolium*) *Rnps.* Embers, *R. monachica* × *Vdps. lissochiloides,* 'Joyce' is a tall growing plant. The influence of *Ascocentrum* has reduced the plant size of *Gotterara* (Sidran, 1984).

In 1977, Sidran registered *Lutherara* Claire Bradshaw (*Rnthps.* Osceola × *Rhy. gigantea*). Unlike its parents, the hybrid blooms freely throughout the year. The clone 'Rosebud' has many star-shaped orange flowers and 'Red Magic', and 'Margaret Chenoweth', are also very colourful clones.

8.28.0 Rhynchostylis Bl.

It is a small genus containing about four species, more or less allied to *Aerides, Saccolabium* and *Vanda.* Because of the densely multiflorous racemes, *Rhynchostylis* are popularly known as 'Fox-tail Orchids'. Plants are usually stout and rather short stemmed epiphytes. Flowers are small and crowded in pendulous or erect racemes, brightly coloured, very attractive. The sepals are broad, obtuse; the spreading petals are similar but smaller than petals. Lip is adnate to the foot of column; spur forming a sac. Pollinia 2.

Species: *Rhynchostylis coelestis, R. gigantea, R. retusa and R. violacea.*

Hybrids: *Aeridostylis* (× *Aerides*), *Renanstylis* (× *Renanthera*), *Vandacostylis* (× *Vanda*), *Rhynchocentrum* (× *Ascocentrum*), *etc.*

8.29.0 Rodriguezia R and P.

This is a genus of about 30 species of very attractive epiphytes or lithophytes, in the American tropics allied to *Lonopsis*. Plants are small having prominent pseudo-

bulbs usually sheathed at base with large leaf-like bracts. These pseudobulbs produce one or more erect or arching racemes bearing many large size flowers of different colours.

Important species: *Rodriguezia decora, R. granadensis, R. maculata, R. secunda, R. venusta, R. candida, R. compacta* and *R. pubescens*.

Hybrids: A few but very handsome hybrids have been registered in which *Rodriguezia* figures. These are *Rodrassia* (× *Brassia*), *Rodrettia* (× *Comparettia*), *Rodridenia* (× *Macradenia*), *Rodritonia* (× *Miltonia*) and *Rodricidium* (× *Oncidium*).

8.30.0 Sophronitis Ldl.

Although as mall genus comprising about six valid species, it has played an important role in the improvement of orchids. These epiphytic orchids endemic to Brazil and dwarf, pseudobulbous and have flat leaves. Flowers are either small in clusters or proportionately very large and solitary ones.

Only two species of *Sophronitis*, *i.e.*, *S. cernua* and *S. coccinea* are in cultivation today.

Hybrids: *Sophronitis coccinea* has widely been utilized by the breeders to impart its vivid scarlet colouration to the progeny in many instances, when crossed with such allied genera as *Cattleya* (*Sophrocattleya*), *Laelia* (*Sophrolaelia*), *Epidendrum* (*Epiphronitis*), *etc.* Many of the finest of orchid hybrids contain *Sophronitis* in their makeup in greater or lesser degrees (Hawkes, 1978).

8.31.0 Spathoglottis Bl.

This is a genus of about 40 terrestrial species allied to *Calanthe* and *Phaius*. The species are widespread from southern China to the Himalayas, throughout Malaysia, Indonesia, the centre of development is New Guinea. The plants bear prominent pseudobulbs giving rise to a few folded leaves. Inflorescences are tall and slender and arise from a basal leaf-axil. Flowers are large and very attractive. Floral colour ranges from brilliant yellow and pure white to crimson and several shades of magenta, purple and mauve. The lip is 3-lobed, with narrow lateral lobes which are oblong and uncurved. The end of the lip-blade is more or less widened and sometimes cleft. The column is slender, curved as footless.

Important species : *Spathoglottis affinis, S. aurea, S. ixioides, S. plicta, S. grandifolia, S. hardingiana, S. kimballiana, S. vieillardii, etc.*

Hybrids: *Spathoglottis* Aureo-Vieillardii (*S. aurea* × *S. vieillardii*) and *S.* Wiganae (*S. kimballiana* × *S. vieillardii*) are two important interspecific hybrids. The genus has also been bred with *Phaius* resulting in the hybrid known as *Spathophaius*.

8.32.0 Thunia Rchb. f.

It is a genus of about 8 valid species of very handsome terrestrial or semi-epiphytic orchids. It has tall, cane-like stem, without any pseudobulb, furnished with lanceolate, opposite leaves which fall off after flowering. Flowers are produced in clusters with petals and sepals strap-shaped. The lip is cornet-shaped. Pollinia 8.

Species: *Thunia alba, T. bensoniae* and *T. marshalliana* are the important species.

Hybrids: *Thunia* Gattonensis (*T. marjorensis* × *T. winniana*) and *T.* Veitchiana (*T. bensoniae* × *T. marshalliana*) are two fine hybrids known to be registered.

8.33.0 Vanda Jones (Papilionanthe Schltr.)

Vanda, a genus of more than 70 species, stands among the most handsome and popular of all orchids, being grown widely around the world. *Vandas* are found throughout the Eastern Hemisphere, with the highest concentration in Southeast Asia. While some grow in mountainous areas, most are lowland plants that live in warm and humid conditions with bright sunlight. *Vandas* can be large or small plants with flowers of corresponding size in a variety of colors, many with a powerful fragrance. They are long-lasting and bloom several times throughout the year. Strictly monopodial, epiphytic in habit, although some lithophytes and true terrestrials are also known. There are two kinds of plants, one with terete leaves and another with strap like leaves. Roots prominent and stout and most of them prefer to remain in the air, rather than enter the compost. Inflorescences are simple, erect, appear from near the upper part of the plant but not from its apex. Flowers are faily large, showy, with sepals and petals having delicate texture. Sepals and petals are usually similar in size, narrowed toward the base, rounded toward the apex, normally flat and spreading. Lip attached to the foot of the column, spurred at its base, with fleshy and erect side lobes, the middle lobe spreading. Colours include white, yellow, rose, purple, blue and brown.

Important species: *Vanda amesiana, V. bensonii, V. coerulea, V. coerulescens, V. cristata, V. denisoniana, V. hookeriana, V. teres, V. tessellata, etc.*

Hybrids: *Vanda coerulea* is very important for breeding blue Vanda hybrids. It was considered necessary to improve *V. coerulea* and a clone 'Buranasin' resulted from breeding of two plants of that species collected from natural habitat. The following hybrids are widely used from breeding cultivars of desired colours. *Vanda* Bhimayothin is used for pink colour, *V.* Madame Rattana for pink and blue vandas, *V.* Pranerm Ornete for brownish yellow, *V.* Piyaporn and Prakpetch for purplish red and blue black, *V.* Gordon Dillon and Siam Ruby for pink and purplish red and *V.* Rassi for breeding yellow vandas (Sophonsiri, 1984). A large number of superior hybrids have been produced using *Vanda coerulea, V. teres, V. hookeriana, V. bensonii, etc.* as parents. The first recognized *Vanda* hybrid *Vanda* Miss Joaquim (*V. hookeriana* × *V. teres*) is floriferous and is still very popular in South East Asiatic countries. Some other outstanding hybrids are *V.* Deva (*V.* Crimson Glory × *V.* Thospol), *V.* Nam Phung (*V.* Thananchai × *V.* Pont Tong), *V.* Rasri (*V.* Thananchai x *V.* Pranerm Ornete), *V.* Kultana Gold (*V.* Pong Tong × *V.* Rasri), *V.* Prakpetch (*V.* Sun Tan × *V.* Laurel Yap), *V.* Sumon Sophonsiri (*V.* Deva × *V.* Gordon Dillon), *V.* Kasem's Delight (*V.* Su Tan x *V.* Thospol), *V.* Pissamai (*V.* Thananchai × *V. laotica*), *V.* Wirat (*V.* Madame Rattana × *V. coerulea*).

Aranda × *Arachnis*), *Vandanthe* (× *Euanthe*), *Liusanda* (× *Luisia*), *Ascocenda* (× *Ascocentrum*), *Christieara* (× *Ascocentrum* × *Aerides*), *Renantanda* (× *Renanthera*), *Opsisanda* (× *Vandopsis*), *Aeridovanda* (× *Aerides*), *Vandofinetia* (× *Neofinetia*), *Trichovanda* (× *Trichoglottis*), *Nakamotoara* (× *Ascocentrum* × *Neofinetia*), *etc.*, are important intergeneric hybrids.

8.34.0 Vandopsis Pftz. (Fieldia Gaud.)

The genus *Vandopsis* has about 12 species of robust and heavy, epiphytic or lithophy-tic orchids. The name of the genus means like *Vanda* except that the lip has no spur. Leaves heavy, leathery, unequally two-lobed at the apex. Inflorescence is arching to erect; flowers large, heavy-textured, long-lasting, colourful.

The important species are *Vandopsis gingantea, V. lissochiloides* and *V. parishii,* var. *Mariottiana* is more frequent in cultivation than the type. The first two species have far been utilized by the breeder with spectacular results.

Hybrids: When crossed with *Vanda, Opsisanda* is the name of the progeny, with *Renanthera, Renanopsis* and with *Arachnis, Vandachnis.*

Vandopsis parishii combined with the upright stem habit of *Rhy. retusa,* created *Opsistylis* Ruben, registered by Ruben in 1983. The flower is cream to white with red spotting, on compact plants.

In 1973, Wilkins created a new genus *Wilkinsara* (*Wknsra.*) by hybridizing *Vandopsis parishii* with Ascda. Red. Gem, *Wknsra*. Gemini. *Wknsra*. Mem. Clyde Knapp (*V. parishii* × *Ascda.* Tammy Wagasugi) has larger flower than *Wknsra*. Gemini and ranged in colour from pink to orange (Sidran, 1984).

8.35.0 Zygopetalum Hook

This genus comprises about eighteen species, most of these are terrestrials (rarely epiphytic or lithophytic) orchids native in Brazil. Plants pseudobulbous and have attractive glossy foliage. The flower stalk bears sparse clusters of three to ten blossoms.

Zygopetalum brachypetalum, Z. crinitum, Z. intermedium, Z. bolivianum, Z. graminifolium, Z. maxillare, etc., are some of the species.

Hybrids: *Zygopetalum* has been utilized in the production of such group as *Chondropetalum* (× *Chondrorhyncha*), *Zygobatemania* (× *Batemania*), *Zygocaste* (× *Lycaste*), *Zygocolax* (× *Colax*) and *Zygonisia* (× *Aganisia*).

Zygocolax veitchii is a natural hybrid between *Z. crinitum* and *Colax jugosus.* The flowers are large, fragrant, about 7 cm in diameter, yellow-green, spotted and blotched with brown purple.

8.36.0 Important Cultivars in Cut-Flower and Pot Plant Trade

The international trade of cut-flower and pot plants of orchids are dominated by hybrids of *Arachnis, Aranda, Asconcenda, Cattleya, Cymbidium, Dendrobium, Miltonia, Mokara, Odontoglossum, Oncidium, Paphiopedilum, Phalaenopsis, Vanda.*

Cultivars of Orchid

Aranthera × 'Anne Black'

Arachnis 'Maggie Oe'

Aranda 'Christine Alba'

Ascocenda 'Udom Chai'

Cattleya 'Tainun city'

Cymbidium × 'Beauty Fred'

Dendrobium × 'Madame Pompadour'

Militonia Augres 'Trinity'

Mokara 'Khaw Phaik Suan'

Oncidium × 'Josephine'

Phalaenopsis 'Happy Valentine'

Vanda ' Gordon Dillon'

Newly Registered Hybrids of Orchid

***Lycaste* ×**
'Autumn Glow'

***Vanda* ×**
'James Sutton'

***Renanthera* ×**
'SCBG Fire'

***Phalaenopsis* ×**
'Blue Ridge Bubblegum'

***Paphiopedilum* ×**
'Alpine Pleasure'

***Dendrobium* ×**
'Snow White'

***Bulbophyllum* ×**
'Manchind'

***Cattleya* ×**
'Cosmic Chili'

Vanda
'Princess Blue'

Though there are hundreds of hybrids in each group, many are grown in different parts of the world by amateur and professional growers. Some of the outstanding hybrids which are in demand in the trade are given below:

Aeridopsis × 'Chienlung Bouquet', 'Chienlung Little Giant'

Aeridachnis × 'Roga'

Angulocaste × 'Helgrove', 'Hitriere'

Arachnis × 'Maggie Oei.'

Aranthera × 'James Storie', 'Anne Black'.

Aranda × 'Wendy Scott', 'Noorah Alsagoff', 'Christine', 'Lilac', 'Majulah'.

Ascocenda × 'Udomchai', 'Dechjum Yip Sum Wah', 'Rakpaibulsombat, 'Meda Arnold', 'Sauvanee'.

Bulbophyllum × 'Gabriel Hoy', 'Dezherae's Delight', 'Tree Frog', 'Manchind'.

Calanthe × 'Annelise and Paul', 'Mana Banana', 'Marmorata'

Cattleya × 'George Angus', 'Dorset Gold', 'Rain Gold' 'Yue', 'Esmeralda Heritage', Greenwich 'Killarney', 'Glorious Sunset', 'Chinese Beauty', 'Tainan', 'Cosmic Chili', 'Dusk Fire'.

Cymbidium × 'Beauty Fred', 'Sensation Chianti', 'King Arthur', 'Sparkle', 'Tracy Reddaway' 'Sungold', La costa 'Pink Lady', 'Pearlbel Pink', 'Beautiful Spirit', 'Cherry Red', 'Gary Cooper,' 'Kiwi Fay', 'Magical Cleo.'

Dendrobium × 'Mme Pompadour', 'Tomi', 'Spellbound', 'Ceaser Alba', 'Jaquelyn', 'Thomas', 'Jaq. Hawaii,' 'Kulltana', 'Jaq. Concert', 'Kanjana Snow', 'Ekapol No. 1', 'Lovely Virgin Angel', 'Stardust Chiyomi', 'Deep Blush', 'Kasem White', 'Snow White'.

Kagawara × '' Yoon Weng × Low, 'Viroonchan Ruby', 'Boon Rab Thonglor'.

Miltonia × 'Angres Trinity' 'Gordon Hoyt 'Dolores', Playgirl 'Cha-Cha', Saint Heiler 'Red Jewel', Zorro 'Yellow Delight'.

Mokara × 'Khaw Phaik Suan', 'Mak Chin On', 'Bangkok Gold.'

Odontoglossum × 'Maronis', 'Elegans', 'Crispania', 'Kalkarstern', 'Geyser Gold', 'Purback Gold', 'Fiona Isler'.

Oncidium × 'Josephine', 'William Thurston', 'Noel Schoenrock', 'Florida Gold', 'Saulsee', 'Red Velvet', 'Oro Quimbaya'.

Phalaenopsis × 'Musashino No. 1', Su's Red Lip 'Tanny', 'Happy Valentine', 'Taipei Gold', 'Blue Ridge Bubblegum', 'Dragon's Gem',

Paphiopedilum × 'Leeanum', 'Arthurianum', 'Harrisianum', 'Charles Richman', 'Winnifred', 'Alma's Moth', 'Amarone Dream', 'Alpine Pleasure', 'Guilty Nay'

Vanda × 'Thananchai', 'Rothschildiana', 'Kultana Gold', 'Gordeon Dillon', 'Kasem's Delight', 'Sreepai', 'Princess Blue', 'Amphai', 'Thonglor', 'TMA', 'April Rose Motes', Kalyana, 'Motes Sparkle', 'William Motes', 'James Sutton', 'Bill Sutton'.

9.0 Genetics and Breeding

9.1.0 Genetics

Cytological studies indicated that the probable basic chromosome number for the family is 20 (Hoffmann, 1930; Duncan, 1959; Blumenschein, 1960). The number 19 is also very prevalent among orchids. Miduno (1938) reported that in the subtribe Limodorinae, the number 20 is derived from16 by fragmentation of chromosomes. He considered the probable basic chromosome number as 8 from which 16 and later 20 were derived. Garay (1960) stated that different groups of orchids have different basic number. In Orchideae, the genera *Habenaria* and *Platanthera* appeared to have the basic number 7. In this group itself, chromosome numbers seemed to have been derived also from 6. The subtribes Limodorinae and Listerine were reported to have a basic number of 8 and in Spiranthinae it was suggested to be 5. The haploid number ranges from 10-21, with a few species having 23, 27 and 29. Of these, the prevalent numbers are 19, 20, 21 in the order.

Changes in chromosome number may have a definite influence on the development of phenotypes. Polyploid orchids have been reported to have many desirable characters valued for cut-flowers (Niimoto and Randolph, 1958; Kamemoto *et al.*, 1961). Flower size was often improved and they tended to be rounder (Wimber and Wimber, 1968). Diploids and triploids are found to produce more number of flowers than higher polyploids. Many diploid hybrids of *Dendrobium* and *Vanda* dominate the field of commercial cut-flowers. In *Cattleya* and *Cymbidium* tetraploid plants possess quality, diploids are floriferous and triploids have both properties. Triploids are partially or completely sterile and seldom successful as parents. Although many excellent plants are polyploids, not all polyploids are excellent plants. Genes are the ultimate determinants and an increase in number of poor genes due to chromosome duplication may even worsen the quality of already undesirable plants (Vajrabhaya, 1977).

The range of variation in orchids which is of potential value in the development of new hybrids with desirable characters, is very great but most of it remained unex-ploited. Genetical considerations are necessary for meaningful utilization of this vast material by planning crosses to avoid unproductive hybrids and result in economy of time, effort and resources. Orchids have received little attention in the hands of gene-ticists and the knowledge about the mode of inheritance of characters in this group of plants is still very inadequate. Generally, interspecific hybrids are intermediate between the parental forms and intraspecific hybrids tend to show dominance or recessiveness (Lenz and Wimber, 1959). Polyploid form of the parents may skew the range of morphological characters of the hybrid toward the polyploid parent. This has been well illustrated in genera like *Vanda, Dendrobium, Phalaenopsis, Cymbidium, Paphiopedilum, Cattleya, Laelia, etc.* (Breddy, 1949; Kamemoto, 1950; Mehlquist, 1953). These polyploid forms are known to have greater influence on the progeny than diploids of the same apparent qualities.

There are many dominant characters among *Cattleya* and its relatives which can be detected in their hybrids. The bifoliate type of growth, with its slender, jointed stems and two or three leaves is dominant over the thick pseudobulbs and single leaf of the labiata group. So also is the spade lip dominant over the labiate lip. If the large-sized *Cattleya* species is crossed with small-flowered type, the hybrid will always produce flowers of intermediate type. The heavy substance of bifoliate flowers is transmitted to their hybrids. The characteristic veins in the lip of *Cattleya dowiana* and the mottling in the lip of *C. mossiae* both persist in their hybrids. *C. labiata* contributes its day-length sensitivity to its hybrids (Northen, 1970).

The reed-stemmed epidendrums are reported to be dominant both for their growth habit and character of their flowers over several genera such as *Cattleya, Laelia, Diacrium, Sophronitis* and members of pseudobulbous *Epidendrum* alliance (Hurst, 1925; Kamemoto, 1950). The majority of the substances responsible for different colours are anthocyanins, anthoxanthins and plastid pigments. The anthocyanins are responsible for scarlet, red, blue and magenta colours and the anthoxanthins for pale ivory to deep yellow colours. The plastid pigments vary from yellow to orange or even red and many produce colours identical with those produced by anthoxanthins. The inheritance patterns of flower colours in orchids have received more attention than has any other character in orchid breeding. As early as 1892, Henri Lecoq (cited by Fenton, 1951) reported that in order to have brown colour in the orchid flower it was necessary to have blue, red and yellow combined between the parents. He also suggested that crosses between white and red orchids will produce hybrids of a more dilute colour and that brilliant colour reappeared when the hybrid was back-crossed to one of its parents or when F2 population was grown.

Charles Chamberlain Hurst was probably the first to study flower colour inheritance in orchids (Lenz and Wimber, 1959). He applied the principles of Mendel to orchid breeding and investigated the genetics of albinism (Anon., 1912; Foote, 1938) in several alba forms of *Paphiopedilum, Cattleya, Dendrobium* and *Laelia* (Hurst, 1913, 1925). Inheritance of albinism was also investigated by others in *Dendrobium* and *Laelia* (Thwaites, 1912), *Dendrobium nobile* (Anon., 1912), *Laeliocattleya* (Anon., 1914b), *Cypripedium insigne* (Anon., 1914a), *Lycaste skinneri* (Twiss, 1914) and *Paphiopedilum* (Wilson, 1923, 1927; Black, 1933; Lenz and Wimber, 1959). The patterns of flower colour inheritance in *Sophronitis, Cattleya* and *Brassavola* hybrids have received much attention (Curtis and Duncan, 1942; Storey, 1946; Storey and Kamemoto, 1960; Woodward, 1964, 1965; Sauleda, 1966). Investigations were also made in *Spathoglottis* (Storey, 1950, 1958). The results of investigation on each genus are summarized below.

9.1.1 Cattleya

Cattleya walkeriana is one of the most beautiful flowers joining the small sized plant with medium large and heavily scented flowers (Picolo *et al.*, 2016). It is widely known and appreciated by its beautiful clones and it has much to offer to breeders because their plants have besides other attributes as small habit and big flowers, many colour variations, form and precocity, becoming flower only four years in *ex vitro* culture. However, in some of the original places it is becoming a red listed species. Notwithstanding, very little is known about the genetics of these flowers and the variability in the species that is widespread in the Brazilian territory (Picolo *et al.*, 2016).

Flowers' color variation in Cattleya is often linked to assemblages of a number of species and varieties (Rego *et al.*, 2009).Cattleya breeders must have been somewhat baffled to discover that crosses between two white flowers often resulted in colored hybrids. The rosy-purple color in *Cattleya* is due to simultaneous presence of two complementary color factors: C and R (Lenz and Wimber, 1959). The true albino rises when one or both of these factors are absent. Some *Cattleya* forms produce flowers with white sepals and petals, but colored lips. The term semi-alba is used to describe these combinations (Mehlquist, 1958). He proved the existence of a gene *p* and its effect in Cattleya color (Rego *et al.*, 2009).

Flower-colour inheritance in *Cattleya* received more attention than any other member of the Orchidaceae. Early *Cattleya* breeders discovered that crosses between two albinos often resulted in coloured hybrids and those between coloured forms produced white progeny (Hurst, 1925; Lenz and Wimber, 1959). It was suggested that two genes, *C* and *R* were responsible for expression of colours. If one or both of these genes were recessive, the result was a true albino with no trace of purple sap. The colour was produced only when those two factors were present in dominant condition, either *CcRr* or *CCRR*. Parents of alba forms, homozygous for the colour factors, produced coloured progeny having the constitution of *CcRr*. When the parents were alba forms but heterozygous for the colour factors, *i.e.*, *Ccrr* or *ccRr*, one fourth of progeny was coloured and the rest white. The crosses between two coloured forms with *CcRr* constitution resulted in some progeny with white colour while others coloured. The ratio of coloured to white may be expected as 9:7 as presented below (Table 4).

Table 4: Ratio of Coloured to White

	CR	*Cr*	*cR*	*cr*
CR	*CRCR*	*CrCR*	*CRcR*	*CRcr*
	Coloured	Coloured	Coloured	Coloured
Cr	*CrCR*	*CrCr*	*CrcR*	*Crcr*
	Coloured	White	Coloured	White
cR	*CRCR*	*CRCr*	*CRcR*	*CRcr*
	Coloured	Coloured	White	White
cr	*crCR*	*crCr*	*crcR*	*crcr*
	Coloured	White	White	White

* Lenz and Wimber (1959).

Some *Cattleya* forms, popularly called as semi-alba, produce white flowers with coloured lip. According to Mehlquist (1958), a gene *P* is responsible for the distribution of colours in these orchids. It was suggested that *CCRRPP* or *CcRrPp* were fully coloured, CCRRpp and CcRrpp white with coloured lips (semi-alba) and ccrrPP, ccrrPp and2 ccrrpp were pure white.

Yellow colour in *Cattleya* which is probably not due to anthocyanins, was reported to be controlled by genes Y1, Y2. y1 and y2 but recessive to purple. A cross between *Cattleya gigas* (purple) and *Cattleya dowiana* (yellow) resulted in all purple offspring (Northern, 1970; Arditti and Fisch, 1977).

9.1.2 Cymbidium

Flower colour in *Cymbidium* was reported to be due to chlorophyll (Arditti, 1966; Matsumato, 1966), anthocyanins and other pigments (Lenz and Wimber, 1959). The presence of anthocyanins is apparently due to a single dominant gene (Strauss and Arditti, 1972). In species like *Cymbidium lowianum* 'Concolor', *C. insigne* 'Album', *C. eburneum, C. sinense* and *C. ensifolium* anthocyanins were noted to be absent (Arditti and Fisch, 1977). Concolor and other pure white forms when inter-crossed produced similarly coloured hybrids. Concolors were also obtained from crosses involving one coloured parent and even two coloured parents. The cross between C. Blue Smoke and C. Vale of Cashmir produced 38 coloured and 9 concolor hybrids in a ratio of 3:1 (Lenz and Wimber, 1959). Colour and marking pattern variations on the lip of flowers were reported to be controlled genetically, the illustration of which is complicated by the existence of intraclonal variations (Lenz *et al.,* 1955; Wimber *et al.,* 1956).

9.1.3 Dendrobium

Several investigations with this genus revealed that flower colour was inherited in the same manner as in *Cattleya*. It was obsered that *Dendrobium nobile* 'Virginale', when selfed yielded only white offspring. The mating of *D. nobile* 'Virginale' with *D. fundlayanum* 'Album' produced all albino offsprings suggesting that both had the same genotype and was either CCrr, ccRR or ccrr. The cross between *D. wardianum* 'Album' x*D. nobile* 'Album', on the other hand, produced only coloured progeny (Lenz and Wimber, 1959) and it was suggested that one of the white parents was ccRR and the other CCrr. In some hybrids, yellow colour was noted to be dominant over purple which is just reverse to that observed in case of *Cattleya*. *D. nobile* (purple) x*D. signatum* (pale yellow) produced D. Wiganiae having buff-yellow flowers. *D. aureum* (yellow) x *D. nobile* (purple) produced *D.* Ainsworthii having usually yellow flowers and occasionally white ones. *D. nobile* 'Burfordiens'e, a peloric cultivar, crossed with *D. findlayanum* has flowers having a blotch on the lower sepals as found in the *D. nobile* 'Durfordiense' parent (Lenz and Wimber, 1959). McConnell and Kamemoto (1993) studied the floral characteristics, meiotic beha-viour and sporad formation in the selfed progency of 3 tissue culture-derived *Dendrobium* tetraploids (amphidiploids) and their diploid counterparts (Jaquelyn Thomas 0550-3-2N, Jaquelyn Thomas 0580-8-2N and Neo Hawaii Y972-2N). Amphidiploid flowers were larger than those of diploids and had fewer reflexed petals, while petals and sepals were larger in size. The preponderance of bivalents in metaphase I and tetrad formation at the end of meiosis in the amphidiploids, in spite of high level of pairing between the genomes in the diploids suggests that some mechanism prevents multivalent formation in the amphidiploids. A regular peloria appeared in a *Dendrobium* hybrid registered by D.B. Nursery of Australia as *D.* D'Bush Pansy in 1988. This cultivar was crossed to several *Dendrobium* plants and also backcrossed. It was concluded that pansy lip is controlled by a single recessive gene pair (Amore and Kamemoto, 1996). Kobayashi and Kamemoto (1989) studied the inheritance of floral necrosis in *Dendrobium* caused by cymbidium mosaic virus. Crosses between susceptible plants produced susceptible offspring, while crosses between resistant plants, produced susceptible offspring with the exception of one cross (D 155-1 xK 135-2) which segregated into 14 susceptible and 10 resistant plants. Expression of necrosis seemed to be resistant.

9.1.4 Laelia

In this genus, yellow and orange colour appeared to be dominant over purple (Lenz and Wimber, 1959; Northen, 1970).

9.1.5 Paphiopedilum

The inheritance of flower colour and albinism in this genus also followed the *Cattleya* pattern (Hurst, 1925; Wilson, 1927; Black, 1933; McQuade, 1949; Lenz and Wimber, 1959; Arditti and Fisch, 1977). *Albino Paphiopedilum* contributed much in the development of present day hybrids. These plants are all green-flowered forms of otherwise coloured species except for *P. bellatulum* var. *Album* which is pure white and *P. insigne* var. *sanderianum* which is yellow. Hurst (1925) suggested that in *Paphiopedilum* two complementary pairs of genes were responsible for albinism. These were designated as C and R and the presence of either or both in the homozygous recessive state produce albinism. Intercrossing among *P. callosum* var. *sanderae, P. lawrenceanum* var. *hyeanum, P. insigne* var. *sanderianum* and *P. curtisii 'sanderae'*, produced all albinos, whereas outcross to *P. bellatulum* var. *album* (*CCrr*) resulted in all coloured progeny only (Lenz and Wimber, 1959). Inheritance of the colour and pattern markings of the dorsal sepals of *Paphiopedilum insigne, P. spicerianum* and *P. boxallii* was of a complex nature dividable into base or ground colour, markings on the ground colour and median band on the dorsal sepal and two lateral petals. Each of these characteristics was noted to be governed by incompletely dominant allelmorph (Hurst, 1925; Lenz and Wimber, 1959; Arditti and Fisch, 1977). By selfing *Phaphiopedilum insigne* var. *sanderae,* a number of plants were produced having flowers of different shades of colour. One of them *P. insigne* 'Purity' had no spotting at all and resembled. *P. insigne* var. *sanderianum* very closely (Lenz and Wimber, 1959).

9.1.6 Sophronitis

Sophronitis has been very important in contributing red colour to many modern *Sophrolaeliocattleya* and *Potinara* hybrids. The gene for red is reportedly dominant, and when present in a homozygous condition it gives a beautiful clear red colour (Lenz and Wimber, 1959; Northern, 1970). According to Ilsley (1961), however, truly red hybrids are still hard to obtain and this was probably due to the inheritance of additional anthocyanins from other parents. Pelargonidin-3-bioside was identified in *Sophronitis grandiflora* and found responsible at least in part for its red-orange-red colour. The cross *Sophrocattleya* x°*Cattleya* or *Laeliocattleya* resulted in the development of flowers with red-violet tones (Sauleda, 1966).

9.1.7 Spathoglottis

Investigations on the inheritance of flower colour and breeding behaviour of *Spathoglottis plicata, S. plicata* var. *Alba* and *S. grandiflora* revealed that three independent gene pairs control the colour (Storey, 1950, 1958).

P × Dominant allele for pink colouring of sepals and petals and red spots on calli and base of labellum.

P × Recessive allele. No colour or spots.

T × Dominant allele for pale pink colour, producing tinging of sepals and petals.

t × Recessive allele. No colour or pale tinging.

B × Dominant allele for bright lemon-yellow colour on calli and base of labellum.

B × Recessive allele producing only pale sulphur-yellow colour.

9.2.0 Breeding

The creation of new orchid hybrids is the most fascinating subject of orchidology. With the development of Lewis Knudson's (1925) asymbiotic method of orchid seedling culture, the production of hybrids received much attention and today orchid breeding stands at the forefront of ornamental horticulture. John Dominy became the first person to bloom a man-made hybrid in the year 1856. This hybrid was named as *Calanthe* × *Dominii* obtained by a cross between *Calanthe masuca* and *C. furcata*. Since then tremendous advances have been made in the production of hybrids all over the world. Thousands of new hybrids and cultivars are regularly produced each year to suit the varied likes and dislikes of the consumers. The Royal Horticultural Society, London is the International Authority for Registration of orchid hybrids which are published in the periodicals like Orchid Review and the American Orchid Society Bulletin.

Historical perspectives on Phalenopsis breeding was described by Griesbach, 2002). According to Griesbach (2002), the first major advance in breeding was made in 1920 by Dr. Jean Gratiot from France when he registered P. Gilles Gratiot. This hybrid, the result of a cross between the typical form of *P. amabilis* and the *rimestandiana* form, had extremely heavy substance. The second major advance in breeding came in 1927 when Henri Lecoufle at Vacherot and Lecoufle Company in France registered the hybrid between the *rimestandiana* and *grandiflora* forms of *P. amabilis*. This hybrid, P. Elisabethae, had the heavy substance of the *rimestandiana* from and the improved flower shape of the *grandiflora* form.

Through extensive in-breeding and out-crossing to *P. aphrodite* Rchb. superior hybrids were created, the most important of which was P. Doris. This hybrid was registered in 1940 by Duke Farms in New Jersey. P. Doris produced long, arching inflorescences on plants that were large and vigorous. The flowers were pure white, huge, and flat (Griesback, 2002). Their heavy substance made them ideal for cut-flower production.The popularity of *Phalaenopsis* naturally led to the creation of many artificial hybrids. The first hybrid (P. Intermedia) was created in 1875 when John Seden at Veitch and Son's Nursery in England crossed *P. amabilis* (L.) Bl. with *P.equestris* Rchb. (Veitch, 1886 cited by Griesbach, 2002). The seed was sown at the base of the mother plant and only a single seedling survived which flowered a decade later in 1886. By 1900, an additional 13 primary hybrids were created and flowered.

John Seden was responsible for creating all but one of these hybrids. The emphasis in hybridization at this time was to create exotic parlor plants (Griesbach, 2002).

9.2.1 Breeding Objectives

The idea behind the development of hybrids may vary according to the genus and species. The requirements of the consumers are also changing. Therefore, breeder's goal should be to produce larger and commercially important hybrids liked by the consumers. In general, the breeding programmes should be drawn out with the following objectives:

1. Improvement in colour, size and substances of flower.
2. In the case of flowers having more than one colour, the correct blending of the colours in sepals, petals and lip.
3. Creation of round full form of sepals and petals with a minimum of fenestration and twists.
4. To increase the length of inflorescence and number of flowers per inflorescence.
5. Development of hybrids showing correct mode of display of flower.
6. Extension of blooming periods.
7. Development of orchids suitable as pot plants.
8. Development of orchids resistant to climatic adversities and diseases, particularly to those caused by viruses.

Breeding for the mass-market was very different than breeding for the hobbyist or cut-flower markets (Griesbach, 2002). In the hobbyist and cut-flower markets, emphasis was placed exclusively on floral traits; while in the mass-market, vegetative characteristics were of nearly equal importance with floral traits (Griesbach, 2002). The *Phalaenopsis* varieties used for breeding are usually divided into two groups - the standard big flower group and the novelty group. The standard big flower group includes the white, pink as well as the varieties with stripes, being derivatives of the white *Phal. amabilis* and the pink *Phal. schilleriana*. The varieties of the novelty group are usually small flowers with special coloration; some have special fragrances, *i.e.* if *Phal. violacea* is involved in the pedigree. Other parental varieties in this group are *Phal. amboinensis*, *Phal. venosa*, *etc.*(Tang and Chen. 2007).

The pot varieties which have small but plentiful flowers have become a new market trend. *Phal. equestris* and *Phal. stuartiana* are the common parental varieties of this group. the breeding programs are designed to improve the size and color of the flowers as well as other characteristics such as,longevity, stalk length, leaf shape, ease of cultivation, disease resistance (Tang and Chen. 2007).

9.2.2 Methods of Breeding Orchids

Most of the genetic diversity in orchids has been the result of plant exploration, collection, introduction, evaluation and conservation. Collection of important species, especially usual forms from their natural habitats and selection of desirable ones from the wild collected plants contributed much to the improvement of these plants. Hybridization proved to be a very potential method of orchid breeding. It involves crossing of two different kinds of plants and occurs both naturally and artificially. Producing hybrids of tried and untried combinations and raising the seedlings in flasks through asymbiotic means have become almost a household art all over the world. Induction of mutation through ionizing radiation and chemical mutagens may also be used. With the new technique of genetic engineering or somatic cell hybridization, genes or sections of chromosomes from an unrelated species or hybrids might be incorporated into others to develop exciting hybrids.

Stulzer *et al.* (2019) described the breeding procedures in Cattleya as follows. Plants of *Cattleya forbesii* and *C. loddigesii,* with vigorous growth and no symptoms of phytosanitary problems were selected in the orchid garden of the Londrina State University, to be used as parents. Pollen of *Cattleya loddigesii* was artificially transferred to the stigma of recently-opened *C. forbesii* flowers by hand pollination. Capsules containing mature seeds were collected eight months after cross-pollination. Seeds were germinated *in vitro,* in half-strength MS culture medium (Murashige and Skoog, 1962), under addition of 1.0 g/l activated charcoal, 30.0 g/l sucrose and 7.0 g/l agar and pH correction to 6.2. After germination, the plantlets were sub-cultured *in vitro* in the same culture medium for 12 months, under a 16-h light photoperiod, 52 µmol m-2 s-1 irradiance and a temperature of 25 ± 2 °C. The plantlets were left to grow in plastic trays with sphagnum moss as substrate for approximately 2 months. After this acclimatization period, the plants were transplanted separately into black polypropylene pots (diameter 10.2 cm; height 7.8 cm; volume 415 mL). A sieved (1.5 and 0.5 cm sieves) mixture of pine bark and wood charcoal (1:1; v:v) was used as substrate. After 24 months of initial growth, the plants were transferred to larger pots (diameter 13 cm; height 9.8 cm; volume 1.000 mL), filled with the above substrate, and left to grow until the beginning of the first flowering (Stulzer *et al.,* 2019). During the acclimatization and growing periods, the plants were maintained in an environment protected with agricultural canvas and a 75 per cent shade net. Irrigation was applied manually and according to the plant needs during each cultivation phase and season. Fertilization was applied fortnightly after acclimatization, consisting of 50 mL NPK 20-20-20 (Peters®) at 3 g/L.

9.2.3 Natural Hybridization

The occurrence of natural hybrids among orchids has been noticed from the earliest days of orchid collection and cultivation. *Phalaenopis intermedia*, collected from the Philippines, is one of the oldest natural hybrids between *P. aphrodite* and *P. rosea*. *Masdevallia splendida* and *M. parlatoreana* were recognized as hybrids of *M. veitchiana* and *M. barlaeana* which existed in the Andes of Peru. Natural hybrids were also traced among *Odontoglossum*, *Laelia* and *Miltonia*. Natural hybridization within the genus *Platanthera* in North America occurred between the closely related species *P. dilatata*, *P. hyperborea*, *P. limosa*, *P. sparsiflora* and *P. stricta* (Schrenk, 1978). A population of *Oncidium* near Santa Cruz was identified as being a natural hybrid between *O. stacyi* and *O. jonesianum* (Ragan and Sauleda, 1982). There are also records of intergeneric hybrids evolved under natural conditions and *Laeliocattleya elegans* is one of such examples.

It is the flower structure which prevents easy fertilization. Some species have such specialized structure that they are dependent upon particular type of insects for pollination. Birds, butterflies, moths, beetles, ants, mosquitoes, fungus, gnats, syrphids, various other flies, and social and solitary bees and wasps are known pollinators of orchids (Ackerman, 1984). Mosquitoes and ants are only known to pollinate temperate species (Thien and Utech, 1960; Jones, 1975). Several studies on orchid pollination revealed fantastic floral mechanisms. In the genus *Listera*, pollinarium release is triggered by an explosive, touch-sensitive rostellum which then covers the stigma for sometime and renders the flower protandrous (Ackerman and Mesler, 1979; Nilson, 1981).

Van der Pijl and Dodson (1966) estimated that about one third of orchids are nectar-less and most of these have developed deceptive means of attracting pollinators. Pseudo-copulation, a deceptive pollination method, is a phenomenon only known to occur in the Orchidaceae. Flowers of this syndrome produce sex pheromone-like odours which attract particular species of male insects. The flower morphology similar to mimicking morphological characteristics of females further stimulates the male insects to attempt copulation. In repeated copulatory efforts, pollen is transferred among flowers and cross pollination is accomplished (Ackerman, 1984). The genus *Trichoceros* was reported to be pollinated by the copulation movements of male flies (Van der Pijl and Dodson, 1966).

A review of comparative levels of reproductive success among nectariferous and nectarless orchids worldwide was compiled by Neil and Wilcock (1998) from a comprehensive survey of fruit set from 117 orchid species in the literature and from field studies. It confirms the hypothesis that nectariferous orchids are more successful in setting fruit than are nectarless species. Overall fruit set figures for nectarless and nectariferous orchids were, respectively, 19.5 and 49.3 per cent for North America, 27.7 and 63.1 per cent for Europe, 41.4 and 74.4 per cent for the temperate southern hemisphere and 11.5 and 24.9 per cent for the tropics, demonstrating that the dichotomy is consistent across all geographical areas.

Most deceptive mechanisms are related to nutrition and involve empty promises of pollen, nectar or insect brood sites. Coloured labellum hair of the genera *Arethusa, Pogonia, Calopogon* and *Cephalanthera* and thought to represent pollen which attract and deceive foraging bees (Thien and Marcks, 1972; Dafni and Ivri, 1981). Genera like *Polystachya, Maxillaria* and *Eria* also possess pseudopollen in the form of papillae or multicellular hair. It was suggested that this material is actually collected and consumed (Van der Pijl and Dodson, 1966). Nierenberg (1972) described a case where a few *Oncidium* species appeared to mimic flowers of sympatric malpighiaceae which produced an oil and collected by female *Centris* bees.

Generally insects enter into the flower lip, search for nectar, release the cap, and pollen masses are lodged on the head. When visiting another flower, the pollinia are transferred to the stigma. Male bees pollinate many orchids like *Catasetum, Orchis, Gongora etc.* while butterflies and moths help in pollination of *Epidendrum secundatum*. Self pollination occurs in *Cypripedium schlimii, Neottia* sp., *Phaius grandi-florus, etc.* because of their inherent structural arrangement which facilitates pollination and fertilization (Bose and Bhattacharjee, 1980).

There are unusual means and circumstances by which orchids deceive their pollinators. *Epipactis consimilis* was reported to mimic the shape and colour of aphids which attracted aphidophagous hover flies (Ivri and Dafni, 1977). Flowers on *Oncidium* and *Ophrys* are attacked and pollinated by territory defending male *Centris* bees (Dodson and Frymire, 1961; Nierenberg, 1972).

Disa, a large genus in South Africa are pollinated by all major groups of pollinating insects. Nineteen different specialized pollination systems were reported in 27 species (Johnson *et al.*, 1995). *Orphys* spp. is mostly pollinated by male bees which are lured to the orchids by the visual cues and semiochemicals (Ayasse *et al.*, 1997). It was reported by Howcroft (1996) that *Amegilla cingulata*, a solitary bee is the most likely pollinator and is attracted by deceit to the callus of the flower which resembles pollen mass. *Pteroglosapsis ruwenzoriensis*, a native of Argentina is pollinated by beetle (*Euphoria lurida*) because of the morphological character of the flowers (Singer and Cocucci, 1997).

Floral biology and phenology were studied by Borba and Semir (1998) in the orchids *Bulbophyllum weddelli*, *B. involutum* and *B. ipanemense* in southeastern Brazil. The three species are pollinated by females of *Pholeomyia* spp. with relatively high fidelity. In these three species, wind with a velocity between 1.5 and 2.0 m/s is necessary for pollination to occur, which requires shaking of the labellum, thus trapping the insect on the column.

9.2.4 Artificial Hybridization

Development of new hybrids and commercial production of cut flowers in orchids are expanding rapidly in the U.S.A. Europe, Thailand, Malaysia and Singapore. There is immense scope for improving orchids in India, because large numbers of species are native to this country and many of them have already proved to be important parent plants and contributed in the production of several outstanding hybrids in the world. Due to the diversity of environmental condition in India, it is possible to grow all types of orchids in suitable places without the control of environment.

The artificial hybridization in orchids was started at a much later date than in other angiospern families. One of the major difficulties was lack of understanding of the method of pollination due to the complexity of the flower structure in orchids. Even before the time of John Dominy, Rev. William Herbert, Dean of Manchester, was probably the first person to attempt orchid hybridization. He succeeded in producing pods out of a cross between *Orchis* and *Ophrys* but the seedlings died after a few days (Lenz and Wimber, 1959). In 1863, the first bigeneric hybrid was produced by Dominy. After this, more and more people took part in hybridization and there has been a commendable achievement in this field. Apart from crossing between two species in one genus, even crossing between plants of two or more genera has success-fully been made and hybrids produced (Curtis, 1950; Dugger, 1975; Moir, 1975; Wimber and Kenney-Norberg, 1984). A large number of hybrids have already been registered and thousands are being added to this every year.

9.2.4.1 Selection of Parents

Selection of parents depends upon the aim of breeding such as creation of new colour, improvement in size and shape of flower, resistance to diseases, *etc.* The extent of genetical affinity of two species or hybrids used as parents is another important factor. The incompatible or genetically distant species belonging to different genera having different chromosome numbers usually cannot be crossed to produce the hybrids. Therefore, knowledge of species affinity, compatibility and chromosome number and behaviour is very important for obtaining successful hybrids (Lenz and Wimber, 1959; Wilfret and Kamemoto, 1969; Kamemoto *et al.*, 1972; Ando, 1982). It is also advisable to choose parents with identical flowering season. Selection of good and healthy plant and flower also counts to a great extent. Very young plants or seedlings blooming first time should not be selected as female plants (Bose and Bhattacharjee, 1980).

9.2.4.2 Pollination

This is a simple process involving transfer to pollinia from one flower to the stigma of another flower. Following are the steps:

The pollinia from the male parent are removed and collected on a clean paper. For this, a matchstick, toothpick or clean stainless steel needle is pushed below the edge of anther cap and it is gently lifted to take out the pollinia which stick to the tip of matchstick, *etc.* by its gland and comes out of the anther cap. Similarly, the pollinia are removed from the female parent. A separate toothpick or matchstick should be used for this purpose. Now the pollinia of the male parent are transferred over the stigmatic surface thereby effecting pollination. After this process is over, the flower should be bagged with butter paper bag piercing tiny holes so that pollinia from other flowers may not get deposited on the same stigma.

9.2.4.3 Storing Pollen

In the case female parent flowers later, one has to store the pollen in viable condition for future use, the pollens are placed in small pieces of tissue paper and after folding carefully they are kept in small tubes, placed in a desiccator having calcium chloride. These desiccators are then kept in a refrigerator. By this way, pollens can be stored for at least six months (Northen, 1970). Dehydrants were found to reduce pollen viability. Pollens stored at 45°F (7.2°C) germinate better than those stored at 72°F (22.2°C). It was also recorded that viability varies with species.

9.2.4.4 Post-pollination Phenomena

Pollination not only shortens the life of flowers, but also induces numerous and remarkable changes in morphology and colouration. If the cross is compatible, the column begins to swell and ovary enlarges within 2-4 days. The perianth lobes shrivel and arch over the stigma indicating successful crossing. At this time, bags should be removed and withered parts of the flower may carefully be clipped. Wilting of the perianth in some orchids is one of the post-pollination phenomena while in other orchids pollination can cause parts of the perianth to become green and to survive until the capsule ripens. Other postpollination phenomena include

Intergeneric Hybrids

Aeridocentrum
(*Aerides* × *Ascocentrum*)

Aranthera
(*Arachnis* × *Renanthera*)

Angulocaste
(*Anguloa* × *Lycaste*)

Laeliocattleya
(*Laelia* × *Cattleya*)

Renanthopsis
(*Renanthera*× *Phalaenopsis*)

Epicattleya
(*Epidendrum* × *Cattleya*)

Renanstylis
(*Renanthera* × *Rhynchostylis*)

Aeridovanda
(*Aerides* × *Vanda*)

Miltonidium
(*Miltonia* × *Oncidium*)

Vascostylis
(*Ascocentrum* × *Rhynchostylis* × *Vanda*)

Trichocidium
(*Oncidium* × *Trichocentrum*)

Rhynchonopsis
(*Phalaenopsis* × *Rhynchostylis*)

Intergeneric Hybrids

Rhyncholaelio cattleya
(*Rhyncholaelia* Schltr. × *Cattleya*)

Chlorogavilea **'Máxima'**
(*Chloraea crispa* Lindl., × *Gavilea longibracteata* (Lindl.)

Doritaenopsis
(*Doritis* × *Phalaenopsis*)

Asconopsis
(*Ascocentrum* × *Phalaenopsis*)

'*Yapara* Tariflor Pink Fairy'
(*Phalaenopsis* × *Rhyncostylis* × *Vanda*)

'Uptonara'
(*Phalaenopsis* × *Rhyncostylis* × *Sarcochilus*)

'Sapphire' Ascocenda
(*Ascocentrum* × *Vanda*)

Rhyndoropsis
(*Doritis* × *Phalaenopsis* × *Rhyncostylis*)

Waironara **'Tango Fire'**
(*Aerides* × *Renanthera* × *Rhyncostylis* × *Vanda*)

Interspecific Orchid Hybrids

***Bulbophyllum.* Fascination (*B. longissimum* × *B. fascinator*)**

***Cattleya.* Hardyana (*C. dowiana aurea* × *C. gigas*)**

***Calanthe.* Veitchii (*C. rosea* × *C. vestita*)**

***Vanda*. Rasri (*V.* Thananchai × *V.* Pranerm Ornete**

***Ophrys* × *arachnitiformis* (*O. fuciflora* × *O. sphegodes*)**

***Dendrobium* × *usitae* (*D. bullenianum* × *D. goldschmidtianum*)**

***Cattleya* × *elegans* (*Cattleya purpurata* Van den Berg × *Cattleya tigrina* A.Rich.)**

Lycaste Frank Hoyt (*Lycaste costata* × *Lycaste skinneri*)

***Phalaenopsis.* Golden Dawn (*P.* 'GoldenLouis' × *P.* Palm Beach)**

stigmatic closure, increase in fresh and dry weight of ovaries and gynostemia, hormone production, synthesis and/or destruction of pigments, deresupination, nastic movements, new biochemical pathways, cessation of scent evolution (Arditti and Knauft, 1969; Dijkman and Burg, 1970; Arditti *et al.*, 1971a, 1971b, 1973; Harrison and Arditti, 1976; Arditti and Harrison, 1979; Arditti *et al.*, 1984).

Pollination of flowers initiates postpollination development in orchid (*Doritaenopsis hybrid*) flowers, including perianth senescence, stigma closure, and ovary development. Ethylene is thought to play a key role in coordinating these developmental changes. Zhang *et al.* (1999) studied the temporal and spatial carboxylic acid (ACC) synthase and ACC oxidase following pollination-associated factor treatments in orchid flowers. Both ACC synthase and ACC oxidase mRNA accumulation in the various parts of the flowers is induced by auxin, and ethylene, but not by emasculation. The patterns of both ACC synthase and ACC oxidase mRNA accumulation are similar in all floral organs following auxin and ethylene treatments. Further, *in situ* hybridization analysis indicates that the ACC oxidase mRNA is localized in epidermal and parenchyma cells of the stigma after auxin and ethylene treatments.

In flowers of cymbidium, Tanase *et al.* (1999) reported that *o*-phenanthroline (PA) inhibited pollination-induced labellum colouration and flower senescence by suppre-ssing ethylene evolution through chelation of Fe, an essential cofactor of ACC oxidase.

9.2.4.5 Harvesting of Fruits

Harvesting of the capsules should be done before they dehisce. When the capsule starts becoming brownish or yellowish it is a sign of maturity. Generally, it takes 4-10 months for a capsule to mature and ripen. Time taken for maturity of pods also depends upon the habit of species crossed. According to Hegde (1984) it may vary as follows:

Calanthe	4-5 months
Cymbidium	3-6 months
Dendrobium	9-17 months
Paphiopedilum	7-13 months
Phaius	6-9 months
Vanda	5-7 months

Abraham and Vatsala (1981) suggested toharvest pods earlier which prevented contamination of the seeds with fungal or bacterial spores and resulted in good germination. The highest percentage of germination obtained in *Bletilla striata, Calanthe discolor, C. furcata, C. cardiosa* and *Phaius minor* was with seeds harvested when the embryos were almost mature (Nagashima, 1982a, 1983). The pods, after harvesting should carefully be wrapped in clean tissue paper and stored, following the method described earlier for the storage of pollen.

9.2.4.6 Seed Storage

Seeds of three species *Dactylorhiza juchsii, Dendrobium anosmum* and *Eulophia gonychila* as well as those of *Dactylorhiza majalis* subsp. *praetermissu* and *Paphiopedilum rothschildianum* generally survived 12 months storage at a range of sub-zero temperatures, including _196°C (*i.e.*, cryopreservation). However, *P. rothschildianum, D. fuchsii* and *D. majalis* subsp. *praetermissa* displayed significantly lower germination when stored at either 30°C or 50°C compared to other sub-zero temperature (Pritchard *et al.*, 1999).

9.2.4.7 Seed Germination and Care of Seedlings

Orchid pods contain millions of extremely minute seeds which are non-endospermic and do not contain nutrients. So, for germination they require symbiotic association of suitable mycorrhiza. That is how they germinate in nature. Seeds are germinated and seedlings are grown in culture media containing agar and all necessary nutrients. Ten to twelve months after sowing, the seedlings are ready to transplant in community pots. The compost used for seedlings in the community pots should be a mixture of equal parts of finely chopped tree fern and crushed bark or moss. Later on, they are transferred to larger pots. The details of seed germination, care of seedlings, *etc.* are discussed later in the chapter on propagation.

The hybrids thus obtained are to be evaluated and the best ones sorted out and multiplied through tissue culture method. Nomenclature of these hybrids should be in accordance with the International Code of Nomenclature for Cultivated Plants. Later, these should be registered with the International Registration Authority.

Intergeneric Hybrids

Several Indian species of *Aerides, Ascocentrum, Dendrobium, Cymbidium, Paphiopedilum, Phalaenopsis* and *Vanda* have extensively been used in producing outstanding hybrids of the world today. Many bigeneric, trigeneric and quadrigeneric hybrids have been evolved by artificial hybridization. Some important multigeneric hybrids are listed below:

Hybrids	*Parental Combination*
Bigeneric hybrids:	
Aeridachnis	*Aerides* ×*Arachnis*
Aeridopsis	*Aerides* × *Phalaenopsis*
Aeridovanda	*Aerides* × *Vanda*
Arachnopsis	*Arachnis* × *Phalaenopsis*
Aranda	*Arachnis* × *Vanda*
Ascocenda	*Ascocentrum* × *Vanda*
Brassocattleya	*Brassavola* × *Cattleya*
Brassoepidendrum	*Brassavola* × *Epidendrum*
Laeliocattleya	*Cattleya* × *Laelia*
Vandaenopsis	*Phalaenopis* × *Vanda*
Renanthopsis	*Renanthera* × *Phalaenopsis*

Hybrids	*Parental Combination*
Trigeneric hybrids:	
Brassolaeliocattleya	*Cattleya* × *Laelia* × *Brassavola*
Holttumara	*Arachnis* × *Vanda* × *Renanthera*
Sophrolaeliocattleya	*Sophronitis* × *Laelia* × *Cattleya*
Trevorara	*Arachnis* × *Paraphalaenopsis* × *Vanda*
Tanakara	*Phalaenopsis* × *Aerides* × *Vanda*
Quadrigeneric hybrids:	
Iwanagara	*Brassavola* × *Cattleya* × *Diacrium* × *Laelia*
Potinara	*Brassavola* × *Cattleya* × *Laelia* × *Sophronitis*
Yamadara	*Brassavola* × *Cattleya* × *Epidendrum* × *Laelia*
Kirchara	*Sophronitis* × *Laelia* × *Cattleya* × *Epidendrum*
Beaumontara	*Schomburgkia* × *Brassavola* × *Cattleya* × *Laelia*
Pentageneric hybrid:	
Hasegawaara	*Cattleya* × *Brassavola* × *Broughtonia* × *Laelia* × *Sophronitis*

Hexageneric Hybrids

A great step forward in the history of orchid hybridization occurred when in 1982 *Brilliandeara* Gary was registered. It is the first hexageneric hybrid not only produced in the orchid family but also in the whole plant and animal kingdom. It is named in the honour of Gary Brilliande, a famous orchid breeder. The Forgetara Mexico (*Brapasia* Serene x*Miltonia* Fortaleza) is a combination of plants of three genera having 60 chromosomes. *Burrageara* Sambu River (*Miltonidium* Aztee Gold × *Oda*. Isabella) is a combination of two different bigenerics. *Mtdm*. Aztec Gold (*Oncidium* Powelli × *Mtdn*. Aristocrat) is a *stellata* Oncidium of the southern group and *Mtdm*. Aristocrat is the first registered *Miltonidium* (*Milt. Schroederiana* × *Onc. Leucochilum*) (Moir, 1983).

Brilliandera Gary

Forgetara Mexico   *Burrageara* Sambu River

Brapasia Serene   *Miltonia* Fortaleza *Miltonidium* Azetec Gold   *Odontioda*Isabella

Apasia   *Brassia*   *Miltonia Oncidium Odontoglossum Cochlioda*

The new hybrid genera reported in 2001 are Aerangaeris [*Arg.*] *Aerangis* × *Rangaeris;* Agasepalum [*Agsp*] *Aganisia* × *Zygosepalum;* Arachnocentron [*Act.*] Arachnis × *Cleisocentron;* Engkhiamara [*Ekma*] *Aerides* × *Arachnis* × *Ascocentrum* × *Renanthera* × *Vanda;* Holcocentrum [*Hctm.*] *Ascocentrum* × *Holcoglossum;* Neostylopsis [Nsls.] *Neofinetia* × *Phalaenopsis* × *Rhynchostylis;* Priceara [*Pri*] *Anguloa* × *Colax* × *Lycaste* × *Promenaea* × *Zygopetalum;* Promelli (Pmla.) *Ansellia* × *Promenaea;* Promosepalum [*Prsm.*] Promenaea × Zygosepalum; Rehfieldara [Rfda.] Ada × Odontoglossum × Oncidium; Stenosarcos (*Stsc.*) *Sarcoglottis* × *Stenorryncbos;* Stephenara (*Stph*) *Aganisia*

× *Batemannia* × *Neogardeneria* × *Otostylis* × *Zygopetalium* × *Zygosepalum* × Tanara (*Tanara*) *Aerides* × *Ascocentrum* × *Renanbera* × Rynchostylis × Vanda (Anon., 2001).

9.2.4.8 Reciprocal Hybridization

Many investigations on hybridization showed that reciprocal crosses between species or cultivars were sometimes difficult or impossible to accomplish. Barriers recorded might be structural, functional or genetic. One of the most common causes was probably physical barrier. Hybridization between long-styled and short-styled plants was sometimes only successful if the long-styled plants were used as the staminate parent. The pollen tubes of short-styled species might not be able to traverse a long style. The sensitivity of one of the parents to the environment could block the genes while it would not affect the other and as a result, development of capsule could be checked. Plants that are infertile due to irregularities at meiosis may produce viable egg cells whereas the pollen could be non-functional (Lenz and Wimber, 1959).

The weak reproductive barriers in the genus Dendrobium provided an edge to evolve many interspecific hybrids and intersectional hybrids. Results of reciprocal crosses between *Dendrobium nobile* and 8 other species which flowered at the same time and of crosses between *D. moniliforme* and 21 other pollinating species, showed that *D. nobile* was reciprocally compatible only with *D. haniffii*, *D. linawianum* and *D. moniliforme* each of which resulted in 100 per cent seed set when used as a pollinator. With *D. moniliforme*, however, all crosses were fertile, the seed set varying between 17 and 100 per cent (Ando, 1982). McConnell and Kamemoto (1983) also studied the characteristics of four sets of reciprocal crosses in *Dendrobium*. With crosses involving 2 accessions of *D. canaliculatum* (D 173-2 and D-129), the off springs with cytoplasm of D. 173-2 were taller, produced more pseudobulbs and gave a higher flower yield than with D-129. *D. canaliculatum* offspring with D 173-2 cytoplasm produced taller plants and more pseudobulbs in reciprocal crosses between *D. canaliculatum* and *D. strebloceras*. Reciprocal crosses of *D. schulleri* x *D. sunset* were different in chromosome number and flower quality. *D. schulleri* x *D.* Sunset produced triploid offspring, while the reci-procal produced diploid offspring. Reciprocal matings of 2 amphidiploid *D.* Jaquelyn Thomas selections did not differ in height, yield, number of pseudobulbs or floral characteristic. There were 1040 interspecific hybrids have been developed till 2019. Many of them have proved beneficial breeding lines for developing commercial cultivars. Among them Den Jaquelyn Thomas, *Den* Caesar, *Den* Ellen, *Den* Janice Tanaka, *Den* Ursula, *Den* Cassiope, *Den* Sunglow *etc. Den* Jaquelyn Thomas was evolved by crossing *D. gouldii* and *D. phalaenopsis* in 1949. It bears about 20 flowers on a spike measuring approximately 6.5 cm across. This hybrid has been used to evolve 146 first-generation hybrids, where it has been used 70 times as the female parent and 76 times as the male parent. The *Den* Caesar proved to be an excellent female line for the breeding of Dendrobium orchids. It was evolved in 1937 by crossing *D. phalaenopsis* and *D. stratiotes,* and since then 114 primary hrybrids have been developed using this hybrid where it has been used 104 times as the female parent and only ten times as the male parent. *Den* Ellen is another primary hybrids evolved by crossing *D. kingianum* x *D. tetragonum* in 1928.

This hybrid has been used for developing 76 primary hybrids; 40 times, it has been used as the female parent and 36 times as the male parent.

9.2.4.9 Backcrossing and Selfing

Selfing and backcrossing techniques are of immense value in breeding for superior hybrids. These techniques are regularly used in most genetic research and even in practical plant and animal breeding but they have largely been avoided by orchid breeders (Lenz and Wimber, 1959). The reason for eschewing the techniques is that inbreeding leads to inferior forms or a depression in vigour. Undoubtedly, individual seedlings may be poor but in the long run for every inferior species there is one that is superior.

A pure white and several coloured cultivars were produced from the same capsule obtained by selfing *Cattleya gaskelliana* (Lenz and Wimber, 1959). The long continued inbreeding of *Phalaenopsis sanderiana* is a famous example leading eventually to superior clones, one of which in the seventh generation received a gold medal award of American Orchid Society.

Phalaenopsis Princess Kaiulani is a hybrid of *P. violacea* and *P. amboinensis*. When it was crossed back to *P. violacea* it made the very variable *P.* Princess Violet. The best example of variability in backcross progency that exceeds the extremes in the two original species is found in *P.* Frank McClain which is a hybrid of *P.* Amblearis and *P. amboinensis*. The overall mahogany flower having waxy sheen is one extreme in this case (Walbrunn, 1984).

9.3.0 Polyploidy and Mutation Breeding

Genetic diversity which is the backbone of crop improvement may be introduced deliberately by employing chemical mutagens like colchicine and by ionizing radiation.

Polyploidy tends to increase the size and improve the form of flowers. Mehlquist (1974) observed that chromosome doubling of sterile hybrids may produce fertile amphidiploids.

Yield (total number of racemes) and morphology of 11 named diploid *Dendrobium* hybrids and their corresponding colchicine-induced amphidiploids from intersectional combinations of Ceratobium × Eleutheroglossum, Phalaenanthe × Ceratobium and Phalaenanthe × Eleutheroglossum were compared. The amphidiploids were less floriferous and produced fewer and shorter racemes than their corresponding diploids (Amore and Kamemoto, 1992).

Similarly, Lee and Mowe (1983) produced polyploid plants by using meristem culture in *Aranda*. Improvement in flower size and lasting quality was obtained by applying 1 per cent colchicine in lanolin paste at various times and at different bud positions in cymbidiums (Parker, 1979). In 1985, Griesbach produced fertile hexaploids from triploid *Phalaenopsis* orchids by using naturally occurring unreduced gametes and by artificially inducing polyploidy.

Devi (1989) reported that relatively low colchicine concentration of 50 mg/l for 5 days was found to be suitable for inducing polyploidy in *Phaius tankervilliae* since it produced highest percentage (30 per cent tetraploid and 10 per cent aneuploids) of polyploids.

The polyploidy has been an important factor in the development of improved commercial varieties and hybrids of the *Cattleya* orchids (Kamemoto, 1950). Colchicine levels of 0.05 and 0.10 per cent on the protocorm-like bodies of *Cattleya intermedia L.* (*Orchidaceae*) cultured in vitro appeared to be effective on the production of mixoploids and tetraploids and could be further used in breeding programs. The suitability of stomatal area and stomata density as criteria for the distinction between diploids and tetraploids was tested. The results suggested that tetraploid plants could be identified with a fair amount of certainty when the screening was based on the density of stomata (Silva *et al.*, 2000).

Polyploidy played an important role in the development of *Phalaenopsis* varieties. Many popular commercial hybrids are tetraploids. However, analysis of the pedigree of these hybrids indicated that their genetic base was quite narrow (Chen *et al.*, 2010). The problems in using these tetraplois are (i) it is difficult to transfer favorable genes of the diploid wild species to the commercial hybrids because of the difference in the ploidy levels, (ii)The chromosome sizes of *Phalaenopsis* species varied tremendously, (iii). Hybrids developed from the crosses between these species were usually sterile due to the problem of genome homology. Chen *et al.* (2010) while listing the above problems suggested the use of chromosome doubling technique to restore the fertility of these hybrids and to develop simple and effective methods to detect the nuclear DNA content by flow cytometry. They also suggested todouble the chromosome of *Phalaenopsis* species and varieties by the use of tissue culture without application of anti-microtubule agents. The nuclear DNA contents of the native species and some of the related hybrids of *Phalaenopsis* in several germplasm banks and orchid nurseries in Taiwan were determined. Based on this information, *Phalaenopsis* species were grouped into low, medium and high nuclear DNA content which reflects the chromosome sizes of the species. This will provide useful information to orchid breeders for the selection of parental varieties for hybridization. Through chromosome doubling techniques, a conversion program to develop a series of tetraploid native species of *Phalaenopsis* is in progress (Chen *et al.*, 2010).

Endopolyploidy was observed in the protocorms of diploid *Phalaenopsis aphrodite* subsp. *formosana* with ploidy doubling achieved by in vitro regeneration of excised protocorms, or protocorm-like bodies (PLBs) (Chen *et al.*, 2009). Thirty-four per cent of the PLBs regenerated from the first cycle of sectioned protocorms were found to be polyploids with ploidy doubled once or twice as determined by flow-cytometry. The frequency of ploidy doubling increased as the sectioning cycles increased and was highest in diploid followed by the triploid and tetraploid (Chen *et al.*, 2009).

It was reported that three years after radiation treatment, the use of ^{60}Co. gamma rays for trearment had conspicuously influenced to growth, mutation frequencies in mmorphological and physiological reits of *Phalaenopsis* varieties. Survival rate and growth vigour of treated plants reduced while total variation ratio increased by higher doses. The useful variations were the highest in 20 Gy and next to 40 Gy radiation treatments. In this study, LD_{50} dose in radiation treatment for Phalaenopsis orchid variety was at dose 40 Gy. Suitable dose in radiation treatment which made wide diversity for Phalaenopsis orchid at dose 20-40 Gy (Tan *et al.*, 2011).

Table 5: Flower Characteristic of Selected Mutants and Control

Mutant	*Description*	*Horizontal Measure-ment (cm)*	*Vertical Measure-ment (cm)*	*Petal Measure-ment (cm) (L/W/T)*	*Sepal Measure-ment (cm)*	*Lip Measure-ment (cm)*
Control	Red purple (71 A) petals, sepals and lip. Narrow petals. Long and narrow lip with curled at the edges.	8.0	7.3	L: 4.7 W: 3.0 T: 0.075	L: 4.0 W: 1.6 T: 0.062	L: 3.7 W: 2.0 T: 0.067
DS 35-C	Purple (78B) narrow petals with curled at theedges. Smudge of purple (78B) sepals and purple(78A) long and narrow lip with curled at the edges.	**9.0**	**8.2**	L: 5.0 W: 3.5 T: 0.077	L: 4.6 W:1.7 T:0.067	L:4.0 W:1.9 T:0.053
DS 35-1	Purple (78C) narrow and elongated petals. Petalcolour is paler around the edges and more intense towards the center region of the petal. Smudge of purple (78C) sepals and purple (78B) long and broad lip.	**9.5**	**9.6**	L: 6.2 W: 3.6 T: 0.079	L:4.2 W:1.4 T:0.074	L:4.2 W:2.3 T:0.075
DS 35-R	Purple (78B) petals, sepals and lip. Narrow petals. Long and narrow lip.	6.7	6.5	L: 4.5 W: 2.7 T: 0.063	L:5.4 W:1.7 T:0.074	L:3.0 W:1.4 T:0.044
DS 35-M	Purple (78C) petals, sepals and lip. Narrow and broad petals. Veinous petals and sepals. Long and broad lip with curled at the edges.	7.4	**7.1**	L: 4.4 W: 3.3 T: 0.068	L:4.2 W:1.4 T:0.061	L:4.0 W:3.9 T:0.064
DS 35-U	Purple (78B) petals and lip. Smudge of purple(78B) with green tip sepals. Short and broad lip with curled at the edges.	6.5	5.6	L: 3.7 W:3.3 T: 0.070	L:4.0 W:1.7 T:0.056	L:3.0 W:2.1 T:0.074
DS 35-N	Smudge of purple (76B) petals and sepals.Rounded petals. Purple (76A) short and broad lipwith curled at the edges.	5.5	5.7	L:3.1 W: 3.0 T: 0.064	L:2.9 W:1.8 T:0.055	L:2.7 W:2.0 T:0.049
DS 35-White	Pure white petals, sepals and lip. Narrow petals.	7.3	6.7	L:5.0 W:3.4 T:0.063	L:4.5 W:1.7 T:0.051	L:3.8 W:1.7 T:0.058

Mutant	*Description*	*Horizontal Measure-ment (cm)*	*Vertical Measure-ment (cm)*	*Petal Measure-ment (cm) (L/W/T)*	*Sepal Measure-ment (cm)*	*Lip Measure-ment (cm)*
DS 35-135	Purple (78A) veinous, broad and elongated petals. Smudge of purple (78B) sepals. Purple (78 A) long and broad lip with curled at the edges.	8.2	7.2	L:4.5 W:3.7 T:0.052	L:3.9 W:1.7 T:0.044	L:3.6 W:2.4 T:0.047
DS35-1/B	Purple (78B) broad petals. Petals is white at the centerbut gradually becomes purple towards the edges.Smudge of purple (78D) sepals. Purple (78B) narrowand curled backward lip.	8.5	7.5	L:4.7 W:3.5 T:0.068	L:4.3 W:1.8 T:0.056	L:3.5 W:1.9 T:0.059
DS 35-1/0	Red-purple (71 A) petals, sepals and lip. Roundedpetals. Short and broad lip	7.0	6.5	L:3.8 W:3.5 T:0.060	L:3.5 W:1.9 T:0.050	L:3.5 W:2.4 T:0.060
DS 35-2/L	Purple (78B) petals, sepals and lip. Narrow and curledbackward petals. Veinous sepals. Short and broad lip.	6.2	6.5	L:4.0 W:3.1 T:0.065	L:3.5 W:1.6 T:0.052	L:3.5 W:2.1 T:0.057

The benefits of induction of polyploidy in cymbidium orchids were realised when a clone of grex *Cym* Alexander 'Westonbirt' flowered. Clone Westonbirt was a chance tetraploid and had large flower size, intense colour, full shape and heavy substance than its diplod form. Similarly, clone 'Pinkie' of grex *Cym* Rosana was found to be tetraploid. These two clones 'Pinkei' and 'Westonbirt' might developed due transfer of unreduced gametes. The efforts were made to produce tetraploids by applying colchicines on shoot developing from the pseudobulbs (Menninger, 1963). The chromosomes of *Cym* Silky were doubled by treating the PLBs with 0.05 per cent colchicines concentration for a week (Kim *et al.*, 2003). The doubling of chromosome is also useful in restoring of fertility. Cymbidium Peter Pan 'Greensleeves' (*Cymbidium ensifolium* x Cymbidium Miretta) proved to be outstanding parent for hybrisation of cymbidium after restoring the fertility through doubling the chromosome number (DuPuy and Cribb, 2007). Majority of modern cultivars are tetraploids and transfer of characters to these progenies from wild species or primary hybrids (2n) leads to production of triploids that cannot be used further in breeding program. Thus conversion of desired species or primary hybrids is useful for transferring the desired traits from species or primary hybrids in improving the cultivars. According to Dunn (1979), tetraploid *Cymbidium* plants are far superior to their diploid counterparts in respect of colour, substances, spike habit, size and shape. Many award winning plants are polyploids, and of these, the majority are tetraploids. Wimber and VanCott (1967) treated protocorms in colchicine solution and observed that approximately 40 per cent of the treated protocorms of *Cymbidium* developed into nonchimeric tetraploids. Griesbach (1981) reported that when young protocorms were treated in liquid culture with 50 mg/l colchicine about 50 per cent of the protocorms developed into tetraploids.

A wide spectrum of induced mutation breeding programme was initiated to explore the possibilities of obtaining new colour and forms of Dendrobium orchid variety 'Sonai kai' (Ariffin and Basiran, 2000). Matured seeds from self pollination were cultured and irradiated at 35 Gy at the protocorm-like bodies (PLBs) stage. Selection of induced mutations was done after the first flowering of the plants regenerated from the irradiated protocorms. Results showed changes in flower colour, shape and size. Most of these changes are expressed in different combinations in the petals, sepals and lip of the flowers.

Another aspect of orchid mutation breeding was focused to produce resistant or tolerant mutant to insect infestation on commercial hybrids or species of *Dendrobium* orchids; namely *Dendrobium* Sonia 17 Red from Thailand, *Dendrobium* jayakarta from Indonesia and *Dendrobium mirbellianum* from Malaysia. Tissue culture orchid materials (protocorm-like-bodies or PLBs) were exchanged among the participated countries early in the project. The Malaysian research team had irradiated these PLBs with two ionizing mutagens; gamma rays and ion beams (JAEA). Following irradiation, regenerated plantiets were randomly selected from each dose and subjected to *in vitro* infestation with mites and thrips, to analyse their resistance towards these pests and to pre-select potential mutants for secondary screening at flowering stage. Potential insect resistant orchid mutants were then selected and subsequently planted in the glasshouse. Secondary screenings were carried out at flowering stage by challenge-infestation with the target insects. This project has successfully generated two *D. jayakarta* mutants tolerant to thrips, and one *D. mirbellianum* tolerant to both mite and thrips (Ahmad *et al.*, 2010).

Wild type and peloric flower buds of *Phalaenopsis* hybrids derived from flower stalk nodal culture were used for cDNA-RAPD and cDNA suppression subtractive hybridization analyses was experimented by Chen *et al.*, 2005 in order to study their genetic difference in terms of expressed sequence tags. A total of 209 ESTs from normal flower buds and 230 from mutants were sequenced. These ESTs sequences can be grouped into several functional categories involved in different cellular processes including metabolism, signal transduction, transcription, cell growth and division, protein synthesis, and protein localization, and into a subcategory of proteins with unknown function. Cymbidium mosaic virus transcript was surprisingly found expressed frequently in the peloric mutant of *P.* Little Mary. Real-time RT-PCR analysis on selected ESTs showed that in mutant flower buds, a *bZIP* transcription factor (TGA1a-like protein) was down-regulated, while up-regulated genes include auxin-regulated protein kinase, cyclophilin, and *TCP*-like genes. A retroelement clone was also preferentially expressed in the peloric mutant flowers. On the other hand, ESTs involved in DNA methylation, chromatin remodeling and posttranscriptional regulation, such as DNA methyltransferase, histone acetyltransferase, ERECTA, and DEAD/DEAH RNA helicase, were enriched in normal flower buds than the mutants. The enriched transcripts in the wild type indicate the down regulation of these transcripts in the mutants, and vice versa.

9.4.0 Somaclonal Variation

It is well known that genetic variation occurs in undifferentiated cells, isolated protoplasts, calli, tissues and in differentiated plants while grown in *in vitro* conditions.

There are few reports on utilization of somaclonal variation in orchids. Laneri (1990) observed two somaclones in the plantlets developed from protocorms produced by *in vitro* culture of immature flower buds of cymbidium Oriental Legend 'Cinnamon'. These somaclones produced racemes of flowers with a slightly larger labellum and more pointed yellow brown rather cinnamon petals.

10.0 Growth and Flowering

For successful cultivation of orchids, care should be taken to ensure that conditions identical to those of their natural environment are provided. There are several factors influencing their growth and flowering. The genotype is the basic factor determining the growth and development. It takes more than 2 years for popular orchid cultivars with high commercial value (*e.g.*, *Phalaenopsis*) to switch from the vegetative to the reproductive phase (Lu, 2007). Environmental conditions play a key role in growth and flower production and these can hardly be separated from nutrients and growing media. *D. nobile* requires vernalization for flowering whereas *D. phalaenopsis* flowers at high temperatures. Spiking of *Phalaenopsis aphrodite* subsp. *formosana* is significantly inhibited under warm conditions while the natural flowering period of *P. sanderiana* is through the summer in the Philippines (Wang *et al.*, 2017). Induction of flowering by manipulating growth rate is also possible. Similarly, the role of certain growth substances is well-known. Nutrient management and wate-ring to provide moisture have been dealt with in later chapter and discussion in this chapter will be restricted to other factors.

10.1.0 Genetic or Inherent Factors

Striking variations in growth and flowering of different genera and species have been observed which are attributed to their genetic make-up (Withner, 1959). Several tropical lowland species are free flowering and produce flower throughout the year. This freedom of flowering is said to be a genetic character (Goh, 1984).

Tang and Chen (2007) reviewed the genetics of flower colour in Phalaenopsis. According to them, it was reported that several independent genes control the colors of the lip of *Phal. equestris* through the expression of both anthocyanins and carotenoids(Christense, 2001). By crossing between the white and red forms of *Phal. equestris*, Fu *et al.* (1996) cited by Tang and Chen (2007) reported that the pink floral color was controlled by a single dominant gene. This gene acts on the coloration of petals, sepals, the mid-lobe and the apex area of the side-lobe of the lip. Also,it is expressed as a brownish color through a pleiotrophic effect on the coloration of the floral stalk and the spot of the callus. Compared with the petals and sepals, the color inheritance of the lip is more complicated. Two dominant duplicated genes which are independent of the above mentioned red gene, control the yellow color of the base of the mid-lobe and side-lobe as well as the callus of the lip.

10.2.0 Environmental Factors

Most orchid species have definite growth and flowering seasons and these are more pronounced in species and hybrids from temperate regions (Curtis, 1954; Dunsterville and Dunsterville, 1967; Quisumbing, 1968). Amongst the tropical lowland species, many show peak flowering period even if they flower throughout the year. Plants of the same species may flower at different times of the year in different countries or in different localities in the same country (Goh, 1984). West African orchid species also exhibit peak flowering periods (Sanford, 1971). Thus, growth and flowering in tropical as well as temperate orchids are regulated by some environmental factors. Light, temperature, humidity, aeration, CO_2 concentration, *etc.*, are important environmental factors which regulate the growth and flowering processes. The optimum require-ments of temperature, humidity and light of some important orchid genera are given in Table 6.

Table 6: Some Features of Prominent Orchid Genera Popular in the Cut-Flower Industry*

Genera	*Native Country*	*Culture Temp. °C*	*Relative Humidity (Per cent)*	*Light Requirement (f.cdl.)*
Arachins and its hybrids	S.E. Asia	18	60-75	2400-3600
Cattleya	C and S American tropics	15-18	40-70	2400-3600
Cymbidium	Asia and the Philippines	10-21	40-70	Partial shade to full sun
Dendrobium	Western Pacific Basin	10-21	40-70	2400-3600
Oncidium	C and S American tropics	15-21	40-70	2400-3600
Paphiopedilum	S.E. Asia	18-21	65-70	1800-2400
Phalaenopsis	Asia, Philippines and Indonesia	21	40-70	1500
Vanda	S.E. Asia	15-21	40-70	2400-3600

* Sheehan and Sheehan (1979).

10.2.1 Light Intensity

Light is essential for plant growth. This is true in the case of orchids also. They, however, require less light intensity compared to common garden plants and hence are grown under greenhouse conditions. Insufficient light, on the other hand, promotes spindly and succulent growth and the plant may fail to flower. Yoneda and Sasaki (1978) observed an adverse effect on growth and flowering of *Epiendrum radicans* grown under more than 50 per cent shade. The optimum light requirement varies in different genera. *Arachnis, Cattleya, Oncidium,* most species of *Dendrobium* and *Vanda* grow and flower well at a light intensity ranging from 2400 to 3600 foot-candle. *Cymbidium* prefers higher light intensity. The optimum requirement of light for *Paphiopedilum,* in general, varies between 1800 and 2400 foot-candle while *Phalaenopsis* shows satisfactory growth and flowering at 1500 foot-candle light intensity (Sheehan and Sheehan, 1979).

Goh (1984) reported that many species of *Arachnis, Ascocentrum, Renanthera* and *Vanda* require full sun for free flowering and any shading delayed or suppressed the flowering process. Many of their hybrids such as *Aranda* and *Aranthera* are known to behave in the same manner. Shade-loving orchids such as *Dendrobium, Oncidium* and *Phalaenopsis* do not tolerate direct exposure to tropical full sun and they would be scorched within hours if exposed to the strong mid-day sun directly. Even under partial shade condition, the growth and flowering behaviour of *Oncidium* Goldiana was shown to have a 'negative correlation' 2 months before harvest (Ding *et al.,* 1980). In Hawaii, the seasonal flowering of *Vanda* Miss Joaquim was found to be correlated with sunlight availability. Reduction in sunlight resulted in late commencement of flowering and lower yields (Murashige *et al.,* 1967). Similarly, when the plants were exposed to varying periods of direct sunlight in Singapore, it was observed that inflorescence production was dependent on the length of exposure to direct sunlight. Those plants exposed longest, flowered most profusely while those exposed to only 4 hours of direct sunlight daily did not flower at all (Goh and Wan, 1974). In *Aranda* Wendy Scott, plants receiving 3 hours of direct sunlight (plus 8 hours of diffused and reflected light) remained vegetative whereas those exposed for 8 hours of direct sunlight (plus 3 hours of shaded light) produced inflorescences regularly (Goh, 1982).

The source of artificial light was also reported to influence the growth and flowering of some orchids. The additional fluorescent lighting from nightfall to 11-30 pm and from 6-30 am until daybreak was found to favour the vegetative growth of 2-year-old *Phalaenopsis* seedling (Springer, 1962). *Cattleya* hybrids grew best under Luxor, Gro-Lux-WS, Ho-WW and VHO-WW lamps, and least well under Power-Groocee. All 5 fluorescent lamps were better than Lucalux high intensity discharge lamp and natural greenhouse light for *Phalaenopsis* species (Poole and Seeley, 1977). *Laeliocattleya* seedlings illuminated by Gro-Lux lamps were more vigorous than seedlings grown under natural conditions. The growth promoting effect of the lamps appeared to be due to their particular light spectrum (Lawrence and Arditti, 1964). Perata (1981) grew some orchids under artificial light and observed that *Phalaenopsis* species and *Wilsonara* responded best to artificial light whereas *Cattleya* species flowered very poorly. True-Light lamps, when used with Gro-Lux, gave good results.

Regulation of light proved to be a key to successful commercial culture of orchids. The procedure called light gardening is now very popular among orchid growers. Extra amount of illumination is supplied to orchids either to supplement sunlight or to replace it entirely, with the result that greater amount of flowers are achieved. There are several kinds of illuminants for this purpose. For proper growth and flowering fluorescent as well as incandescent lights are necessary. Only low intensities are needed and this illumination is to be given only for short periods. The ratio between the two types of illumination received by the plants is also important in the production of flowers and this ratio varies with species. Powell

(1964) suggested using 25 watts of incandescent to every 40 watts of fluorescent. Walker and Abernathie (1964) recommended addition of 3 watts of fluorescent to every 1 watt of incandescent. Generally, the best growth and flowering were recorded by using a combination of red and blue fluorescent bulbs with 5 to 10 per cent visible radiation from incandescent bulbs.

The effects of light intensity (no shading, one 50 per cent shading net or two 50 per cent shading nets) and photoperiod on flowering rate and flower quality of *Oncidium* cultivars Gower Ramsey and Aloha Iwanaga was investigated in greenhouse conditions (Lin *et al.*, 1998). Both *Oncidium* cultivars showed the best inflorescence quality (longest flower stalks, flower stalk branches and florets) under one 50 per cent shading net, which gave a light intensity of 230-420 μmol/m^2/s at mid-day. At this degree of shading, flowering rates of Gower Ramsey and Aloha Iwanaga were 83.3 and 100 per cent respectively, and soluble sugar and starch contents of pseudobulbs were at their highest. Kim *et al.* (1999a) reported that flowering in *Phalaenopsis* was accela-rated by a photosynthetic photon fluxes (PPFs) of 123-210 μmol/m^2/s; the light saturation point was at 130-180 μmol/m^2/s. Higher PPFs (239-371 μmol/m^2/s) increased inflorescence length and diameter, and the number of inflorescence nodes and flowers. Low PPFs (64-119 μmol/m^2/s) decreased flower quality. The effect of photosynthetic photon fluxes (PPFs; upto 180 μmol/m^2/s) on flowering of *Doritae-nopsis* cv. Happy Valentine (grown at an alternating temperature of 23/28°C, day/night) was also investigated on clones and seedling plants (Kim *et al.*, 1999b). Increasing PPF increased the number of flowers and inflorescences in both clone and seedling plants. Higher axillary bud development and flowering rate were observed at higher PPFs in clones compared with seedlings. Days to flowering were also shortened.

10.2.2 Photoperiod

Although, many orchids are day-neutral and their flowering is not affected by day length, there are several species which are controlled by day length. *Cattleya* and its hybrids are some of the important examples of such orchids. *Cattleya* trianae normally initiates flower buds in March and April in Ohio. The buds reach the sheath stage in June-July, elongate in October-November and flower in December-January. By artificially increasing the length of day, flowering was inhibited until the plants were again exposed to short days. In this way it was possible to delay the crop until Easter (Holdson and Laurie, 1951).

In the genus *Cattleya* itself, there are both short-day and long-day plants. *C. warsce-wiczii, C. gaskelliana, C. warneri, etc.* are long-day plants while *C. labiata, C. mossiae, C. trianaei* are short-day plants. *Cattleya* hybrids were observed to produce almost 100 per cent more pseudobulbs per plant by long-day (supplementary lighting from 10 pm to 2 am) treatment than short (shading from 4 pm to 8 am) or normal days (Sheehan *et al.*, 1965). The flowering of *C. labiata* and its hybrids could be timed for any date one wishes (Northen, 1970). In *Cattleya bowringiana* and *C. trianaei*, flowering was hastened by 47-49 days by short-day (8 h) treatment, but with *C. bowringiana*, this treatment reduced the flower number per spike (Bhattacharjee, 1980). Eleven *Cattleya* hybrids showed variation in the season of flowering. Exposure to long-day (16 h/day) for 170 days flowered in July when the hybrids do not flower under natural day length (Cocozza *et al.*, 1995).

Effects of photoperiod on growth and flowering were also studied in some other orchids. Bose and Mukherjee (1977) recorded earlier flowering (42-49 days earlier than in plants held in long-days) in *Aerides multiflorum, Renanthera imschootiana* and *Rhynchostylis retuse* by short-day (9 hours light) treatment. Long-day (15 hours light) treatment, on the other hand, did not even produce flowers in *Phaius tankervilliae.* Flowering in *Dendrobium phalaenopsis, Phalaenopsis amabilis* and *P. schilleriana* was hastened by 50-59 days under short-day (8 hours light) treatment. However, leaf area index of both *Phalaenopsis* species was reduced by the treatment. The flower spike length and number of flowers per spike increased by long-day (16 hours light) treatment (Bhattacharjee, 1979). Short days advanced flowering date in both juvenile and adult plants of *Phalaenopsis* at 23 and 22°C. At 28°C, no flowers were initiated. Plants grown under short days (8h) had longer stalks and more flowers than those grown under natural day length. Most flower stalks were borne at the 4th and 5th nodes below the uppermost leaf-node (Yoneda *et al.*, 1991). The correct ratio between the light and dark periods also initiated the production of phytochrome. It occurs in two forms, one absorbing light in the red and the other in the far red region (Arditti, 1966b).

Day lengths of 16 and 10 h gave flowering rates for *Onicidium* cv. Gower Ramsey of 100 and 50 per cent respectively. Flowering rate was promoted when plants grown under a 10-h photo-period were moved to a 16-h photoperiod, but flowering rate was reduced when plants were moved from a 16- to a 10-h photoperiod. The cultivar Aloha Iwanaga was not as responsive as cv. Gower Ramsey to daylength. Only flower stalk length and floret number of the flower quality parameters increased under a long daylength (Lin *et al.*, 1998).

10.2.3 Temperature

Temperature markedly regulates the growth and flowering of orchids (Rotor 1950, 1952; Richardson, 1955; Milletti, 1967; Holman, 1976). Although various species vary in their individual requirement of optimum temperature, most of the cultivated orchids thrive in a day temperature varying from 15.5° to 21.0°C and night temperature of 10.0° to 15.5°C. Depending upon their preference for a particular range of temperature, the cultivated orchids are categorized as cool, intermediate and warm species. Sheehan and Sheehan (1979) suggested that the optimum range of temperature for successful growing of *Cymbidium* and *Dendrobium* species is between 10 and 20°C. *Arachnis, Cattleya, Oncidium, Paphiopedilum, Phalaenopsis* and *Vanda* require a temperature between 15 and 21°C during the growing period. The optimum temperature requirement also depends on the environmental conditions of the habitat of a particular species and not always for the genus as a whole occurring naturally at different locations.

For optimum growth and flowering, night temperature should be lower than day temperature, preferably by a margin of about 5.0°C. *Cattleya labiata* does not flower when night temperature is higher than 70°F (21°C) (Northen, 1970). Most *Cymbidium* and *Cypripedium* species need a temperature of 12.5°-15.0°C during nights for proper growth and flowering (Wilcox, 1960; Helton, 1961). Sakai *et al.* (1981) studied the influence of temperature conditions during winter on flowering of *Cymbidium* cvs. Sazanami, Otome, Sazanami Champion and Sweetheart also. Plants were brought to the greenhouse at the time of new pseudobulb appearance and the night temperatures were maintained at a minimum of 8°, 12° or 6°C or ambient (unheated) from 5th October to 30th March after which the plants were kept at the same temperature until flowering. It was recorded that higher the night temperature, faster was the growth and flowering of new pseudobulbs, but heating to 6°C had no noticeable advantage over ambient temperature. Heating was most effective with cv. Sazanami Otome, an early flowering cultivar. Under higher night temperature flower abortion frequently occurred where flower stems appeared before July. A two-year comparison of 2 day/night temperature regimes (21/21° or 21/18°C) for growing *Phalaenopsis* showed that lowering the night temperature by 3°C delayed flowering date in the first and second year by 1 and 2 months, respectively. The number of flowers per inflorescence and the total number of inflorescence were unaffected (Schenk and Brundert, 1982).

The temperature between 18° and 32°C was found to be the most suitable range for successful cultivation of *Phalaenopsis* (Quyu, 1959). Many species of orchids are induced or promoted to flower by low temperature (Kosugi, 1952; Leffring, 1972; Kronenberg, 1976). *Phalaenopsis schilleriana* required 2-3 weeks period of low temperature for floral induction (Rotor, 1952; Vries, 1953). Adult *P. amabilis* plants grown continuously, at 28°C produced leaves but no flowering stalks. A temperature less than 25°C was required for more than 12 h/day to produce reproductive growth. Flowering stems emerged after 6-8 weeks' exposure to low temperature (Sakanishi *et al.*, 1980). Schenk and Brundert (1983) reported that chilling at night for 6 weeks during summer advanced flowering in potted *Phalaenopsis* seed-ling while at both night and day, it was delayed. He suggested that the plants should be chilled when the leaf area reaches 150 cm^2. For Christmas market, plants potted in February should be chilled at night in June-July for 4-6 weeks at 14°-15°C.

Cut flower crops of *Vuylstekeara cambria* 'Plush' mainly flowers during summer. Os and Mourik (1991) obtained 34 and 71 per cent increase in flower yield by growing at 14°C rather than 17°C in the first and second seasons, respectively. Growing at higher temperature adversely affected yield and quality.

Hirai and Mori (1996) observed that cooling of *Zygopetalum mackayi*, *Dendrobium* hybrid and *Phalaenopsis* spp. in 1-m-high tunnel covered with a heat insulating material and maintaining a temperature of 4°-7°C lower than that of in ordinary greenhouse resulted in flowering 1 month earlier than the controls.

Kim *et al.* (1999c) studied the effect of temperature treatments on flowering of *Doritaenopsis* cv. Valentine propagated from leaf segments (clone) and or from seeds (seedling). Inflorescence length and diameter showed no difference between alternate temperature (23 ± 3/18 ± 2°C, day/night) and constant temperature (23 ± 3°C) treatments. Flower number/inflorescence was higher in the alternate temperature treatment, and better growth and flower quality were observed in clonal plants compared with seedling plants.

Cooling of *Phalaenopsis* plants covered with heat insulating material and cooled around 25°C resulted in early flowering and good quality flowers (Hirai and Mori, 1997).

Mesophyll cell collapse was reported to occur in *Phalaenopsis* plants during winter months (Sheehan and McConnell, 1978). One or some leaves became yellow to tan then to black, finally looking like a virus infection. The next leaf, however, developed normally, indicating it was not a virus infection. This disorder occurred on cold nights having a temperature below 7°C or when cold water is applied to leaves that are almost mature. The malady could be prevented by maintaining the temperature of greenhouse above 15°C and by using water having suitable temperature (Sheehan, 1980).

In Japan, Komori and Yoneda (2001) cultivated Miltonia 'Second Love' in greenhouses from autumn to spring by raising night temperature. The emergence of flowering stalk was less than 90 per cent when the minimum temperature was kept at 5°C before the night temperature was raised, while at 15°C, it was elevated to 100 per cent irrespective of the time of raising temperature and also produced early flowering.

Bose *et al.* (1999) suggested the following temperature regime broadly suitable for different groups of orchids.

Cool Orchids

This group is represented by *Cymbidium*, plain leaf *Paphiopedilum*, some *Miltonia, Odontoglossum* and *Dendrobium*. The temperature range of 15.5°-21°C during day and 10°-12.5°C at night is suitable for proper growth and flowering of orchids belonging to this group.

Intermediate Orchids

This group includes *Cattleya, Laelia, Brassovala, Oncidium, Epidendrum, Miltonia,* few species of *Odontoglossum, etc.* The temperature range of 18° to 21°C during day and 15.5-18°C at night is ideal for these orchids.

Warm Orchids

Orchids of this group is represented by *Phalaenopsis, Vanda* and *Dendrobium*. Temperatures between 21° and 29°C during day, from 18° to 21°C at night are suitable for the growth of these orchids.

10.2.4 Light and Temperature

In an investigation on the effect of light and temperature on the growth and flowering of *Phalaenopsis*, Kubota and Yoneda (1990) concluded that high light intensity substantially increased total plant dry weight, sugar content, and N absorption. The differential in assimilation and N uptake under high light intensity resulted in larger plants with lower N contents. High light intensity preceding 23°/18°C treatment increased the number of inflorescences, but the 28/23°C treatment brought the time of flower emergence in plants subjected to high light intensity forward to that of plants exposed to lower light intensities.

Kubota *et al.* (1997) reported that in light at 20°C, *Phalaenopsis* produced few spikes 21.3 days after treatment. No buds developed on plants kept in 20°C in dark after 36 days but when transferred from the dark to the light they spiked 18.6 days later. When *Epicattleya* cv. Siam Jade with red spotted splashed petals and yellowish green sepals flowered in light under high temperature (32° /28°C day/night), chlorophyll and carotenoid contents were low in sepals and petals, whereas anthocyanin contents were high in petals and lips. Darkness also reduced chlorophyll and carotenoid contents but increased more of anthocyanin (Matsui and Tanase, 1995).

In *Dendrobium nobile* cultivars (Malones Fantasy, Golden Blossom Kogane, Snow Flake, Red Star, Yukidaruma, a King) plants grown under 7 different day/ night temperature with 30°C day temperature and 15, 20, 25, 30°C night temperature and vice versa, 250/50 mol/m²/s natural light for 9 h photoperiod. The time of flag leaf appearance was influenced by ADT (average of daily temperature) and they appeared earlier as ADT increased (Chae *et al.*, 1998). *Dendrobium kingianum* photosynthesized only when the PPF was 63 or 125 µmol/m²/s and the temperature was 10 or 15°C. The optimum temperature for photosyn-thesis of *Dendrobium* cv. Permos Glory was 25 °C, above which photosynthesis decreased sharply. Photosynthesis also increased as PPF increased upto 63 µmol/m²/s At a PPF of 12 or 25µmol/m²/s (Paek *et al.*, 1998).

Time of flowering and flower quality in cymbidium orchids is influenced by the temperature (Li *et al.*, 2001). Transferring the plants of *Cymbidium* Sazanami after the initiation of flower buds in three environments *viz* cold room (8-28 °C), warm room (15-30°C and hot room (20-30 °C) delayed the flowering in plants which were transferred in cold room whereas the flowering was advanced by one in plants transferred to hot room as compared with plants subjected to warm room. Warm room provided optimum temperature for flowering of cymbidium. Normal growth in vegetative buds occur at 30/25 °C and low temperature (2°C) and high light intensity during winters encourages leads and accumulation of assimilates (Komori and Yoneda, 2002). Studies on the initiation and development of flower buds in *Cymbidium* revealed that the highest percentage (33 per cent) of plants with visible flower initials occurred in those grown at the lowest possible temperature using full ventilation (Klougart, 1973). In *Cymbidium ensifolium* var. misericors day/ night temperature (30/25°C and 25/20°C) accelerates pseudobulb development. The lead production in Cymbidium cv. Lovely Angel 'The Two Virgins' was affected by temperature and light. Cultivating Lovely Angel 'The Two Virgins' at 2°C under full sunlight nearly doubled the production than the cultivated at the same temperature under 50 per cent shade (Komori and Yoneda, 2002). Low light intensity reduces fresh and dry weight of shoots and roots and also causes severe reduction in flower stalk emergence (88.89 per cent). For proper growth and flowering of cymbidium orchids the humidity range between 50-80 per cent is necessary. The excess humidity in growing environment increases the incidence of fungal and bacterial diseases. The excess humidity should be managed by providing proper ventilation for air circulation and the low humidity can be managed by installing misting or fogging devices in the green houses. The areas lying between 1400-1800 meter altitude in North Eastern Himalayas offer congenial climate for growing of cymbidium orchids. In these areas these can be cultivated with slight manipulation in growing environment.

Zheng and Matsui (1998) studied the effects of dark treatment of pseudobulbs on sprouting of lateral buds and subsequent growth were investigated in 3 cultivars of *Cymbidium* Miss Taipei, Elf's Castle, 'China Peak' and Hong Gui. Effects of dark treatment became obvious 4 or 5 months later after the start of the treatment because it inhibited defoliation from 2- and 1-year old pseudobulbs and stimulated healthy root development. The increased number of roots could be responsible for new shoot development. Hong Gui developed flower stalks from back pseudobulbs and bloomed in the dark and light. The rate of flower stalk development was slow in plants which produced numerous shoots. These findings indicate that dark treatment of pseudo-bulbs could hasten clonal multiplication of *Cymbidium* by division.

The photosynthetic rate of Cymbidium cv. Water King peaked at 25°C and increased with photosynthetic photon flux (PPF) from 25 to 125 $\mu mol/m^2/s$. Photosynthetic rate decreased sharply at temperatures above 30°C, irrespective of the PPF. Respiration was observed in darkness and even at a PPF of 12 $\mu mol/m^2/s$, and increased as temperature increased (Paek *et al.*, 1998). An additional supply of CO_2 at high rate of 2000-2300 ppm also increased the growth of Cymbidium hybrid seedlings, as compared to those grown without additional CO_2 (Berg, 1965). Humidity is also important under orchid houses. Ohsawa *et al.* (2000) reported that flowering in Cymbidium was earlier under fog cooling system in greenhouse than in the open.

Some bulbs of *Miltonia* were exposed to 15°C, 16 h photoperiod and a light intensity of 20 $\mu mol/m^2/s$ for 40 days whereas others kept at room temperature under natural photoperiod and light intensity. Growth and flowering of *Miltonia* cultivars were accelerated by artificially chilling of the plants for 40 days (Matsui and Yoneda, 1997).

The effect of light intensity and temperature on photosynthesis and respiration was investigated by Paek *et al.* (1998a) on orchids of several genera. The photosynthetic rate of *Cattleya* cv. Spring Mount was highest at 20°C, but a similar photosynthetic rate was achieved upto 25°C at a PPF of 63 $\mu mol/m^2/s$. Respiration occurred at temperatures greater than 30°C, irrespective of the PPF. The photosynthetic rate of *Phalaenopsis* cv. Lipstick responded to temperature and PPF in a similar fashion to *Cattleya*, except photosynthesis was observed at 30° C when PPF was 25 $\mu mol/m^2/s$. Respiration was observed only at 35° C, irrespective of the PPF.

In *Doritaenopsis* cv. Happy Valentine, Kim *et al.* (2000c) reported that chlorophyll content was lower at low temperature (15/10°C day/night) and high light intensity (150 $\mu mol/m^2/s$). Chlorophyll content was highest at day/night temperatures of 25/20°C. Photosynthesis was the highest at PPFD 100 $\mu mol/m^2/s$ and 25/20°C day/night, and decreased at 50 $\mu mol/m^2/s$. Changes in stomatal conductance and trans-piration were similar to that of photosynthesis. Intercellular CO_2 concentration increased at low light intensity (50$\mu mol/m^2/s$). CO_2 uptake showed typical CAM type photosynthesis. Sugar and starch contents increased in the day/night treatment of 25/20°C, but decreased at high temperature and light intensities.

When plants achieve a certain size after a specific period of vegetative growth, they bloom under suitable temperature and light conditions. For example, most Phalaenopsis species and hybrids must be exposed to relatively cool temperatures, *i.e.*, <28°C, to trigger elongation of the spike (Lee and Lin, 1984). Cypripedium flavum, an alpine orchid, requires two years from floral bud formation to flowering. In the first year, two new buds are formed at the lateral base of the two-year-old bud. They then differentiate seven to nine spires before becoming dormant in winter. In the second year, one of those younger buds develops into a floral bud that then produces flowers and fruits in the third year (Weng *et al.*, 2002).

10.2.5 Light, Temperature and CO_2

Effect of photosynthetic photon flux (PPF) day/night temperatures and CO_2 concentration on CO_2 exchange for two *Phalaenopsis* hybrids were determined by Lootens and Heursel (1998) (Table 7). At 20°C, the saturating PPF for photosynthesis was 180 μmol/m²/s. At this PPF and ambient CO_2 level (380 μl/l), a day/night temperature of 20/15°C resulted in the largest daily CO_2 uptake. Higher night temperatures probably increased the respiration rate and lowered daily CO_2 uptake in comparison with 20/15°C. An increase in the CO_2 concentration from 380 to 950 μl/l increased daily CO_2 uptake by 82 per cent.

Table 7: Main Effects of Cultivar and Day/Night Temperature Regime on Cumulative CO_2 Uptake during 24-h (Pn) by Leaves of Two Phalaenopsis Cultivars*

Factor	P_n^{-1} *(mmol $CO_2 m^{-2}$)*
Cultivar	
70	167 b
L	114 a
Temperature	
20/15°C	159 my
20/20°C	133 n
25/20°C	139 n
25/25°C	131 n
Significance	
Cultivar (CV)	***
Temperature (TR)	***
CV x TR	NS

Determined at a PPF of 180 μmol $M^{-2}s^{-1}$, a 12-h photoperiod, and a relative humidity of 65 per cent.

y Mean separation among cultivars by ANOVA and among temperature regimes by Duncan's multiple range test at p = 0.05 (m. n).

NS, ***Non-significant or significant at $P \leq 0.001$, respectively.

*Lootens and Heursel (1998).

Kim *et al.* (2001) studied the effect of different shading treatments on the CO_2 uptake and growth and development of *D. phalaenopsis*. The chlorophyll content was lower at higher light intensities (35000-58000 lx) compared with the other light treatments, while CO_2 uptake, stomatal conductance, and transpiration rate were higher. The intercellular CO_2 concentration increased at low light intensities (4000-10000 lx). One year-old seedlings had higher leaf and node numbers and fresh weight at 25000-31000 lx, whereas 3-year-old seedlings had higher flower numbers and flower production rate at higher light intensities from March to June and at medium light intensities from July to December.

10.2.6 CO_2

Carbon dioxide concentration in the atmosphere also influences the growth of orchids (Frackowiak, 1933; Tsuchiya, 1935; Meyer, 1943; Stockdale, 1950). Cattleya seedlings grown in an atmosphere with an additional supply of 0.1 or 1 per cent carbon dioxide were found greener, stouter and had thicker leaves than those grown in normal atmosphere (Stockdale, 1950).

10.2.7 Humidity

Orchids, in general, prefer high humidity for their proper growth and flowering. The optimum requirement, however, varies depending on the species. According to Quyu (1959), humidity for the successful culture of *Phalaenopsis* should be about 70 per cent. For *Odontoglossum*, a relative humidity of 75 per cent was reported to be satisfactory (Richardson, 1955). Monopodial types, *e.g.*, *Vanda*, *Phalaenopsis*, *etc.* require high humidity (70-75 per cent) than sympodial types, *e.g.*, *Cattleya*, *Laelia* requiring 40-55 per cent. High humidity between 75 and 80 per cent was also reported to promote growth and flowering in orchids (Bose and Bhattacharjee, 1980). Mukherjee (1983) reported that humidity should not be less than 30 per cent at night and above 80 per cent during day time. Almost similar suggestion was made by Yadav and Bose (1985). Microclimate of the orchid house was more favourable by spraying at intervals and mechanical ventilation (Chen *et al.*, 1999). The temperature for *Oncidium* can be reduced to 28°C with ventilation at 1.2 m/s. Spraying for 30s every 8 minutes was the most efficient evaporative cooling system.

10.2.8 Air

Orchids require fresh air throughout the year for normal growth and flower production. As most of the commonly cultivated species are epiphytic, free circulation of fresh air is essential. Many tropical orchids like *Vanda*, *Phalaenopsis*, *Aerides*, *etc.* with aerial roots were found to grow well when exposed to current of fresh air. Perfect ventilation which maintains an uniformity of temperature inside the house and also reduces the occurrence of diseases and physiological disorders in orchids is important for green or glasshouse management (Milletti, 1967; Bose and Bhattacharjee, 1980). Flap ventilation is good, although automatic operation of ventilation equipment is necessary for large commercial green or glasshouse.

10.3.0 Growth Rate and Flowering Gradient

In some species, the differentiation of flower primordia in the meristems occurs when stem growth ceases.

Many *Aranda* hybrids exhibit a flowering gradient along the stem axis (Goh, 1975, 1977a). Decapitation was found to induce flowering in these hybrids (Goh and Seetoh, 1973). When the cut was near the apex, the plants produced either inflorescences or mixture of inflorescences and vegetative shoots. When the cut was low, only vegetative shoots were produced (Tables 8 and 9). Investigations with fully grown plants showed that in *Aranda* Lucy Laycock as well as *A.* Mei Ling, decapitation at the 15th or the 16th node (counting from shoot apex basipetally) still caused the production of inflorescences, in *A.* Hilda Galistan and *A.* Nancy, this level was found to be around the 18th to 20th node. In *A.* Deborah, inflorescences were still produced when decapitated at the 25th node (Goh, 1975, 1977a, 1978). Almost similar observations were made with *A.* Peter Ewart, *A.* Christine No.1 (Koay and Chua, 1979) as well as *Aranthera* James Storie and *Holttumara* Maggie Mason (Koay and Chua, 1980).

Naturally occurring hormones play an important role during the process of plant growth and development. In fact, involvement of auxin in the regulation of flowering in orchids was suggested as early as 1953 (Vries, 1953). Exogenous application of growth regulators may induce or promote flowering, prevent or delay it. Plants of *Phalaenopsis schilleriana* sprayed with auxin solutions were less abundant in flowering and it was suggested that there seemed to be an excess auxin which inhibited the production of inflorescence (Vries, 1953). The flowering intensity of *Vanda* Miss Joaquim was found to be inversely correlated with the auxin level in the shoot apex. Plants exposed longest to direct sunlight and producing profuse flowering had the lowest level of auxin, whereas those given 4 hours of direct sunlight remained vegetative and had the highest level of auxin (Goh and Wan, 1974).

Table 8: Frequency of Flower Production of *Aranda* Deborah during 1-month Period after Transplantation on 10 January, 1970*

Treatment	*No. of Plants Treated*	*No. of Inflorescences Produced*	*Average No. of Inflorescences per Plant*	*Percentage of Plants which Flowered*
Control				
Mature plant	18	7	0.39	33.3
Top cuttings	12	3	0.17	16.7
Decapitated				
Mature plant	22	73	3.32	100.0
Top cuttings	12	38	3.17	100.0

* Goh and Seetoh (1973).

Table 9: Effect of Decapitation on *Aranda* Deborah _ Flowering Responses during the 3-week Period after Decapitation*

Treatment	*No. of Plants Treated*	*No. of Inflorescences Produced*	*Average No. of Inflorescences per Plant*	*Percentage of Plants which Flowered*
Control	10	3	0.3	20.0
Decapitated	10	39	3.9	100.0

* Goh and Seetoh (1973).

10.4.0 Role of Growth Regulators

Decapitation was found to induce flowering in *Aranda* Deborah but this response could be inhibited by continuous supply of 10^{-4} M IAA solution (Goh and Seetoh, 1973). Application of cytokinin (benzyladenine) caused production of multiple inflorescences (Table 10). Similarly, antiauxins and growth retardants were found effective in stimulating flowering in *Aranda* Deborah (Goh, 1977b).

Table 10: Effects of Benzylaminopurine (BAP) and Gibberellic Acid (GA_3) on Floral Initiation and Floral Development in Aranda Deborah*

Treatment (5 days)	*No of Plants Treated*	*Time Taken for Floral Initiation (Parenthesis indicate no of initials produced)*	*Total No. of Floral Initials Produced*	*No. of inflorescences Developed to Maturity*	*Proportion of Plants Responded*
Control	5	12(1), 30(1)	2	1	1/5
BAP 10^{-3} M	5	4(8), 7(5), 12(4), 15(1), 17(1)	19	12	5/5
BAP 2 x 10^{-3} M	5	5(4), 7(8), 9(4), 15(2), 19(1)	19	9	5/5
BAP 10^{-3} M+					
GA 10^{-4} M	5	5(7), 8(2), 10(1), 14(4), 16(7), 19(3), 24(1), 27(1)	26	11	5/5
BAP 10^{-3} M+					
GA 10^{-3} M	5	5(8), 8(4), 10(3), 12(1), 14 (4), 16 (2), 19 (4), 22 (6)	32	16	5/5

* Goh (1977b).

Spraying with ethrel at 1000-4000 ppm, 1-3 times at intervals, has been found to hasten the flower induction in Golden Shower *Oncidium* (Widyastuti and Soedjono, 1979).

Seedlings from cultivars of *Brassolaeliocattleya, Laeliocattleya, Dendrobium, Phalaenopsis* and *Epidendrum radicans*, after 3-6 months of greenhouse growth, were treated with dichlorophenoxy triethylamine (DCPTA) at 30 μM. The treated plants produced a 2- or 3-fold increase in shoot growth compared with the controls (Keithly *et al.*, 1991). Application of 30 μM DCPTA during the transfer of seedlings of *Phalaenopsis* from sterile agar culture to greenhouse pots produced a 2- to 3-fold increase in root and leaf growth. After 2 years in a greenhouse, total plant dry weight of treated plants doubled. Flowering after 11 or 18 months of growth increased significantly following treatment of seedlings with 30 or 300 μM DCPTA. The number of racemes produced during a 2-year-growth period doubled in plants given 30 μM DCPTA at the seedlings stage compared with untreated controls (Keithley and Yokoyana, 1990).

Flower buds of intact or defoliated plants of *Calanthe sieboldii* when treated with GA_3 at 250, 500, 1000 mg/l advanced flowering, the effect was more pronounced on intact plants (Sawa, 1989). Treatment with 0, 5, 10 or 20 mg each of ABA and GA_3 at shoot initiation or as new shoots developed, stimulated root and shoot growth, particularly when applied in summer. Winter treatment stimulated early production of flowers (Zheng *et al.*, 1998).

The effects of growth regulators and low temperature treatment on growth and flowering of *Doritaenopsis* cv. Happy Valentine during summer were investigated by Kim *et al.* (2000b). In case of flower induction by low temperature (day/night 23/18°C), higher BA concentrations (100-400 mg/l) increased inflorescence number, but decreased inflorescence length. Number of flowers per inflorescence and flowering rate increased most at BA rates of 100-200 mg/l compared with other treatments and the control. Ethephon (100, 200 and 400 mg/l) application induced earlier flowering and decreased the rate of physiological disorders, but caused defoliation of lower leaves, and decreased flower number and flowering rate.

Treatment with BA at 0-1 per cent alone or in combination with GA_3 at 0.25-1 per cent on secon-dary flowering of *Phalaenopsis* after cutting of inflorescence was investigated by Kim *et al.* (1999d). The best flowering was observed following treatment with BA at 0.25 per cent. Application of BA in combination with GA_3 decreased the percentage of flowering. Higher BA concentrations increased inflorescence number.

In an investigation, the stimulatory effect of exogenous cytokinins and ethylene on root-to-bud conversion in *Catasetum fimbriatum* seemed to be mediated by the establishment of an endogenous auxin : cytokinins balance favourable to shoot formation (Peres *et al.*, 1999). Chung *et al.* (1999) reported that in *Bletilla striata* uniconazole was the most effective retardant for retarding leaf growth in seedlings *in vitro*. Paclobutrazol, ancymidol and N-dimethyl succinamic acid also retarded leaf growth.

11.0 Propagation

Orchids can be propagated by both seeds and vegetative means. The creation of hybrids and raising the seedlings in flasks are being practised by commercial growers throughout the world. Plants raised from seeds, however, take long time to bloom. The technique of meristem culture is used for rapid and large scale multiplication (Ilsley, 1965, 1966; Bertsch, 1967; Marston and Voraurai, 1967) and in fact it has revolu-tionized the orchid propagation. Multiplication by traditional methods of vegetative propagation such as by division, cutting, air layering, *etc.* are followed, particularly by amateurs.

11.1.0 Aseptic Culture of Orchids

Detailed account is presented under biotechnology under tissue culture. Aseptic culture method involves an elaborate process and is generally employed by commercial nurserymen and orchid breeders. This is a very useful method for fast and large scale multiplication of orchids which are multiplied very slowly by conventional methods of propagation. A million-fold increase in the rate of multiplication over traditional methods is not unrealistic now. Seeds, meristems, leaf tips, stem segments, embryos, *etc.*, all can successfully be cultured. The fusion of protoplasts of genetically distinct species and achievement of somatic hybrid plants through protoplast cultures has opened a new horizon in the field of plant breeding.

11.2.0 Seed Propagation

As stated earlier, orchid seeds do not possess any stored food material or endosperm, and for their successful germination it is necessary to protect them from fungal infection and also to supply food. There are two methods of seed culture. One is symbiotic and the other asymbiotic.

11.2.1 Symbiotic Seed Germination

This method is based on the principle that orchid seed requires fungal association for their germination (Bernard, 1899, 1903; Schaffstein, 1938). Following this discovery, a method was developed in which specific fungus was cultured first and then seeds were sown on it. Seeds were also cultured by inoculating the test tubes with both orchid seeds and fungi (Bernard, 1905, 1909; Burgeff, 1932; Irpa, 1965). The fungus grew by itself and encouraged the germination and good growth of the seedlings. The relationship between orchid seed and fungus was found to be symbiotic (Arditti and Ernst, 1974). Most of the mycorrhizal fungi of the orchids were identified as *Rhizoctonia*, the major species being *R. repens, R. mucoroides* and *R. languinosa* (Arditti, 1981).

Immature embryos as well as seeds of *Listera ovata* were successfully inoculated with a symbiotic fungus (*Epulorhiza* sp.). Symbiotic protocorms developed from embryos in all stages, while asymbiotic culture of embryo only succeeded if the capsule were almost mature (Rasmussen *et al.*, 1991).

Epulorhiza repens and *Epulorhiza* sp. (syn. *Rhizoctonia globularis*) were isolated as mycorrhizal fungi in association with *Microtis parviflora* (Perkins *et al.*, 1995). Symbiotic seed germination of *Platanthera clavellata* was investigated *in vitro* by inoculating micorrhizal fungi (*Epulorhiza* spp.). A single fungal isolate (*E. inquilina*) obtained from the roots of *P. calvellata* promoted seedling growth (Zettler and Hoffer, 1998).

11.2.2 Asymbiotic Seed Germination

This method of seed culture involves germinating seeds and growing seedlings under aseptic conditions in an agar medium supplemented with nutrients required by the plant. While studying the influence of carbohydrates on green plants, Knudson (1921) observed that a symbiotic fungus produces sugars for the initial development of an orchid embryo. Based on this, he successfully germinated *Cattleya, Laelia* and *Epidendrum* seeds on agar medium containing sugar and concluded that fungus was not necessary for orchid seed germination (Kundson, 1925). He also grew *Laeliocattleya* hybrid asymbiotically from seed to flower. This was followed by aseptic culture of *Dendrobium* by Tanaka (1957) and *Oncidium* by Livingston (1962). The validity of asymbiotic seed germination was firmly established, and the original medium (Knudson, 1946) was improved to germinate the other species of orchid (Arditti, 1967a). Inorganic salts, carbohydrates, organic substances like vitamins, amino acids, growth hormones, water and proper pH along with optimum light and temperature were necessary for success in germination. The procedure of asymbiotic seed culture and factors influencing the germination of seeds and seedling growth are discussed below.

In an endangered species, *Vanda coerulea,* seeds from capsules (pods) at various stages of development [180-260 days after pollination (DAP)] were cultured asymbiotically *in vitro*. Germination percentage incrased with increasing capsule are upto 270 DAP, then decreased. Seeds obtained from 270 DAP pods exhibited the highest germination (90.13 per cent) on Knudson C medium. A study was conducted by Kim *et al.* (1998) to investigate effective media and optimal days after pollination for germination of seeds which are crossed by selfing or reciprocal pollination in *Cymbidium* spp. Cross combinations expressing high capsule number and viability showed good germination when they were sown *in vitro*. Capsules collected 150-180 days after pollination showed good germination performance and rapid germination. Hyponex medium was the most effective for protocorm development and germination.

11.2.2.1 Culture Media

Several standardized media have been used for the germination of orchid seed. The earliest one which is still widely used is Knudson C medium, the composition of which is given below :

Monopotassium acid phosphate	0.25 g
Calcium nitrate	1.00 g
Ammonium sulphate	0.50 g
Magnesium sulphate	0.25 g
Ferrous sulphate	0.025 g
Manganese sulphate	0.0075 g
Sucrose	20.00 g
Agar	17.50 g
Water	1000 ml

Apart from Knudson C, media used in tissue culture of orchids include Murashige and Skoog (MS), Nitsch and Nitsch (NN), Vacin and Went, White, Gumborg, Bergeff, *etc.* Commercially available formulations of culture media are also used.

The choice of medium depends much on the type of orchid whose seeds are to be sown for germination. It was reported that seeds of *Dendrobium, Cattleya, Epidendrum, Odontoglossum, Phalaenopsis* and *Vanda* germinated and seedlings grew well in Knudson C medium while the seeds of *Aerides multiflorum* and *Phaius wallichi* failed to germinate in this medium. Ichihashi (1979) studied the seedling growth of several orchids in 6 different media, *i.e.*, Knudson C, a modified Vacin and Went, Kano, Thompson and 2 media devised for *Bletilla striata*. It was observed that *B. striata* made satisfactory growth in all media except Knudson's medium. Seedlings of *Laelia anceps* grew normally on Kano's medium but the germination was inhibited in Thompson's medium. *Doritaenopsis* seedlings, however, grew best in Thompson's medium.

Arditti *et al.* (1981) observed that seeds of *Calypso bulbosa, Epipactis atrorubens* and *Goodyera tesselata* germinated faster and in higher proportion in media of Norstog and modified Curtis containing urea and $CaCO_3$ instead of $Ca(NO_3)_2$ and NH_4NO_3. Nagashima (1982b) reported that hyponex was a better germination medium than Murashige and Skoog or Knudson's C medium. Lee *et al.* (1982) however, reported that percentage of *Laelia briegeri* seeds germinating in Murashige and Skoog's medium was greater than in Knudson C or hyponex medium. Vacin and Went's medium was found best for germination and growth of *Spathoglottis* plicta and *Phaius wallichii* (Nagabhusana, 1982). Numerous investigations were carried out to improve these media by supplementing with various carbohydrates, organic nitrogen sources, growth substances, *etc.* Sugars had been the most common carbon source in different media and most of the orchids utilize disaccharides such as sucrose, cellobiose, maltose, lactose and polysaccha-rides such as starch. Galactose was found toxic to most of the orchid seeds which inhibited the germination and growth of seedlings even at low level.

Arditti (1967a) studied several nutritional factors for orchid seed germination and reported that low concentrations of nitrogen promoted seed germination and further growth of the seedlings whereas high concentrations retarded these processes. Nitrates, ammonia and urea were well-utilized by orchid seeds. Several species, however, were unable to use nitrate during early stage of germination. *Cattleya* seedlings utilized nitrates only after 60 days from germination. According to Mitra (1970), NH4NO3 was the best nitrogen source for orchid seed germination and seedling growth. Growth of *Dendrobium* and *Phalaenopsis* seedlings was found to be inhibited by using urea in the nutrient media while the development of *Laeliocattleya* and *Vanilla planifolia* was enhanced. Orchid seeds were also found sensitive to phosphate and it was observed that low concentration of phosphate in a medium increased the germination (Wynd, 1933). Potassium played an important role in the germination of *Galeola septentrionalis*, a terrestrial orchid lacking chlorophyll. Soaking of seeds of this species in solutions of several potassium salts increased germination 3.0-3.5 fold (Nakamura, 1962, 1963, 1964). Similarly, excellent seed germination of several species was recorded in media low in calcium. Microelements like boron, iron, manganese, iodine, copper, cobalt and molybdenum were found to induce faster growth in orchid seedlings (Arditti, 1967a; Thompson, 1974, 1977) and therefore, most of the modern media contain these elements which, if not helpful, are not harmful.

From a number of investigations on the effects of vitamins on germination and growth of seedlings, it appeared that certain species had specific vitamin requirements (Arditti, 1967a). Ascorbic acid was found beneficial in enhancing seed germination and seedling development in *Cattleya autumnalis, Cymbidium, Oncidium pulvinatum* and *Vanilla planifolia* (Pollacci and Bergamaschii, 1940; Bouriquet, 1948; Torikata *et al.*, 1965). Thiamine was also reported to enhance orchid seed germination (Mariat, 1948; Hegarty, 1955; Arditti and Harrison, 1977). Niacin was another vitamin which proved very effective in seed germination and development in several orchids like *Phalaenopsis* hybrids, *Vanda coerulea, Epidendrum cochleatum, Cattleya* hybrids, *Laeliocattleya* and *Cypripedium* (Schaffstein, 1941; Withner, 1943; Harley, 1951; Hegarty, 1955; Arditti, 1965, 1967b, 1977).

Certain growth substances also proved effective in enhancing the germination of seeds and subsequent growth of seedlings. In majority of cases, auxins-IAA, IBA and NAA improved seed germination and seedling growth (Withner, 1974; Chennaveeraiah and Patil, 1975; Nagabhushana, 1982). Although, addition of gibberellic acid to the medium improved the growth of *Cattleya* seedlings compared to the control, the percentage of germination was poor (Hirsch, 1959).

A large number of additives like coconut water (Lawrence and Arditti, 1964; Vanstaden and Drewes, 1975), banana pulp (Anderson, 1967; Arditti, 1968), tomato juice (Rao and Avadhani, 1963; Ong, 1969), pineapple juice (Hoyt, 1960; Arditti, 1967a; Northen, 1970), potato extract (Harvais, 1974, 1982; Sahid, 1980), peptones (Withner, 1953; Tsukamoto *et al.*, 1963), fish emulsion (Griffith and Link, 1957), honey (Thomale, 1954), yeast extracts (Burgeff, 1954), *etc.*, were found beneficial in germination of seeds and seedling growth. The most useful among these are coconut water, banana pulp and juices of tomato and pineapple (Lawrence and Arditti, 1964; Arditti, 1968). Of these banana has been found consistently beneficial. Addition of 2.6 ml of ground banana pulp in this basal medium stimulated the growth of *Cattleya* seedlings (Anderson, 1967). Beneficial effects of coconut water on *Vanilla* and *Cypripedium* were reported by Hegarty (1955). The optimum concentration of coconut water for seed germination was found to be 10-15 per cent. Hoyt (1960) suggested addition of 15 per cent pineapple juice as substitute for water in Chang's medium for early germination and healthy growth of seedlings. Tomato juice was found the most effective additive both for seed germination and seedling development of *Vanda* Miss Joaquim (Rao and Avadhani, 1963; Ong, 1969). Synergistic effects of coconut water, banana extract, tomato juice and NAA or IBA on seed germination and protocorm development were reported in *Cattleya, Dendrobium, Phalaenopsis* and *Vanda* (Valmayor, 1974).

Complex organic additives, such as extracts of taro, potato, etiolated germinating soyabeans, and green pods of kidney bean as well as banana homogenate enhanced the germination and growth of *Cattleya* seedlings. Inhibitory effect was noted with higher concentrations (Islam *et al.*, 2000).

Seeds of *Eulophia cucullata, E. streptopetala* and *E. petersii* were placed in aseptic culture and germinated on MS medium supplemented with 3 per cent sucrose and 0.01 per cent *myo*-inositol. *E. cucullata* and *E. petersii* germinated after 3 months whereas *E. streptopetala* took 6 months to germinate. (McAlister and Staden, 1998).

11.2.2.2 Pretreatment of Seeds

Before sowing, seeds should properly be cleaned. They may carry on their surface spores of fungi and bacteria and, therefore, must be disinfected (Wilson, 1915; Laurie *et al.*, 1958; Arditti, 1967a; Northen, 1970). Calcium hypochlorite or the commercial bleaching powder, suggested by Wilson (1915) is still the most commonly used seed sterilant. Redlinger (1964) reported that 2.6 per cent solution of chlorox (sodium hypochlorite) was a better sterilant than bleaching powder. Jessel

(1962) recommended 3 per cent hydrogen peroxide as an orchid seed sterilizing agent. Waves and Geest (1983) reported that mature seeds sterilized in calcium hypochlorite solution for 10-15 minutes gave 75-80 per cent germination in modified Thomale's medium.

Thomasen (1964) suggested soaking of seeds for 4 hours in a mixture of 4 per cent honey in water at a temperature of 36.4°C. The honey stimulated the fungal and bacterial spores to germinate while the seeds remained intact. Decantation of the solution and addition of calcium hypochlorite solution or hydrogen peroxide killed the germinated spores quicker with a weaker solution of the disinfectant than in case of the ungerminated spores. Luks *et al.* (1977) recorded 100 per cent germination in seeds of *Bletilla striata* sterilized with alcohol. The use of certain fungicides and bactericides like 1-10 ppm amphotericin, 5-50 ppm benomyl, 5-50 ppm gentamicin, quitozene, 5-50 ppm vancomycin, either singly or in combination for killing spores of fungus and bacteria was also suggested by Brown *et al.* (1982). These substances were found to have no deleterious effects on seed germination of various species of *Cattleya, Vanda, Cymbidium* and *Zygopetalum*. Miyoshi and Mii (1995) soaked the seeds of *Calanthe discolour* for 7 days in solutions of NAA (1-100 mg/l), ethephon (1000 mg/l) or BA (10 and 100 mg/l) prior to germination in a growth regulator-free medium resulted in 1.5-2.9 fold increased, in protocorm formation compared with control cultures (Table 11).

Table 11: Effects of Phytohormone Pre-treatment on Seed Germination and Proto-corm Formation of *Calanthe discolor* Lindl. Seed*

Phytohormone	*Concentration (mg l^{-1})*	*No Germination*	*Pre-germination*	*Germination*	*Protocorm*
NAA	0	8.0	54.5	30.3	7.3
	0.01	8.8	55.2	26.6	9.4
	0.1	8.3	51.8	30.1	9.8
	1	10.9	38.7	31.7	18.7***
	10	7.8	36.9	34.2	21.1**
	100	8.0	37.4	39.8	14.9**
BA	0	6.4	52.3	33.4	7.9
	0.01	4.9	57.4	27.6	10.2
	0.1	5.7	54.2	28.8	11.3
	1	4.8	54.6	30.5	10.1
	10	7.4	48.0	32.6	12.0
	100	41.2***	15.1***	25.5	18.1***
Ethephon	0	11.6	48.1	31.7	8.7
	10	11.0	44.7	32.7	11.6
	100	10.9	44.6	32.1	12.3
	1000	13.0	35.0	31.4	20.7

Values followed by asterisks in each phytohormone treatment within the same column are significantly different from control (0 mg l^{-1}), using Student's t-test, at : ** 5 per cent level; *** 1 per cent level.

* Miyoshi and Mii (1995).

11.2.2.3 Flasking

Flasking of seeds, the most important process, should be done very carefully. The hands should properly be washed in a solution of sterilant. A pair of sterilized gloves may also be used. The surface on which this operation is to be done should be wiped with cotton dipped in sterilizing solution. Open the container, flame the neck over a Bunsen burner or spirit lamp, pick up the wire loop, heat it, and then pick up a loop full of seeds and transfer them to the flask. Flame the neck of the flask and put back the stopper. Gently swirl the flask so that the seeds are evenly distributed over the surface of the solidified medium. The flask is then kept on a shelf and maintained in suitable light and temperature.

11.2.2.4 Germination and Seedling Growth

Depending on the genus, the inoculated seeds will turn green after only a few days or a few months. The seeds enlarge in size assuming the shape of a spherule. It develops hairs and a depression on top, where the first leaf primordium appears after 2-3 months. Light period of 12-18 hours and intensity of about 2,000-3,000 lux were reported to be suitable for seed germination in most orchid species (Arditti, 1967a; Mukherjee, *et al.*, 1974). According to Lee *et al.* (1982) light increased seed germination in *Laelia briegeri.* Cultured seeds of *Cypripedium reginae* held in light did not germinate (Reyburn, 1978). Arditti *et al.* (1982) observed that seeds of *Goodyera oblongifolia* also germinated well both in light and dark whereas seeds of *G. tesselata* germinated only in light. In an experiment on the effect of light on germination and protocorm growth of *Calanthe* cv. Satsuma, Fukai *et al.* (1997) reported the highest germination and development of protocorms in the dark. Islam *et al.* (1999) found that red, yellow and blue light significantly enhanced the germination of *Cattleya walkeriana* hybrid seeds and increased both fresh and dry weight of seedlings *in vitro* on solid Kyoto medium.

Temperature extremes for germinating seeds of orchids were reported to be 6° to 40° C, the optimum range being 22-29°C. According to Harvais (1973) and Mukherjee *et al.* (1974), the optimum temperature for the germination of most of the species was 20-25°C.

A high production of protocorms from seeds of *Phalaenopsis* was obtained by the use of a modified medium with 2 g peptone and 50 g banana juice/litre and this medium, with or without the banana juice, gave a high proportion of green protocorms. A high rate of production of green protocorms was also given by a medium with 100 g potato juice/litre in place of the banana juice (Kimura and Kurihara, 1991). *In vitro* propagation by seeds of rare species of orchids in Kenya including *Ansellia africana, Angracium dives, Solenangis wakefieldii, Polystachya bella* and *Arangis confusa* has beeen reported by Bytebier *et al.* (1994).

11.2.2.5 Transplanting of Seedling

Seedlings should remain in the flask for about 10 to 12 months before they are transplanted to community pots. Sometimes, depending on the genus, seedlings are reflasked into a fresh flask with medium resulting in rapid growth. Seedlings are taken out carefully one by one with the help of a looped wire, put in water till agar is washed, and then dipped in a mild solution of fungicide. A wooden flat having either moss alone or moss and finely cut osmunda or fern fibre in equal proportions may be used as transplanting beds. After transplanting, the flats should be kept away from direct sunlight preferably under glass. Seedlings in these flats are allowed to grow for a year or so and then planted in individual small pots. Later, they are transferred to larger pots. Half strength MS medium supplemented with 2.4 mg/l IBA, 3 per cent sucrose and 60 $\mu mol/m^2/s$ PPFD were found as suitable conditions for optimal induction of roots. In vitro regenerated *Cattleya* plantlets were successfully transferred to potting medium containing bark:perlite:peatmoss (1:1:1) and perlite:peatmoss (1:1), and 98.3 and 80.7 per cent were survived respectively (Dewir *et al.*, 2015). Four substrates were compared (peat, no. 2 gravel only, mixture of no. 2 gravel and peat at a 3:1 ratio, and xaxim) for use in *ex vitro* growth of *Cattleya forbesii* (Junior and Venturieri, 2011). A substrate x species interaction was observed for the variables vigor and height, showing that each species has its own specificity for substrate. For *C. forbesii*, xaxim placed first in two of the four assessed parameters. Next came the gravel:peat substrates and peat in the second position for two parameters (Junior and Venturieri, 2011).

11.3.0 Vegetative Propagation

Though easier to follow, these are very slow processes and only few plants can be multiplied from one plant in a year. However, one can maintain the uniformity in a particular type. It is also difficult for a common grower to raise plants from seeds which require good amount of experience and knowledge.

11.3.1 Division

In fact, before the development of techniques of aseptic culture of orchids, the major means of clonal propagation in case of a large number of species was the division of plants. The method involved in dividing large clumps into smaller units can be adopted satisfactorily for orchids belonging to sympodial group. It was however, suggested not to divide the plants unless there were 8-12 pseudobulbs of good size (Northen, 1970; Arora *et al.*, 1975). *Coelogyne, Dendrobium* and *Oncidium* should be allowed to develop into sizeable clumps with 6-8 shoots.

This method of propagation is suitable for genera like *Cattleya, Laelia, Cymbidium, Dendrobium, Coelogyne, Zygopetalum, Epidendrum, etc.* Multiplication of *Phaius* by division of pseudobulbs was reported as the most suitable and popular method which produced flowering sized plants earlier than any other method (Mukherjee, 1979). Generally after flowering, when the new shoots are healthy, robust and with a cluster of fresh roots, they should be separated with a clean knife without damaging the roots.

Treatment of pseudobulbs with several chemicals proved very effective in promoting root formation. Backbulbs of *Cattleya* treated with IBA at 2000 ppm induced better rooting than the untreated ones (Withner, 1951). Increase in shoot and root growth of *Cymbidium* (Kofranek and Barstow, 1955) and improved rooting in *Dendrobium aggregatum* (Nagabhusana, 1982) were also reported with the application of IBA at 100-200 ppm. Another chemical found beneficial for root formation, was

NAA. Delizo (1974) recorded 86 per cent rooting in pseudobulbs of *Cattleya* hybrid treated with NAA at 90 ppm as compared to only 14 per cent in untreated materials.

11.3.2 Offshoots or Keikis

Orchids with long cane-like stems, as *Dendrobium* and *Epidendrum*, sometimes produce little plants with roots at the nodal region of their pseudobulbs or stem. These offshoots (keiki) can be separated from the parent plant and set in small pots con-taining compost recommended for mother plant. Offshoots are also produced in *Phaius* and *Phalaenopsis* if their flower stems are cut off and laid horizontally in moist sand, sphagnum or other medium (Laurie *et al.*, 1958; Hartmann and Kester, 1972; Bose and Bhattacharjee, 1972, 1980). Emergence of offshoots was also reported in *Cymbidium* (Arora *et al.*, 1975) and *Paphiopedilum* (Sampolinski, 1983).

11.3.3 Cutting

Monopodial species are difficult to propagate vegetatively. Cuttings, preferably with a few roots, are sometimes successful with *Vanda, Renanthera, Arachnis, Phalaenopsis, etc.* Stems should neatly be cut in sections of 3-4 nodes and placed in moist sand or damp sphagnum moss, kept in a cool and dry place for root formation. Flower stalk cuttings were also used for the multiplication of some orchids like *Phaius, Thunia, Calanthe* and *Phalaenopsis* (Mukherjee, 1979). Treatment of cuttings with growth substances stimulates rooting. Hartono (1978) reported that application of NAA or IBA at 5000 ppm enhanced rooting in terminal cuttings of *Renanopsis* Lena Rowold. Goh (1983) recorded 100 per cent rooting in *Aranda* cuttings treated with NAA (10^{-4} M) as compared to 20-25 per cent in the untreated ones.

11.3.4 Air-layering

Vanda and some other monopodial orchids may also be propagated by air-layering (Hartmann and Kester, 1972; Bose and Bhattacharjee, 1972, 1980). The method suggested was to give a slant cut partially through the stem, about 20-30 cm below the apex and wrap with sphagnum moss or osmunda fibre around the cut portion. The rooting media should be kept moist and when the root formation is completed the layer is removed from the mother plant and planted in pot having compost used for growing the mother plant.

11.3.5 Other Methods

Grafting was also tried to propagate *Vanilla* orchids. Though graft union was successful, plants survived for a period of only two years (Muzik, 1958). In a British orchid, *Hammabya paludosa*, new plants are produced from the nodules appearing on the edge of leaves under moist conditions. Similarly, roots of *Listera* ovata produce knob-like thickenings which are used for the production of new plants (Summerhayes, 1968). In orchids like *Peristylus stenostachyus, Nervilia* spp., *etc.*, roots are produced from above the tubers, each of which ends in a tubercle. The small tubers were reported to produce new plants. Rhizomes of *Goodyera* produce special lateral bran-ches which turn up and form aerial shoots. When properly rooted, they are detached from the mother plant and established separately (Yadav and Rao, 1986). Cells of a suspension culture of *Doritaenopsis* cv. New Toyohashi strain 643 were placed in a mixture of 2 M glycerol and 0.4 M sucrose for 15 minutes at room temperature and then dehydrated with a vitrification solution (PVS2) for 1-3 h on ice and plunged into liquid nitrogen. The highest viability (64 per cent by 2.3, 5-triphenyltetrazolium chloride stainability was obtained when the cells were precultured in liquid New Dogashima medium with 0.1 M sucrose and 1.0 mg abscisic acid/l for 1 week at 25°C in the light. Protocorm-like bodies were induced from cryopreserved cells without morphological variations (Tsukazaki *et al.*, 2000).

11.3.6 Micropropagation

For a long time, the major means for propagating orchids clonally was the division of plants. As mentioned earlier, this is a very slow process for obtaining large number of plants and it is also very difficult to multiply monopodial species by such method. It was, thus considered necessary to develop method of vegetative propagation which will produce large number of plants from single individual. Morel (1960) first tried to culture the meristem tips of virus infected *Cymbidium* and succeeded in obtaining plantlets free of infection. The technique was further improved by him and hundreds of plants were obtained from a single bud. Following this, tissue culture methods for the clonal propagation of orchids were introduced by several workers using various plant parts and a number of media (Morel, 1963, 1964a,b, 1965a,b; Wimber, 1963; Bergman, 1972a,b).

11.3.6.1 Explant

Various plant parts like shoot-tip or meristem, leaf and leaf segments, stem segments, pseudobulbs, floral parts, aerial roots, *etc.*, have successfully been used for *in vitro* propagation of orchids. Among these, meristem and leaf-tip culture are, however, more popular and have extensively been used for the propagation of commercial species and hybrids. Tissue culture methods for the clonal propagation of many species of orchids and their hybrids have been standardized using various plant parts and number of media.

Goh (1990) described the media and culture conditions required for shoot-tip and axillary bud culture of nearly 30 *Aranda* hybrids. Arditti and Ernst (1993) summarized micropropagation protocols of hybrids and varieties of 83 orchid genera which include *Arachnis, Aranda, Aranthera, Cattleya, Cymbidium, Dendrobium, Holttumara, Lycaste, Paphiopedilum, Miltonia, Odontoglossum, Phaius, Mokara, Oncidium, Phalaenopsis, Spathoglottis, Vanda* and many others.

Shot Tip and Meristem

Protocorm-like body formation from shoot-tip explants of *Oncidium* was best in Vacin and Went medium supplemented with 15 per cent coconut milk and plantlet differentiation in MS medium supplemented with 200 mg casein hydrolysate/litre (Bagde and Sharon, 1997). The highest percentage of success had been with axillary buds weighing 2 g and 4.9 cm long (Kim *et al.*, 1970). Morel (1970) suggested using the buds from the base of the young shoots for the propagation of *Cattleya* from bud meristem *in vitro*. The growing points excised from dormant buds on old pseudobulbs of *Cattleya* produced protocorm-like bodies which could be propagated like protocorms grown from meristem of young shoots (Vajrabhaya, 1978). Working with *Cattleya*, Ishii *et al.* (1979) reported that fully mature lateral buds excised from new pseudobulbs, 9-12 cm long, caused better survival and adventitious bud formation than buds of younger pseudobulbs. Rhizome (pseudobulb) explants of 4 species and hybrids of *Cymbidium* were cultured containing homogenates of banana and tomato fruits and potato or coconut milk. Shoot growth was greatest when explants were cultured on a medium containing 10 mg/l for 10 days then transferred to a medium containing 0.5 mg BA/litre (Paek *et al.*, 1989). Vij and Abhilasha Dhiman (1997) reported that modified Ichihastri medium supplemented with BA at 2 mg/l favoured development of multiple shoot-buds from pseudobulb segments of *Bletia striata*. The best result in the propagation of *Brassocattleya* was obtained by using the third bud from 15 cm long, newly growing shoots (Arditti, 1977). In case of *Cattleya* he recommended the use of enlarging, non-dormant axillary shoots. According to Sagawa and Kunisaki (1982), the apical shoot or axillary bud was the best source of explants for the successful clonal propagation of orchids.

Torres and Mogollon (1998) described clonal propagation of *Cattleya luddemanniana* and developed protocol in 4 different phases : (i) culture initiation, from the caulinary apex to active shoot growth on filter paper bridges, with Huang's modification of MS medium, (ii) shoot multiplication in the same medium on a polypropylene membrane raft at 2800 lux, with 3.0 mg BA + 13.2 µM TDZ/l, without coconut water; (iii) rooting in half-strength MS salts + 0.7 per cent agar, without BA; and (iv) acclimatization of plants in a mist system.

Variation in response to tissue culture by using type of buds as explants was also recorded by Widiastoety (1986). Apical or axillary buds of *Aranthera* (*Arachnis* x *Renanthera*) cv. James Storie were placed on Vacin and Went medium modified by adding coconut milk and 0 (control), 0.1, 0.2, 0.3, 0.4 or 0.5 ppm each of NAA and BA. The time required for the explants to form protocorm-like bodies (PLBs) was shortest on the control medium, with 52 per cent of the apical-bud explants forming PLBs in 61 days and 10 per cent of the axillary-bud explants forming PLBs in 81 days.

The technique of etiolation is widely used in the micropropagation of various plant species. de Jesus Rodrigues *et al.* (2017) evaluated growth and development in shoots of *Cattleya labiata* grown *in vitro*, as a function of growth regulators and the presence and absence of light. Stem segments, approximately ± 1.0 cm in height, were inoculated in test tubes containing 15.0 mL of MS culture medium supplemented with different concentrations of BAP (0.0, 2.0 and 4.0 mg/l) and NAA (0.0, 1.0 and 2.0 mg/l), and in different growth environments (growth room, darkness and a photoperiod of 16 hours) in a completely randomised design. At 150 days, the following were evaluated: a) number of etiolated shoots; b) number of nodes per etiolated shoot; c) length of stem segment (cm); d) number of roots; and e) total seedling dry weight (g). The highest mean values were achieved with the use of 4.0 mg/l BAP and 1.0 mg/l NAA for the number of etiolated shoots, and 2.0 mg/l NAA for the number of nodes per etiolated shoot in the absence of light. The length of stem segment was greater in the absence of light; in contrast, the number of roots and the total seedling dry weight were greater in a bright environment, irrespective of the growth regulator added to the culture medium. The etiolation of stem segments is a viable technique, as it provides features that are advantageous in the *in vitro* multiplication stage of the species (de Jesus Rodrigues *et al.*, 2017).

Leaf

Investigations carried out by several workers indicated that culure of leaf-tip or whole leaf was a promising possibility without sacrificing the important portion of a plant, as done in the case of meristem culture (Arditti, 1967a; Harrison, 1968; Harrison and Arditti, 1970; Churchill *et al.*, 1970). They were able to obtain plantlets from the leaf-tips of the young seedlings and matured plants of *Dendrobium*, *Epidendrum* and *Laeliocattleya*. Tanaka *et al.* (1975) studied the formation of protocorms in *Phalaenopsis* and *Vanda* by using leaf tissue, leaf segments (8-12 mm) and entire leaves (8-10 mm long). It was observed that protocorm-like bodies were formed earlier on leaf segments, and in both the species the proximal tissue gave better result than the distal ones. Similar studies in *Aranda, Cattleya, Dendrobium* and *Ascocenda* cultivars and hybrids were made by Lay (1978). He reported that leaf-tips did not survive but leaf parts including the base proliferated depending on the cultivar and formed calli which differentiated into plantlets. It was also recorded that the entire young leaves were better than the leaf sections in respect of plantlet production (Lay, 1979). The success of foliar explants for *Ascocenda* cv. 50th State Beauty depended upon the juvenility of tissue, use of BA and IAA in the nutrient medium, and optimization of surface sterilization procedure for the explants of *in vivo* grown plants. A treatment having BA and NAA promoted callus based development of multiple shoots capable of rooting (Vij and Kaur, 1999).

Protocorm-like bodies (PLBs) obtained through *in vitro* culture of seedling leaf segments of *Phalaenopsis* were cultured on media containing peptone or tryptone at various concentrations. Segments from dorsal parts of PLB had a tendency to form plantlets, while segments from basal parts produced no plantlet (Amaki and Higuchi, 1989). Nodal and leaf explants of *Spathoglottis apticata* cultured on charcoal amended MS medium supplemented with NAA and BA produced plantlets (Teng *et al.*, 1997). Leaf-tip of several species of orchid was also propagated *in vitro* by Tom and Weatherhead (1991). The best medium for leaf-tip culture of *Pholidota chinensis* was Heller's medium containing 1 mg BA and NAA/l. In *Acampe rigida*, largest number of shoots was produced Ichihashi and Yamashita medium containing 1 mg BA and 0.2 mg NAA/l. Only a few shoots were produced from leaf-tips of *Cleisostoma fordii* on modified MS medium with 1 mg BA/l. *Oncidium* cv. Aloha Iwanaga shoots 0.6, 2.0, 3.2 or 4.8 cm long derived from PLBs were cultured on full- or half-strength MS medium or VW medium. Full-strength MS medium gave the best growth, and Gellan Gum gave better results than agar, PLBs were formed only on 0.6 and 2.0 cm shoots (He *et al.*, 2001).

Myint *et al.* (2001) obtained the highest percentage of shooting (95 per cent) and rooting (95 per cent) when PLBs from leaf segments of *Phalenopsis* were cultured on ND medium with 1.0 mg NAA/l, 1.0 mg BA/l, 30 g potato homogenate/l and 20 g sucrose/l with a cotton support.

Stem Segment

Stem segments, each having a bud, were successfully cultured for the propagation of *Phalaenopsis* (Sagawa, 1961; Scully, 1965, 1966). Buds enlarged in two weeks followed by leaf and root development in 6-20 weeks and plantlets were ready for transfer within thirty weeks. Development of plantlets in *Vanda* Miss Joaquim in 2-3 months was also possible by using similar technique (Sagawa and Sehgal, 1967).

In 1995, Prakash and his co-workers developed a technique for rapid regeneration of orchid hybrids which utilizes thin sections (TS) obtained from shoot-tips as explant. With this procedure, they could produce more than 80,000 plantlets from a single shoot-tip in a year. Bui *et al.* (1999) used transverse thin cell layer (TCL) method for the micropropagation of *Rhynchostylis gigantea*. tTCL explant (0.3-0.5 mm) excised along the stem from the base to the shoot tip of 1-year-old plantlets were cultured on MS medium supplemented with BA, NAA, thidiazuron (TDZ) and 3 per cent sucrose. The combination for higher bud regeneration was 3 μM TDZ, giving rise to 11.7 buds per tTCL.

Torres and Mogollon (2000) developed a protocol of *Cattleya mossiae* through micropropagation. Clonal propagation was achieved in 4 phases : (i) culture initiation, starting from caulinary apices of dormant vegetative buds the latest mature growth flux on filter paper bridges, with Hang's modification (1984) of MS medium; (ii) shoot multiplication medium containing 0.7 per cent agar and 3.0 mg BA/l, without coconut water. Shoot elongation and quality were reduced above 2800 lux. Over 3500 shoots/explant were produced in 5 subcultures using modified plant tissue culture techniques.

Flower Stalk and Inflorescence

The use of flower stalks having well-developed buds was found economical and safe for the propagation of *Phalaenopsis* (Rotor, 1949; Tanaka and Sakanishi, 1977). Schmude (1984) cultured the segments of flower stalk of *Phalaenopsis*, each having dormant bud. The buds in liquid culture developed protocorms and after transferring into solid media, produced plantlets. In a similar study, Tse *et al.* (1977), however, recorded poor growth in *Phalaenopsis* compared with *Cattleya*. Young inflorescence bud (1-3 mm) from cultivars of *Aranda, Dendrobium, Mokara, Aranthera* and *Renantanda* developed plantlets when cultured in liquid Vacin and Went medium containing 1 mg BA and 0.1 mg NAA/l and in medium containing 1 mg kinetin and 0.1 mg 2,4-D/l (Lamchan and Lee, 1996).

Inflorescence tips of *Mokara* cv. Chark Kuan produced plantlets *in vitro* when cultured in liquid Vacin and Went medium supplemented with 200 ml coconut water/l and proliferated 2-month-old PLB transferred to solid Vacin and Went medium (Abdul Karim *et al.*, 1992). Clonal propagation of *Aranda* 'Deborah' was successful through *in vitro* culture of inflorescence tip explants. The initiation medium was Knudson, enriched with 15 per cent coconut water, 2 per cent sucrose and BA 1 mg/l. Growth and proliferation was better in Vacin and Went medium with sucrose and coconut water (Goh and Wong, 1990).

Explants (0.8-1.0 cm long segments) from inflorescence of *Rhynchostylis gigantea* from the distal region with undifferentiated buds regenerated better than those from the distal part of the inflorescence (Vij *et al.,* 1994). Scape nodes of *Oncidium* cv. Gower Ramsey, and various *Phalaenopsis* hybrids were cultured on Vacin and Went medium (*Dendrobium* and *Oncidium*) or modified MS (*Phalaenopsis*). Both media were supplemented with 2 mg BA. Shoot formation was higher from younger scapes (fully formed flower buds) than from scapes from full-bloom flower stalks (Nuraini and Mohd. Sahib,1992). Floral parts of *Cymbidium* Harunoumi were cultured by Kim and Kako (1984) and it was observed that column, ovary and flower stalk explants were better than sepal, petal and tip explants for adventitious bud formation and chlorophyll synthesis.

Segments taken from flower-stalk internodes of *Oncidium* cv. Sweet Sugar formed somatic embryos and shoot buds directly from wound surfaces or via nodular mass proliferation within 1.5 months, when cultured on a Gelrite gelled 1/2-MS basal medium supplemented with TDZ (O.1-3 mg/l) in darkness (Chen and Chang, 2000a). Supplementing with 0.1-1 mg/l NAA enhanced embryo formation, but retarded proliferation of shoot buds and yellowish abnormal structures.

Rhizome

The effect of culture media on PLB [protocorm-like body] formation and multiplication and shoot regeneration was studied by Park *et al.* (1998) to establish a micropropagation system for *Phalaenopsis*. The highest frequency of PLB formation was observed on VW medium containing 1,2-times ion concentration, 1 per cent sucrose, 1.5 g/l PVP or 2.5 g/l active charcoal, apple and potato extract and 4 g/l gellan gum. The highest ratio of PLB multiplication was obtained on VW medium with 2 per cent sucrose, apple and potato extract and a cotton-based medium matrix. Shoot regeneration was most effective on hyponex medium with 3 per cent apple, 3 per cent potato and 4 per cent banana extract. The RPLBs proliferated and produced numerous rhizome tips in the same medium and environment (Takahashi and Kondo, 1998).

Root According to Beechey (1970), aerial roots of orchids could also be used for *in vitro* propagation. If the attempts were successful, there would be tremendous scope for growing useful orchid hybrid without losing, either the main apex or lateral buds of mother plants. Vegetative multiplication of *Catasetum* hybrid was achieved by cultu-ring 1.5 mm long root-tip and protocorm-like bodies regenerated directly from the explants (Kerbauy, 1984). Plantlets were produced from root-tip of *Bletilla striata* and *Cleisostoma fordii* cultured in modified MS medium containing 1 mg NAA and 0.2 mg BA/l (Tom and Weatherhead, 1991).

Holters and Zimmers (1991) recorded shoot formation from root-tips from species of *Catasetum, Clowesia, Cycnoches, Cyrtopodium, Galeandra* and *Momordes*, cultured in growth regulator-free medium. *Grobya galeata* produced only a few protocorm- like bodies. Tips of isolated roots of *Mormodes histrio* produced 1-2 protocorm-like bodies with several shoot primordia, but only one of these primordia grew to a whole plantlet (Holters and Zimmer, 1990). Apical root segments from *Catasetum pileatum* were grown *in vitro* and placed in MS medium containing 1 g peptone per litre and 1 g charcoal/l in addition to vitamins. Protocorm-like bodies were formed which gave rise to seedlings (Krans and Monteiro, 1989). Root tip explants of *Cattleya* hybrid were cultured in modified Vacin and Went medium supplemented with various combinations of 2,4-D, NAA, IBA and BA. Regeneration of protocorm-like bodies only occurred on media containing NAA (Kerbauy, 1991). Root apex conversion of *Catasetum fimbriatum* into protocorm-like bodies can occur in the absence of plant growth regulator. But BA caused marked acceleration in the process of root-tip coversion and on the number of PLBs formed (Colli and Kerbauy, 1993).

Chang and Chang (2000) obtained seven shoot buds in 45 days when callus-derived rhizomes of C. *ensifolium* was cultured in 1/2 MS basal liquid medium supplemented with 0.17μM 2iP, 0.17 μM TDZ and 1.5 μM NAA under 10 mol/ m^2/s artificial light and agitated at 60 rpm on a rotary shaker.

11.3.6.1 Mericlone Culture and Production of Virus-Free Plants

Apart from propagation of large number of plants in a relatively short period of time, tissue culture technique has been effectively used for obtaining virus-free clones. Systemic mosaic virus infects roots, bulbs, leaves and flowers of mature plants of many orchids. In some hybrids, infected plants show greatly reduced vegetative growth within two years of cultivation. The flowers were smaller, fewer in number and sometimes deformed (Lawson and Branningan, 1986; Zettler *et al.*, 1990). These flowers were unacceptable for the cut-flower industry, resulting in great economic loss. Cymbidium Mosaic Potex virus (CyMV) commonly infects many members of the Orchidaceae. When expressed, foliar symptoms of CyMV include necrotic flecks, colour break and flower distortion (Lawson, 1995; Wisler, 1989). CyMV can thus reduce plant vigour and cause irregular blooming. This virus is mechanically transmitted. Seed-transmission of the virus has not been documented.

Lim *et al.* (1993) obtained high percentages of virus-free plantlets of *Mokora* Char Kuan 'Pink' from thin section cultures of infected plantlets and protocorm-like bodies with ribavirin treatment. Regenerance from thin section cultures without ribavirin treatment were also found to be free from *Cymbidium* mosaic virus and odontoglo-ssum ring spot.

Tissue cultured plantlets of *Oncidium* Goldiana (Clone 018) infected with cymbidium mosaic virus (CyMV) were treated with ribavirin (syn. virazole) which is known to be active against different plant DNA and RNA viruses (Yap *et al.*, 1999). Both the meristems and callus tissues were treated with ribavirin at 10, 20 and 40 mg/l to eradicate the CyMV, 10 mg/l of ribavirin for 30 days was sufficient to eradicate CyMV from 43 per cent of meristem plantlets compared with 31 per cent eradication for plantlets derived from callus tissues treated with 20 mg/l for 60 days.

11.3.6.2 Regeneration and Flowering In vitro

Regeneration and flowering *in vitro* in *Cymbidium* were reported by Wang (1998). Shoot meristems and axillary buds of *C. ensifolium* and *C. goeringii* were cultured on MS or White media supplemented with 10 per cent coconut water + 5 mg NAA/ litre. The explants formed protocorm-like bodies (PLBs) within 4-6 months. When

Propagation of Orchid

Division Method of Propagation

Cutting

Off-Shoot or Keiki

Tissue Culture

Shoot Regeneration in micropropagation

Potting and Repotting of Orchid

Potting Mixture

Orchid Nursery

Cultivation under Protected Structures

the PLBs were divided and subcultured in liquid medium containing the same additives, they formed new bright green rosettes of PLBs within a month. Plantlets from the PLBs in a liquid medium supplemented with 1 mg BA+0.1 mg/l NAA produced flower *in vitro* within a few months. A method to induce early flowering in the temperate orchid, *Cymbidium niveomargi-natum in vitro* was also reported by Kostenyuk *et al.* (1999). Combined treatment of BA, restricted N supply with P enrichment, and root excision (pruning), induced the transition from vegetative to reproductive stage. Nearly 100 per cent of plants flowered within 90 days after application of this combined treatment. When root excision and/or BA were omitted, flower induction was significantly reduced.

Psygmorchis pusilla is an interesting model to study orchid flower formation *in vitro* (Vaz and Kerbauy, 2000). A low concentration of N stimulated flower formation and this effect was more pronounced than that observed for the lowest total salt level used. Presence of NH_4^+ in the medium was essential for plant growth and development, while absence of NO_3^- depressed flowering. Flower formation was also enhanced by P and K reduction to a half-strength level.

11.3.6.3 Somatic Embryogenesis

Somatic embryos can be induced directly from the explant or through the callus phase. These embryos can be utilized for mass propagation as well as for conservation of germplasm.

In *Phalaenopsis*, a friable and embryogenic callus was developed by Sagawa and his associates at the University of Hawaii. The callus could rapidly multiply in Vacin and Went medium and produced embryoids when starved of sucrose (Sagawa, 1990). Callus induction and somatic embryogenesis in *Phalaenopsis* cv. Santa Cruz were reported by using Vacin and Went medium with sucrose. PLBs were formed in the medium without sucrose and PLBs were somatic embryos (Ishi *et al.*, 1998). Ichihashi and Hiraiwa (1996) reported that calli of *Phalenopsis* and *Doritaenopsis* grew better in medium solidified with gelrite, developed into friable and yellow or pale green structures in sucrose-rich substrate and turned green and generated somatic embryos when sucrose was eliminated from, or replaced with sorbitol or mannitol in the nutrient pool.

Chen and Chang (2000b) obtained compact and yellow-white embryogenic calli of *Oncidium* 'Gower Ramsey' formed from root tips and cut ends of stem and leaf segments on 1/2 MS basal medium supplemented with TDZ (0.1-3 mg/l), 2,4-D (1-3 mg/l) and peptone (1 g/l) for 4-7 weeks. Initiation of somatic embryogenesis and development upto the protocorm-like bodies from callus cultures was achieved on hormone-free basal medium. Regenerants were recovered from somatic embryos after transfer to the same medium and showed normal development.

12.0 Cultivation

12.1.0 Orchid House

Orchids are cultivated in greenhouse. The greenhouse for growing orchids is of two types – uncooled and controlled. Uncooled greenhouse provides mainly shade from direct sun which also reduces the inside temperature and increases the humidity, thus creating a favourable environment for the orchids to grow and flower. In tropical climate, tropical or warm orchids are usually grown in such greenhouse where the top is covered with shade-net, made of high density polythene or first covered with transparent plastic and then shaded with shade-net. The sides usually remain open for free circulation of air which is essential for orchids, particularly the epiphytes are covered with shade-net to reduce the light intensity and increase the humidity, when necessary.

Another type of greenhouse, with arrangements for controlling temperature, light and humidity, is constructed to provide a favourable environment for growing commercially important orchids, that is controlled greenhouse, which is mainly used in developed countries for growing high-value orchids. The production and quality of flowers produced in controlled greenhouse is much better than in uncooled greenhouse.Usually, cool and intermediate orchid groups are commercially grown in such greenhouse, because warm orchids will require heating in temperate climate which is expensive and the price of such orchids grown in uncooled greenhouse in warm climate is less than the other groups. Irrigation and feeding in controlled greenhouse, in most cases, are regulated by automatic devices.

12.1.1 Cooling

Cooling of greenhouse is necessary in places, where the outside temperature goes beyond 30°C and also when cool-season orchids are to be grown. One of the most common cooling systems used all over the world, to reduce the greenhouse temperature, is the fan and pad system of cooling where ventilation air is drawn into the greenhouse through a wet pad, where evaporation and air cooling take place (Mears, 1991). Once the greenhouse is covered, the inside temperature can be brought down through evaporative cooling method using horizontal draft fans on one side andcooling pads on the other side. The capacity of the fan and the size of the pad depend on the length and width of the greenhouse and also the elevation and other parameters. This method is very effective where the ambient humidity is low. Cooling can also be effected by providing vents in the greenhouse so as to facilitate free air circulation in the greenhouse.

12.1.2 Heating

In tropical and subtropical regions, although the climatic condition during most parts of the year is congenial for growing warm and intermediate orchids, heating is required in places where the winter temperature is very low. Similarly, in places where the climate is extreme (too cold or too warm), both cooling and heating are required. At higher elevations, where temperatures do not normally go above 30°C, cooling may not be necessary; only providing proper ventilation will serve the purpose. However, these places may require heating during winter. Solar heating of greenhouses occurs naturally. Greenhouses can be heated with the help of oil burners, hot water or steam, and or propane (or electric) heaters or gas-lit heaters.

12.2.0 Selection of Plants

A careful selection of proper species and hybrids is necessary for the successful cultivation of orchids. There are many beautiful commercial orchids belonging to different climatic zones and all of them cannot be grown in a particular climatic condition. Therefore, choice of orchids depends upon the climatic conditions of the place where they are to be grown. It should be kept in mind that tropical orchids like warm-humid conditions with frequent watering during summer, subtropical ones intermediate temperature and humidity and temperate ones cool and humid conditions. The area or space available and choice of the grower is other important criteria for selecting orchids. Further, orchids should be so chosen that one gets flowers throughout the year. Plants selected should be healthy and free from pests and diseases.

Assessment of species and hybrids for their performance in a particular area and selection of promising ones are very important for the commercial exploitation of orchids. Species and hybrids suitable for growing under warm conditions were listed by Bose and Bhattacharjee (1972). Singh (1978) scored out certain species suitable for cultivation under Bangalore conditions. Investigations were also carried out and the promi-sing orchids were suggested for cultivation under the conditions of Yercaud in the state of Tamil Nadu (Rao, 1982). Similarly, a large number of orchid species were assessed for their performance and suitable ones were recommended for cultivation under the plains of West Bengal (Yadav and Bose, 1986a). Many tropical, subtropical and temperate orchids including species of *Aerides, Vanda, Dendrobium, Paphiopedilum, Cymbidium, etc.*, grow and flower well in ordinary greenhouses at Kalimpong in Darjeeling district at different altitudes.

12.3.0 Containers

Clay pots, wooden or galvanized wire baskets, tree-fern blocks, logs of wood, *etc.*, are the different types of containers used for growing orchid. Clay pots provided with several cuts and holes to facilitate adequate aeration and proper drainage are very popular in different places. Davidson and Bennett (1961) compared the suitability of standard clay, silicon-treated clay, white and green plastic pots for growing *Laeliocattleya* seedlings and recorded 27 per cent more growth in the ordinary clay pots than in the white plastic pots. The requirement of water was, however, more in ordinary clay pots. The size of the pots may vary depending upon the size of the plant. Bigger pots have been found to encourage vegetative growth whereas plants grown in small pots produce better flowers.

Ground orchids can be grown in ordinary earthen pots, like other plants. Wooden baskets allowing sufficient aeration and drainage are the most suitable containers for epiphytic orchids producing pendulous inflorescence. Tree fern block makes an excellent container because of its fibrous nature and porosity. Wooden logs or blocks with the bark still on them are also suitably used for growing many epiphytic orchids. About 10 cm in diameter and 25 cm long logs are good enough. Similarly, coconut shells can be split into halves, bored, and used as containers.

In recent years, plastic containers are rapidly giving way to others and have become popular throughout Europe and the USA. Denmark is another country where almost similar types of containers are extensively used with good results. Plastic pots are light in weight, easy to transport and do not accumulate algae or salts on their sides. Besides, they retain moisture longer than clay pots, and therefore, the requirement of water is also less. Davidson and Bennett (1961) reported that plants in white plastic pots required only 49 per cent of the water needed by those in the ordinary clay pots.

12.4.0 Potting Media

The type of the medium will vary according to the habit of the orchids. The media suitable for growing epiphytic orchids should hold moisture but not remain soggy and wet. For many years the potting medium used all over the world was osmunda fibre either as such or in combination with moss which helps in increasing the moisture holding capacity. It lasts longer and contains 2 to 3 per cent nitrogen which is released slowly as the fibre decomposes (Sheehan, 1980). The supply of osmunda fibre has become scarce and growers had to turn to the roots and trunk fibres of various species of tree fern. In their various forms and grades tree fern fibre makes an excellent medium for growing orchids. Since the last decade, particularly in the USA, the orchids have successfully been grown in the potting media of bark chips, by-products of lumber industry. Probably the most widely used type of potting material in recent years has been fir bark or the many bark-based mixes. 'Off Mix', a commercial mixture developed at Rutgers University, is basically a combination of fir bark, peat and perlite which has been improved and modified many times by various commercial and amateur orchid growers. 'Servo' is the trade name of a medium which is manufactured from the residue of the sugar industry. To date it is still rather localized in its usage (Henderson, 1984). Fire-clay products, such as, 'Solite' are used by some growers where plants are exposed to frequent rain, such as, in South Florida (Sheehan, 1980). Cork is being used in much the same way as fir bark, either alone or as part of a mixture. In many of the tropical countries, orchids have successfully been grown in coconut husk and pieces of brick and charcoal. Apart from these, perlite, gravel, moss peat, sawdust, rice husks, stone chips, peanut shells, branches and twigs of various trees and several more have successfully been tried individually and in combination with other materials.

Over a 2-year period the best plant growth and flowering were recorded in plants grown in medium containing moss peat. Davidson (1956) suggested the use of equal parts of coarse peat moss, dried and undecomposed fibre for successful cultivation of *Cattleya* and its hybrids.

The demand of orchid growers for fibre of the tree fern started denuding the natural vegetation in parts of New Zealand. While searching for suitable substitute of tree fern, Hunter (1958) reported that excellent growth of *Cattleyas* could be obtained in bark and cypripediums and cymbidiums in bark, peat and sand mixture, without using tree fern fibre. The media consisting of fine pine bark + 30 per cent coarse poultry peat or a mixture of 30 per cent oak leaves, 50 per cent bark and 20 per cent palco wool also gave much better results in cymbidiums than pure green sphagnum moss (Berg, 1965). A mixture of 40 per cent each of pine bark pieces and sphagnum along with 20 per cent dry birch or oak leaves or that of 30 per cent each of polypodium, sphagnum and pine bark along with 10 per cent beech or oak leaves was recommended for growing most of the orchids (Elle, 1960, 1962). Sheehan (1960, 1962) reported that *Cattleyas* and *Phalaenopsis* Pink Chiffon grown in ease and evory barks produced more flowers on large stems than those grown in vita bark.

Studies of Poole and Sheehan (1977) indicated that both terrestrial and epiphytic orchids can be grown in a mixture of peat and perlite (1:1) with excellent results. *Aerides multiflorum* was found to perform well in a substrate of hardwood charcoal. Most of the orchid cultivars have been found to grow and flower with coconut husk dust, husk chips of coir fibre in the media depending on the types of orchids (Nash, 1997).

The maximum growth and flowering in terrestrial orchids were noted when plants were grown in a compost containing one part each of loamy soil, river sand, leafmould and half part each of charcoal dust and old morter. Loamy soil or leafmould alone or in combination as potting compost, however, responded poorly (Bhattacharjee and Mukherjee, 1981). Bhattacharjee (1984) suggested that terrestrial Eulophias should be grown in a medium consisting of loamy soil, sand, leafmould and tree fern fibre in equal parts, with some charcoal and bonemeal.

Epiphytic species grow well on fibre or tree fern blocks with sphagnum moss. Experiments revealed that a potting medium containing equal parts of charcoal, brick pieces, coconut fibre and leafmould recorded the best growth and flowering in several epiphytic orchids like *Aerides, Cattleya, Dendrobium, Vanda, etc.* In case of terrestrial orchids compost containing loamy soil, river sand, leafmould, charcoal dust and old morter, 1 part each, has been found very useful in improving the growth and flower production (Yadav and Bose, 1985).

Rajamani *et al.* (1999) grew in brick pieces + coir dust and recorded highest number of leaves (7.66), inflorescences (6.33) per plant, flowers (9.33) per inflorescence, flower diameter (4.03 cm) and the longest inflorescence (25.20 cm), compared to plants grown in brick pieces, charcoal, coir dust, gravel pieces and fern bark + moss.

Under natural conditions, terrestrial orchids are found growing in forest soils composed of rich humus, accumulated from the falling leaves over a period of time. For a potting medium stimulating this natural humus, a mixture of equal parts of leaf-mould, garden soil and coarse river sand will be suitable (Pradhan, 1985). In fact, any well-drained potting mixture containing 50 per cent of more organic matter can be used. From the results of an experiment on the effect of substrate on growth of *Cymbidium*, Kim *et al.* (1997) concluded that physical characteristics of substrate, especially higher bulk density, water holding rates and porosity are important factors promoting the growth of the plants.

Several investigations were carried out to select suitable and inexpensive media for different types of orchids. The responses of *Cymbidium* to the different potting media including clay soil, moss peat, sedge peat, sawdust and manure alone or in various mixtures and the standard mixture of osmunda fibre, leaf- mould and manure were studied by Mott (1954 b,c). *Cymbidium sinense*, *C. goeringii* (6 cultivars), *C. faberi* (1 cultivar), *C. kanran* (1 cultivar) and *C. ensifolium* (4 cultivars) were cultivated in 5 different substrates (Yu *et al.*, 1998). Growth rate, chlorophyll content and substrate macronutrient content were recorded in each substrate. A substrate containing leaf mould of pine needles and vermiculite was the most effective substrate for *Cymbidium* cultivation. The logarithmic growth phases of *C. sinense* and *C. goeringii* lasted from mid-July to early September, and from early July to mid-September, respectively. It was concluded that the growth of *Cymbidium* plants may be promoted and prolonged by applying inorganic fertilizers and manures, and by controlling the temperature, relative humi-dity and light intensity before and after the logarithmic growth phase.

Cymbidium goeringii, C. kanran cv. Jeju × *C. kanran* cv. Dosakwan, *C. niveo-marginatum* and *C. gyokuchi* were grown in 8 different types of potting media (Paek *et al.*, 1998b). Shoot and root growth of *Cymbidium* spp. were greater when planted in bark, leafmould, habitat soil or carbonized rice hull than in sphagnum moss, which was favourable for growth of *C. kanran* cv. Jeju × *C. kanran* cv. Dosakwan. Shoot and root N contents were highest when plants were potted in leaf mould or habitat soil, and P content was higher than K content when grown in a sphagnum moss or carbo-nized rice hull potting medium. Leaf and root analyses indicated that optimum growth of temperate *Cymbidium* spp. creates a large demand for K.

12.5.0 Potting and Repotting

Potting orchids is an important item of work and the knowledge of growth and flowering behaviour of plants is very essential to perform such operation. As a general rule, the period soon after the flowering season is the best for potting or repotting orchids.They should not be disturbed during flowering or at dormant stage. Care should also be taken that pots and all tools used in potting are perfectly clean. Crocks should thoroughly be washed and dried before use. These will minimize the spread of diseases. The amount of crocks to be used varies from about one-third the depth of the pot in terrestrial orchids to a greater depth in epiphytic orchids. It also depends on the type of potting material. When osmunda fibre or other porous material are used, the quantity of crocks will be less (Bose and Bhattacharjee, 1980).

Whatever material is used, the ingredients of the media should be mixed in suitable proportion. It should be heaped in the pot making a small mound on the top. The plant to be potted is placed on the top of this mound with the roots spread outwards and downwards along the slopes of the mound. Holding the plant in place with one hand, the space between the roots and wall of the pot is filled with more medium, till the pot is filled to an inch from the rim of the pot and packed properly. The medium should cover only the roots and the rhizome, but never the pseudobulbs (Abraham and Vatsala, 1981). While potting vandaceous orchids, care should be taken so that the aerial roots are not buried inside the compost. The taller orchids might need a support, and therefore, a stake is provided with and the plant loosely tied to it.

Orchids remain at their best when left undisturbed. Repotting is advised only in case where the plant has filled the pot and there is no room for further growth. The physiological status of the plants at repotting plays an important role on their establishment and subsequent growth and flowering. Plants should be repotted when it has entered into a stage of active vegetative growth. While repotting, the whole plant is taken out of the pot without causing any root damage. The rhizome is cut with a clean knife without injury to the pseudobulbs into as many portions as is desired. Dead backbulbs and roots are cut away, and each portion is potted in a fresh pot. Spring is the best time to undertake any repotting, when the new growths are getting underway and the new roots soon to be made can penetrate immediately into the fresh compost (Rittershausen and Rittershausen, 1970). Orchids like *Cymbidium* react favourably when repotted every other year or so. *Pleone*, *Calanthe*, *Thunia* and *Phaius*, *etc.*, need annual repotting. Vandaceous orchids tolerate very little disturbance (Yadav and Bose, 1985).

12.6.0 Watering

Orchids like spray of water, unlike other plants which are watered by pouring. Majority of the commercial growers employ some form of overhead watering in their greenhouses though they may alter their watering procedure at certain times of the year because of cut-flower production. There are several forms of the overhead system but they are usually some modifications of two basic types, either a stationary mist head or oscillating sprinkler type. Some growers put water by hand and it is no doubt better for the individual plant, it becomes impractical in large production. Syrovatka (1980) suggested the use of fibre glass netting in wicks or some other form for capillary watering of pot-grown orchids. The main advantage of using fibre glass rather than a cotton wick is the durability of the material. Irrigation thus achieved is continuous, and any salt deposit on the foliage avoided. Another system that has been tried is individual pot watering through a spaghetti tube system. But this appeared to have only limited success in the orchid industry (Henderson, 1984).

Like other flower crops, orchids should be watered thoroughly, and until the surface of the potting mixture begins to dry further watering should not be done. The frequency of watering depends on several factors, such as, climatic conditions, type of growing medium, type and size of container, growth habit of orchid, *etc.* Atmospheric humidity influences evaporation of moisture from the compost, and orchids in general, prefer high relative humidity. In a dry and well-ventilated atmosphere, damping of the floor of the greenhouse and frequent overhead sprinkling of water will increase the humidity. But in high humid atmosphere, watering should be less frequent. Small pots dry out faster than large ones, plastic pots hold water longer than clay ones. Similarly, osmunda and coconut fibre retain moisture longer than either tree fern or bark. Epiphytic orchids cannot tolerate under or overwatering but terrestrial ones need more moisture. Mott (1953) reported that plants of *Cattleya labiata* Amesiana watered daily were large and produced more pseudobulbs and flowers than plants watered less frequently. Almost similar observations were recorded in plants of *Cattleya brunoyensis* grown in osmunda (Mott, 1954a). In an investigation on the effect of watering frequency on the growth of cattleya, cymbidium and phalaenopsis, Poole and Seeley (1977) noted that *Cattleya* plants grew best when watered 5 or 7 times per week but *Phalaenopsis* plants had larger plant diameter when watered 3 times a week. Plants of genera like *Cypripedium*, *Cymbidium*, *Anoectochilus*, *etc.*, require ample moisture around their roots at the time of flowering otherwise the flower quantity is reduced (Curtis, 1950).

Newly potted orchids require less water than those well-established. They should not be watered very frequently but the compost should be kept moist by fine spraying. As new roots emerge and growth starts, the plants need more frequent watering and it should be decreased again after flowering. Actively growing plants need more water than resting plants, and the orchids require less water in winter than in summer.

The quality of water is as important as the quantity of water applied. Cattleyas were reported to tolerate a wide range of water quality. Plants supplied with moderately hard water with a pH ranging from 5 to 9 made significantly better growth than plants given very soft water. Water with specific conductance values between 25 and 75×10^{-5} mhos was considered suitable for orchid culture (Davidson, 1960; Northen, 1970). Hard water, however, should not be used in overhead sprinkler system because the leaves will be covered with a thin film of calcium crystals, resembling white salt on the leaves (Sheehan, 1980). Water having a pH of not lower than 5.6 and not higher than 6.8 was found most suitable for the culture of cymbidiums (Duckitt, 1984). The use of very cold water should be avoided as it may damage the leaves causing mesophyll collapse, resembling a virus infection (Curry, 1975). According to Curtis (1950) and Beckner (1979), rain water is the best for watering orchids such as *Calopogon, Pogonia, Spiranthes* and *Blephariglottis*. The level of soluble salts in water is another important factor to be considered before watering. Water having soluble salts less than 125 ppm was reported to be excellent for use, 125 to 500 ppm was good, 500 to be used with caution and salts above 800 ppm was suggsted not be used (Sheehan, 1980).

12.7.0 Fertilization

Like other plants, the growth and flowering in orchids are markedly improved by proper supply of nutrients in liquid forms (Carnes, 1953; Penningsfeld and Fast, 1960; Sheehan, 1960, 1962; Gething, 1974; Bose and Bhattacharjee, 1980; Sakai *et al.*, 1982). Inadequate nutrition, on the other hand, limits the growth and development (Davidson, 1960; Northen, 1970; Yadav and Bose, 1986b) and also causes various disorders leading to the decline of plants (Poole and Sheehan, 1970; Sheehan, 1980). The major mineral elements required in large amount are nitrogen, phosphorus, potassium, calcium, sulphur and magnesium. Iron, boron, copper, manganese, molybdenum, zinc and other micronutrients are also necessary but in extremely small quantity. Under natural conditions, orchids get their necessary inorganic nutrients from the soil or decayed organic matter on the bark upon which they are growing and also from the atmosphere washed down by rain. The organic matters are supplied by the decaying vegetable and animal debris. When under cultivation, all these nutrients should be supplied to the orchids regularly for their growth and flowering.

12.7.1 Role of Nutrients

12.7.1.1 Nitrogen

The beneficial effects of nitrogen in improving the growth and flowering of orchids have been reported by many workers (Sheehan, 1960, 1962; Penningsfeld and Fast, 1962a,b; Sagarik and Siripong, 1963; Poole and Seeley, 1978; Bhattacharjee, 1981; Yadav and Bose, 1986b). Sheehan (1960) studied the effects of three levels of nitrogen (0, 50 and 100 lb/acre) on the growth of *Phalaenopsis* seedlings and flowering *Cattleya* and recorded a significant increase in leaf growth of *Phalaenopsis* with each increment of nitrogen. In *Cattleya*, N level upto 100 lb/acre also enhanced the number of flowers per plant. In another trial with *Phalaenopsis* Pink Chiffon, he (1962) noted that control plants produced only 2.7 flowers per spike and those receiving the highest level of nitrogen (1000 ppm) produced 4.6 flowers. Flower spike length ranged from 24.2 cm in plants treated with the lowest dose to 37.5 cm at the highest level of nitrogen. Flower diameter also increased with increasing levels of this nutrient. Increase in growth and flowering of *Cattleya* with the application of higher dose of nitrogen was also recorded by Penningsfeld and Fast (1962a). The quality of flowers was, however, better at 2N than at 3N level. Gething (1974) studied the response of *Odontoglossum* seedlings to different doses of nitrogen. The average fresh weight noted after 20 months was found to be 11, 22 and 43 g with the treatment of 0, 50 and 100 ppm, respectively. *Aerides multiflorum*, an elegant and attractive monopodial epiphytic orchid also responded well to similar treatment with nitrogen (Table 8). The plants treated with 1000 ppm nitrogen grew best and produced 116.6 and 67.8 per cent more spikes and flowers, respectively, than those under control (Yadav and Bose,1986b).

The ratio of NO_3 -N : NH_4 -N on flowering and plant inorganic elements of *Oncidium* was studied by Lin *et al.* (1999). The greater the proportion of NH_4 -N the highest were the N and P contents in plant tissues. However, K, Ca and Mg contents were decreased. The N content was lowest when the ratio of NO_3-N : NH_4 -N was 1:0. The flowering quality of Gower Ramsey and Aloha Iwanaga was best when the NO_3 -N : NH_4 -N ratio was 6 :1.

When nitrogen was deficient, the plants were stunted and matured too early (David-son, 1960; Yadav and Bose, 1986b). Older leaves turned yellow and dropped off. Too much nitrogen produced excessive growth and delayed flowering (Northen, 1970). Penningsfeld and Fast (1962a) reported that increased dressing of nitrogen was responsible for rotting of bulbils and roots of *Cattleya* orchids. Nitrogen deficiency caused marked suppression of vegetative growth, reduction of dry weight and early leaf fall. The concentration of chlorophyll decreased but no decreae in flower-stalk occurred. (Yoneda *et al.*, 1997).

12.7.1.2 Phosphorus

Like nitrogen, application of phosphorus was also found beneficial in improving the growth and flowering of orchids (Poole and Sheehan, 1970; Wong and Chua, 1975; Bhattacharjee, 1981; Yadav and Bose, 1986b). Gething (1974) reported that this nutrient alone could not influence much the growth and flowering of *Odontoglossum* but fresh weight of plant was considerably higher when P2O5 was added to the nitrogen. Improvement of growth and flowering of an important hybrid *Aranda* Wendy Scott was also recorded by Wong and Chua (1975), by using P2O5. Bhattacharjee (1981) obtained the largest number of spikes and flowers in *Dendrobium moschatum* with the application of phosphorus at 500 ppm. In *Aerides multiflorum*, the effect of this nutrient was most favourable at 1000 ppm which significantly increased the leaf number and length of spike and number of flowers per spike (Yadav and Bose, 1986b) (Table 12). Sheehan (1960), however, did not observe any effect of this nutrient on growth of *Phalaenopsis* and flowering of *Cattleya*.

Phosphorus deficiency leads to stunting, but the leaves, instead of turning yellow, become dark green (Northen, 1970). Davidson (1960) observed a severe reduction in the growth of *Cattleya* due to the deficiency of this nutrient.

12.7.1.3 Potassium

Potassium, is another important nutrient, regulating many physiological processes in plants. Its beneficial effects on growth and flowering of orchids were recorded by several workers. Sheehan (1962) reported that treatment with this nutrient at 500 ppm significantly increased the stem length in *Phalaenopsis* Pink Chiffon. Production of leaves in *Cattleya* was also found to be influenced by the application of this nutrient and the N content varied with the level of K (Poole and Sheehan, 1970).

Table 12: Growth and Flowering of *Aerides multiflorum* at different Levels of Nitrogen, Phosphorus and Potassium**

Treatments	*Leaves*		*Number of*		*Length of*	
	Number (Average)	*Length (cm)*	*Flowering Spikes per Plant*	*Flower per Spike*	*Flowering Spike (cm)*	*Flowers (cm)*
Nitrogen						
0 ppm	7.7	15.2	1.2	26.1	19.1	3.0
1000 ppm	13.4	29.9	2.6	43.8	30.1	4.1
1500 ppm	11.1	24.8	2.2	39.6	27.2	3.7
F-test	*	*	*	*	*	*
C.D. at 5 per cent	0.00	0.42	0.07	0.63	0.50	0.12
Phosphorus						
0 ppm	9.8	21.5	1.8	34.2	24.9	3.6
1000 ppm	12.2	25.6	2.2	38.6	29.5	3.8
1500 ppm	10.4	22.8	1.9	37.8	25.1	3.4
F-test	*	*		*	*	
C.D. at 5 per cent	0.30	0.42	N.S.	0.63	0.50	N.S.
Potassium						
0 ppm	9.6	21.9	1.9	35.2	24.6	3.5
1000 ppm	12.0	24.3	2.1	37.7	27.6	3.7
1500 ppm	11.9	23.3	2.0	36.5	26.2	3.6
F-test	*	*		*	*	
C.D. at 5 per cent	0.30	0.42	N.S.	0.63	0.50	N.S.

N.S. = Not significant.

*Significant. **Yadav and Bose (1986b).

In *Aranda* Wendy Scott, the vegetative growth and flowering improved with the treatment of a solution containing 36 mg potassium. Inflorescence yield showed a positive correlation with stem length increment and production (Wong and Chua, 1975). The most effective dose of this nutrient was found to be 500 ppm in *Dendrobium moschatum* (Bhattacharjee, 1981) and 1000 ppm in case of *Aerides multiflorum* (Yadav and Bose, 1986b). Gething (1974), on the other hand, noted that application of potassium had no effect on the growth of *Odontoglossum*.

Potassium deficiency results in dwarfness with the edges of the leaves frequently scorched and dead. Chin (1966) noted reduction in plant growth and drying of foliage in both *Dendrobium* and *Phalaenopsis* hybrids. Potassium-deficient plants showed reduction in leaf area (Yoneda *et al.*, 1997). The deficiency of this element can be corrected with the treatment of potassium sulphate, potassium nitrate, or potassium chloride.

12.7.1.4 Calcium

Sulphur is necessary for cell wall formation and for regulation of cell activities. If this nutrient is deficient, the new growths are stunted and distorted (Poole and Sheehan, 1970). The discolouration of leaves starts from the margin and extends the midrib. It develops due to the loss of chlorophyll. The veinlets become tinged with purple colour while the rest of the tissues in the affected areas generally change from greenish yellow to dark brown or black. Mature leaves may turn completely black and drop off. In advanced stage, the tip of the plant and tips of upper leaves die back and the young pseudobulbs turn black. The deficiency can be corrected by the application of monocalcium phosphate, calcium nitrate or calcium chloride.

12.7.1.5 Sulphur

Sulphur is also an ingredient of proteins and the deficiency of this element may affect the root growth (Northen, 1970).

12.7.1.6 Magnesium

It helps in producing leaves with dark green colour and promotes uptake and translocation of phosphorus and other elements. In nutrient culture, growth of hybrid *Phalaenopsis* was optimal with 25 ppm Mg and of *Cattleya* with 50 ppm. The nutrient at 1000 ppm decreased growth in all the 3 genera in comparison with 50 ppm (Poole and Seeley, 1977). The deficiency of magnesium causes chlorosis between veins of older leaves and plants do not thrive. This can be prevented by the treatment with magnesium sulphate.

12.7.1.7 Manganese

It acts as a catalyst and regulates growth and flowering. Deficiency of manganese causes chlorosis between the veins which can be prevented by manganese sulphate.

12.7.1.8 Iron

This is essential for chlorophyll formation and is useful for maintaining the normal growth. Iron is seldom deficient. However, the deficiency of this element causes the younger leaves to become yellow. Application of ferrous sulphate will correct the deficiency of iron.

Very little is known about the specific activities of boron, molybdenum, manganese, copper and zinc in orchid nutrition which are also necessary for the better growth and flowering, but in very minute quantity.

12.7.2 Effect of Fertilization

Fertilizer for orchids should contain all the necessary nutrients soluble in water. The proportion of one element to another may vary according to the age and growth rate of the plant and growing medium used. A combination of NPK in the ratio 10:12:10 was found the best for growth and flowering of a large number of species and hybrids suitable for growing under warm humid conditions (Bose and Bhattacharjee, 1972). Wong and Chua (1975) considered that the plants of *Aranda* Wendy Scott should be fed on a 10-day round with a solution containing 72 mg each of N and P and 36 mg of K. Growth of hybrids of *Phalaenopsis* was found optimal with the application of 100 ppm N, 50-100 ppm K and 25 ppm Mg and of *Cattleya* with 50 ppm each of N, K and Mg (Poole and Seeley, 1978). According to Sheehan (1980) the ratio of fertilizer depends on the growing medium used. He suggested that osmunda, tree fern, terrestrial mixtures, 'Holite' and 'Solite' should be fertilized with a 1:1:1 fertilizer, such as, 10:10:10 (NPK) at the rate of 453 g/393 litres of water applied to 43.3 m2 of growing area. Fir or other barks require additional nitrogen and should be fertilized with a 3:1:1 fertilizer, *i.e.*, 30:10:10 (NPK) at the rate of 453 g/393 litres of water. Application of N, P, K in the ratio of 20:20:20, used every week for 3 successive weeks followed by a 10:30:20 (NPK) mixture were successfully used by Abraham and Vatsala (1981) for different orchids growing under South Indian conditions. They further suggested that during the main growing season, the proportion of nitrogen should be increased and NPK ratio of 30:10:10 used for three weeks. The above NPK ratio to be used once a week was also suggested by Duckitt (1984) for growing cymbidiums in South Africa. In an experiment on NPK requirement of *Vanda* Miss Joaquim, a commercial cut flower cultivar, grown on trunk sections of tree fern, Higaki and Imamura (1987) observed that plant height was maximum with 200 kg P + 300 kg N/ha and effect of N was most pronounced, but K and P were also limiting.The highest flower yields were obtained with 200 kg P + 275 kg K at all levels of N, being the maximal at 300 kg N. Flower size was largest when all 2 elements were applied, P having most effect and N the least.

In a greenhouse experiment, *Phalaenopsis* hybrids Syllea, Nopsya and Abylos were grown in peat substrate and supplied with 150, 275, 400 or 650 mg N/plant from a soluble fertilizer (N:P:K 16:4:18). For all three cultivars, increasing fertilizer rates resulted in larger plants and developed more inflorescence and flower (Amberger-Ochsenbaner, 1997).

The effects of 8 kinds of commercial liquid or solid fertilizers with different ratios of NPK on the growth and mineral content of shoots and roots of 4 species of temperate *Cymbidium* were investigated by Kim *et al.* (1998). Shoot and root growth were favoured when NPK ratio of liquid fertilizer was 5-10-5. Leaf analysis showed that N content was the highest when ratio of liquid fertilizer was 30-10-10 and the lowest when it was 6-40-6. Among the major nutrients analysed, K had the highest content. K content was higher in roots than in leaves. N content of leaves was higher than that of roots. P content of roots was higher than that of leaves with the exception of *C. goeringii*. Effects of macroelement concentration on growth, flowering and nutrient absorption in mericlone seedlings of an *Odontoglossum* hybrid (*Odontioda* cv. Lovely Morning 'Sayaka') were studied in sphagnum moss culture. N, P and K were supplied via fertigation (once every 5 irrigations) at concentrations which ranged from zero to twice the control concentration for each nutrient separately. Leaf number and area varied little. However, leaves became paler, bulb diameter smaller and roots fewer as N rate decreased. Root length was less when N and P were not supplied. Low N rates also resulted in shorter and thinner stalks, fewer flowers, and advanced flowe-ring date. Flowers became narrow with low K rates (Yoneda *et al.*, 1999) (Tables 13 and 14).

Table 13: Concentrations of Major Elements in Butrient Solutions (mM). Solutions were adjusted to pH 5.8 with HCL*

Element	*Treatment*			
	0	*0.5*	*1*	*2*
N	0.00	7.50	15.00	30.00
P	0.00	0.50	1.00	2.00
K	0.00	5.00	10.00	20.00

* Yoneda *et al.* (1999).

Table 14: Effects of Nutrient Concentration on Flower Stalk Emergence and Flowering of *Odontiodaa**

Treat-ment	*Element (Times[b])*	*Flower Stalk Emergence Conc.no.*	*Flower Stalk Length Total*	*Diameter of Flower Stalk per cent*	*First Flowering (cm/ stalk)*	*No. of Florets (mm/ stalk)*	*Width of Floret (Date)*	*(cm/ floret)*
N	0	16ac	100	25.4c	3.6c	16 July	4.0c	7.5a
	0.5	16a	100	30.6b	4.0c	13 July	5.2b	7.6a
	1	15a	93.8	33.8a	4.7b	17 July	6.6a	7.5a
	2	14a	87.5	23.9c	5.4a	23 July	6.6a	6.0b
P	0	15a	93.8	34.4a	4.4a	19 July	6.2a	7.2a
	0.5	16a	100	35.0a	4.7a	11 July	6.6a	7.4a
	1	15a	93.8	33.8a	4.7a	17 July	6.6a	7.5a
	2	16a	100	32.6a	4.5a	14 July	6.5a	7.4a
K	0	16a	100	20.3c	5.4a	26 July	5.7a	5.8c
	0.5	14a	87.5	29.0b	5.4a	21 July	6.9a	7.1b
	1	15a	93.8	33.8a	4.7b	17 July	6.6a	7.5a
	2	15a	93.8	32.6a	4.4b	21 July	6.1ab	7.4b

* Yoneda *et al.* (1999). Values are the means of 16 plants.

a: *Odontioda* Lovely Morning 'Sayaka'; b: The control treatment used 15.0 mM N, 1.0 mM P, 10.0 mM K. The concentrations of N, P and K were changed independently to give concentrations 0, 0.5, 1 and 2 times the control; c: Mean separation within columns by Duncan's multiple range test, 5 per cent level.

In an experiment on the effect on N and P on *Phalaenopsis*, Wang (2000) observed that high P treatments, regardless of the frequence of application, had no effect on the date of emergence of the flowering stem (spiking), anthesis or flower size (Tables 15 and 16). All plants treated with the high P fertilizer had fewer flowers (15-19) than controls (24 flowers). Continuous application of adequate N appeared to be more important than low N and increased P for best flowering.

Table 15: Effects of Continuing High N Fertilizer or Application of a low N, High P and K Fertilizer at Various Frequencies (At every irrigation, PP, or at every fourth irrigation, PW) on Performance of the *Phalaenopsis* orchids. Flower longevity was recorded for the basipetal first flower on each inflorescence*

Fertilizer Treatment[z]	*Date of*		*Flower*		*Leaf No.*	*Total Flower No.*
	Spiking	*Anthesis*	*Diam (cm)*	*Longevity d)*		
Continuous high N	12 Oct.ay	23 Jan.a	10.5a	93a	6.5a	25.2a
PP	6 Oct.a	16 Jan.a	10.7a	85a	5.0b	18.6b
PW	8 Oct.a	15 Jan.a	10.8a	85a	4.5b	15.0c

z: High N = Peters 20 N-8.7P-16.6K soluble fertilizer at 100, 43 and 83 mg.L^{-1} of N, P and K respectively. P = low N, high P and K fertilizer at 30, 390 and 506 mg.L^{-1} of N, P and K respectively. W = water with no fertilizer.

y: Mean separation within columns by Tukey's HSD, $P \leq 0.05$.

* Wang (2000).

Table 16: Effects of Terminating Fertilization at Various Dates or Developmental Stages on Performance of *Phalaenopsis* Orchids. Flower longevity was recorded for the basipetal first flower on each inflorescence.*

Fertilizer Termination Date[z]	*Date of*		*Flower*		*Flower No.*	*Leaf No.*
	Spiking	*Anthesis*	*Longevity (d)*	*Diam (cm)*		
By date						
1 Sept.	30 Sept.ay	17 Jan.a	91b	10.7a	16.0c	4.1c
29 Sept.	30 Sept.a	13 Jan.a	96ab	10.4a	20.4b	4.8bc
27 Oct.	3 Oct.a	17 Jan.a	97ab	10.8a	18.0bc	4.6bc
24 Nov.	2 Oct.a	14 Jan.a	95ab	10.4a	24.8a	5.0b
Continuous	30 Sept.a	12 Jan.a	103a	10.6a	24.4a	5.7a
By developmental stage						
At spiking	1 Oct.a	9 Jan.a	96a	10.5a	19.0b	5.0b
10-cm spike	9 Oct.a	12 Jan.a	92a	10.4a	21.0ab	4.9b
1st flower opening	1 Oct.a	9 Jan.a	92a	10.5a	20.8ab	5.3ab
Continuous	30 Sept.a	12 Jan.a	103a	10.6a	24.4a	5.7a

z: The 20N-8.7P-16.6K fertilizer was used at 0.5 g.L^{-1}.

y: Mean separation within columns and dates/stages by Tukey's HSD, $P \leq 0.05$.

* Wang (2000).

Several formulae for fertilizer solutions have been recommended by various workers.

Some of the popular ones are Ohio W. P., Davidson, Sagarik and Siripong. Composition of Ohio W. P. solution is given below.

Ohio W.P. Solution (to be used half-strength)

Potassium nitrate	2.63 g
Ammonium sulphate	0.44 g
Magnesium sulphate	2.04 g
Monocalcium phosphate	1.09 g
Calcium sulphate	4.86 g
Iron sulphate (Ferrous sulphate)	0.50 g
Manganese sulphate (1 per cent)	2.50 ml
Water	1.00 gal

Commercial orchid growers often prepare formulations of nutrient solution based on their experience for different groups of orchids.A mixture of 1.0 g calcium nitrate, 0.5 g ammonium nitrate, 0.25 g monopotassium phosphate, 0.25 g magnesium sulphate, 0.02 g phosphate of iron, dissolved in four gallons of water is also reported to suit most of the orchids (Bose and Bhattacharjee, 1980). Ohio W.P. solution was found very effective in improving the growth and flowering of a large number of species and hybrids. It was suggested that the pH of the prepared solution should range between 5.5 and 6.5 and it should be sprayed once a week during the growing season and once a fortnight during other seasons (Yadav and Bose, 1985).

Generally, orchids are fed with nutrients in liquid forms. To prepare the solution, each salt is dissolved completely in water before adding the next salt. Feeding of plants during their growing season is very useful for obtaining better growth and flowering. According to Rittershausen and Rittershausen (1970), orchids should not be fed during the winter months. However, a weak solution applied at longer intervals is bene-ficial. Penningsfeld and Fast (1960) suggested that during growing season fertilizer solution should be sprayed every 2-3 weeks and in winter every 4-6 weeks.

It is a good practice to first water the plant thoroughly and then apply the fertilizer solution. Both plants and the growing media should be sprayed with the solution. After every three or four consecutive applications of fertilizer solution the compost should thoroughly be leached with water to remove the salts accumulated in it and also on clay pots (Pradhan, 1985). Combination of organic and inorganic fertilizers was found to promote growth and flowering in *Cyprepedium macranthum* (Kim *et al.*, 1996).

12.7.3 Chemical Composition and Nutrient Uptake

The chemical composition of plant helps in determining the nutritional requirement of different elements. Experimental evidences in this line are meagre in case of orchids. Poole and Sheehan (1973) reviewed the orchid composition in relation to nutrition. Plant parts were analyzed for dry matter, N, P, K, Ca, Mg, Fe, Mn, Zn and Cu contents and the results indicated that tissue analyses of the upper half of 1- to 2-year-old *Cattleya* leaves provide adequate information on the nutritional status of the plants. The uptake of nutrient elements by the entire plant of *Cattleya* and *Laelia* species, determined by Carlucci *et al.* (1980) was found to be 384.1, 522.9, 654.2, 875.1 and 888.3 mg for *L. crispilabia, L. cinnabarina, C. loddigesii, L. flava* and *C. intermedia*, respectively. Increased application rate of macro- and micro-elements generally led to an increase of individual elements in the plant part. There were positive correlations between elements, such as, Ca and Mo and negative correlations between others. The deficiency of one major nutrient resulted in increase of other elements in the ash. Determination of elements from different parts of orchids by neutron activity analysis was suggested to be a rapid and sensitive method for Na, K, Mg, Ca, Mn, Zn, Fe, Cl and Br. Among the leafless shoots, an increase in Br content and a decrease of Fe and Zn were recorded with the age of the shoot (Yamaguchi, 1979). Growth and flowering in *Dendrobium* were found to increase with higher concentration of N at 50-300 ppm and P at 25-150 ppm but flowering was delayed at high N levels. Leaves of plant receiving 100 ppm N contained 1.5 per cent N, 0.5 per cent P, 3.2 per cent K2O and 0.9 per cent Ca on dry weight basis (Uesato *et al.*, 1987). In several species of epiphytic orchids, Awasthi *et al.* (1995) reported that stem leachates from the host plant (*Schima wallichii*) were the main source of ammonium-N and nitrate-N for uptake by the orchids, while organic N from stem flow leachates was not utilized by them.

13.0 Diseases and Pests

Like other cultivated plants, orchids are also attacked by a number of pests and diseases. Most of the diseases are infectious and spread from plant to plant. The grower should always be careful and necessary preventive measures are to be followed in time. It is better to prevent diseases from its initial development and it is possible if a grower uses good cultural practices unfavourable for disease infection or pest attack (Burnett, 1965). Proper sanitation in the greenhouse and application of preven-tive sprays are very important to keep the plants free from various pests and diseases.

13.1.0 Diseases

Like pests, diseases can be a serious problem with orchids. Fungi, bacteria and viruses infect orchids causing various diseases. The most serious at present are the viruses of which reliable control measures are available. Although much research has been done and is continuing in the field of virus removal from infected plants, there is still little one can do except to utilize every technique available to prevent spread of viruses. The most common orchid diseases and their control measures are discussed below.

13.1.1 Fungal Diseases

13.1.1.1 Leaf Spot

This disease is very common and can be found to almost all cultivated species of orchids. It is caused by species of Gloeosporium, Colletotrichum, Cercospora and Phyllostictina. The disease has been reported from Germany, England and other countries by Simon (1938) and Dodgte and Rickett (1943). Within a few days of infection sunken spots appear at any place on the leaves which latter turn brown. Most leaf spot organisms do not kill the leaves but formation of spots make the leaves unsightly (Sheehan, 1980). Warm humid weather and lack of light encourage the spread of disease.

Spraying of Bordeaux mixture (6-6-50 or 4-4-50) will effectively control the disease (Ark, 1959). Dithane M-45 (0.2 per cent) and bavistin (0.1 per cent) are also effective.

13.1.1.2 Pythium Black Rot

This disease affects mostly seedlings in community pots and is caused by Pythium ultimum. Affected plants turn black and leaves start falling. The pseudobulbs also start to rot within. Since in community pots the plants are close to each other, the disease spreads from plant to plant at a very rapid rate. Orchids such as Cattleya and Epidendrum species were found to be susceptible to *Pythium ultimum*.

Withholding of watering for a few days and shifting the plants to less humid part of the greenhouse help to check the disease. Infected plants should be removed and the remaining healthy ones sprayed with fungicide like mancozeb @ 2.5 g/l. Irrigation with a mild solution of a copper fungicide once a month till the plants are tender, is an effective preventive measure. Copper oxychloride (0.4 per cent) and metalaxyl (0.1 per cent) control the malady.

13.1.1.3 Heart Rot

Cattleya, *Phalaenopsis* and *Vanda* are the important orchid genera susceptible to the disease which is caused by Phytophthora palmivora. The leaves of the affected plants turn yellow and drop off. Pseudobulbs have dark rotted areas inside.

The infected pseudobulbs should be removed and the plants repotted in fresh media after dipping in systemic fungicide. The infection spreads rapidly in low light and high humidity. Dithane M-45 and metalaxyl have been successfully used to control heart rot of orchids.

13.1.1.4 Brown Speck and Blight of Flowers

The disease affects a number of orchids like Cattleya, Phalaenopsis, Dendrobium, Oncidium and Vanda. It is caused by Botrytis cinerea. When infection is mild, spots appear on the flowers. In severe cases, the entire flower may be covered with spore masses and a blighting effect is produced.

All infected flowers should be cut off and destroyed and the plants be kept in a place where humidity is low, temperature is high and there is good ventilation. For prevention of Botrytis, it is important to keep the plants dry during cool weather. The best control may be achieved by careful spraying with systemic fungicide and by elimination of host plants near the orchid house (Jefferson and others, 1954). A black spot disease on blooms of Vanda tricolor caused by Glomerella sp. is controlled by spraying dithane M-45 and bavistin. Feder (1952) also reported fungal complex causing rot of Vanda flowers.

13.1.1.5 Root Rot

The fungus Fusarium is responsible for this disease which causes the destruction of the root tissue. Pellicularia filamentosa is also said to cause root rot. The disease kills the seedlings and retards the growth of mature plants. Affected plants should be treated with dithane M-45.

13.1.1.6 Orchid Wilt

This disease is caused by Sclerotium rolfsii which affects mostly Cymbidium and Paphiopedilum plants. The leaf base initially becomes yellow later turning brown. Warm humid weather encourages spread of the disease. Affected leaves should be removed and plants repotted in fresh medium after treatment with protective fungicide. Biological preparations of *Trichoderma* has proved effective.

13.1.1.7 Rust

In the United States and other parts of the world, Cattleya and other orchids are subject to the attack of a rust fungus, Hemileia americana. The disease is characterized by yellow pustules covered with fine powdery masses of spores (Ark, 1959). Treatment with copper oxychloride has been suggested to control rust.

13.1.2 Bacterial Diseases

13.1.2.1 Bacterial Soft Rot

It is a serious disease of Cattleya orchids and is caused by Erwinia carotovora. The disease starts at the upper end of the leaf as a small, water-soaked and some-what darker than normal green spot. Pseudobulbs of infected plants turn soft and pulpy and become yellow in colour. A foul smelling liquid sometimes starts to ooze from the bulbs. It is difficult to control the disease, but spraying with agrimycin has certain beneficial effects (Ark, 1959).

Phalaenopsis, *Dendrobium* and *Oncidium* are susceptible to soft rot caused by Erwinia chrysanthemi. Antibiotics such as achroplant 30.3 per cent S.P. and atakin 68.8 per cent W.P. inhibited soft rot (Wey, 1991). Solution of tetracycline (200 and 300 ppm) and gentamycin (100 ppm) also minimises the infection.

13.1.2.2 Bacterial Brown Rot

This disease is serious on *Paphiopedilum*. It is caused by a rod-shaped bacteria Erwinia cypripedii as described by Hori (1911). The organism attacks the plants through wounds and cracks. Yellowing and browning of leaves indicate the presence of the disease. The infection spreads into the crown and causes death of the plant. The disease could be suppressed by using antibiotic like agrimycin.

13.1.2.3 Bacterial Brown Spot

This is a disease of both seedlings and full grown plants and is caused by Xantho-monas (Phytomonas) cattleyae. In the beginning, a small water-soaked spot appears on any part of the leaves. The lesion enlarges and soon the leaf becomes soft and dies. The disease may progress into the growing point often killing the plant. Bacterium oncidii causing yellow spots on Oncidium and the bacterium which causes the scorch in Miltonia are other bacterial pathogens which damage the cultivated species. Treatment with agrimycin/tetracycline has been suggested.

13.1.3 Viral Diseases

Viruses are the worst enemies of orchids. Jensen (1959) listed 32 virus diseases to occur in orchids. But it does not mean that there were 32 different kinds of viruses. The same virus may produce more than one disease symptoms in different species of orchids. Cymbidium mosaic virus (CyMV), tobacco mosaic virus-o (TMV-O) and odontoglossum ring spot virus (ORSV) are the most common among orchid viruses. Most of the cultivated orchids are susceptible to these viruses. ORSV is a rigid rod-shaped virus, 18 nanometers in diameter and 300 in length (Zettler *et al.*, 1984). CyMV is a flexuous rod, 18 nanometers in diameter and 500 nanometers in length, and described by Gold (1950), Gold and Jensen (1951) and Jensen and Gold (1955). TMV-O is a straight short rod about 300 nanometers in length. All these viruses are obligate parasites, their wellbeing is dependent upon the hosts (Zettler *et al.*, 1984). At least 25 viruses have been identified from orchids. Of these cymbidium mosaic potex virus and odontoglossum ring spot tobacco virus received most attention (Zettler *et al.*, 1990).

13.1.3.1 Symptoms

Viruses induce chronic rather than lethal diseases and, therefore, while they may discolour, deform or stunt the affected plants, they do not reduce the life span of plants. Although they generally invade the entire plant, symptoms are mostly evident on the foliage or flowers where they may induce uneven and unpredictable colour patterns variously known as mosaic, ring spot or colour break. In severe cases malformation of leaves and flowers, crinkling and twisting may result. Growth and quality of infected plants are reduced, resulting them inferior to their healthy counterparts.

CyMV produces mosaic mottles on young leaves of *Cymbidium* which later turn black and necrotic. Leaves having severe necrotic areas drop prematurely (Jensen, 1950, 1951; Bellardi, 1983). Genera like *Cattleya, Phalaenopsis* and *Spathoglottis* are also susceptible to this disease and the symptoms reported are necrosis and mosaic (Murakishi, 1952, 1954; Jensen, 1959; Jensen and Gold, 1955).

A virus disease of *Odontoglossum grande,* caused by ORSV, was described by Jensen and Gold (1955) which is characterized by single or concentric necrotic rings of green to black tissue appearing on leaves. On the leaves of *Cymbidium*, it produces the diamond mottle and in *Cattleya* and mild form of colour breaking (Jensen, 1955). TMV-O produces chlorotic streaks on the leaves of *Cymbidium* which later turn brown. In *Cattleya*, it causes colour breaking in flowers.

In field experiments, Tan *et al.* (1999) assessed the growth performance of cymbi-dium mosaic virus inoculated plants of two *Dendrobium* clones (clones 013 and 014) were compared with virus-free plants of the same clones. In the first year, the virus-free plants for both clones grew faster (45.5 per cent average increase in stem length and 22.5 per cent increase in stem), flowered earlier and produced an average of 75.0 per cent more sprays, longer inflorescences (9.2 per cent longer) and taller flowers (4.0 per cent taller) than virus infected plants. No significant difference was noted for vase-life for both clones. An average of 61.9 per cent more sprays were produced in two years by virus-free plants of both clones than virus-infected plants.

13.1.3.3 Diagnosis

Several techniques are used for detection of viruses. These include host plant bio-assay, electron microscopy, various forms of serology and light microscopy (Burnett, 1965; Lawson and Ali, 1975; Brannigan, 1983; Wisler *et al.*, 1983). However, no one technique alone is fully reliable and even in combination, some infections can not be detected. Once a particular virus has been well studied, one or several of these detec-tion techniques can be used exclusively with reasonable accuracy. According to Bodnaruk *et al.* (1978), immunodiffusion technique proved reliable, convenient and definite for detecting the CyMV and ORSV in a wide range of orchids.

13.1.3.4 Transmission

Two forms of transmission of orchid viruses are probably found, that by man or that by some agents other than man. The two most prevalent viruses, ORSV and CyMV seem to have no insect vector (Francki, 1970; Paul, 1975), and transmission is accomplished by transfer of sap from one infected plant to a wounded area of healthy one (Bodnaruk *et al.*, 1978). This transfer is generally made while dividing plants or harvesting flowers. Other viruses known to infect orchids are likely to have specific vectors. In general, aphids are the most important vectors although other insects such as white flies, leaf hoppers and thrips also transmit viruses.

Diseases of Orchid

Leaf Spot

Heart Rot

Botrytis Blight

Wilting

Viral Diseases

Bacterial Soft Rot

13.1.3.5 Control

Viruses are difficult to control when compared with other types of pathogens because there is no reliable means of chemical control. Once the plants are infected, they cannot be made free of virus. So, care should be taken to prevent the disease. Following are the most important measures of virus control.

13.1.3.6 Sanitation

Proper sanitation in the greenhouse and use of sterilized tools and pots can enable orchid growers to entirely eliminate the detrimental effects of viruses (Corbett, 1959). Recommendations prescribed by Lawson and Ali (1975) may be followed for sanitary measures.

13.1.3.7 Viricide

Many viruses are transmitted by specific vectors, such as aphids, and hence a routine pest control programme is advisable. A proper weed control programme is also recommended as a precaution since some viruses could harbour on them. Many attempts have been made over the years to find an antiviral compound, but to date none was recommended. Attempts with such chemicals as ribavirin (virazole) have been made but with mixed results (Zettler *et al.*, 1984).

13.1.3.4 Tissue Culture for Elimination of Virus

This procedure proved to be very useful in rescuing valuable clonal orchids infected with virus. However, it is not applicable for all orchids or for all viruses. Meristemming is one of the ways of getting rid of the virus in *Cymbidium* and many other orchids. CyMV is easier to get rid of than TMV-O (Morel, 1960). According to Jensen (1959) the most important aspects of virus control are (i) selection of virus-free propagation, (ii) use of seedlings since virus is not kown to be transmitted through the seeds of orchids, (iii) destruction or isolation of diseased plants, (iv) protection of healthy plants from vectors, and (v) precautions to avoid infection of healthy plants. Heat therapy is useful for some plant virus infections, but CyMV and ORSV are so heat stable that high temperature required to inactivate the virus will kill the orchids (Wisler and Zettler, 1984).

Odontoglossum ring spot tobacco virus (ORV) was eradicated from *Cymbidium* by chemotherapy based on incorporation of ribavirin (as virazole) into the *in vitro* culture medium of protocorms. Applications of the virustatic agent for several consecutive subcultures freed protocorms of the virus. Acclimatized plantlets regene-rated from these protocorms were healthy as determined by Elisa (Toussaint *et al.*, 1993). Kim *et al.* (1997) reported elimination of this virus in *Odontoglossum* by combined treatment of 80 mg/l amantadine and using alternating temperature of 30 and 35°C. In an attempt to eradicate viruses from tissue-cultured plants of *Oncidium* Goldiana (clone 018) infected with cymbidium mosaic virus (CyMV) were treated with ribavirin (syn. virazole) which is known to be active against different plant DNA and RNA viruses (Yap *et al.*, 1999). Both the meristems and callus tissues were treated with ribavirin at 10, 20 and 40 mg/l to eradicate the CyMV, 10 mg/l of ribavirin for 30 days was sufficient to eradicate CyMV from 43 per cent of meristem plantlets com-pared with 31 per cent eradication for plantlets derived from callus tissues treated with 20 mg/l for 60 days.

13.2.0 Pests

Following are some of the most troublesome insects and pests which have been recorded to cause damage to orchids in different parts of the world.

13.2.1 Mites

These are minute insects which suck juice from the cells of leaves, leaving them speckled with white. In heavy infections, webbing appears on the plants. In Brazil, during the dry season, red spider mite (Tenuipalpus sp.) was recorded to be a serious pest of orchids, especially in Laelia and Cattleya species and hybrids. Sulphur, 3 per cent wettable or colloidal, satisfactorily controlled the nymphs and adults, but a double concentration was required to kill the eggs. The omnivorous mite, Brevipalpus australis originally known as a citrus pest, was reported to attack orchids. B. californicus has been recorded from many types of orchids under glasshouse condition in various parts of the United States.

The oncidium mite (*B.oncidii*) is a serious pest of Oncidium and Odontoglossum in England and California. Tenuipalpus pacificus, the phalaenopsis mite, is known from Australia, Indonesia, the Philippines, the Canal Zone and the United States. It is a serious pest of phalaenopsis and some other orchids. Aramite was reported to be the best material for spider mite control. A single thorough application was found suffi-cient for several months (Pritchard, 1959). Kelthane effectively controls mites.

13.2.2 Thrips

Thrips are very small insects which feed on the foliage or remain hidden in the flower buds. The epidermis is punctured, causing silvery areas on the plant. The flower buds are disfigured and sometimes they even fail to open. Many species of thrips have been recorded from orchids. *Anaphothrips orchidaceus*, the English orchids thrips, is known to feed on a number of orchids in England, Ireland and the United states, and *A. orchidii*, the California orchid thrips, is recorded from orchids in the United States and Hawaii. *Taeniothrips xanthius*, the yellow orchid thrips, earlier reported from Trinidad was subsequently found on orchids in Hawaii and the United States. Both *Heliothrips haemorrhoidalis* and *Hercinothrips femoralis* were found trouble-some to orchids in the United States, Hawaii and England, while *Parthenothrips dracaenae*, the dracaena thrips, was recorded to attack orchids in the United States and Europe. Onion thrips (*Thrips tabaci*) and flower thrips (*Frankliniella* sp.) also affect the orchids and damage the flowers (Pritchard, 1959).

The control of *Thrips tabaci* could be obtained by (i) misting with 100 ml deltamethrin, 40-50 ml permethrin or 100-150 ml pirimiphosmethyl/1000 m_2, (ii) spraying with 25-50 g or 25-50 ml permethrin or 50 ml deltamethrin, 100 g propoxur, 50 g acephate or 100 ml pirimiphosmethyl, and (iii) dusting with 300 g propoxur or 200-300 g parathion (Timmerman and Vrie, 1982). Biweekly spraying with 0.1 per cent malathion or metasystox will also control this pest.

13.2.3 Scales

Both armoured and soft scales attack orchids. They attach themselves to leaves and suck the juice from the cells. Armoured scales are amongst the most serious and persistent pests of orchids. They not only extract plant juice but also have a toxic effect upon the plant. About 100 species of armoured scale insects have been recorded from orchids and thirtyone species have been reported from Florida alone (Dekle and Kuitert, 1968). The boisduval scale is the common armoured scale found on many orchid genera, especially *Cattleya* and *Cymbidium*. It damages both leaves and pseudobulbs. Orchid scale (*Furcaspis biformis*), originally recognized in Jamaica and Trinidad, is known from *Cattleya* orchids. *Genaparlatoria pseudaspidiotus* was found on *Vanda*, and *Lepidosaphes mackiana* and *L. noxia* were observed on *Dendrobium* species in California. *Cymbidium* orchids in the United States, Europe and the Orient were noted to be the host of *L. machili*. Other widely distributed scale insects that occur on orchids are *Melanaspis aliena, Parlatoria proteus, Leucaspis cockerelli, Fiorinia fioriniae, Hemiberlesia lataniae, H. cyanophylli, Aspidiotus hederae* and *Chrysomphalus distyospermi*. Soft scales are small, elliptical and shiny in colour. *Coccus pseudohesperidum*, a species of soft scale, is known to infest *Cattleya* and *Dendrobium* in Europe, the United States and Hawaii. *C. hesperidum* and *Pulvinaria phaiae* are other species of soft scales causing severe damage to orchids (Pritchard, 1959; Sheehan, 1980).

To control the scale insects, Holttum (1964) recommended the use of the root extract of *Derris*, a plant which is common in Malaysia. Because of their scales, these insects get protection from the insecticides. Only the crawler stage which occurs just after the eggs are hatched in susceptible to insecticides is necessary. Spraying of methyl parathion is effective in eradicating this pest (Abraham and Vatsala, 1981). The use of pyrotox was also suggested against larvae of two species of scale insects (*Saissetia coffeae* and *Aspidiotus nerri*) which were reported to attack *Cymbidium* in Hungary (Ordogh and Jakacs, 1983).Treatment with 0.1 per cent metasystox also controls scales.

13.2.4 Plant Bugs

Bugs injure orchids by sucking sap and causing pale spots on the leaves. *Tenthecoris orchidearum, T. bicolor, T. vestitus* and *T. figueiredoi* are some of the common species of plant bugs that feed on orchids. *T. figueiredoi*, unlike the wellknown *T. bicolor*, attacks *Dendrobium, Oncidium* and *Newtonia* but has not been observed on *Laelia* or *Cattleya*. Good control could be obtained with 0.02 per cent parathion or metasystox (Abraham and Vatsala, 1981).

13.2.5 Aphids

Aphids suck the sap from plants and excrete honeydew on the new spikes and foliage. They also transmit certain virus diseases of orchids. *Cerataphis lataniae* and *Macrosiphum luteum* are two important species of aphids recorded on orchids. *Myzus persicae, Neomyzus circumflexus* and *Aphis gossypii* are other species found in the orchid greenhouses. Certain organic phosphates, parathion and malathion give good control of aphids.

13.2.6 Mealy Bugs

The mealy bugs are easily identified by its soft and filamentous body with a covering of white powdery wax. Underneath the covering they may look pink or yellow. These pests suck sap, secrete honeydew and attract ants. In large infestations, sooty mold may develop on the leaves. *Pseudococcus adonidum*, the long-tailed mealy bug, is a pest of *Phalaenopsis*, and *Pseudococcus maritimus*, the Baker mealy bug, attacks orchids on North American greenhouses. It is also known from orchids in India. *Planococcus citri, Ferrisiana virgata* and *Nipaecoccus nipae* are other mealy bugs. Spraying with systemic insecticides like 0.02 per cent parathion, metasystox will control these bug species found on orchids (Pritchard, 1959; Abraham and Vatsala, 1981).

13.2.7 Beetles

Several species of beetles, *i.e., Mordellistena cattleyana, M. epidendrana, Diaxenes dendrobii, D. phalaenopsis, etc.*, are reported to feed on orchids, the grubs feeding within the pith and the adults being found on the leaves or flowers. Organophosphorus insecticides may be used for the control of these pests.

13.2.8 Orchid Weevils

Weevils are serious pests of orchids. The adults feed on the flowers or puncture the flower buds and the grubs bore and feed within the plant tissues. The common species found on orchids are *Orchidophilus* spp., *Cholus* spp., *Eucactophagus weissi* and *Acythopeus aterrimus*. Orchid weevil larvae and adults have chewing mouthparts. The adult females chew feeding holes into pseudobulbs or other plant parts, crawl in, and deposit eggs. After hatching, the grubs continue feeding, creating frass/ fiber chambers suitable for pupation. Pupae do not feed. adults eventually eat their way out of the pseudobulb. While feeding may not kill the pseudobulbs, bulbs often stop growing and fail to produce flowers. Color break, a white streaking of Vanda flowers, is associated with larval infestation of pseudobulbs or stems (Mau, 1983; Hara and Mau, 1988). Spraying with 0.2 per cent thiodan emulsion at 3 weeks intervals for 3 months was suggested to control *Acythopeus aterrimus* attacking *Paphiopedilum callosum* (Helle, 1958).

13.2.9 Cattleyafly

In Hawaii, cattleyafly, a tiny wasp (*Eurytoma orchidearum*) was found to be a serious pest of *Cattleya*. The larvae burrow within the small flower buds or pseudobulbs. Removal of infested shoots and spraying with 0.1 per cent malathion proved effective against this pest.

13.2.10 *Cattleya midge*

The larvae of this pest form galls on the roots of *Cattleya* species. *Parallelodiplosis cattleyae* is an important pest of *Cattleya* and several other genera of orchids in Europe and North America. Malathion is effective to control gall midge.

Pests of Orchid

Spider Mite

Thrips

Scale

Mealy Bug

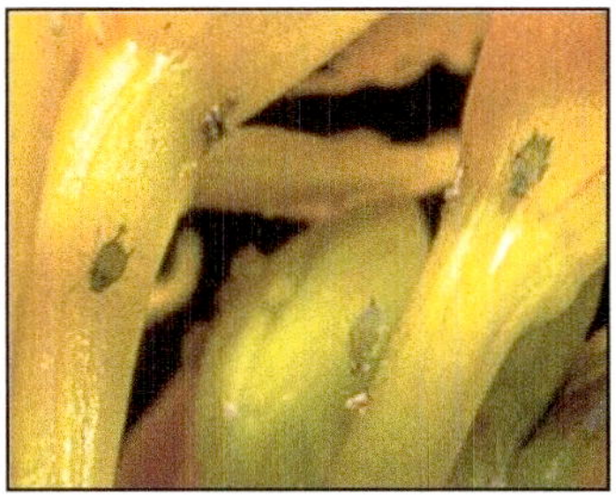

Aphid

Orchid Weevil

Nematode

Blossom Midges

13.2.11 Root Miner

In several orchid collection in Brazil it was noticed that leaves and pseudobulbs of many orchid plants were wilting. The trouble was traced to infestation by root miner which was identified as *Agromyza orchidearum*. Systemic insecticides will control the pest.

13.2.12 Slugs and Snails

Slugs and snails are common greenhouse pests where the humidity is high. They feed on the young leaves, roots, flower buds and even open flowers. Slim trails leading to the plant injury are indicative of the presence of these pests and need to be brought under control. Metaldehyde is the best material for control of snails. Alicata (1957) recorded that two ground sprays of 1 per cent metaldehyde at an interval of 3 weeks proved effective to control slugs (probably *Deroceras laeve*). This spray along with application of 3 per cent metaldehyde powder had also been effective in killing slugs in pots. The giant African snail (*Achatina fulica*) might also be controlled by metaldehyde sprays. No phytotoxic effect was observed with the application of these chemicals (Davidson, 1962).

13.2.13 Nematodes

A leaf nematode (*Aphelenchoides ritzema-bosi*) caused great loss in the cultivation of *Vanda* Miss Joaquim, the most widely grown orchid in Hawaii. Crippled flowers, yellow buds and blighted racemes are the symptoms (Sher *et al.*, 1953). Feder (1951) also recorded that up to 50 per cent flowers of *Vanda* was affected by *A. ritzema-bosi* or a closely related form. Application of carbofuran is effective to control leaf nematode.

14.0 Harvesting and Yield

14.1.0 Harvesting

This is a very important operation and the growers should have thorough knowledge about the flowering behaviour of the orchids used for cut-flower production. Proper time, stage and method of harvest determine the quality of the produce. In general, orchid flowers do not mature until 3 to 4 days after they open. Flowers cut prior to their maturity may wilt before reaching the wholesaler. Harvesting should preferably be done in the evening. All the tools should be sterilized. Flowers harvested in the heat of the day can be stressed because of high temperature. Dark coloured flowers may be as much as 5.5°C warmer than white flowers during mid-afternoon (Rij *et al.*, 1979). When individual flower stalks are cut, the peduncle should immediately be inserted in a tube of water. In Hawaii and Singapore, some growers immerse the entire sprays of flowers in water for 15 minutes before packing and shipping (Sheehan, 1980).

14.2.0 Yield

The yield of spike and number of flowers/spike varies with cultivars, growing environment, production technology including media, irrigation, nutrition and plant protection.

15.0 Postharvest Management

15.1.0 Grading and Packaging

There are no standard grades for orchid flowers and their prices are fixed according to the size. In case of cattleya flowers both colour and size are considered while pricing. If the flower size is equal, a white flower will cost more than a purple flower or a white flower with purple lip. Certain attempts have also been made to standardize the flower grades. *Aranthera* James Storiei is graded as a single spray or a branched spray, but no limits are put on the size of spray. Grading is done mainly on length of the flower spike, flower number and size and arrangement of flowers on the spike. Sometimes the number of lateral branches on the inflorescence is also taken into consideration (Sheehan, 1980).

Packaging is another important aspect in the flower trade. If it is not done properly, the flowers may wither or suffer mechanical injury during transit. An ideal package should be airtight, water proof, strong enough to withstand handling and small in volume. Many ways are followed to pack orchid flowers.

15.2.0 Storage

Since most orchid flowers are long-lived on the plants, they should not be harvested until needed. If these are to be cut they should be stored at 5-7°C. At this temperature most orchid flowers can be stored for 10 to 14 days (Sheehan, 1980). The correct use of modified atmosphere can often usefully substitute for the low temperature advantage. Plastic film storage is attractive and can be utilized (Nair, 1984).

15.3.0 Physiology of Flower Senescence

In cut-flower industry major chemical environmental pollutant affecting senescence is ethylene (Halevy and Mayak, 1979). Orchid flowers are very sensitive to ethylene (C_2H_4) and proper management and control of C_2H_4 and C_2H_4-like pollutants are important for the success of the industry (Arditti, 1979). He also suggested that the rostellum is the probable site for C_2H_4 production in orchid flowers. Any disturbance of the rostellum-viscidium interface which occurs during the dislodgement of pollinia can trigger off a wound response and enhances C_2H_4 production as recorded in a number of orchid flowers including Vanda (Burg and Dijkman, 1967). It is, therefore, important to ensure that the pollinium is not dislodged while handling the cut-flower.

Lizada and Rimando (1982) reported a marked increase in ACC (1-aminocyclopropane-1-carboxylic acid) content in pollinated Phalaenopsis flowers where column tissue showed the highest concentration among various floral parts observed. In Aranthera James Storiei it was observed that exogenous application of ACC (1 ml) on to the column and adaxial surface of the labellium of mature flower induced wilting within 24 hours. However, this exogenous supply of C_2H_4 had no visual effect on the newly opened flowers (Nair, 1984).

Lin (1999) found that in Oncidium cv. Gower Ramsay pollinia cap dislodgement of florets resulted in a decline in water uptake, reduced vase life and increased ethylene production and electrolyte leakage. Ethylene-forming-enzyme (EFE) activity was detected (at 0.45 nl/g per h) 8 h after pollinia cap dislodgements. Pulsing with silver thiosulfate at 1.5-2.0 mM at 15°C for 24 h extended vase life and promoted flower quality.

15.3.1 Ethylene Biosynthesis

Investigations on the expresson pattern of three 1–aminocyclo propane-1-carboxyase (ACC) genes in pollinated phalaenopsis flowers revealed the basis for an expanded model of interorgan regulation of three ACC synthase genes that respond to both primary (Phal - ACS2 and Phal- ACS3) and secondary (Phal ACS1) pollination signal (Bui and Oneill, 1998).

15.4.0 Vase Life

Orchid flowers, though long-lasting, should properly be handled to ensure maximum shelf life of cut flowers. The optimum harvesting stage of commercial orchids is fully open and mature. The hybrids of *Dendrobium*, *Vanda* and *Mokara* remain perfect from 7 days to 30 days (De *et al.*, 2014). The flowers of *Cattleya* and *Phalaenopsis* remain fresh for 1 to 4 weeks whereas *Aranda* lasts for 18 to 28 days. Higher sugar levels of flowers improve longevity of cut flowers. In *Cymbidium* hyb. 'PCMV', harvest at two buds opened stage had maximum vase life (66.8 days). Ethylene is the main factor responsible for early senescence. In Cymbidium hybrid 'Red Princess'pulsing with 5 per cent sucrose increases vase life upto 56 days. Pulsing with 4 mM STS for 10 minutes in Aranda and 0.5 mM STS for 24 hours in Phalaenopsis blocks the deleterious effect of ethylene. In tropical orchids like Dendrobium and Oncidium, AgNO3 (10-30 ppm) and HQS (50-100 ppm) extends vase life and bud opening of cut flowers. In Cymbidium, 1-MCP and AVG are superior to STS in prolonging the vase life of cut flowers. In Cymbidium 'PCMV', highest per cent of fully opened buds (75 per cent) and maximum vase life (45 days) were recorded with the chemical combination of sugar 4 per cent + salicylic acid 200 ppm (De *et al.*, 2014). Immediately after arrival, the lower 0.75 cm of the peduncle is cut off, and the flower is inserted into a fresh tube of water containing preservative. Sometimes, cut spikes are inserted in tube containing water or water with preservatives and bunch of 5 or more or individual spikes are placed inside the CFB box in alternate fashion. Cool growing orchids are stored at lower temperature even at 5°C in cold chambers whereas tropical and subtropical orchids are stored at 7-10°C and 90-95 per cent relative humidity (De *et al.*, 2014). In case of spray-type orchids, the basal 2.5 cm of the stem is cut upon arrival, placed in warm water at 38°C with a preservative and hardened off at 5°C (Sheehan, 1980). Foliar application of aluminium chloride at 500 ppm, ammonium molybdate at 100 ppm or boric acid at 1000 ppm lengthened the vase-life of Oncidium Goldiana (Ong, 1982). In an investigation on Oncidium Goldiana, inflorescences were cut when 0-20, 20-40, 40-60, 60-80 or 80-100 per cent of the flowers were open. Vase-life was extended by a mixture of 8-hydroxyquinoline, ascorbic acid, aspirin and sucrose or hydroxyquinoline alone for flowers cut at all stages. Flowers that were not yet open when the inflorescences were cut with more than 40 per cent bloom did not develop further in the solutions (Young and Ong, 1979). In another investigation, cut-flowers harvested at the 30 per cent bloom stage, were pre-treated with silver nitrate or distilled water and placed in holding solution with various concentrations of sucrose, 8-hydroxyquinoline and silver nitrate and citric acid. Hydroxyquinoline resulted in additional bloom opening of the flowers and also increased the vase-life (Dong and Lim, 1983). Treatment with 8-hydroxyquinoline citrate + 5 per cent

Postharvest Handling of Orchid

Spikes Inserted in Tube

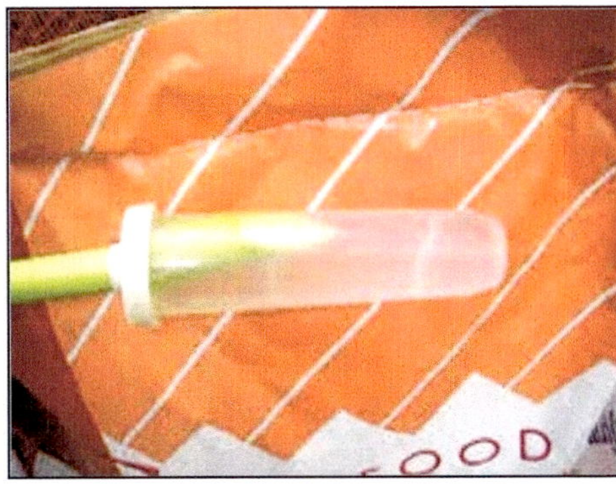

Water with Preservatives

Pulsing of Orchid

Grading of Orchid

Drying and Resting Prior to Export

Packaging for Export

International Market

South-east Asian Market

Japan Market

Orchids in International Market

Orchids in Local Market

sucrose improved the flower quality and vase-life of many orchid flowers (Nowak and Vacharotayan, 1980). The vase-life of Phalaenopsis hybrid was prolonged by using chrysal VB and sucrose mixture (Carow, 1981). Phalaenopsis cut flowers with about a third of the floret at bud stage, when pulsed with silver thiosul-phate (STS) after harvest and stored at 20°C in sucrose + 8 hydroxyquinoline sulphate (8-HQS) doubled the vase-life than those stored at 11°C (Huang *et al.*, 1995). Pretreatments with BA (upto 3 mM for 24 hours) significantly suppressed ethylene production and prolonged the vase-life of cut florets of Laeliocattleya and Brassocattleya hybrids. A combination of BA pretreatment and 0.4 mM silver thio-sulphate spray had synergistic effect on the vase-life of cattleya hybrid (Yamane *et al.*, 1997).

16.0 Biotechnology

Extensive investigations on clonal propagation *in vitro* have been carried out on many orchid species and cultivars using different explants and the methods standardized for commercial exploitation. *In vitro* propagation of orchid has been briefly summarized under propagation.Results of researches on protoplast culture and fusion, biochemical and molecular markers, genetic transformation and *in vitro* conservation in orchids are given hereunder.

16.1.0 Protoplast Culture and Fusion

Protoplast was isolated from protocorms, roots, leaves and flowers of various species and hybrids of *Aranda, Renantanda, Cattleya, Phalaenopsis, Dendrobium, Paphiopedilum, Vanilla, Odontonia, Maxillaria, Vanda, Calanthe, Cymbidium and Oncidium* after suitable enzyme treatment (Price and Earle, 1984); (Loh and Rao, 1985). Oshiro and Steinhart (1991) described a method for preparation of protoplasts from flowers, roots, and leaf tissue of the genera *Cymbidium, Cattleya, Epidendrum, Ascocentrum* and *Vanda*. A slow gentle process of cellulase and macerase digestion as well as sucrose floatation/sedimentation stem for protoplasts purification seemed to be the critical feature in the procedure to produce satisfactory yields. Isolation of protoplast of *Spathoglottis plicata* was reported by Devi *et al.* (1997).

In *Phalaenopsis*, 70-90 per cent of the protoplasts showed budding and about 1 per cent showed autofusion. Protoplasts of *Cattleya, Paphiopedilum* and *Renantanda*, however, did not show budding. It proved possible to fuse protoplast from different plants within the genus, *e.g.*, *Phalaenopsis, Dendrobium* and *Renantanda*. In the presence of 28 per cent polyethylene glycol the frequency of fusion was increased (Teo and Neumann, 1978). Intergeneric fusion of terrestrial orchid protoplasts induced by different fusion promoting agents was also found possible by Pais *et al.* (1983).

Devi and Deka (1999) reported use of mesophyll and petal derived protoplasts for fusion with PEG and protoplast fusion occurred at a frequency of 5-10 per cent in different combinations of *Vanda, Dendrobium, Phalaenopsis, Cattleya, Ascocenda*. Isolation and electrofusion of protoplast of *Phalaenopsis* were also done by Chen *et al.* (1991).

16.2.0 Biochemical and Molecular Markers

16.2.1 Isozymes

Klier *et al.* (1991) analyzed 44 morphological characters and 21 allozyme loci from 17 populations of yellow lady's slipper orchid (*Cypripedium pubescens*), nine populations of small white lady's slipper (*C. candidum*), and five sites that appeared to contain hybrid swarms, to assess the direction and magnitude of interspecific exchange of genes. The data demonstrated the existence of bidirectional gene flow whenever the two species are sympatric.

Electrophoresis of 5 isoenzymes and total proteins of 5 taxa of *Calanthe* was performed by Kim *et al.* (1990). In their study, the intermediate superoxide dismutase pattern of *C. discolor* var. *bicolor* suggested that it was a hybrid between *C. discolor* and *C. sieboldii.*

16.2.2 Molecular Markers

Knowledge of genome size has application in the fields of diverse interest such as cell and molecular biology, ecology, phytogeography and systematics. Genome size has also been used to confirm taxonomic assignments.

Genetic relationships between orchids of the genus *Vanda* and the related genus *Ascocentrum* were evaluated by Lim *et al.* (1999) using RAPD analysis. The results obtained indicated that strap-leaved *Vanda* species (including *Vanda sanderiana*) and *Ascocentrum miniatum* were more closely related to each other than to the terete-leaved *Vanda* species studied.

Chen *et al.* (1999) investigated the application of the PCR-based fingerprinting technique AFLP in orchids. The optimal AFLP patterns were determined using primer combinations of EcoRI + 4 and MseI + selective nucleotides. Genomic variations among different cultivars of vandaceous orchid hybrids were successfully determined by AFLP analysis. More than 10 per cent of the AFLP bands were polymorphic DNA when siblings derived from the same original crosses (two cultivars of *Aranda* Christine, five cultivars of *Mokara*) were used.

16.2.2 Phylogenetics and Diversity

The variability within the Orchidaceae family is perceived in a large number of genera and species, as well as in the diversity within the same species, as found in *Cattleya violacea*. Few plants surpass the orchids in their distribution throughout the world, variation in growth habits, and the magnificent spectrum of colors produced by their flowers and leaves. Rego *et al.* (2009) analyzed the genetic variability among eleven *C. violacea* varieties from the Amazonian Region (three Brazil Amazonian States and one Venezuela Amazonian region) with different flower coloration: *Cattleya violacea* cv. *caerulense*, *Cattleya violacea* cv. *maravilha*, *Cattleya violacea* cv. *semi-alba*, *Cattleya violacea* cv. *semi-alba estriata*, *Cattleya violacea* cv. *semi-alba flamea*, *Cattleya violacea* cv. *rubra*, *Cattleya violaceae* cv. *caerulea*, *Cattleya violacea* cv. *semi-alba fantasia*, *Cattleya violacea* cv. *amesiana*, *Cattleya violacea* cv. *bicolor* and *Cattleya violacea* cv. *alba*. RAPD analysis indicated that the varieties were clustered in ten different groups. The most similar varieties were *Cattleya violaceae caerulense* and *Cattleya violacea caerulea* that had blue flowers and were collected in Mato Grosso state and Roraima state. The most dissimilar were *Cattleya violacea rubra* and *Cattleya violacea semi-alba estriata* from the same region of Roraima state and they were similar with white and purple flower coloration. The genetic diversity analysis showed the necessity for conservation of the biologic diversity of the genus *Cattleya* and Orchidaceae family in Amazon Region. Random Amplified Polymorphic DNA (RAPD) was found to be good enough to estimate the variability in *C. walkeriana*. The selected primers were able to define colour group, especially the *coerulea*. Inbreeding will occur in crosses of clones with the same colour (Picolo *et al.*, 2016).

Cattleya intermedia is native from the South Atlantic rainforest. Neto and Vieira (2011) investigated the variation of wild and bred *Cattleya intermedia* accessions using RAPD to clarify the genetic relationships between bred lines and accessions collected in the wild. A total of 65 polymorphic bands were used to generate a genetic distance matrix. No specific groupings were revealed by the cluster analysis as bred materials were not different from wild plants. The genetic differentiation (FST = 0.01626) was very low indicating a high gene flow in *C. intermedia* due to artificial crosses and a high differentiation between populations. The genetic variability available within this species is high enough to allow genetic progress in flower shape and size (Neto and Vieira, 2011).

Cattleya elongata, is one of the most beautiful species of orchids and is the only species of the genus that is exclusively rupiculous. Its distribution is restricted to the Chapada Diamantina in Brazil (Cruz *et al.*, 2003 cited by Cruz *et al.*, 2011). Nine natural populations of *C. elongata* for morphological and genetic variability,covering the whole distribution area of the species, using allozymes and ISSR markers and morphometric multivariate analyses (Cruz *et al.*, 2011). Genetic variability in allozimes was relatively high (He = 0.12–0.25), and unexpectedly higher than the values based on ISSR (He = 0.16–0.19). Genetic similarity among the populations was high in both markers, in spite of the discontinuity of the outcrops of the region. The differences observed between the two genetic markers and the various morphological markers examined indicated that the isolated use of any single parameter of these different populations for conservation planning or management would not consider all of the variability to be found in the species.

Pinheirom *et al.* (2012) provided the first data on genetic diversity and structure of wild populations in the genus *Cattleya*, in particular *C. labiata*, using RAPD and intersimple sequence repeat (ISSR) markers. 130 individuals, 117 belonging to *Cattleya labiata* and 13 from 10 other species in the same genus were studied, Data generated from 12 ISSR and 12 RAPD primers were used to determine genetic variability. The marker data indicated that *C. labiata* has a high level of polymorphism, and five reconstructed populations were identified by Structure (Pinheirom *et al.*, 2012)

16.2.3 Species Identification

Jiménez *et al.* (2017) evaluated the feasibility of DNA barcode regions, namely ITS1, ITS2 and *rpoC1*, to discriminate between *Cattleya walkeriana* and *C. loddigesii* species. DNA barcode regions were successfully amplified using primers designed to amplify plants. They included sequences from public databases in order to test if these regions were able to discriminate *C. walkeriana* and *C. loddigesii* from other *Cattleya* species. These regions, and their combinations, demonstrated that the ITS1+ITS2 had the highest average interspecific distance (11.1 per cent), followed by *rpoC1* (1.06 per cent). For species discrimination, ITS1+ITS2 provided the best results. The combined data set of ITS1+ITS2+*rpoC1* also discriminated both species,but did not result in higher rates of discrimination. These results indicate that ITS region is the best option for molecular identification of these two species and from some other species of this genus.

16.3.0 Genomics

In an exhaustive review on genomics of orchids, Tsai *et al.* (2017) opined that the first milestone was the sequencing of the *Phalaenopsis equestris*, a tropical epiphytic orchid and recurrently be used as parent species for orchid breeding. The *P. equestris* genome was whole-genome shotgun and sequenced by Illumina technology.The genome size was estimated to 1.16 Gb containing 29,431 predicted protein-coding genes. Cai *et al.*(2015) presented the genome sequence of the *Phalaenopsis equestris*, a frequently used parent species for orchid breeding. *P. equestris* is the first plant with crassulacean acid metabolism (CAM) for which the genome has been sequenced. The assembled genome contains 29,431 predicted protein-coding genes. It was observed that contigs are likely to be underassembled, owing to heterozygosity, are enriched for genes that might be involved in self-incompatibility pathways. The results indicated that that gene duplication might have contributed to the evolution of CAM photosynthesis in *P. equestris*. Finally, they found expanded and diversified families of MADS-box C/D-class, B-class *AP3* and *AGL6*-class genes, which might contribute to the highly specialized morphology of orchid flowers. Analysis of the *P. equestris* genome showed that the majority of the genome (about 62 per cent) was occupied by repetitive DNAs, mostly transposable elements (TEs). Furthermore, Tsai *et al.* (2017) concluded that the *Phalaenopsis* genome sequence was practical to identify MYB genes controlling floral pigmentation patterning (Hsu *et al.*, 2015) and TCP genes participated in the ovule development (Lin *et al.*, 2016).

16.4.0 Transcriptomics

Most ESTs were revealed as being related to carbohydrate metabolism, and regulatory function, biosynthesis of mannose, pectin and starch, stress-related, and transportation (Tan *et al.*, 2005). To illustrate expressed genes in reproductive organs of *Phalaenopsis*, *P. equestris* mature flower buds were collected and 5593 ESTs were sequenced and annotated (Tsai *et al.*, 2006). In addition, 2359 ESTs were sequenced from scented *P. bellina* flower buds cDNA library to deduced ESTs involved in scent biosynthesis pathway (Hsiao *et al.*, 2006 cited by Tsai *et al.*, 2017). According to Tsai *et al.* (2017), 454 technology was independently applied to generate 8233 contigs and 34,630 singletons sequenced from the mixed tissues of three *Phalaenopsis* species (Tsai *et al.*, 2013), and 50,908 contig sequences released from six different tissues of *Oncidium* Gower Ramsey (Chang *et al.*, 2011). These data set expansively increased information of expressed genes in *Phalaenopsis* and *Oncidium* and speed identifying sets of ESTs associated with a broad range of biological processes (Chang *et al.*, 2011; Hsiao *et al.*, 2011; Huang *et al.*, 2015 cited by Tsai *et al.*, 2017).

Even though the whole-genome sequence of *Phalaenopsis equestris* is published (Cai *et al.*, 2015), the current scaffold number of *P. equestris* genome is still high (236,185 scaffolds), and the assembly of *P. equestris* scaffolds remains to be improved. A saturated genetic linkage map is regarded as an effective tool to facilitate scaffolds assembly. Besides, genetic mapping is referred to a stepin genome mapping, which can integrate with a physical map and a cytogenetic map to access genes in genome (Tsai *et al.*,2017). Previously, 950 SSRs from *P. equestris* have been identified, and among them 206 SSRs primer sequences were developed for *Phalaenopsis* genetic mapping (Hsu *et al.*, 2011; Tsai *et al.*,2017). *P. equestris* contains 38 chromosomes (2N = 2x = 38) and its genome size is 1600 Mb per haploid genome. It is a relatively large genome compared to other crops, like rice or sorghum.

Hence, development of abundant molecular markers for *Phalaenopsis* genome mapping was felt necessary (Tsai *et al.*,2017). In addition, a genetic linkage map can be adopted for useful trait analysis and applied for quantitative trait loci (QTL) mapping. Combining the phenotypic data of the mapping population, QTLs may be able to be mapped on the saturated genetic linkage map (Tsai *et al.*, 2017).

Gawenda *et al.* (2012) analysed the genetic diversity of a broad collection of *Phalaenopsis* varieties and selected wild species by means of molecular markers. The marker data were used to obtain genetic distances, estimates of the degree of linkage disequilibrium and population structure for the genotypes under study. With a total of 492 markers, the genotypes clustered according to their horticultural classification (for example, old hybrids vs. more recent hybrids) but not according to their origin, indicating extensive exchange of germplasm among breeders. Linkage disequilibrium was found to decrease relatively slowly, most likely due to the small number of generations that have occurred since the first hybrids were generated. Based on the most likely estimates for the population structure (ranging from 10 to 12 subpopulations), associations between ornamental traits like flower size, flower colour, flower type, flower texture, stem length and leaf shape were calculated. These results can now serve as starting points for detailed analyses of the genetic architecture of these traits.

16.5.0 Genetic Transformation

Transformation in orchid has been attempted using protoplasts (Chen *et al.*, 1991; Oshiro and Steinhart, 1991). Griesbach and Hammond (1993) reported incorporation of *gus* gene into orchids through eletrophoresis. Griesbach (1994) improved the original eletrophoretic transformation method described by Ahokas (1989) and was able to obtain transgenic plants in orchids. He used one-month-old protocorms of *Calanthe sedenii* and a plasmid pBI-121 containing the *gus A* gene under control of the cauliflower mosaic virus 35 S promoter. Introduction of DNA leading to stable transformation occurred at 0.5 mA constant current after 10 min of electrophoresis. Marker gene was expressed in about 55 per cent of the surviving seedlings.

Belarmino and Mii (2000) genetically transformed a hybrid [*Doritaenopsis* Coral Fantasy × *Phalaenopsis* (Babay Hat × Ann Jessica)] after cocultivation of cell clumps with *A. tumefaciens* strain LBA4404 (pTOK223) and EHA101 (pIG121Hm) that harbour *gus* gene and hygromycin resistance. Regeneration of transgenic plantlets was enhanced by incubating PLBs on NDM medium supplemented with 0.1 mg/l abscisic acid, followed by partial desiccation for 10-30 min. More than 100 hygromycin resistant plantlets were produced in about 7 months following infection of the cell aggregates.

Zhang *et al.* (2010) cocultivated protocorm-like bodies (PLBs) of *Cattleya* orchid CM2450 were with *Agrobacterium tumefaciens* strain EHA101 carrying either plasmid pIG121-Hm harboring genes coding for neomycin phosphotransferase II, hygromycin phosphotransferase, and β-glucuronidase (*GUS*) or plasmid pBBR*acdS* harboring these same genes along with a gene coding for 1-aminocyclopropane-1-carboxylate (ACC) deaminase. PLBs were maintained in a liquid New Dogashima (ND) medium and then added to a bacterial suspension culture ($OD_{600} \approx 0.6$) yielding dilution ratios of either 1:10 or 1:40, and these were incubated for either 30 min, 3 h, or 6 h. Hygromycin-resistant secondary PLBs were induced after 4 weeks of culture on an ND medium containing 1.0 mg/l naphthaleneacetic acid (NAA), 0.1 mg/l benzyladenine (BA), 10 mg/l hygromycin, 20 mg/l meropenem, 10 g/l sucrose, and solidified with 2.5 g/l gellan gum. The highest frequency of transformation was obtained when PLBs were inoculated with 1:10 *Agrobacterium* liquid culture for 3 h. The transformation of hygromycin-resistant plantlets regenerated from different sites of inoculated PLBs was confirmed by histochemical GUS assay, polymerase chain reaction (PCR) analysis, and Southern blot hybridization.

Qin *et al.* (2011) cited several studies on successful *Agrobacterium*-mediated transformation of *Phalaenopsis* (Belarmino and Mii, 2000; Chai *et al.*, 2002; Chan *et al.*, 2005; Mishiba *et al.*, 2005; Semiarti *et al.*, 2007), among which an effective *Agrobacterium*-mediated transformation system for *P. amabilis* was carried out as well.

Genetically transformed plants of a phalaenopsis orchid [*Doritaenopsis* Coral Fantasy×*Phalaenopsis* (Baby Hat×Ann Jessica)] were regenerated after cocultivation of cell clumps with *Agrobacterium tumefaciens* strains LBA4404 (pTOK233) and EHA101 (pIG121Hm) that harbored genes for β-glucuronidase (GUS) and hygromycin resistance (Belarmino *et al.*, 2000). The efficiency of transformation was markedly increased by 10 h cocultivation of cell clumps with *A. tumefaciens* that had been induced with 200 µM acetosyringone, and by inclusion of 500 µM acetosyringone in the cocultivation medium. Hygromycin-resistant cell clusters (0.5–3 mm in diameter) were selected from the infected cell clumps after 4–6 weeks of culture on agar (8 g/l)-solidified new Dogashima medium (NDM) containing 20 g/l sucrose, 0.1 mg/l naphthaleneacetic acid, 1.0 mg/l benzyladenine (BA), 50 mg/l hygromycin and 300 mg/l cefotaxime. The cell clusters proliferated 4 weeks after transfer onto the same medium. To induce callus greening, the carbon source was changed from sucrose to maltose. The green calli obtained produced protocorm-like bodies (PLBs) after 4 weeks of culture on phytohormone-free NDM medium. Regeneration of transgenic plantlets was enhanced by incubating PLBs on NDM medium supplemented with 0.1 mg/l abscisic acid, followed by partial desiccation for 10–30 min. Successful transformation was confirmed by histochemical GUS assay, PCR analysis and Southern hybridization of transformants. With this transformation system, more than 100 hygromycin-resistant phalaenopsis plantlets were produced about 7 months following infection of the cell aggregates.

Embryogenic callus of *Phalaenopsis amabilis* derived from leaf tissue was cocultivated with *Agrobacterium tumefaciens* strain LBA4404 harboring a plant cloning vector (Qin *et al.*, 2011). The vector carried the lipid transfer protein (LTP) encoding gene cloned from cold tolerant Brazilian upland rice cv. IAPAR 9. The highest transformation efficiency (12.16 per cent) was obtained when 1–2 mm calli were infected and cocultivated with 0.4 (OD_{600}) *A. tumefaciens* for 20 min. Transgene integration of *kan*-resistant plants was confirmed through polymerase chain reaction analysis and Southern hybridization. Four hundred seventy transgenic plants, each derived from an independent protocorm-like body, were obtained. The expression of rice cold-inducible *LTP* gene in transgenic *P. amabilis* improved its adaptive responses to cold stress. The examination of transgenic plants revealed that enhanced cold tolerance was most likely due to the increased accumulation of several compatible solutes such as total soluble sugars, proline, antioxidant superoxide dismutase, decreased accumulation of malondialdehyde, and maintained electrolytes within the membrane compared with controls (Qin *et al.*, 2011).

16.6.0 *In vitro* Conservation

Development of artificial seeds has been reported in orchids. The encapsulation technique offers a method of preserving desirable genotypes and an easy and novel delivery system for orchid propagates (Devi and Deka, 1999).

Artificial seeds were produced in *Phaius tankervilliae* by encapsulation of protocorms with sodium alginate. Malemnganba *et al.* (1966) also reported that 96 per cent of freshly prepared encapsulated protocorms differentiated into shoots and roots when cultured on Nitsch medium. Chetia (1996) prepared artificial seeds in three orchid species, *viz.*, *Cymbidium longifolium*, *Renanthera imschootiana* and *Agrostophyllum myrianthum* by encapsulating protocorms cultured in Nitsch medium and 0.7 per cent agar water and observed higher germination percentage compared to non-encapsulated protocorms in both the media. Cells of a suspension culture of *Doritaenopsis* cv. New Toyohashi strain 643 were placed in a mixture of 2 M glycerol and 0.4 M sucrose for 15 minutes at room temperature and then dehydrated with a vitrification solution (PVS2) for 1-3 h on ice and plunged into liquid nitrogen. The highest viability (64 per cent by 2.3, 5-triphenyltetrazolium chloride stainability) was obtained when the cells were precultured in liquid New Dogashima medium with 0.1 M sucrose and 1.0 mg abscisic acid/1 for 1 week at 25°C in the light. Protocorm-like bodies were induced from cryopreserved cells without morphological variations (Tsukazaki *et al.*, 2000).

17.0 References

Abdul Karim, A.G., Hairani Haris and Nazarnddin, U. (1992) *Lindleyana*, **7** : 7-10.

Abhilasha, D. (1997) *J. Orchid Soc. India*, **11** : 93-97.

Abraham, A. and Vatsala, P. (1991) *Introduction of Orchids*, Tropical Botanic Garden and Research Institute, Trivandrum.

Ackerman, J.D. (1984) *Proc. 11th World Orchid Conf.*, Florida, pp. 98-101.

Ackerman, J.D. and Mesler, M.R. (1979) *Amer. J. Bot.*, **66** : 820-824.

Agrawal, K.S. and Dey, A.C. (1969) *G. K. J. Sci. Res.*, **1** : 37-38.

Ahmad, Z., Ariffin, S., Ramli, R.A.A., Basiran, M.N. and Abu Hassan, A. (2010) 7th R and D Seminar, Research and Development Seminar; Bangi (Malaysia).

Alicata, J.E. (1957) *Hawaii Fm. Sci.*, **6** : 11.

Alphonso, A.G. (1975) *Proc. First Asean Orchid Cong.*, Bangkok, pp. 1-5.

Alphonso, A.G. (1978) *Proc. Symposium on Orchidology*, Singapore, The Orchid Society South-East Asia, Singapore, pp. 70-76.

Amaki, W. and Higuichi, H. (1989) *Sci. Hort.*, **39** : 63-72.

Amberger-Ochsenbaner, S. (1997) *Acta Hortic.*, **450** : 105-112.

Ames, O. (1948) *Orchids in Retrospect*, Bot. Mus., Harvard University.

Amore, T.D. and Kamemoto, H. (1991) *Lindleyana*, **6** : 117-167.

Amore, T.D. and Kamemoto, H. (1992) *Lindleyana*, **7** : 162-167.

Amore, T.D. and Kamemoto, M. (1995) *Lindleyana*, **12** : 12-15.

Anderson, L. (1967) *Amer. Orchid Soc. Bull.*, **36** : 304-338.

Ando, T. (1982) *Tech. Bull. Fac. Hort., Chiba Univ.*, No. 30, pp. 1-6.

Anonymous (1912) *Orchid World*, **2** : 122.

Anonymous (1914a) *Orchid World*, **4** : 97.

Anonymous (1914b) *Orchid World*, **4** : 194.

Anonymous (1949) *The Charaka Samhita*, Vol. 1-6, Ayurvedic Society, Jamnagar, India.

Anonymous (2000) The Orchid Review, **108** : 3-13

Anonymous (2001)The Orchid Review, **109** : 3-15

Arditti, J. (1965) *Orchid Dig.*, **29** : 432-433.

Arditti, J. (1966a) *Cymbidium Soc. News*, **20** : 10-11.

Arditti, J. (1966b) *Amer. Orchid Soc. Bull.*, **37** : 385-387.

Arditti, J. (1967a) *Bot. Rev.*, **33** : 1-97.

Arditti, J. (1967b) *Amer. J. Bot.*, **54** : 291-298.

Arditti, J. (1968) *Amer. Orchid Soc. Bull.*, **39** : 112-126.

Arditti, J. (1976) *Proc. 8th World Orchid Conf.* pp. 528-533

Arditti, J. (1977) In : *Orchid Biology – Reviews and Perspective*, I (Ed. J. Arditti ed), Cornell University Press, Ichaca and London, pp. 203-293.

Arditti, J. (1979) *Adv. Bot. Res.*, **7** : 421-455.

Arditti, J. (1981) *Adv. Bot. Res.*, **9** : 422-438.

Arditti, J. and Ernst, R. (1974) *Proc. 7th World Orchid Conf.*, Medellin, pp. 299-307.

Arditti, J. and Ernst, R. (1993) *Micropropagation of Orchids*, John Wiley, New York.

Arditti, J. and Fisch, M.H. (1977) In : *Orchid Biology – Reviews and PerspectivesI, I* (Ed. J. Arditti), Cornell University Press, Ithaca and London.

Arditti, J. and Harison, C.R. (1979) *Bot. Gaz.*, **140** : 133-137.

Arditti, J. and Harrison, C.R. (1977) In : *Orchid Biology – Reviews and Perspectives*, I (Ed. J. Arditti), Cornell University Press, Ithaca and London.

Arditti, J. and Knauft, R.L. (1969) *Amer. J. Bot.*, **56** : 620-628.

Arditti, J., Arditti, M. and Ernst, R. (1984) *Proc. 11th World Orchid Conf.*, Florida, pp. 102-105.

Arditti, J., Flick, B.H. and Jeffrey, D.C. (1971a) *New Phytol.*, **70** : 334-341.

Arditti, J., Healey, P.L. and Ernst, R. (1972) *Amer. Orchid Soc. Bull.*, **41** : 503-510.

Arditti, J., Hogan, N.M. and Chadwick, A.V. (1973) *Amer. J. Bot.*, **60** : 883-888.

Arditti, J., Jeffrey, D.C. and Flick, B.H. (1971b) *New Phytol.*, **70** : 1125-1141.

Arditti, J., Oliva, A.P. and Michaud, J.D. (1982) *Amer. Orchid Soc. Bull.*, **51** : 394-397.

Ariffin, S. and M. Nazir. (2005) "MY0101588".

Ark, P.A. (1959) In : *The Orchids – A Scientific Survey* (Ed. C.L. Withner), Ronald Press Co., New York.

Ark, P.A. and Snyder, W.C. (1951) *Plant Dis. Reptr.*, **35** : 43-44.

Arora, Y.K. (1983) *Indian Hort.*, **28** : 11-12, 16.

Arora, Y.K., Rao, A.S. and Mukherjee, A. (1975) *Indian Hort.*, **20** : 23-27.

Ashokas, H. (1989) *Theor. Appl. Genet.*, **77** : 469-472.

Atwood, J.T. (1984) *Proc. 11th World Orchid Conf.*, Florida, pp. 106-110.

Awasthi, O.P., Sharma, E. and Palni, L.M.S. (1995) *Ann. Bot.*, **75** : 5-11.

Ayasse, A., Schiestt, F.P., Panlus, H.F., Erdmann, D. and Frenche, W. (1997) *Proc German Society for General and Applied Entomology*, Germany, 18-22 March.

Bagde, P. and Sharon, M. (1997) *Indian J. Plant Physiol.*, **2** : 10-14.

Bailes, C. (1994) *Amer. Orchid Soc. Bull.*, **63** : 238-249.

Baker, C. and Baker, M. (1995) *Amer. Orchid Soc. Bull.*, **64** : 1102-1107.

Baker, C.(1995) Awer orchid Soc. Bull., 64 : 1102-1107

Barendse, L.V.J. (1979) *Vakblad voor de Bloemisterij*, **33** : 20-21.

Barendse, L.V.J. (1980) *Vakblad voor de Bloemisterij*, **35** : 25.

Barry, K., Hu, J. S., Kuhenle, A.R. and Sughii, N (1996) J. Phytopathology, **144** : 179-186.

Beckner, J. (1979) *Amer. Orchid Soc. Bull.*, **48** : 556-560.

Beechey, N. (1970) *Amer. Orchid Soc. Bull.*, **39** : 1085-1088.

Begum, A.A., Tamaki, M., Tahara, M. and Kako, S. (1994) *J. Japanese Soc. Hort. Sci.*, **63** : 419-427.

Belarmineo, M.M. and Mii, M. (2000) *Plant Cell Rep.*, **19** : 435-442.

Belarmino, M., Mii, M. (2000) *Plant Cell Reports*, **19**: 435–442.

Bellardi, M.G. (1983) *Sunformatore Fitopatologico.*, **33** : 31-35.

Bentham, G. (1881) *Jour. Linn. Soc. Bot.*, **18** : 281-360.

Berg, F. (1965) *Amer. Orchid Soc. Bull.*, **34** : 899-902.

Bergman, F.J. (1972a) *Amer. Hort.*, **51** : 20-23.

Bergman, F.J. (1972b) *Amer. Hort.*, **51** : 41-44.

Bernard, N. (1899) *C.R. Acad. Sci.*, Paris, **128** : 1253-1255.

Bernard, N. (1905) *C.R. Acad. Sci.*, Paris, **140** : 1272-1273.

Bernard, N. (1909) *Ann. Sci. Nat. Bot.*, **9** : 1-196.

Bertsch, W. (1967) *Amer. Orchid Soc. Bull.*, **36** : 32-38.

Bess, H.A. (1953) *Hawaii Fm. Sci.*, **1** : 3.

Bhattacharjee, S.K. (1979) *Sci. and Cult.*, **45** : 293-295.

Bhattacharjee, S.K. (1980) *Singapore J. Prim. Ind.*, **7** : 90-92.

Bhattacharjee, S.K. (1981) *Die Gartenbauwiss*, **46** : 178-181.

Bhattacharjee, S.K. (1984) *Amer. Orchid Soc. Bull.*, **53** : 390-393.

Bhattacharjee, S.K. and Mukherjee, T.P. (1981) *Haryana J. Hort. Sci.*, **10** : 58-66.

Bik, R.A. (1982) *Bedrjsontwikkeling*, **13** : 562-563.

Bingham M.G. (2009) Chikanda, an unsustainable industry.*Pollinia* **7**: 23–25.

Bivin, J.L. (1970) *Amer. Orchid Soc. Bull.*, **39** : 1005-1006.

Bivins, J.L. (1968) *Amer. Orchid Soc. Bull.*, **37** : 385-387.

Black, J.M. (1933) *Orchid Rev.*, **41** : 69-72.

Blume, C.L. (1825) *Bijdragen tot de Flora Van Nederlandssch Indie*, Stuk, 6, 7, 8 met Vitt Tabellen, 376.

Blumenschein, A. (1960) *Publ. Cien. Univ. Sao Paulo Inst. Genet.*, **1** : 47-48.

Bobisud, C.A. and Kamemoto, H. (1982) *J. Amer. Soc. Hort. Sci.*, **107** : 1024-1027.

Bodnaruk, W.H., Hennen, G.R. and Zettler, F. (1978) *Amer. Orchid Soc. Bull.*, **47** : 1113-1116.

Borba, E.L. and Semir, J. (1998) *Lindleyana*, **13** : 203-218.

Bose, T.K. and Bhattacharjee, S.K. (1980) *Orchids of India*, Naya Prokash, Calcutta, India.

Bose, T.K. and Bhattacharyee, S.K. (1972) *Indian Hort.*, **17** : 25-27.

Bose, T.K. and Mukherjee, T.P. (1976) *Orchid Rev.*, **84** : 107-108.

Bose, T.K. and Mukherjee, T.P. (1977) *Orchid Rev.*, **85** : 245-247.

Bose, T.K., Bhattacharjee, S.K., Das, P. and Basak, U.C. (1999) *Orchids of India*, (Revised edition), Naya Prakash, Calcutta, India.

Bose, T.K., Maiti, R.G., Dhua, R.S. and Das, P. (1999) *Floriculture and Landscaping*, Naya Prakash, Calcutta, India.

Bouriquet, G. (1948) *L' Agron. Trop.*, **2** : 105-164.

Brandange, S. and Luning, B. (1969) *Acta Vhem. Scand.*, **23** : 1151-1153.

Brannigan, M. (1983) *M. S. thesis*, University of Maryland, USA.

Breddy, N.C. (1949) *Orchid Rev.*, **57** : 128-132.

Brown, D.M., Groom, C.L., Cvitanik, M., Brown, M., Copper, J.L. and Arditti, J. (1982) *Plant Cell Tissue Organ Cult.*, **1** : 168-180.

Bruyns, P. (1989) *South African J. Bot.*, **55** : 492-497.

Bui Van Le, Hang Phuong, N.T., Anh Hong, L.T. and Tran Thanh Van, K. (1999) *Plant Growth Regulation*, **28** : 179-185.

Burg, S.P. and Dijkman, M.J. (1967) *Planta*, **150** : 439-442.

Burgeff, H. (1932) *G. Fisher Verlag*, Jena, pp. 249.

Burgeff, H. (1954) *G. Fisher, Stuttgart*, 4. 48.

Burnett, H.C. (1965) *Fla. Dept. Agric. And Consumer Services Div., Plant Ind. Bull.*, **10** : 50-60.

Bytebier, B., Simiyu, S.W. and Pearce, T.R. (1994) *Proc. 14th AFTFAT Congress, 22-27 August*, Wagingen, Netherlands.

Cai J, Liu X, Vanneste K, Proost S, Tsai WC, Liu KW, Chen LJ, He Y, Xu Q, Bian C,et al. (2015) *Nat Genet* **47**: 65–72.

Cameron, K.M. and Chase, M.W. (1998) *Lindleyana*, **13** : 148-169.

Carlucci, M.V., Hoag, H.P. and Bello, A.F.J. (1980) *Solo*, **72** : 27-34.

Carnes, A.E. (1953) *N.J. Agric.*, **35** : 8-9.

Carnevali, G. and Romero, G.A. (1991) *Novon*, **4** : 86-93.

Carow, B. (1981) *Deutscher Gartenbau*, **35** : 320-322.

Carpenter, M.O. (1988) *Amer. Orchid Soc. Bull.*, **57** : 1360-1369.

Chadwick, A.V., Hogan, N.M. and Arditti, J. (1980) *Bot. Gaz.* **141** : 422-427.

Chae Soo Cheon, Sonkicheol Yun Jae Gill (1998) *J. Korean Soc. Hort. Sci.*, **39** : 60-65.

Chai ML, Xu CJ, Senthil KK, Kim JY, Kim DH (2002) *Sci Hortic* **96**: 213–224.

Chan YL, Lin KH, Liao LJ, Chen WH, Chan MT (2005)*Transgenic Res* **14**: 279–288

Chang YY, Chu YW, Chen CW, Leu WM, Hsu HF, Yang CH (2011) *Plant Cell Physiol* **52**: 1532–1545.

Chang, C. and Chang, W.C. (2000) *In Vitro Cellular Dev. Biol.*, **36** : 517-520.

Chang, W. (1984) *Proc. 11th World Orchid Conf.*, Florida, pp. 136-137.

Charansari, U. and Kamemoto, H. (1978) *Amer. Orchid Soc. Bull.*, **47** : 701-705.

Cheamuangphan, A., C. Panmanee, and R. Tanusuchat. (2013) Value Chain Analysis for Orchid Cut Flower Business in Chiang Mai. Business and Information, Chiang Mai, Thailand.

Chen, C.C., Sheng, D.C. and Lin, R.S. (1999) *J. Agri. and For.*, **48** : 31-44.

Chen, J.T. and Chang, E.C. (2000a) *Plant Cell Tissue Organ Cult.*, **62** : 95-100.

Chen, J.T. and Chang, W.C. (2000b) *Plant Science*, **160** : 87-93.

Chen, J.Y., Cheng, J.S. and Zhu, Y. (1998) *J. Beijing Forestry University*, **20** : 32-35.

Chen, Jian Yuan, Pan Rui Chi and Wen Zhas Qins (1994) *J. Trop. Subtrop. Bot.*, **2** : 70-76.

Chen, W.H., Tang, C.Y. and Kao, Y.L. (2009a) *Plant Cell Tiss Organ Cult* **98,** 229–238.

Chen, W.H., Tang, C.Y. and Kao, Y.L. (2010) *Acta Hortic.* **878**: 133-138.

Chen, W.H., Tsai, W.T., Wu, C.C. and Hsich, R.M. (1991) *Taiwan Sugar J.*, March, pp. 14-18.

Chen, Y.C., Chang, C. and Chang, W.C. (2000) *In Vitro Cellular and Developmental Biology - Plant*, **36** : 420-423.

Chen, Y.H., Chen, W.H., Chyou, M.S., Fu, Y.M., Lin, Y.s., Hsiau, T.W., Lin, K.Z. and Hong, L.S. (2000) *J. Chinese Soc. Hort. Sci.*, **46** : 147-156.

Chen, Y.H., Tsai, Y.J., Huang, J.Z. and Chen, F.C. (2005) *Cell. Res.*, 15(8) : 639-657.

Chennaveeraiah, M.S. and Patil, S.J. (1975) *Curr. Sci.*, **44** : 68.

Chetia, S. (1996) *M.Sc.Thesis*, Assam Agricultural University, Jorhat, India

Christenson, E.A. (1992) *Amer. Orchid Soc. Bull.*, **61** : 243-247.

Chin, T.T. (1966) *Amer. Orchid Soc. Bull.*, **35** : 549-554.

Christense EA. (2001) *Phalaenopsis: A Monograph*. Timber Press, Inc.,Portland, Oregon

Christenson, E.A. (1992) *Amer Orchid Soc. Bull.*, **61** : 243-247.

Christenson, E.A. (1995) *Amer. Orchid Soc. Bull.*, **64** : 392-394.

Chung, J.D. and Chun, C.K. (1983) *J. Korean Soc. Hort. Sci.*, **24** : 236-242.

Chung, J.D., Lee, J.H., Jee, S.O. and Kim, C.K. (1998) *Korean J. Horti. Sci. Technol.*, **16** : 239-241.

Chung, J.D., Park, Y.K., Kim, H.Y., Jee, S.O. and Koh, J.C. (1999) *J. Korean Soc. Hort. Sci.*, **40** : 485-488.

Chung, M.Y. and Chung, M.G. (2000) *Plant Biology*, **2** : 77-82.

Churchill, M.E., Ball, E.A. and Arditti, J. (1970) *Orchid Dig.*, **34** : 271-273.

Cocozza Talia, M.A., Lucia, B.De and Marini, P. (1995) *Colture Protette*, **24** : 77-82.

Colli and Kerbauy, G.B. (1993) *Plant Cell Tissue Organ Cult.*, **33** : 39-44.

Cooray, D.A. (1940) *Orchidologia Zeylanica*, **11** : 3, 73.

Cootes, J. and Banks, D.P. (1997) *Orchids*, **68** : 800-807.

Corbett, M.K. (1959) *Proc. Fla. St. Hort. Sci.*, **72** : 398-403.

Cribb, P. and Robbins (1993) *Kew Magazine*, **10** : 5-8.

Croix, L.A. (1994) *Amer. Orchid Soc. Bull.*, **63** : 1130-1137.

Cruz, D.T. da., Schnadelbach, A. S., Lambert, S. M., Ribeiro, P. L. and Borba, E. L.. (2011) *Plant Syst. Evol.*, **294**: 87–98.

Cruz, D. T., Borba, E. L. and van den Berg, C. (2003) *Sitientibus ser. Ci. Biol.* 3: 26–34.

Curry, R.D. (1975) *Amer. Orchid Soc. Bull.*, **44** : 497-498.

Curtis, C.H. (1950) *Orchids : Their Description and Cultivation*, Putnam and Co. Ltd., London.

Curtis, J.T. (1954) *Bull. Torrey Bot. Club*, **81** : 340-352.

Curtis, J.T. and Duncan, R.E. (1942) *Amer. Orchid Soc. Bull.*, **10** : 283-286, 304-307.

da Silva, G.M., T. L. Varella, I. V. Karsburg *et al.* (2017) *Cytologia*, **82**(2): 137–140.

Dafni, A. and Ivri, Y. (1981) *Plant Syst. Evol.*, **137** : 229-240.

Dahlgren, R.M.T. (1980) *Bot. J. Linn. Soc.*, **80** : 91-124.

Damen, P. (1980) *Vakblad voor de Bloemisterij*, **35** : 42-44.

Dauncey, E.A. and Cribb, P. (1993) *Kew Bull.*, **48** : 545-576.

Davidson, B. (1994) *Amer. Orchid Soc. Bull.*, **63** : 638-645.

Davidson, O.W. (1956) *N. J. Agric.*, **38** : 3-6.

Davidson, O.W. (1960) *Proc. 3rd World Orchid Conf.*, London, pp. 224-233.

Davidson, O.W. and Bennett, D.E. (1961) *N.J. Agric.*, **43** : 13-15.

Davidson, R.H. (1962) *Down to Earth*, **18**(2): 2, 23-24.

De Figueiredo, E.R. Jr and Lepage, H.S. (1951a) *Biologico*, **17** : 139-141.

De Figueiredo, E.R.Jr. (1950) *Arq. Inst. Biol. S. Paulo*, **20** : 249-253.

De Figueiredo, E.R.Jr. and Lepage, H.S. (1951b) *Biologico*, **17** : 196-199.

de Jesus Rodrigues, A.A., E. de Oliveira Santos, R. J.Takane *et al.* (2017) *Revista Ciência Agronômica*,. **48**: 296-302.

De, L.C. (2020) *J Agric Forest Meteorol Res*, **3**(5): 402-404.

De, L.C. (2020) *J.Agril. and Forest Meteorology Res.*, **3**(5): 402-404.

De, L.C., Pathak, P., Rao A.N.and Rajeevan, P.K. (2019) Global Orchid Industry. *In* : Commercial Orchids.De Gruyter Open, Warsaw, Polland. pp.300.

De, L.C., Vij, S.P. and Medhi, R.P. (2014) *J Hort.*, **1**: 102, doi: 10.4172/ horticulture.1000102.

Dekle, G.W. and Kuitert, L.C. (1968) *Fla. Dept. Agric. Bull.*, No. 8, pp. 1-28.

Delizo, T.C. (1974) *Arenata Res. J.*, **20** : 153-164.

Dematte, M.E.S.P. and Graziano, T.T. (2000) *Acta Horticulturae*, **511** : 265-270.

Devi, C.G., Damayanti, M. and Sharma, G.J. (1998) *J. Orchid Soc. India*, **12** : 83-87.

Devi, J. (1989) *Ph.D. Thesis*, Assam Agricultural University, Jorhat, India.

Devi, J., Borthakur, B. and Deka, P.C. (1997) *J. Orchid Soc. India*, **11** : 19-21.

Dewir, Y.H., M. E. El-Mahrouk, H. N. Murthy, and K. Y. Paek. (2015) *Hort. Environ. Biotechnol.* **56**(1): 89-93.

Dijkman, M.J. and Burg, S.P. (1970) *Amer. Orchid Soc. Bull.*, **39** : 799-803.

Dillon, G.W. (1971) *Amer. Orchid Soc. Bull.*, **40** : 408-410.

Ding, T.H., Ong, H.T. and Yong, H.C. (1980) *Proc. 3rd Asean Orchid Congress*, Ministry of Agriculture, Malaysia, pp. 65-78.

Dodge, B.O. and Rickett, H.W. (1943) *Diseases and Pests of Ornamental Plants*, The Jaques Cattell Press, Lancaster, Pa.

Dodson, C.H. and Atwood, J.T. (1992) *Amer. Orchid Soc. Bull.*, **61** : 1218-1219.

Dodson, C.H. and Frymire, G.P. (1961) *Missouri Bot. Gard. Bull.*, **49** : 133-152.

Dourado, F.M. (1992) *Amer. Orchid Soc. Bull.*, **61** : 226-229.

Dressler, L.R. (1974) *Proc. 7th World Orchid Conf.*, Medellin, pp. 259-279.

Dressler, L.R. and Dodson, C.H. (1960) *Ann. Miss. Bot. Gard.*, **47** : 1-67.

Duckitt, F. (1984) *Proc. 11th World Orchid Confi.*, Florida, pp. 159-161.

Dugger, R. (1975) *Orchid Rev.*, **83** : 171-174.

Duncan, R.E. (1959) In : *The Orchids : A Scientific Survey* (Ed. C.L. Withner), Ronald Press, New York.

Dunn, M.E. (1979) *Aust. Orchid Rev.*, **44** : 145-149.

Dunsterville, G.C.K. and Dunsterville, E. (1967) *Amer. Orchid Soc. Bull.*, **36** : 790-797.

Dycus, A.M. and Knudsoh, L. (1957) *Bot. Gaz.*, **119** : 78-87.

Easton, A. (1992) *Amer. Orchid Soc. Bull.*, **61** : 340-341.

Elle, A. (1960) *Gartenwelt*, **60** : 79.

Elle, A. (1962) *Die Orchidee.*, **13** : 48.

Emmart, E.W. (1940) *An Aztec Herbal of 1552*, John Hopkins Press, Baltimore.

Ernst, R. (1967) *Amer. Orchid Soc. Bull.*, **36** : 694-704.

Ewing, J. (1984) *Proc. 11th World Orchid Conf.*, Florida, pp. 162-165.

Faragher, J., Borochov, A. and Halevy, A.H. (1982) *Proc. 21st Int. Hort. Congr.*, Hamburg, Abstr. No. 1735

Feder, W.A. (1951) *Phytopathology*, **41** : 938-939.

Feder, W.A. (1952) *Phytopathology*, **42** : 373-376.

Felix, L. P. and Guerra, M. 2010. *Bot. J. Linn. Soc.* **163**: 234–278.

Fenton, K.J. (1951) *Amer. Orchid soc. Bull.*, **20** : 519-523, 587-591.

Fiske, M.D. (1982) *Amer. Orchid Soc. Bull.*, **51** : 43-46.

Foote, G.B. (1938) *Orchideologia Zeylanica*, **5** : 145-146.

Forero M.G., Beltrán C.E., Troncoso A., González-Santos C. (2020) Classification of *Cattleya Trianae* and Its Varieties by Using Colorimetry. In: Figueroa Mora K., Anzurez Marín J., Cerda J., Carrasco-Ochoa J., Martínez-Trinidad J., Olvera-López J. (eds.) Pattern Recognition. MCPR 2020. Lecture Notes in Computer Science, vol 12088. Springer, Cham. https: //doi.org/10.1007/978-3-030-49076-8_4

Frackowiak, E. (1933) *Amer. Orchid Soc. Bull.*, **2** : 34.

Francki, R.I.B. (1970) *Commonw. Mycol. Inst. Ser.* **3,** No. 27.

Freitas-Astua, J. and Rezende, J.A.M. (1998) *Fitopatologia Brasileira*, **23** : 158-160.

Freson, R. (1969) *Bull. Soc. Royale Bot. Belgique*, **102** : 205-209.

Fu, Y.M., Chen, W.H., Tsai, W.T. *et al.* (1996) Studies on floral color heredity of *Phalaenopsis equestris*. Report of the Taiwan Sugar Research Institute **152**: 35–49. (In Chinese with English abstract).

Fukai, S., Fujiwara, K., Okamoto, K., Hasegawa, A. and Goi, M. (1997) *Lindleyana*, **12** : 169-171.

Gangaprasad, A.N., Decruse, W.S., Sooriamuthn Seeni and Sarojini Menon (1999) *Lindleyana*, **14** : 38-46.

Garary, L.A., Hamer, F. and Siegerist, E.S. (1994) *Nordie J. Bot.*, **14** : 609-646.

Garary, L.A., Hamer, F. and Siegerist, E.S. (1995) *Lindleyana*, **10** : 174-182.

Garay, L.A. (1960) *Bot. Museum Leaflet*, Harvard University, **19** : 57-96.

Garay, L.A. (1972) *J. Arnold arb.*, **53** : 202-215.

Gawenda, I., Schröder-Lorenz, A. and Debener, T. (2012) *Mol Breeding* **30,** 305–316.

Gething, P.A. (1974) *Orchid Rev.*, **82** : 133-135.

Gilbert, J.C., Mainland, G.B. and Lohman, M.L. (1950) *Bull. Pac. Orchid Soc.*, **8** : 289-293.

Goh, K. (2013) National flower series – South East Asia 6 – Myanmar. Available at: https: //kwgls.wordpress. com/2013/04/30/98/

Goh, C.J. (1975) *Ann. Bot.*, **39** : 931-934.

Goh, C.J. (1976) *Ann. Bot.*, **40** : 645-646.

Goh, C.J. (1977a) *Ann. Bot.*, **41** : 1061-1063.

Goh, C.J. (1977b) *Ann. Bot.*, **41** : 763-769.

Goh, C.J. (1978) *Orchid Rev.*, **86** : 121-122.

Goh, C.J. (1979a) *New Phytol.*, **82** : 375-380

Goh, C.J. (1979a) *Proc. 10th. International Conf. on plant growth Substances, Madisen, Wiscensin,* USA, Abs. No. 29

Goh, C.J. (1982) *Malayan Orchid Rev.,* **16** : 24-31.

Goh, C.J. (1983) *Ann. Bot.,* **51** : 145-147.

Goh, C.J. (1984) *Proc. 11th World Orchid Conf.,* Florida, pp. 166-173.

Goh, C.J. (1990) In : *Hand Book of Plant Cell Culture* (Eds. P.V. Ammirato, D.A. Evans, W.R. Sharp and Y.P.S. Bajaj), McGraw-Hill, New York, pp. 598-637.

Goh, C.J. and Seetoh, H.C. (1973) *Ann. Bot.,* **37** : 113-119.

Goh, C.J. and Wan, H.Y. (1974) *Plant Growth Substances* (Ed. Y. Sumiki), Hirokawa Publishing Co., Tokyo.

Goh, C.J. and Wong, P.F. (1990) *Scientia Hort.,* **44** : 315-321.

Goh, C.J. and Young, A.L. (1978) *Plant Sci. Letter,* **12** : 287-292.

Gold, A.H. (1950) *Phytopathology,* **40** : 365.

Gold, A.H. and Jensen, D.D. (1951) *Amer. J. Bot.,* **38** : 577-578.

Griesbach, R.J. (1981) *Plant Cell Tissue Cult.,* **1** : 103-107.

Griesbach, R.J. (1994) *Plant Sci.,* **102** : 81-89.

Griesbach, R.J. and Hammond, J. (1993) *Acta Horticulturaei,* **336** : 165-169.

Griffith, E. and Link, C.B. (1957) *Amer. Orchid Soc. Bull.,* **26** : 184-191.

Halevy, A.H. and Mayak, S. (1979) *Hort. Rev.* (Ed. J. Janik), AVI Publishing, Westport, Conn., Vo. I, pp. 204-236.

Hamilton, R.M. (1965) *Amer. orchid Soc. Bull.,* **34** : 429-430

Hara, A.H. and Mau, R.F.L. (1988) *Proc. Haw. Ent. Soc.,* **26**: 71-75.

Harley, J.L. (1951) *Amer. Orchid Soc. Bull.,* **20** : 291-293.

Harn, C. (1968) *Amer. Orchid Soc. Bull.,* **37** : 1066-1068.

Harrison, C.R. (1968) *Amer. Orchid Soc. Bull.,* **39** : 715-716.

Harrison, C.R. and Arditti, J. (1970) *Orchid Dig.,* **39** : 199-204.

Harrison, C.R. and Arditti, J. (1976) *Amer. J. Bot.,* **63** : 911-918.

Hartmann, H.T. and Kester, D.E. (1972) *Plant Propagation : Principles and Practices,* Prentice Hall of India Ltd., New Delhi.

Hartono, S.H. (1978) *Gard. Bull.,* **31** : 222-228.

Harvais, G. (1973) *Canadian J. Bot.,* **51** : 327-332.

Harvais, G. (1974) *Canadian J. Bot.,* **52** : 451-460.

Harvais, G. (1982) *Canadian J. Bot.,* **60** : 2547-2555.

Hawkes, A.D. (1943) *Amer. Orchid Soc. Bull.,* **11** : 412-415.

Hawkes, A.D. (1944) *Amer. Orchid Soc. Bull.,* **13** : 56-58.

Hawkes, A.D. (1978) *Encyclopaedia of cultivated orchids*, Faber and Faber Ltd., London.

He, J., Kho, G.H. and Hew, C.S. (1998) *Environ. Bot.*, **40** : 255-264.

He, S.L., Jutori, M., Kong, D.Z., Zhang, Q.X. and Ranaka, M. (2001) *J. Beijing Forestry Univ.*, **23** : 29-31.

Hegarty, C.P. (1955) *Amer. Orchid Soc. Bull.*, **24** : 547-564.

Hegde S.N. (2020) Status of Orchid Industry in India. In: Khasim S., Hegde S., González-Arnao M., Thammasiri K. (eds) Orchid Biology: Recent Trends and Challenges. Springer, Singapore. https: / /doi.org/10.1007/978-981-32-9456-1_2

Hegde, S.N. (1981) *Indian Hort.*, **25** : 7-9.

Hegde, S.N. (1984) *Orchids of Arunachal Pradesh*, Forest Department, Itanagar, Arunachal Pradesh, India.

Hegde, S.N. (2016) *J. Orchid Soc. India*, **30**: 43-49.

Helle, W. (1958) *Jversl Proefst. Bloem.*, Aalsmcer, pp. 83-84.

Helton, O.M. (1961) *Proc. 3rd Word Orchid Conf.*, London, pp. 254-257.

Henderson, J.L. (1984) *Proc. 11th World Orchid conf.*, Florida, pp. 221-224.

Heyes, J.A. and Johnston, J.W. (1998) *New Zealand J. Crop Hort. Sci.*, **26** : 319-324.

Higuchi, H. and Sakai, K. (1977) *Aichi-Ken Agril. Res. Cent. Res. Bull.*, B., No. 9, pp. 79-81.

Him, Tae Joung, Pack Kee Yoeup, Yun Tae, Cho Jin Tae and Jcong In Myeong (1996) *J. Korean Soc. Hort. Sci.*, **37** : 435-441.

Hinsley *et al.* (2018) *Botanical Journal of the Linnean Society*, **186**: 435–455.

Hirai, H. and Mori, G. (1996) *Environment Control in Biology*, **34** : 295-301.

Hirai, H. and Mori, G. (1997) *J. Soc. High. Tech. Agric.*, **9** : 193-198.

Hirsch, D.H. (1959) *Amer. Orchid Soc. Bull.*, **28** : 342-344.

Hoffmann, W. (1930) *Planta*, **10** : 523-595.

Holdson, J. and Laurie, A. (1951) *Proc. Amer. Soc. Hort. Sci.*, **57** : 379-380.

Holman, R.T. (1976) *Amer. Orchid Soc. Bull.*, **45** : 415-422.

Holters and Zimmer (1990) *Gartenbauwissenchaft*, **55** : 201-205.

Holters, J. and Zimmer, K. (1991) *Gartenbauwissenchaft*, **56** : 114-117.

Holttum, R.E. (1964) *Review Flora of Malaya*, 3rd Edition, Govt. Printing Office, Singapore.

Hori, S. (1911) *Centralbl. Bakt. U. Infektionkr. Abt.* 2, **31** : 85-92.

Howcroft, N.H.S. (1996) *Science in New Guinea*, **21** : 159-175.

Hoyt, G.M. (1960) *Amer. Orchid Soc. Bull.*, **29** : 597-599.

Hsiao, Y.Y., Chen, Y.W., Huang, S.C., Pan, Z.J., Fu, C.H., Chen, W.H., Tsai, W.C. and Chen, H.H. (2011) *BMC Genomics* **12**: 360.

Hsiao, Y.Y., Tsai, W.C., Kuoh, C.S., Huang, T.H., Wang, H.C., Wu, T.S., Leu, Y.L., Chen, W.H. and Chen, H.H. (2006) *BMC Plant Biol.*, **6**: 14.

Hsing, H., Lin, Y., Tong, C. *et al.* (2016) *Bot Stud* **57,** 30. https: //doi.org/10.1186/ s40529-016-0146-6.

Hsu,C.C., Chung, Y.L., Chen, T.C., Lee, Y.L., Kuo, Y.T., Tsai, W.C., Hsiao, Y.Y., Chen, Y.W., Wu, W.L. and Chen, H.H. (2011) *BMC Plant Biol.*, **11**: 3.

Hua, H.S. (1999) *Orchids*, **68** : 242-243.

Huang JZ, Lin CP, Cheng TC, Chang BC, Cheng SY, Chen YW, Lee CY, Chin SW, Chen FC (2015) *PLoS ONE* **10**: e0123474.

Hunter, J.A. (1958) *N.Z. Plants and Gdns.*, **2** : 342-345.

Hurst, C.C. (1913) *J. Roy. Hort. Soc.*, **38** : 412-429.

Hurst, C.C. (1925) *Experiments in Genetics*, Cambridge Univ. Press, Cambridge.

Hyun, M.R., Choi, J.Y., Suh, J.N., So, I.S. and Lee, J.S. (1998) *Korean J. Hort. Sci. Technol.*, **17**: 346-351.

Ichihashi, S. (1979) *J. Japanese Soc. Hort. Sci.*, **48** : 345-352.

Ichihashi, S. (1982) *Japanese Assoc. Pl. Tissue Cult.*, pp. 181-182.

Ichihashi, S. and Hiraiwa, H. (1996) *J. Orchid Soc. India*, **10** : 81-88.

Ilsley, P. (1961) *Proc. 4th Hawaiian Orchid Conf.*, pp. 43-45.

Ilsley, P. (1965) *Orchid Rev.*, **73** : 371-376.

Ilsley, P. (1966) *Orchid Rev.*, **79** : 69.

In: W.H. Chen and H.H. Chen (Eds.).*Orchid Biotechnology*. World Scientific Publishn Co., P.1-22.

Irpa, L. (1965) *Die Orchidee*, **16** : 54-55.

Ishi, Y., Takamura, T., Goi, M. and Tanaka, M. (1998) *Plant Cell Rept.*, **17** : 446-450.

Ishii, M., Uemoto, S. and Fujieda, K. (1979) *J. Japanese Soc. Hort. Sci.*, **48** : 199-204.

Islam, M.O. and Ichihashi, S. (1999) *J. Japanese Soc. Hort. Sci.*, **68** : 1124-1131.

Islam, M.O., Matsui, S. and Ichihashi, S. (1999) *J. Japanese Soc. Hort. Sci.*, **68** : 1132-1138.

Islam, M.O., Matsui, S. and Ichihashi, S. (2000) *Lindleyana*, **15** : 81-88.

Ito, I. (1965) *Japanese Orchid Soc. Bull.*, **11** : 4-15.

Ivri, Y. and Dafni, A. (1977) *New Phytol.*, **79** : 173-177.

Jauti, S. (1998) *J. Orchid Soc. India*, **12** : 43-45.

Jefferson, R.N. and Others (1954) *Amer. Orchid Soc. Bull.*, **23** : 729-743.

Jensen, D.D. (1950) *Phytopathology*, **40** : 966-967.

Jensen, D.D. (1951) *Phytopathology*, **41** : 401-414.

Jensen, D.D. (1955) *Amer. Orchid Soc. Bull.*, **24** : 756-766.

Jensen, D.D. (1959) In : *The Orchids – A Scientific Survey* (C.L. Withner Ed.), Ronald Press Co., New York.

Jensen, D.D. and Gold, A.H. (1955) *Phytopathology*, **45** : 327-334.

Jessel, W.H. (1962) *Amer. Orchid Soc. Bull.*, **30** : 954.

Jiménez, H.R., Rossini, B.C., E. V. Tambarussi *et al.* (2017) *Acta Scientiarum. Biological Sciences*, **39**: 45-52.

Johansen, B. (1990) *Bot. J. Linnean Soc.*, **103** : 165-96.

Johnson, L.P. (1995) *Lindleyana*, **10** : 1.

Jones, D.L. (1975) *Ann. Bot.*, **39** : 585-589.

Júnior, D.S. and G. A. Venturieri. (2011) *Acta Scientiarum. Agronomy* **33**: 97-103.

Kamemoto, H. (1950) *Amer. Orchid Soc. Bull.*, **19** : 366-373.

Kamemoto, H. (1950) *American Orchid Soc. Bull.*, **1**: 366-373, 1950.

Kamemoto, H. and McConnell, J. (1984) *Proc. 11th World Orchid Conf.*, Florida, pp. 220-223.

Kamemoto, H., Mecyot, W. and Takeshita, M. (1972) *J. Amer. Soc. Hort. Sci.*, **97** : 3-5.

Kamemoto, H., Tanaka, R. and Kosaki, K. (1961) *Bull. Hawii agric. Exp. Stat.*, pp. 127.

Kamemoto, H., Thammasin, K., Marutani, M. and Kobayashi, R.S. (1987) *J. Orchid Soc. India*, **1** : 7-10.

Kano, K. (1968) *Amer. Orchid Soc. Bull.*, **37** : 690-698.

Kaputo MT. (1996) *Food Chemistry* **55**: 115–119.

Kaushik, P. (1983) *Ecological and Anatomical Marvels of the Himalayan Orchids*, Today and Tomorrow's Printers and Publoishers, New Delhi, India.

Keithly, J.H. and Yokoyana, H. (1990) *Plant Growth Regulation*, **9** : 19-26.

Keithly, J.H., Jones, D.P. and Yokoyama, H. (1991) *HortScience*, **26** : 1284-1286.

Kerbauy, G.B. (1984) *Z. Pflanzenphysiol.*, **113** : 287-291.

Kerbauy, G.B. (1991) *J. Plant Physiol.*, **138** : 248-251.

Ketsa, S. and Kosonmethakul, N. (2001) *Acta Horticulturae*, **543** : 41-44.

Ketsa, S. and RugKong, A. (1999) *J. Hort. Sci. Biotech.*, **74** : 608-613.

Ketsa, S. and RugKong, A. (2000) *J. Hort. Sci. Biotech.*, **75** : 149-153.

Kim Doug Won, Cho Kuen Ho, Pak Chun Ho, Kang, Hun, So Insup and Kwek Beyong Hwa (1997) *J. Korean Soc. Hort. Sci.*, **38** : 749-753.

Kim, B. C. Kim, M.H. and Oh, M.Y. (1990) *Korean J. Plant Taxonomy*, **20** : 53-64

Kim, H.Y. (1998) *Korean J. Hort. Sci. Technol.*, **16** : 40-41.

Kim, H.Y. and Chung Jaedong (1999) *Korean J. Hort. Sci. Technol.*, **17** : 344-345.

Kim, H.Y., Koh, J.C., Chung, J.D., Liu, D.W. and Son, B.G. (2000a) *J. Korean Soc. Hort. Sci.*, **41** : 96-100.

Kim, J.Y. and Lee, J.S. (1998) *Korean J. Hort. Sci. Technol.*, **16** : 30-32.

Kim, K.K., Kunisaki, J.T. and Sagawa, Y. (1970) *Amer. Orchid Soc. Bull.*, **39** : 1077-1080.

Kim, K.W. and Kako, S. (1984) *J. Korean Soc. Hort. Sci.*, **25** : 65-71.

Kim, M.S., Kim, J.Y. and Eun, J.S. (1999) *J. Korean Soc. Hort. Sci.*, **40** : 607-610.

Kim, M.S., Lee, Y.R., *et al.* (2011) *Hort. Sci. and Technol.*, **29**: 503-506

Kim, M.S., Lee, Y.R., Won, J.Y., Kim, J.Y., Kim, B.H. and Eun, J.S. (1998) *Korean J. Hort. Sci. Technol.*, **16** : 236-238.

Kim, T. J., Kim, J. H., Lee, J. W., Lee, C. H., Lee H. D., Choi, K. S. and Paek, K. Y. (1999a) *J. Korean Soc. Hort. Sci.*, **40** : 611-614.

Kim, T., Peak, K.Y., Yung, T., Cho, J.T. and Jcong, I.M. (1996) *J. Korean Soc. Hort. Sci.*, **40** : 435-441.

Kim, T.J., Choi, J.I., Shin, K.S. and Paek, K.Y. (2000c) *J. Korean Soc. Hort. Sci.*, **41** : 221-225.

Kim, T.J., Kim, J.H., Lee, J.W., Lee, C.H. and Paek, K.Y. (2001) *J. Korean Soc. Hort. Sci.*, **42** : 111-115.

Kim, T.J., Kim, J.H., Lee, J.W., Lee, C.H., Choi, K.S., Shim, S.W. and Paek, K.Y. (1999b) *J. Korean Soc. Hort. Sci.*, **40** : 615-618.

Kim, T.J., Lee, C.H. and Paek, K.Y. (2000b) *J. Korean Soc. Hort. Sci.*, **41** : 101-104.

Kim, T.J., Lee, J.W., Lee, C.H., Kim, H.H., Choi, K.S. and Paek, K.Y. (1999c) *J. Korean Soc. Hort. Sci.*, **40** : 375-378.

Kim, T.J., Lee, J.W., Lee, C.H., Kim, H.H., Choi, K.S. and Paek, K.Y. (1999d) *J. Korean Soc. Hort. Sci.*, **40** : 739-880.

Kim, T.J., Seon, J.H. and Paek, K.Y. (1998) *J. Korean Soc. Hort. Sci.*, **39** : 469-474.

Kimura, Y. and Kurihara, N. (1991) *Agril. and Hort.*, **66** : 951-958.

King, G. and Pantling, R. (1898) *The Orchids of Sikkim Himalaya*, Ann. Royal Bot. Garden, Calcutta, **8** : 1-342.

Kirtikar, K.R. and Basu, B.D. (1975) *Indian Medicinal Plants*, Vol. IV, Bishen Singh Mahendra Pal Singh, Dehra Dun, India.

Klier, K Leoschke, M. J. and Wendel, J.F. (1991) *J. Hered.*, **82** : 305-318

Klougart, A. (1973) *Gartner Tidende*, **89** : 405-408.

Knudson, L. (1921) *Bot. Real.Soc. Espanola Hist. Nat.*, **21** : 250-260.

Knudson, L. (1925) *Bot. Gaz.*, **79** : 345-379.

Knudson, L. (1946) *Amer. Orchid Soc. Bull.*, **15** : 214-217.

Knyazev, M.S., Kulikov, P.V., Knyazeva, O.I. and Semer Ikov, V.L. (2000) *Botanicheskii Zhurnal*, **85** : 94-102.

Koay, S.H. and Chua, S.E. (1979) *Singapore J. Prim. Ind.*, **7** : 51-61.

Koay, S.H. and Chua, S.E. (1980) *Singapore J. Prim. Ind.*, **8** : 76-84.

Kobayashi, R. and Kamemoto, N. (1989) *HortScience*, **24** : 499-500.

Kofranek, A.M. and Barstow, G. (1955) *Amer. Orchid Soc. Bull.*, **24** : 751-753.

Kohl, H.C.Jr. (1962) *Amer. Orchid Soc. Bull.*, **31** : 117-120.

Komori, T. and Yoneda, K. (2001) *Environ. Control Biol.*, **39** : 67-70.

Koopowitz, H. and Hasegawa, N. (1984) *Proc. 11th World Orchid Conf.*, Florida, pp. 227-228.

Kostenyuk, I., Oh, B.J. and So, I.S. (1999) *Plant Cell Rep.*, **19** : 1-5.

Kosugi, K. (1952) *J. Hort. Ass., Japan*, **21** : 179-182.

Krans, J.E. and Monteiro, W.R. (1989) *Ann. Bot.*, **64** : 491-498.

Kronenberg, H.G. (1976) *Amer. Orchid Soc. Bull.*, **45** : 513.

Kubota, S. and Yoneda, K. (1990) *Bull. College of Agriculture and Veteninary Sci., Nihon University*, **47** : 75-80.

Kubota, S., Hisamatsu, T., Ichimura, K. and Koshioka, M. (1997) *J. Japanese Soc. Hort. Sci.*, **66** : 581-585.

Kuehnle, A.R. and Sugii, N. (1992) *Plant Cell Rep.*, **11** : 484-488.

Kulkarni, A.R. and Surwase, B.S. (1998) *J. Orchid Soc. India*, **12** : 47-51.

Kull, T., 1999. *J. Ecol.*, **87**: 913-924.

Kurup, P.N.V. (1977) *Hand Book of Medicinal Plant*, Vol. I, Central Council for Research in Indian Medicine and Homoeopathy, New Delhi, India.

Lamchan, L.T. and Lee, S.M. (1996) *Singapore J. Prim Indust.*, **24** : 25-29.

Laneri, U. (1990) *Acta Horticulturae*, **280** : 445-453.

Laurie, A., Kiplinger, D.C. and Nelson, K.S. (1958) *Commercial Flower Forcing*, McGraw-Hill Book Co., Inc., New York, Tooronto, London.

Lawler, L.J. and Slaytor, M. (1969) *Phytochemistry*, **8** : 1959-1962.

Lawler, L.J. and Slaytor, M. (1970) *Proc. Linn. Soc. N.S.W.*, **94** : 237-241.

Lawrence, D. and Arditti, J. (1964) *Amer. Orchid Soc. Bull.*, **33** : 948.

Lawson, R.H. (1995) In : *Orchid Pests and Diseases*, Amer. Orchid Soc., pp. 74-104.

Lawson, R.H. and Ali, S. (1975) *Handbook on Orchid Pests and Diseases*, American Orchid Soc. Inc., Cambridge, Mass.

Lawson, R.H. and Branningan, M. (1986) In : *Handbook on Orchid Pests and Diseases*, Revised edition, Amer. Orchid Soc., pp. 2-49.

Lay, F.F.M. (1978) *Gard. Bull.*, **31** : 132-138.

Lay, F.F.M. (1979) *Orchid Rev.*, **87** : 343-346.

Leal, B.S.S.,C. J.N.Chaves, S.Koehler and E.L. Borba. (2016) *Botanical J. Linnean Society*, **181**: 621–639.

Leander, K. and Luning, B. (1967) *Tetrahedron Letters No* : 3477-3478.

Lee, J.K., Paek, K.Y., Lee, H.S. and Sin, S.H. (1982) *J. Korean Soc. Hort. Sci.*, **23** : 163-171.

Lee, J.S. and Kim, J. Y. (1999) *J. Korean Soc. Hort. Sci.*, **40** : 389-391

Lee, N. and G.M. Lin. (1984) *J. Chin. Soc. Hortic. Sci.* **30**: 223-231

Lee, Y.H. and Mowe, B.L. (1983) *Singapore J. Prim. Indust.*, **11** : 46-48.

Lee, Y.R., Kim, J.Y. and Kim, B.H. (1999) *J. Korean Hort. Sci. and Technol.*, **17** : 755-757.

Leffring, L. (1972) *Vakblad Voor de Boloemisterij*, **27** : 13.

Lenz, L.W. and Wimber, D. 1959. Hybridization and inheritance in orchids, p. 261-314. In C. L. Witner (ed), The orchids; a Scientific survey. Ronald Press, New York.

Lenz, L.W. and Wimber, D.E. (1959) In : *The Orchids – A Scientific Survey* (Ed. C.L. Withner), Ronald Press Co., New York.

Lenz, L.W., Wimber, D.E. and Dodson, C.H. (1955) *Cymbidium Soc. News*, **10** : 17-20.

Leon, C. and Lin, Y.L. (2017) *Chinese medicinal plants, herbal drugs and substitutes: an identification guide*. Kew: Kew Publishing.

Leur, C.A. (1991) *Novon*, **1** : 165-176.

Leur, C.A. (1995) *Lindleyana*, **10** : 232-237.

Lewis Kundson (1925) *Bot. Gaz.*, **79** : 345-379.

Liang, Xu Ye and Pan Rui Chi (1994) *J. Trop. Subtrop. Bot.*, **2** : 65-70.

Lim, S.H., Lee, Y. H. and Goh, C. J. (1999) *Aun. Bot.*, **83** : 193-196

Lima, L.H. (1993) *Amer. Orchid Soc. Bull.*, **62** : 262-265.

Lin, R.S. (1999) *J. Agric. and Forest.*, **48** : 63-83.

Lin, R.S., Shyu, H.E. and Lai, S.F. (1999) *J. Chinese Soc. Hort. Sci.*, **45** : 43-52.

Lin, R.S., Shyu, H.F., Lai, S.E. and Chen, C.C. (1998) *Chinese J. Agromet.*, **5** : 193-202.

Lin, Y.F., Chen, Y.Y., Hsiao, Y.Y., Shen, C.Y., Hsu, J.L., Yeh, C.M., Mitsuda, N., Ohme-Takagi, M., Liu, Z.J. and Tsai, W.C. (2016) *J. Exp Bot.*, **67**: 5051–5066.

Lin, Y.H., Chang, C. and Chang, W.C. (2000) *Plant Cell Tissue Organ Cult.*, **62** : 21-25.

Linder, H.P., Johnson, S.D. and Liltved, W.R. (1998) *Novon*, **8** : 405-407.

Lindley, J. (1840) *The Genera and Species of Orchidaceous Plants* (Reprinted in 1963), A. Asher and Co., Amsterdam.

Liu, Z.J. and Zhang, J.N. (1998) *J. South China Agri. University*, **19** : 114-118.

Livingston, R.B. (1962) *Amer. Orchid Soc. Bull.*, **31** : 230-231.

Lizada Ma, C.C. and Rimando, T.J. (1982) *Proc. Fourth Asean Orchid Congress*, Manilla.

Loh, C.S. and Rao, A.N. (1985) *Malayan Orchid Rev.*, **19** : 34-37.

Lootens, P. and Heursel, J. (1998) *HortScience*, **33** : 1183-1185.

Lu, H.C., Chen, H.H., Tsai, W.C., Chen, W.H., Su, H.J., Chang, C.N. and Yeh, H H. (2007) *Plant Physiol.*, **143** : 558-69.

Luer, C.A. (1998) *Lindleyana*, **13** : 138-147.

Luks, Y., Yu, A. and Shevchenko, S.V. (1977) *Byull. Gosuda. Nikits. Botanich. Sada*, No. 1, pp. 30-35.

Luning, B. (1964) *Acta Chem. Scand.*, **18** : 1507-1516.

Luning, B. (1967) *Phytochemistry*, **6** : 857-861.

MacLeod, R.A. (1947) *Amer. Orchid Soc. Bull.*, **16** : 336-337.

Malemnganba, H. (1994) *M.Sc. thesis* Assam Agricultural University, Jorhat, India.

Malemnganba, H., Ray, B.K., Bhattacharya, S. and Deka, P.C. (1996) *Indian J. Exp. Biol.*, **34** : 802-805.

Manglitz, G.R. (1953) *J. Econ. Ent.*, **46** : 116-119.

Mansfeld, R. (1937) *Notizblatt des Botanischen Gartens und Museums zu Berlin-Dahlem. 13 Band* No. 116-120, pp. 666-676.

Mao, B.Z., Lin, W.H., Qian, X.H., Fu, H.D. and Li, D.B. (1998) *J. Zhejiang Agric. Univ.*, **24** : 66-68.

Mariat, F. (1948) *Rev. Gen. Bot.*, **55** : 229-243.

Marston, M.E. and Voraurai, P. (1967) *A Review of the Literature*, Univ. of Nottingham, Dept. Hort., Misc. Pub. No. 17, pp. 1-8.

Marv and Ragan, J. (1973) *Amer. Orchid Soc. Bull.*, **42**; 981.

Matsui, N. and Yoneda, K. (1997) *J. Japanese Soc. Hort. Sci.*, **66** : 597-605.

Matsui, S. and Kannero, Y. (1994) *Res. Bull. Fac. Agric. Gifu. Univ.*, **59** : 49-55.

Matsui, S. and Tanase, K. (1995) *Res. Bull. Fac. Agric. Gifu Univ.*, **60** : 173-179.

Matsumato, K. (1966) *Cymbidium Soc. News*, **29** : 11-14.

Mau, R.F.I. (1984) *Proc. Haw. Ent. Soc.*, **24** : 298-297.

McAlister, B.G. and Staden, J.van (1998) *South African J. Bot.*, **64** : 264-266.

McConnell, J. and Kamemoto, H. (1983) *J. Amer. Soc. Hort. Sci.*, **108** : 1003-1006.

McConnell, J. and Kamemoto, H. (1993) *HortScience*, **28** : 935-937.

McQuade, H.A. (1949) *Hort. Ann. Missouri Bot. Gardi.*, **36** : 433-472.

Mehlquist, G.A.L. (1953) *Proc. 7th Int. Bot. Congr.*, Stockholm, pp. 335.

Mehlquist, G.A.L. (1958) *Proc. 2nd World Orchid Conf.*, Honolulu, pp. 200-2009.

Mehlquist, G.A.L. (1974) *The Orchids : Scientific Sutides* (C.L. Withner Ed.), Wiley Interscience, New York.

Mehlquist, G.A.L. 1958. Genetics and orchid breeding. Proceedings of 2nd World Orchid Conference Honolulu. 1957, pp. 200-209.

Menezes, L.C. (1988) *Nat. Nov. Brasil Florestal*, **15** : 16-18.

Menninger, E.D. (1963) *Amer. Orchid Soc. Bull.*, **32** : 885-887.

Meyer, J.R. (1943) *Biologico*, **9** : 401-406.

Miduno, T. (1938) *Cytologia*, **8** : 505-514.

Milletti, G. (1967) *Frutticoltura*, **29** : 313-328.

Mishiba, K., Chin, D.P. and Mii, M. (2005) *Plant Cell Rep.*, **24**: 297–303.

Mitra, G.C. (1970) *Indian J. Exp. Biol.*, **9** : 79-85.

Mitra, J.N. (1958) *Flowering Plants of Eastern India*, Vo. 1. Monocotyledons, Calcutta.

Miyazaki, S. and Nagamatsu, T. (1975) *Agric. Bull. Saga Unit.*, No. 21, pp. 131-149.

Miyoshi, K. and Mii, M. (1995) *Scientia Hort.*, **63** : 263-267.

Moir, W.W.G. (1975) *Orchid Rev.*, **83**; 191-194.

Moir, W.W.G. (1983) *Orchid Rev.*, **91** : 131.

Monkhouse, S.A. (1984) *Proc. 11th Orchid Conf.*, Florida, pp. 251-254.

Moon, A. (1989) *Orchid Rev.*, **97** : 159-162.

Moore, E.T. (1947) *Amer. Orchid Soc. Bull.*, **16** : 512.

Mora Retana, D.E. and Garcia Castra, J. (1990) *Nuevas Combinaciones de Orchidaceae de Costa Rica Brenesia*, **33** : 123-126.

Morel, G.M. (1960) *Amer. Orchid Soc. Bull.*, **29** : 495-497.

Morel, G.M. (1963) *C.R. Acad. Sci.*, Paris, **256** : 4955-4957.

Morel, G.M. (1964a) *Amer. Orchid Soc. Bull.*, **33** : 473-478.

Morel, G.M. (1964b) *Rev. Cytol. Veg.*, **27** : 304-314.

Morel, G.M. (1965a) *Cymbidium Soc. News*, **20** : 3-11.

Morel, G.M. (1965b) *Die Orchidee*, **16** : 165-176.

Morel, G.M. (1970) *Die orchidee*, **20** : 433-443.

Moses, J.R. (1994) *Amer. Orchid Soc. Bull.*, **63** : 1000-1008.

Mott, R.C. (1953) *Bull. N.Y. St. Flower Grs.*, No. 94, pp. 3-4.

Mott, R.C. (1954a) *Bull. N.Y.St. Flower Grs.*, No. 105, pp. 2.

Mott, R.C. (1954b) *Bull. N.Y.St. Flower Grs.*, No. 106, pp. 2-3.

Mott, R.C. (1954c) *Bull. N.Y.St. Flower Grs.*, No. 107, pp. 1-2.

Mourik, N.M. Van and Os, P.C. Van (1991) *Vakblad Voor de Bloemisterij*, **46** : 48-49.

Mukherjee, A. (1979) *Indian Hort.*, **24** : 23-28.

Mukherjee, S.K. (1983) *Orchids*, Indian Council of Agricultural Research, New Delhi, India.

Mukherjee, T.P., Roy, S. and Bose, T.K. (1974) *Indian Hort.*, **18** : 21-24.

Murakishi, H.H. (1952) *Phytopathology*, **42** : 517.

Murakishi, H.H. (1954) *Bull. Pacific Orchid Soc.*, Hawaii, **12** : 20-22.

Murashige, T., Karnemoto, H. and Sheehan, T.J. (1967) *Proc. Amer. Soc. Hort. Sci.*, **91** : 672-679.

Muzik, T.J. (1958) *Science*, **127** : 82.

Myint, K.T., Chung, M.Y., Chung, J.D. and Kim, C.K. (2001) *J. Korean Soc. Hort. Sci.*, **42** : 1-5.

Nagabhushana, S.R. (1982) *Thesis Abstract*, UAS, Bangalore, **8** : 383-388.

Nagano, Y. (1960) *Proc. 3rd World Orchid Conf.*, London, pp. 50-55.

Nagashima, T. (1982a) *J. Japanese Soc. Hort. Sci.*, **51** : 82-93.

Nagashima, T. (1982b) *J. Japanese Soc. Hort. Sci.*, **51** : 94-105.

Nagashima, T. (1983) *J. Japanese Soc. Hort. Sci.*, **52** : 65-77.

Nagashima, T. (1998) *J. Japanese Soci. Horti. Sci.*, **67** : 792-797.

Nair, H. (1984) *Proc. 11th World Orchid Conf.*, Florida, pp. 255-260.

Nair, H., Idris, Z.M. and Arditti, J. (1991) *Lindleyana*, **6** : 49-58.

Nakamura, S.I. (1962) *Zeitschr. Fur Botanik*, **50** : 487-497.

Nakamura, S.I. (1963) *Int. Symp. Physiol. Ecol. Biochem.*, Germination, A 111, **9** : 1-4.

Nakamura, S.I. (1964) *Mem. Coll. Agr. Kyoto Univ.*, **86** : 1-48.

Nan, G.L., Tang, C.S. and Kuehnle, A.R. (1995) *In Vitro Cell Dev. Biol. Plant*, **31** : 131-136.

Nan, G.L., Tang, C.S., Kuehnle, A.R. and Kado, C.I. (1997) *Physiol. Mol. Plant Path.*, **51** : 391-399.

Nash, N. (1984) *Proc. 11th World Orchid Conf.*, Florida, pp. 261-263.

Nash, N. (1997) *Orchids*, **66** : 1268-1277.

Nayak, N.R., Chand, P.K., Rath, S.P. and Patnaik, S.N. (1998) *In Vitro Cell. Dev. Biol. Plant*, **34** : 185-188.

Neil, M.R.M. and Wilcock, C.C. (1998) *Amer. J. Bot.*, **85** : 1657-1671.

Neto, N.B.M and Vieira, L.G.E. (2011) *Braz. Arch. Biol. Technol.*, **54**(5): 939-946.

Nierenberg, L. (1972) *Amer. Orchid Soc. Bull.*, **41** : 873-882.

Niimoto, D.H. and Randolph, *I.E.* (1958) *Amer. Orchid Soc. Bull.*, **27** : 157-162.

Nilson, L.A. (1981) *Nord. J. Bot.*, **1** : 461-480.

Nishida, T. and Boyle, W.W. (1957) *Hawaii Fm. Sci.*, **6** : 7-8.

Nishikawa, K. and Hirata, Y. (1968) *Tetrahedron Letters*, pp. 6289-6291.

Nongdam, P. (2014) *Research Journal of Medicinal Plants*, **8**: 126-139.

Northen, R.T. (1970) *Home Orchid Growing*, 3rd Edition, Van Nostrand Reinhold Co., New Work.

Norton, H.P., Raven-Riemann, C. and Fighetti, C.F. (1995) *Amer. Orchid Soc. Bull.*, **64** : 470-477.

Nowak, J. and Vacharotayan, S. (1980) *Prace Instytutu Sadownictwa I Kwiarciarstwa, Skierniewicach*, B **5** : 83-93.

Nuraini and Mohn Sahib (1992) *Acta Horticulturae*, **292** : 169-172.

Oakeley, H.F. (1991a) *Amer. Orchid Soc. Bull.*, **60** : 222-231.

Oakeley, H.F. (1991b) *Amer. Orchid Soc. Bull.*, **60** : 316-326.

Oakeley, H.F. (1991c) *Amer. Orchid Soc. Bull.*, **60** : 426-433.

Oakeley, H.F. (1991d) *Amer. Orchid Soc. Bull.*, **60** : 634-644.

Oakeley, H.F. (1991e) *Amer. Orchid Soc. Bull.*, **60** : 738-747.

Obara-Okeyo, P. and Kano, S. (1998) *Euphytica*, **99** : 95-101.

Obara-Okeyo, P., Fujii, K. and Kano, S. (1998) *Hort.* Science, **33** : 133-135

Ohsawa, U., Fukuda, M. and Utunomiya, H. (2000) *Research Bulletin Aichi-ken Agricultural Research Centre*, **32** : 175-182.

Onaka, T., Kamata, S. Maeda, T., Kawazoe, Y., Natsume, M., Okamoto, T., Uchimaru, F. and Shimizy, M. (1965) *Chem. Pharm. Bull.*, **13** : 745-747.

Onal, K. (1999) *Turkish J. Agric. and For.*, **23** : 1057-1064.

Ong, H.T. (1982) *Orchid Rev.*, **90** : 264-266.

Ordogh, G. and Jakacs, A. (1983) *Novenyvedelem*, **19** : 417-419.

Os, P.C. Van and Mourik, N.M. Van (1991) *Vakblad voor de Bloemisterij*, **46** : 66-67.

Oshiro, M.A. and Steinhart, W.L. (1991) *Lindleyana*, **6** : 36-41.

Paek, K.Y., Kim, T.J. and Seon, J.H. (1998a) *J. Korean Soci. Hort. Sci.*, **39** : 597-604.

Paek, K.Y., Seon, J.H. and Kim, T.J. (1998b) *J. Korean Soc. Hort. Sci.*, **39** : 593-596.

Paek, K.Y., Shim, G.B. and Kim, J.J. (1989) *J. Korean Soc. Hort. Sci.*, **7** : 192-193.

Pais, M.S., Barroso, J., Fevereiro, P. and Oliveira, M. (1983) *Protoplast*, Birkhauser Verlag, pp. 74-75.

Palma, M., Y.J.Chen,C.Hall *et al.* (2010) *HortTechnology*. **20**. 10.21273/ HORTTECH.20.1.239.

Park, J.S., Chung, J. D and Chung, M,S. (1990) *J. Korean Soci Hort Sci.*, **31** : 176-183

Park, K. and Kim, S. (1995) *FRI J. Forest Sci. (Seoul)*, **51** : 64-73.

Park, M.J., Park, S.J. and Kim, D.H. (1998) *Korean J. Hort. Sci. Technol.*, **16** : 42-44.

Parker, N. (1979) *Orchid Rev.*, **87** : 274-275.

Paul, H.L. (1975) *Commonw. Mycol. Inst.*, No. 155.

Penningsfeld, F. and Fast, G. (1960) *Mitt. Kloslernenburg, Ser. B.*, **10** : 113-120.

Penningsfeld, F. and Fast, G. (1962a) *Gartenwelt*, **62** : 5-7.

Penningsfeld, F. and Fast, G. (1962b) *Zierpflanzeub*, No.1, pp. 3.

Penningsfeld, F. and Fast, G. (1970) *Gartenwelt*, **70** : 205-208.

Perata, P. (1981) *Informatore di Ortoflorofruticoltura*, **22** : 15-18.

Peres, L.E.P., Amar, S., Kerbauy, G.B., Salatino, A., Zaffari, G.R. and Mercier, H. (1999) *J. Plant Physiol.*, **155** : 551-555.

Perkins, A.J., Masuhara, G. and McGree, P.A. (1995) *Austr. J. Bot.*, **43** : 85-91.

Petroski, R. (1984) *Proc. 11th World Orchid Conf.*, Florida, pp. 281-284.

Pfitzer, E.H.H. (1989) *Dipraturlichen pflanzenfa millien Pt-2,* Abt. 6, pp. 52-220.

Picolo,M.R., C.C.Custódio and N.B.M.Neto. (2016) *Aftican J. Biotech.,* **15**: 2613-2619.

Pindel, A. and Miczynski, K. (1996) *Folia Horticulture,* **8** : 95-105.

Pinheiro, L.R., A. R. C. Rabbani, A. V. C. da Silva *et al.* (2012) *Plant Syst Evol.* DOI 10.1007/s00606-012-0682-9.

Pollacci, G. and Bergamaschi, M. (1940) *Boll. Soc. Ital. Biol. Sper.*, **15** : 326-327.

Poole, H.A. and Seeley, J.G. (1977) *Amer. Orchid Soc. Bull.*, **46** : 923-928.

Poole, H.A. and Sheehan, T.J. (1970) *Proc. Fla. St. Hort. Soc.*, **83** : 465-469.

Poole, H.A. and Sheehan, T.J. (1973) *Amer. Orchid Soc. Bull.*, **42** : 889-895.

Poole, H.A. and Sheehan, T.J. (1977) *Amer. Orchid Soc. Bull.*, **46** : 155-160.

Porter, K.G., Kuehnle, A.R. (1997) *Hort. Technol.*, **7** : 161-164.

Porter, K.G., Kuehnle, A.R. and Hu, J.S. (1996) *Lindleyana,* **11** : 211-213.

Powell, T. (1964) *Amer. Orchid Soc. Bull.*, **33** : 184-186.

Pradhan, G.M. (1985) *Indian Orchid J.*, **1** : 114-116, 128.

Pradhan, U.C. (1975) *Indian Orchids – Guide to Identification and Culture,* Vol. 1, Kalimpong, India.

Prasad, R.N. and Mitra, G.C. (1975) *Indian J. Exp. Biol.*, **13** : 123-126.

Prayitno, S. (1979) *Bull. Penelitian Hortikultura,* **7** : 19-25.

Price, G.R. and Earle, E.D. (1984) *Amer. Orchid Soc. Bull.*, **53** : 1035-1043.

Pritchard, A.E. (1951) *Calif. Agric.*, **5** : 9-11.

Pritchard, A.E. (1959) In : *The Orchid – A Scientific Survey* (Ed. C.L. Withner), Ronald Press Co., New York.

Pritchard, M.W., Poynter, A.L.C. and Seaton, P.T. (1999) *Lindleyana,* **14** : 92-101.

Puri, H.S. (1970) *Amer. Orchid Soc. Bull.*, **39** : 723.

Qin, X., Liu, Y., Mao, S. *et al.* (2011) *Euphytica,* **177,** 33–43.

Quisumbing, E. (1968) *Araneta J. Agric.*, **15** : 195-212.

Quyu, B.S.H. (1959) *Trop. Agriculturist,* **115** : 257-281.

Ragan, M.E. and Sauleda, R.P. (1982) *Amer. Orchid Soc. Bull.*, **51** : 388-390.

Raghunathan, K. (1976) *Pharmacopoeial Standards for Ayrvedic Formulations,* Central Council for Research In Indian Medicine and Homoeopathy, New Delhi.

Rajamani, K., Jawaharlal, M., Muthumanickam, D. and Balakrishnamurthy, G. (1999) *South Indian Hort.*, **47** : 348-350.

Rao, A.N. (1996) *J. Econ. Taxon. Bot.*, **20** : 707-709.

Rao, A.N. and Avadhani, P.N. (1963) *Proc. 4th World Orchid Conf.*, Singapore, pp. 194-202.

Rao, A.S. (1979) *Orchids of India*, National Book Trust, New Delhi, India.

Rao, A.V.N. (1982) *Indian Hort.*, **30** : 168-170.

Rasmussen, H.N., Johansen, B. and Anderson, T.F. (1991) *Lindleyana*, **6** : 134-139.

Redlinger, J.R. (1964) *Amer. Orchid Soc. Bull.*, **33** : 759-761.

Rego, E.R. do, M.M. do Rêgo and B.C. Souza. (2009) *Acta Hort.* **813**: 413-420.

Reybrun, A.N. (1978) *Amer. Orchid Soc. Bull.*, **47** : 798-802.

Richard, L.C. (1818) *Memoires du Museum d'Histoire Naturelle*, **4** : 50.

Richardson, D. (1955) *Gdn. J. N. Y. Bot. Gdn.*, **5** : 54-57.

Rij, R.E., Thompson, J.E. and Farnham, D.J. (1979) *USDA-SEA. Adv. Agr. Tech. West. Ser.*, **5** : 1-26.

Rittershausen, B. and Rittershausen, W. (1970) *Popular Orchids*, David and Charles Ltd, Newtop Abbot.

Robbins, S. (1991) *Amer. Orchid Soc. Bull.*, **60** : 234-235.

Rodrigues, J.F., van den Berg, C., Abreu, A.G. *et al.* (2015) *Plant Syst Evol.*, **301:** 1345–1359.

Rogerson, W.P. (1991) *Amer. Orchid Soc. Bull.*, **60** : 673-682.

Romero, G.A. and Carnevali, G. (1991) *Amer. Orchid Soc. Bull.*, **60** : 770-774.

Romero, G.A. and Carnevali, G. (1992) *Amer. Orchid Soc. Bull.*, **61** : 354-360.

Rossi, W., Contorni, M. and Lutti, A. (1990) *Webbia*, **44** : 323-327.

Rotor, G.B.Jr. (1950) *Flor. Exch.*, **114** : 15-17.

Rotor, G.B.Jr. (1952) *Cornell Univ. Agric. Exp. Sta. Bull.*, 885.

Rotor, G.B.Jr. (1958) *Proc. 2nd World Orchid Conf.*, Honolulu, pp. 159-170.

Sagarik, R. and Siripong, S. (1963) *Amer. Orchid Soc. Bull.*, **32** : 299-302.

Sagawa, Y. (1961) *Amer. Orchid Soc. Bull.*, **30** : 808-809.

Sagawa, Y. (1990) In : *Hand Book of Plant Cell Culture*, (Eds. P.V. Ammirato, D.A. Evans, W.R. Sharp and Y.P.S. Bajaj), pp. 638-653.

Sagawa, Y. and Kunisaki, J.T. (1982) *Japanese Assoc. Pl. Tissue Cult.*, pp. 683-684.

Sagawa, Y. and Sehgal, O.P. (1967) *Bull. Pacific Orchid Soc.*, Hawaii, **25** : 17-18.

Sagawa, Y., Shoji, T. and Shoji, T. (1966) *Amer. Orchid Soc. Bull.*, **35** : 118-122.

Sahid, C.A. (1980) *Bull. Penelitian Hortikult.*, **8** : 37-41.

Sakai, K., Osuga, M., Youemura, K. and Higuchi, H. (1981) *Res. Bull. Aichi0ken Agric. Res. Cent.*, **14**: 219-227.

Sakai, K., Osuga, M., Youemura, K. and Higuchi, J. (1982) *Res. Bull. AichiOken Agric. Res. Cent.*, **14**: 187-192.

Sakanishi, Y., Imanishi, H. and Ishida, G. (1980) *Osaka Univ. Perfect. Bull.*, B **32** : 1-9.

Sakharam, R.J. (1953) *Sci. and Cult.*, **19** : 97-99.

Salisbury, E. (1953) *Proc. Roy. Inst. G.B.*, **35** : 293-298.

Sampolinski, J. (1983) *Orchid Rev.*, **91** : 1097.

Sanford, W.W. (1971) *Bot. J. Linn. Soc.*, **64** : 163-181.

Sanguthai, O., Sanguthai, S. and Kamemoto, H. (1973) *Hawaii Orchid J.*, **2** : 12-16.

Sanguthai, S. and Sagawa, Y. (1973) *Hawaii Orchid J.*, **2** : 17-19.

Sauleda, R.P. (1966) *Amer. Orchid Soc. Bull.*, **35** : 630-632.

Sawa, Y. (1989) *Report Kochi Univ. Agril. Sci*, Japan, **38** : 9-18.

Schaffstein, G. (1938) *Jahrb. Wiss. Bot.*, **86** : 720-752.

Schaffstein, G. (1941) *Jahrb. Wiss. Bot.*, **90** : 141-198.

Schenk, M. and Brundert, W. (1982) *Deutscher Gartenbau*, **36** : 10-11.

Schenk, M. and Brundert, W. (1983) *Deutscher Gartenbau*, **37** : 1786-1788.

Schlechter, R. (1926) *Notizblatt des Botanischen Gartens u. Museums Berlin-Dahlem*, **10** : 563.

Schmidt, G., Zotz, G. (2002) *J. Veg. Sci.*, **13**: 527-534.

Schmude, N.F. Hass-Von (1984) *Proc. 11th World Orchid Conf.*, Florida, pp. 311.

Schrenk, W.J. (1978) *Amer. Orchid Soc. Bul.*, **47** : 429-437.

Schultes, R.E. and Pease, A.S. (1963) *Generic Names of Orchids : Their Origin and Meaning*, Academic Press, New York.

Scully, R.M. (1965) *Bull. Pacific Orchid Soc.*, **28** : 823-829.

Scully, R.M. (1966) *Amer. Orchid Soc. Bull.*, **35** : 40-42.

Semiarti E, Indrianto A, Purwantoro A, Isminingsih S, Suseno N, Ishikawa T, Yoshioka Y, Machida Y, Machida C (2007) *Plant Biotechnol* **24**: 265–272.

Sharma, A., Tandon, P. and Kumar, A. (1992) *Indian J. Exp. Biol.*, **30** : 747-748.

Sheehan, T.J. (1960) *Proc. 3rd World Orchid Conf.*, London, pp. 211-218.

Sheehan, T.J. (1962) *A.R. Fla. Agric. Exp. Stat.*, **35** : 6.

Sheehan, T.J. (1980) In : *Introduction to Floriculture* (Ed. R.A. Larson), Academic Press, New York.

Sheehan, T.J. and McConnell, D.B. (1978) *Proc. 9th World Orchid Conf.*, Bangkok.

Sheehan, T.J. and Sheehan, M. (1979) *Orchid Genera Illustrated*, Van Nostrand Reinhold Co., New York.

Sheehan, T.J., Murashige, T. and Kamemoto, H. (1965) *Amer. Orchid Soc. Bull.*, **34** : 228-232.

Shefferson, R.P. (2006) *Oikos* **115**: 253-262.

Sher, S.A., Kamasaki, H. and Murakishi, H. (1953) *Hawaii Fm. Sci.*, **2** : 2.

Sheviak, C.J. (1992) *Amer. Orchid Soc. Bull.*, **61** : 546-559.

Shi, Y.Z., Pan, R.C., Wang, X.J., Ye, Q.S. and Guo, L.R. (2000) *Chinese J. Appl. Environ. Biol.*, **6** : 326-330.

Shimasaki, K. and Kemoto, S. (1991) *Plant Cell Tissue Organ Cult.*, **25** : 49-52.

Shukla, B.K., Singh, P. and Chauhan, A.S. (1998) *J. Orchid Soc. India*, **12** : 53-76.

Shushan, S. (1959) In : *The Orchids – A Scientific Survey* (Ed. C.L. Withner), Ronald Press Co., New York.

Sidran, C.B. (1984) *Proc. 11th World Orchid Conf.*, Florida, pp. 312-318.

Silva,P.A.K.X. M., S. C.Jacques and M.H.B.Zanettini. (2000) *Ciência Rural, Santa Maria*, **30**: 105-111.

Simon, E. (1938) *Gartenwelt*, **62** : 254-256.

Singer, R.B. and Cocucci, A.A. (1997) *Botanic Acta*, **110** : 338-342.

Singh, F. (1978) *Indian Hort.*, **22** : 14-17, 24.

Singh, H. and Sagawa, Y. (1972) *Hawaii Orchid J.*, **1** : 9.

Sinoda, K., Suto, K., Hara, M. and Aoli, M. (1988) *Bull. National Research Institute of Vegetables*, Ornamental Plants and Tea, **A(2)** : 279-290.

Sophonsiri, T. (1984) *Proc. 11th World Orchid Conf.*, Florida, pp. 319-321.

Sosa, V. and Luma, D.E.D. (1998) *Plant Systematics and Evolution*, **214** : 185-213.

Soysa, E. (1943) *Orchid Culture in Ceylon*, Caxton Press, Colombo, pp. 153.

Springer, A. (1962) *Die Orchidee*, **13**; 64.

Stearn, W.T. (1960) *Proc. 3rd World Orchid Conf.*, London, pp. 26-41.

Stewart, J. (1992) *Nature and Environment series*, (1992), No. 57, pp. 64, ISBN, 92T, 871-2001.

Stewart, J. and Button, J. (1977) *Amer. Orchid Soc. Bull.*, **46** : 415-418.

Stockdale, A. (1950) *Amer. Orchid Soc. Bull.*, **19** : 302-305.

Storey, W.B. (1946) *Bull. Pacific Orchid Soc. Hawaii*, **4** : 1-10.

Storey, W.B. (1950) *Bull. Pacific Orchid Soc. Hawaii*, **8** : 327-331.

Storey, W.B. (1958) *Bull. Pacific Orchid Soc. Hawaii*, **16** : 7-13.

Storey, W.B. and Kamemoto, H. (1960) *Amer. Orchid Soc. Bull.*, **29** : 24-29.

Strauss, M.S. and Arditti, J. (1972) *Cymbidium Soc. News*, **7** : 170-171.

Stulzer, G.C.G., R.T.Hoshino, A.B. P. Suzuki, J.C. B. de Paula. (2019) *Crop Breeding and Applied Biotechnology* **19**: 3, 364-367.

Su, W.R., Chen, W.S., Koshioka, M., Mander, L.N., Hung, L.s., Chen, W.H., Fe, Y.M. and Huang, K.L. (2001) *Plant Physiol. Biochem.*, **39** : 45-50.

Summerhayes, V.S. (1968) *Wild Orchids of Britain*, Collins, London.

Sushmita, B., Khan, H.A. and Reddy, P.V. (1998) *Orissa J. Hort.*, **26** : 114-116.

Suzuki, H., Keimarsu, 1 ano 1 to, M. (1934) *J. Pharm Soc. Japan*, **54** : 802-819.

Suzuki, H., Keimarsu, I. and Ito, M. (1932) *J. Pharm. Soc. Japan*, **52** : 1049-1060.

Swartz, O. (1800) *Kongl. Vetenskaps Academiens Nya Handlingar*, **21** : 235.

Synder, G. (1992) *Amer. Orchid Soc. Bull.*, **61** : 28-33.

Syrovatka, T. (1980) *Amer. Orchid Soc. Bull.*, **49** : 1258-1259.

Takahashi, C.T. and Kondo, K. (1998) *Lindleyana*, **13** : 284-291.

Tamer, C., Karaman, B. and Copur, O. (2006) *Food Review International* **22**: 43–50.

Tan, J., Wang, H.L. and Yeh, K.W. (2005) *Biotechnol Lett* **27**: 1517–1528.

Tan, K.K.T., Yik, C.P., Lee, S.M., Yap, M.L., Kueh, J. and Loh, H.M. (1999) *Singapore J. Prim. Indus.*, **27** : 20-21.

Tan, V.H.,H.P.Tran, L. V. Hong *et al.* (2011) Vietnam Atomic Energy Association, Hanoi (Viet Nam); Vietnam Atomic Energy Institute, Hanoi (Viet Nam), pp. 736-742.

Tanaka, H. (1957) *Na Pua Okika O. Hawaii Nei*, **7** : 100-102.

Tanaka, H., Hasegawa, A. and Gopi, M. (1975) *J. Japanese Soc. Hort. Sci.*, **44** : 47-58.

Tanaka, M. and Sakanishi, Y. (1977) *Amer. Orchid Soc. Bull.*, **46** : 733-737.

Tanase, K., Ohno, H. and Matsui, S. (1999) *Lindleyana*, **14** : 147-151.

Tang, C.Y. and W.H. Chen (2007) Breeding and Development of New Varieties in Phalaenopsis

Teng, W.L., Nicholson, L. and Teng, M.C. (1997) *Plant Cell Rep.*, **16** : 831-835.

Teo, C.K.H. and Neumann, N.H. (1978) *Orchid Rev.*, **86** : 186-189.

Teoh, E.S. (2016) *Medicinal orchids of Asia*. Cham: Springer

Thammasiri, K. (2015) *Acta Hort.*, DOI: *10.17660/ActaHortic.2015.1078.2.*

Thien, L.B. and Marcks, B.G. (1972) *Canadian J. Bot.*, **50** : 2319-2325.

Thien, L.B. and Utech, F. (1960) *Amer. J. Bot.*, **57** : 1031-1035.

Thomale, H. (1954) *Engen Ulmer Stuttgart*, p. 189.

Thomasen, H.B. (1964) *Amer. Orchid Soc. Bull.*, **33** : 111.

Thompson, P.A. (1974) *J. Roy. Hort. Soc.*, **99** : 117-122.

Thompson, P.A. (1977) *Orchids from Seeds*, Royal Bot. Gardens, Kew.

Thwaites, R.G. (1912) *Orchid World*, **3** : 52-55.

Timmerman, D. and Vrie, M. Van De (1982) *Vakblad voor de Bloemisterij*, **37** : 38-39.

Torikata, H., Sawa, Y. and Sisa, M. (1965) *J. Japanese Soc. Hort. Sci.*, **34** : 63-70.

Torres, J. and Mogollon, N. (1998) *Proc. Interamerican Society of Tropical Horticulture*, **42** : 92-98.

Torres, J. and Mogollon, N. (2000) In : *Proce. of the Interamerican Society for Tropical Horticulture* (Ed. J. Crane), **42** : 87-92.

Toussaint, A. Kummert, A., Maroquin, C., Lebrun, A. and Roggemans, J. (1993) *Plant Cell Tissue Organ Cult.*, **32** : 303-309.

Treub, M. (1879) *Verh. Akad. Amesterdam,* Naturk, **19** : 1-50.

Tsai, W.C., Fu, C.H., Hsiao, Y.Y., Huang, Y.M., Chen, L.J., Wang, M., Liu, Z.J. and Chen, H.H. (2013) *Plant Cell Physiol.*, **54**: e7.

Tsai, W.C., Hsiao, Y.Y., Lee, S.H., Tung, C.W., Wang, D.P., Wang, H.C., Chen, W.H. and Chen, H.H. (2006) *Plant Sci* **170**: 426–432.

Tsai, W. C., Dievart, A., Hsu, C. C., Hsiao, Y. Y., Chiou, S. Y., Huang, H., and Chen, H. H. (2017) Post genomics era for orchid research. *Botanical studies, 58*(1), 61. https: / / doi.org / 10.1186 / s40529-017-0213-7

Tse, A.T.P., Smith, R.J. and Hackett, W.P. (1977) *Amer. Orchid Soc. Bull.*, **40** : 807-810.

Tsuchiya, I. (1935) *Na Pusa Okika O Hawaii Nei*, **4** : 11-16.

Tsukamoto, Y., Kano, K. and Katsuura, T. (1963) *Amer. Orchid Soc. Bull.*, **32** : 354-355.

Tsukazaki, H., Mii, M., Tokuhara, K. and Ishikawa, K. (2000) *Plant Cell Rep.*, **19** : 1160-1164.

Twiss, R. (1914) *Orchid World*, **4** : 145

Uesato, K., Yagi, N. and Ddo, Si. (1987) *Soi. Bull. College of Agril. Univ. Ryukus*, **34** : 11-19.

Vajrabhaya, M. (1978) *Orchid Rev.*, **78** : 256-257.

Vajrabhaya, M. and Vajrabhaya, T. (1984) *Proc. 11th World Orchid Conf.*, Florida, pp. 335-338.

Vajrabhaya, T. (1977) In : *Orchid Biology, Reviews and Perspectives*, 1 (Ed. J. Arditti), Cornell Univ. Press, Ithaca, New York, pp. 117-201.

Valmayor, H.L. (1974) *Proc. 7th World Orchid Conf.*, Medellin, Colombia, pp. 211-229.

Van Den Berg, C. (2014) *Phytotaxa* **186**(2): 75–86.

Van der Pijl, L. and Dodson, C.H. (1966) *Orchid Flowers – Their Pollination and Evolution*, University of Miami Press, Coral Gables.

Vanstaden, J. and Drewes, J.E. (1975) *Plant Physiol.*, **32** : 347-352.

Vaz, A.P. and Kerbauy, G.B. (2000) *Acta Horticulturae*, **520** : 149-156.

Veitch, H.J. (1886) *J. Royal Hort. Soc.* **7**: 22–49.

Vermeulen, P. (1966) *Acta. Bot. Neer.*, **15** : 224-253.

Vij, S.P. and Abhilasha Dhiman (1997) *J. Orchid Soc. India*, **11** : 79-84.

Vij, S.P. and Parminder, Kaur (1999) In : *Proc. of the National Academy of Science India, section B, Biological Science*, **69** : 317-321.

Vij, S.P., Kondo, K., Promila Pathak (1994) *J. Orchid Soc. India*, **8** : 19-23.

Vries, J.T. De (1953) *Annales Bogor.*, **1** : 61-76.

Walker, J.N. and Abernathie, J.W. (1964) *Amer. Orchid Soc. Bull.*, **33** : 391-396.

Wallbrunn, H.M. (1984) *Proc. 11th World Orchid Conf.*, Florida, pp. 378-384.

Wallis, F.M., Joubert, J.J. and Schlosser, I.P. (1975) *Phytophylactia*, **7** : 125-130.

Wang, H.M., Tong, C.G. and Jang, S. (2017) *Plant Signaling and Behavior*, **12**(5) : e1322245.

Wang, X. (1998) *Lindleyana*, **3** : 184-189.

Wang, Y., Ma, Y., Dai, S. (2010). *Chin. Bull. Bot.*, **45**: 641-653.

Wang, Y.T. (2000) *HortScience*, **35** : 60-62.

Watrous S.B. and Wimber, D.E. (1988) *Lindleyana*, **3** : 177-183.

Waves, J. Van and Geest, S.De (1983) *Revue del' Agric.*, **36** : 1415-1432.

Webb, A. and Webb, M. (1990) *Amer. Orchid Soc. Bull.*, **62** : 964-973.

Webb, A. and Webb, M. (1993) *Amer. Orchid Soc. Bull.*, **62** : 1016-1017.

Webb, M. (1988) *Amer. Orchid Soc. Bull.*, **57** : 138-143.

Webb, M. and Webb, A. (1992) *Amer. Orchid Soc. Bull.*, **61** : 12-19.

Weng, E.S., H. Hu,S.Y. Li S.Y., *et al.* (2002) *Acta Bot. Yunnanica*, **24**: 222-228.

Wey, G.C. (1991) *Report of Taiwan Sugar Research Institute*, **133** : 1-10.

Widiastoety, D. (1986) *Buletui Peneletin Horticultural*, **13** : 1-8.

Widyastuti, D. and Soedjono, S. (1979) *Penoli. Hortikult. Bull.*, **7** : 3-6.

Wilcox, E.C. (1960) *Proc. 3rd World Orchid Conf.*, London, pp. 346-352.

Wilfret, G.J. and Kamemoto, H. (1969) *Amer. J. Bot.*, **56** : 521.

Wilson, G. (1923) *Orchid Rev.*, **31** : 299-300.

Wilson, G. (1927) *Orchid Rev.*, **35** : 40-41.

Wilson, J.K. (1915) *Amer. J. Bot.*, **2** : 420-440.

Wimber, D.E. (1963) *Amer. Orchid Soc. Bull.*, **32** : 105-107.

Wimber, D.E. (1965) *Cymbidium Soc. News*, **20** : 7-10.

Wimber, D.E. and Kenney-Norberg, J. (1984) *Orchid Rev.*, **92** : 260-262.

Wimber, D.E. and Van Cott, A. (1967) *Proc. Vth World Orchid Conf.*, Long Beach, pp. 27-31.

Wimber, D.E. and Wimber, D.R. (1968) *Amer. Orchid Soc. Bull.*, **37** : 572-576.

Wimber, D.E., Lenz, L.W. and Dodson, C.H. (1956) *Amer. Orchid Soc. Bull.*, **25** : 153-158.

Wisher, G.C. (1989) In : *The Complete Guide Book*, Maupin House Pub., Gainesville, Florida.

Wisler, G.C. and Zettler, F.W. (1984) *Proc. 11th World Orchid Conf.*, Horida, pp. 402-406.

Wisler, G.C., Zettler, F.W. and Sheehan, T.J. (1983) *Amer. Orchid Soc. Bull.*, **52** : 255-256.

Withner, C.L. (1943) *Amer. Orchid Soc. Bull.*, **11** : 261-263.

Withner, C.L. (1951) *Amer. Orchid Soc. Bull.*, **20** : 276-278.

Withner, C.L. (1953) *Orchid J.*, **2** : 473-477.

Withner, C.L. (1959) In : *The Orchid – A Scientific Survey* (Ed. C.L. Withner), Ronald Press Co., New York.

Withner, C.L. (1974) *The Orchids : Scientific Studies*, Wiley Inter Science, New York.

Wong, K.M. and Chan, C.L. (1993) *Sandakania*, **2** : 85-89.

Wong, Y.K. and Chua, S.E. (1975) *Singapore J. Prim. Indust.*, **3** : 75-106.

Wood, J. (1992) *Kew Magazine*, **10** : 9-12.

Wood, J. (1993) *Kew Magazine*; **10** : 9-12

Wood, J. and Comber, J.B. (1988) *Orchid Rev.*, **96** : 51-53.

Wood, J. and Comber, J.B. (1993) *Lindleyana*, **8** : 115-117.

Woodward, G.P. (1964) *Amer. Orchid Soc. Bull.*, **33** : 569-571.

Woodward, G.P. (1965) *Amer. Orchid Soc. Bull.*, **34** : 116-122.

Wynd, F.L. (1933) *Ann. Miss. Bot. Gard.*, **20** : 223-237.

Xia, H.X., Zhu, L.Q., Zhang, M., Zhang, Y.J. and Bie, Z.L. (1999) *Acta Horticulturae Sinica*, **26** : 275-276.

Yadav, L.P. and Bose, T.K. (1985) *Farmer and Parliament*, **20** : 19-21, 40.

Yadav, L.P. and Bose, T.K. (1986a) In : *Biology, Conservation and Culture of Orchids* (S.P. Vij Ed.), Affiliated East-West Press Private Ltd., New Delhi, India.

Yadav, L.P. and Bose, T.K. (1986b) In : *Biology, Conservation and Culture or Orchids* (S.P. Vij Ed.), Affiliated East-West Press Private Ltd., New Delhi, India.

Yadav, L.P. and Rao, D.P. (1986) In : *Propagation of Tropical and Subtropical Horticultural Crops* (T.K. Bose, S.K. Mitra and M.K. Sadhu Eds.), Naya Prokash, Calcutta, India.

Yam, T.W. and Weatherhead, M.A. (1991) *Lindleyana*, **6** : 147-150.

Yamaguchi, S. (1979) *J. Amer. Soc. Hort. Sci.*, **104** : 739-742.

Yamamura, S. and Hirata, Y. (1964) *Tetrahedron Letters* : 79-87.

Yamane, K., Kuchii, E., Fujishige, N. and Minegishi, N. (1997) *J. Japanese Soc. Hort. Sci.*, **65** : 801-807.

Yang, J., Lee, H.J., Shin, D.H., Oh, S.K., Seon, J.H., Paek, K.Y., Han, K.H. (1999) *Plant Cell Rep.*, **18** : 978-984.

Yap, M.L., Lam-Chan, L.P., Yik, C.P., Kueh, J., Lee, S.M., Loh, H.M. (1999) *Singapore J. Prim. Indust.*, **27** : 11-19.

Yates, R.C. and Curtis, J.T. (1949) *Amer. J. Bot.*, **36** : 390-396.

Yee, K.K. (1983) *Pl. Propagator*, **29** : 3-4.

Yoneda, K. and Sasaki, H. (1978) *Bull. Coll. Agric. and Vet. Med. Nihan Univ.*, No. 35, pp. 80-87.

Yoneda, K., Momose, H. and Kubato, S. (1991) *J. Japanese Soc. Hort. Sci.*, **60** : 651-657.

Yoneda, K., Sakamoto, T. and Sasaki, H. (1979) *Bull. Coll. Agric. and Vet. Med. Nihan Univ.*, No. 36, pp. 64-73.

Yoneda, K., Suzuki, N. and Hasegawa, I. (1999) *Scientia Hort.*, **80** : 259-265.

Yoneda, K., Usin, M. and Kubota, S. (1997) *J. Japanese Soc. Hort. Sci.*, **66** : 141-147.

Young, H.C. and Ong, H.T. (1979) *Orchid Rev.*, **87** : 282-285.

Yu, P.Z., Bai, L.J., Li, H.Y. and Sun, J. (1998) *Acta Horticulturae Sinica*, **25** : 374-378.

Yukawa, T. and Nishida, M. (1992) *Lindleyana*, **7** : 374-378.

Zettler, F.W., Ko, N.J., Wisler, G.C., Elliot, M.S. and Wong, S.M. (1990) *Pl. Disease*, **74** : 621-626.

Zettler, F.W., Wisler, G.C., Sheehan, T.J. and Ko, N.J. (1984) *Proc. 11th World Orchid Conf.*, Florida, pp. 407-412.

Zettler, L.W. and Hoffer, C.J. (1998) *Environ. Exp. Bot.*, **39** : 189-195.

Zhang, F.P., J.L. Huang and S.B. Zhang. (2016) *Plant Signal. Behav.* **11**, e1149668. https: / /doi.org/

Zhang, F.P.,J.J. Zhang, Y. Yan *et al.* (2015) *Seed Sci. Res.*, **25**: 395-401.

Zhang, G., Liu, K., Li, Z. *et al.* (2017) *Nature*, **549:** 379–383.

Zhang, L., Chin, D.P. and Mii, M. (2010) *Plant Cell Tiss Organ Cult.*, **103,** 41–47.

Zhang, S., Yang, Y., Li, J., Qin, J., Zhang, W., Huang, W. and Hu, H. (2018) *Plant Diversity*, **40** : 196 -208.

Zhang, X.S., Zheng, C.C. and O'Neill, S.D. (1999) *Acta Botanica Sinica*, **41** : 468-471.

Zheng Xin Shu, Kamuro, Y. and Matsin, S. (1998) *J. Soc. High Tech. Agric.*, Japan, **10** : 27-33.

Zheng, X.S. and Matsui, S. (1998) *J. Japanese Soc. Hort. Sci.*, **67** : 939-944.

Zhu, Y.Q. and Wang, X.G. (1999) *J. Zheijiang Forestry College*, **16** : 164-169.

3

Narcissus

V.A. Parthasarathy, Rajive Kumar, C. Aswath and Sk Masudul Islam

1.0 Introduction

Narcissus is an ornamental bulbous plant, naturally grown in the area of Mediterranean Sea, also in China and Japan (Hanks, 1993). The genus of *Narcissus* L. belonging to the sub-family Amaryllidoideae, is one of the fifteen genera of the Amaryllidaceae family, which contains 80 to 100 wild species of perennial bulbiferous geophytes (Berkov *et al.*, 2009; Hanks, 2002). The genus is widely distributed in the southwest of Europe, with the Iberian Peninsula diversity center (Spain and Portugal) and North Africa (Ito *et al.*, 1999). Few species of *Narcissus* have been developed in France and Italy, and some are found in the Balkans and the Eastern Mediterranean regions (Hanks, 2002). The genus members are commonly known as daffodil, daffadowndilly, narcissus, and jonquil throughout the world. *Narcissus tazetta* L. is an indigenous ornamental and fragrant plant species that is distributed from Portugal to Turkey.

2.0 Importance and Uses

2.1.0 Aesthetic Uses

The genus *Narcissus* (Amaryllidaceae family), including the large-cup cultivar 'Carlton', occupies an important position in the worldwide bulbous plant industry. These are spring flowering bulbous perennial plants of temperate region. They are grown for their decorative qualities and used as garden flowers, cut flowers, beds, borders, hanging basket as well as pot flowers. The crop is well suited to different landscape design use, including shrubbery, under tree and in rock or perennial gardens. The narcissus is perceived in the West as a symbol of Vanity, in the East as a symbol of wealth and good fortune.

2.2.0 Therapeutic and Other Uses

Throughout history, the scent of narcissus has been an important ingradient of perfume, a quality that comes from essential oils rather than alkaloids. Also, because of the pharmacologically active alkaloids (for the treatment of Alzheimer's disease and anti-cancer therapy) and lectins that narcissus produce, they are also a potential material for the pharmaceutical industry (Malik and Bach, 2016). Plants are important for new pharmaceuticals because of the complexity of their secondary metabolic compounds. Indeed, one third of the approximately 1000 new pharmaceuticals registered in the past 25 years originated from or were inspired by natural products. The global market in plant-derived drugs in 2008 was estimated to be worth about £12bn from both prescribed and over-the-counter sales (Jones and Pulman, 2011).

Ornamental flower growers know that placing a cut daffodil (a.k.a. narcissus) in a vase with other flowers has a negative effect on the quality of those flowers and significantly shortens their vase life (Kornienko and Evidente, 2008). Furthermore, a common horticultural practice for the cultivation of narcissus flowers involves the introduction of cuts on the bulbs before immersing them into water. The mucilage that leaches out from the cuts is constantly removed by frequent changing of water and this leads to sprouting. These observations raise speculation that specific components in the mucilage of the narcissus bulbs may have powerful growth-inhibitory effects (Kornienko and Evidente, 2008).

Daffodils have been used regularly in folk and traditional medicine, in addition to the well-documented sickness-inducing properties, extracts can produce numbness, hallucinations, convulsions and cardiac effects, as well as being used for centuries in cancer treatments. Applying juice or aqueous extracts to burns or open wounds can result in numbness and these effects explain why there has been so much interest in identifying their chemical and pharmacological bases (Jones and Pulman, 2011).

3.0 Origin and History

3.1.0 Origin

To investigate the origin of cultivated trumpet daffodils the genome size (2C-value) of more than 100 old and new cultivars was measured (Zonneveld, 2010). A large number of concolored yellow tetraploid cultivars with large coronas seem to be derived from a doubling of chromosomes of a hybrid of the two species *N. hispanicus* Gouan and *N. pseudonarcissus*. This can also still be recognized by the presence of a black spot at the top of the anthers in about 15 per cent of the tetraploid cultivars. Assuming *N. pseudonarcissus* as one of the parents of the allotetraploid cultivars, species of trumpet daffodils of section *Pseudonarcissi* investigated earlier were compared. Doubling of the nuclear DNA content of 23.8 pg of *N. pseudonarcissus* falls short of the values found for the trumpet daffodils. Species of trumpet daffodils of section *Pseudonarcissi* with a complementing higher amount of nuclear DNA like *N. hispanicus* with 25.8 pg, *N. poeticus* L. with 26 pg and the hexaploid *N. pseudonarcissus* ssp. *bicolor* (L.) with 67.7 pg (Zonneveld, 2008) and other species were evaluated. *N. hispanicus* with 25.8 pg clearly differs from *N. pseudonarcissus* with 23.8 pg and is accepted here also as a species. The nuclear DNA content of *N. bujei* with 30 pg fits with the (ancient) hybrid origin for *N. bujei* between *N. longispathus* (36 pg) and *N. hispanicus* (25.8 pg). The tetraploids with white tepals and yellow corona's can be obtained by crossing a diploid *N. pseudonarcissus* with the hexaploid and bicolored *N. pseudonarcissus* ssp. *Bicolor* (L.) (Zonneveld, 2010).

According to Nuenz *et al.* (2003) the 'Ajax Group' (*i.e., Pseudonarcissus*) is one of the most important ancestors of modern daffodils cultivars. The manner in which these plants were introduced into the English, French and Dutch gardens appears relatively obscure since most are endemic to the Iberian Peninsula. Therefore, it was necessary to determine how their introduction into cultivation and domestication occurred. This study primarily utilized the comparison of data from Arab texts of agriculture, European Renaissance, and Prelinnaean ancient texts and illustrations, with the morphological characteristics of the currently known wild taxa from the territories of Spain and Portugal and primitive cultivars, which are the ancestors of the modern hybrid trumpet daffodils.

Uses of Narcissus

In Pot

As Specimen in Garden

As Border in Garden

Uses of Narcissus

In Flower Arrangement

In Flower Vase

Species of Narcissus

Narcissus asturiensis

Narcissus alpestris

Narcissus bulbocodium

Narcissus calcicola

Species of Narcissus

Narcissus canaliculatus

Narcissus cyclamineus

Narcissus tazetta

Narcissus jonquilla

Narcissus elegans

Narcissus juncifolius

Species of Narcissus

Narcissus minor

Narcissus nanus

Narcissus poeticus

Narcissus pseudonarcissus

Narcissus rupicola

Narcissus triandrus

3.1.0 History

Narcissus is a classical Greek name in honor of a beautiful youth who became so entranced with his own reflection that he pinned away and the gods turned him into this flower. Although the name daffodil is often applied only to the larger trumpet-flowered cultivars, with the short-cupped and multi-headed cultivars referred to as narcissi, breeders and other enthusiasts refer to all kinds as daffodils (https://daffseek.org/daffodil-history/).

Daffodils were introduced into gardens in about 300 BC. The Greek botanist and philosopher Theophrastus listed and described many of the earliest known kinds of narcissus in his nine-volume '*Enquiry into Plants*'. Daffodils were brought to Britain by the Romans who thought that the sap from daffodils had healing powers (Raven, 2020).

4.0 Distribution

Narcissus species are found in a variety of habitats in Europe and North Africa ranging from sea level to subalpine meadows, woodlands and rocky places. Spain hosts the greatest variety of species, but they can also be found in Morocco, Portugal, western France, Italy, and other countries (https://daffseek.org/daffodil-history/).

The genus *Narcissus* is estimated to comprise generally between 35-70 species (Barrett *et al.*, 1996), the majority of which are considered native to the western Mediterranean area, with the center of diversity located in the southern Iberian Peninsula and northern Morocco (Heidi *et al.*, 1997). The phylogenetic relationships within *Narcissus* are little known, and both the status and boundaries of many of the taxa are unclear. This is due in part to the group's long history of cultivation as an ornamental (including extensive artificial hybridization), which makes it often difficult to ascertain whether populations, and even taxa, growing in the wild are native or represent naturalized escaped ornamentals (Webb, 1980). The frequent occurrence of natural hybridization and polyploidy complicates the picture even further (Heidi *et al.*, 1997).

5.0 Taxonomy

Narcissus is a genus of geophytes (bulbous plants) belonging to the family Amaryllidaceae. It is well known because a number of the species are traditionally used in gardening and horticulture (Jimenez *et al.*, 2015). This genus is mainly located in the western Mediterranean basin, its chief diversification occurring in the Iberian Peninsula and North Africa with number of species between 60 and 80 occurring, most of which are narrow endemics, with few morphological and/or genetic differences (Jimenez *et al.*, 2015). The taxonomy of this group has traditionally been controversial mainly due to its biological features, such as recent phylogenetic divergence, rapid speciation without complete morphological differentiation, hybridization and polyploidy (Díaz-Lifante and Andrés-Camacho, 2007; Hodgings and Barret, 2007; Díaz-Lifante *et al.*, 2009; Medrano *et al.*, 2014).

The taxonomy of all species of *Narcissus* (Amaryllidaceae) has been investigated recently by flow cytometry (Zonneveld, 2008; Zonneveld, 2010). The somatic nuclear

DNA contents (2C) were shown to range from 14 to 38 picogram for the diploids. The total number of *Narcissus* species was determined as 36, nine more than in *Flora Europaea* and they were divided up in two subgenera and eleven sections. Section *Pseudonarcissi* was much more heterogeneous in nuclear DNA content than expected. Sixty five accessions of *N. pseudonarcissus* possessed, with 23.8 pg, similar amounts of DNA. However, several species from this section were clearly distinctive in nuclear DNA content. It runs from the diploid *N. primigenius* Fern. Casas with 21.7 pg to the also diploid *N. nevadensis* with 38.2 pg. Also *N. cyclamineus, N. abscissus* and *N. moleroi* are with about 26 pg clearly different from *N. pseudonarcissus.* For the first time, in eleven accessions, hexaploidy was found in *N. pseudonarcissus* ssp. *bicolor*. A new section *Nevadensis* Zonn. with 30-39 pg of nuclear DNA was split off from the section *Pseudonarcissi* with now 21-27 pg (Zonneveld, 2010).

6.0 Species and Cultivars

6.1.0 Species

***Narcissus asturiensis* (*N. minimus*)** is a native in the mountainous regions of Spain growing to 7.5 cm high bearing weak stems, leaves minute and grass-like, and flowers deep golden-yellow with some 2.5 cm long, yellow and dentate trumpet. It belongs to trumpet-flowered narcissus in miniature. It fits the division I and is most suitable for growing in pots and rock gardens. Flowering occurs in February. It requires protection against rough weather.

N. alpestris is a native of Pyrenees grows only to 10 cm in height, bearing narrow and deeply channelled leaves, white flowers appearing in April with twisted segments which droop towards the slender and little cut-corona, little flanged corona is small and the anthers are bright yellow.

N. bulbocodium is a native of SW France, Spain, Portugal and NW Africa growing up to 15 cm high with green, slender, thread-like, cylindrical and 10-30 cm long leaves and bears solitary, all-yellow and funnel-shaped wide-open flowers some 2.0-3.5 cm long having 2.5 cm long, funnel-shaped and pale to deep yellow trumpets over narrow and pointed petals and curved stamens and style. It flowers from February to March.

N. calcicola is a native of Portugal growing up to 15 cm with linear leaves, jonquil-like clustered deep yellow and scented flowers, each some 2 cm across with flattened corona and appear in March and April. It is also sometimes considered as the taller form of *N. juncifolius*. It is suitable for pot-growing. It fits roughly to division VII.

N. canaliculatus, a native of Canary Islands and S. Europe is a dwarf form of *N. tazetta*. It grows about 20 cm tall and has all the characteristics of that species though flowers later from February to April in bunches of up to six per stem. Its flowers are highly fragrant, the petals are white but corona is yellow. It is a most suitable species for rock gardens and for growing in pots.

N. cyclamineus is the only species of its group. A native of Spain and Portugal growing up to 20 cm in height with dark green narrow-linear leaves having a deeply grooved keel, slender scape and most beautiful rich gold pendent mini-flowers having some 2-4 cm long petal-coloured trumpets with crenate mouths though petals are entirely swept back similar to cyclamen. It can be grown in open or partial shade best in moist soils at a cool and shady location though is not frost-hardy. It is the parent of many hybrids and varieties fitting into division VI.

N. elegans (*N. autumnalis*), a native of Algeria, S. Italy and the Mediterranean islands are a non-frost hardy autumn-flowering (September-October) species which grows up to 25 cm in height. The leaves are linear and flat, some 25 cm long and 6 mm wide, channelled and appear with the flowers. The stems are erect and may bear solitary or more blooms some 5 cm across where petals are widely spaced, thin and wavy and greenish-white while central corona is pale-orange and almost flat. It likes sunny location.

N. jonquilla is a native of Algeria with distribution in Portugal, Spain, North Africa and Balearic Islands growing up to 45 cm in height, bearing cylindrical stems and dark green rush-like leaves up to 30 cm long and strongly scented deep yellow flowers in a cluster of 2-6, each measuring 3.7-5.0 cm wide, and small central corona. Its double variety is 'Queen Anne's Jonquil'. *N. jonquilla* with *N. poeticus* produced *N. gracilis* growing up to 38 cm in height and bears linear mid-green leaves and scented bright pale-yellow flowers some 3.7-5.0 cm wide with short corona which appear in May though others flower in April. *N. jonquilla* with *N. pseudonarcissus* (a trumpet narcissus) has produced an April-flowering hybrid *N. odorus* known since 16th century, growing up to 45 cm in height with linear leaves and 2-3 clustered, fragrant and pale-yellow flowers some 6.25 cm across with bell-shaped corona. *N. odorus* 'Campernelli Plenus' is its double form. These prefer warmer climate and deep rich soil with alkaline reactions. These all the varieties fit into division VII.

N. juncifolius is a native of SW Europe (Spain, Portugal and France) growing up to 15 cm in height with cylindrical leaves and May-borne deep yellow some 2.5 cm across flowers with flattened corona. This fits division VII. It is suitable for a cool location and in rock gardens.

N. minor, a native of Portugal growing up to 20 cm in height with narrowly strap-shaped leaves, is a variable species flowering in March having pale-yellow flowers but slightly deeper corona which is 3.75 cm long. It fits into division I.

N. nanus is a native of Europe and grows to 20 cm or more in height. The leaves are grey-green, flowers are upright with segments and corona almost equal in size, *i.e.*, some 3.5 cm long, perianth segments are flat, pointed and soft yellow though corona is deep yellow.

N. poeticus, a native of Mediterranean region (S. Europe, Spain to Greece) is a variable species, growing to a height of 45 cm bearing 2-edged scape, glaucous green and narrow strap-shaped leaves, and solitary, 4.5-7.5 cm across, pure white and highly fragrant flowers in April with crisped and frilled, flat and bright red corona. This has many wild forms and named varieties found on wet mountain meadows so requires moist conditions for their growth.

N. pseudonarcissus is a native of France, Switzerland, N. Italy, N. Spain and United Kingdom, growing up to 30 cm in height with grey-green, strap-shaped and 1.5 cm wide leaves. It is hardy and very variable in shape, size and colour of the flowers which are scented, nodding and near-white to light yellow with 5.0-6.25 cm long and lemon-yellow trumpets, which appear in April. This is found growing in the wild in the open and damp meadowlands.

N. rupicola is a native of Spain and Portugal growing up to 13cm high, bearing grey-green, narrow, triangular and some 20 cm long leaves, and fragrant, clear yellow solitary flowers some 1.8-2.5 cm across with nearly flat 6-lobed corona. It flowers in April and May and fits into division VII.

N. scaberulus is a native of Portugal, grows up to 13 cm high and bears slightly glaucous green and cylindrical leaves. Flowers up to 13 mm across are yellow (deeper than *N. rupicola*). It is best grown at cooler locations. It fits into division VII.

N. tazetta, a native of Canary Islands, S. Europe (Spain) to Iran, Kashmir, China, Japan and North Africa, is half-hardy and grows up to 45 cm high bearing strap-shaped some 30 cm long leaves. This variable species is divided into three categories, *viz*. albae, bicolours and luteae. In albae, the flowers are pure white, in bicolours the perainth is white and corona yellow while in luteae the flowers are yellow. *N. tazetta* has several (4-10) white flowers on one scape, each flower some 2.5-3.8 cm across and a small cup-like white to yellow corona, generally strongly scented and earlier to flower (January to February). It does not require cold for its growing (Bailey and Bailey, 1976). It is most suitable for growing in pots. Among all the three ancestral species involved in breeding programme, *N. tazetta* occupies the third place. With *N. poeticus* it has produced numerous poetaz varieties.

N. triandrus is a native of Spain, Portugal and Isles des Glenans. The scape is very slender and about 30 cm long. Its leaves are grey-green, few, rush-like, some 15 cm long and channelled. The flowers are white, nodding, 3 cm across, petals swept-back and in a bunch of 1-6. The corona is 1 cm long and broad, and cup-shaped. The seeds are produced freely. They also need plenty of moisture, but combined with a well-drained sandy, peaty soil which partly dries out in summer. It fits to division V. *N. t. albus* from Spain and Portugal grows up to 10 cm high bearing pendent cream-white flowers with reflexed and back-swept petals and cup-shaped corona some 1.25 cm long, and the flowers appear in March April. *N. t. concolor* from Spain and Portugal is a golden-yellow and sweetly scented version of *N. t. albus*.

N. viridiflorus, a native of Gibralter, S. Spain and Morocco is autumn-flowering (October-November) which grows up to 20 cm in height and prefers sun. Leaves are filiform, the flowers are 1-4 per stem of dull olive-green colour with about 5 cm in diameter. Petals are narrow, twisted, widely-spaced and star-like.

N. watieri, a native of N. Africa and the Atlas Mountains is not hardy and grows to 15 cm high with narrowly strap-shaped, erect and some 15 cm long leaves, solitary pure white and sweetly scented flowers some 1.9 to 3.75 cm wide in April possessing an almost flat cup. It does well in pots in cold houses. It fits in division VII.

6.2.0 Classification of Cultivars

All daffodils are classified into one of thirteen divisions by Royal Horticultural Society. Each daffodil cultivar or garden hybrid belongs to one of the first twelve divisions. Those daffodils distinguished solely by botanical name belong in Division 13. Below are the definitions for each Division.

Division 1: Trumpet Daffodil Cultivars

One flower to a stem; corona (trumpet) as long as, or longer than, the perianth segments (petals). This division has four sub-divisions based on their colour combinations.

Sub-Division IA: Height 35-45 cm and perianth and corona coloured but corona not paler than perianth such as 'Burgemeester Gouverneus' (golden-yellow), 'Dutch Master' (golden yellow with central trumpet dentate), 'Garron' (primrose-yellow), 'Goldcourt' (lemon-yellow, cup deeper), 'Golden Harvest' (golden-yellow), 'Irish Luck' (golden-yellow), 'King Alfred' (golden-yellow), 'Kingscourt' (lemon-yellow), 'Moonrise' (lemon-yellow), 'Moonstruck' (lemon-yellow), 'Mulatto' (sulphur-yellow, central trumpet white inside), 'Unsurpassable' (golden-yellow), *etc.*

Sub-Division IB: Height 30-48 cm, perianth white and corona coloured (bicolour trumpets) such as 'Celebrity' (cream-white, trumpet yellow), 'Foresight' (milky-white, trumpet yellow), 'Music Hall' (perianth white, trumpet yellow), 'Oklahoma' (white, trumpet dark golden-yellow), 'Preamble' (perianth white and smooth, corona bright yellow), 'Queen of Bicolours' (white, trumpet canary-yellow), 'Spring Glory' (white, trumpet yellow), 'Trousseau' (white, trumpet pale-yellow), 'World's Favourite' (white, trumpet pale-yellow), *etc.*

Sub-Division IC: Height 30-45 cm, perianth and corona white (sometimes yellow in bud) such as 'Angel's Wings' (large silver-white), 'Beersheba' (white, trumpet narrow), 'Broughshane' (cream but later white), 'Cantatrice' (white, trumpet narrow), 'Mount Hood' (perianth rounded and white, trumpet cream-white), 'Mrs. E.H. Krelage' (white, trumpet ivory-white), 'Vigil' (pure white), *etc.*

Sub-Division ID: Height 40-45 cm. Any colour combination nor covered above such as 'Mrs. R.O. Backhouse' (ivory-white, trumpet dark pink), 'Pink Isle' (white, trumpet pink), 'Rose Caprice' (white, trumpet pink, base suffused green), 'Salmon Trout' (best pink, perianth white, trumpet salmon-pink), 'Spellbinder' (yellow, trumpet turning white), 'Toscanini' (white but reverse apricot-pink, trumpet ivory-white), *etc.*

Division 2: Large-cupped Daffodil Cultivars

One flower to a stem; corona (cup) more than one-third but less than equal to the length of the perianth segments (petals).

Sub-Division II A: Height 35-55 cm. Perianth segments coloured in shades of yellow, orange and red and cup usually darker such as (i) yellow-cupped 'Carlton' (yellow, broad-frilled cup), 'Galway' (deep yellow), 'Golden Torch' (golden-yellow), 'St Keverne' (rich gold), *etc.*; and (ii) orange or red-cupped 'Armada' (cup orange),

Division of Narcissus

Div 1 Trumpet Daffodil

Div 2 Large-cupped Daffodil

Div 3 Small-cupped Daffodil

Div 4 Double Daffodil

Div 5 Triandrus Daffodil

Div 6 Cyclamineus Daffodil

Div 7 Jonquilla and Apodanthus Daffodil

Div 8 Tazetta Daffodil

Div 9 Poeticus Daffodil

Div 10 Bulbocodium Daffodil

Div 11 Split Corona Daffodil

Div 12 Miniature Daffodil

Cultivars of Narcissus

Kingscourt

Mulatto

Carlton

Golden Torch

Kissproof

Doubtful

La Riante

Bridal Crown

Double Campernelle

Liberty Bells

Cassata

Palmeres

Cultivars of Narcissus

February Gold

Golden Echo

Chromacolor

Congress

Pride of Cornwall

White Lion

Actaea

Petticoat

Angel's Tears

Van Sion

George Washington's

New Baby

'Aranjuez' (cup golden with bright red rim), 'Fortune' (orange to red cup) and 'Scarlet Elegance' (cup orange-red), *etc.*

Sub-Division IIB: Height 33-50 cm. Segments white and corona coloured in varying shades of yellow and orange such as (i) yellow-cupped 'Chinook' (perianth white, corona yellow), 'Coverack Perfection' (petals star shaped and cream-white, cup rich yellow), 'Duke of Windsor' (petals white, cup apricot-orange), 'Green Island' (segments cream-white, corona variable green-yellow), 'Hyperion' (tepals and cup yellow), 'Kissproof' (segments and cup yellow), 'Polindra' (petals cream-white, cup mimosa-yellow), 'Tudor Minstrel' (segments white and overlapping, cup orange-yellow and frilled), *etc.*; and (ii) orange or red-cupped 'Arbar' (tepals white, cup orange-red), 'Eddy Canzony' (tepals white, corona flat and bright orange but centre yellow), 'Fermoy' (petals white, corona orange-red), 'Flower Record' (tepals white, corona red), 'John Evelyn' (tepals white, corona orange and expanded), 'Kilworth' (tepals cream with yellow base, corona red), 'Roseworthy' (tepals white, corona pink), 'Salome' (tepals white, corona salmon); *etc.*

Sub-Division IIC: Height 35-45 cm. Segments white, corona white but not paler than the segments such as 'April Snow' (white), 'Ave' (white, segments broad and pointed, corona rounded), 'Castella' (corona lemon, ageing cream), 'Gypsy Moth' (creamy-white), 'Ice Follies' (white, corona wide, flattened, yellow fading to white), 'Tenedos' (cream, corona yellow turning cream-white), 'Truth' (rounded and slightly frilled), *etc.*

Sub-Division IID: Any other combination other than IIA, IIB or IIC such as 'Bethany', 'Binkie', 'Charter', 'Day-dream', 'Gleeful', 'Rushlight', *etc.*

Division 3: Small-cupped Daffodil Cultivars

One flower to a stem; corona (cup) not more Than one-third the length of the perianth Segments (petals).

Sub-Division IIIA: Height 43-45 cm. Segments and corona coloured such as 'Alcida' (yellow and orange), 'Ardour' (yellow, corona orange-scarlet), 'Chungking' (flowers circular, yellow, corona orange-red), 'Doubtful' (predominantly red), 'Edward Buxton' (yellow and orange), 'Irish Coffee' (yellow, corona brown-red), 'Perimeter' (red), 'Seraglio' (yellow, corona green-yellow with red rim), 'Varna' (lemon yellow, corona orange-red), *etc.*

Sub-Division IIIB: Height 35-45 cm. Perianth white and corona coloured such as 'Barret Browning' (tepals white, corona orange-red), 'Bishmills' (white, corona red), 'Blarney' (cream-white pointed, corona open-fluted and rich apricot), 'La Riante' (white, corona vivid orange), 'Merlin' (white, corona orange), 'Rockhall' (white, corona orange-red), 'Valhalla' (white, corona orange-red), 'Verger' (flowers white, base yellow, corona flat and orange-red), 'Yalta' (cream, corona orange-yellow and rim red, *etc.* Other show cultivars without red colouration are 'Aircastle', 'Green Hills', 'Lady Kesteven', 'Silken Sails', 'Syracuse', *etc.*

Sub-Division IIIC: Height 35-40 cm. Perianth white, corona short and white and not paler than perianth such as 'Angel' (white with a green and yellow eye), 'April Clouds', 'Chinese White', 'Cushendall', 'Polar Ice', 'Sacramento'Tern', 'Verona', *etc.*

Sub-Division IIID: Any colour combination not described above but such varieties are very rare such as 'Reversa'.

Division 4: Double Daffodil Cultivars

One or more flowers to a stem, with doubling of the perianth segments or the corona or both. The combination of doubled perianth and corona segments can vary widely between cultivars, and there may be predominantly one, rarely more flowers per stem, also varying by cultivar such as 'Bridal Crown' (tepals white, centre yellow), 'Camellia' (soft yellow, no visible trumpet), 'Cheerfulness' (1-3 flower on a stem, outer segments cream-white, cup orange-rimmed and filled with more cream petals), 'Dick Wilden' (yellow), 'Golden Ducat' (deep yellow), 'Flower Dream' (yellow), 'Indian Chief' (sulphur-yellow, striped orange), 'Inglescombe' (primrose-yellow), 'Irene Copeland' (white and yellow), 'Mary Copeland' (cream-white, some central petals lemon and orange), 'Tahiti' (yellow-orange), 'Telemonius Plenus' syn. 'Van Sion' (golden-yellow, with or without a complete trumpet), 'Texas' (yellow and orange-red), 'Twink' (pale-primrose with orange centre), 'Valencia' (yellow and orange), 'White Cheerfulness' (blooms fragrant creamy-white, 3-4 blooms per stem), 'Yellow Cheerfulness' (3-4 yellow fragrant blooms per stem), *etc.* Other show cultivars are 'Acropolis', 'Double Event', 'Gay Challenger', 'Gay Time', 'Hawaii', 'White Lion', *etc.*

Division 5: Triandrus Daffodil Cultivars

Characteristics of *N. triandrus* predominant: usually two or more pendent flowers to a stem; perianth segments reflexed.

Sub-Division VA: Corona not less than two-thirds the length of the perianth segments such as 'Liberty Bells' (creamy-yellow), 'Rippling Waters' (2-3 white flowers on a stem), 'Thalia' (pale-cream, corona white), 'Tresamble' (cream-white), *etc.* One of its varieties 'Arish Mell' is very showy cultivar.

Sub-Division VB: Corona less than or equal to two-third the length of the perianth segments such as 'April Tears' (deep yellow), 'Hawera', 'Silver Chimes', 'Tincleton', 'Tuesday's Child', *etc.*

Division 6: Cyclamineus Daffodil Cultivars

Characteristics of *N. cyclamineus* clearly evident: one flower to a stem; perianth segments significantly reflexed; flower at an acute angle to the stem, with very short pedicel ("neck").

Sub-Division VIA: Corona not less than two-thirds the length of the segments such as 'Baby Doll' (yellow with little deeper corona), 'Bartley' (golden reflexed flowers), 'Charity May' (yellow, corona little deeper), 'Dove Wings' (tepals white, corona yellow), 'February Gold' (yellow with slightly deeper corona), 'February Silver' (tepals white, corona yellow), 'Jack Snipe' (cream-white and a bit reflexed segments, corona orange-yellow with pale margin), 'Jumblie' (back-reflexed yellow tepals, corona turmeric-yellow), 'March Sunshine' (yellow), 'Peeping Tom' (shining yellow), *etc.* Other show cultivars are 'Jenny', 'Tête à Tête' (yellow), 'Woodcock', *etc.*

Sub-Division VIB: Corona less than two-third the length of tepals such as 'Beryl' (cream and orange), 'Kitten', 'Little Imp', 'Lofty', 'Roger', *etc.*

Division 7: Jonquilla and Apodanthus Daffodil Cultivars

Characteristics of Sections Jonquilla or Apodanthi clearly evident: one to five flowers to a stem; perianth segments spreading or reflexed flowers usually fragrant.

Sub-Division VIIA: Corona not less than two-thirds the length of the segments such as 'Desert Song' (golden), 'Golden Guinea' (golden), 'Golden Perfection' (golden), 'Golden Sceptre' (golden yellow, corona long), 'Shah' (golden), 'Sweetness' (canary yellow, corona small), *etc.*

Sub-Division VIIB: Corona less than two-thirds the length of the perianth segments such as 'Chérry' (pale-cream, corona dull yellow turning pink), 'Finch' (yellow), 'Lanarth' (yellow), 'Lintie' (yellow, tepals rounded, corona orange-edged and small), 'Orange Queen' (orangish-yellow), 'Pipit' (deep yellow), 'Stratosphere' (orange, corona darker), 'Suzy' (segments yellow, corona dep orange), 'Trevithian' (buttercup yellow), *etc.* Other outstanding show cultivars are 'Nancegollan', 'Sweet Pepper', 'Tittle Tattle', *etc.*

Division 8: Tazetta Daffodil Cultivars

Characteristics of the *N. tazetta* group predominant: usually three to twenty flowers to a stout stem; leaves broad; Perianth segments spreading, not reflexed; flowers fragrant. True tazettas though do not survive outdoors but are forced in bowls for winter flowering such as 'Avalanche' (segments white flushed yellowish, corona small and lemon-yellow), 'Grand Primo Citronière' (milky-white, corona lemon-yellow, strongly sweet scented), 'Minnow' (2-4 small creamy-yellow flowers per stem, corona lemon, and only 15 cm high), 'Paper White' (whiter, strongly scented), 'Pride of Cornwall' (white, corona small and deep orange), 'Scarlet Gem' (dull yellow, corona small and orange), 'Silver Chimes' (beautiful light yellow, corona little darker), *etc.* Poetaz which are hardy, bear fewer and larger flowers per stem but flower later than true tazettas and the varieties in this group are 'Cragford' (white, corona red), 'Geranium' (pale-cream, corona orange), 'Grand Soleil d'Or' (yellow, corona darker, strongly scented), 'Martha Washington', 'Soleil d'Or' (golden-yellow, deliciously and strongly scented), 'St. Agnes', *etc.*

Division 9: Poeticus Daffodil Cultivars

Characteristics of the *N. poeticus* group and related species clearly evident; Perianth segments pure white; corona very short or disc shaped, not more than one fifth the length of the perianth segments; corona usually with a green and or/yellow centre and red rim, but sometimes wholly or partly of other colours; anthers usually set at two distinct levels; flowers fragrant. Varieties are 'Actaea' (tepals glistening white, corona very small, yellow and red-edged), 'Cantabile' (white, corona green and edged red), 'Sarchedon' (tepals white, base yellow, corona yellow but edged red), 'Sonata' (white, corona flat, green-yellow and red-edged), *etc.* Other show cultivars are 'Milan', 'Sea Green', *etc.*

Division 10: Bulbocodium Daffodil Cultivars

Characteristics of Section Bulbocodium clearly evident; usually one flower to a stem; perianth segments insignificant compared with corona; filament and style are usually curved. *N. b. obesus* is a deep yellow form bearing thick leaves and some 3.75 cm long flowers which appear in March and April and the corona is large and deep yellow. *N. b. romieuxii* is a pale-yellow form where flowers appear from December to February. *N. b. vulgaris conspicuus* is deep yellow and *N. b. citrinus* is lemon-yellow, both flowering in March and April. *N. b. monophyllus* (*syn. N. cantabricus*) from Spain, France and NW Africa grows up to 10 cm in height bearing thread-like leaves, 2-3 transparent-white and funnel-shaped flowers per stem, with widely expanded corona and very prominent stamens, flowering in December to January and is suitable for pot-growing.

Division 11: Split Corona Daffodil Cultivars

Corona split-usually for more than half its length; In some cases, the segments may be broad enough to underlap and overlap alternating perianth segments such as 'Cassata', 'Golden Orchid', 'Joli Coeur', 'Lemon Beauty' and 'Parisienne'. though flowers are most often borne one to a stem, there are cultivars with multiple flowers per stem. Division split into a and b types, Collar and papillon.

a) *Collar Daffodils:* daffodils with the corona segments opposite the perianth segments; corona segments usually in two whorls of three such as 'Ahoy', 'Evolution' and 'Gold Collar'.

b) *Papillon Daffodils:* Split-corona daffodils with the corona segments alternate to the perianth segments; the corona segments usually in a single whorl of six such as 'Donna Bella' and 'Elfhorn'.

Division 12: Other Daffodil Cultivars

Daffodil cultivars which do not fit the definition of any other division. Miniature daffodils are not an official ADS Division but can occur in each of the other divisions and possess the same descriptive characteristics. However, the flowers are 3.8 cm or less in diameter, and are borne on proportionately smaller plants.

Division 13: Daffodils Distinguished Solely by Botanical Name

All species and wild or reputedly wild variants and hybrids. The genus *Narcissus* is complex taxonomically having about 50 species and three sub-genera, *i.e.*, Ajax, Corbularia and Narcissus. All the narcissi described up to division IX are of garden origin, but all those narcissi and their naturally wild occurring forms (ecotypes and natural hybrids) are described under this division such as *Narcissus asturiensis* (*N. minimus*), *N. bulbocodium*, *N. b. subsp. bulbocodium*, *N. b. cantabricus*, *N. b. conspicuus*, *N. b. var. citrinus*, *N. b. monophyllus* (*N. cantabricus monophyllus*), *N. b. nivalis*, *N. b. obesus*, *N. b. romieuxii* (*N. romieuxii*), *N. b. tenuifolius*, *N. campernellii* (*N. canaliculatus*, *N.odorus*, *N. tazetta*), *N. cantabricus* (*N. bulbocodium cantabricus*, *N. clusii*), *N. cantabricus foliosus*, *N. cyclamineus*, *N. elegans* (*N. autumnalis*), *N. jonquilla*, *N. juncifolius* (*N. assoanus*), *N. j. rupicola* (*N. rupicola*), *N. lobularis* (*N. obvallaris*), *N. minor* (*N. nanus*, *N. pumilus*), *N. moschatus*, *N. odorus regulosus*, *N. poeticus*, *N. p.*

verbenensis, N. pseudonarcissus, N. pseudonarcissus gayi (N. gayi), N. p. moschatus, N. p. obvallaris, N. poeticus var. recurvus, N. romieuxii, N. scaberulus, N. tazetta canaliculatus, N. t. papyraceus, N. triandrus, N. t. albus, N. t. concolor, N. viridiflorus, N. watieri, etc. Most popular ones are being described here with (Meyer, 1966).

7.0 Genetics and Breeding

7.1.0 Genetics

7.1.1 Cytology

Many reports are available on the cytology of the genus *Narcissus* (Philp 1934; Darlington and Wylie, 1955). This genus is especially important because of the variability it presents in its species and species complex. Not only it is constituted of diploids, aneuploids and polyploids but even supernumerary B chromosomes have been claimed to exist in certain populations of some species (Wylie, 1952). The species of this genus have been classified by Wylie (1952) under three groups, *viz.*, sexually reproducing, clonal hybrids and fertile hybrids. Sharma and Sharma (1960) studied different varieties of *Narcissus tazetta,* described in the text, show the presence of different numbers of chromosomes, such as 14, 24, 28, 30 and 32 in their somatic complements. Of these, only the variety Chastity, having thirty chromosomes in the diploid set, may belong to the same line as the series with n=10 chromosomes, reported previously. Fourteen and twenty-eight chromosomes in the somatic sets of varieties 'Ponterius' and King daffodil, respectively are unusual for this species in the sense that, the absence of this number in *N. tazetta,* according to Wylie (1952) distinguishes it from the other species of this genus (Sharma and Sharma, 1960). A significant feature observed in all the varieties of *N. tazetta* studied by Sharma and Sharma (1960) was the occurrence of chromosomal abnormalities in the somatic cells. Such irregularities include both structural and numerical alterations of chromosomes. This behavior is specially characteristic of those species which propagate through vegetative means. It may be noted that though *N. tazetta* is sexually fertile, yet all the varieties in their collection propagate at the same time profusely through vegetative means.

7.1.2 Genetic Diversity

There are thousands of cultivars of Narcissus registered during the past century. Among them, many are natural hybrids or natural interspecific hybrids. Many thousands of named cultivars of garden origin are known (Kington, 2002), suggesting that some of the discriminating characters could be very plastic and their variability may be influenced by environmental and/or developmental factors (Jiménez *et al.,* 2009). After a long history of cultivation, complex breeding programs of daffodils over the past 150 years have resulted in more than 30,000 registered cultivars (Kington, 2014), showing rich genetic diversity in this genus. One of the taxonomically most complex groups is *Narcissus* sect. *Pseudonarcissi* (the trumpet daffodils), which has its centre of diversity in the Western Mediterranean basin (Iberian Peninsula and southern France). This section includes up to 30 variable species, though the exact number differs according to various authors. Many of them are narrowly-distributed endemics that are endangered and composed of only a few populations (Jiménez *et al.*, 2009). As the members of the genus are so popular in cultivation and have been hybridized over such a long period, a complex range of hybrid forms occur with unknown state of the parents' genetic characteristics (Li *et al.*, 2017). It is now difficult or impossible to elucidate the exact parents for most existing hybrids (Marques *et al.*, 2010; Ribeiro *et al.*, 2007). Therefore, artificial hybridization with distinct breeding objective on *N. pseudonarcissus* is still relatively backward development for the following reasons (Li *et al.*, 2017).

1. The chromosome number is usually 2n=14 or 28 (Philp, 1934) for the species *N. pseudonarcissus*, but there is considerable variation throughout the genus, and aneuploidy and polyploidy are quite common. In addition to some diploids, most types face the certain problem of fertility decrease in various degrees or even sterility (Brandham, 1992).
2. *N. pseudonarcissus*, as a perennial bulbous geophyte, mainly reproduces by developing vegetative daughter bulblets. The seedling plant is usually very small during the first 4 years and only has vegetative growth and first flowers in the fifth or sixth season of growth, if survived from insect visitation every year (Caldwell and Wallace, 1955). This makes breeding quite inefficient, expensive, and time-consuming.
3. Breeding is also being impeded because of the synchronous nature of flowering in *N. pseudonarcissus*. To conduct successful crosses, it is necessary to select genotypes on their individual values and to plant them in a way that will enable flowering at almost the same time.
4. The cultivars chosen as parents are always not homozygous in most genetic loci. Genetic segregation will be comprehensive in the unbeknown genetic background and molecular breeding still has a far way to go. Moreover, low fertility and disease also affect the obtaining of artificial hybrids. All those mentioned previously hindered the efficiency of crosses to obtain new *N. pseudonarcissus* cultivars with desirable traits. Although hybridization, polyploidization, and adaptive radiation occur within the genus along with evolutionary processes, there were no artificial genetic populations with clear genetic background, and no F_1 hybrids generation has been reported (Li *et al.*, 2017).

7.2.0 Breeding

Despite the importance of *Narcissus* to ornamental horticulture, there have been no population genetic studies of wild species, many of which have narrow distributions. Barrett *et al.* (2004) measured selfing rates and levels of genetic diversity at allozyme loci in six populations of *Narcissus longispathus*, a self-compatible daffodil endemic to a few mountain ranges in south eastern Spain. The populations were distributed among four distinct river valleys encompassing two main watersheds in the Sierra de Cazorla Mountains. Selfing rates averaged 0.37 (range 0.23-0.46), resulting in significant inbreeding coefficients for the progeny ($f \leq 0.324$). In contrast, estimates of inbreeding in parental genotypes were not significantly different from zero ($f \leq 0.001$), indicating that few selfed offspring survive to maturity because of inbreeding depression. Marques *et al.* (2007) evaluated the influence of the breeding system in the reproductive success of *Narcissus cavanillesii* (*Amaryllidaceae*), a species listed under Annexes II and IV of European Community Habitat and Species Directive. This taxon, which occurs in only two populations in Portugal, is of conservation interest in this country due to its low area of occupancy and population size, and fragmentation. Hand-pollinations were carried out to evaluate the performance of self and cross-pollination offspring. Results showed that xenogamous pollinations produce more fruits and seeds than autogamous ones. However, no significant differences were detected neither in seed weight nor in seed germination percentage between treatments.

Narcissus longispathus (Amaryllidaceae) is a bee-pollinated, perennial herb endemic to a few mountain ranges in south eastern Spain (Moreno Saiz and Sainz Ollero, 1992). It is related to the widely distributed, highly polymorphic, self-incompatible *N. pseudonarcissus*; the most commonly cultivated trumpet daffodil of horticulture (Barrett *et al.*, 2004). In a study of the floral biology of *N. longispathus* in the Sierra de Cazorla mountains, Herrera (1995), as quoted by Barrett *et al.* (2004), demonstrated that *N. longispathus* is self-compatible producing equivalent amounts of seed following experimental self and cross-pollinations. This contrasts with other species in the genus most of which are moderately to strongly self-sterile (reviewed in Barrett *et al.*, 1996, 2004; Sage *et al.*, 1999).

Natural hybrids between rare and common sympatric species are commonly eradicated to avoid the potential extinction of the rare species, although there is currently no clear predictive framework to quantify this risk (Marques *et al.*, 2007). As hybrids can have intrinsic value as new evolutionary pathways, further knowledge on the factors controlling hybridization is needed. They evaluated the role of pollination patterns in hybridization events in two sympatric populations of *Narcissus cavanillesii* and *N. serotinus* in Portugal. *Narcissus cavanillesii* is a rare species, while *N. serotinus* is widely distributed across the Mediterranean. The hybrid, *N. perezlarae*, is quite frequent in southeastern Spain but is scarce in Portugal. Reciprocal manual crossings confirmed compatibility between the two species, although hybridization was more successful when *N. cavanillesii* participated as female.

7.2.1 Resistance Breeding

Basal Rot

Basal rot of *Narcissus* caused by *Fusarium oxysporum* f. sp. *narcissi* frequently results in major losses in UK commercial bulb crops. Until recently, resistance breeding programmes have used as parents only tetraploid varieties, none of which have complete resistance to basal rot. The diploid species *N. jonquilla*, which has been reported to be completely resistant, was used as the pollen parent in crosses with two varieties ('Golden Harvest' and 'St. Keverne') that display opposite extremes of 'field resistance' and the seedling progenies were clonally propagated (Carder and Grant, 2002). The hybrids were screened for resistance by planting bulbs in compost infested with chlamydospores of the pathogen and recording bulb survival one year later. Progeny from self-pollinated *N. jonquilla* were also propagated and tested to study the inheritance of resistance in an F_1 generation from the diploid species. All of the *N. jonquilla* 'selfs' displayed similar very high survival rates indicating homozygous resistance in this species. A continuous distribution of survival rates from 0 to 100 per cent was observed in both the hybrid progenies but more lines with 100 per cent survival were found when 'St. Keverne', a variety with high field resistance, was the maternal parent than when 'Golden Harvest', a more susceptible variety was used. A detached-scale puncture-inoculation technique has been developed as a rapid *in vitro* assay for basal rot resistance in *Narcissus*. Varying amounts of rot were recorded for eight varieties and five species that correlated well with their known resistance to basal rot (Carder and Grant, 2002).

8.0 Growth and Flowering

The low-temperature flowering-response pathway, used as an inductive stimulus to induce flowering in plant species from temperate regions in response to cold temperature, has been extensively studied. However, limited information is available on the flower transition of several bulbous species, which require high temperature for flower differentiation. *Narcissus tazetta var. chinensis* (Chinese narcissus) exhibits a 2-year juvenile phase, and flower initiation within its bulbs occurs during summer dormancy. The genetic factors that control flower initiation are mostly unknown in Chinese narcissus (Li *et al.*, 2013). A high storage temperature is necessary for flower initiation. Flower initiation was advanced in bulbs previously exposed to extended high temperature. The heat accumulation required for flower transition was also determined. High temperature treatment rescued the low flower percentage resulting from short storage duration under natural conditions. In addition, extended high storage temperature was found to increase the flowering percentage of 2-year-old plants, which can be applied in breeding. Narcissus FLOWERING LOCUS T1 (NFT1), a homolog of the *Arabidopsis thaliana* gene FLOWERING LOCUS T, was isolated. NFT1 transcripts were abundant during flower initiation in mature bulbs and were up-regulated by high temperature. The genetic experiments, coupled with an expression profiling assay, suggest that NFT1 possibly takes part in flower transition control in response to high temperature (Li *et al.*, 2013).

Floral display (the size, number, and arrangement of open flowers) influences pollinator visitation to animal pollinated plants and should be an important determinant of reproductive success. Worley *et al.* (2000) examined variation in the size and number of open flowers in wild daffodils (*Narcissus*). Analysis of published data on 45 taxa showed that flower number varied negatively with flower diameter among *Narcissus* species, which supports the widespread assumption that there is a trade-off between these traits. In contrast, field measurements indicated a positive relation between flower number and diameter within two populations of *Narcissus dubius*, and no relation was evident due to variation in bulb size.

9.0 Propagation

9.1.0 Seed Propagation

Seed propagation requires that pollinated and fertilized flowers be left on the plant so the seed capsules can ripen. Collected seed can be sown in the fall in a prepared seedbed for spring germination. Seedlings require several years to reach flowering size, also they show variation in many characters as they are not homozygous.

9.2.0 Vegetative Propagation

9.2.1 Bulbs

Narcissi, like other bulbous plants, may be increased by two methods: division and seed. Bulbs multiply underground naturally by division. Division occurs as the 'mother' bulb matures, thus creating a number of new 'daughters' bulbs. Bulbs resulting from division always produce flowers identical to those of the 'mother' plant (Dana and Lemur, 2016). A bulb consists of successive generations of bulb units, each unit comprising bulb scales, leaf bases and flower (in larger bulbs). The vegetative buds develop in the axils of the fleshy leaves. The apical meristems in the leaf axils of mature bulbs in each season produce 2, 3 or 4 scale leaves, 2-3 leaves and then a flower bud. These new bulbs enlarge and cause gradual disintegration of the previously intact outer bulb scales. With the growing of the buds in the axils of the fleshy leaves (Rees, 1972), the slabs of small and flattened bulbs are formed so when such bulbs are planted after separation develop into round, double-nosed and triple-nosed and mother bulb stages, the latter stage requiring about three years to develop from a slab. The life span of a bulb unit extends over about four years. Therefore, the mother bulbs have a number of noses and offsets (bulbils), and are usually replanted after removing any loose offset. These provide a high proportion of rounds and many large bulb offsets. These are only rounds which are used for flowering and early forcing if they are large enough though second and third grade rounds are used for later forcing. Top size bulbs are known as 'wares' and these develop only when every year lifting and planting are effected. Double- and triple nosed bulbs do not have proper-sized rounds, hence are used as planting stock for second or third year forcing after these once or twice have been grown after separation. Mother bulbs on replanting may produce large rounds, small rounds, double or triple-nosed bulbs and many offsets. These offsets develop directly in rounds within shorter duration (1 year) as compared to the flat bulbs (2 years). Bulb production in a 2-year period is roughly 2.5 times (Misra *et al.*, 2013).

Scooping is a process where the whole basal plate (root disc) is carefully sliced out (not too deep into the scales) through a specially designed curved-bladed knife in a manner that neither the potential portion of the scale bases are removed nor any part of root disc is left after scoring as both the cases are harmful. The main aim is to kill the apically dominant bud and to provide full access to scale bases to form new bulblets after callusing (Rees, 1972). Scooping process should be initiated some 15-17 weeks before final planting. The scooped bulbs are immediately dipped in clean hydrated lime or in hypochlorite solution and then placed in propagating chamber at 21°C, and this temperature may be raised up to 30°C for two weeks but after about 15-20 days of scooping. When newly formed bulblets on the scale bases of the planted bulbs have started forming roots, these are planted out. Through scooping, one can obtain some 30 bulblets per parent bulb, and such bulblets will take up to five years to attain flowering size.

Chipping (cutting or cross-cutting) is the conical incision on the basal plate of bulbs measuring 10-18 cm circumference, to kill the shoot for breaking apical dominance so that numerous daughter bulbs along the cut edges of the scales are formed. Chipping may be effected into halves, quarters, 1/8 and 1/16 and incubated in plastic bags and planted in a 50:50 grit plus peat medium which was reported better than twin-scaling and quicker method for obtaining bulbs without wasting any bulb material.

Twin-scaling is a further exploitation of chipping, where many (8-10) such incision through apical portion to basal plate is made in 12-14 cm grade bulb, and then attached twin-scales are gently removed, disinfected in Thiram or Captafol, planted in moist sand or vermiculite and incubated at 15-23°C for 8-10 weeks in darkness, these regenerate into daughter bulbs with better bulbil formation in the outer scales as compared to inner ones though bulbil size is better in the inner scales (Sanyat Misra and Misra, 2013). The effects on yield of propagating narcissus cv. 'Fortune' bulbs of different size grades by chipping and twin scaling were studied in field and frost-protected greenhouse experiments. After the first year, the average number of bulbs per propagule cut from offset-derived bulbs varied between 0.51 and 1.50, with propagule size being a major determining factor. Total bulb weight yield, derived from a single initial bulb was dependent on sectioning treatment. Up to an initial bulb weight of 25 g, uncut bulbs gave the highest yield, but sectioning above that by chipping gave higher total yields. There was a linear relationship between bulb weight per chip and initial chip weight; allowing target bulb weights to be achieved after the first year by adjusting the cutting rate relative to initial bulb grade (Fenlon, 1990).

9.2.2 Micropropagation

A workable protocol for micropropagation of narcissus is in Langens-Gerrits and De Klerk (1999). One of the earliest attempts to multiply Narcissus by tissue culture was reported by Seabrook *et al.* (1976). A new technique is reported here for the rapid propagation of *Narcissus* cultivars involving the induction of shoot and root apices from small pieces of leaf base tissue, inverted scape sections, and from ovaries, in axenic culture (Seabrook *et al.*, 1976). Shoot and root apices were obtained on the above explants of all nine *Narcissus* cultivars tested. The optimal medium contained modified Murashige and Skoog (1962) inorganic salts, the organic constituents recommended by Ziv *et al.* (1970), and higher than normal growth regulator levels. Proliferating shoot apices were induced on a medium containing 4.4×10^{-5} *M* (10 mg/l) 6-benzylaminopurine (BAP) and 5.3×10^{-6} *M* (1 mg/l) naphthaleneacetic acid (NAA). Plantlets grew optimally if transferred to a medium containing 9.0×10^{-6} *M* (2 mg/l) BAP and 2.0×10^{-7} *M* (2 mg/l) NAA. The induction of roots on plantlets grown *in vitro* required a half-strength salt and sucrose solution without growth regulators. In 5 months, 2620 plantlets were produced from two leaf base explants. This technique is considerably more efficient than any known conventional methods of propagating *Narcissus*. Root apices were optimally produced on a medium containing 2.0×10^{-6} (0.5 mg/l) to 2.0×10^{-5} *M* (4 mg/l) BAP and 1.0×10^{-4} (20 mg/l) to 2.0×10^{-5} (4 mg/l) NAA. Callus was induced on leaf base explants, inverted scape sections, and on ovaries; the ovary explants required the

Propagation of Narcissus

Mother Bulb

Seedlings

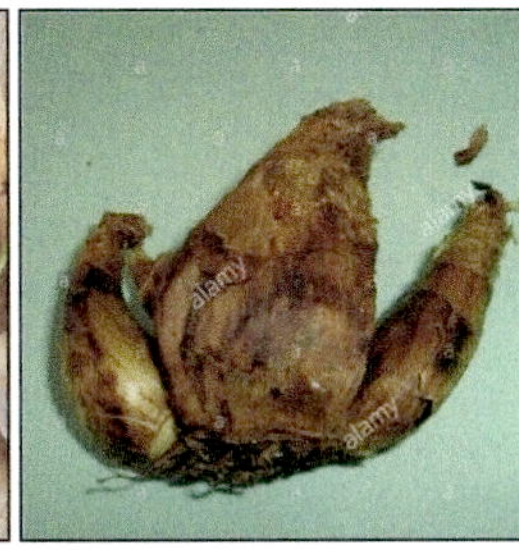

Offsets

Twin Scaling

Seedlings Arise from Bulb in Pots

Bulb Chipping

Micropropagation

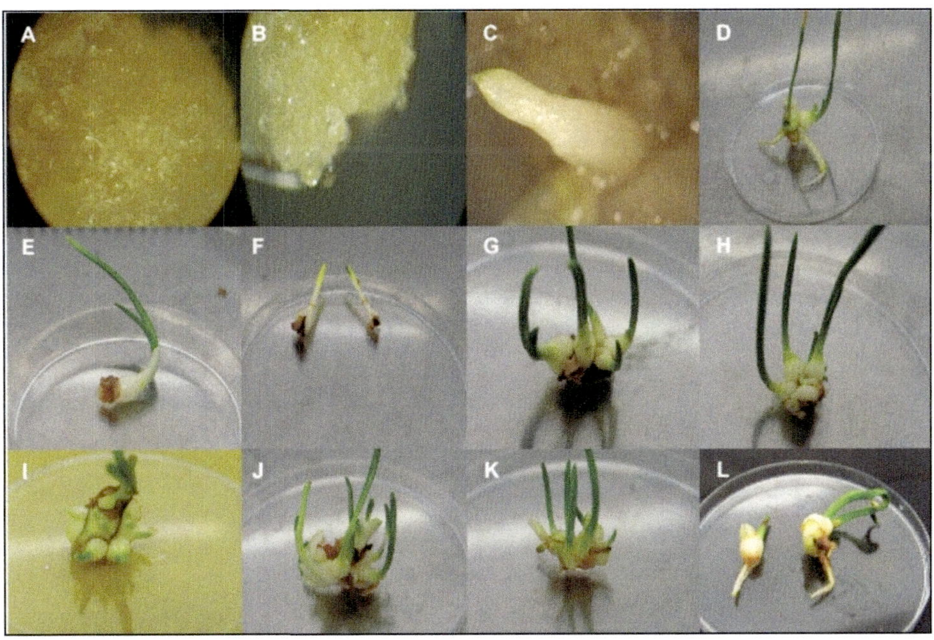

Hosseini, R., Moradnejad, M., Nezami-Alanagh, E., Ashrafi, S., Golmohammadi, F. G. (2013). Somatic embryogenesis and bulblet production in Narcissus papyraceus cv. Shirazi: effect of plant growth regulators, light intensity, sucrose concentration, methyl jasmonate and anti-gibberellins. *Iranian journal of genetics and plant breeding*, 2:1

highest levels of auxin for the induction of callus and were most responsive. Thus, the levels of growth regulators required to induce a response *in vitro* are higher than so far reported for plants that have been propagated by tissue culture. Although the levels of various growth regulators used were found to be important in obtaining apices, the relative ratio of cytokinin to auxin was also found to be critical. Although considerable differences in clonal responsiveness were noted, cultures were obtained from all nine *Narcissus* cultivars tested. The induction of adventitious meristems of leaf base explants is a particularly promising method for the propagation of virus-free material and for the rapid propagation of valuable horticultural material (Seabrook *et al.*, 1976).

Adventitious shoots were induced on leaf, scale and stem explants taken from the basal plate region of flowering-size bulbs on media containing 2-16 mg/l 6-benzylaminopurine and 0·25-4·0 mg/l 1-naphthal-ene acetic acid (NAA). Only the levels of NAA had a significant effect on the numbers of shoots produced. When trimmed and split in half, 6 mm or more diameter *in vitro* shoots regenerated further adventitious shoots which in turn grew in size suitable for splitting within 16 weeks (Hussey, 1982).

9.2.3. *In vitro* Bulblet Formation

The effect of photoperiod (16/8 h light/dark and 24 h dark) and various concentrations of NAA, BAP, and GA_3 on the *in vitro* mass bulblet regeneration of *N. tazetta* was studied (Khonakdari *et al.*, 2020). The GAL production ability of the regenerated bulblets was assessed by HPLC-UV-MS. Light treatments significantly affected the number of bulblet and leaf, the ratio of bulblet/leaf, and leaf length. The maximum number of bulblet (31.0±1.58) and leaf (13.3±1.33) was recorded from the cultures fortified with NAA and BAP (2 mg/l) kept in 16/8 h light/dark, while the maximum leaf length (2.1±0.92 cm) was measured on the MS medium containing 0.5 mg/l NAA and 2 mg/l BAP incubated in the same photoperiod. The average ratio of bulblet proliferation per explant was significantly different between studied photoperiod (1.1±0.86) and 24 h dark (0.62±0.31). The regenerated bulblets contained 40 and 20 µg/g DW GAL underexposed photoperiod and 24 h dark, respectively. This information could be useful in the commercial production of GAL as a valuable anti-AD compound through *in vitro* mass bulblet proliferation of *N. tazetta* (Khonakdari *et al.*, 2020).

10.0 Soil and Climate

10.1.0 Soil

According to Dana and Lerner (2016), narcissi grow adequately in many soil types. They recommend a well-drained, deep, fertile soil containing abundant organic matter. If natural drainage is poor, break the sub-soil with a pickaxe. If the site is low, create a raised bed several inches above the surrounding soil level. Incorporate large amounts of coarse organic matter (5 cm or more) to loosen very heavy clay soil. Good sources of organic matter are well-rotten manure, undecomposed peat moss, or compost. Narcissi, like most bulbous plants, thrive in slightly acid soils (pH 6.0-6.5). Species will, of course, favour a soil type close to that of their habitat, *e.g.*, acid to neutral soils having a pH range of 5.5-7.0 for most forms of *Narcissus bulbocodium*, *N. triandrus*, *N. asturiensis* and *N. cyclamineus*, and the slightly alkaline pH, *i.e.*, 7-8 for *N. jonquilla* and *N. tazetta*.

10.2.0 Climate

10.2.1 Temperature

A high summer temperature signal is often used to advance narcissus flowering. Chinese narcissus (*Narcissus tazetta* var. *chinensis*) bulbs sprout in October to November (when soil temperature drops), grow throughout the winter and flower in January to February. The above-ground parts of the plants begin to senesce in late spring. Chinese narcissus exhibits a 2-year juvenile phase. Florogenesis is initiated within large-sized bulbs during summer dormancy. Timing of flower initiation in its dormant bulbs varies, depending on where they were cultivated (Li *et al.*, 2013).

Hanks (1996) reviewed different meteorological factors which are likely to affect the growth and development of narcissus. The important factors *viz.* temperature and water availability are considered the most important.

Bock *et al.* (2015) recorded the first flowering dates of 26 Narcissus cultivars over a 27-year period on the island of Guernsey, within the English Channel. They analysed flowering trends over time and relationships with climate variables. The study revealed that earlier flowering cultivars advanced most and were more variable than later flowering *Narcissus*. A strong relationship between flowering and climate variables, with temperature appearing to be the main driver was reported.

10.2.2 Light

More than 6-hour sunlight is suitable for quality flower production in majority of the cases (Miles, 1963). Planting in partial shade ensure a longer flowering period. Flower regulation in narcissus is not dependent on photoperiod because flowers can be initiated at any time of the year (Rees, 1972) though for harvesting long-stemmed flowers, long days are essential (Stevens, 1977). However, the overall performance with regard to bulb and flower production is certainly better under full sunlight.

11.0 Cultivation

11.1.0 Planting

October is generally best for bulb planting. This enables the bulbs to develop strong roots in the fall, which supports flowering in the spring. The proper depth to plant narcissus bulbs is governed by the size of the bulbs and the soil type. As a general rule, plant bulbs twice as deep as their greatest diameter, in medium to heavy soils and about three times their depth in sandy soils (Dana and Lemer, 2016).

11.2.0 Spacing

Spacing of bulbs will depend upon the size and type of bulb. Minimum spacing will vary from 8 cm for small bulbs to 12-16 cm for large bulbs. Planting too close resulted in poor development of foliage and flowers. The closer the bulbs are planted together, the more massive the color display, but it will be necessary to dig the

Cultivation of Narcissus

Bulbs

Planting of Bulbs

Sprouting of Bulbs

In Greenhouse

Open Cultivation

bulbs and divide them sooner to keep up flowering. Narcissi do multiply in time, so wider spacing will eventually result in a dense stand of plants.

There is little need for cultivating other than to control weeds. In fact, deep cultivation can be harmful. A 5 cm winter mulch of straw, peat moss, leaves, or similar materials is beneficial the first year after planting. It is particularly useful for bulbs planted late in the season or for bulbs planted in southern exposures, where the soil warms early and the growing tips may be damaged. The mulch provides insulation as well as prevents heaving during periods of freezing and thawing (Dana and Lemer, 2016)

11.3.0 Manuring and Fertilization

Dana and Lemer (2016) recommend the use of commonly available bulb 'food' or 'bulb booster' fertilizers are useful when preparing soil for narcissus planting. These products are relatively low in nitrogen, but contain enough to support healthy plant growth.

Bone meal is a traditional fertilizer for bulbous plants that releases phosphorus slowly, but it is usually too low in nitrogen to be used alone. An additional source of nitrogen must be used together with bone meal. Wood ash is an alternative way to supply potassium. For average situations, apply 1.5 2.0 kg of bone meal and 2.5- 3 kg of wood ash to each 100 square feet of soil (Dana and Lemer, 2016). The fertilizers should be thoroughly with the soil before planting. The nutrients need to be located in the soil in the vicinity of and just below the bulb for best uptake by the plant. Generally, a yearly application of fertilizer should be made immediately after flowering (Dana and Lemer, 2016). Application of nitrogen in two splits at 1st week of March and April significantly improved the growth characters and registered 9.88 and 21.56 per cent higher bulb yield than single split of nitrogen at 1st week of March, respectively (Khan *et al.*, 2013).

11.4.0 Pot Culture

Most of the smaller species, *e.g.*, *N. assoanus, N. cantabricus, N. romieuxii, N. rupicola* and *N. watieri* are best grown in pots in an unheated glasshouse or bulb frame where their miniature beauty can be appreciated and there is more control of their growing conditions. *N. serotinus* needs a real baking to induce them to flower. *N. bulbocodium* variants are also successful in pots but they can be tried in the open ground on a rock garden or in fairly fine grasses not forming tufts. These are usually seen on sloppy sandy acidic soils. Some clustered species are less frost-hardy and require a warm dry summer for flowering such as *N. papyraceus, N. tazetta* and their relatives. In cold areas these will certainly require to be planted against a warm sunny wall. The selections and hybrids of these types are used for early forcing indoors. The varieties which are best for pot growing are miniatures such as 'February Gold', 'Jetfire' and 'Tete a Tete'. When growing in pots, it should be ensured that there is plenty of crocks in the bottom of the pot for good drainage. Bulbs can of course be planted touching each other if required. Daffodils are well suited to pot cultivation, either for exhibition or for any other decorative purpose. Bulbs and medium in and around bulbs should be pressed firmly. Watering should be done with Watering Can and soil should be kept moist but not too wet. If plastic pots are used, care should be taken that pots do not get overheated due to direct sunlight. In no case the medium should be allowed to dry out at any time from planting to harvesting. Two parts high fertility soil-based medium mixed with one part coarse grit or fine chippings is suitable for most species. Small bulbs can be planted 7 cm deep in the 12-cm pots (Misra *et al.*, 2013).

12.0 Diseases, Pests and Disorders

The following are the serious pests and diseases described by The Daffodil Society (https://thedaffodilsociety.com/a-guide-to-dafodils/pests-diseases/).

12.1.0 Diseases

12.1.1 Fungal Diseases

12.1.1.1 Basal Rot (Fusarium oxysporum f. sp. narcissi)

This is the most serious daffodil bulb disease which although was described a hundred years ago as a serious threat, is growing in intensity as a result of hotter summers. The initial symptoms is the premature dying back of the foliage which when investigated, or at lifting time, reveals a soft or mummified bulb containing a chocolate brown rot spreading upwards from the base plate. At this point recovery of the bulb usually becomes impossible. The bulbs can become infected in storage or after planting and spores become widespread and are viable for over ten years in the soil. Control is very difficult even commercially. Resistant cultivars such as St. Kevern are grown, but even these are not immune. The avoidance of fresh manure

Diseases and Pests of Narcissus

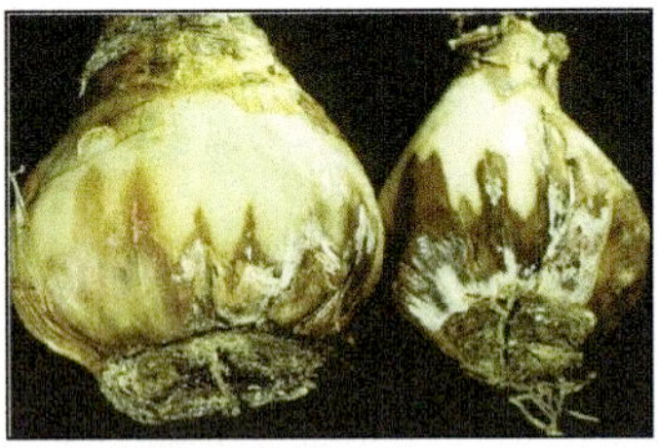
Basal Rot

Leaf Scorch

Narcissus Mosaic Virus

Narcissus Yellow Stripe Virus

Narcissus White Stripe Virus

Bulb Rot

Sun Scorch

Flower Bud Death

Large Narcissus Fly

Garden Swift Moth

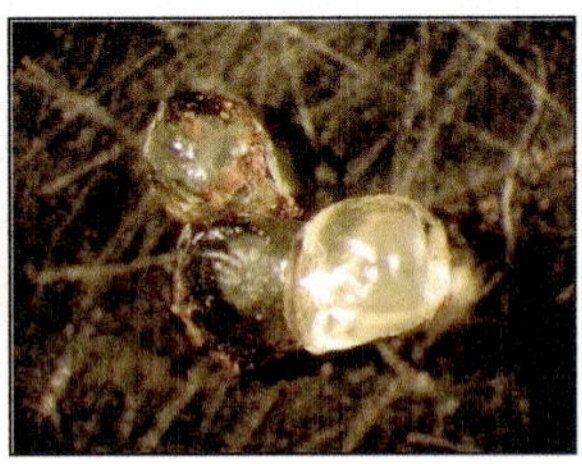
Slugs and Snails

Root-lesion Nematode

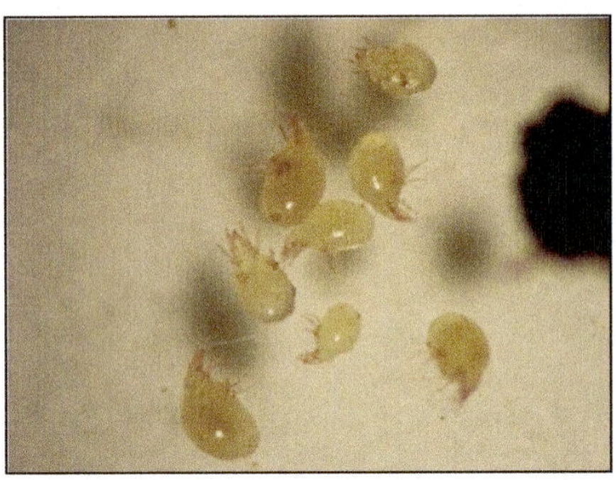
Bulb Scale Mite

Stem Nematode

or excessive nitrogen is essential and the early lifting of the bulbs is preferred. These should immediately be sprayed with a suitable fungicide and dried rapidly in a good air-flow by using electric fans. Storage should be done at low temperature (17-18°C) and planting should be undertaken in late September or early October when soil temperatures are lower. Bulbs in storage should be inspected regularly and soft ones be destroyed.

12.1.1.2 Neck Rot (F. oxysporum f. sp. narcissi and others)

Neck rot (or top rot) has been recognised since the 1930s, but has caused serious concern only since the 1980s, perhaps because it was seen more often as a result of pre-export bulb inspections. Having usually started after harvest, neck rot may be seen on the remains of the old flower stalk. The rot can progress downwards and eventually invade the base plate, the tissues becoming dark-chocolate to reddish-brown. Eventually the new shoot is attacked, and seriously affected bulbs will be blind or produce weak shoots. Neck rot appears to affect a wider range of cultivars than basal rot. Neck rot is usually associated with *F. oxysporum* f. sp. *narcissi*, and recently the general term 'Fusarium rots'has been preferred to cover both base and neck rot since both are considered different expressions of the same disease. Neck rot can be induced by inoculating the bases of senescing leaves with *F. oxysporum* spores or water, the latter implying there is some predisposition to the condition. However, other fungi-including *Penicillium hirsutum*, *Botrytis narcissicola* and *Stagonospora curtisii* as well as 'greasy skin' condition and mites, have also been associated with neck rot (Hawker, 1935; Price, 1982; Davies *et al.*, 1998; Davies *et al.*, 1994; Carder *et al.*, 2003; Carder *et al.*, 1999; Chastagner *et al.*, 2010).

12.1.1.3 Grey Bulb Rot (Rhizoctonia tuliparum)

Grey bulb rot is characterised by the distal ends of leaves developing grey-brown areas in early spring, becoming crooked and withered. The disease is usually mild in narcissus, and the bulbs are not damaged.

12.1.1.4 Soft Rot (Rhizopus sp.)

Soft rot affects bulbs that are stored in unsuitable conditions, usually at a relative humidity >80 per cent and a temperature >30°C. The bulbs rot to a mushy, grey-white mass covered with coarse mycelial growth. Soft rot can occur in warm storage, in the incubation of chips, and in transit, and can be corrected though only if early enough by rectifying storage conditions. In propagation, the temperature of chips should be maintained at 20°C or lower. *Rhizopus* species are common and it may be normally present on bulbs without causing damage, rapidly causing damage if conditions deteriorate.

12.1.1.5 Penicillium Rots (Penicillium sp.)

P. corymbiferum occurs on damaged tissues after lifting, for example where the outer scales are torn or where offsets have recently been removed. Aggressive strains may spread rapidly under damp conditions, causing a dark-brown rot, though often the symptoms are limited to the damaged outer bulb scale. The outer scales may show blistery markings. Grey-green mould can be found under the tunic; a 'puffer' is a completely rotted bulb under an intact skin, that, when broken, releases a cloud of spores. In some cases, *P. corymbiferum* has been linked with basal rot-like symptoms and has caused necrosis when inoculated to healthy bulbs. *Penicillium* is a particular problem in propagation (chipping), and the dwarf cultivars such as 'Tete-a-Tete' are prone to *Penicillium* and other fungal pathogens. Control is generally through the use of appropriate fungicides and storage conditions (Plate *et al.*, 1967; Nicholson *et al.*, 1989; van der Weijden, 1989).

12.1.1.6 Black Slime (Sclerotinia bulborum)

Black slime has been observed only in the Netherlands and following a heavily infested crop of another host, such as hyacinth or *Muscari*. The underground parts of the leaves become soft and black, and the bulb is superficially or completely attacked, with the development of large sclerotia. Other *Sclerotinia* species have been reported on narcissus.

12.1.1.7 White Root Rot (Rosellinia necatrix)

There used to be occasional but significant outbreaks of white root rot in the Isles of Scilly, but it has not been seen for many years. Affected bulbs have black, rotted outer scales and roots and white fungal strands near the base plate. Soil disinfection is the only control measure reported (Tompsett, 2000).

12.1.1.8 Dry-scale Rot (Stromatinia gladioli)

This is reported to cause a dry rot with sharp margins and black sclerotia. It occurs in heavy, wet soils and control is through hot water treatment (HWT) and planting on a better drained site (Chastagner *et al.*, 1985).

12.1.1.9 Honey Fungus (Armillaria mellea)

Honey fungus can cause a bulb rot, particularly on newly cleared ground; bulbs are covered with white mould and rhizomorphs may be present (Chastagner *et al.*, 1985).

12.2.1.10 Crown Rot (Corticium rolfsii)

Crown rot occasionally causes a rot of the outer scale of narcissus bulbs in the US Pacific Northwest. The rot is initially brown and wet, later dry and woody. The bulbs may have a white mycelial mat and small reddish-brown sclerotia (Chastagner *et al.*, 1985).

12.1.1.11 Leaf Scorch (caused by Stagonospora sp.)

The symptoms are leaf tips that become reddish brown with a yellow border. The flowers may become spotted and there is usually premature die back. Control is by application of an appropriate foliar fungicide spray and Hot Water Treatment.

12.1.2 Viral Diseases

Yellow stripe virus is the most serious disease. The symptoms are yellow stripes on the green foliage which is more apparent as the foliage emerges and which often disappears as the season progresses. Other common viruses are cucumber mosaic virus, white streak virus and tobacco rattle virus. In some cases, the flowers are also affected with "breaking" or light patches on the petals or dark streaks. In most cases, even though the flowers are not affected there will be a loss of vigour and reduced yield in the plants involved. There are several vectors that transmit the various viral diseases including aphids above ground and nematodes and millipedes below ground. The spread of the disease may be slow or rapid throughout the collection but the only solution is to rigorously remove any obviously infected plants. It should be assumed that any very old cultivars are likely to be infected by viral diseases of this type and therefore, it is probably wise to not grow them together with a modern collection. Some growers try to convince themselves the symptoms which they see are not viral but as a result of the plant being stressed. However, it is when the plant is stressed that the virus will normally be able to be most easily seen and it is safer to assume the worst rather than risk infecting the whole of one's collection. Control is by destroying any infected bulbs and by controlling the agents that cause the spread of the disease by spraying regularly throughout the growing season to kill aphids and HWT for the control of nematodes.

Table 1: Viruses Reported for Narcissus in 1995, with the more Significant Viruses

Virus and Abbreviation	*Occurrence*	*Vector*	*Symptoms*
Broadbean wilt (BBWV)	Tazetta in Japan	Aphids	-
Carnation latent (CarLV)	Israel	Aphids	-
Cucumber mosaic (CMV)	Tazetta cultivars usually; wide host range	Aphids	Occurs with other viruses, symptomless or mosaic
Narcissus degeneration (NDV)	Tazetta cultivars	Aphids	Chlorotic leaf stripes, chlorosis, decline, potentiallyserious
Narcissus late season yellows (jonquil mild mosaic) (NLSYV)	Common, widespread, jonquils	Aphids	Mild chlorotic streaking or mosaic late in season.May be responsible for white/ silver streaks attributedto NWSV
Narcissus latent (NLV)	Widespread narcissus, other bulbs	Aphids	Symptomless or mild chlorosis alone, exacerbateseffects of other viruses when together
Narcissus mosaic (NMV)	Widespread	Not known (mechanical)	Mild mosaic leaf symptoms or dark brown ovalspots, late season
Narcissus tip necrosis (NTNV)	Widespread	Not known (mechanical)	Symptomless; or late-season leaf tip chlorosisleading to leaf senescence
Narcissus white (or silver) streak (or white stripe or silver leaf) (NWSV)	Widespread	Aphids	When >18°C (high temperature and light intensity),green to purple leaf streaks, later white, yellowishor grey streaks, coalescing; rapid prematuresenescence, seriously reduced yield
Narcissus yellow stripe (or grey disease or mosaic) (NYSV)	Ubiquitous	Aphids	Plant stunted, distorted green, grey or yellowish leafstreaks or mottling early in season, flower breaking,rapid premature senescence, seriously reduced yield

Virus and Abbreviation	Occurrence	Vector	Symptoms
Raspberry ringspot (RRSV)	Scotland	Nematodes	-
Tobacco rattle (TRV)	Widespread; especially 'Golden Harvest' and 'Carlton'	Nematodes	Symptomless; or yellow streaks in leaf bases, leafdistortion, mild flower breaking
Tomato ringspot (TomRSV)	Japan	Nematodes	Mild mosaic leaf symptoms

Source: Brunt *et al.* (1995).

12.2.0 Pests

12.2.1 Large Narcissus Fly (*Merodon equestris*)

The fly, which resembles a small bumble bee, is easily recognised by the high pitched whine it emits in flight and it is only active when the temperatures are in excess of 20°C on windless days. Eggs are laid on the daffodil foliage near to ground level and hatch in seven days. At this point the larvae crawl down and enters the bulb by chewing through the base plate, where it remains, increasing in size by eating the centre of the bulb.

Narcissus bulb flies are difficult to control. Cultural control of this pest includes mowing leaves as soon as they dry in the spring. Lightly tilling the soil after mowing destroys the soil openings that the adult females use to locate the bulb for egg laying. A control strategy used by some gardeners includes using floating row covers as a barrier to egg-laying (Gorsuch *et al.*, 2020).

12.2.2 Small Narcissus Fly (*Eumerus* spp.)

Several small maggots are found in the bulb than the single one in the case of the large fly. Cultural control of this pest includes mowing leaves as soon as they dry in the spring. Lightly tilling the soil after mowing destroys the soil openings that the adult females use to locate the bulb. When bulbs are dug for dividing, any soft bulbs should be destroyed. Treating slightly damaged bulbs by placing them in hot water (43°C to 44°C) for 3 hours can control these pests. Be sure to keep an eye on the temperature and not let it go over the recommended temperature range as higher temperatures can damage the bulbs (Gorsuch *et al.*, 2020).

12.2.3 Garden Swift Moth (*Hepialus lupulinus*)

The garden swift moth is an occasional pest of narcissus. Eggs are dropped in flight over weedy ground and the caterpillars emerge and attack the roots of many plants, especially grasses. They can make large holes in the sides and bases of narcissus bulbs. There are no recognised controls but, if garden swift moth is known to be a problem locally, attention should be paid to keeping the areas around narcissus crops weed-free. The ghost swift moth (*H. humuli*) may also be a pest. Management is mainly through HWT, though this controls only larvae already in the bulbs; adding chlorpyrifos to the tank gives protection during the year after planting.

12.2.4 Narcissus Leaf Miner (*Norellia spinipes*)

The leaf miner does not appear to have been reported in the UK, though it has spread to the Netherlands from Eastern Europe and survives in sheltered areas such as woods. Leaf mining results in wilt and yield loss, but the leaf miner does not attack the bulbs.

12.2.5 Bulb Scale Mite (*Stenotasonemus laticeps*)

It is a light-coloured mite which is only just visible to the naked eye and is likely to found only on daffodils in storage above 17°C, where it attacks the top third of the bulb. It is killed by HWT and may control by an insecticide approved for that purpose.

12.2.6 Stem Nematodes (Eelworm-*Dityenchus dipsaci*)

The most devastating pest of daffodils and most feared by growers. It is not visible to the naked eye and the first symptoms are 'spickels', small yellow raised and lumpy lesions on the edges of leaves or stems. It will usually be accompanied by large areas of the beds with weak growth, stunted plants, or even where no growth is seen. Bulbs cut across will show brown rings where the individual scales have been attacked by the nematodes, Infected plants should be destroyed and care taken not to transfer the infection on boots or clothing. There is no approved chemical treatment of eelworm in the ground and infected areas should not be reused for the growing of daffodils. Hot water treatment (HWT) will kill the nematodes but this requires a high temperature, very careful temperature control and the use of approved chemicals within the solution.

12.2.7 Root-lesion Nematode (*Pratylenchus penetrans*)

Most frequently occurring in the Isles of Scilly where it is not easy to find un-cropped land for new plantings, *P. penetrans,* in association with the fungus *Cylindrocarpon destructans* (also known as *Nectria radicicola*), causes a root rot, 'decline', 'replant disease' or 'soil sickness'. Especially in cultivars with fine roots, the fungus enters through wounds on the roots caused by the nematode, spreads and leads to areas of stunted plants with severe rotting, water stress and premature senescence. The controls include HWT, soil disinfection, growing *Tagetes* (marigolds) and pre-planting pesticide soil application and bulb fumigation (Apt *et al.,* 1961; Miller *et al.,* 1969; Wood, *et al.,* 1982).

12.2.8 Other Pests

Slugs and Snails

It is evident that slugs (*e.g., Arion hortensis* and *Deroceras reticulatum*) and snails can cause extensive damage to narcissus both above and below ground, especially in mild damp weather on heavy soils. The coronas of *Narcissus poeticus* varieties are especially prone to attack. Wild populations are damaged extensively by slugs. Despite this, molluscs are not generally listed amongst the serious pests of narcissus. Slug pellets can be used, though this is not usual and they are effective only for molluscs on the surface. Good soil structure is important in denying slugs and snails hiding places, so cultivation is a key factor (www.slugcontrol.rothamsted.ac.uk/SlugsBrochure.pdf).

12.3.0 Disorders

Narcissus crops may be affected by a number of physiological disorders. These are often due to damage from weather (frost, waterlogging and sun scorch), chemicals (herbicides, and formerly formalin), HWT or mechanical damage. The disorders 'chocolate spot', 'physiological rust', 'soft rot', 'root rot' and 'physiological neck rot' may have the appearance of fungal diseases, but no pathogen has been isolated from their lesions. Many of the damage symptoms might be difficult to distinguish from basal rot, but only the latter will show a white/pink mycelium. There are also numerous disorders of flower development.

12.3.1 Sun Scorch

Exposure of bulbs to strong sunlight can damage tissues several scales deep, turning them grey, shrivelled, hard and prone to attack by *Botrytis*. Flowers and leaves can also be damaged, even if there is no external damage, with withered perianth tips and thickened areas near the leaf tips; in extreme cases plants may be stunted and blind. Sun scorch is one danger of drying lifted bulbs naturally on the soil surface ('windrowing') before harvesting.

12.3.2 Grassiness

Also known as 'horses teeth' or 'crowns', this occurs in bulbs where apical dominance has been lost and many offsets are produced, giving a 'grassy' plant with many small leaves, usually no flower, and producing many small bulbs. This may be due to damage to the growing point by large narcissus fly, bulb-scale mite, HWT or viruses. No pathological cause has been identified, but persistence of the condition for many years has been reported (Industry discussions, 2007).

12.3.3 Chocolate Spot

This condition is unrelated to 'chocolate spot' of beans. *Narcissus* leaves develop elongated spots the colour of dark chocolate, from which no pathogen has been isolated. It has been suggested that the condition results from rising ambient temperatures or other environmental conditions. It is seen in many cultivars in the field, and especially in 'Dutch Master' and 'Mount Hood', but does not appear to occur in forced crops (Mowat, 1983).

12.3.4 Bullhead

In this condition up to 10 per cent of the flowers may fail to emerge from the spathe, which remains dry and membranous. Single-flowered blooms of the double cultivar 'Cheerfulness' are often affected, with extra perianth segments (often green in colour), dissected corona and petaloid styles fused with the corona, poor ovary development and non-elongating stem. Marked plants produce bullhead flowers the next year (Rees, 1972).

12.3.5 Flower Bud Death

The bud may die and remain within a dry spathe, resembling a drumstick. The disorder is common in *N. poeticus* 'Flore Pleno', where it occurs most on warmer slopes and when there are temperature extremes, hot dry growing seasons and wet autumns. It may be related to poor rooting and adverse water relations, though irrigation, mulching and shading treatments failed to prevent it in trials (Rees, 1972, Tompsett, 1971).

12.3.6 Fasciation

Fasciation is a 'doubling-up' of the stem and sometimes flower-heads is seldom severe in narcissus, but has been reported in a few per cent of a stock.

13.0 Harvesting and Postharvest Management of Flowers

13.1.0 Harvesting

Narcissus has number of beautiful and fragrant flowers appearing as bunch on a leafless peduncle and is used mainly as cut flowers. However, there is no detailed postharvest study on these beautiful delicate flowers. Cut narcissus, like tulips has a very short vase life. The life of different cultivars in vase, however, can vary significantly. Therefore, choosing cultivars with a long vase life is an important step in assuring quality of cut narcissus.

Flowers are harvested when the buds have opened (the papery spathe has already split and the tips of the petals are showing). Flowers harvested at an earlier stage may not open. Harvesting is usually followed with leaves along with the flowers.

13.2.0 Postharvest Management

One large bulb produces 1-2 flower stems depending upon the number of noses present. Trumpet and cupped narcissi should be cut at the fat goose-neck stage when the perianth segments are just showing colour (Stevens, 1977) and polyanthus types at the time when the first two flowers in the bunch are opening (Waters, 1977; De Hertogh, 1996). The cut scapes can be stored dry for local markets but immediately after reaching there these should be kept with their cut ends in water. Wet storage for 2-4 hours or overnight in water at 0-2°C immediately after cutting will help in proper opening and prolonging the life. Cut stems can be held for two weeks at 1-2°C and 90 per cent RH, and then can be rehydrated in warm water. Tulips and daffodils should not be kept in same vase solution because cut stems of narcissus exude a poisonous sap. These are graded for quality and packed in the bundles of tens just by putting ne rubber band below the flowers and one near the cut ends (Misra and Misra, 2017). These are now stored either in water or dry in polythene bags or in newspapers and held vertically in wooden or metallic boxes but for distant marketing, these are kept flat in wooden or corrugated boxes. Whether these are stored dry or wet but they are to lose less than a quarter of their potential life. Lower temperatures prevent flower opening while higher storage temperatures reduce the vase life, but a temperature regime of 10-16°C is best for flower opening. Vase life of narcissus varies greatly with the cultivar, and it ranges from 5-11 days, with an average life of 7 days at 15.6-16°C and 65 per cent RH. 8-HQC at 200 ppm and sucrose at 2 per cent enhance the vase life. Shipment of potted trumpet narcissus is done when buds are in the pencil stage of development, *i.e.* when scape is erect.

Store the flowers in an upright position in a 0–1.6°C cooler with a relative humidity of more than 90 per cent. Rapid cooling of the flowers is essential for a good product. Make sure that the stems remain upright to avoid bending. Flowers can be stored up to 1 week.

Before selling the flowers, trim 1-2 cm off the stems and then place them in 10-15 cm of warm water (37°C). Allow the flowers to harden for at least 2 hours. A word of caution is that you must never place other fresh cut flowers, especially tulips, in the same bucket as narcissus during the first 24 hours. Experiments have demonstrated that the sap from cut narcissus is 'toxic' to tulips and significantly reduces the vase life of the flowers. After the 24-hr period, cut narcissus can be used in a bouquet without negatively affecting the postharvest quality of the other flowers.

13.2.1 Senescence and Vase Life

Enhanced expression of protease genes is one of the earliest senescence related gene changes to be identified in various ornamental cut flowers pointing to the fact that protein cleavage is a prime step during petal senescence (Jones *et al.*, 2005; van Doorn and Woltering, 2008). The interactions between proteases and their inhibitors have been linked to modulation of cell death processes in plants and in certain cut flowers. Inhibition of protease action chemically modulates the onset of cellular senescence thereby developing a new strategy of delaying flower senescence and improving the postharvest performance in ornamental flowers (Shahri and Tahir, 2010; Gul *et al.*, 2012). Longevity of flowers can therefore be enhanced with application of specific protein synthesis inhibitors. A study was made to investigate differential responses of petal senescence and postharvest performance at varying concentrations of Cycloheximide (CHI) (a protein synthesis inhibitor at translational level) in cut spikes of *Narcissus tazetta* cv. Kashmir Local (Gul and Tahir, 2013). Cycloheximide at 0.01 and 0.05 mM concentrations delayed senescence. Above 0.05 mM concentrations CHI prevents flower opening and promotes senescence. Senescence delay by CHI points to the synthesis of some specific proteins (enzymes) responsible for execution of cell death programme in flower petals. Cycloheximide at lower concentrations (0.01 and 0.05 mM) enhanced longevity, maintained a sustained rate of flower blooms, delayed senescence and optimized postharvest performance. Pulse treatment of spikes with CHI concentrations at 0.01 and 0.05 mM concentrations maintained high fresh and dry mass of flowers and lowered electrical conductivity of leachates. The content of sugars and proteins decreased, whereas that of a-amino acids and total phenolics increased in the petal tissues with CHI treatment; besides improving postharvest performance. Pre-treatment of flowers with 0.01 or 0.05 mM CHI concentrations for 1 h enhanced vase life and improved postharvest performance in narcissus.

Narcissus carving is a unique craft in the world. Carving narcissus bulbs is possible in different ways. The shapes of flowers, leaves and whole narcissus are full of artistic and ornamental value. After long term storage and transportation in dark, artistically carved

Narcissus bulbs will develop yellow leaves and mouldy, and the calyx will wither. However, using water culture with 50 mg/l VitaC at for 14 days, fumigating with 1 ppm 1-MCP for 12 h after carving the bulbs, packing with polyethylene bags with storage and transportation at 4°C remarkably decreased the loss of water and chlorophyll, and reduced the content of MDA and extended the postharvest life. The marketable proportion after 36 days of simulated storage and transportation was up to 94.7 per cent (Wang *et al.*, 2013).

14.0 Bulb Lifting, Storage, Grading and Forcing

14.1.0 Lifting

Lifting of narcissus bulbs is done when entire foliage has senesced and 95 per cent have died down and that is from mid-May to late July, or when required especially for forcing or for export, the desiccant spray generally of diluted sulphuric acid will kill the non-senesced foliage. In New Zealand they lift these during December-January (Stevens, 1977). The dormancy in narcissi is little and of tulip-type, and when leaves have completely senesced, the bulbs seem to be dormant, but in fact, inside there is a very hectic activity of primordial initiation and growth of daughter bulb units, shoots and roots. Poeticus and Poetaz have short resting period and in these cases generally roots begin appearing before senescence of leaves. Through fork the soil is loosened and gently the bulbs are cleaned, roots removed and bulbs collected without injuring them and giving any chance to narcissus fly to lay its eggs over them. Any loose scale if is there, that should be taken out. After lifting the bulbs are divided, graded and kept in trays at a ventilated place for drying and curing.

14.2.0 Storage of Bulbs

Experiments are described on storage treatments for forcing three cultivars of narcissus (Rees and Hanks, 2015). For early lifted, high-temperature treated bulbs in a range of temperatures and durations for cool storage were investigated. A third variable examined was the time of starting cool storage, relative to the completion of flower differentiation within the bulb; between high-temperature treatment and cool storage bulbs were stored for variable durations at 17°C. Successful, rapid flowering was achieved after a minimum of 15 weeks of cool storage, and a shorter duration in the glasshouse could be achieved by extending the duration of cool storage by 1-2 weeks. Two weeks extra cold resulted in a later flowering date. The optimum temperature was 9°C; at 5°C the stems were shorter. Earlier blooms (early-to late-November, depending on cultivar and experiment) could be obtained following shorter cool storage (12-14 weeks), but these were associated with shorter stems, smaller flowers and an extended glasshouse period. There were no indications of benefits to flower quality from a storage period at 17°C, to allow the completion of flower differentiation (Stage Pc), after high-temperature treatment and before starting cool storage. Satisfactory flowering was obtained following starting cool storage as early as Stage PI (first whorl of perianth segments visible). The use of accumulated day-degrees below 30°C was a useful guide to assess the satisfaction of the cold requirement, with 2000 units needed to obtain flowers after 30 days at 16°C, and 2250 for flowers in 21 days (Rees and Hanks, 2015).

14.3.0 Bulb Size and Grade

Bulb and flower yields are largely determined by the grade of bulbs planted, their planting rate and the length of the growing cycle (one, two or more years). In the UK, narcissus bulbs are usually graded by cm circumference using long (slotted) riddles. Under this system flat bulbs may have a high weight in proportion to their grade. In current commercial UK practice the bulb grades used have become standardised to even numbers, such as 10-12 cm (often written 10/12 cm), 12-14 cm, *etc.* Prior to that, bulbs of trumpet and large-cup cultivars were traditionally graded to 12-15, 15-17 and >17 cm, and other (generally smaller) bulbs to 12-14 cm, *etc.* But, with the adoption of higher planting rates, bulbs are now generally smaller, and even-numbered grades have been adopted for all types. For most types the minimum bulb size (circumference) required for flowers to be formed is about 10 cm. In the following list, the bulb grades refer to cultivars with 'standard sized' bulbs, not small-bulbed types (Strojny, 1975; le Nard *et al.*, 1993)

14.3.1 Small Bulbs (<10 cm)

Small bulbs are usually replanted, though there is a good trade in such 'planters. Bulbs of 7-10 or 8-10 cm have a high growth rate and can make good planting stocks, and many will not flower. The smaller the bulbs, the more difficult it will be to clean and inspect them effectively. Very small bulbs are sometimes discarded.

14.3.2 Small Flowering-size Bulbs (10-12 cm)

These give a high percentage weight increase and make ideal planting stock. As a result of their small size, they are also favoured for sales in pre-packs, and they produce more flowers for their weight compared with larger bulbs. Some growers aim to maximise yields in this grade.

14.3.3 Middle Size Bulbs (Particularly 12-14 cm)

The middle sizes are in demand for dry bulb sales and bulb forcing.

14.3.4 Large Bulbs (mainly 14-16 cm)

Large bulbs are usually sold, but they may produce relatively fewer flowers per tonne. However, these larger bulbs are in demand by some retailers who regard them as of high quality: large bulbs will be more expensive, but may emerge sooner and produce larger and more flowers.

14.3.5 Very Large Bulbs (>16 cm or 16 cm +)

Very large bulbs are not in demand, and are often replanted along with small bulbs, a practice traditionally said to 'increase vigour'in a stock. Dwarf and some other cultivars that are naturally small bulbed may be sold in smaller grades, such as 8-10 cm.

Bulbs of *Narcissus* species are often much smaller, and the minimum bulb size that will flower may be as small as 5 cm-circumference.

In the Netherlands bulbs are more usually graded on round, not long-slot, riddles, so, for example, a Dutch 12-14 cm bulb may be about equivalent to a British 10-12 cm bulb. Often, packers require a specific count of bulbs per tonne, so growers will sometimes vary the riddle size to achieve this, resulting in less usual grades such as 9-12 cm. Sometimes, also, Dutch suppliers may use an older system of grading double-nosed (DN) bulbs in increasing sizes as DN III (roughly 12-14 cm), DN II (14-16 cm) and DNI (>16 cm), which corresponds approximately to 375, 275 and 175 bulbs/25 kg. This grading system may be found in retail outlets or in the mail order trade.

The best flowers are generally produced in the second, third, or fourth year. When the flowers gradually become smaller and less in number, usually due to crowding, it is better to lift and replant the bulbs. Dig the bulbs as soon as the tops begin to die back. Digging before the tops have fully disappeared helps locate clumps and makes digging easier (Dana and Lemur, 2016).

14.4.0 Bulb Forcing

Daffodil bulbs are used for glasshouse forcing in North America and Western Europe, mainly for cut flowers together with potted plants though they are also usually grown outdoors. In India, however, the crop is grown outdoors for bedding and cut flower production both under temperate and sub-temperate climates. At harvest, narcissus has an almost completely initiated flower (Hartsema, 1961) inside which had started taking place shortly after flowering of mother bulbs in April-May. The bulbs of *Narcissus* require warm-cool-warm temperature sequence for growth and development (Rees, 1972). Normally, round or double-nosed bulbs are used for forcing having bulb weight of at least 25 g (De Hertogh, 1977). When the bulbs are lifted from the field, they have partially formed flower parts. Hence, while the initial warm temperature helps to complete flower development, the subsequent low temperatures are required for stalk and leaf elongation and warm temperature for flowering (De Hertogh, 1974). For early forcing, the bulbs are harvested in July and kept at 34°C for one week, followed by 17-20°C until pre-cooled in August at 9°C (Gordon *et al.*, 1984). Bulbs for cut and potted daffodils are handled separately after planting in early October. Cut daffodils are rooted and cooled continuously at 9°C, which is optimum temperature for shoot elongation. After accumulation of 15-16 weeks of cold period, forcing takes place at a 13-15°C, whereas pot daffodils are rooted at 9°C but as soon as rooting takes place the bulbs are shifted to 5°C. They are forced in the greenhouse at 16-18°C after 15 cold weeks. The bulbs of daffodils are harvested in late July or August and stored at 17-20°C until planting for mid-season and late forcing. Bulbs are planted by providing 17-18 cold weeks for cut daffodils and 14-16 cold weeks for pot daffodils. These are subjected to rooting at 9°C and then cooled at 5°C or 2°C. The growth of the bulb shoot should not be more than 10 cm in the rooting room. Forcing of cut and potted daffodils is done at 13-15°C or 16-17°C in the greenhouse. 'Paper White' tazetta narcissus leads bulb production in Israel, with more than 20 million bulbs exported annually for dry sales and flower

Harvest and Postharvest Management of Flowers and Bulbs

Narcissus in Full Bloom

Harvesting of Flowers

Freshly Harvested Flowers

Harvesting and Storage of Bulbs

production. Recent release of new narcissus cvs 'Ariel', 'Inbal' and 'Nir' suitable for pot-plant production, developed by Agricultural Research Station, Volcani Center of USA necessitates to device the methods for flowering manipulations. Under natural conditions in Israel, 'Paper White' flowers in December-January. In order to advance flowering to October-November, three forcing strategies were applied to 'Ariel', 'Inbal' and the well-known cultivar 'Ziva'. Following the determination of the stage of flower development, the bulbs kept in different storage regimes, *viz.* open shed, temperatures 18-22/25-30°C (night/day), constant 25°C, 25°C followed by 2 weeks at 13°C, and 25°C followed by 2 weeks at 9°C, storage at 25°C accelerates intra-bulb flower differentiation and shortens forcing time by 2 weeks in'Ziva' and 'Inbal' and by 4 weeks in 'Ariel', and storage at 9°C and 13°C for 2 weeks prior to planting causes stalk elongation inside the bulbs. In all the cultivars, storage at 25°C followed by a short period at lower temperatures of 9°C or 13°C results in anthesis after 37-40 days of growth. Thus, the time to anthesis is reduced by 16-18 days. Storage at 25°C advances anthesis by 9-10 days. In 'Ariel' and 'Inbal', lower storage temperatures significantly increase the number and diameter of individual flowers, while the opposite result is obtained for 'Ziva'. Storage at 25°C significantly increases the number of flower stalks per bulb in 'Inbal', but in 'Ziva' the increased number of flower stalks is observed after storage in the open shed. It is so inferred that physiological requirements for 'Ariel' and 'Inbal' are different from those of 'Ziva', and that a precise pre-planting protocol is needed for each cultivar.

Forcing techniques were developed in the Netherlands in the 1930s (van Slogteren,1938) and programme consists of several elements: (a)early bulb lifting – allowing forcing programmes to begin earlier, (b)intermediate temperature storage – holding the bulb until it reaches a developmental stage where it can respond to subsequent warm and cool treatments. (c)warm storage – speeding the bulb's response to cool treatment, (d)cool storage – providing the winter cold that spring flowering bulbs need before they can grow-on quickly to flowering, the different cultivars varying in the length of cold storage required, (e)low-temperature holding period if necessary, holding bulbs until their glasshouse slot is available, (f)flower production in the glasshouse.

15.0 Biotechnology

15.1.0 Somatic Embryogenesis (SE)

There are several developed protocols for SE of *Narcissus*, in which different initial explants and different growth regulators were used (Selles *et al.*, 1999; Sage *et al.*, 2000; Sage and Hammatt, 2002; Chen *et al.*, 2005; Anbari *et al.*, 2007; Malik, 2008; Malik and Bach, 2016). The narcissus plant regeneration systems that are known are based on two major *in vitro* pathways: organogenesis or somatic embryogenesis. Shoot organogenesis was induced on bulb scales, flower stems, leaves and basal plates as well as mature seeds, while cytokinin BA in combination with auxin NAA was most frequently used for the initiation of the culture (Seabrook *et al.*, 1976; Chen, 1982; Hussey, 1982; Kozak, 1991; Bergonon *et al.*, 1996; Selles *et al.*, 1997; Sochacki and Orlikowska, 1997). Somatic embryogenesis was obtained on the bulb scale, leaf, flower stem, ovary, anther explants as well as mature and immature embryos excised from mature seeds (Selles *et al.*, 1999; Sage *et al.*, 2000; Chen *et al.*, 2005; Malik, 2008). Studies carried out so far on narcissus somatic embryogenesis only generally demonstrate that somatic embryos formed in callus cultures obtained under the influence of 2,4-D or Picloram auxins applied alone or in combination with cytokinin BA.

Somatic embryos (SEs) have been produced from bulb and shoot culture leaf explants of *Narcissus pseudonarcissus* cvs. Golden Harvest and St. Keverne (Sage and Hammatt, 2002). The highest yields of direct SEs were obtained from scape explants orientated with their basipetal surface away from the MS-based medium. SEs converted to plantlets most efficiently with a pre-treatment of 24°C for 10-20 weeks on a medium with IBA, followed by transfer to 4°C. These plantlets readily transferred to *ex vitro* conditions. Nodular callus was produced from scape explants of 'Golden Harvest' and 'St. Keverne'. This regenerated most successfully on a medium with TDZ in combination with either NAA or 2,4-D. Preliminary studies of the growth of callus in a liquid medium were promising (Sage and Hammatt, 2002).

15.2.0 Cryopreservation

Somatic embryos of narcissus are obtained and maintained in controlled and sterile conditions, as *in vitro* cultures. The maintenance of *in vitro* collections is expensive, labour-intensive and threatened with contamination and somaclonal variation (Sakai and Engelmann, 2007). Moreover, the process of narcissus SE induction followed by somatic embryo development takes at least twelve weeks. It is, therefore, necessary to develop a protocol for the long-term preservation of the embryos, which simultaneously will be useful to preserve the natural biodiversity. An efficient cryopreservation protocol for the geophyte giant narcissus that guarantees high rate plant regeneration was reported by Macœlanka *et al.* (2016). Globular somatic embryos (1-, 2-, and 3-mm diameter) were picked up from the embryogenic callus. The embryos were treated with plant vitrification solution (PVS2) for 10, 20, 30, 45, and 60 min. For 1 mm embryos the survival rate does not exceeding 76.7 per cent, irrespective of PVS2 exposure duration. In case of 2 mm and 3 mm embryos, low survival (73.3-86.7 per cent) was observed only after 10 and 20 min of PVS2 exposure. In other cases, the survival rate ranged between 93.3 and 100 per cent. During post-rewarm regeneration, the highest growth rate amounting to 26.7 per cent was achieved after 45 min treatment of PVS2, in case of the 2 mm embryos. The best re-growth among 1 mm embryos, at level 20 per cent, was achieved after 60 min treatment of PVS2. Large sized embryos did no regenerate at all. The regenerated globular embryo formed a few secondary embryos. All the secondary embryos formed bulblets, which exhibited the same morphology as that of non-cryopreserved, control bulblets (Macœlanka *et al.*, 2016).

15.3.0 Genetic Transformation

Despite *Narcissus* being a monocotyledonous genus, it has been shown that at least some members are amenable to transformation via *Agrobacterium tumefaciens*. Regeneration from nodular callus and successful callus growth in liquid medium sets the scene for potential *Narcissus* mass production in bioreactors (Sage and

Hammatt, 2002). Transformation via *Agrobacterium tumefaciens* may be used for the introduction of desirable genes to *Narcissus* cultivars (Sage and Hammatt, 2002).

Phytoene synthase (PSY), as a key regulatory enzyme for carotene biosynthesis, plays an important role in regulating color formation in many species (Lu *et al.*, 2007). A protocol was developed for the transformation *Narcissus tazzeta* var *chinensis* using *Agrobacterium tumefaciens* strain LBA4404 harboring a binary vector pCAMBIA1301 plasmid which contained an antisense phytoene synthase gene, a reporter β-glucuronidase gene and a selectable marker hygromycin phosphotransferase gene (Lu *et al.*, 2007). Effects of some factors on efficiency of transformation and regeneration were examined. Preculture of the explants for 6 days before inoculation enhanced the transient GUS expression. The addition of acetosyringone (AS) at 100 μmol/l for inoculation and a period of 3 days co-cultivation yielded efficient transient GUS expression. Transformants were obtained through selection on MS medium containing 5 mg/l 6-benzylaminopurine (BA), 0.1 mg/l α-naphthalene acetic acid (NAA) and 40 mg/l hygromycin. The transformation frequency was 1.24 per cent based on PCR analysis of gus gene. One or two copies of transgene were demonstrated in different transformations by Southern blotting analyses. Northern blotting results confirmed that the transcription of the endogenous psy gene in transgenic plants was inhibited or silenced. The method reported here provides new opportunities for improvement of quality traits of *Narcissus tazzeta* via genetic transformation (Lu *et al.*, 2007).

15.4.0 Markers

15.4.1 Population Structures

Determination of species boundaries is essential for understanding and preserving biological diversity, yet it remains a difficult task for many plant lineages. Population genetics approaches explicitly taking into account the geographic context of processes driving population divergence and speciation may help to identify species boundaries in complex groups. Medrano *et al.* (2014) adopted this approach to investigate genetic boundaries in an endemic group of trumpet daffodils (*Narcissus*, section *Pseudonarcissi*) whose taxonomic distinctiveness remains controversial. They analyzed amplified fragment length polymorphism markers from a total of 36 populations (526 individuals) spanning the group's entire distribution range in the southern Iberian Peninsula. To identify the most likely number of distinct genetic groups, model- and non model-based methods (Bayesian, principal coordinates, and neighbor-joining classification) were applied. Effects of long-term historical divergence were dissected from more recent or local differentiation processes using simple and partial Mantel tests major genetic split, consistently supported by the three analytical methods used, differentiated all populations generally ascribed to *Narcissus bujei* in traditional taxonomic treatments from the rest, which included all populations generally designated as *Narcissus longispathus* and *Narcissus nevadensis* (Medrano *et al.*, 2014). These findings emphasize the usefulness of molecular data and population genetics approaches in a geographic context to delineate morphologically cryptic species in complex lineages.

15.4.2 Diversity Analysis

Analysis of the genetic diversity of narcissus still needs to be conducted in a systematic and quantitative manner (Nunez *et al.*, 2003; Santos-Gally *et al.*, 2012). AFLP markers were used to distinguish two taxa of narcissus (Tucci *et al.*, 2004), evaluate the effect of radiation on regeneration of Chinese Narcissus (*N. tazetta* var. chinensis), and analyze the genetic variation (Lu *et al.*, 2007). ISSR and internal transcribed spacer markers were also used for species identification in narcissus (Jiménez *et al.*, 2009). In addition, Hodgins and Barrett (2007) investigated cpDNA sequence and nuclear microsatellite variation among populations of the wild daffodil (*Narcissus triandrus*).

15.4.3 Hybrid Analysis

Narcissus pseudonarcissus, also known as daffodil, is a world-famous ornamental flower. Li *et al.* (2017), for the first time, carried out cross-pollinations between two widely cultivated *N. pseudonarcissus* varieties 'Slim Whitman' and 'Pinza'.

After eight consecutive years of cultivation, 27 reciprocal hybrids with different genotypes survived; 15 hybrids in 'Slim Whitman' 3 'Pinza' and 12 in 'Pinza' 3 'Slim Whitman'. The hybrids showed great variation in most morphological characters compared with the parents, especially in leaf and flower characters. Hybrids SP03, SP04, SP05, SP12, SP13, PS04, PS06, PS07, PS08, and PS11 had evident growth advantage in some aspects compared with both parents. Of these hybrids, SP04 got novel flowers with white petals and a yellow-orange corona, and had the potential to become a new popular *N. pseudonarcissus* cultivar. Hybrids SP01, SP03, SP05, PS04, PS06, and PS07 also possessed a great ornamental value.

15.4.4 Sexual Polymorphism

The genus *Narcissus* provides a model system to study the evolution and maintenance of sexual polymorphisms (Barranco *et al.*, 2018). Microsatellite markers were used studies on *N. dubius*, *N. cuatrecasasii*, *N. assoanus* and *N. rupicola* for genetic diversity and paternity analyses to investigate the stability of stylar dimorphism. 40 new primer pairs from a genomic library of *N. papyraceus* and 12 microsatellite markers characterized for *N. papyraceus* including those from previous study (52 primer pairs overall). Twenty markers amplified, but their transferability and variability were different among species. Polymorphism was tested at least on 74 individuals and one population per species. The number of polymorphic loci per species ranged from four to eight.

15.4.5 Transcriptomics and Metabolomics

Singh and Penix (2017) presented the first comprehensive metabolome and transcriptome study from *N. pseudonarcissus* 'King Alfred' providing invaluable resources for metabolic engineering and biotechnological applications. Amaryllidaceae alkaloids (AAs) represent a diverse class of plant specialized metabolites and many display potent pharmacological activities. The AA metabolic pathway is poorly understood and resources are minimal. To enable AA pathway elucidation and novel biosynthetic enzymes discovery, they generated comprehensive metabolomic and corresponding transcriptomic datasets from different tissues of *Narcissus pseudonarcissus* 'King Alfred'. They performed untargeted UPLCQTOF-MS metabolite analysis from different tissues, which generated exhaustive list of compounds, including several AAs, most predominant and diverse in bulbs.

16.0 References

Abdel-Rahman, H.M., Al-Ansary, A.M.F., Rashed, K.N. and Rizkalla, A.A. (2017) *J. Appl. Life Sci.*, **14** : 1-11.

Abdulla, M.F. and Çelikel, F.G. (2019) *Acta Hortic.*, **1263** : 455-460.

Anbari, S., Tohidfar, M., Hosseini, R. and Haddad, R. (2007) *Plant Tissue Cult. Biotech.*, **17** : 37-46.

Apt, W.J. and Gould, C.J. (1961) *Plant Dis. Report.*, **45** : 290-295.

Bailey, L.H and Bailey, E.Z. (1976) International Flower bulb Information Centre, Hillegom, The Netherlands. pp. 69-93.

Barrett, S.C.H., Cole, W.W. and Herrera, C.M. (2004) *Heredity*, **92** : 459-465.

Berkov, S., Georgieva, L., Kondakova, V., Atanassov, A., Viladomat, F., Bastida, J., Codina, C. (2009) *Biotech. Biotechnol. Equip.*, **23**(2) : 1170-1176.

Bock, A., Sparks, T.H. and Estrella, N. (2015) *Int. J. Biometeorol.*, **59** : 477-480.

Bowes, S.A. (1990) *Euphytica*, **48** : 275-278.

Brandham, P.E. (1992) *Plantsman*, **14** : 133-168.

Brunt, A.A., Loebenstein, G. and Lawson, R.H. (1995) Wiley, Chichester, UK. pp. 322-334

Caldwell, J. and Wallace, T.J. (1955) *J. Ecol.*, **43**(1) : 331-341.

Carder, J.H. and Grant, C.L. (2002) *Acta Hortic.*, **570** : 255-262.

Carder, J.H. (1999) Report on MAFF Project HH2717SBU.

Carder, J.H. (2003) Report on Defra Project HH1748TBU.

Chastagner, G.A. and DeBauw, A. (2010) *Acta Hortic.*, **2** : 1

Chastagner, G.S. and Byther, R.S. (1985) *Praeger Scientific*, pp. 447-506

Chen, L., Zhu, X., Gu, L. and Wu, J. (2005) *Plant Cell Rep.*, **24** : 401-407.

Chen, Z. (1982) Laboratory of Breeding Floricultural Plant Tissue.

Dana, M.N. and Lerner, B.R. (2016) https: //www.purdue.edu/hla/sites/ yardandgarden/wp-content/uploads/sites/2/2016/10/HO-11.pdf.

Darlington, C. D. and Wylie, A. P. (1955) Chromosome atlas of flowering plants. George Allen and Unwin Ltd., London.

Davies, J.M.L.I (1994) Final Report on Project BOF 28. HDC, Petersfield, UK.

Davies, J.M.L.I., Dickens, J.S.W., Inman, A.J., Jones, O.W., Reed, P.J. and Wilson, D.G. (1998) *J. Horti. Sci. Biotech.*, **73** : 245-250.

De Hertogh, A.A. (1974) *Scient. Hort.*, **2** : 313-355.

De Hertogh, A.A. (1977) Netherlands Flower bulb Institute, The Netherlands.

De Hertogh, A.A. (1996) In: Holland Bulb Forcer's Guide (5th ed.). International Flower Bulb Centre, Hillegom, The Netherlands.

Díaz-Lifante, Z. and Andrés-Camacho, C. (2007) *Bot. J. Linn. Soc.*, **154** : 237-257.

Díaz-Lifante, Z., Andrés-Camacho, C., Viruel, J. and Cabrera-Caballero, A. (2009) *Bot. J. Linn. Soc.*, **159** : 477-498.

Fenlon, J.S., Jones, S.K. Hanks, G.R. and Langton, F.A. (1990) *J. Hort. Sci.*, **65**(4) : 441-450.

Fernandes, A. (1951) *Bol. Soc. Brot.*, **25** : 113-190.

Gordon, R., Hanks, G.R. and Rees, A.R. (1984) *Scient. Hort.*, **23**(3) : 269-278.

Gorsuch, C.S., Scott, J.M. and Williamson, J. (2020) Flowering Bulb fact Sheet. Clemson University

Gul, F., Tahir, I. and Rasool, I. (2012) *Int. J. Bot.*, **8** (1): 22–30.

Gul, Fahima and Tahir, Inayatullah (2012) *J. Saudi Soc. Agril. Sci.*, **12**: 39–46.

Hanks, G. R. (1993) Elsevier, pp. 463-558.

Hanks, G. R. (1996) *J. Hortic. Sci.*, **71**(4) : 517-532.

Hanks, G.R. (1992) *Elsevier*, pp. 463-558.

Hanks, G.R. (2002) CRC Press, Boca Raton.

Hartsema, A.M. (1961) Springer-Verlag, Berlin and New York. pp. 123-167.

Hawker, L.E. (1935) *Ann. Appl. Biol.*, **22** : 684-708.

Heidi, E.M., Dobson, H.E.M., Arroyo, J., Bergstrom, G. and Groth, I. (1997) *Biochem. Syst. Ecol.*, **25** : 685-706.

Herrera, C.M. (1995) *Ecol.*, **76** : 218-228.

Hodgins, K.A. and Barrett, S.C.H. (2007) *Mol. Ecol.*, **16** : 2317-2322.

Hussey, G. (1982) *Ann. Bot.*, **49** : 707-719.

Industry discussions (2007) personal communications

Ito, M., Kawamoto, A., Kita, Y., Yukawa, T. and Kurita, S. (1999) *J. Plant Res.*, **112**(2) : 207-216.

Jiao, C., Yizhu, X. and Jian, W. (2005) *Plant Cell Rep.*, **24** : 401-407.

Jones, M.L., Chaffin, G.S., Eason, J.R. and Clark, D.G. (2005) *J. Exp. Bot.*, 56 : 2733–2744.

Jones, M. and Pulman, J. (2011) *Chem. Ind.*, **4** : 21.

Khan, I.M., Khan, F.U., Salmani, M., Khan, M.H., Mir, M.A., and Hassan, A. (2013) *Afr. J. Agric. Res.*, **8**(31) : 4189-4193.

Khonakdari, M.R., Rezadoost, H., Heydari, R. and Mirjalili, M.H. (2020) *Plant Cell, Tissue and Organ Cult.*, **142** : 187-199.

Kington, S. (2002) The international daffodil register and classified list 1998. Available at: www.rhs.org.

Kington, S. (2014) The daffodil register and classified list (2008) 6th supplement. Royal Horticultural Society, London, UK.

Langens, M.G. (1996) *Acta. Bot. Neerl.*, **45**(4) : 578-579.

Langens-Gerrits, M.M. and De Klerk, G.J.M. (1999) *Plant Cell Culture Proto.*, pp. 141-147.

le Nard, M. and De Hertogh, A.A. (1993) *The Physiology of Flower Bulbs*, pp. 29-43.

Li, X., Tang, D. and Shi, Y. (2017) *HortScience.*, **52**(12) : 1668-1675.

Li, X.F., Jia, L.Y. and Xu, J. (2013) *Plant Cell Physiol.*, **54**(2) : 270-281.

Lin, H.X, Lin, R. and Xie, L. (1997) *Plant Quarantine Sin.*, **10** : 227-229.

Lu, G., Zou, Q. and Guo, D. (2007) *Plant Cell Rep.*, **26** : 1585-1593.

Malik, M. (2008) *Plant Cell, Tissue and Organ Cul.*, **94** : 337-345.

Marques, I., Feliner, G.N., Munt, D.D., Martins-Loucxao, M.A., and Aguilar, J.F. (2010) *J. Evol.*, **64**(8) : 2353-2368.

Marques, I., Rossello-Graell, A., Draper, D., and Iriondo, J.M. (2007) *Am. J. Bot.*, **94**(8) : 1352-1359.

Maœlanka, M., Panis, B. and Malik, M. (2016) *Prop. Ornam. Plants.*, **16** : 28-35.

Meyer, F.G. (1966) *The Amer. Hort. Mag.*, **45**(1): 47-76.

Miles, B. (1963) The Wonderful World of the Bulbs. D. Van Nostrand Co., Inc, London, England.

Miller, P.M. and Ahrens, J.F. (1969) *Plant Disease Reporter*, **53** : 642-644.

Misra, S. and Misra, R.L. (2013) *Westvelle Publishing House, New Delhi.* pp. 109-124.

Misra, R.L. and Misra, S. (2017) Kruger Brentt Publishers UK. LTD. pp. 373-388.

Moreno Saiz, J.C. and Sainz Ollero, H. (1992) Atlas corológico de las monocotiledóneas endémicas de la Península Ibérica y Baleares. ICONA, Ministerio de Agricultura: Madrid.

Mowat, W.P. (1983) *Scottish Crop Research Institute Annual Reportn.*, pp.187.

Nicholson, P and Ingram, D.S. (1989) *Mycological Research*, 92 : 359-382.

Nunez, D.R., De Castro, C.O., Ruiz, S.R. and Ariza, F.A. (2003) *Sci. Hort.*, **98** : 307-330.

Philp, J. (1934) *Daffodil Year Book.* Royal Hort. Soc., London, U.K.

Plate, H.P. and Schneider, R. (1967) *Gartenwelt*, **67** : 229-230.

Price, D.J. (1982) *Glasshouse Crops Research Institute Annual Report 1981*, pp. 130-131.

Raven, S. (2020) https: //www.sarahraven.com/articles/a-brief-history-of-daffodils.htm

Rees, A.R. (1972) Academic Press, London, UK.

Rees, A.R., Hanks, G.R. (1984) J. Hortic. Sci. **59** : 229-239

Rees, A.R.1972. The Growth of Bulbs. Academic Press. London and New York.

Ribeiro, P.M., Paiva, J.A., and Freitas, H. (2007) *Anales Jard. Bot. Madrid.*, **641** : 43-46.

Rosselló-Graell, A., Marques, I., Draper, D. and Iriondo, J.M. (2007) *Bocconea*, **21** : 359-365.

Sage, D. and Hammatt, N. (2002) *Acta Hortic.*, **570** : 247-249.

Sage, D.O., Lynn, J. and Hammatt, N. (2000) *Plant Sci.*, **150** : 209-216.

Sage, T.L, Strumas, F., Cole, W.W. and Barrett, S.C.H. (1999) *Am. J. Bot.*, **86** : 855-870.

Sakai A. and Engelmann, F. (2007) *Cryoletters*, **28** : 151-172.

Santos-Gally, R., Vargas, P. and Arroyo, J. (2012) *J. Biogeogr.*, **39** : 782-798.

Seabrook, J.E.A., Cumming, B.G. and Dionne, L.A. (1976) *Canadian J. Bot.*, **54** : 814-819.

Selles, M., Viladomat, F., Bastida, J., Codina, C. (1999) *Plant Cell Rep.*, **18** : 646-651.

Shahri, W.and Tahir, I. (2010) *Pak. J. Bot.*, 42 (5) : 3577–3585.

Singh, A. and Penix, I.I D. (2017) *Sci. Rep.*, **7**: 17356.

Slug control and control of slug damage in horticultural crops, available at: www.slugcontrol.rothamsted.ac.uk/SlugsBrochure.pdf

Stevens, R.B. (1977) Min. Agric. and Fish., Levin, New Zealand.

Strojny, Z. (1975) *Prace Instytutu Sadownictwa w Skierniewicach B.*, **1** : 115-127.

Tompsett, A.A. (1972) *Rosewarne EHS Annual Report 1971*, pp. 15-62.

Tompsett, A.A. (2000) *personal communication.*

Van der Weijden, B. (1989) *Bloembollencultuur*, **100**(15) : 16-18.

Van Slogteren, E. (1938) *Laboratorium voor Bloembollen-onderzoek Lisse Publicatie 59.*

van Doorn, W.G.m and Woltering, E.J. (2008) *J. Exp. Bot.*, 59 : 453–480.

Wang, S.F., Li, X.M., Liu, S.X., Wu, S.B., Lan Y.Y. and Chen, Y.Y. Hong, Z.F. Guo X.F., Wu, Y.B. and Dong, J.L. (2013) *Acta Hort.*, **970**: 129-140.

Waters, R.A.S. (1977) In: *Narcissus Production*, 11 pp. Ministry of Agriculture and Fisheries, New Zealand.

Webb, D.A. (1980) Narcissus. In *Flora Europaea*, eds.

Wood, F.H. and Foot, M.A. (1982) *New Zealand J. Experi. Agri.*, **10** : 439-441.

Wylie, A. P. (1952) *Heredity*, **6** : 137-156.

Zonneveld, B.J.M. (2010) *Anales del Jardín Botánico de Madrid.*, **67**(1) : 29-39.

4

Freesia

V.A. Parthasarathy, Rajiv Kumar, C. Aswath and S. Bal

1.0 Introduction

Freesia hybrida is an important cut flower, especially in America and Europe (Gao *et al.*, 2010). It is an ornamental bulbous plant from the Iridaceae family that has recently become one of the most popular cut flowers due to its fragrance, long vase life and the wide colour range (Anderson, 2007). Market research shows that over 110 million freesia stems are sold in the UK each year (Gao *et al.*, 2009). The modern freesia, *Freesia*×*hybrida*, is derived from crosses among several *Freesia* species.

Holland is the largest producer of freesia with over 280 ha under cultivation (Ali *et al.*, 2011) and In the Netherlands, the sales amount of freesia as cut flower was approximately €50 million in 2010 (Uwagaki *et al.*, 2015). The diversity of floral pigmentation patterning and abundant flavonoids accumulation make it an ideal model system to study the regulation of flavonoid biosynthesis in Iridaceae family, which has numerous ornamental genera (Ju *et al.*, 2018; Li *et al.*, 2016, 2017).

2.0 Importance and Uses

2.1.0 Aesthetic Uses

Freesia is ideal for both indoor decoration and outdoor cultivation. Freesia can be grown in garden as well as in pots. It is suitable of hanging baskets and also in flower arrangements. Freesia is primarily used as cut flowers and can last at least a week in a vase solution. The flowers are used as an elegant, stylish and sophisticated flowers for indoor decoration in small designer pots, can be kept in dining table, study table, in balcony *etc.*

2.2.0 Therapeutic and Other Uses

The flowers of freesia are well is known for its fragrance, which has many applications and uses. Freesias are widely used in aromatherapy as a means for stress relief. Since the fragrance from freesia is so pleasantly sweet, the species *Freesia refracta* is widely cultivated as a fragrance for use in shampoos, body lotions and perfumes.

3.0 Origin and History

The freesia originally came from the Cape Colony in South Africa. It is also known as the 'Cape lily-of-the-valley'. The flower received its name around 1830 from botanist cum Ecklon, a collector of plants in South Africa. He named the crop after a friend of his, the German doctor F. Freese (http://www.unicumfreesia.nl/en/herkomst). No flower has become more popular in recent years than the fragrant freesia. It appears that the bulbs were first brought to Britain from the Cape of Good Hope, about the year 1850 as the species *Freesia refracta*, which has sweetly perfumed flowers, white with a deep orange throat (https://www.houseplantsguru.com/freesia-history).

4.0 Distribution

Freesia refracta Klatt was introduced into England early in the nineteenth century (probably in 1816 it was first described in the Botanical Register.) Apart from its attractive fragrance, it offered little floricultural promise with its small, tubular flowers of a dingy, greenish yellow. At a later date, a freesia possessing a 'pinkish-mauve' flower was found at Humansdorp, South Africa, by a land agent named Armstrong. A white form *F. refracta* alba was introduced in England in 1878 and soon became popular as a florist's flower (Morgan, 1929).

5.0 Species and Cultivars

5.1.0 Species

There are impressive 16 different species of freesia flowers in existence, each one boasting equal measures of beauty and fragrance (https://www.serenataflowers.com/pollennation/all-about-freesia/). These variants include:

i. *Freesia caryophyllacea*
ii. *Freesia andersoniae*
iii. *Freesia alba*
iv. *Freesia fergusoniae*
v. *Freesia fucata*
vi. *Freesia grandiflora*
vii. *Freesia viridis*
viii. *Freesia laxa*
ix. *Freesia corymbosa*
x. *Freesia sparrmannii*
xi. *Freesia marginata*
xii. *Freesia leichtlinii*
xiii. *Freesia occidentalis*
xiv. *Freesia refracta*
xv. *Freesia speciosa*
xvi. *Freesia verrucosa*

5.2.0 Cultivars

Freesia Patent 98 (Morgan, 1934)

New freesia was produced by the crossing of an unnamed lavender seedling of unknown parentage and Elder's Giant White. In the large size of the blossoms, upright growth and small size of the corms my improvement resembles the Elder's Giant White. It has large blossoms, a pleasing 'light lavender-violet' color, tall stems sufficiently strong to hold the blossom. Spike upright, with dark green foliage whose tips rise to the base of the flower spike, with unusually robust constitution and other desirable and useful characters which make it suitable for florists' cut-flower use. Many patents were made in thirties and forties but they were just named 'Freesia' with no varietal name.

Varared (van Andel, 1992)

A new and distinct Freesia cultivar distinguished by its very tall stems bearing a single red flower at its upper end and immediately preceding a jointed rachis which extends almost at right angles to the stem and has from six to nine longitudinally spaced buds of which several may be open at the same time, the flowers being of medium size and blooming successively on strong spikes with substantially uniform spacing along the rachis to its end. This freesia cultivar originated as a seedling of a

Uses of Freesia

In Pot

As Specimen in Garden

As Border in Garden

In Flower Arrangement

In Flower Vase

As Cut Flowers

Species of Freesia

Freesia caryophyllacea

Freesia andersoniae

Freesia alba

Freesia fergusoniae

Freesia fucata

Freesia grandiflora

Species of Freesia

Freesia marginata

Freesia viridis

Freesia laxa

Freesia corymbosa

Freesia leichtlinii

Freesia occidentalis

Freesia refracta

Freesia speciosa

Freesia verrucosa

Freesia sparrmannii

Cultivars of Freesia

Golden Amor

Varayel

Varatwisno

Propagation of Freesia

Freesia Bulbs

Planting of Bulbs in Pots

After 30 and 60 Days Sowing

Application of Water

Freesia Flowers after 130 Days of Sowing in Pots

Cultivation of Freesia

In Greenhouse

Growth of Seedling in Open

Growth of Freesia under Greenhouse

hybrid produced by crossing two varieties selected from a collection of freesia being maintained at nursery at Aalsmeer, Holland for breeding purposes with the object of producing freesia hybrids that can be grown faster and at higher temperatures than ordinary freesia hybrids. This seedling was particularly selected because of its height of the main stem and its apparent growth habits, medium red flowers and numerous buds.

Vaposma (van Andel, 1995a)

A new and distinct variety of freesia which originated by crossing two unpatented, unnamed varieties of at the Van Staaveren b.v. Nursery in Aalsmeer, the Netherlands. Among the features which distinguish the new variety from other presently available and known commercial cultivars are the following combinations of characteristics: attractive pale yellow flowers, a habit of growth which distinguishes the new variety as a plant suitable for growing in pots and referred to as a pot freesia variety reflecting its natural dwarf characteristics without need of growth retardants, and its ability to flower, independent of soil temperature up to about 23°C in about 70 to about 100 days after planting corms.

Vapopey (van Andel, 1995b)

A new variety of pot freesia which is suitable for growing indoors as a potted plant, produces light yellow flowers, which originated by crossing two unpatented, unnamed varieties at the Van Staaveren Nursery in Aalsmeer, the Netherlands. The female parent is identified as 86311-AP1 and the male parent is identified as 86317-AP1. The varietal denomination of the new variety is Vapopey.

Vaposu (van Andel, 1995c)

It is a new variety of pot freesia suitable for growing indoors as a potted plant, which originated by crossing two unpatented, unnamed varieties of this creation at the Van Staaveren b.v. Nursery in Aalsmeer, the Netherlands. The female parent is identified as 86429-AP1 and the male parent is identified as 86411-AP1. The varietal denomination of the new variety is VAPOSU. Among the features which distinguish the new variety from other presently available and known commercial cultivars are the following combinations of characteristics: attractive off-white flowers, a habit of growth which distinguishes the new variety as a plant suitable for growing in pots and referred to as a pot freesia variety reflecting its natural dwarf characteristics without need of growth retardants, and its ability to flower independently without requirement of soil temperature up to about 23°C. in about 80 to about 110 days.

Varablue (van Andel, 1998a)

It is a new plant variety of freesia with early flowering response and is suitable for growing as cut flower. This invention relates to a new and distinct variety of Freesia which originated by crossing two unpatented, unnamed varieties of this creation at the Van Staaveren Nursery in Aalsmeer. The female parent is identified as 3208-AT-1 and the male parent is identified as 841-J. Among the features which distinguish the new variety from other presently available and known commercial cultivars are attractive flowers and early flowering response after planting which distinguishes the new variety to be Suitable as a cut flower variety. Asexual reproduction of the new variety by corms and cormlets in Aalsmeer, The Netherlands has established that the distinctive characteristics can be transmitted through succeeding generations.

Varatwisno (van Andel, 1998b)

It is a new variety of freesia suitable for growing as a cut flower under glass. This invention relates to a new and distinct variety of Freesia which originated by crossing two unpatented, unnamed varieties at the Van Staaveren Nursery in Aalsmeer. The female parent is identified as 884135 AT3 and the male parent is identified as 72242H5. Among the features which distinguish the new variety from other presently available and known commercial cultivars are an attractive flower and fast flowering after planting which distinguishes Varatwisno as suitable for use as a cut flower variety.

Varayel (Van Andel, 1998c)

It is a plant variety of freesia suitable for growing as a cut flower under glass. This invention relates to a new and distinct variety of Freesia which originated by crossing two unpatented, unnamed varieties at the Van Staaveren Nursery in Aalsmeer. The female parent is identified as 873.16-AT1 and the male parent is identified as 87310AT4. Among the features which distinguish the new variety from other presently available and known commercial cultivars are attractive flowers and early response to flowering after planting which distinguishes the new variety as suitable for use as a cut flower variety. Asexual reproduction of the new variety by corms and cormlets has established that the distinctive characteristics are transmitted through succeeding generations. The varietal denomination of the new variety is 'Varayel.

Golden Amor (Cho *et al.*, 2009)

It is a double type freesia having golden colour (*Freesia hybrida* Hort.). 'Golden Amor' was developed from a cross between 'Golden Crown' and 'White Wing' followed by seedling and line selections at the National Horticultural Research Institute, Rural Development Administration (RDA) in 2005. Growth and flowering characteristic tests were conducted from 2000 to 2005. 'Golden Amor' has large double type flower with bright golden color (RHS 12B). It has multiple florets, good flower shape, upright leaves and stalks. The average yield of 'Golden Amor' was 6.0 cuttings per plant in greenhouse carried out from 2003 to 2005. It has good production of cormlets, vigorous growth and resistance to *Fusarium* and virus.

6.0 Genetics and Breeding

In an interesting presentation on the history of breeding of freesia, Morgan (1929) graphically described the story of freesia breeding. According to him, several hybridists recognized the possibilities of the freesia. Prior to 1900, work had been mainly with pale-tinted selections from seedlings of *F. refracta* alba, also crosses with *F. leichtlinii*, a large yellow variety discovered by Max Leichtlin among some neglected seedlings in an Italian garden (about 1874). One of the first to use

F. armstrongii in crosses with *F. refracta* alba was Dr. A. Ragionieri of Castello, Italy. At around the same time similar crosses were made at Kew Gardens, England, and by C. G. Van Tubergen of Haarlem, Holland. Results of these and subsequent crosses were several distinct strains of freesia with colors ranging from light yellow to orange, pink to deep red, brown, mauve to deep bluish-lavender, and variegated.

In recent past many new freesia cultivars suitable for cut flower production have been developed. Cho *et al.* (2013a) developed a freesia cultivar 'Dancing Flame' for the cut flower from a cross between a seedling 'Vulcano' and 'Sailor', which is scarlet double flowering and purple single flowering, respectively, and 'Figaro' which is bright red color and semi-double flowering. Dancing Flame is red color (RHS R44B) and single flower with yellow center color (RHS Y9A). The growth of the plant is vigorous and the average height is 93.4 cm. The average flower width is 60 mm, number of floret per stalk is14.3, and stalk is 13.2 cm length. The average yield, 7.8 cuttings per plant, is 2.5 stems per plant. The average days to first flowering is 133.

Cho *et al.* (2013b) also developed a new freesia cultivar 'Grace Girl' from selfing lines between 'Athene' and 'Vienna', followed by seedling and line selections. 'Grace Girl' has double flowering florets with bright white color (RHS 155D). It has good production of cormlets, vigorous growth and low virus infection. It takes approximately 135 days after planting for flowering and its vase life is 7.0 days. Based on greenhouse yield trials, the average yield is 8.3 flower stems per plant.

Similarly, Choi *et al.* (2014) developed a freesia cultivar 'Shiny Bell' for the cut flower from a cross between 'Teresa' and 'Rossini', which is white, semi-double flowering and red double flowering, respectively. 'Shiny Bell' is red color of RHS W155C, and double flower. The growth of the plant is vigorous and the average height is 109 cm. The average flower width is 5.8 mm, number of floret per stalk is 11.3, and stalk length is 8.5 cm. The average yield, 7.3 branches per plant, is 3 stems per plant. The average days to first flowering of 'Shiny Bell', is 143.3, and the average yield is 4.0 cormlets per plant.

7.0 Growth and Flowering

Endogenous carbohydrates were investigated during development of florets of *Freesia hybrida* 'Polaris' from inflorescences attached to the plant as well as from cut inflorescences placed in water (van Meeteran *et al.*, 1995). During development at the plant the amount of glucose, fructose and sucrose increased about 15–20 times; maximum was reached at anthesis. For detached inflorescences, at anthesis the amount of sugars in the 5th floret (from the base of the inflorescence) was only about 20 per cent as compared to attached inflorescences, however, opening of the florets at position 5 was normal. The increase of sugars during development of buds of cut inflorescences could not only be accounted for by starch hydrolysis. Decreasing the stem length of cut inflorescences greatly reduced the increase in dry weight and in total carbohydrates of the buds. However, there was no influence of stem length on the number of flowering florets/inflorescence (van Meeteran *et al.*, 1995).

Corms used for freesia reproduction must be regenerated every few years due to their ageing process, decreasing vigour and increasing problems associated with diseases and pests (Bach, 1992; Startek *et al.*, 2005). The rate of senescence and quality deterioration depend both on cultivar characteristics and growing conditions as well as corm reproduction methods. The production of healthy corms, their preparation in order to obtain even, abundantly flowering, good quality plants as soon as possible are of utmost importance in freesia cultivation (Imanishi, 2002; Startek *et al.*, 2002, 2005).

The effect of corm storage treatments and gibberellic acid foliar spray on plants and their interaction on some morphological and physiological traits, aiming to plant it in a large scale, El-Bably and Samia (2016) found that plant height, fresh and dry weights of leaves were significantly increased as a result of corm cold storage treatments. Moreover, all cold storage treatments tended to induce a steady and significant increase in flowering precocity, spike length, as well as fresh and dry weights of spike, rachis length, number of florets per spike, vase life, cormels number and fresh and dry weights of the daughter corm by prolonging cold storage period in both seasons as compared to the control plants. The percentages of nitrogen, phosphorus, potassium and total carbohydrates in the leaves increased as a result of these treatments. Also, cold storage treatments tended to induce a steady and significant increase in photosynthetic pigments such as chlorophyll (a), chlorophyll (b) and carotenoids as well as most of the studied characteristics which were significantly increased due to application of gibberellic acid. Referring to the interaction between the two factors, data showed the superiority of storing corms for 6 weeks at 5°C coupled with spraying plants whith gibberellic acid at 200 ppm, 30 days after planting corms, three times at 3 weeks interval. In brief, for obtaining plants of *Freesia refracta* with good morphological and physiological traits, it is recommended to store corms for 6 weeks at 5°C coupled with gibberellic acid foliar spray at 200 ppm, 30 days after planting corms, three times at three weeks interval (El-Bably and Samia, 2016).

8.0 Propagation

8.1.0 Seeds and Vegetative Propagation

There are two main methods of propagating freesias: by seed and by corm division. Both have high success rates. Freesias grown from seed usually take 8 to 12 months to bloom, while plants grown from divided corms will take a few years. The other main method of freesia propagation is corm division. Freesias grow from corms, which are similar to bulbs. These are called cormels, and each can be grown into its own new freesia plant. (https://www.gardeningknowhow.com/ornamental/bulbs/freesia/propagating-freesia-plants.htm).

In the Northern India plains corms can be planted during October-November and February-April in hills. In the western countries corms are planted from February to May, 12 cm deep and in a sunny, free draining position. Early planting will allow the corms to establish, but will not result in particularly early flowering as soil temperatures must fall below 17°C before the corms will set flowers, and this does not happen until later in autumn (https://nzbulbs.co.nz/media/ecom/factsheets/freesias.pdf).

The effects of the storage temperature of corms, the corms size, differentiated soil moisture and mulching; on some morphological traits of the plants, and the yield coefficient and the quality of corms were studied by Startek *et al.* (2002). Two sizes of corms were used in 10 cultivars and the corms were stored in a special climatic chamber at the temperature 28-30°C, air humidity 85 per cent; and in a store room where the maximum temperature was 18° C, and air humidity 55- 65 per cent. Corms were always planted in May in a plastic tunnel and in open ground in the field. In both the places the beds can be covered with sawdust or without sawdust cover. Two different levels of soil moisture and the same level of fertilization were maintained. There were differences between the results obtained in particular years of the experiment. The greatest effect on the developmental stages and quality of freesia were from the storage conditions, and the course of the weather, soil temperature in particular, the growing site and the size of corms. Soil moisture and soil cover had the least effects. There were also considerable differences between the cultivars, especially between 'Rapid' group and the others.

Seed vernalization of the variety Buttercup resulted in reduced germination. Shading of young freesia plants had little effect on growth. A soil dressing of chaff had a good effect on the growth of Buttercup. Reduced soil temperatures resulted in earlier flowering, but flower quality was better with higher soil temperatures. Corm weight in Buttercup was improved by soil coverings of reeds or chaff. Cross-pollination between Buttercup and *F. refracta alba* produced more seed than self-pollination (Kragtwijk, 1954).

8.2.0 Micropropagation

True-to-type plantlets of *Freesia* × *hybrid*, cultivar 'Royal' were generated from callus after 27 months of sub-culture in liquid medium (Stimart and Ascher, 1982). Callus was initiated from young flower pedicels cultured on semi-solid Linsmaier—Skoog (LS) medium supplemented with 5 mg/l of 2, 4-D and 0.5 mg/l kinetin, and transferred to the same medium in liquid form without hormones and thereafter sub-cultured every 7–10 days. Liquid cultures with 2.4–4.3 g of callus per 25 ml medium produced largest increases in callus fresh weight. Callus generated the most shoots when cultured on LS medium supplemented with 5 mg/l of kinetin and incubated in the light, while fewer plantlets were produced when no growth regulator or GA_3 or PBA were used. Callus cultures incubated in continuous darkness did not form shoots (Stimart and Ascher, 1982). A tissue culture method has been developed to successfully establish a rapid *in vitro* propagation system for virus-indexed freesia (*Freesia refracta*). Using 6 freesia cvs., apical shoot-tips (0.5–0.7 mm) were established (Stage I) on an agar medium containing Murashige-Skoog salts, 2.0 mg/1 kinetin, and 0.1 mg/1 naphthylacetic acid (Foxe, 1991). Plantlets were transferred to a similar medium containing 4.0 mg/1 kinetin and no auxin 6–7 weeks later for shoot proliferation (Stage II). Plantlets were subcultured on Stage II medium every 4–5 weeks as needed and yielded approximately 12 axillary shoots per culture. Roots developed readily in this medium suggesting that it may not be necessary to transfer the shoots to a specific root-inducing medium. Freesia cvs. were successfully transferred to the field and have produced true-to-type plants during a three year trial. Virus indexing was accomplished using several methods, including electron microscopy and enzyme-linked immunosorbent assay.

Ali and Morteza (2015) studied the effects of explant sources, plant growth regulators, sucrose concentrations, and light conditions on in vitro cormlet formation of freesia (*Freesia ×hybrida Bailey* 'Argenta'). Interestingly, it was observed that the pupae had lower contamination levels compared to mother corms. Using 40 per cent sodium hypochlorite solution for 40 min, contamination levels of daughter corms (DC) and mother corms reduced to 19.80 and 46.40 per cent, respectively. Moreover, DC showed the highest regeneration frequency. In the first experiment, 6.67 cormlets were directly produced per pupa (cold storage-produced corm) on Murashige and Skoog (MS) medium supplemented with 6 mg/l 1-naphthaleneacetic acid (NAA), 1 mg/l 6-benzylaminopurine (BA), and 60 g/l sucrose, when cultures were stored in the dark. In the second experiment, on average, 5.67 shoots were proliferated per pupa explant in the presence of 4 mg/l BA and 2 mg/l Kinetin (Kin).

For efficient regeneration of this flower from young inflorescence and rachillae in tetraploid, Gao *et al.* (2010) developed a simple *in vitro* micropropagation protocol. Explants of Freesia hybrida can regenerate plantlets through somatic embryogenesis via two kinds of pathways, that is, directly from the epidermal cells or indirectly from an embryonic callus, depending on the exogenous plant growth regulators (PGRs) used in the culture media. In direct embryogenesis, when the explants were cultured on Murashige and Skoog (MS) medium supplemented with 11.43 μM indole acetic acid (IAA) and 4.44 μM 6-benzylaminopurine (6-BA), the induction rate was 84 per cent for young inflorescence and 100 per cent for rachillae. After the multishoots were subcultured on the rooting MS medium containing 1.08 μM α-naphthalene acetic acid (NAA), the rooting rate was close to 100 per cent. In indirect embryogenesis, embryonic calluses were formed when the culture medium contained 22.22 μM 6-BA and 4.52 μM 2,4-dichlorophenoxy acetic acid (2,4-D), and the induction rate was 92.4 per cent for young inflorescence and 100 per cent for rachillae. After the embryonic calluses were transferred to the medium supplemented with 11.43 μM IAA and 13.33 μM 6-BA, they could develop into plantlets with roots.

Doi *et al.* (1992) observed efficient shoot regeneration when the primordia were cultured on modified MS agar medium with 1/10 strength NH_4NO_3 and 1ppm BA under 16 hr day length. When the regenerated shoots were transplanted on hormone free medium and cultured under 16 hr day length and 0.1 per cent CO_2 enriched conditions, root system developed well and shoots grew vigorously. Plantlets cultured under CO_2 enriched conditions were easily acclimatized to *ex vitro* as compared with the plantlets cultured under ambient atmospheric conditions. Consequently, one corm and three cormlets were produced per plant, indicating that more than 500 corms were obtained per year by originating from 1g shoot primordia. Corm formation was also observed under in vitro conditions, if regenerated shoots were exposed to chilling at 15°C for 10 weeks and then moved to 23°C (Doi *et al.*, 1992).

8.3.0 Storage of Corms

The bulb production phase consists of five steps (1) harvesting, grading of the bulbs into planting stock and marketable bulbs, and pre-planting storage; (2)

planting, rooting and low temperature mobilization for flowering and/or bulbing; (3) leaf and flower stalk growth; (4) flowering; and (5) increase in bulb size and/or numbers (De Hertogh, 1980; Koksal *et al.*, 2010).

In general, bulbs and corms of ornamental plants are highly recommended to store in ambient temperature for certain periods of time depending on the production strategies (Koksal *et al.*, 2010). During the storage period, many physiological changes occur in bulbs and corms. The corms of *Freesia reflecta* 'Polaris' and 'Aladin' were stored in either non-cold (18-25°C and 55-65 per cent RH) or cold (5±1°C and 70-85 per cent RH) storage conditions. Total soluble sugars and loss of weight in corms stored in both conditions were evaluated for 120 days with 40 days intervals. Loss of weight of corms increased linearly in both the storage conditions. Significant varietal differences were detected between the cultivars. No significant difference between storage conditions was detected in corms of freesia cultivars (Koksal *et al.*, 2010).

9.0 Soil and Climate

9.1.0 Soil

Freesia can be grown in slightly sandy soil that has been amended with moisture retaining organic material. It prefer a location in full sun but can tolerate light shade. When freesia is actively growing and blooming, the soil should be kept moist.

9.2.0 Climate

9.2.1 Temperature

The growth rate of freesia is influenced by temperature. The results show that there is a trend of speeding growth at higher temperature. The growing speed and temperature are relative particularly from sowing to emergence and in anthesis. The growing period will be shortened by higher temperature by 0.32 and 1.04 days/degree (Jin *et al.*, 1994). Leaf height can also be affected by temperature. Mansaur (1968) reported that flower height decreased with decreasing day temperatures between 21 and 12°C. He also found that final leaf height was reduced when flower inducing conditions (10°C) were provided soon after planting and that vernalization of corms at 5°C induced early flowering and significantly shorter flowering spikes (Wulster and Gianfagna, 1991).

Fifteen-day-old seedlings of *Freesia hybrida* 'Shangnong Jinhuanghou' were exposed to 30°C for 144 hr to investigate the physiological response to high temperature stress (Yuan *et al.*, 2011). The results showed that high temperature stress caused a significant decrease in chlorophylls content and a significant increase in malondialdehyde (MDA) content and electrolytic leakage (EL) in *Freesia* seedlings. Under high temperature stress, accumulation in proline content was observed, while a slight decrease in soluble sugar content occurred in the first 6 hr exposure to stress, followed by a constant increase. Superoxide radical (O_2^-) producing rate increases in the first 2 hr exposure and then significantly decreases up to 72 hr, followed by an increase at the end of stress.

9.2.2 Light

According to Startek *et al.* (2002), *Freesia* x *hybrida* is photoperiodically indifferent plant, with low thermal requirements and short production cycle but very sensitive to little light and too high soil temperature (Berghoef *et al.*, 1986; Berghoef and Zevenbergen, 1990; Doorduin and van Winden, 1986; De Hertogh and Le Nard, 1993).

In two winter seasons the possibilities of photosynthetic lighting in a winter crop of freesia cv. 'Blue Heaven' were investigated (Doorduin, 1992). The equipment used consisted of SON-T lamps. The effects of the daily light sum (unlit, 155, 215 and 310 Wh/m^2/day), lighting period (16 and 24 hours per day) and starting time of lighting on growth, yield, quality and vase life were studied. Light-integral-based and current-level-based lighting resulted in higher stem and corm production, stem quality and flowering as the daily light sum was higher. Starting at a bud length of 2 cm gave the same results as starting 3 weeks later at a bud length of 15 cm. Lighting for 24 hours per day had adverse effects on stem production, compared to 16 hours lighting (8 hours at night), whereas the effects on corm and cormlet production were favourable. Twenty-four hours lighting gave less stem yield when started at a bud length of 2 cm than at a bud length of 15 cm. As the daily light sum increased the percentage of visible leaf necrosis also increased. It was concluded that light-integral-based and current-level-based lighting till a lightsum of 200 – 225 Wh/m^2/day offered most prospects for freesias grown in the winter. To start with lighting was only useful at a bud length from 15 cm. Twenty-four hours lighting in an early stage was disadvantageous.

A greenhouse study was carried out in Belgium to investigate the effects of supplementary light (3000, 4500 and 6000 lux) on the stem length and weight of *Freesia* (*cvs.* Elysee Ricasee and White Lace). Bulbs were planted in September and supplementary light was used from 20 December onwards when outdoor light levels were less than 50 W/m^2 (Blindeman, 2000). The cultivar Elysee Ricasee reacted more to supplementary light than cv. White Lace. Supplementary light resulted in longer stems, improved flower quality, and heavier plants and bulbs. Range of supplementary light between 20 December and 29 February vary from 201 BF/m^2 (3000 lux) to 302 and 398 BF/m^2 (4500 and 6000 lux, respectively).

10.0 Cultivation

10.1.0 Greenhouse Cultivation

It has been demonstrated for many glasshouse crops that increasing the CO_2 concentration lead to improved quality and yield, while concentrations below the ambient level lead to loss of quality and yield. Relatively few experimental data are known for freesia in this area and on freesia holdings CO_2 enrichment is applied only to a limited extent (Doorduin, 1990). For this reason in an experiment 4 CO_2 concentrations were combined with 4 plant densities to investigate the effect of CO_2 on quality and yield of the cultivar 'Blue Heaven'. The CO_2 concentrations realised were 265, 360, 560 and 860 ppm. Plant densities were 57, 78, 100 and 121 corms per net square metre. The concentration of 265 ppm resulted in a 20 per cent lower yield of stems, corms and cormlets and a reduction of stem quality as compared to 360 ppm. Increasing the concentration to 560 ppm resulted in a 20 per cent higher yield and an improved quality as compared to 360 ppm. Levels exceeding 560 ppm did not improve yield and/or quality. Higher plant densities did not give higher yields but evidently reduced quality. Vase life was not affected by differences in CO_2 concentrations or plant densities (Doorduin, 1990). Greenhouse conditions only accounted for a very low percentage of the large variance in firmness of stems from the different nurseries, with low soil- and air temperatures as favourable conditions (Slootweg, 2005).

10.1.1 Hydroponics

A study was conducted to determine the effects of various growth substrates on the growth and flowering of hydroponically grown *Freesia hybrid* 'Gold Rich'. Perlite, peat moss and a perlite: peat moss mixture (1:1 ratio, v/v) were used as the growing media. The greatest plant height before flower bud differentiation was attained using mixed medium compared to the others (Lee, 2017). Among the substrates tested, it was found that use of mixed perlite: peat moss medium (1:1 ratio, v/v) for hydroponic culture of freesia to be the best, as the use of this medium improved the physical properties of the plants, producing the best results in terms of plant growth and cut-flower quality.

10.2.0 Manuring and Fertilization

10.2.1 Foliar Feeding

In natural conditions in temperate areas, freesia flowers bloom in August and September, due to vernalization completed during the winter season. It is possible to achieve this effect to advance flowering, with a cold treatment of corms. However, the quality of flowers decays. The effect of applying a foliar fertilizer once a week, to corms previously treated with cold forcing, on the quality of the stems was studied. The measured variables were length of rod and inflorescence and number of flowers per inflorescence. The application of a foliar fertilization promotes a delay in flowering. The facts also showed that there is no improvement in the quality of the cut flowers in most varieties and measured variables (Piovano *et al.*, 2017).

10.2.2 Organic Supplements

Recently in agriculture, there has been a growing interest in new proecological technologies whose main aim is the protection of biological material against negative effects of direct contact with the environment. The effect of spraying liquid coconut (0, 100, 200 ml/l) and seaweed extract (0, 2.5, 5 ml/l) on flowering growth and vase life in plant freesia was studied (Alasadi, 2016). The results showed superiority of treatment with liquid coconut 200 ml/liter on the rest of concentrations in all traits.

The use of humic fertilizers instead of mineral fertilizers is one of the means to reduce the pollution caused by the use of mineral fertilizers. Humus is a rich source of nitrogen and phosphorus and contains the acids of the humic and folvoulic (Abdalla, 2019). The effect of spraying foliar of the biostimulants Izomen and humus on the growth and flowering of Freesia plants was reported by Abdalla (2019). The corms in similar size were planted in pots with a diameter and height of 25 cm, which filled with 2.5 kg of sterilized loam soil. After 50 days of planting, the plants sprayed with humus at 0, 1.5, 2.5 ml/l and after five days sprayed with Izomen at 0, 1.5, 2.5 ml/l. The different concentrations of biostimulants were sprayed three times, the period between one spray and another 15 is days. The results showed that the spraying of humus at 2.5 ml/l significantly increased the plant height, the number of leaves and the leaf content of chlorophyll recorded (29.56 cm, 8.33 and 58.43 per cent) respectively.

Abbass *et al.* (2020a) studied the effect of spraying organic acid (Lag-Humus) and chelated iron (Fe 6 per cent, EDTA) on the growth and flowering indicators of freesia. The results showed that spraying organic acid at a concentration of (6 ml/l) with chelated iron at a concentration of (40 mg/l) significantly increased the dry weight of the total vegetative, the leaves content of total chlorophyll, floret diameter, the length of the flower inflorescence, vase life and iron content.

10.3.0 Irrigation

Water-saving irrigation management strategies are among the options available to horticultural growers to reduce water consumption and improve water use efficiency (WUE). Possible strategies include deficit irrigation (DI), regulated deficit irrigation, and partial root-zone drying (Abdel-Fattah *et al.*, 2020). Four irrigation intervals of 3, 6, 9 and 12 days with four irrigation levels of 100 (control), 120, 80, and 60 per cent cumulated pan evaporation (CPE) were tested. Results showed that irrigation scheduling significantly influenced the stated items. The mean seasonal irrigation water for the highest economical yield of fresh weight of spike was 12.58 cm, while it was 14.05 cm for new corm under 9 days and 60 per cent or 12 days and 80 per cent, respectively with mean value of 13.32cm.

In protected cultivation, intensive fertilizers and chemicals used for nutritional purposes and irrigations practiced unconsciously causes salt accumulation in the root zone resulting decrease in quality and quantity of production. It was found that saline irrigation water levels decreased the corm, cormel and flower yield, flower stem length and diameter, flower numbers and flower spike length in all freesia varieties studied. The threshold values for Athena, Oberon, and Cordula varieties were determined as 2.7, 1.0, and 3.0 dS/m, respectively (Aydinsakir *et al.*, 2010).

10.4.0 Growth Regulators

Freesia hybrida is grown extensively in Europe and Japan as a cut-flower crop. Its vividly colored and fragrant inflorescence would also make freesia a potential container-crop commodity if stem length could be reduced without adversely affecting flower quality. Gianfagna and Wulster (1986) have reported that height can be reduced in freesia by use of the growth regulators ancymidol (A-Rest) and paclobutrazol), although flowering was delayed by these treatments (Wulster and Gianfagna, 1991).

Growth and flowering of *Freesia hybrida* for the container-plant market can be controlled chemically using growth retardants and environmentally by cold storage of corms at 5°C for 2 to 6 weeks before planting. Corms stored at 5°C for 4 weeks flowered 20 days earlier than corms not stored at 5°C. Preplanting at 5°C, storage life of corms also reduces leaf and flower height. An ancymidol soil drench (3 mg) reduced leaf height and flower height by more than 50 per cent and delayed flowering by 9 days. Combining growth regulator application with cold storage of corms produced the greatest reduction in leaf height and flower height. Moreover, plants flowered earlier than controls when corms were stored for at least 4 weeks, regardless of growth regulator treatment (Wulster and Gianfagna, 1991).

The effect of polyamines on growth, flowering and corm production of 'Golden Wave' and 'Blue Sea' cultivars of freesia was studied (Rezvanypour *et al.*, 2016). For this purpose, freesia plants were sprayed at 35, 70 and 100 days after planting with concentrations of 0.5 and 1 mM of putrescine, spermidine and spermine and distilled water (control). The results showed that spermine treatment delayed flowering time. But spermidine treatment accelerated flowering and also increased leaf length, flowering stem height, inflorescence height, diameter and length of florets and number of lateral stems in both cultivars. Maximum number of corms was obtained in spermidine treatment. The results revealed that polyamines, and especially spermidine and putrescine, could increase yield and improved quality of flower and corm of freesia.

11.0 Diseases and Pests

11.1.0 Diseases

11.1.1 Bacterial Soft Rots, Leaf Spots, Blights, Wilts (*Erwinia, Pseudomonas, Xanthomonas* spp.)

Pathogens causing spots, blights, and soft rots can also cause vascular wilt if the infecting bacteria become systemic. Above ground plant parts become yellow, droop, wilt, and die.

To control this, avoid using infected corms, soil solarization may be useful. As bacteria enter through wounds, it is better to handle the corms with care.

11.1.2 Dry Rot (*Stromatinia gladioli*)

The initial infection attacks corms either in storage or in the soil. After the initial infection, decay spreads up to the leaf bases, dry rot causes leaves to yellow and die. Plants are stunted and fail to bloom, leaf bases or stems rot near the soil and tissue appears shredded. Bulbs develop dark brown, sunken lesions with raised margins.

Use of pathogen free corms is the solution and avoid growing freesia in infected soil.

11.1.3 Fusarium Wilt—*Fusarium oxysporum*

Fusarium wilt causes foliage to become yellow, curve, wilt, then turn brown and die. Older leaves usually die first in infected plants, followed by death of the entire plant, young infected plants often dry. Soil solarization before planting is recommended, provide proper sanitation and cultural care to reduce plant susceptibility to infection and damage. Also avoid overwatering, provide good drainage and refrain from using excessive fertilizer.

11.1.4 Root, Stem, and Crown Rots (*Fusarium, Pythium, Phytophthora,* and *Rhizoctonia* spp.)

Dull-colored foliage or wilting followed by yellowing of plants are often the first above ground symptoms of root and crown disease. Plants may be stunted and can eventually die.

Control measures involve good cultural practices and sanitation is critical control measures for all root and crown rots. Avoid excess moisture in the root zone and minimize other plant stresses. Use only pathogen-free plants or bulbs and plant them in well drained soil.

11.1.5 Viral Diseases (Bean Yellow Vein Mosaic Virus and Cucumber Mosaic Virus)

Viruses can slow plant growth and change the appearance of foliage, flowers, and fruits. Virus-infected leaves can become spotted, streaked, mottled, distorted, or stunted. Veins may lose their color or develop outgrowths. Flowers can be dwarfed, deformed, streaked, or faded, or they can remain green or develop into leaf like structures.

Dig out and destroy virus-infected plants. Prevent spreading viruses by using good sanitation and cultural practices. Use only virus-free plant material and consider growing virus-resistant cultivars.

11.1.6 Petal Specking (*Botrytis cinerea*)

Freesia flowers grown in the Netherlands are frequently rejected by UK wholesalers and retailers because of *Botrytis cinerea* infection-induced petal specking. Such rejections result in loss of goodwill as well as substantial economic losses to both growers and importers (Darras *et al.*, 2004). Postharvest petal specking caused by *Botrytis cinerea* is a major concern for freesia growers and sellers in Holland and the UK. Control of *Botrytis* disease can be achieved by pre and/or post-harvest chemical applications. However, disease control with conventional fungicides carries the risk of development of fungicide resistant strains (Darras *et al.*, 2005). Treatment of cut freesia var. Cote d'Azur flowers with methyl jasmonate (MeJA) vapour suppressed petal specking caused by *Botrytis cinerea* infection. MeJA efficacy was concentration and incubation temperature dependent. Disease severity, lesion numbers and lesion diameters decreased with increasing MeJA concentration from 0.025 to 0.1 µl MeJA/l (Darras *et al.*, 2007).

11.2.0 Pests

11.2.1 Green Peach Aphid (*Myzus persicae*)

Green peach aphids can attain very high densities on young plant tissue, causing water stress, wilting, and reduced growth rate of the plant. The major damage caused by green peach aphid is through transmission of plant viruses. Indeed, this aphid is considered by many to be the most important vector of plant viruses throughout the world.

Diseases and Pests of Freesia

Root/Crown Rot

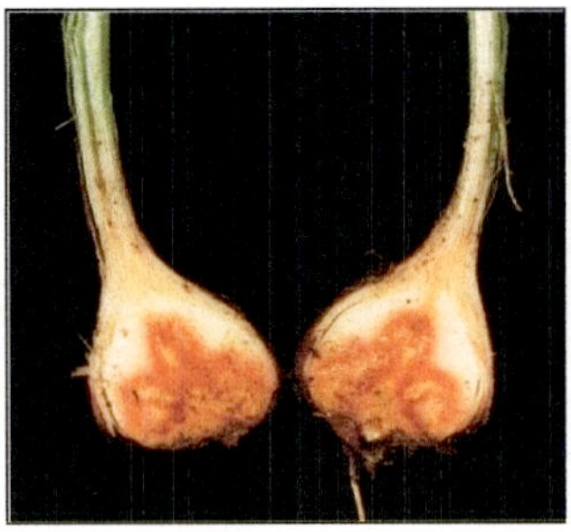

Fusarium Wilt

Bacterial Soft Rot

Aphids

Snail

The overwintering behavior of green peach aphid, which in many areas is restricted to *Prunus* or other relatively restricted sites, has fostered research on techniques to reduce aphid abundance and disease transmission to vegetables, by either removing the overwintering site or by eliminating the aphids before they disperse to freesia. Application of effective insecticide should be done before flowering.

11.2.2 Tulip Bulb Aphid

Bulbs infested with bulb mites may rot and fail to produce new growth, or new growth may be off color, stunted, and distorted. Although the bulb mite is not considered a primary pest of bulbs, it is often responsible for serious losses; the slightest injury to a bulb will allow bulb mites to enter and become established. Bulb mites damage bulbs by penetrating the outer layer of tissue and allowing rotting organisms to gain entry. Plant growth may be reduced and bulbs may rot in storage or in the field. Infested fleshy bulb scales commonly turn reddish brown, bulb mite-infested roots, rhizomes, or basal stems can become soft and decayed.

To control this, inspect bulbs carefully before planting for signs of bulb mites, and destroy any that are infested, decayed, or are soft when squeezed. Avoid injuring bulbs during digging, handling or storage. After harvest, clear all vegetation to allow organic matter to decompose; this reduces populations of the mite. Flood irrigation or heavy rains during the winter may reduce mite levels in the soil.

12.0 Harvesting and Postharvest Management

12.1.0 Harvesting

Harvesting longer stems lead to a longer vase life with a maximum of 2 extra days. All other factors such as soil type, fertilisation, plant density, days until harvest, crop protection and production had very little or no effect on vase life and bud opening. After transport simulation of the stems, the correlation between vase life and bud opening and greenhouse conditions was much lower than without the simulation, but the data pointed in the same direction. The first and second lateral stems reacted in the same way as the main stems (Slootweg, 2005).

12.2.0 Postharvest Management

Postharvest trials showed marked differences in the length of vase life and in bud opening of *Freesia* stems of the same cultivar, obtained at a similar time from various nurseries (Slootweg, 2005). The factors that cause these differences in vase life and bud opening are yet unknown. The main factors affecting vase life and bud opening were temperature, light level, CO_2 level, and relative humidity. The values of these factors during the last two weeks of production explained about half of the variance in bud opening and vase life. The results of the investigation provide useful advice for growers on how to optimise greenhouse conditions in order to get the best possible postharvest quality of their product (Slootweg, 2005).

12.2.1 Storage of Flowers

Cut freesia flowers have a relatively short vase life (5 days) and two methods have been developed for the storage of these cut flowers; namely, 'wet' and 'dry' storage. Wet storage is carried out in containers containing water or preservative solutions and the flowers are usually kept at temperatures between 0.5 and 4°C. In this method of storage, the flowers do not require packaging, however, they occupy more space in the storage room, compared with dry storage. During wet storage, nutrition breakdown occurs at higher rates than for flowers stored dry (Halevy and Mayak 1981). During wet storage, bud development and senescence of flowers occurs more rapidly compared to flowers held dry. The flowers could be stored wet at 0-0.5°C for up to14 days, and dry for 21 days. Longer-term storage resulted in a

Harvesting and Marketing of Freesia

Harvesting Stage in Greenhouse

Cut Flowers

Freesia in Market

reduction in vase life and poor quality flowers. Dry storage is the more commonly used method for long-term storage of cut flowers. For dry storage, the cut flowers are graded after harvest, they may also be pulsed and placed into sealed plastic bags or cardboard boxes (Zencirkiran, 2002).

12.2.2 Vase Life

Some physiological changes of cut Freesia inflorescences (*cv.* Oberon and Ballerina) were evaluated in distilled water at room temperature with or without STS (silver thiosulphate) treatment for 24 hours prior to placing them in distilled water (Amariutei *et al.*, 1994). The ultrastructure of cells from freshly opened floret petal tissues was compared to that of open floret petal tissues from inflorescences of Oberon cultivar held 8 days in distilled water continuously or treated with STS, held 7 days in distilled water. The enhancement of senescence of cut flowers are generally related to the several factors, such as depletion of carbohydrates and other nutrient reserves, increase of ethylene production and action, and reduction of water absorption by flowers (Borochov and Woodson, 1989). Flower preservative solutions are composed of sucrose, bactericides, weak acid, anti-ethylene agent, and growth regulator. Jin *et al.* (1999) studied the effects of pulsing treatment and to develop techniques for extending vase life and improving flower quality in cut freesia. Thirty minutes dipping treatment of STS 2 mM followed by 20 hr pulsing in sucrose 10 per cent + BA 10 ppm + 8-HQS 300 ppm solution showed the best results in vase life and flower quality of cut freesia when kept in vase water. This pretreatment extended vase life by 24.7 per cent than control, and improved quality of cut freesia significantly in flower diameter, per cent flowering (35.4 per cent), fresh weight, water uptake, and carotenoid content, and depressed ethylene production and respiration rate.

Inflorescences of both cultivars treated with STS had greater values for fresh weight, water uptake, conductivity of holding water, floret opening and vase life, than those held continuously in distilled water, but smaller values for transpiration, respiration and membrane permeability measured as electrolyte leakage from inflorescences. Floret petal tissues of Oberon cultivar from STS-treated inflorescences held 7 days in distilled water displayed fewer ultrastructural changes than those observed in the petals of florets from inflorescences not treated with STS, when compared with the ultrastructure of fresh petal tissues.

Recent work indicates that the use of nano-silver particles (because of a high area/volume ratio) shows antimicrobial effects and has been proven as a beneficial agent in preservative solutions of several cut flowers (Hajizadeh, 2016). Hence, the effects of different concentrations of nano silver particles in the vase water (5, 10 and 15 ppm) combined with 3 per cent sucrose on cut freesia vase life were studied. Distilled water was used as a control. Measurements included the water uptake, number of opened buds, petal membrane stability index, lipid peroxidation and vase life. Results showed a positive effect of nano silver on cut freesia flower vase life. Flowers placed in 10 ppm nano silver and 3 per cent sucrose had the longest vase life, as compared to controls having a vase life of 5 days. The number of opened florets of that treatment was also improved, as compared to the control.

Petal membrane stability index decreased during vase life, but no effect of nano silver was noted. The 10 ppm nano silver combined with 3 per cent sucrose showed the highest flower fresh weight. Hajizadeh (2016) concluded that using of 10 ppm nano silver combined with 3 per cent sucrose in vase solution of freesia flowers was the best treatment in preserving the best quality of cut freesia flowers. The longest spike vase life was also obtained in the 1-MCP (4 nl/l + 3 hours) treatment as 9.06 days, spike vase life of control group flowers was 6.33 days (Zencirkiran, 2010).

13.0 Biotechnology

13.1.0 Protoplast Culture

It has potential to be a model plant to unravel the molecular basis of plant specialized metabolites biosynthesis in family Iridaceae, which are crucial to genetically modify ornamental traits such as flower color and flower scents. However, the lengthy life cycle propagated from tissue culture and extremely low transformation rate severely decelerate functional characterization of genes through transgenic method in this iridaceous plant. Using calluses induced from young flower buds from one popular cultivar of *Freesia hybrida*, 'Red River®', an efficient protoplast isolation method was established and optimized, which was further employed in the gene transient expression system. Results demonstrated that the newly explored transient system was versatile and could be applied in numerous aspects, such as protein immunoblotting, enzyme activity, protein subcellular localization, protein–protein interaction, protein–DNA interaction and regulatory gene transregulation property assays. Moreover, calluses induced from another cultivar of *Freesia hybrida*, 'Ambiance' could also be used to isolate protoplasts for gene transient transfection analysis, indicating the wide applicability of this system. Results in this study provided new methods for the elucidation of gene function in *Freesia in vivo*, which might also be inspirational in other species with long life cycles or hardly to be transformed, such as plants in family Iridaceae.

13.2.0 Somatic Embryogenesis

In one of the earliest attempts on somatic embryogenesis, Bajaj and Pierik (1974) isolated segments of corm, stem, leaves, flower-buds and anthers from flowering *Freesia* cv. 'Ballerina' plants have been induced to form callus on a modified Murashige and Skoog's medium supplemented with NAA and PBA. The best callus was obtained from young flower-buds kept at 25°C in darkness. Callus could be subcultured in darkness on a medium containing auxin and cytokinin. Complete plantlets were obtained by transferring the callus to light on an auxin-free medium containing kinetin or PBA. Plantlets have also been regenerated from young anthers. Adventitious organ formation in explants and callus was strongly determined by the auxin/cytokinin ratio in the medium. Generally, auxins promoted rooting, whereas cytokinins enhanced budding. Light, in comparison to darkness promoted both root and bud formation while at the same time callus growth was retarded. Later, in the same laboratory (Pierik and Steegmans, 1975), flower-buds of 10 *Freesia* cultivars were grown *in vitro* and were induced to regenerate adventitious buds on a modified Murashige and Skoog medium, containing PBA and IAA. The formation of adventitious buds was strongly enhanced by growing the explants first in darkness and subsequently in light. Subcultured shoots, grown on a medium with IAA, could be rooted easily and viable plants were obtained.

13.3.0 Cormlet Regeneration

Ali and Khui (2015) studied the effects of explant sources, plant growth regulators, carbohydrates and light conditions on indirect cormlet regeneration and the induction of embryogenic callus of freesia (*Freesia × hybrida* Bailey 'Argenta'). Sections of two different types of explants, corms and pupae (cold storage-produced corms), were placed on Murashige and Skoog (MS) media containing different concentrations of plant growth regulators. The results showed that the highest percentage of callus induction (100 per cent), the highest callus growth (15 mm diameter) and the best type of calli were achieved for pupa explants grown on the medium that contained 4 mg/l 1-naphthaleneacetic acid (NAA) and 2 mg/l 6-benzylaminopurine (BAP) in the dark. Increasing BAP up to 3 to 4.5 mg/l resulted in the maximum number of regenerated cormlets from 1 cm^2 calli (2 cormlets) under light conditions.

13.4.0 Bioreactors

Kang *et al.* (2017) studied the bioreactor culture conditions for rapid production of freesia 'Shiny Gold' shoots. Young shoots (< 1 cm) obtained from freesia corm section *in vitro* were used as plant materials. As a basic experimental environment, 20 young shoots were inoculated into a 5 L balloon type bubble reactor which contained 1 L 1/2 strength MS medium supplemented with 30 g sucrose (3 per cent), and the aeration was 0.1 vvm (vessel volumes per minute). The bioreactors were placed in a growth room with 23°C temperature, 60 per cent relative humidity and 60 μmol/m^2/s light condition (16 hr/8 hr, day/night). The concentrations of MS media were set with 1/4, 1/2, full strength, medium volume 10, 20, 40 per cent, sucrose concentration 3, 6, 9 per cent, and aeration 0.1, 0.2, 0.4 vvm. After 4 weeks of cultivation, the growth indexes including the fresh and dry weight, and plant height were evaluated. At the same time, the consumption, pH, and EC of medium were estimated 4 weeks after incubating. The best results were achieved when 40 young shoots were incubated in a bioreactor in which 1 L of 1/2 strength MS medium supplemented with 6 per cent sucrose was used for the rapid production of freesia shoots. The shoots were 17 cm in plant height and 1.0 g in fresh weight only 4 weeks after incubation which could be a good plant material suitable for corm enlargement in vitro. No correlation was observed between the growth of freesia shoots and the consumption, pH or EC of medium (Kang *et al.*, 2017).

13.5.0 Genetic Transformation

A method for *Agrobacterium*-mediated transformation of *Freesia × hybrida* was described for the first time in 2015 (Uwagaki *et al.*, 2015). Cormlet-derived calli of two cultivars, 'Mosera' and 'Ishikawa f3' were co-cultivated with *Agrobacterium tumefaciens* strain EHA105 harboring the binary vector pIG121-Hm, which included hygromycin phosphotransferase gene and an intron-containing β-glucuronidase gene in the T-DNA region. Callus pieces were co-cultivated with *A. tumefaciens* on the callus proliferation medium [Murashige and Skoog (MS) medium containing 1 mg/l thidiazuron, 1 mg/l dicamba, 20 mg/l 32,52 -dimethoxy- 42 -hydroxyacetophenone, 1 per cent (w/v) glucose, 3 per cent (w/v) sucrose, and 0.2 per cent (w/v) Gelrite]. Then, they were cultured on the callus proliferation medium containing 300 mg/l cefatoxime and 10 mg/l hygromycin B. Hygromycin-resistant lines of both cultivars regenerated into plantlets after transfer onto MS medium containing 2 mg/l 1 3-indoleacetic acid and 3 mg/l 6-benzyl aminopurine and/or plant growth regulator-free MS medium. Transgenic plants were identified by β-glucuronidase assay and verified by Southern blot analysis. Two transgenic plant lines were obtained from 475 callus pieces of 'Mosera' and one transgenic plant line was obtained from 290 callus pieces of 'Ishikawa f3'.

13.6.0 Markers

Markers have been used for variety of purposes as detailed below.

13.6.1 Genetic Fidelity

In assessing the two regeneration pathways of embryogenesis, in terms of genetic and epigenetic fidelity of the regenerants, two kinds of molecular markers [amplified fragment length polymorphism (AFLP) and methylation-sensitive amplified polymorphism (MSAP)] were employed (Gao *et al.*, 2010). The AFLP analysis used 20 primer pairs that yielded 916 scorable bands among the donor plant and 11 regenerants from direct embryogenesis, of which 8 (0.87 per cent) were polymorphic. The regenerants from indirect embryogenesis had 1075 clear bands of which 3 (0.27 per cent) were polymorphic scorable bands from 18 primer pairs.

13.6.2 Transcriptomics

Based on Illumina platform, the transcriptome profiling of Freesia hybrida 'Pink Passion' was conducted by de novo sequencing method to reveal basic information and provide data on regulatory mechanism of flower color formation in freesia (Tang *et al.*, 2018). Totally, 49,503,460 short reads, corresponding to total 4.46 GB nucleotides, were yielded. These short reads were then classified into 74,192 unigenes, of which 42,934 were annotated in several databases, including Nr, Nt, Swiss-Prot, KEGG, COG, and GO. A total of 43,594 coding sequences were obtained and 25,409 unigenes were allocated to 128 KEGG pathways. The "metabolic pathways" (6642 counts, 26.14 per cent) were present as the largest category. The freesia transcriptome results revealed 205 unigenes involved in the flavonoid biosynthesis pathway and 18 unigenes in anthocyanin biosynthesis pathway.

13.6.3 Chalcone Synthase

The range of flower colours available in the freesias includes red, pink, yellow, white, blue, lavender, purple and bicolors. Thereby, the freesias are a good system for studying the secondary metabolism of flowers with different colors. Furthermore, the characterization of *CHS* gene from *F. hybrida* may help the studies of the duplication and evolution of the CHS gene family considering that there are only a few reports of *CHS* gene from Iridaceae (Sun *et al.*, 2015). Chalcone synthase (*CHS*) catalyzes the first committed step in the flavonoid biosynthetic pathway. The cDNA (FhCHS1) encoding CHS from *Freesia hybrida* was successfully isolated and analyzed.

13.6.4 Terpene Synthase

The development of flower scents was a crucial event in biological evolution, providing olfactory signals by which plants can attract pollinators. Bioinformatics, metabolomics, and biochemical and molecular methodologies were integrated to investigate the candidate genes involved in the biosynthesis of volatile components in two cultivars of *Freesia* x *hybrida*, Red River® and Ambiance, which release different categories of compounds (Gao *et al.*, 2018).

Despite fragrance being an important floral trait, little is known of its associated molecular mechanisms in *F. hybrida*. Huang *et al.* (2018) constructed cDNA libraries for three floral developmental stages in *F. hybrida*. A total of 74,660 unigenes were obtained from RNA sequencing and de novo assembly, of which 72.20 per cent were annotated by seven public protein databases. Approximately 12,420 differentially expressed genes were identified during flower development. Gas chromatography–mass spectrometry analysis detected a total of 34 floral volatile compounds, primarily volatile monoterpenes, which accounted for approximately 90 per cent of the volatiles. Using a system analysis-based approach, 36 candidate genes related to volatile terpenes were identified, in which the majority of up-regulated genes associated with the biosynthesis of monoterpenes and the majority of down-regulated genes associated with the biosynthesis of sesquiterpenes. Of these, FhDXS2A, FhGPPs and FhTPSs were considered to be important in floral scent formation. This study provides an in-depth assessment of floral scent and a reference for the future molecular breeding of floral scent in ornamental plants (Huang *et al.*, 2018).

14.0 References

Abbass, J.A., Al-Zurf, M.T.H. and Ajami, A.T. (2020a) *Euphrates J. Agric. Sci.*, **12**: 31- 41.

Abbass, J.A., Al-Zurfi, M.T.H., Al-Abbasi, A.M., Swadi, M.J. and Kadhim, E.S. (2020b) *Plant Arch.*, **20**: 574-578.

Abdalla, N.A. (2019) *QJAS Al-Qadisiyah J. Agric. Sci.*, **9**: 240-246.

Abdel- Fattah, I.M., Attia, E.A. and El- Banna, G.M. (2020) *Env. Biodiv. Soil Security* **4**: 59-71.

Ahmad, I., Tanveer, M.U., Liaqat, M. and Dole, J. M. (2019) *Sci. Hort.*, **254**: 21-25.

Alasadi, Z.N. (2016) *Basrah J. Agric. Sci.*, **29**: 608-618.

Ali, P. and Khui, M.K. (2015) *Folia Hort.*, **27/2**: 123-134.

Ali, P. and Morteza, K. K. (2015) *Int. J. Hortic. Sci. Tech.*, **2**: 97-109.

Ali, T., Khattak, A.M., Amin, N.U. and Khan, M.A. (2011) *Sarhad J. Agric.*, **27** (1).

Amariutei, A., Burzo, I. and Craciun, C. (1994) *Acta Hortic.*, **368**: 586-597.

Anderson N.O. (2007) Flower Breeding and Genetics. Springer, the Netherlands.

Aydinsakir, K., Tepe, A. and Buyuktas, D. (2010) *Akdeniz Üniversitesi Ziraat Fakültesi Dergisi*, **23**(1): 65-72.

Bajaj, Y.P.S. and Pierik, R.L.M. (1974) *Neth. J. Agric. Sci.*, **22**: 153-159.

Berghoef, J. and Zevenbergen, A.P. (1990) *Acta Hort.*, **266**: 169-176.

Berghoef, J., Melcherts, J., Mourits, J.W.F. and Zevenbergen, A.P. (1986) *Acta Hort.*, **177**: 636.

Blindeman, L. (2000) *Verbondsnieuws.*, **44**: 28-30.

Borochov, A. and W.R. Woodson. (1989) *Hort. Rev.*, **11**: 15-43.

Capinera, J.L. (2020) http: //entnemdept.ufl.edu/creatures/veg/aphid/green_peach_aphid.htm

Cho, H.R., Lim, J.H., Rhee, H.K., Kim, M.S., Choi, S.R., and Shin, H.K. (2009) *Korean J. Breed. Sci.*, **41**: 346-348.

Cho, H.R., *et al.* (2013a) doi : 10.7744/cnujas.2013.40.4.305.

Cho, H.R., Rhee, J.H., Lim, J.H. and Kim, M.S. (2013b) *Korean J. Breed. Sci.*, **44**: 633-636.

Choi, Y.J., *et al.* (2014) doi: 10.7744/cnujas.2014.41.4.299.

Darras, A.I., Joyce, D.C. and Terry, L.A. (2004) *Aust. J. Exp. Agric.*, **44**: 103–108.

Darras, A.I., Terry, L.A. and Joyce, D.C. (2005) *Postharvest Biol. Technol.*, **38** : 175–182.

Darras, A.I., Joyce, D.C., Terry, L.A., Pompodakis, N.E. and Dimitriadis, C.I. (2007) *Australas. Plant Path.*, **36**: 332–340.

De Hertogh, A.A. (1980) Bulbous plants. In: R.A. Larson (ed.), *Introduction to Floriculture.* Academic Press, New York. pp. 215-235.

De Hertogh, A.A. and Le Nard, M. (1993) The Physiology of flower bulbs: a comprehensive treatise on the physiology and utilization of ornamental flowering bulbous and tuberous plants. Amsterdam; Elsevier.

Doi, M., Hamatani, S., Hirata, T., Imanishi, H., and Hisamoto, T. (1992) *Acta Hortic.*, **319**: 249-254.

Doorduin, J.C. (1990) *Acta Hortic.*, **268**: 171-178.

Doorduin, J.C. (1992) *Acta Hortic.*, **325**: 85-90.

Hajizadeh, H.S. (2016) *Acta Hortic.*, doi: 10.17660/ActaHortic.2016.1131.1.

Doorduin, J.C. and C.M.M.va Winden, van. (1986) *Acta Hortic.*, **177**: 331-340.

El–Bably and Samia, M.Z. (2016) *Middle East Journal of Agriculture Research,* **5**(4): 507-521.

Foxe, M.J. (1991) Rapid *in vitro* propagation of virus-indexed Freesia. In: Prakash J., Pierik R.L.M. (eds). *Horticulture — New Technologies and Applications. Current Plant Science and Biotechnoloy in Agriculture*, vol. 12. Springer, Dordrecht. https: //doi.org/10.1007/978-94-011-3176-6_32.

Gao, X., Yang, D., Cao, D., Ao, M., Sui, X., Wang, Q., Ao, M., Sui, X., Wang, Q., Kimatu, J.N. and Wang, L. (2009) *J. Plant Growth Regul.*, **29**: 257-267.

Gao, F., Liu, B.F., Li, M., Gao, X., Fang, Q., Liu, C., Ding, H., Wang, L., and Gao, X. (2018) *J. Exp. Bot.*, **69** : 4249–4265.

Gianfagna, T.J. and Wulster, G.J. (1986) *Hort Sci.*, **21**: 263-264.

Halevy, A.H. and Mayak, S. (1981) *Hort Reviews*, **3**: 60-112.

Huang, M., Fan, R., Ye, X., Lin, R., Luo, Y., Fang, N., Zhong, H. and Chen, S. (2018) *Plant Growth Regul.*, **86**: 93–104.

Imanishi, H. (2002) *Acta Hort.*, **570**: 95-99.

Jin, K.H., Hwang, M.J. and Kim, K.S. (1999) *Kor. J. Hort. Sci. Technol.*, **17(3)**: 361-364.

Jin, W.M., Qin W.Y. and Lin, Y.X. (1994) *J. Shanghai Agrl. Coll.*, 01.

Ju, Z., Sun, W., Meng, X., Liang, L., Li, Y., Zhou, T., Shen, H., Gao, X. and Wang, L. (2018) *Plant Cell Tissue Organ Cult.*, **135**: 99–110. doi: 10.1007/s11240-018-1447-0.

Kang, J.S., Jeong, K.J., Choi, Y.J. and Yun, J.G. (2017) *Korean J. Plant Res.*, **30**(6): 699-706.

Koksal, N., Gulen, H. and Eris, A. (2010) *Acta Hort.*, **877**: 1791-1799.

Kragtwijk, C. (1954) *Jaarverslag van het proefstation voor de bloemisterij te Aalsmeer* . pp. 63-77.

Piovano Li, Y., Shan, X., Zhou, L., Gao, R.,Yang, S., Wang, S., Wang, L. and Gao, X. (2019) *Front. Plant Sci.*, **9**: 1935. doi: 10.3389/fpls.2018.01935.

Mansaur, B.M.M. (1968) *Landbouwhogeschool Wageningen.*, **68**(8): 1-76.

Morgan, W.P. (1934) US Patent Number: 98. Plant Date of Patent: May 22, 1934.

Morgan, W.P. (1929) *Proc. the Indiana Academy of Science*, USA.

Pierik, R.L.M. and Steegmans, H.H.M. (1975) *Neth. J. Agric. Sci.*, **23**: 334-337.

Piovano, M.V.G. Pisi, Luna, M.G. and Mata, D. (2017) *Horticultura Argentina*, **36**: 19-25.

Rezvanypour, Sh., A. Hatamzadeh, S.A. Elahinia and H.R. Asghari. (2016) *J. Sci. Tech. Greenhouse Cult.*, **7**: 63-75.

Scagel, C. (2003) *J. Environ. Hortic.*, **21**: 196-205.

Shan, X., Li, Y., Zhou, L., Tong, L., Wei, C., Qiu, L., Gao, X. and Wang, L. (2019) *Plant Cell, Tissue and Organ Culture*. https: //doi.org/10.1007/s11240-019-01649-9.

Slootweg, G. (2005) *Acta Hort.*, **669**: 297-302.

Startek, L., Mynett, K., and Zurawik, P. (2002) *Acta Hort.*, **570** : 377-383.

Stimart, D.P. and Ascher, P.D. (1982) *Sci. Hort.*, **17**: 153-157.

Tang, D., Sun, Y., Li, X. *et al.* (2018) *Acta Physiol. Plant.*, **40** : 168. https: //doi. org/10.1007/s11738-018-2739-z.

Uwagaki, Y., Matsuda, E., Komaki, M., Murahama, M., Otani, M., Nishizawa, N. and Hamada, T. (2015) *Plant Biotechnol.*, **32**: 165–168.

van Andel, J. (1992) US Patent Number: Plant 7,887. Date of Patent: Jun. 16, 1992.

van Andel, J. (1995a) US Patent Number: Plant 9,294. Date of Patent: Sep. 19, 1995.

van Andel, J. (1995b) US Patent Number: 9293. Plant. Date of Patent: Sep. 19, 1995.

van Andel, J. (1995c) US Patent Number: 9295. Plant. Date of Patent: Sep. 19, 1995.

van Andel, J. (1998a) US Patent Number: 10,470 Plant. Date of Patent: Jun. 30, 1998.

van Andel, J. (1998b) US Patent Number: 10,469. Plant. Date of Patent: Jun. 30, 1998.

van Andel, J. (1998c) US Patent Number: 10,468. Plant. Date of Patent: Jun. 30, 1998.

van Meeteren, U., H. van Gelder and A.C. van de Peppel. (1995) *Acta Hortic.*, doi: 10.17660/ActaHortic.1995.405.14.

Wulster, G.J. and Gianfagna, T.J. (1991) *Hort Sci.*, **26**: 1276-1278.

Yuan, Y., Tang, D.Q. and Shi, Y.M. (2011) *J. Shanghai Jiaotong Univ. Agric. Sci.*, **29**: 30-36.

Zencirkiran, M. (2002) *N. Z. J. Crop Hortic. Sci.*, **30**: 171-174.

Zencirkiran, M. (2010) *Sci. Res. Essays*, **5**(17): 2409-2412.

5

Tuberous Begonia

V.A. Parthasarathy, C. Aswath and Debmala Mukherjee

1.0 Introduction

Begonias are mainly native to moist tropical and subtropical climates, and are mostly distributed in Asia, America and Africa. It is one of the most attractive ornamentals with both beautiful foliage and flowers. In "Year of the" crops programme hosted by the National Garden Bureau of USA, begonia was voted the annual of 2016. *Begonia* x *tuberhybrida* Voss (syn: *Begonia* × *tuberhybrida* var. *grandiflora* Voss. *Begonia tuberhybrida* Group, *Begonia tuberosa* Hort.) belong to the family Begoniaceae. The common names are Hybrid Tuberous Begonia, Tuberous Begonia. Approximately 2.75 million *Begonia* plants, mainly *Begonia x hiemalis, Begonia rex, Begonia semperflorens,* and *Begonia tuberhybrida,* were grown in commercial facilities in Norway in 2015. *B. tuberhybrida* made up about 10.8 per cent (298 000 units) of the total number of *Begonia* plants produced (Sandgrind, 2017).

Begonia elatior, tuberous begonia, is the product of the cross of *B. socotrana* x *B. tuberhybrida* and is known as *B. hiemalis* and is considered a flower with great potential because of the great acceptance in local consumption, but with good management in marketing. It can be considered for the international market. In the domestic market of Mexico, it is the predominant one since sales indicate that it is around 90 per cent and possibly its demand increases.

2.0 Importance and Uses

2.1.0 Aesthetic Uses

Genus *Begonia* are mostly succulent plants with colourful flowers or leaves. Many of the species and cultivars are used as pot plants in indoors or as garden plants in outdoor. These are well suited in flower bed, border along garden roads, floral carpet, hanging basket *etc.* Tuberous begonia is one of the most beautiful flowering plants in sub-tropical and mild temperate climate having large number of cultivars varing in flower size and colour.

2.2.0 Therapeautic and Other Uses

Additionally, there are various reports from several countries from Mexico to India where Begonia species have been used for medicine, vitamin C and in some instances as a source of food (Jain and Dam, 1979; Lans, 2006; Sajem and Gosai, 2006). *B. lempuyangensis* is used as a palliative medicine for coughs. Similarly, *B. lombokensis* and *B. longifolia* are also eaten and/or used medicinally (Girmansyah, 2009). The sap has been put on wounds, and concoctions used for cough, consumption, and fever. Shoots have been used to treat upset stomachs and enlarged spleens. The acid-based sap has been used to treat toothaches and gum aliments as it is considered anti-bacterial. It is also proven to have anti-tumor activity and commonly used in the West Indies by herbalists to treat cancer (Jordon, 2008). Flower infusions used to promote blood circulation and elimination of toxins from the body. Poultice of flowers used for burns and toxic sores. *Begonia picta* juice of the plants is drunk to use relieve headaches. Roots of the juice used for conjunctivitis. The tuberous roots and fruits of *Begonia grandis evansiana* are anodyne, antiphlogistic and antispasmodic.

Uses of Tuberous Begonia

In Border of a Garden

In Pot

In Hanging Basket

In Garden Oathway

In Garden Design

As Specimen in Garden

In Bed of a Garden

Beaufying the Pergola

Decoction used in the treatment of traumatic pain, haematemesis, gonorrhoea, amenorrhoea and snakebites.

3.0 Origin, History and Distribution

3.1.0 Origin

The first cultivar mentioned in the literature is 'John Heal' from 1883. This cultivar was a result of a cross between *B. socotrana* and *B.* x *tuberhybrida* 'Viscountess Doneraile'. Since then a number of cultivars has been introduced and cultivated. Names as 'Ensign', 'Elatior', 'Exquisite' and 'Baardse' belong to the hybrid-group

B. x *hiemalis*, and some of them have also been used, and are still in use as a name for the whole group. Elatior-begonia and Baardse-begonia are names commonly used in Europe. In USA the name Holland-begonia is used.

Begonia tuberhybrida is a complex group of cultivars developed from hybridization of several Andean species such as *B. boliviensis* with pink flowers or *B. pearcei* with yellow flowers. Other parental Andean species used in hybridization include *Begonia veitchii* and *B. davisi*. Tuberous begonias are cultivated and not found wild.

Doorenbos and Legro (1968) traces the history of two species *B. dregei* and *B. socotrana* used in the interspecific hybridization for developing modern *B.* x *tuberhybrida*. According to them '*Begonia dregei*' was sent to the Botanical Garden at Berlin from the Cape of Good Hope by J. F. Drége. In 1836, it was sent from Berlin to Edinburgh, apparently still unnamed as it was described later, in 1840, as *B. parviflora* by Graham (Curtis' Botanical Magazine 66, t. 3720).

Neal *et al.* (2006) made a comprehensive review on the evolution of diversity in Begonia. According to them, the genus is perhaps best known horticulturally. Species have been in cultivation for at least 600 years, *B. grandis* having been recorded in Chinese gardens in 1400. However, it was not until the plant collecting expeditions of the late 1700's and the 1800's that begonias arrived in Europe and North America. The species *B. vetchii* for example was introduced into England from South America in 1867. It is still available and has been crossed extensively by breeders to create numerous hybrids. The huge amateur interest in begonia cultivation and hybridization represents a great resource of both knowledge and experience with hybrids. There is an economically valuable market for hybrids as both bedding plants and potted plants. In northern Europe in particular, there is a large market for hybrids derived from crosses involving the winter flowering species *B. socotrana*. These hybrids form two main groups. The Cheimantha group contains all crosses and back crosses between *B. socotrana* and the South African species *B. dregei*. The Hiemalis group contains all crosses and back crosses between *B. socotrana* and floriferous South American tuberous species. In 1990, the turnover at auction from Heimalis hybrids in the Netherlands was approximately Dfl 35 million (€15.88 million; Kroon 1993). Semperflorens hybrids, derived from *B. cucullata* crossed with *B. schmidtiana* and the *B.* x *tuberhybrida* cultivars, are also popular as bedding plants. In the US in 2003, the market for these plants was estimated to be worth over $60 million (€50.49 million; Tebbit, 2005). It is in the US that the greatest use is made of Begonia's diversity. Interest is not limited to bedding plants but includes many hybrids and species cultivated for their vegetative form. Official records of hybrids and distribution of seeds is controlled by the American Begonia Society (ABS).

3.2.0 Distribution

Begonia is a genus of about 1550 described species, placing it in the top ten most speciose angiosperm genera. The distribution of *Begonia* diversity is uneven throughout tropical regions, with the greatest diversity in America and Asia (>600 species each), whilst being relatively species poor in Africa (160 species) and absent in Australia. The genus is thought to have originated in Africa, while South American and South East Asian species are the results of parallel radiations over the last million years.

According to Dewitte *et al.* (2011), the mean age estimates for the *Begonia* crown group diversification indicate an Oligocene origin (c. 23-34 Ma), although large confidence ranges have to be considered. The earliest divergent clades in the *Begonia* phylogeny are African, suggesting an early diversification on the African continent long after the Gondwanan break-up. From Africa, *Begonia* dispersed independently to both America and Asia, and the early Miocene has been inferred as the most likely time frame for these dispersal events.

4.0 Classification, Species and Cultivars

4.1.0 Classification

Only two genera are currently recognized in the Begoniaceae: the monotypic genus *Hillebrandia,* and morphologically diverse genus *Begonia.* A third genus, *Symbegonia,* was previously included in Begoniaceae, and separated from *Begonia* by floral characters (syntepalous perianth and a monadelphous androecium).

The genetic and morphological diversity of the genus *Begonia* has been exploited through cultivation to produce over 10,000 cultivars (Dewitte *et al.*, 2011). These are horticulturally divided into 5 classes as below:

a) The tuberous begonias (*B.* x *tuberhybrida*), a complex group derived from crosses between species such as *B. boliviensis* or *B. pearcei.*
b) Elatior begonias (*B.* x *hiemalis*), a cross between tuberous begonias and *B. socotrana.*
c) Lorraine begonias (*B.* x *cheimanta*), a cross between tuberous hybrids and *B. dregei.*
d) Semperflorens begonias (*B. semperflorens*-cultorum), with *B. cucullata* and *B. schmidtiana* as important ancestors.
e) Begonias grown for their ornamental foliage (*B. rex*-cultorum), which is asiatic in origin.

4.2.0 Species

The number of species in Begonia is increasing over time. Okuno and Nagai (1953) reported 400 to 500 species. But, according to Tian *et al.* (2018) Begonia, one of the most diverse plant taxa and the fifth or sixth largest angiosperm genus, consists of over 1800 accepted species. The number of species recognized within this genus has greatly increased over the past 20 years, rising from 80 to 200 species in China alone. They predict that the number of species may actually in total between 2000 and 2500 (Tian *et al.*, 2018, 2020).

There are numerous garden varieties, most of which are grown in the greenhouse. However, there is one wild plant *B. evansiana* Andr. This is widely distributed throughout Japan and commonly known as "Shukaido" (Okuno and Nagai, 1953). However, only relatively a few of the species have made an important

Species of Begonia

Begonia tuberhybrida

Begonia cucullata

Begonia josephii

Begonia boliviensis

Begonia socotrana

Begonia dregei

Begonia × malipoensis

Begonia maculata wightii

Begonia wui-senioris

contribution to the current commercial cultivars. *Begonia boliviensis, B. clarkei, B. pearcei,* and *B. veitchii,* for example, have been the main parents of the *B. tuberhybrida* group (the tuberous hybrids). Similarly, *B. socotrana* and *B. dregei* were important parents of the *B. cheimantha* group, and *B. socotrana,* when crossed with members of the *B. tuberhybrida* group, resulted in the *B. hiemalis* group. Furthermore, *B. cucullata* and *B. schmidtiana* were founding parents of the *Begonia* Semperflorens-Cultorum Group, while *B. rex* and a wide variety of other species were important in the development of the *Begonia* Rex-Cultorum Group, the latest of which are mainly cultivated for their beautiful leaves (Haegeman, 1979; Tebbitt, 2005).

Recently, Tian *et al.* (2020) reported that *Begonia dioica* and *B. flagellaris,* both from Xizang (Tibet) are new to China; *Begonia lipingensis, B. muliensis* T.T.Yu and *B. sizemoreae* are synonyms of *B. circumlobata, B. taliensis* Gagnepain and *B. longiciliata,* respectively; and *Begonia* × *lancangensis* and *B.* × *malipoensis* are natural hybrids.

4.3.0 Cultivars

'Nonstop' was the first seed-propagated cultivar of *Begonia tuberhybrida* introduced during the 1970s by Benary Seeds of Germany (Ewart, 1985) in USA. Many additional seed-propagated cultivars are now produced commercially. There are a large number of commercial varieties of tuberous begonias which have been patented. Few of them are Encore, Fantasia, Jane, Karita, Lemon Rose, Manderijn, Manifesto, NZCTWO, Ninon, Northern Sunset, NZC Three, Peach, Plum Paisly, Puff Pink, Soft Pink, Solenia Orange, Solenia Rose, Solenia Rot, Star David, Tess dark, TNBEGDY, Yabon, Yates, Yamina, Yamoon, Yamour, Yantana, Yasped, Yasmon, Yasprink, Yaspwit, Yaspyell *etc.* are in the market as patented varieties.

Begonia x *tuberhybrida* hybrids have been classified in 13 groupings based on plant habit or flower color:

1. (S) Single-large single flowers, 4 usually flat tepals (flower part undistinguishable as sepal or petal).
2. (Fr) Frilled, Crispa-large single flowers, tepal margins frilled or ruffled.
3. (Cr) Cristata, Crested—large single flowers, frilled or tufted center of tepals.
4. (N) Narcissiflora, Daffodil-flowered large more or less double flowers, central tepals form 'trumpet'.
5. (C) Camellia, Camelliflora-large double flowers resembling camellias, unruffled, solid colors.
6. (RC) Ruffled Camellia-camellia flowers, ruffled on edges.
7. (R) Rosebud, Rosiflora-large double flowers with rose bud-like center.
8. (Car) Carnation, Fimbriata Plena-large double carnation-like flowers, tepals fringed on margins.
9. (P) Picotee-large usually double flowers like camellias, tepals with different color on margin blending with other color.
10. (M) Marginata-like Picotee, only distinct non-blending line of color on margins.
11. (Mar) Marmorata, Marbled-like Camellia but rose-colored, blotched or spotted with white.
12. (HB) Hanging basket, pendula-stems trailing or pendant, large to small flowers single or double.
13. (Mul) Multiflora-low, bushy, compact plants with many small single or double flowers.

Table 1: Details of Cultivars Based on Flower Colour

Cultivars	*Group*	*Flower Color*
Allan Langdon	C	Dark red
Bernat Klein	C	White
Bridesmaid	P	White, red edge
Buttermilk	C	Cream, apricot
Cascade series	HB	4 colors
Charisma series (Rieger)	C	Coral, salmon
Clips series	C	Mix
Fairy Light	P	White, pink edge
Illumination series	HB	Pink, orange
Jean Blair	P	Yellow, red edge
Lace Red	P	Red, white edge
Marmorata series	Mar	Mix
Midnight Beauty series	C	5 colors
Musical series	HB	5 colors
Nonstop series	C	9 colors
Pin-Up	M	White, rose edge
Prima Donna'	P	Yellow, red edge

Source: Leonard Perry, University of Vermont.

The following are the details of patented varieties.

Ellen (Schelbeck, 1985): Ellen is characterized by its large flower size and compound inflorescence, brilliant and uniform red flower color, fast propagation, rapid flowering, and its foliage which is infused with maroon in margins and underside. Flower color is the most brilliant red known to date in commercial *hiemalis begonias*, being a more brilliant red rather than the somewhat tinted red of Dorthe. Ellen is less susceptible to flower bud drop in the winter months

Dorthe (Schelbeck, 1983): This variety, botanically known as *Hiemalis begonia*, has double deep bright red sterile flowers, medium size firm waxy foliage, upright growth easily controlled for plant height, outstanding keeping qualities, and being capable of flowering in all seasons for commercial purposes. The flowers are double red while Socotrana flowers are single and light pink. Dorthe's flowers are true red, whereas the flowers of Connie have a tint of blue giving the flowers a rose-red coloration. The flower keeping quality is very good in all seasons.

Barbara (Schelbeck, 1984b): A begonia plant named Barbara having large double and compound bright pink flowers contrasting with dark green firm glabrous foliage; vigorous in growth habit, and easily manipulated for year round flower production, especially for larger containers and hanging baskets for interior decorating without flower fading under abnormal conditions.

Connie (Schelbeck 1984a): This is a hybrid, botanically known as *Hiemalis begonia*, named Connie having rose-red, double sterile flowers up to 5 cm in diameter, medium size slightly convex dark green glabrous leaves, upright growth, and having excellent flowering and keeping qualities for year round production in various size containers. The flowers are double rose-red while the flowers of *socotrana* are single (4 tepals) and light pink. Connie's flowers have a tint of blue giving a rose red coloration. whereas, Dorthe's flowers are true red. The racemes are tightly clustered in the early flowering stages and may have up to 8-10 flowers in the inflorescence.

Edinburgh (Soren, 1996a): 'Edinburgh' is characterized by a tall and reasonably branching growth habit; simple, alternate, medium green, double-serrated foliage that is generally orbicular and purple-red flowers on a sturdy peduncle with 12 or more tepals. The new cultivar 'Edinburgh' is the product of a planned breeding program. "Edinburgh' has purple-red flowers unlike any other known hiemalis begonia. It is purpler than 'Cheers'. The flowers of 'Edinburgh' are not ruffled like those of 'Cheers'.

Berlin (Soren, 1996b): A new cultivar of Begonia named 'Berlin' characterized by a compact growth habit with excellent branching; simple, alternate, dark green, double-serrated foliage that is ovate to orbicular and pink flowers on a sturdy peduncle with 15 or more tepals.

Bonfire (Cross, 2004): A new and distinct cultivar of Begonia named 'Bonfire' (*Begonia boliviensis*). It is comprised of hairs on both the upper and lower leaf surface; bright orange-red flower color; and red margin on serrated leaf edge. Bright orange-red flower color, RHS 40A.

Brothglow (Booth and Brown, 2013): A cultivar of tuberous Begonia, Begonia Brothglow, characterized by its upright growth habit with good basal branching, its single male flowers that are apricot in color, and its evenly formed leaves that are dark olive green bronze in color.

Cheers (Mikkelsen, 1987): The cultivar 'Cheers' is particularly characterized as to uniqueness by the combined characteristics of rose-red flower color, strong and upright petioles which support flowers above foliage, double flower form, floriferous habit, fast growth with good branching, and by its adaptability to 10 or 15 cm pots. Cheers has a bright rose flower, while Aphrodite Cherry Red is duller in color.

Desire (Drewlow, 1990): It has double flowers orange in color, relatively large flower diameter, red pigmentation on upper and lower leaf surfaces, early flowers, tepal margin entire, long lasting flowers, and an ability to be propagated by stem or leaf cuttings. Desiré has orange flowers which are brighter in color than the lighter orange St. Helena and have more red pigmentation that the orange-flowered Lido (not patented) and Schwabenland Improved Orange.

Enchanted Evening (Pavlich, 2015): A new and distinct form of *Begonia* characterized by attractive dark chocolate brown leaves with green veins, numerous large coral pink flowers, an upright habit when young, becoming decumbent with maturity, an extremely long bloom time, a caudiciform base, and excellent vigor.

Cultivars of Tuberous Begonia

Allan Langdon

Bridesmaid

Enchanted Sunrise

Nonstop Series

Cascade Series

Illumination Series

Marmorata Series

Enchanted

Evening Rose

Enchanted Sunrise (Pavlich, 2016): A new and distinct form of Begonia characterized by attractive dark hocolate brown leaves with green veins, numerous large salmon flowers, upright habit when young, becoming decumbent with maturity, an extremely long bloom time, a caudiciform base, and excellent vigor.

Yabos (Yates, 2007) A new cultivar of *Begonia boliviensi*. It is characterized by its compact, freely branched, pendulant plant habit combined with red-orange flowers and ease of propagation. It was selected as unique for its compact pendulous plant habit, its abundant production of side shoots and its red-orange flowers.

5.0 Genetics and Breeding

5.1.0 Genetics

5.1.1 Cytology

The earliest report on cytology on the genus Begonia was undertaken by Matsuura and Okuno (1943). They investigated 20 species of begonia and progenies of their inter-specific and intra-specific crosses in *Begonia semperflorens*. They concluded that there are three basic chromosome numbers in *Begonia*, namely, 6, 7 and 13. Thirteen may be of secondary origin, synthesized from the primary numbers, 6 and 7. Within each series, there may exist several different genomes. Matsuura and Okuno, 1943 reported that this plant has 26 chromosomes in root-tip cells, but observations on meiotic chromosomes have not been made (Okuno and Nagai, 1953).

Twenty six somatic chromosomes were analysed morphologically. From prometaphase to metaphase, trabants have been frequently observed in two paired chromosomes but unfortunately such figures do not always exhibit to the best advantage. In *B. evansiana*, meiotic chromosomes were first demonstrated in the present study. The secondary association of meiotic chromosome has also been described. The following various types of secondary association were ascertained. (i) 6(2)+1(1), (ii) 5(2)+3(1), (iii) 4(2) +5(1), (iv) 3(2)+7(1), (v) 2(2)+9(1), (vi) 1(2)+11 (1) and (vii) 13(1). Six groups of two bivalents and one bivalent was the maximum type. *B. evansiana* may be a secondary balanced allopolyploid based on n=6 and 7 which is the primary basic number.

Legro and Doorenbos (1969) determined the somatic chromosome numbers of 100 *Begonia* species, of which 72 have been determined for the first time. Eighteen different numbers have been established, ranging between 22 and 156, fifteen of which could be fitted into a tentative scheme of five polyploid series. The most common numbers were 22 (14 species), 28 (35 species) and 56 (17 species).

Genomic *in situ* hybridization (GISH) has been employed to assess the genome composition in eleven 'Elatior'- begonias hybrids and their ancestor genotypes (Marasek-Ciolakowska *et al.*, 2010). Genomic DNA of tuberous begonia was sonicated to 1–10-kb fragments, labelled by nick translation with digoxigenin-11-dUTP and used as a probe whereas *B. socotrana* DNA was autoclaved to 100 bp fragments and used as block. The genome of tuberous begonia was clearly pronounced in 'Elatior'-begonias when the probe concentration was 3.75 ng/ll (150 ng/slide), with 30 times the excess of *B. socotrana* blocking DNA and stringency of post hybridization washings at 73 per cent (0.19 SSC at 42°C). In 'Elatior'-begonias hybrids GISH distinguished two groups comprising short (0.6–1.03µ in length) and relatively longer chromosomes (1.87–3.88 µ) which represent *B. socotrana* and tuberous begonia genomes, respectively. The number of chromosomes derived from tuberous begonia ranged from 14 to 56 and for *B. socotrana* from 7 to 28 which suggest the presence of different ploidy levels in analyzed 'Elatior'- begonia hybrids. Intergenomic recombination has not been detected through GISH in hybrids analyzed. Genomic *in situ* hybridization turned out to be useful to identify the genome constitution of 'Elatior'-begonia hybrids and thus gain an insight into the origin of these cultivars. This knowledge on the ploidy level and genome composition is essential for further progress in breeding begonias.

Genomics

The genome sizes of a *Begonia* collection comprising 37 species and 23 hybrids of African, Asiatic, Middle American, and South American origin were screened using flow cytometry. Within the collection, 1C values varied between 0.23 and 1.46 pg DNA. Genome sizes were, in most cases, not positively correlated with chromosome number, but with pollen size. A 12-fold difference in mean chromosome size was found between the genotypes with the largest and smallest chromosomes. In general, chromosomes from South American genotypes were smaller than chromosomes of African, Asian, or Middle American genotypes, except for *B. boliviensis* and *B. pearcei.* Cytological chromosome studies in different genotypes showed variable chromosome numbers, length, width, and total chromosome volume, which confirmed the diversity in genome size. Large secondary constrictions were present in several investigated genotypes. These data show that chromosome number and structure exhibit a great deal of variation within the genus *Begonia*, and likely help to explain the large number of taxa found within the genus (Dewitte *et al.*, 2009).

5.2.0 Breeding

5.2.1 Hybridization

Despite the large variation in chromosome numbers, begonias cross-hybridize quite easily, which explains the wide variation in begonia hybrids that can be obtained (Anderson, 2007). Eugene created the modern base of large- flowered tuberous Begonia with breeding of Begonia 'Erecta Super' variety in 1878.

In their detailed analysis of evolution of begonia, Neal *et al.* (2006) traced the reasons for prevalence of diversity. According to them, the diversity within species may be a result of the prevalence of out crossing in begonias. This is encouraged by a number of features including separation of male and female inflorescences, dioecy and pollination syndromes. A series of papers by Schemske and co-workers demonstrated that the prevalence of protandry in the genus may be linked to deceit pollination, commonly seen in *Begonia* (Le Corff *et al.*, 1998 cited by Neal *et al.*, 2006). Many *Begonia* species readily hybridise, a feature widely exploited by horticulturalists. This, combined with the relatively short generation times and copious seed production, makes this genus a good candidate for classical genetics and segregation analysis. They performed some crosses to investigate the genetic basis of a number of traits. One example is that between short day flowering *B. socotrana* and long day flowering *B. dregei*. This cross recreates the original 1891 Cheimantha group hybrid, Gloire de Lorraine. Crosses have also been made to investigate the genetic basis of differences in leaf morphology in the section Gireoudia. *B. nelumbiifolia* crossed to a deeply lobed species *B. heracleifolia* showed lobing to be semi-dominant. A further cross between *B. nelumbiifolia* and a species with simple leaves, *B. lindeyana*, showed peltateness also to be semi-dominant. Selfing of these hybrids and the production of an F_2 generation will segregate and will determine the number of loci involved including their relative strengths.

According to Marasek-Ciolakowska *et al.* (2010) interspecific hybridization of various tuberous begonia species with *Begonia socotrana* results in so-called 'Elatior'-begonias hybrids (*B.* x *hiemalis*). The current commercial assortment of begonias comprise mostly of 'Elatior-hybrids' (*B.* x *hiemalis*), over 100 varieties, representing around 88 per cent of the total begania production (Kroon 1993 cited by Marasek-Ciolakowska *et al.*, 2010). The early varieties of 'Elatior'- begonias resulted from crosses between winter flowering *Begonia socotrana* and 'Viscoutess Doneraile', a very early hybrid tuberous begonias (*B.* x *tuberhybrida*) (Arends 1970; Doorenbos 1973; Marasek-Ciolakowska). However, the majority of modern varieties of 'Elatior'-begonias are the result of crossing various tuberous begonia species hybrids (*B.* x *tuberhybrida*) and *B. socotrana* (Gleed, 1961). Some varieties of 'Elatior-hybrids' have also been obtained from the following crosses: 'Elatior'-begonia x *B. socotrana* and 'Elatior'-begonia x tuberhybrid (Arends 1970; Marasek- Ciolakowska, 2010). A new era in development of the Elatior begonia began with the introduction of the first hybrid in 1955 by Otto Rieger of Germany. Its strongest assets are absence of bud drop, relative resistance to mildew and easy propagation from both leaf and stem cuttings (Molnar, 1974).

5.3.0 Polyploidy

Begonia chromosome numbers range from 16 to 156. Within this range the chromosome numbers are extremely variable. This may be a result of hybridization, polyploidisation and aneuploidy. It is also possible that there has been frequent chromosome fission and fusion leading to dysploid variation (variation without pattern - not doubling or trebling) in basic numbers. However, chromosomes in begonia are very small and this variation may to some extent reflect the difficulties of chromosome counts. Published chromosome numbers can vary enormously for the same species. Chromosome counts are also incomplete for the genus, but there have clearly been several polyploidy events during the evolution of begonia. Gene and genome duplication is thought to have been a important in evolutionary change and diversification, with duplicates acting as an important source of raw material on which evolutionary change can operate (Kellogg, 2004). There have been both genome duplications and reductions in begonia and the variation in chromosome number is suggestive of underlying genetic instability which may be associated with the rapid creation of genetic variation and the evolution of morphological diversity (Neale *et al.*, 2006).

In *Begonia*, 10 genotypes that produce unreduced pollen have recently been identified (Dewitte *et al.*, 2009). In all these genotypes, meiotic aberrations resulted mainly in the formation of dyads. Unreduced gametes are the driving force for the polyploidization of plants in nature, and are also an important tool for ploidy breeding. The final heterozygosity of a 2n pollen grain depends on the cytological mechanism behind 2n pollen formation.

5.4 Mutation Breeding

The occurrence of mutants in plants of four clones of *Begonia* x *hiemalis* grown from leaf cuttings irradiated with 0, 1500, 2000 or 2500 rad of X-rays was studied by Doorenbos and Karper (1975). Irradiation reduced the number and slowed down the development of adventitious shoots, but to a different degree in the four clones used. The percentage of mutated shoots was as high as 18–35 per cent in the non-irradiated shoots. Irradiation raised it to more than 80 per cent. In the control plants, almost exclusively flower colour mutants were found. Irradiation caused a wider range of flower colour mutants and also caused mutations of growth habit, colour and size of the leaves, and size and shape of the flower. Almost all mutants were 'solid', *i.e.* non-chimeral.

6.0 Growth and Flowering

Tuberous begonias bloom throughout the summer, thriving in shady spots where few other plants with long bloom periods and showy flowers can grow. With up to 20 cm, showy flowers that bloom from summer to fall, tuberous begonias provide all-season color when planted in containers and beds. Tubers started in February should begin to bloom in June; those started in March or early April should bloom in July.

7.0 Propagation

7.1.0 Seed Propagation

Tuberous begonias can be propagated from seed. At a constant temperature of 21°C, germination usually takes about ten days, but it can take as long as three weeks, especially if the temperature of the potting soil is lower than 21°C. Thinning of seedlings is required to 5 cm apart. Bright light such as fluorescent light is helpful in producing sturdy seedlings. When they become crowded, transplanting is to be done in larger pots, using a porous, well-drained potting mix. After hardening, the plants are transferred in hanging baskets, in larger containers, or directly in field.

The relatively long germination time makes the growing of *Begonia X tuberhybrida* more difficult for the raising of seedlings. It takes about 3 to 5 months and in the winter months is necessary supplemental light, too (Kisvarga *et al.*, 2015). Under additional artificial light plants develop more and bigger flowers.

7.2.0 Vegetative Propagation

Begonia x *tuberhybrida* is propagated mainly by seed, because vegetative propagation by cuttings or tissue culture is difficult. However, seedlings generally exhibit a wide range of variation due to their high level of heterozygosity, and also the pollination period is limited. In case of leaf cuttings, BA may be effective for inducing adventitious bud formation of *B.* x *tuberhybrida* (Shimada *et al.*, 2005).

Propagation by stem cuttings or by tuber division is also possible. Stem cuttings are made from surplus shoots which arise from the tuber. Stem tips may also be used. The cuttings about 8 cm long are cut off just below a node. Sand may be used as the rooting media. Keep cuttings out of direct sun and in temperatures between 15°C and 18°C. Rooting occurs in five weeks. Tuber division is the other way to propagate the plants. Divide the tuber so each division has a bud and use a fungicide to prevent rot. Allow the pieces to dry for several days, then place them 3 cm deep in sand. MacKenzie and Moncada (2018) suggest the use of forsythe pots

Propagation of Tuberous Begonia

Seed Pod and Seeds

Cuttings

Potted Cuttings

Tubers

Shoot Regeneration from T-thin Cell Layer System

Cultivation of Tuberous Begonia

Cultivation in Open

Cultivation in Greenhouse

for rooting and to be kept in warm place with bright indirect light. This pot works like water reservoir, so proper humidity is being maintained here. After two to four weeks, cuttings become well rooted and harden them off and transplant in bigger pots or outside. By September, small tubers are formed and new plants develop more rapidly with bottom heat (Gilman and Heatley, 2014).

It was found that the method used in multiplication influences significantly both the quality of planting material and the quality of the obtained plants. The whole tubers and half-sectioned one promote vegetative growth, flowering per cent and the flowers quality, but the best multiplication coefficient was obtained in the case of the tubers sectioned in quarters. The three years old tubers have better results compared with the younger ones. The multiplication by apical shoot cuttings is superior to the leaf cuttings multiplication, and the best period for this operation proved to be in June.

For the larger leaved plants, begonia propagation can begin with a single leaf. Here, cutting of mature leaf just below the petiole and burying the petiole only will develop new growth. Rooting of begonias this way will produce whole new plant grown from the roots that develop at the end of the petiole. It can be used for a windowsill garden or for next spring's outdoor planting and both these two method of propagation through root cuttings and leaf cuttings are comparatively cheaper.

The effects of day length and temperature on rooting and growth of Elatior begonia cuttings was studied. The temperature treatment was more effective for the growth of cuttings than the day length treatment. Temperature treatments had a statistically significant effect on the development and growth of roots, whereas there was no significant difference among the day length treatments. However, continuous day length of 16 h and 24°C temperature showed the best shoot and root growth of Elatior begonia cuttings (Shim *et al.*, 2005).

7.3.0 Micropropagation

Tuberous begonias are traditionally propagated by seeds, except those belonging to the 'multiflore' group, which are propagated by cuttings. Hence, a protocol for the *in vitro* micropropagation of *Begonia* x *tuberhybrida* and regeneration of plants from petiole calluses was standardized (Viseur and Lievens, 1987). The complete cycle, from *in vitro* induction of calluses to the flowering of regenerated potted plants, took 6 to 7 months (less than with the conventional methods). Mean multiplication rate was about 10 per month (with high values of 30) and the rate of successful establishment of plants in peat containers amounted 90 per cent. The first flowers opened about 60 days after the plants were transferred to soil. Flowering was very homogeneous, and no variant phenotype was observed within 297 regenerated plants. After flowering, tuberisation was normal, the average diameter of the tubers being approximately 4 cm.

The efficiency of commercial micropropagation programs for *Begonia* x *hiemalis* depends on the production of large adventitious shoots for easy handling and on effective rooting and acclimatization procedures (Simmonds, 1984). Maximum induction of adventitious buds on petiole segments occurred in response to NAA (0.1 mg/l) and BA (0.5 mg/l), but continued shoot growth was limited. With a lower concentration of BA (0.1 mg/l) fewer shoots were produced but shoot growth was enhanced. With a combined agar/liquid culture program the low BA (0.1 mg/l) medium produced 50 per cent more shoots larger than 1 cm than did the high BA (0.5 mg/l) medium. *In vitro* rooted explants developed weak root systems and acclimatization losses occurred during adaptation to greenhouse conditions. Adventitious shoots treated with commercial rooting powder and placed directly in mist frames produced much stronger root systems and could be adapted to greenhouse conditions without loss. The elimination of the *in vitro* rooting stage also simplifies the micropropagation program.

In another study, Nakano *et al.* (1999) induced adventitious shoots elongating to over 10 mm in length (8±9 elongated shoots per explant) from both leaf and petiole segments of the 'rose-formed' strain of hybrid tuberous begonia on MS media containing 0.54 mM NAA and 0.44 mM BA. Shoots were efficiently multiplied by shaking shoot-forming leaf segments in a liquid medium containing 0.54 mM NAA and 0.44 mM BA, while shoot growth was stimulated by shaking them in a liquid medium without plant growth regulators. Plantlets were obtained by rooting the elongated shoots on half-strength MS media containing 0.54 mM NAA and solidified with 8 g/l agar or 2 g/l gellan gum, and successfully transferred to the greenhouse. Regenerated plants grew into the flowering stage and showed no apparent morphological alterations.

In vitro regeneration of four *Begonia* genotypes, *B. semperflorens*, *B. rex*, *B. × elatior*, and hybrid of *Begonia* with unknown parents 'Tiger' was carried out starting from leaf and petiole segments as explants (Espino *et al.*, 2004). Five Murashige and Skoog2 s derived media were tested, three of them supplemented with α-naphthaleneacetic acid (NAA) and 6- benzyladenine (BA), and the other two with NAA and kinetin (KIN) in different concentrations. Shoot regeneration was preferentially induced on the BA containing media, quantitative differences being observed among explants and genotypes.

7.4.0 Shake Culture

Takayama and Misawa (1981) stimulated plantlet growth in submerged culture by Murashige-Skoog's medium with 10 or 30 g/l sucrose, but was inhibited by a lower (×1/4) concentration of basal medium. Root development was superior in the lower strength medium. The greater amount of sucrose, 30 g/l, stimulated root growth more than the 10 g/l did. Buds required kinetin at a concentration of 0.3 to 1 mg/l, but did not require auxin. Plantlets reached maximum growth after 3 weeks of cultivation in 300-ml Erlenmeyer flasks containing 30 to 150 ml of medium that were shaken at 180 rpm. Cultivation in jar and bubble column fermentors appears to be advantageous for the large scale propagation of begonia plantlets (Takayama and Misawa, 1981).

7.5.0 Thin Cell Layer System

Transverse thin cell layer (tTCL) technology were cut from *Begonia tuberhybrida* petioles about 0.8-1.0 mm thick and cultured on half-strength Murashige and Skoog (MS) medium supplemented with various concentrations of BA, TDZ, and different combinations of BA or TDZ and NAA. The highest shoot bud formation was obtained on medium with 0.2 mg/l TDZ and 0.2 mg/l NAA, giving rise to 100 per cent shoot formation after 8 weeks. TDZ at low concentrations (0.1-0.2 mg/l) were effective for caulogenesis from begonia tTCLs (Nhut *et al.*, 2005).

8.0 Soil and Climate

8.1.0 Soil

Tuberous begonia being a bulbous plants, prefers porous substrate rich in organic matter (see cultivation).

8.2.0 Climate

Tuberous begonias thrive best in a mild cool summer climate and are totally intolerant of high temperatures or very high humidity levels or frost. The ideal conditions for tuberous begonias are areas where evening temperatures do not fall below 15°C. Tuberous begonias reportedly produce tubers under short days (SD) and low temperatures but in contrast with *B. hiemalis* (hiemalis begonia, rieger begonia, or elatior begonia), continuous long days (LD) and temperatures >17°C are required for proper plant development and flowering.

8.1.0 Light

Begonia x *hiemalis* form flowers when the length of day is between 12 and 14 hours. Roodenburg states "The main flower-forming period is situated in late summer and the early autumn. It is the result of the shortening of the days in August and September, for day lengths of 14 to 12 and less are favourable for flower bud initiation in these begonias. Later on, in the much shorter days of November and December, the growth gradually comes to a standstill".

To prevent this standstill during winter months, artificial lengthening of the days has been practicised on mother plants in order to provide cutting-material for propagation. On the other hand, shortening the days in the late spring and summer is used to initiate flowers on plants for sale.

Short-day conditions may result in delayed development and dormancy, although seed-propagated tuberous begonias may respond differently to photoperiod than cultivars propagated by tubers (Karlsson. and Werner, 1999). It was reported by Waines in 1972 (Molnar, 1974) that *Begonia socotrona* under short day conditions exhibits a rapid floral initiation and development and plants can be kept in a non-flowering state at 26.6°C with long days. In case of formation of lateral breaks, the plants must be grown in long photoperiods at a temperature of 18–21°C (Djurhuus, 1985).

Tuber formation is inhibited at photoperiods of 14 h or longer and at temperatures above 18°C. At a night temperature of 18°C, the formation of tubers was independent of day temperature. So, the critical photoperiod for tuber development is 10–14 h. (Djurhuus, 1985). The root/shoot dry-weight ratio within cultivars was independent of day length (Karlsson and Werner, 1999).

9.0 Cultivation

9.1.0 Planting

Tubers are planted in February or March at temperatures of 18°C to 20°C on a layer of peat. When new growth is three to four inches tall, repotting and covering the tuber is practiced. The tubers are planted in a flat of peat moss or vermiculite. If the stored tubers show no signs of disease, the same medium in which they were stored can be used. Otherwise fresh peat moss or vermiculite is needed. Plant the tubers right-side up in the flat, so the peat moss or vermiculite just covers them. Place the flats where the temperature is around 18°C, and they will receive indirect light. Supplemental lighting, such as grow lights, is not necessary during this stage of growth (MacKenzie and Moncada, 2018).

In recent years, soilless substrates have been more widely used for growing pot plants because of several advantages that include facilitated root penetration, reduced levels of pathogenic microorganisms, absence of weeds, light weight, high water capacity, and high water retention capacity (Londra *et al.*, 2012).

Among organic materials used as substrates, peat is the most widely distributed worldwide with many desirable characteristics such as both high cation exchange capacity and water retention capacity and resistance to decomposition. Similarly coir (coconut dust/pith) is also found to be very effective substrate. During irrigation of organic substrates (*i.e.* peat, coir), as the amount of water increases, root aeration decreases as a result of the large water retention capacity of these substrates, which render them extremely difficult in retaining a sufficient water–air balance. The water–air balance of four soilless substrates: 75 per cent sphagnum peat–25 per cent perlite (Ps75: P25), 50 per cent sphagnum peat–50 per cent perlite (Ps50:P50), 50 per cent coir–50 per cent perlite (C50: P50), and a fortified substrate with 60 per cent sphagnum peat–30 per cent black peat–10 per cent perlite (Ps60:Pb30:P10) (in a volumetric proportion) was investigated under two different irrigation methods (drip and sub-irrigation), and its effect on the growth of *Begonia elatior* 'The President' was studied. The bulk density, particle size distribution, and water retention curve of the substrates were determined. Furthermore, the water profiles, oxygen (O_2) concentration, and O_2 diffusion rate of all substrates were determined during a 16-week cultivation period. Plant height, flower production, and both shoot and root dry weights as well as per cent growth increase of plants were measured at the end of the experiment. The substrate water profiles showed that the water content was greater and air content was less in substrates of pots irrigated with drip irrigation than with sub-irrigation. The O_2 concentration in all substrates irrespective of the irrigation method was high. The O_2 diffusion rate values of sub-irrigated substrates were greater than those drip-irrigated, and Ps60: Pb30: P10 showed the greatest values. Shoot and root dry weights and per cent growth increase of drip-irrigated plants were greater than that of sub-irrigated plants (Londra *et al.*, 2012).

9.1.1 Thinning of Shoots

Sizes of the plants depend on the size of the tubers but flower size is not affected by it. Flower size is dependent by the number of shoots on each tuber. For bigger flowers, thin the shoots when they are about 8 cm tall. In case of small tubers, two or three stems should be kept while for large tubers three to six stems are kept, extra stems are pinched off. Some varieties, such as those used in hanging baskets, are grown for quantity of flowers, not for large blooms, shoots are thinned. These plants grow as bushy and bear many smaller flowers, especially if the tips of the stems are pinched in May or early June.

9.1.2 Hardening Off

Hardening off the plants is accustomed about two weeks before the last frost date in outdoor condition. This is practiced for pot plants only. The plants are placed outdoors on a warm day to a shaded, protected place and brought them back in before temperatures cool off at night. Same process is followed for consecutive days. Finally, these are moved out into brighter light and exposed to more wind, but continued to bring them in at night. Once the weather has settled, the soil is warm, and all danger of frost is past transplant them to hanging baskets on the porch or out in the garden, or move containers to their permanent place on the deck or patio (MacKenzie and Moncada, 2018).

9.2.0 Fertilization and Irrigation

According to MacKenzie and Moncada (2018), tuberous begonias are fairly heavy 'feeders', hence they suggest fertilizing with a dilute balanced fertilizer throughout the growing season for best bloom and growth. Although tuberous begonias must be grown in well-drained soil, they are not drought tolerant, so it's important to water them. If they are growing in the shade of trees or in containers under a roof, attention must be paid to keeping the soil properly moist. When the soil in containers becomes slightly dry they must be watered. Moss-lined wire hanging baskets work particularly well to provide adequate drainage.

Application of nutrients to plants has varied functions from growth, flowering to their role in stress conditions. Plant performance and appearance in deficient and toxic levels of nutrients are well characterized. However, less is known about the potential subtleties of plant growth, form, development, nutrient uptake, and biotic stress tolerance in the broad tolerable range. *Botrytis cinerea* is a ubiquitous pathogen that infects leaves, stems, and flowers. Symptoms of the disease appear as water soaked lesion spots, which quickly coalesce affecting the entire tissue, causing major economic loss. Susceptibility to *B. cinerea* disease has been reported to decrease with NO_3^- form but increase with NH_4^+ form (Pitchay *et al.*, 2007).

The effect of different watering frequencies and fertilizer amounts on the biomass, plant height, crown diameter, flower number and days of blooming of *B.* × *elatior* was studied by pot culture and quadratic general rotary design (Li and Xiang, 2011). The regression models were established by taking the biomass, plant height, crown diameter, flower number and days of blooming of *B.* × *elatior* as objective function. The effect of water and fertilizer on the growth index and quality indexes was significant by inspection. The results indicated the effect of fertilizer dose was greater than that of watering frequency, and the value of watering frequency and fertilizer amount matching the optimal indexes was conducted as well. The action law of water and fertilizer on the growth index and quality indexes of *B.* x *elatior* was obtained by single factor analysis. It was concluded that the optimal project for potted *B.* x *elatior* in growing medium V (sawdust): V (vermiculite)=3:1 was watering when the water content of medium was 57.58 per cent -70.00 per cent, the water amount was 375.00 ml per pot, and the fertilizer amount was 561.26-625.00 mg per pot (Li and Xiang, 2011).

10.0 Diseases and Pests

10.1.0 Diseases

Moorman (2016) has given details of the Begonia diseases.

Disease	*Symptoms*	*Pathogen*	*Management*
Bacterial Leaf Spot and blight	Water-soaked areas surrounded by yellow halos develop on leaves. Plants slowly die one leaf at a time.	*Xanthomonas campestris pv. begoniae*	Planting of disease free plants. Discarding infected plants; particularly Rieger-types which are systemically infected. Leave the leaves dry while irrigation. Destroy crop debris.
Botrytis Blight	Cuttings rot at their base. Tan spots develop on leaves. Established plants rot at the crown. Infected tissue is covered with dusty gray fungal growth.	*Botrytis cinerea*	Maintain low humidity by spacing plants and venting to improve air circulation. Remove dead and dying flowers and leaves. copper based fungicide help to protect plants.
Powdery Mildew	White, mealy fungal growth develops on leaves, flowers, and stems. Tissue beneath the fungus may die.	*Erysiphe cichoracearum*	Examine plants carefully and frequently to detect the onset of disease. Apply any sulfur based fungicide to control this.
Pythium Rot	Seedlings die. Shiny blackened areas develop on the stems and petioles of established on the stems and petioles of established plants at and just above the soil line as plants collapse and die.	*Pythium sp.*	Steam or chemically sterilize soil. Observe strict sanitary measures. Drench plants with systemic fungicide. Do not drench very young seedlings.
Rhizoctonia cown rot	Stems rot at the soil line. Plants collapse and die. A fine webbing of fungal growth may be seen between the infected stems and nearby soil particles.	*Rhizoctonia* spp.	Plant in pasteurized potting media. Apply any systemic fungicide to protect plants.
Viruses	Mosaic, malformed leaves, pale yellow ring spots, severe chlorotic mottling of leaves or plant stunting may occur.	Impatiens necrotic spot, tobacco ring spot, broadbean wilt, cucumber mosaic, tobacco necrosis virus.	Disease free planting material is must. Discard infected plants. Plant in pasteurized potting media. Maintain good aphid and thrips control to reduce spread.

Source: Moorman, G.W. (2016). https://Extension.Psu.Edu/Begonia-Diseases

Besides the above, begonias are also susceptible to Armillaria root rot (*Armillaria mellea*), anthracnose (*Gloeosporium* sp.), black root rot (*Chalara elegans*), cottony rot (*Sclerotinia sclerotiorum*), crown gall (*Agrobacterium tumefaciens*), foliar nematode (*Aphelenchoides olesistis*), leaf spot (*Phyllosticta* sp.), Rhizoctonia stem rot (*Rhizoctonia solani*), root knot nematode (*Meloidogyne* spp.), soft rot (*Erwinia carotovora*), and *Verticillium* wilt (Koike and Wilen, 2020).

10.2.0 Pests

10.2.1 Thrips

Thrips are tiny, slender insects with fringed wings. They feed by puncturing the epidermal (outer) layer of host tissue and sucking out the cell contents, which results in stippling, discolored flecking, or silvering of the leaf surface. Thrips feeding is usually accompanied by black varnish like flecks of frass (excrement). Pest species are plant feeders that discolor and scar leaf, flower, and fruit surfaces, and distort plant parts or vector plant pathogens. Thrips can readily move long distances floating with the wind or transported on infested plants, and exotic species are periodically introduced. For example, myoporum thrips causes severe galling of *Myoporum laetum* and *M. pacificum* (Bethke *et al.*, 2014).

Western flower thrip, *Frankliniella occidentalis* (Pergande), is among the most economically important agricultural pests globally, attacking a wide range of vegetable and horticultural crops. They cause irregular reddish brown lines on the upper sides of the leaves. Spots form on the undersides of the leaves, especially along the main veins. The leaves may be deformed. In addition to causing extensive crop damage, the species is notorious for vectoring destructive plant viruses, mainly belonging to the genera Orthotospovirus, Ilarvirus, Alphacarmovirus and Machlomovirus. Once infected by orthotospo viruses, thrips can remain virulent throughout their lifespan and continue transmitting viruses to host plants when and wherever they feed (He *et al.*, 2020).

Superior horticultural oils which are also considered nontoxic and are less likely to harm beneficial insects. The oil suffocates the aphid insects. These oils are highly refined and under proper conditions, can be applied to plants in foliage without damage. Insecticides containing pyrethrins, a plant-derived insecticide, are effective and more benign than other chemical pesticides.

10.2.2 Whiteflies

Whiteflies, tiny pests with white wings, leave a dusty, white powder on plant. They damage begonia by feeding on the bottom of the leaves, causing mottled and yellowing foliage. These are flying around the plant when the leaves are disturbed. Remove infested leaves and control the whiteflies with yellow sticky traps. Insecticide sprays might also be useful in removing whiteflies.

10.2.3 Scales

Begonia plants covered with crusty, brown bumps or white, cottony lumps are likely infested with scales, an unusual insect that can appear inanimate. Scales cause begonia leaves to wilt, turn yellow or drop from the plant. Scales might also cause discoloration and deformities on the leaves. Remove scales with a toothbrush dipped in soapy water and spray both sides of the leaves with horticultural oil to get rid of any lingering pests.

10.2.4 Aphids (*Aphis gossypii*)

Aphids suck moisture and nutrients from begonia plant, causing the leaves to curl, turn yellow and distort. Aphids can affect growth as well. Rinse begonia plant under running water to remove the pests and spray any systemic insecticide or horticultural oil.

10.2.5 Mealy Bugs (*Planococcus citri*)

The mealy bugs cause slow growth and leaf damage. Prevent mealy bugs by keeping your begonia well watered and healthy. Remove mealy bugs by spraying your plant with water. Separate infested plants from those not affected. It can be controlled by spot treatment with Isopropyl Alcohol: On small infestations on houseplants, a 70 per cent or less solution of isopropyl (rubbing) alcohol in water may be dabbed directly on mealy bugs with a cotton swab to kill them or remove them. Test the solution out on a small part of the plant 1 to 2 days before hand to make sure it does not cause leaf burn (phytotoxicity). In some cases, a much more diluted solution may be advisable. Where infestations are extensive, a 10-25 per cent solution of isopropyl alcohol can be applied with a spray bottle. This spray is needed to repeat every week until the infestation is gone (Flint, 2016).

10.2.6 Foliar Nematode (*Aphelenchoides fragariae*)

Plants become stunted due to foliar nematode infestation. Excessive red pigment develops in infected leaves. Bronzed or water-soaked areas develop on leaves of some cultivars. Fibrous rooted cultivars have small brown leaf spots. Some cultivars exhibit no symptoms despite heavy infection. To control this nematode infestation, nematode-free plants should be planted or purchased, should avoid overhead irrigation and splashing. Spraying with nematicide and discarding the infected plants as and when seen is beneficial to spread the nematode in field.

10.2.7 Mites

10.2.7.1 Spider Mites (Tetranychus urticae)

Spider mites can cause damage if they aren't controlled immediately. These pests are extremely tiny and hard to detect. They cause stippled and yellowed leaves by sucking the moisture from the soft tissues of the flower. Hold a piece of white paper under the leaves of your begonia and tap them gently. If plant is infested, tiny dark dots will drop to the paper. To eradicate spider mites, take the plant outside and spray it with a garden hose. This knocks the mites from the begonia and prevents them from spreading to other plants. (Bethke *et al.*, 2014).

10.2.7.2 Tarsonamid Mite or Broad Mite (Polyphagotarsonemus latus)

These are particularly nasty infestations. Too small to be seen by the naked eye their effects can rapidly become devastating if left unattended. Symptoms are corky deposits on the stem and leaf, also discolouration and withering of the

Diseases and Pests of Tuberous Begonia

Powdery Mildew

Bacterial Leaf Spots

Botrytis Blight

Foliar Nematode

Botrytis

Pythium Root Rot

Impatiens Necrotic Spots

White Fly

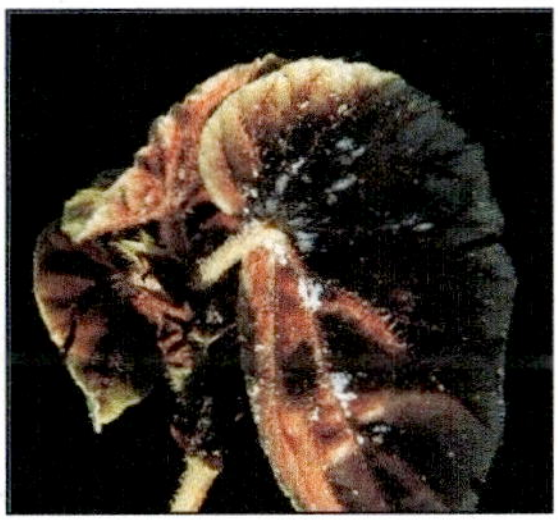
Mealy Bug

growing tip which takes on a blackened knurled effect. Any plant that is suspected of being affected should be isolated before the problem has a chance to spread. A preventative treatment practiced by some growers is to soak tubers in a solution containing 10 per cent Domestos for 15 minutes followed by 15 minutes in fresh water. This treatment has not however been proven. Gaining popularity is a biological preventative measure in the form of tiny predators that attack the mite, *Amblyseius cucumeris* has proved very effective in combating this awful insect that left untreated will rapidly destroy an entire collection.

It can be controlled by periodically wash down of house plants with a strong spray of water and be sure to get good coverage of the lower leaf surfaces. Use insecticidal soap for house plants or broad-spectrum miticide can give good control.

11.0 Harvesting and Postharvest Management

11.1 Harvesting

Effects of climate conditions on postharvest quality were investigated for cyclamen, begonia and Poinsettia by Bulle (2005). The postharvest life of begonia was better when RH during the last weeks of the growth period was lower. The mean RH-levels of the lowest and highest group in the cluster analysis of begonia were 68 per cent and 78 per cent respectively. Light levels during cultivation also caused different effects depending on the crop. The results of Bulle's (2005) research with begonia show that postharvest life can be improved with high light levels in the beginning of the growth period. Probably because it was in February, a dark period of the year, in which light is often an important limiting factor. Mean levels of relative humidity of approximately 60 per cent showed the best postharvest life of begonia (Fjeld, 1986). The numbers of yellow leaves and leaf drop were less when light levels were high. In other research no effects from relative humidity on postharvest life of begonia were found (Mortensen, 2000).

11.2 Postharvest Management

11.2.1 Shelf Life

The spectacular effects of silver thiosulphate (STS) in overcoming the deleterious effects of ethylene in potted flowering plants (Cameron and Reid, 1981) have led to widespread commercial use of this chemical (Veen, 1983). Because STS contains

Harvesting and Marketing of Tuberous Begonia

Harvesting Stage in Greenhouse

Harvesting of Tuberous Begonia

Tuberous Begonia in Market

Cut Flowers in Market

Marketing of Potted Begonia

silver, which is seen as a potential environmental pollutant, its use for potted flowering plants has been restricted in some countries (Serek and Reid, 1993). Researchers have therefore been searching for alternative strategies for preventing ethylene damage in potted plants, including using ethylene biosynthesis and binding inhibitors (Serek *et al.*, 1994a). Serek *et al.* (1994b) observed a 6-hour fumigation of flowering *Begonia ´elatior hybrida* Fotsch. 'Najada' and 'Rosa', *B. ´tuberhybrida* Voss. 'Non-Stop', *Kalanchoe blossfeldiana* Poelln. 'Tropicana', and *Rosa hybrida* L. 'Victory Parade' plants with 1-MCP, (formerly designated as SIS-X), a gaseous non-reversible ethylene binding inhibitor, strongly inhibited exogenous ethylene effects such as bud and flower drop, leaf abscission, and accelerated flower senescence. The inhibitory effects of 1-MCP increased linearly with concentration, and at 20 nl/l this compound gave equal protection to that afforded by spraying the plants with a 0.5 STS mM solution.

Elatior begonia (*Begonia* × *hiemalis*) is considered as an ethylene sensitive plant. Transportation circumstances produce ethylene and accelerate the abscission of flower buds and open flowers, yellowing and abscission of leaves, thus reduce display life in begonia (Kim *et al.*, 2007). They determined the main stress factor during transportation and the effects of 1-MCP using three concentrations (0, 5, 25, and 125 mg/l), and two treatment durations (0, 6, and 12 hr) on 'Blitz' and 'Carneval' begonia. Transportation conditions in simulated export container with darkness, vibration with continuous shaking (150 ± 20 RPM) on a rotary lab shaker, and low temperature (12°C) were held for 4 days. After simulated shipment and placement in an interior environment, a considerable percentage of flower buds and flowers abscised, which could be attributed to elevated ethylene concentrations inside the boxes. All levels of chemical protected the display life of plants, however 1-MCP was the most effective at concentrations of 125 nl/l /12 hr for 'Blitz' and 25 nl/l /12 hr for 'Carneval'. Thus, 1-MCP pre-treatment prior to packing begonia for shipment is recommended to increase the length of display life and to reduce abscission of open flowers and floral buds (Kim *et al.*, 2007).

Subsequently they (Kim *et al.*, 2012) studied the effect of 1-methylcyclopropene (1-MCP) on ethylene production induced by simulated transport stress in *Begonia* × *hiemalis* 'Blitz' and 'Carnival' to improve the display life in potted plants. The simulated transportation conditions were imposed for 4 days in simulated export containers with darkness, vibration with continuous shaking (150±20 rpm) on a rotary lab shaker, and low temperature (12°C). Plants were treated with 1-MCP at three concentrations (5, 25, or 125 nl/l) and for three different periods (0, 6, or 12 hours) before undergoing the simulated transport stress treatments. Treatment with 25 or 125 nl/l 1-MCP inhibited the abscission of open flowers by more than 40 per cent as compared to the untreated plants.

11.3.0 Storage of Tubers

Tuberous begonias enter a dormant period both in greenhouse or outdoor growing, so they are not suitable as year-round houseplants. (Some others, such as Rieger begonias, are a better choice as indoor flowering plants.) During the dormant period, begonia tubers must be stored indoors. If left in the ground during winter, they would get killed. Following a reduction in blooming and yellowing leaves, the leaves and stems drop off. Some gardeners may choose to remove any flower buds that form in September to encourage dormancy and the storing of energy in tuber. Begonia tubers are dug by cutting the stems about five inches and drying at room temperature until the remaining stem pieces are dry and loose. The tubers are washed and allowed to dry before storing. Dried tubers are covered with peat or sand and stored in refrigerator between 7°C and 15°C (Gilman and Heatley, 2014).

12.0 Biotechnology

12.1.0 Somatic Embryogenesis

Direct somatic embryogenesis induction of *Begonia* x *hiemalis* (Elatior Begonia) was initiated from two different explants *i.e.* leaves and petioles (Awal *et al.*, 2008). Both explants were cultured on MS medium supplemented with different concentrations of benzylaminopurine (BAP) and 2,4-dichlorophenoxy acetic acid (2,4-D). The results showed that combinations of 0.5-1.0 mg/l BAP and 0.1 mg/l 2,4-D produced direct somatic embryogenesis from leaf and petiole explants. Different concentrations of casein hydrolysate were also tested to optimize somatic embryo induction. The results showed that 100 mg/l casein hydrolysate could produce 53.08 nodular callus and 24.16 green embryogenic callus, whereas, 500 mg/l casein hydrolysate produced 30.83 nodular callus and 23.75 green embryogenic callus. The embryogenic callus was then transferred to MS medium supplemented with 0.5 mg/l gibberelic acid (GA_3) with 0.2 g/l activated charcoal for further embryogenesis development and further regeneration.

12.2.0 Somaclonal Variation

Somaclonal variation is also a major obstacle in biotechnological breeding techniques, such as genetic engineering and the production of haploid plants. The extent of somaclonal variation is usually assessed by determining the percentage of plants aberrant for a specific trait (*e.g.* phenotypical characteristics, chromosome number, and electrophoresis pattern of proteins). de Klerk *et al.* (1990) have assessed the extent of variation in micropropagated plants of *Begonia* x *hiemalis* cv. Schwabenl and Red in another way. In three somaclonal populations the value of the standard deviation (SD) of a quantitative trait was determined, namely the SD of the natural logarithm of the ratio of the lengths of two leaf ribs. Plants, regenerated from callus, induced and maintained at a high concentration of 2,4-dichlorophenoxyacetic acid (2,4-D), had a significantly larger SD, whereas the mean was not affected at all. The observed increase of the SD is likely to have been caused by somaclonal variation, but epigenetic variation cannot be ruled out. It is concluded that differences in the values of the SD may be used to assess differences in the extent of variation.

12.3.0 Characters of Micropropagated Plantlets

In order to study the variation of *Begonia* × *hiemalis* after *in vitro* propagation, one plant of 'Aphrodite Pink' and one of 'Schwabenland Red' were vegetatively propagated in four ways (Westerhof *et al.*, 1984). One group of plants was obtained by propagation by means of cuttings. For the second group, the bacterial elimination system, as developed by Hakkaart and Versluijs (1983), was applied which includes an *in vitro* propagation step. The third and fourth groups were obtained by adding one and two cycles, respectively, of *in vitro* propagation after the bacterial elimination, so that one, two and three propagation cycles could be compared. One cycle, followed by bacterial elimination, gave nearly uniform offspring. However, when this was followed by one or two cycles of *in vitro* propagation, the variation increased. Variation also increased when the size of the plantlets obtained *in vitro* decreased. Thus, an important source of variation can be avoided by eliminating the smallest *in vitro* plantlets. It is recommended to flower and select the plants before propagation by conventional cuttings, whether or not a further *in vitro* propagation is applied after the bacterial elimination (Westerhof *et al.*, 1984).

12.4.0 *In vitro* Conservation

The historical perspectives of *in vitro* conservation of begonias at Glasgow Botanical Garden make an interesting reading. Bowes and Curtis (1991) wrote a beautiful brief on this in their article. According to them, in 1983, the *Begonia* collection in the Glasgow Botanic Gardens was designated as the National Collection and it currently holds approximately 200 species plus 100 cultivars. In 1987, a joint expedition, from the Botany Department of Glasgow University and the Glasgow Botanic Gardens visited Papua New Guinea (Dickson, 1987) and sent back a number of species of *Begonia* new to cultivation. Several were collected on the expedition as miniature leaf cuttings which were cultured *in vitro* on a Knop's mineral salts medium under non-sterile conditions utilizing a technique developed for cultivars of *B. rex* (Bowes, 1990). Because sucrose is excluded from the medium, growth of microbial contaminants is greatly reduced and does not normally interfere with the spontaneous regeneration which occurs from leaves in a number of begonias (Bowes and Curtis, 1991).

Regeneration was investigated in non-sterile leaf explants taken from 87 *Begonia* lines after culturing for up to 10 weeks *in vitro* on Knop's mineral salts agar medium with no exogenous growth substances present (Bowes and Curtis, 1991). For many lines, only very limited supplies of leaf samples were available; nevertheless, in 66 lines *de novo* regeneration of roots and buds occurred and potted plants were recovered from 49 hues. Additionally, in 13 *Begonia* lines, buds were excised from stock plants and cultured under non-sterile conditions on Knop's agar. In seven of these the buds rooted and gave rise to potted plants. In six lines, organogenic cultures (derived from sterile leaf explants) were established in long-term culture on Murashige and Skoog medium supplemented with growth substances and potted plants have been recovered. Sterile explants from a further 12 lines showed regeneration and were subcultured. The establishment of such sterile cultures, with continued regenerative capacity, is a significant step in the development of a programme for the conservation of rare and endangered begonias.

12.5.0 Genetic Transformation

An *agrobacterium*-mediated transformation system for commercial *Begonia* species was first reported by Kiyokawa *et al.* (1996). The leaf explants of *Begonia semperflorens, Begonia* x *hiemalis* and *B. tuberhybrida* were inoculated with

Agrobacterium tumefaciens LBA4404 harboring a binary vector pBI121 which contains rol A, B and C genes of an agropine type Ri plasmid (pRiA4b). Kanamycin resistant shoots of *B. tuberhybrida* were obtained on MS agar medium supplemented with 0.1 mg/l NAA, 0.5 mg/l BA, 500 mg/l claforan and 100 mg/l kanamycin. These shoot exhibited GUS activity and Southern analysis showed a single copy insertion into the genome. When the transgenic plants were transferred to soil, they displayed the phenotype specific to the transgenic plants by *A. rhizogenes* such as dwarfness, delay of flowering, and wrinkled leaves and petals.

Kishimoto *et al.* (2002) reported a method for *Agrobacterium*-mediated transformation of Elatior Begonia (*Begonia*×*hiemalis*). Young leaf discs were infected with *Agrobacterium tumefaciens* strains AGL0 and LBA4404. Each strain has a binary vector plasmid, pIG121Hm that includes the β-glucuronidase (GUS) gene with an intron as a reporter gene, and both the neomycin phosphotransferase II (NPT II) and the hygromycin phosphotransferase genes as selection markers. Explants were cultured on modified MS medium supplemented with 1.0 mg/l BA, 0.5 mg/l IAA, 300 mg/l ticarcillin, and either 100 mg/l kanamycin and 5 mg/l hygromycin, or 300 mg/l kanamycin for selection and regeneration. Out of 500 explants infected with AGL0, 16 plantlets were regenerated, and out of 628 explants infected with LBA4404, two plantlets were regenerated after 4 months of culture. Transformation was confirmed by Southern blot analysis of the GUS gene and by histochemical assays of GUS activity in plant tissues. Ten *in vitro* transgenic plants were obtained from AGL0 infected explants only (Kishimoto *et al.*, 2002).

12.6.0 Markers

According to Dwitte *et al.* (2011), DNA sequence data from non-coding regions plays an important role in plant classification and barcoding (CBOL Plant working group, 2009), and has widely been used to resolve relationships at the species and sectional level in *Begonia*. A framework phylogeny of *Begonia* based on analyses of 13000 bases of plastid and mitochondrial DNA of 30 *Begonia* species (Goodall-Copestake *et al.*, 2010 cited by Dewitte *et al.*, 2011) indicated that African taxa form the earliest divergent clades in the *Begonia* phylogeny and that both Asian and American *Begonia* lineages are derived from African ancestors. The phylogenetic relationships within the relatively small group of African *Begonia*, which comprises around 160 species subdivided into 17 sections (De Wilde and Plana, 2003; Doorenbos *et al.*, 1998).

Species-specific microsatellite markers were developed for endemic *Begonia socotrana* (Hughes *et al.*, 2002a), and *Begonia sutherlandii* Hook (Hughes *et al.*, 2002b). Later, universal DNA markers from chloroplast *trnL* intron and nuclear ribosomal internal transcribed spacer (ITS) were used for reconstruction of molecular phylogeny of African begonias (Plana *et al.*, 2004).

13.0 References

Arends, J.C. (1970) *Meded Landbouwh Wageningen*, **70–20** : 1–18.

Booth, C. and Brown, C. (2013) *United States Patent US* PP23.447 P2. dated Mar. 5, 2013.

Bowes, B.G. and Curtis, E.W. (1991) *New Phytol.,* **119** : 169-181.

Bulle, A. (2005) *Acta Hortic.*, **669**: 263-268.

Burkill, I.H. (1935) In: *A Dictionary of the Economic Products of the Malay Peninsula. I (A-Cod)*.

Cameron, A.C. and Reid, M.S. (1981) *HortScience,* **16**: 761–762.

CBOL Plant working group (2009) *Proc. National Academy of Sciences,* **106** : 12794-12797.

Constanta, A., Gh., L. and Simona, P. (2015) *J. Hortic., Forestry and Biotechnology.,* **19** : 16-20.

Cross, R.J. (2004) *United States Patent, US* PP15, 108 P2 dated Aug. 24, 2004.

De Wilde, J.J.F.E. and Plana, V. (2003) *Edinb. J. Bot.,* **60** : 121-130.

Dewitte, A., Eeckhaut, T., Huylenbroeck, Van, J. and Bockstaele, E. Van. (2009) *Euphytica,* **168** : 81–94.

Dickson, J.H. (1987) *The Begonian,* **54** : 63.

Djurhuus, R. (1985) *Sci. Hort.,* **27** : 123-131.

Doorenbos, J. and Karper, J.J. (1975) *Euphytica,* **24** : 13–19.

Drewlow, L.W. (1990) *US Patent Number*: Plant 7260; Date of Patent: Ul.3, 1990.

Espino, F.J., Linacero, R., Rueda, J. and Vázquez, A.M. (2004) *Biol. Plant.,* **48**(1) : 101-104.

Ewart, L.C. (1985) *Tuberous rooted begonia*, pp. 420– 422. In: J.W. Mastalerz and E.J. Holcomb (eds.). *Bedding plants III, a manual on the culture of bedding plants as a greenhouse crop*, 3rd ed. Pennsylvania Flower Growers, Harrisburg, Pa.

Fjeld, T. (1986) *Acta Hortic.*, **181**: 251-255.

Flint, M.L. (2016) *http: //ipm.ucanr.edu/PDF/PESTNOTES/pnmealybugs.pdf.*

Gaur, R.D. (1999) In: *Flora of the District Garhwal North West Himalaya (with ethnobotanical notes*). Trans Media Publication, India.

Gilman, E.F. and Heatley, R. (2014) *https: //edis.ifas.ufl.edu/fp062.*

Girmansyah, D. (2009) *Reinwardtia,* **12**(5): 419–434.

Gleed, C.J. (1961) *J. R. Hort. Soc.,* **86** : 319–322.

Haegeman, J. (1979) *Tuberous Begonias. Origin and development.* J. Cramer, ISBN 376821219X, Vaduz, Liechtenstein. pp. 268.

He, Z., Guo, J.F., Reitz, S.R., Lei, Z.R. and Wu, S. Y (2020) *Insect Sci.,* **27** : 626–645.

Hvoslef-Eide, A.K. and Munster, C. (2007) *Begonia*. History and breeding. In: Anderson, N.O. (Ed.) Flower Breeding and Genetics, Anderson, NO. (ed.),pp, 241-275, Springer, ISBN 9781402044274, Dordrecht, The Netherlands.

Irmscher, E. (1960) Begoniaceae. In: *Parey's Blumengärtnerei,* **2** : 67-91, Berlin.

Jordan, D. (2008) www.eattheweeds.com//Entries/1958/5/25_Begonias

Karlsson, M.G. and Werner, J.W. (1999) *HortScience,* **34**(1) : 62–63.

Kim, Y.J., Kwon, S. and Kim, K.S. (2007) *Acta Hortic.,***755** : 437-444.

Kishimoto, S., Aida, R. and Shibata, M. (2002) *Plant Sci.,* **162** : 697-703.

Kishimoto, S., Aida, R. and Shibata, M. (2002) *Plant Sci.,* **165** : 697-703.

Kisvarga, S., Honfi, P. and Tilly-Mándy, A. (2015) *Bulletin UASVM Hortic.,* **72**(1) : 115-119.

Kiyokawa, S., Kikuchi, Y., Kamada, H. and Harada, H. (1996) *Plant Cell Rep.,* **15** : 606–609.

Koike, S.T. and Wilen, C.A. (2020) *http: //ipm.ucanr.edu/PMG/r280110311. html#ipmpagetop.*

Kroon, G.H. (1993) *Acta Hortic.,* **337** : 53–58.

Le Corff J., Ågren J. and Schemske, D.W. (1998) *Ecol.,* **79** : 1610-1619.

Li, S.X. and Xiang, Z.Q. (2011) *J. Northwest A and F University - Natural Science Edition,* **39**(3) : 168-176.

Lim T.K. (2014) *Begonia x tuberhybrida.* In: Edible Medicinal and Non Medicinal Plants. Springer, Dordrecht. https: //doi.org/10.1007/978-94-007-7395-0_36.

Londra, P.A., Paraskevopoulou, A.T. and Psychoyou, M. (2012) *Hort Sci.,* **47** : 1153–1158.

MacKenzie, J. and Moncada, K. (2018) https: //extension.umn.edu/flowers/ tuberous-begonias.

Marasek-Ciolakowska, A., Ramanna, M. S., Laak, W.A.T. and van Tuyl, J.M. (2010) *Euphytica,* **171** : 273–282.

Matsuura, H. and Okuno, S. (1943) *Cytologia,* **13** : 1-18.

Mikkelsen, J.C.(1987) *US Patent Number: Plant* 6,013; Date of Patent: Sep. 15, 1987.

Moorman, G.W. (2016) Https: //Extension.Psu.Edu/Begonia-Diseases).

Mortensen, L.M. (2000) *Sci. Hort.,* **86**: 299-310.

Murugkarm, A.D. (2006) *Human Ecology Special Issue,* **14**: 83–88.

Nakano, M., Niimi, Y., Kobayashi, D. and Watanabe, A. (1999) *Sci. Hort.,* **79** : 245-251.

Neale, S., Goodall-Copestake, W. and Kidner, C.A. (2006) *The Evolution of Diversity in Begonia.* In: *Floriculture, Ornamental and Plant Biotechnology Volume IV* ©2006 Global Science Books, UK. pp. 606-611.

Okuno, S. and Nagai, S. (1953) *Jap. J. Genet.,* **28** : 132-136.

Pavlich, J.C. (2015) *US Patent Number:* PP 25512 Date of Patent: May, 5, 2015.

Pavlich, J.C. (2016) *US Patent Number*: PP 26810 Date of Patent: June, 7, 2016.

Sajem, A.L. and Gosai, K. (2006) *Journal of Ethnobiology and Ethnomedicine,* **2**: 33, doi: 10.1186/1746-4269-2-33.

Sandgrind, S. (2017) *Master's Thesis* (2017 60 ECTS), Faculty of Biosciences, Dept. Plant Sci., Norwegian University of Life Sciences. Oslo pp.107.

Sandved, G. (1969) *Acta Hortic.,* **14** : 61-76.

Schelbeck, E.W. (1983) *United States Patent. Plant* 5, 162 dated Dec. 20, 1983.

Schelbeck, E.W. (1984a) *United States Patent. Plant* 5, 179 dated Jan. 3, 1984.

Schelbeck, E.W. (1984b) *United States Patent*. Plant 5,179 dated Jan. 17, 1984.

Schelbeck, E.W. (1985) *United States Patent*. Plant 5391 dated Jan. 15, 1985.

Serek, M. and Reid, M.S. (1993) *HortScience,* **28**: 1180–1181.

Serek, M., Sisler, E.C. and Reid, M.S. (1994a) *J. Amer. Soc. Hort. Sci.,* **119**: 572–577.

Serek, M., Sisler, E.C. and Reid, M.S. (1994b) *J. Amer. Soc. Hort. Sci.,* **119**: 1230–1233.

Shim, M.S., Oh, W. and Kim, K.S. (2005) *J. Korean Soc. Hort. Sci.,* **46** : 200-203.

Shimada, Y., Mori, G., Katahara, Y. and Oda, M. (2005) *Acta Hortic.,* **673** : 537-541.

Simmonds, J. (1984) *Plant Cell Tiss Organ Cult.,* **3** : 283–289.

Soren, H. (1996a) *United States Patent*. Plant 9460 dated Feb.27, 1996.

Soren, H. (1996b) *United States Patent*. Plant 9,505 dated Apr. 9, 1996.

Takayama, S. and Misawa, M. (1981) *Plant and Cell Physiology,* **22** : 461–467.

Thomas, D.C., Hughes, M., Phutthai, T., Rajbhandary, S., Rubite, R., Ardi, W.H. and Richardson, J.E. (2011) *Mol.Phylogenet. Evol.,* **60** : 428-444.

Tian, D., Xiao, Y., Tong, Y., Fu, N., Liu, Q. and Li, C. (2018) *Plant Diversity,* **40** : 75-90.

Veen, H. (1983) *Scientia Hort*., **181**: 155–160.

Viseur, J. and Lievens, C. (1987) *Acta Hortic.,* **212** : 705-710.

Westerhof, J., Hakkaart, F.A. and Versluijs, J.M.A. (1984) *Sci. Hort.,* **24** : 67-74.

Yates, F.C. (2007) *United States Patent US 2007/0283473 P1* dated Dec. 6, 2007.

6

Cyclamen

V.A. Parthasarathy, Rajiv Kumar,
C. Aswath and Sk. Masudul Islam

1.0 Introduction

Cyclamen is a genus of 23 species of perennial flowering plants family Primulaceae is one of the most important ornamental crops sold worldwide particularly Western Europe and America as a potted ornamental during winter (Jalali *et al.*, 2012). It was traditionally classified in the family Primulaceae, was reclassified in the family Myrsinaceae in 2000, and finally, in 2009 with the introduction of the APG III system, was returned to the subfamily Myrsinoideae within the family Primulaceae (Kallersjo *et al.*, 2000; Stevens, P.F., 2012). Now over 200 cultivars are known and the *Cyclamen* is the most widely grown pot plant in The Netherlands (Legro, 1959).

2.0 Importance and Uses

2.1.0 Aesthetic Uses

Cyclamen are colourful flower and foliage pot plants traditionally sold during winters in Western countries. They are grown for both indoor and outdoor garden. The plant form beautiful flower with upswept petals and variably patterened leaves. The flower colour ranges from pink, violet, white, red and bicoloured. The leaves are highly patterened which are usually dark green with white or silver shades. It can be successfully grown both in shade and rock gardens. Ivy leaved cyclamen are good house plants besides, borders and rock garden. In temperate areas cyclamen leaves remain attractive throughout the winter months.

2.2.0 Therapeutic and Other Uses

Cyclamen are very attractive ornamental plants while they have been used in medicine as well. The traditional use of cyclamen in medicine was already known by Egyptians, Greeks and Romans. Mazouz and Djeddi (2013) reviewed the use of tincture from cyclamen in Homeopathy. The tincture is usefully employed against the migraines and neuralgia. The followers of homeopathy use homeopathic essence of fresh tubers harvested in the spring, against a variety of ailments: headaches and ear, common cold, toothache gouty origin anorexia, digestive disorders, flatulence, colic attacks, chlorosis due to amenorrhea, rheumatism and gout without febrile events, some neuroses of the heart, itching inoperable, ringworm amenorrhea, dysmenorrhea, worms, cramps, lower limbs, various skin diseases (Fournier, 1999).

3.0 Origin and History

Yesson *et al.* (2009) explored the potential driving influence of palaeo-geographical events on the speciation of cyclamen (Myrsinaceae), a group with most species endemic to the Mediterranean region. Cyclamen species have been shown experimentally to have few genetic barriers to hybridization. The emergence of the *Cyclamen* stem lineage is estimated at 30.1–29.2. The average age of cyclamen species is 3.7 million years. Every pair of sister species have mutually exclusive, allopatric distributions relative to each other. This pattern appears typical of divergence events throughout the evolutionary history of the genus. They concluded that

Uses of Cyclamen

In Garden

In Pot

In Hanging Basket

Surrounding a Tree

In Interior Decoration

As Specimen in Garden

In Rock Garden

In Greenhouse

In Flower Arrangements

In Flower Vase

As Cut Flowers

geographical barriers, such as the varying levels of the Mediterranean Sea, are the most plausible explanation for speciation events throughout the phylogenetic history of cyclamen. The genus demonstrates distributional patterns congruent with the temporally reticulate palaeogeography of the Mediterranean region. A detailed account of the history, taxonomy and nomenclature of Persian cyclamen, *C. persicum*, was described by Doorenbos (1950). This document makes an interesting reading for all those interested in ornamental horticulture. The history of cyclamen in cultivation was traced by collecting all available data from the most important Western European horticultural literature, mostly periodicals. In his openion, it is most important horticultural species of the genus. The others, if grown commercially at all, are never to be found in great quantities, but millions of plants of the Persian Cyclamen are grown yearly in numerous nurseries in Western Europe and the U.S.A. A century of cultivation on a steadily increasing scale has altered the appearance of the plant considerably.

4.0 Morphology

The botanical description of cyclamen was given by Doorenbos (1950). The tubers are flat, with a corky layer fissured into plates or fibers, rooting from the base and the sides. The largest tuber was fifteen centimeters across. Petioles straight (rarely ascending), 7 to 10 cm long. Leaves broadly cordate with a blunt top and rarely overlapping basis lobes, cartilaginously dentate, sometimes slightly lobed. Upper side dull green (very rarely papillate) with a very variable lighter marking: mostly the veins are silver or light green and the leaf has a concentric band of light blotches. Underside green or sometimes purplish. Flowers in winter (in the greenhouse from September to March), on straight peduncles 18 to 30 cm long, often slightly scented. Calyx-lobes with one of the lobes showing many branched nerves, about as long as corolla-tube, ovate, minutely dentate, greenish brown with a white margin. Corolla-tube globose, campanulate (rarely narrow, urceolate), lucid. Corolla-lobes lanceolate and strongly twisted, to obovate and only slightly twisted, mostly white with a purplish base. Pink or lilac forms occur; while the dark base, can be much lighter, or even absent. Anthers densely dotted with black glands. Style about 2 mm longer than corolla-tube.

5.0 Species and Cultivars

5.1.0 Species

Cyclamen species are distributed in parts of Europe, Western Asia and part of North Africa (Davis *et al.*, 1999) (Table 1).

5.2.0 Cultivars

There cannot be a wonderful review and history, presented in a lucid style, the history of Cyclamen breeding than the one presented by Doorenbos (1950). The following is the summary of his exhaustive document. About 1880 large flowered forms existed in most colours known at the time, except in the darkest shades (indeed, these have been lagging behind in this respect to the present day). Famous English varieties of this period are *Mont Blanc* (1878), *Baroness Burdett Coutts* (1879) and *Dame Blanche* (1884), pure white; *Rosy Morn* (1878) and *Duke of Connaught* (1879), rose with dark base; *Brilliant* (1877), *Crimson King* (1879) and *Albert Victor* (1885), purplish crimson; the latter strain was the darkest known at the time. In twenty-five years of nineteenth century, progress had been made.

Table 1: Geographical Distribution of *Cyclamen* Species*

Cyclamen africanum Boiss and Reut	Algeria (N) and Tunisia (NW)
Cyclamen balearicum Willk	France (S), Spain (Balearic island)
Cyclamen cilicium Boiss and Heldr	Turkey (S)
Cyclamen colchicum	Georgia
Cyclamen coum Mill	Armenia, Azerbaijan, Bulgaria (E), Georgia, Iran, Lebanon
Russia, Syria (W)	Turkey
Cyclamen certicum	Greece (Crete, Karpathos)
Cyclamen cyprium Kotschy	Cyprus
Cyclamen graecum Link	Cyprus, Greece, Turkey
Cyclamen hederifolium Aiton	Albania, France (Corsica), Greece (Crete), Italy (Sardinia, Sicily), Switzerland, Turkey (W), Yugoslavia (Old)
Cyclamen intaminatum	Turkey (W and SW)
Cyclamen libanoticum Hildebr	Lebanon
Cyclamen mirabile Hildebr	Turkey (SW)
Cyclamen parviflorum Poped	Turkey (NE)
Cyclamen persicum Mill	Algeria, Cyprus, Greece (E of Crete, Karpathos, E of Aegean IS, Rhodes), Jordan, Lebanon, Syria (W), Tunisia (N), Turkey (S)
Cyclamen pseudibericum Hildebr	Turkey (S)
Cyclamen purpurascens Mill	Austria, Czech Republic, France (E), Germany (S), Hungary, Italy (N), Poland (S), Slovakia, Switzerland,Yugoslavia (Former)
Cyclamen repandum	France (SE of Corsica), Greece, Italy (Sardinia), Switzerland (S), Yugoslavia (Former).
Cyclamen rohfsianum Asch	Libya (N)
Cyclamen somalense Thulin and Warfa (N)	Somalia
Cyclamen trochopteranthum	Turkey (SW)

Note: N: North; E: East; W: west; S: South; IS: Island

Source: Mazouz and Djeddi (2013); * Davis *et al.* (1999).

Cyclamen persicum is the most famous and the mother of the cyclamens, providing the cultivation basis for numerous other varieties and sub-groups. The "aroma cyclamen" and the miniature version, which has been added to the small and large cyclamens, are less well-known. As a rule, winter-hard cyclamen species are garden plants, they are also kept in rooms all year round (https://www.plantopedia.com/cyclamen/).

There are many wonderful cyclamen hybrids available, and since they stay in bloom for a long period (https://www.thespruce.com/how-to-care-for-cyclamen-1402644).

- **Sierra Series:** This has larger flowers (2 to 3 inches) in white, pink, salmon, scarlet, lilac, and purple.
- **Scentsation:** This is an open-pollinated variety with a strong fragrance with pink flowers.
- **Victoria:** This is an open-pollinated variety that has ruffled white flowers with red mouths and margins.

The *Cyclamen* Society is the International Registration Authority for cultivars of all species of Cyclamen except those of *C. persicum*. It has details of all varieties at their website https://www.cyclamen.org/plants/cultivar-list/. Details of all varieties as per at October, 2018 is following

5.2.1 *Cyclamen alpinum*

'Album': white-flowered variety, not completely white, having the characteristic solid purplish stain around the mouth.

'Nettleton White': leaves are marbled; flowers near-white with usual purple stain at the corolla mouth.

'Pink Swirl': leaves are unremarkable; flowers particularly large and elegant, in clear pink.

'Red Devil': leaves two-toned; flowers rather tall, clear deep red.

'Speckles': flowers are quite distinctive, pale pink with darker carmine speckling.

5.2.2 *C. cilicium*

'Album': leaves having variable amounts of marking, flowers are white.

5.2.3 *C. colchicum*

'Deirdre': a completely plain green leaved selection by Jan Bravenboer.

5.2.4 *C. coum* subsp. *coum forma coum* (various coloured flower forms)

'Blush': green-edged pewter-coloured leaves; flowers pale rose-pink with rose-magenta blotch at mouth.

'Crimson King': attractively marked leaves; rich crimson-coloured flowers.

'Dusky Maid': greyish-green leaves; crimson-magenta flowers

'Roseum': leaves are attractively patterned silver; flowers large, rose-pink with magenta markings at the mouth.

'Silver Star': striking silver-coloured leaves with obvious veining; flowers bicoloured pink and white, roughly shuttlecock-shaped.

'Urfa': leaves are rounded with a slightly pointed apex, marked with a striking 'Christmas tree' pattern; flowers rather squat, deep magenta-pink with a deeper coloured blotch at the mouth.

5.2.5 *C. coum* subsp. *coum forma pallidum* (White flower forms with normal marking at the mouth)

'Maurice Dryden': variable silver or pewter-coloured leaves with green margin; flowers are white with a dark purple stain at the mouth.

'White Pewter': white version of 'Pewter Group'; leaves pewter-coloured with a green margin and central vein; flowers are white.

5.2.6 *C. coum* subsp. *coum forma albissimum* (Pure white flower forms with no marking at the mouth)

'Ashwood Snowflake': leaves are variously patterned in silver; flowers sometimes pale pink in bud.

'George Bisson': leaves are patterned in pale green; a pure albino with no red at all on the leaf or flower stalk.

'Golan Heights': plain are green glossy leaves; pure white flowers.

5.2.7 *C. coum* subsp. *caucasicum*

'Album': leaves are heart-shaped with variable markings; flowers white with a basal darker blotch and pink or lavender rim to the mouth (unlike subsp. *coum* in which this zone is white).

'Linnett Charm': leaves are heart-shaped, deep green with a hastate pattern in various shades of pewter; flowers mid-pink with a basal darker blotch and pink or lavender rim to the mouth.

5.2.8 *C. creticum*

'Silvery Hope': bright silvery leaves with white flowers.

5.2.9 *C. cyprium*

'Galaxy': leaves are deep grey-green background, heavily spattered over the entire surface with silver flecks and spots.

5.2.10 *C. elegans*

'Reiko': leaves are heart-shaped, deep green, unremarkably patterned in paler green.

5.2.11 *C. graecum* subsp. *graecum*

'Elizabeth Drew': leaves are green coloured with a thin green margin; flowers large, pure white.

'Rhodopou': leaves are silver with a narrow green border; flowers pink.

5.2.12 *C. hederifolium* subsp. *hederifolium* (Various coloured flower forms)

'Antiochus': leaves are attractively marbled; flowers are deep pinkish-red.

Species of Cyclamen

Cyclamen africanum

Cyclamen balearicum

Cyclamen colchicum

Cyclamen graecum

Cyclamen hederifolium

Cyclamen mirabile

Cyclamen persicum

Cyclamen parviflorum

Cyclamen repandum

Cultivars of Cyclamen

Ericales

Fantacia

Montreal

Cyclaam Verzorgen

Victoria

BSBE 518

Ashwood Snowflake

George Bisson

Silverleaf

Dreamscape

Metis

Tianis

Cultivars of Cyclamen

Super Verano Series

Michler

Rainer Scarlet

Fuchsia

Silverleaf Deep

Verano Neon Pink

Ciclame

Origami

Ciclamino

Sowbread

Rokoko

Indiaka

'Highfield': distinctively glossy foliage; pink-flowered.

'Silver Shield': leaves are deep green with a striking pewter or paler green shield pattern in the centre; flowers pink.

'Stargazer': leaves are unremarkable; flowers upward-facing instead of downward-facing in the usual pattern.

5.2.13 *C. hederifolium* subsp. *hederifolium forma albiflorum* (Pure white flower forms with no marking at the mouth)

'Daley Thompson': leaves are unremarkable; flowers very large, white, with the petals measuring up to 4 cm long.

'Discovery': classic ivy-like foliage in shades of green and grey-green; flowers pure white, up to 3.5 cm long; the first flowers to open are the largest, the later ones more normal size.

'Ellen Corker': flowers are white, suffused with salmon-pink around the mouth.

'Tilebarn Helena': leaves are silver, arrow head shaped; flowers are white.

5.2.14 *C. intaminatum*

'E. K. Balls': leaves are generally deep green with a zone of silvery green blotches; flowers are white.

5.2.15 *C. mirabile*

'Tilebarn Nicholas': leaves are bright silver with a dark green central 'Christmas-tree' shaped pattern, but notable for the silver areas being flushed bright pink.

5.2.16 *C. persicum* var. *persicum*

'Tilebarn Karpathos': leaves are unremarkable for the species; flowers are deep cerise with notably twisted petals, a very dark.

5.2.17 *C.n purpurascens*

'Fatra Form': leaves are plain green and shiny; flowers are deep pink.

'Garibaldi': leaves are bright silver centre and silver veins extending across a broad green margin; flowers pink with short broad petals.

'Green Ice': leaves are mint green; flowers are deep pink.

'Green Lake': leaves are silver-green with veins extending across a broad green margin but with a darker green hastate shape in the centre; flowers are pink.

'Lake Bled Form': leaves are beautifully silvered with a netted pattern of veins towards the margin; flowers are pink.

'Verdume': leaves are plain green; flowers pink.

5.2.18 *C. repandum*

'Amazing Hope': leaves are with a well-defined silver central hastate pattern: flowers pink, typical of the species.

5.2.19 *C. rhodium* subsp. *peloponnesiacum*

'Pelops': leaves are pale green, shiny grey-green background; flowers with narrow twisted petals, pale pink-purple with maroon-red mouth.

'Silver-Leaf Group:' leaves are all over silver; flowers pale pink with a dark nose.

6.0 Genetics and Breeding

6.1.0 Genetics

6.1.1 Cytogenetics

According to Legro (1959), Heitz published for the first time, in 1926, some chromosome numbers of cyclamen. He recorded for the *C. persicum cult. hor.* the haploid number of 42-44 and the diploid number of 88. This cultivar belonged to the 'Splendens-Giganteum' type.

The chromosome numbers reported by De Haan and Doorenbos (1951) are given below.

C. repandum 20 big chromosomes

C. balearicum 20 big chromosomes

C. creticum 22 big chromosomes

C. libanoticum 30 big chormosomes

C. pseudibericum 30 big chromosomes

C. orbiculatum 30 big chromosomes

C. cilicium 30 big chromosomes

C. cyprium 30 big chromosomes

C. europaeum 34 medium-sized chromosomes

C. neapolitanum 34 medium-sized chromosomes

C. africanum 68 medium-sized chromosomes

C. persicum 48 small chromosomes

C. graecum 84 or 85 small chromosomes

Ishizaka (2003) have analyzed cytologically haploids, diploids and tetraploids of *cyclamen persicum* (2n x=24, 2x=48 and 4x=96), *Cyclamen greacum* (2n=84) and their interspecific hybrids (2n=45, 3x=66, 4x=90).

6.2.0 Breeding

As discussed under the sub-heading on History, Doorenbos (1950) has also detailed the history of breeding as follows. During the first period forms are obtained with flowers in colours ranging from pure white via pink and lilac to purplish red. About 1870 forms with corolla-lobes about twice as long as those of the wild forms are found in several nurseries in England as well as in Germany. Fifteen years later these *giganteum-forms* have been obtained in the same colours as the small flowered ones. The third period begins in 1894 with the introduction into the trade of the salmon hue, found in England and afterwards, in 1898, also in Germany. Combination of salmon pink with purplish red leads to a rich real red, another colour entirely new in *Cyclamen*, obtained by several breeders. In the mean time progress was being made as to general appearance of the plant, shape and size of the flower, *etc.*

6.2.1 Interspecific Hybrids

Inter-specific crosses were made to introduce the scent of flowers of *C. purpurascens* into *C. persicum* cultivars and ovule culture was used to rescue the abortive hybrid embryos (Ishizaka and Uematsu, 1995). Cultivars of *C. persicum* diploid (CPD, 2n=2×=48) and *C. persicum* tetraploid (CPT, 2n=4×=96) were the pistillate parents and wild species of *C. purpurascens* (CP, 2n=34) were staminate parents. After pollination, crossed ovaries were collected periodically and examined using paraffin sections. Histological observations suggested that both hybrid ovules of CPD x CP and CPT x CP should be transferred to culture medium 35 days after pollination. Based up on this observation, crossed ovaries were collected 28 days after pollination and ovules with placenta were transferred to MS (1962) medium containing 3 per cent sucrose. These ovules were cultured in the dark at 25° C. The hybrids (2n=41) derived from CPD x CP had the scent of *C. purpurascens*, whereas the hybrids (2n=65) derived from CPT x CP had the scent of *C. persicum*. Although, both hybrids had complete genomes from the parents and produced a few viable pollen grains, they failed to yield viable seeds by self- and cross pollination with fertile pollen grains of *C. persicum* cultivars (Ishizaka and Uematsu, 1995).

Inter-specific crosses were made to introduce the disease resistance of *Cyclamen graecum* into *C. persicum* cultivars and the abortive hybrid embryos were rescued by ovule culture. Diploid and tetraploid cultivars of *C. persicum* (CPD, 2n=2x=48; CPT, 2n=4x=96) were the pistillate parents and wild form of *C. graecum* (CG, 2n=84) were the staminate parents. After pollination, crossed ovaries were periodically collected and examined using paraffin sections. Histological observations suggested that both hybrid ovules of CPD × CG and CPT × CG should be transferred to the culture medium 35 days after pollination. Based upon this observation, crossed ovaries were collected 35 days after pollination, then ovules with placenta were explanted on culture medium and cultured in the dark at 25°C. Plantlets were induced from ovules cultured in MS medium containing 3 per cent sucrose and 10 per cent coconut milk. The hybrids (2n=66) derived from CPD × CG failed to yield viable seeds by self-pollination, although they showed some pollen fertility. In contrast, the hybrids (2n=90) derived from CPT × CG showed high pollen fertility and yielded viable seeds by self-pollination. Furthermore, they were resistant to disease caused by *Fusarium oxysporum* f. sp. *cyclaminis*, *Erwinia herbicola* pv. *cyclamenae* and *Pseudomonas marginalis* pv. *Marginalis* (Ishizaka and Uematsu, 1995).

6.2.2 Mutation Breeding

High-energy ion beams that are generated by accelerators have been shown to have greater biological effects than gamma rays. Fast neutron radiation is also a technique that has not yet been applied to cyclamen but may be a useful form of inducing variants (Tagipur *et al.*, 2016). The effect of $^{12}C_6$ ion beam irradiation on SE from *Cyclamen persicum* callus was examined by Sugiyama *et al.* (2008) using scapes obtained from potted plants. Callus was exposed to a $^{12}C_6$ ion beam at doses ranging between 10 and 80 Gy. The fresh weight of somatic embryos produced was determined. The initial trial showed that mutant callus, somatic embryos, and plantlets could not be obtained. In a second trial, tubers were irradiated at 4–16 Gy to examine the effect of radiation dose on tuber viability. Two petal color mutants were obtained at 12 Gy indicating that mutations induced by heavy-ion beam irradiation can be a useful means of obtaining variation in flower characteristics in a short period of time.

The flower colour of *Cyclamen graecum* gra6 (wild-type) is pink-purple in the main part of the petal, referred to as the 'slip', and deep purple` at the petal base, referred to as the 'eye'. On the other hand, flowers of *C. graecum* gra50 (a white-flowered variant) exhibit a white colour in both the 'slip' and 'eye' regions. The relationship between floral pigmentation and the expression of several anthocyanin biosynthesis genes was investigated in *C. graecum* gra6 and gra50 (Akita *et al.*, 2010). The pigments in the 'slip' and 'eye' regions consist mainly of malvidin 3,5-diglucoside in gra6, suggesting that the difference between the colour of the 'slip' and 'eye' regions is related to the amount of anthocyanin present. White-flowered *C. graecum* gra50 possessed lower amounts of anthocyanins, but higher amounts of flavonols compared to gra6, suggesting a change in metabolism caused by a disruption of anthocyanin biosynthesis.

6.2.3 Polyploid Breeding

In vitro colchicine applied to hybrid ovules (*C. persicum* diploid x *C. hederifolium*) and colchicine applied to the mature hybrids (*C. persicum* diploid x *C. hederifolium* or *C. hederifolium* x *C. persicum* diploid) was used to produce fertile amphidiploids. Many fertile plants were derived from *in vitro* culture of the ovule treated with 0.05 per cent colchicine for 10 and 15 days throughout embryogenic callus formation and several of the fertile plants were detected to be amphidiploids with a chromosome number of 2n=82. On the other hand, hybrids were derived from *in vitro* culture of the ovules without colchicine treatment throughout germination and they were sterile plants with a chromosome number of 2n=41. Viable seeds were barely produced from the mature hybrids of *C. hederifolium* x *C. persicum* diploid which were treated with 0.5 per cent colchicine using tragacanth gum by self pollination with a chromosome number of 2n=82. Although, each of the amphidiploids induced by the two methods and the hybrids exhibited the characteristics of the parents, the amphidiploids showed larger flowers and peduncles than the hybrids and grew vigorously. They showed high pollen fertility and yielded viable seeds self-pollination (Ishizaka and Uematsu, 1994).

The possibility of obtaining tetraploids from diploid yellow-flowered cyclamen 'Kage Yellow' through colchicine treatment and the effects of polyploidization on the characteristics of yellow flowered cyclamen were investigated (Takejiro *et al.*, 1996). A 4-day treatment of tuber segments with 100 mg/l colchicine *in vitro* yielded two tetraploid plants. The petals of the tetraploids were larger and had a greater ability to accumulate chalcone than those of their diploid relatives. Polyploidization may, therefore, represent a useful method for the commercial breeding of deeper yellow-flowered cyclamen (Takejiro *et al.*, 1996).

7.0 Growth and Flowering

7.1.0 Seed and Embryo Development

Embryo development and germination of *Cyclamen persicum* have been characterized for zygotic and somatic embryos with regard to mitotic activity and morphology so as to identify developmental abnormalities in somatic embryogenesis (Schmidt *et al.*, 2006). Zygotic embryo development proved to be highly synchronous with distinct periods of cell division, cell elongation and embryo maturation within a total period of 17 weeks of seed development. Somatic embryo development was accomplished within only 3 weeks, resulting in a mixture of morphologically highly variable embryos. No distinct developmental periods could be identified and no reduction of the mitotic activity was recorded for non-desiccated somatic embryos. Controlled desiccation of somatic embryos severely reduced their germination rate, demonstrating resemblance of somatic embryos to recalcitrant seeds, whereas, zygotic cyclamen seeds could be characterized as typically orthodox (Schmidt *et al.*, 2006).

7.2.0 Flower Development

Flower shape and colour can enhance the value of ornamental flowers. Therefore, the improvement or creation of novel flower traits can provide substantial commercial value. *Cyclamen persicum* (cyclamen) is a commercially valuable, winter-blooming perennial plant. Lyngved *et al.* (2008) cloned two cyclamen orthologues of AGAMOUS (AG), CpAG1 and CpAG2, which are mainly expressed in the stamen and carpel, respectively. Cyclamen flowers have 5 petals, but expression of a chimeric repressor of CpAG1 (CpAG1-SRDX) caused stamens to convert into petals, resulting in a flower with 10 petals. By contrast, CpAG2-SRDX only caused incomplete formation of stamens and carpels. Expression in *Arabidopsis thaliana* showed similar effects on flower organ specification. Simultaneous expression of CpAG1-SRDX and CpAG2-SRDX in cyclamen induced rose-like, multi-petal flowers, a potentially valuable trait in commercial ornamental varieties. Expression of CpAG2-SRDX in a cyclamen mutant lacking expression of CpAG1 more effectively produced multi-petal flowers. They controlled the number of petals in cyclamen by simple genetic engineering with a chimeric repressor. This strategy may be applicable useful for other ornamental plants with two distinct AG orthologues (Lyngved *et al.*, 2013).

8.0 Propagation

8.1.0 Seed Propagation

Traditionally, *Cyclamen persicum* hybrids have been propagated sexually for commercial and private use. It is suggested that during the period of sowing and germination, sowing trays to be covered with vermiculite and keep them for 23 days in full darkness at a temperature of 18°C and with a humidity of 95 per cent. After germination, the seedlings should be kept sufficiently humid to get a well developed first cotyledon.

Studies on the factors that affect germination and tuber development in *C. persicum* have shown that continuous light inhibits germination, at temperatures below 5°C and over 20°C seeds do not germinate, and that germinability is high following dry storage between –30 and 20°C. Moreover, germinability increases with seed maturation and in the presence of gibberellic acid, in particular at 20°C (Neveur *et al.*, 1986). Among the seeds that were sown 5-6 months later (October-November), the germination rate was 35 per cent at 15°C and 47 per cent at 20 °C in October. Germination increased to 80 per cent at 15°C and 87 per cent at 20 °C in November. The highest germination rate was observed in December (95 per cent) and these seeds had been stored for 8 months. Seeds that were stored for 8 months (using different storage methods) were subjected to *in vitro* and *in vivo* germination tests at 15°C and 20°C. Seed storage method (in capsules and out of capsules) did not have a significant effect on *in vitro* germination in MS medium, but storing seeds in capsules had a significant effect on germination on filter paper and *in vivo* (mixtures of peat, perlite, and sand). The highest germination rate was obtained after 8-9 months of storage (*in vitro* average, 94 per cent; *in vivo* average, 83 per cent). A mixture of peat + perlite + sand in equal parts was a suitable *in vivo* germination medium (93 per cent germination) for *C. alpinum* (Burun and Sahin, 2009).

8.2.0 Vegetative Propagation

Cyclamen is most often propagated from its tubers (roots). Tubers are divided into sections with each section having at least one eye and planted in soil.

8.3.0 Micropropagation

Although it is commercially propagated *via* seeds, tissue culture and somatic embryogenesis are important tools in breeding and mass propagation (Jalali *et al.*, 2012). They presented an overview of the tissue culture techniques available for the micropropagation of cyclamen. Micropropagation has been used to propagate some interesting tetraploid yellow flowered cyclamen with "eye" (Takamura *et al.*, 1998). Explants from etiolated petioles of each cultivar (Vuurbaak and Victoria) and from each stock plant showed complete elimination of internal microbial contamination in differentiated shoots (Murasaki and Tsurushima, 1988). Schwenkel and Grunewaldt (1988) used greenhouse-grown flowering plants of diploid and tetraploid genotypes of different origin as well as wild types were the explant sources. Lamina pieces (8 mm) and petiole and peduncle segments (6 mm) were used as explants. Of the 8 genotypes belonging to the *Giganteum* group, only 2 regenerated normal leaves and a few shoot buds. An efficient protocol for callus induction and somatic embryogenesis of *Cyclamen premium* was developed by Naderi *et al.* (2012) who studied the difference between response of two explants type leaf and petiole and two medium salt concentrations MS and ½ MS on callus percentage and somatic embryo number. Media was non-significant on callus production but leaf explants was better than petiole and interaction between explants and media was significant. Schwenkel *et al.* (1990) studied the somaclonal variation in cyclamen after *in vitro* mass propagation. Rooted shoots from *in vitro* culture of 8 diploid genotypes were grown until flowering to give about 6000 plants. Variations in growth habit, flower shape and colour, and chlorophyll pattern and shape of leaves were noted. The frequency of altered plants, which ranged from 4.5 to 36.3 per cent, and the type of alteration depended on the original genotype.

8.3.1 Embryo Culture

Embryo culture has been primarily attempted to produce interspecific hybrids by rescuing the embyos. Hybrids between *C. persicum* diploid (2n=48) and *C. hederifolium* (2n=34) are completely sterile. In order to obtain fertile amphiploids, Ishizaka and Uematsu (1994) applied colchicine treatment *in vitro* to ovules (namely hybrid ovules) embryos of *C. persicum* collected 21 days after pollination with *C. hederifolium* and mature hybrids (*C. persicum* x *C. hederifolium* or *C. hederifolium* x *C. persicum*). Many fertile plants were derived from *in vitro* culture of the hybrid ovules treated with 0.05 per cent colchicine for 10 and 15 days during embryogenic callus formation, and several of them were amphidiploids with a chromosome number of 2n=82. Hybrids derived from *in vitro* culture of the hybrid ovules without colchicine treatment were sterile with a chromosome number of 2n=41. The very few viable seeds produced from mature *C. hederifolium* x *C. persicum* hybrids were treated with 0.5 per cent colchicine following self-pollination. The flowering plants that developed from these seeds were amphidiploids with a chromosome number of 2n=82. Although, the amphidiploids induced by the two methods and the hybrids exhibited parental characteristics, the amphidiploids had larger flowers and peduncles than the hybrids and grew vigorously. They also had high pollen fertility and yielded viable seeds following self-pollination.

8.3.2 Anther Culture

Ishizaka (1998) cultured the anthers of a F_1 hybrid (2n=41) between *C. persicum* (2n=2x=48) and *C. purpurascens* (2n=2x=34) were cultured to produce microspore-derived plants. Embryoids were produced when anthers, containing microspores at the early uninucleate stage of pollen development, were cultured on B5 medium containing sucrose (90 g/l) and NAA (0.1 and 1 mg/l) or 2,4-D (0.1 mg/l) in the dark at 5°C for 4 days, then at 25°C for 60 days. Embryoids usually developed into plantlets when cultured in B5 medium containing sucrose (30 g/l) in the dark at 25°C. At meiosis, the F_1 hybrid, used as source for anther culture, formed some cells with restitution nuclei at telophase and dyads at the tetrad stage, which resulted in the production of viable pollen grains as unreduced gametes. Plants produced by anther culture were grouped into sterile plants with 2n=41 chromosomes and fertile plants with 2n=82 chromosomes.

8.3.3 Suspension Cultures

Winkelmann *et al.* (1998) used as inocula a non-embryogenic and an embryogenic callus line derived from somatic tissue of unpollinated ovules of the same plant for long-term suspension cultures showing stable growth for more than one year. Critical factors for suspension establishment were the starting density, the growth phase of the embryogenic callus, and the dilution and subculture regime. Growth

Propagation of Cyclamen

Seed Pods

Germination of Seeds

Cyclamen Tubers

Division of Tubers

Tagipur, E. M., Seker, G., da Silva, J. A. T., and Mendi, Y. Y. (2016). Somatic embryogenesis, cryopreservation, and in vitro mutagenesis in Cyclamen. In *Somatic Embryogenesis in Ornamentals and Its Applications,* pp. 155-167. Regeneration stages in *Cyclamen persicum.* Leaves (a), petioles (b), ovules (c), and ovaries (d); somatic embryo formation on each explant (a_1, b_1, c_1, and d_1, somatic embryos); a_2, b_2, c_2, and d_2, plantlets.

Cultivation of Cyclamen

Cultivation in Open

Cultivation in Greenhouse

Flowering in Greenhouse

curves for embryogenic and non-embryogenic suspensions are presented. Both types of suspensions were successfully established in bioreactors in initial trials. The regeneration of globular-shaped somatic embryos in high frequencies was observed after plating the fraction 200-500 μm on hormone-free medium. However, the conversion to normal plantlets remains to be optimized, since pronounced root growth was observed in somatic embryos derived from callus and suspension cultures. A yield of about 90000 plantlets from 1 litre of suspension can be achieved representing only a small portion of the embryogenic potential of the cultures.

8.3.4 Thin Cell Layer

Morphogenesis of *Cyclamen persicum* using the thin cell layer technology (TCL) was successfully established (Nhut *et al.*, 2012) The peduncles were sliced into thin cell layers (1 mm thick), which were then cultured on MS medium supplemented with NAA, IBA, 2,4-D at various concentrations (0.1; 0.3; 0.5; 0.7 or 1.0 mg/l) alone or in combination with TDZ (0.2 mg/l). The results showed that the presence of auxin alone in variants of culture medium was not sufficient for callus induction from peduncle TCL explants. MS medium supplemented with 0.2 mg/l TDZ and 1.0 mg/l 2,4-D was found to be the optimal medium for callus induction. The 3 mm initiating explants at the upper position of the peduncle gave 100 per cent callus induction rate and high callus fresh weight. After 8 weeks of culture, calluses which were transferred onto MS medium supplemented with 0.5 mg/l BA and 0.7 mg/l IBA induced the highest number of adventitious shoots (39.4 shoots/explant). Rooting reached 100 per cent after transferring 2-3 cm long shoots to MS medium supplemented with 1.0 mg/l IBA. The results indicated that different kinds of plant growth regulators, the explant size and position are important in determining the morphogenesis from peduncle TCL explants of *Cyclamen persicum* (Nhut *et al.*, 2012).

8.3.5 Protoplast Culture

The first report of protoptlast isolation was described by Morgan (1999). He isolated the protoplasts from *in vitro* grown adventitious shoots and their culture in agarose lenses. After four to 5 days, cell divisions were observed and within 6 weeks calluses developed. However, final plating efficiencies were very low (<0.03 per cent) and only in three calluses shoot regeneration was found, a process which was described by the author to be not reproducible (Morgan, 1999).

Embryogenic cultures of *Cyclamen coum* were established on solid media and in suspension, and their growth characteristics in response to different concentrations of plant growth regulators (PGRs) were evaluated (Prange *et al.*, 2010a). Embryogenic cultures exhibited a high regeneration capacity of 876 somatic embryos per gram fresh mass. Up to 4.24×10^5 protoplasts per gram of fresh mass were isolated from somatic embryos and embryogenic suspension cultures. Protoplasts derived from both embryos and suspension cultures were successfully cultured *in vitro* and regenerated into plants via somatic embryogenesis. Phenotypic analyses and flow cytometric measurements revealed that some regenerated plants were tetraploid. About 20 per cent of the protoplast-derived calluses used for regeneration were tetraploid, while tetraploidy was found in 0.9 per cent of the plants regenerated from the embryogenic cultures.

9.0 Soil and Climate

9.1.0 Soil

Cyclamen can be grown on a wide range of soil. But it prefers slightly acidic pH 5.5-6.0. Well-drained soils rich in humus and containing leaf mould provide a suitable medium for cyclamen corms (Grey-Wilson, 1988). For potted plants the combination of normal potting mix with sphagnum peat is very sutable for its growth and flowering.

9.2.0 Climate

Yesson and Culham (2006) presented a phyloclimatic study that showed the high impact of global climate change on the natural habitat of these species. It is also assumed that the natural habitat of cyclamen will decline and almost all of the wild species, such as *C. mirabile, C. graecum* and *C. alpinum*, will be endangered in about 40 years (Prange *et al.*, 2010b). Cyclamen prefers subtropical to mild temperate climate.

10.0 Cultivation

Cyclamen grows in the wild on wooded low mountain regions or rocky pastures as it loves chalky soil and the humus resulted from leaf decomposition (Preda, 1979). Four cyclamen varieties namely Superba, Dunkelrot, Sylphide and Vuurbach, obtained from seed propagation were grown in a modern greenhouse where the climatic factors are automatic controlled. The quality aspects followed were leaf number on plant, flower number on plant, leaf surface, appearance of first flower and the end of flowering for each variety. It is recommended that cultivation of all varieties is easy as pot plants. Cyclamen is traditionally produced for flowering in winter and the tubers can be kept for flowering year after year (Bala and Berecici, 2010).

10.1.0 Biocontainers

Biocontainers are generally categorized as being plantable or compostable. Plantable biocontainers are containers that allow plant roots to grow through their walls and may be directly planted into the final container, the field, or the planting bed. Compostable biocontainers cannot be planted into the soil because the roots cannot physically break through the container walls and the biocontainers do not break down quickly enough to allow the plant roots to grow through the container walls. Instead, these containers must be removed before planting but can be placed in a compost pile to decompose in a relatively short time (Mooney, 2009).

Beeks and Evans (2013) evaluated the growth of a long-term crop in biodegradable containers compared with traditional plastic containers using a sub-irrigation system. Plastic, bioplastic, solid rice hull, slotted rice hull, paper, peat, dairy manure, wood fiber, rice straw, and coconut fiber containers were used to evaluate plant growth of 'Rainier Purple' cyclamen (*Cyclamen persicum*) in ebb-and flood sub-irrigation benches. The days to flower ranged from 70 to 79 and there were no significant differences between the plastic containers and the biocontainers. The dry shoot weights ranged from 23.9 to 37.4 g. Plants grown in plastic containers had dry shoot weights of 27.6 g. The dry shoot weight of plants grown in containers composed of wood fiber was 23.9 g and was lower than plants grown in plastic containers. The plants grown in the bioplastic, solid ricehull, slotted ricehull, paper, peat, dairy manure, rice straw, and coconut fiber containers had significantly higher dry shoot weights than plants grown in plastic containers.

10.2.0 Irrigation

Grey-Wilson (1988) reported that cyclamen species do not need full watering in summer, but do need some moisture and prefers to be kept a little moist; a weekly watering of the plunge material should suffice. Some experiments on fertilizing for *Cyclamen hederifolium* have been conducted, but water consumption through the whole growing season is needed to achieve a successful cultivation (Altay and Muftuoglu, 2004). Maximum water requirement for optimum morphological and physiological parameters of the bulbs of *C. hederifolium* is 144 mm.

10.3.0 Manuring and Fertilization

Miura (1970) tried the following (a) 1000 mg N + 2000 mg P_2O_5 + 2000 mg K_2O per pot applied in solution; (b) 1000 mg N + 1000 mg P_2O_5 + 1000 K_2O in solution; (c) as (a) but with P applied as superphosphate at potting; (d) a water extract of rotted rape seed waste; (e) composted bone meal, fish waste, rice husks and plant ash; and (f) unfertilized control. He found that of the mineral treatments (a) produced the greatest and (c) the least fresh weight and leaf number by November, and of the organic treatments (d) was more effective than (e). Fresh weight was highly correlated with the absorption of N, P and K but poorly correlated with Ca and Mg absorption.

The effects of five levels of nitrogen (N), phosphorus (P), potassium (K), and lime on potted cyclamen (*Cyclamen persicum*) grown in peat/sand, 1:1 v/v were studied (Gillespie and Thomas, 1983), Nitrogen strongly influenced most aspects of growth and flowering; it was supplied by Osmocote (26 per cent N). Strongest corm and foliage growth occurred at 450 to 600g N/m^3, while early flowering and flowers per plant were promoted at these levels as long as added P was low. Low to medium levels of P and K appeared primarily to be required for flower quality.

Dual inoculation of peat based horticulture substrate with a mixture of four species of arbuscular mycorrhizal fungi and fungal biocontrol agent *Trichoderma harzianum* showed a significant positive effect on the growth and flowering of cyclamen plants. Inoculation substantially decreased plant mortality caused by spontaneous infection by the fungal pathogen *Cryptocline cyclaminis*. Plant mortality was also reduced by separate inoculation with arbuscular mycorrhizal fungi. Both separately inoculated agents positively affected the plant growth, although to a lesser extent (Dubsky *et al.*, 2002).

11.0 Diseases and Pests

11.1.0 Diseases

11.1.1 Fusarium Wilt (*Fusarium oxysporum* f. sp. *cyclaminis*)

The disease is manifested with progressive yellowing and wilting of leaves (older are first). As the disease progress brown discoloration of the vascular tissues in tubers occurs. The tuber remains firm unless secondary bacteria are introduced. The disease urvives as resting spores (chlamydospores) in soil.. The severity of disease is favored by temperatures above 70°F. Some workers have suggested that it might be due to seedborne and therefore seeds should not be collected from infected plants. If the case is severe discard the infected plants and surrounding soil. Good sanitation generally provides adequate control. Drenching with Thiophanate-methyl during early growth period is beneficial and the soil pH should be adjusted to 6.5 to 7.0. Also treat seed with a fungicide before sowing.

11.1.2 Tuber and Root Rot

When the *Cyclamen persicum* is exposed to water logging or excessive much moisture, root rot and subsequent tuber rot may occur. The suspicion may be caused by downward-inclined petioles as well as leaves that are rugged. If the soil can be pressed down more than three centimeters for days, it's probably a case of water logging **(https://www.plantopedia.com/cyclamen/)**.

11.1.3 Tomato Spotted Wilt Virus

This virus is one of the Tospo virus group, a particulate virus. Under the electronic microscope the viral particles have a spherical appearance and are of the order of 70 to 120 nm in diameter, composed of a single strand of RNA covered with a knobbly membrane. The virus had been known in Europe earlier (1932) but had disappeared and was reintroduced in France with the arrival in 1986 of *Thrips frankliniella occidentalis*, one of its most effective vectors. It is a serious threat to many vegetable and flower crops. High temperatures favour its multiplication within a plant. Symptoms and damage are extremely variable from species to species and may take most untypical forms; the viral attack may be so bad as to destroy the infected plant utterly. The most common symptoms are dwarfism, marbling, leaf mosaic patterns due to cell death, discoloration and stains. In the case of cyclamen, the first symptoms are the appearance of a mosaic of dead cells (necrosis) and spots on the leaves and brown signs of cell death on ribs, sometimes mistaken for signs of scorching. The leaves perish, often with yellowing.

11.1.4 Cyclamen Mosaic Virus

This is a virus from the Tobacco virus group (single-strand RNA, without coat). It causes mosaic patterns on leaves. Under the electron microscope it looks like small straight rods 300 nm x 18 nm. In plant tissue, TMV is highly stable and very concentrated.

It is easily transmitted by physical contact, but also by vectors such as the Cotton Aphis (*Aphis gossypii*). The main means of its persistence are the soil, and seeds. A number of strains have been described, differing in their range of host plants and in the severity of their symptoms.

11.1.5 Impatiens Necrotic Spot Virus

This belongs to the Tospo virus group, like the TSWV. The INSV is only transmitted, so far as we know at present, by Thrips (including *Thrips frankliniella occidentalis*). The incubation period within the plant is 2 months. In cyclamen, as opposed to other plants, its symptoms are fairly typical: outward signs are concentric ring-shaped stains of cell death, chestnut or yellow in colour, on the leaves, flower stalks and leafstalks. The roots are not attacked. As with TSWV, when immature leaves are affected, the necrotic zones do not grow along with the others, and leaf deformity results. Flowers also can be misshapen, and the tuber can become elongated, while stripes may be seen in the vascular tissue.

11.1.6 Fig Mosaic Emaravirus (FMV)

FMV belongs to the genus Emaravirus in the newly established family Fimoviridae. Its genome consists of six single-stranded, negative-sense RNA segments which respectively encode the RNA-dependent RNA polymerase (RdRp, RNA-1), a putative glycoprotein (GP, RNA-2), a putative nucleocapsid protein (NCP, RNA-3), a putative movement protein (MP, RNA-4), and two proteins with unknown functions (p5 and p6, respectively RNA-5 and RNA-6). This virus is transmitted in nature by the eriophyid mite vector *Aceria ficus* but not by seed or by mechanical inoculation (Elbeaino *et al.*, 2012; Elbeaino *et al.*, 2009; Slykhuis *et al.*, 1973).

11.1.7 Ring Spot Virus

The ringspot disease of cyclamen is a particularly costly disease because of the long growing period of 6.5 to 9.5 months from seed, and the long latent period of the disease, which may be 4 months or longer. During the latent period, commercial production costs continue as does the undetected spread of the virus. By the time symptoms are evident, most of the plants in a greenhouse may be infected, and those plants in an affected greenhouse that are apparently healthy are seldom sold because of the fear that they will express symptoms later and generate consumer dissatisfaction.

11.1.8 Soft Rot

There are a great many species, many of which are polyphagous and capable of infesting a number of plant species. Along with *Pythium, Fusarium* and *Rhizoctonia solani* it is one of the main causes of damping-off disease in seedlings and root rot. Infection starts with the roots (rootlets). The plants affected start to rot, or show necrotic lesions, in the root system, in the section (hypocotyl) between shoot and radicle (in seedlings and young plants), or at the collar and the base of the stalks. The attacks occur most often after the seedlings have shown above ground, and they kill the plant. When the seeds themselves are attacked this results in a failure to show; the impression is that the seeds have rotted. In adult plants, the roots are blackened. The plant above ground turns yellow and then withers, but this is of course not a distinguishing sign of this disease. *Phytophthora spp.* are vascular fungi, which block the sap-transporting vessels and the plant dies of 'apoplexy'. Dry soil makes these symptoms more likely, since the dead roots can no longer assimilate free water from the soil, and the plant suffers from drought.

11.2.0 Pests

Thrips and Mites are most common pests of cyclamen (https://www.koppert.co.uk/news-item/cyclamen-main-pest-pressures/).

11.2.1 Thrips

Thrips (*Frankliniella occidentalis*), or western flower thrips (WFT), is the most common species found in Cyclamen. If a biological control strategy is not implemented, both larvae and adult thrips will cause plant damage, by piercing cells on the leaf surface, and sucking out the contents. This causes silver-grey spots on the leaves, with dark green spots of frass, and feeding marks to the flowers. Their life-cycle is very temperature-dependent. Although, thrips can be found at lower temperatures of 10°C, the ideal development conditions are 18°C-26°C; above 35°C, the life-cycle stops. They are also very prolific egg-layers, giving rapid rise to the next generation. Thrips larvae can also transmit tomato spotted wilt virus (TSWV) and Impatiens Necrotic Spot Virus (INSV) (https://www.koppert.co.uk/news-item/cyclamen-main-pest-pressures/).

11.2.2 Cyclamen Mite

Cyclamen mite, *Phytonemus pallidus*, is a very tiny mite, only just visible to the naked eye, and is most commonly found in glasshouses. The mites may be concentrated around the buds, or infest the entire plant, resulting in severely stunted plants, with crinkled leaves (https://www.koppert.co.uk/news-item/cyclamen-main-pest-pressures/).

11.2.3 Aphids

Aphid is the most common and pesky pest that can be handled relatively easily if you immediately notice and take all the necessary measures to destroy it. Most often, aphid gets on the plant from the street through the window or window when airing. This is a small wingless insect of 1.4 - 2.5 mm in length. The color of their bodies can be gray, green, orange and black. Aphids feed on cyclamen soft tissue juice.

12.0 Harvesting and Postharvest Management

12.1.0 Harvesting

Cyclamen cut flowers are attractive to florists due to their particular flower morphology and their generally long vase life of more than twelve days and even up to three weeks and more. *Cyclamen* cut flowers should be harvested by picking them from the tubers in order to avoid fungal rot caused by the remains of stems attached to the tubers.

12.2.0 Postharvest Management

The postharvest life of cyclamen can be improved with a low temperature at the end of cultivation. In the nursery comparison the temperature ranged from 16 to 24°C, but Bulle (2005) found that a temperature of 15°C resulted in the best post harvest life, although the greatest number of flowers during post harvest life was

Diseases and Pests of Cyclamen

Tuber Rot

Fusarium Wilt

Botrytis

Tomato Spotted Wilt Virus

Necrotic Spot

Virus Soft Rot

Mite Infestation

Thrips Infestation

Aphids

Harvest and Postharvest Management of Cyclamen

Cyclamen in Full Bloom

Harvesting Stage

Cyclamen Cut Flowers

Harvesting of Greenhouse Cyclamen

Freshly Harvested Flowers

Harvested Tubers

Deadheading a Cyclamen

Plant Cyclamen in Market

Storage of Cyclamen Tubers

Dormant Cyclaman Tubers

Storage of Tubers

Packaging and Marketing of Cyclamen

Bagged Cyclamen

Packed in Decorative Containers

Cyclamen in Floriculture Auction Markets

Exhibition of Cyclamen

Cyclamen in Exhibition Hall

Indoor Exhibition

seen at a temperature of 12 °C. Also a low light intensity and high relative humidity at the end of cultivation resulted in a better postharvest life of cyclamen. Relative humidity had an opposite effect on postharvest life of cyclamen. Plants had a better ornamental value with more flowers when RH was 75 per cent during the last weeks of the growth period. Low levels of fertilization at the end of the growth were found to improve keeping quality of cyclamen (Hendriks, 1993; Sprau, 1997).

A special technique of cutting the flowers by making a longitudinal slit in the stem base enables sufficient water uptake and therefore results in a longer shelf life (Wartenberg, 2000). Moreover, the sugar content of the stems which decreases with lower light supply during cultivation was shown to be positively correlated with the vase life of *Cyclamen* cut flowers (Wartenberg, 1998). Previous studies revealed negative effects of cut flower foods containing sugar on postharvest quality of *Cyclamen* (Wartenberg, 2006). Several authors agreed that the period just after anthesis is the best developmental stage for an optimal time of harvesting (Carow, 1981; Escher, 1996; Horn, 1996).

The effect of the time of harvest during the flowering period, vase water quality and different types of cut flower food on the longevity of *Cyclamen* cut flowers was analysed (Neumaier *et al.*, 2009). Flowers harvested before anthesis as closed buds were compared to those harvested just after anthesis. It was at first remarkable, that seven days after cutting the budded *Cyclamen* cut flowers showed significantly more intact flowers (95 per cent) than the open ones (82 per cent). However, most of the buds did not open further, which led to a bad appearance. The number of flowers with symptoms of wilting was much higher in the variant with Chrysal clear professional 3 (containing sucrose) and in the variant with the cellulose card than in the control vase with plain water.

Cold storage is the common method of storing cut flowers. It facilitates the adjustment of flowers and other planting material supply against the market demand and enables the accumulation of large quantities of flowers. Low temperature treatment during storage or shipment period reduces the entire metabolism in the tissue, slows down the respiration. Flowers should be harvested at the optimal stage of development, should be pre-treated with an appropriate floral preservative and pre-cooled for fast packing. It is necessary to maintain a stable and uniform temperature. In general, cyclamens can be stored at 0-1°C up to 21 days (Senapati *et al.*, 2006).

13.0 Biotechnology

Cyclamen is a fascinating ornamental plant. Even though micropropagation and somatic embryogenesis has been successfully reported, use of embryo culture and anther culture for transferring desirable traits such as fragrance from related species *C. purpurascens* has to be perfected. Since *C. persicum* (2n) x *C. purpurascens* (2n) hybrids form unreduced micropspores which fail to produce progenies, due female sterility, the production and availability of fertile amphidiploid plants (2n=82) is important for cyclamen breeding (Ishizaka, 1998). A transformation system to modify flower colour would be of value. A new colour, pale yellow has been reported (Miyajima *et al.*, 1991) but blue cyclamens donot exist (Aida *et al.*,

1999). Genetic transformation to produce deep yellow or blue cyclamens would be of value. Transfer of rice chitinase gene would enhance the resistance of cyclamen to gray mold (Aida *et al.*,1999).

13.1.0 Somatic Embryogenesis (SE)

Due to inbreeding depression, abortive zygotic embryo formation, and varying ploidy levels, vegetative production is frequently selected as the preferred form of propagation by breeders (Tagipur *et al.,* 2016). Somatic embryogenesis is a highly effective method to produce cyclamen plants vegetatively. Somatic embryo formation in cyclamen is affected by genotype, source of explant, medium content, plant growth regulators, and their concentrations and culture conditions. Somatic embryos can be multiplied efficiently by repetitive or secondary embryogenesis on MS without plant growth regulators. The obtained somatic embryos can be used to produce synthetic seeds for clonal plant propagation and to protect valuable or rare cyclamen genetic resources (Tagipur *et al.,* 2016). Early research on cyclamen SE generally focused on hormonal regulation, and new methods were developed to produce a wider range of *Cyclamen* species using SE. SE studies on cyclamen were initiated in the 1990s (Tagipur *et al.,* 2016). SE in cyclamen usually consists of two culture steps: (1) induction of embryogenic callus and somatic embryos in medium with an auxin or with an auxin and a low concentration of cytokinin and (2) germination of somatic embryos from callus in medium without PGRs (Tagipur *et al.,* 2016).

Cotyledon, petiole, tuber, and root explants from aseptic seedlings of cv. Anneke formed embryoids when cultured on MS medium supplemented with various concentrations of 2,4-D, kinetin, and sucrose (Takamura *et al.,* 1995). The optimal frequency of somatic embryogenesis occurred on MS medium containing 5 µM 2,4-D, 0.5 µM kinetin and 3-6 per cent sucrose. Somatic embryogenesis was also enhanced at higher temperatures ranging from 25 to 30°C. Conversely, light inhibited somatic embryogenesis (Takamura *et al.,* 1995). Kiviharju *et al.* (1992) produced somatic embryos of Cyclamen through a callus phase from juvenile explant material including anthers, ovaries, and zygotic embryos. The auxin 2,4-D (1.0-1.5 mg/litre) and coconut milk (10 per cent v/v) in MS medium were important factors for the induction of somatic embryogenesis. Somatic embryos germinated into plantlets in MS medium without growth regulators. The plants grew well in the greenhouse and flowered normally.

13.2.0 Bioreactor Technology

Propagation in liquid culture, especially bioreactors, is one possible way to produce clonal propagules of *Cyclamen persicum* at a low cost. The current propagation method for *C. persicum* is from expensive hybrid seeds (Lyngved *et al.,* 2008). There are several reports on somatic embryo production systems using bioreactors, but they conclude that the effect of physical environmental factors has not been sufficiently investigated in detail. One potentially important growth factor is the gas composition in the culture vessel. The effect of O_2 partial pressure and CO_2 accumulation on cell proliferation and the subsequent regeneration of somatic embryos of *C. persicum* have been reported for flasks and Applikon bioreactors (Hohe *et al.*, 1999a,b).

13.3.0 Markers

Identification and characterization of gene markers for somatic embryogenesis offer the possibility of determining the embryogenic potential of somatic cells before any morphological changes appear and to provide information on molecular regulation of early somatic embryogenesis (Darvari *et al.*, 2014). Based on evidence reported in the literature, Darvari *et al.* (2014) assessed the reliability of the established SERK, LEC and WUS marker genes as well as the usefulness of mediating epigenetic regulation by DNA methylation, chromatin remodeling and micro RNAs as markers of the induction of somatic embryogenesis. The evidence suggests that the SERK gene marker may be less reliable than has been thought while the germin-like protein encoding genes may, like their protein products, be useful markers of early somatic embryogenesis.

13.4.0 Cryopreservation

Preserving genetic resources by cryopreservation is a beneficial way to protect endangered or wild species. Tagipur *et al.* (2016), quoting the work of Winklemann *et al.* (2014) indicated that only two cyclamen cryopreservation protocols exist. Winkelmann *et al.* (2004) cryopreserved embryogenic suspension cultures of *C. persicum*. Different cryoprotectants (0.09, 0.2, 0.4, or 0.6 M sucrose or 0.4 M sorbitol) were tested during pre-culture, and 0.6 M sucrose resulted in the highest rate of regrowth (75 per cent). Additional pretreatment with 0.6 M sucrose and 10 per cent DMSO for 1 day positively affected regrowth. The optimal time of exposure to 0.6 M sucrose was 2–4 days, and regrowth rates of cryopreserved and unfrozen callus were similar. In another study by Winkelmann *et al.* (2004), with the aim of producing synthetic embryos from globular somatic embryos of *C. persicum*, two encapsulation systems conventional alginate beads and alginate hollow beads) were tested. The percentage germination of alginate beads was as high as that for control embryos (97 per cent), but germination was delayed. Hollow beads showed lower germination percentage (71 per cent). When stored at 4°C, the viability of encapsulated and nonencapsulated somatic embryos was reduced although the speed of germination was increased by incorporating medium into the capsules in both capsule types.

13.5.0 Biochemical and Molecular Markers

Very few reports are available on the biochemical and molecular markers in cyclamen. Zhang *et al.* (1997) examined the use of random amplified polymorphic DNA (RAPD) markers as a means to identify cultivars of cyclamen seeds and to determine the genetic purity within cyclamen seeds. Bulked samples of five cyclamen hybrid cultivars, respectively, produced consistent RAPD marker profiles. Evaluation of individual seeds from a single cyclamen hybrid produced polymorphic banding patterns that were attributed to genetic variability present in the female and male inbred parents. These results show that RAPD markers can be used to quickly assess the genetic purity of selected cultivars of these two flower seed crops.

13.6.0 Genetic Transformation

There has been great scope of use of genetic transformation to transfer traits of value such as flower colour and disease resistance particularly against gray mold (*Botrytis cinerea*). Aida *et al.* (1999) described a method for *Agrobacterium* mediated transformation. Etiolated petiole segments of cv. Anneke were infected with *A. tumefaciens.* Strain AGL0 or LBA4404 harboring the binary vector plasmid pIG121Hm that includes *gus* reporter gene and *nptii* and *hpt* genes as selectable markers. Explants were cultured on MS medium supplemented with TDZ (1.0 mg/l), 2,4-D (1.0 mg/l), ticarcillin (300 mg/l) and hygromycin (5 mg/l) or Kanamycin (100 mg/l) in selection medium. Transformation was confirmed by histochemical assays of GUS activity and Southern blot anlaysis. They could recover 103 transformed plants from 920 explants. This study should further lead to the transfer of desirable genes. Transgenic *C. persicum* have been produced using *Agrobacterium tumefaciens,* and etiolated petioles of aseptic seedlings are used as a target for transformation (Aida *et al.,* 1999; Boase *et al.,* 2002). However, normal (not etiolated) petioles and somatic embryos have not been used as the starting material. There are two essential elements for *Agrobacterium*-mediated transformation: stable introduction into the plant genome and regeneration of transgenic plants. Somatic embryos may be one of the best sources to produce a large number of transgenic Cyclamen plants. Somatic embryogenesis is a very efficient system for clonal mass propagation of plants. Terakawa *et al.* (2008) developed an effective method for somatic embryo-mediated plant regeneration and genetic transformation using *Agrobacterium,* with somatic embryos as the source of plant material. Numerous somatic embryos were formed from the leaf segments of *Cyclamen* cv. "Fragrance Mini" 16 weeks after initiation on a 2,4-D and kinetin-containing medium. The maximum regeneration frequency from the somatic embryos was achieved using a regeneration medium containing 0.1 mg/l benzylaminopurine, 0.01 mg/l α-naphthaleneacetic acid, and 0.2 mg/l gibberellin. After co-cultivating the somatic embryos with *Agrobacterium* strain LBA4404 (the plasmid vector pIG121Hm) that harbors genes for hygromycin phosphotransferase, neomycin phosphotransferase, and b-glucuronidase, hygromycin-resistant clones were selected on a medium containing 5mg/l hygromycin. Hygromycin-resistant plantlets were regenerated from 14 hygromycin-resistant embryos. Using this transformation system, 74 independent plants were obtained from 1 ml packed cell volume, approximately 2,000 embryos.

14.0 References

Aida, R., Hirose, Y., Kishimoto, S. and Shibata, M. (1999) *Plant Sci.,* **148** : 1-7.

Altay, H. and Muftuoglu, N.M. (2004) *International Soil Congress (ISC) on Natural Resource Management for Sustainable Development,* June 7-10, 2004 Erzurum, Turkey.

Beeks, S.A. and Evans, M.R. (2013) *Horttechnology,* **23** : 173-176.

Boase, M.R., Marshall, G.B., Peter, T.A., Bendall, M.J. (2002) *Plant Cell Tiss. Org. Cult.,* **70** : 27–39.

Bulle, A. (2005) *Acta Hortic.,* **669**: 263-268.

Carow, B. (1981) Frischhaltung von Schnittblumen. Verlag Eugen Ulmer, Stuttgart.

Darvari, F.M., Noor, N.M. and Ismanizan, I. (2014) *Plant Cell Tiss. Organ. Cult.* Doi: 10.1007/s11240-014-0615-0.

Davis, G. and Grey-Wilson, M. (1999) CITES Bulb Checklist for the genera: *Cyclamen, Galanthus* and, *Sternbergia*, the Trustees of the Royal Botanic Gardens Kew, pp. 35-46.

De Haan, I. and Doorenbos, J. (1951) *Mededelingen Van De Landbouwhogeschool* Te Wageningen/Nederl and-DEEL51, VERH. **7**, pp. 151-166.

Dubsky, M., Sramek, F. and Vosatka, M. (2002) *Rostlinná Výroba*, **48**(2) : 63-68.

Escher, F. (Hrsg.) (1996) Schnittblumenkulturen. Verlag Eugen Ulmer, Stuttgart.

Gillespie, J.B. and M.B. Thomas. (1983) http: //researcharchive.lincoln.ac.nz/ handle/10182/3005.

Grey-Wilson, C. (1988) The Genus Cyclamen. The Royal Botanic Gardens, Kew in association with Christopher Helm and Timber Press, 9999 sw Wishire Portland, Oregon.

Hendriks, L. (1993) *Deutscher Gartenbau*, **40**: 2541.

Hohe, A., Winkelmann, T. and Schwenkel, H.G. (1999a) *Plant Cell Tissue Org. Cult.*, **59** : 39-45.

Hohe, A., Winkelmann, T., Schwenkel, H.G. (1999b) *Plant Cell Rep.*, **18** : 863-867.

Horn, W. (Hrsg.) (1996) Zierpflanzenbau. Blackwell Wissenschaftsverlag, Berlin.

Ishizaka, H. and Uematsu, J. (1994) *Breed. Sci.*, **44** : 161-166.

Ishizaka, H. and Uematsu, J. (1995) *Euphytica*, **82** : 31–37.

Ishizaka, H., Yamada, H. and Sasaki, K. (2002) *Sci. Hortic.*, **94** : 125-135.

Ishizaka, H. (1998) *Plant Cell Tiss.Org. Cult.*, **54**: 21-28.

Ishizaka, H. (2003) *Plant Syst. Evol.*, **239** : 1-4.

Ishizaka, H. and Uematsu, J. (1994) *Breed. Sci.*, **44** : 161-166.

Ishizaka, H. and Uematsu, J. (1993) *Jap. J. Breed.*, **43** : 207-218.

Jalali, N., Naderi, R., Gharahlar, A.S. and daSilva, J.A.T. (2012) *Sci. Hortic.*,**137** : 11-19.

Kao, K.H. and Michayluk, M.R. (1975) *Planta*, **126** : 105–110.

Kreuger, M., Postma, E., Brouwer, Y. and Van Holst, G.J. (1995) *Physiol. Plantarum*, **94** : 605-612.

Källersjö, Mari; Bergqvist, Gullevi; Anderberg, Arne A (2000) *American Journal of Botany*, **87** (9) : 1325-1341.

Prasad, M., Widmer, R.E. and Marshall, R.R. (1983) *Commun. Soil Sci. Plant Anal.*, **14**(7) : 553-573.

Miyajima, I., Doi, I. and Kage, T. (1990) Science Bulletin of the Faculty of Agriculture, Kyushu University, **45**(1-2) : 83-89.

Miyajima, I., Maehara, T., Kage, T. and Fujieda, K. (1991) *J. Jap. Soc. Hort. Sci.*, **60** : 409-414.

Mooney, B.P. (2009) *Biochem. J.*, **418** : 219–232.

Morgan, E.R. (1999) *Plant Cell Tiss. Org. Cult.*, **55** : 63–65.

Murasaki, K. and Tsurushima, H. (1988) *Acta Hortic.*, **226**(II) : 721-724.

Naderi, R., Jalali, N., Babalar, M. and Mirmasoumi, M. (2012) *Int. J. Agric. Crop Sci.*, **4**(11) : 699-702.

Neumaiera, D., Bausbach, E., Simon, M., Tisch, M.and Winkelmann, T. (2009) *Acta Hortic.*, **847**: 269-274.

Neveur, N., Corbineau, F. and Come, D. (1986) *J. Hortic. Sci.*, **1**(3) : 379-387.

Nhut, D.N., Thu, H.T.M., Vinh, B.V.T., Binh, N.V. and Luan, V.Q. (2012) Propag. Ornam. Plants, **12**(2) : 89-95.

Prange, A.N.S., Serek, M., Bartsch, M. *et al.* (2010a) *Plant Cell Tiss. Organ Cult.*, **101** : 171–182.

Schmidt, T., Ewald, A., Seyring, M. *et al.* (2006) *Plant Cell Rep.*, **25** : 643. https: //doi.org/10.1007/s00299-006-0130-9.

Schwenkel, H.G. and Grunewaldt, J. (1988) *Acta Hortic.*, **226**(II) : 659-662.

Schwenkel, H.G., Grunewaldt, J. and de Jong, J. (1990) *Proc. Symp. Integration of in vitro techniques in ornamental plant breeding.*, 10-14 November 1990, pp. 62-67.

Senapati, A. K., Raj, Dev, Jain, Ritu and Patel, N. L. (2006) In : *Commercial Horticulture* (Eds. N.L. Patel, S.L. Chawla and T.R. Ahlawat), New India Publishing Agency, New Delhi, India, pp. 473-488.

Sprau, G. (1997) *Deutscher Gartenbau*, **32**: 1723-1726.

Sugiyama, M., Saito, H., Ichida, H., Hayashi, Y., Ryuto, H., Fukunishi, N., Terakawa, T. and Abe, T. (2008) *Plant Biotechnol.*, **25** : 101–104.

Stevens, P. F. (2012) Myrsinoideae. Angiosperm Phylogeny Website.

Tagipur, E.M., Seker, G., da Silva, J.A.T. and Mendi, Y.Y. (2016) *Somatic Embryogenesis in Ornamentals and Its Applications*, Springer India. pp.155-168.

Takamura, T., Miyajima, I. and Matsuo, E. (1995) *Plant Cell Rep.*, **15** : 22-25.

Takamura, T., Sugimura, T., Tanaka, M. and Kage. T. (1998) *Acta Hortic.*, **454** : 119-126.

Takejiro, T. and Miyajima, I. (1996) *Sci. Hortic.*, **65** : 305-312.

Tanaka, Y., Oshima, Y., Yamamura, T., Sugiyama, M., Mitsuda, N., Ohtsubo, N., Takagi, M.O. and Terakawa, T. (2013) *Sci. Rep.*, **3**: 2641, doi : 10.1038/srep02641.

Wartenberg, S. (2000) Nachgefragte Blüten – Kultur von Alpenveilchen für den Schnitt. Gärtnerbörse 01/2000 21-23.

Wartenberg, S.(2006) Qualität bei Topf- und Schnittcyclamen. Sächsische Landesanstalt für Landwirtschaft, Pillnitzer Fachseminar am 24.01.2006, www.umwelt.sachsen.de/ de/wu/Landwirtschaft/lfl/inhalt/download/QualitaetCyclamenWartenberg.pdf.

Winkelmann, T., Mobmann, V. and Serek, M. (2004) *Plant Cell Rep.,* **23** : 1–8.

Winkelmann, T. and Serek, M. (2005) *Euphytica,* **144** : 109–117.

Winkelmann, T., Sangwan, R.S. and Schwenkel, H.G. (1998) *Plant Cell Rep.,* **17** : 400-404.

Winkelmann, T., Ratjens,S., Bartsch, M., Rode,C.,Niehaus, K, Bednarz, H. (2015) Plant Sci., 04 August 2015 | https: //doi.org/10.3389/fpls.2015.00597

Yesson, C., Culham, A. (2006) *BMC Evol. Biol* **6**: 72, http: //www.biomedcentral.com/1471-2148/6/72.

Yesson, C., Toomey, N.H. and Yesson, C. (2009) *J. Biogeography,* **36** : 1234-1252.

Zhang, J.H., McDonald, M.B. and Sweeney, P.M. (1997) *HortScience.,* **32** : 246-247.

7

Zantedeschia

M.L. Chaitra, S.P. Sharma and K. Dutta

1.0 Introduction

Zantedeschia is commonly known as calla lily. It is perennial herbs grown for their ornamental corolla like spathe and sometimes for their variegated or spotted foliage. Attributes of *Z. aethiopica* which make it desirable as an ornamental are its ability to propagate vegetatively with prolific offset production and its abundant flowering and seed production (Plummer and Wykes, 1997). It grows very well in warm climate, but can be grown inside greenhouses also.

2.0 Importance and Uses

2.1.0 Aesthetic Uses

The calla lilies (*Zantedeschia aethiopica*) are useful for planting along the ponds, lakes or in bog gardens. These are attractive as pot plants. The spathes are used as cut flowers, while the leaves are very elegant and look impressive in floral arrangements. *Zantedeschia* is a commercially important cut flower industry in New Zealand (including the export of bulbs) (Clemens *et al.*, 1993) and Mexico (Cruz-Castillo and Cardenus-Alonso, 1998). The plants grown in Africa serve as a source of starch and pig feed.

2.2.0 Therapeutic and Other Uses

In Africa, *Z. aethiopica* is used to treat rheumatism, and as a poultice for sores, boils, wounds, insect bites and burns (Sivakumar and Nair, 1992). Maroyi (2019) observed that compounds isolated from the species have antibacterial, antifungal, antioxidant, antihistaminic, antialgal, antithrombotic, and anticoagulant activities. *Z. aethiopica* is an integral part of the traditional pharmacopeia in several countries where the species is indigenous or naturalized, but there is lack of alignment between the known medicinal applications, phytochemistry, and biological activities of the species. Lewis *et al.* (2003) found anthocyanin, cyanidin 3-glucoside and the carotenoids lutein and β-carotene were the main pigments present. Other flavonoids, predominantly 7-methylated luteolin C-glycosides, were also detected by them in the spathe tissue. Variation in spathe colour appears primarily due to changes in the pigment group produced and pigment concentration rather than the production of different individual pigments.

3.0 Origin and History

The common calla lily was long known in the flower trade as *Calla aethiopica* which was changed to *Richardia africana*, but presently it is renamed as *Zantedeschia aethiopica*. Genus of 6 species, found in southern and eastern Africa (Moore and Wykes, 1997; Brickell, 1996). Calla lily belongs to family Araceae. The plants 60-75 cm high, rhizomatous, herbaceous; petioles hairy, or bristly below and spongy throughout; leaf lamina lance-shaped or narrowly to broadly arrow or heart-shaped; peduncles long or longer than leaves; spathes larger, open, with a recurved, wavy tip; white creamy base; spadix is staminate above and pistillate below.

4.0 Species and Cultivars

4.1.0 Species

Zantedeschia aethiopica (Arum lily)

Clump forming, rhizomatous perennial; plants semi-erect; leaves about 40 cm long, arrow shaped, green, glossy; spathe pure white, up to 25 cm long; spadix creamy yellow; Apple Court Babe: plants 60 cm high; Chlidsiana : dwarf and free-flowering; Crowborough: pure white; Candidissima : spathe large, pure white; Devonensis: dwarf, free bloomer, fragrant; Green Goddess: leaves dull green, white centred, spathes bright green, 15-20 cm long; Gigantea: very large; Godfreyana : dwarf, spathe white; Grandiflora: spathe large; Little Gem: dwarf, spathe white, 8-12 cm long; White Sails: spathe open white, 10 cm long.

Z. albomaculata syn. *Z. melanoleuca* (Spotted lily)

Rhizomatous, perennial, semi-erect; leaves arrow-shaped, mid-green, translucent white spots; spathe 12 cm long, white to cream, pale yellow or pale pink, purple mark inside at the base; spadix yellow.

Z. elliottiana (Golden Calla)

Rhizomatous erect, leaves heart-shaped, dark green, covered with translucent white spots; both spathe and spadix golden yellow.

Z. oculata (Yellow calla)

Spathe cup shaped, greenish yellow, black purple at the base.

Z. pentlandii syn. *Z. angustiloba*

Upright, rhizomatous, perennial; leaves oblong-elliptic to oblong-lance, green, rarely spotted; spathe 12 cm long, bright golden to lemon yellow, dark purple mark inside at the base; spadix golden yellow.

Z. rehmannii (Pink or rose calla)

Rhizomatous, perennial, semi-erect; leaves lance-shaped, green; spathe 12 cm long, white to pink or dark purple; spadix yellow.

Z. odorata

It is a sweetly-scented, white-spathed species of restricted habitat in the Western Cape with short stipitate spadix and the green infructescence which bends to the ground on ripening (Perry, 1989).

Karyotypes of 5 species and 3 cultivars of zantedeschia were studied by Yao *et al.* (1994b). Karyotypes were distinct between the evergreen species *Z. aethiopica* (section I) and all the winter-dormant species (section II).

Elliottiana Hybrids

Aztec Gold - Leaves unspotted; spathe golden yellow, at mature stage burnt orange.

Uses of Zantedeschia

In a Garden

Side of a Pond

In Pot

In a Park

Side of a Lake

In Greenhouse

As Cut Flowers

In Flower Arrangement

Species of Zantedeschia

Zantedeschia aethiopica

Zantedeschia albomaculata

Zantedeschia elliottiana

Zantedeschia pentlandii

Zantedeschia rehmannii

Zantedeschia odorata

Cultivars of Zantedeschia

Black Magic

Florex Gold

Majestic Red

Dusty Pink

Marshmallow

Auckland

Red Desire

White Cutie

Flamingo

Gold Rush

Red Alert

Odessa

Propagation of Zantedeschia

Seeds

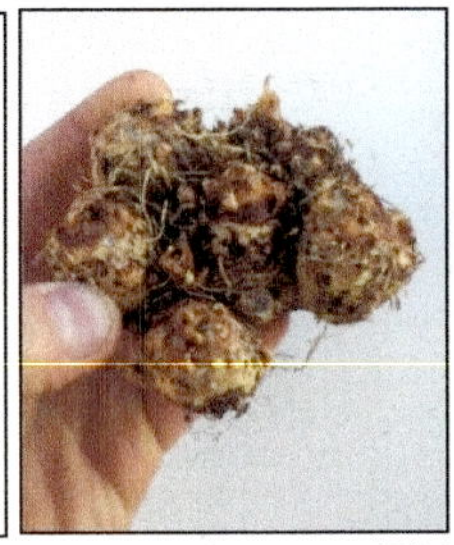

Rhizomes

Germination of Seeds

Growth of Seedling in Pot

Tranplanted in Bed

Micropropagation of Zantedeschia

Danuta, K. (2016) Micropropagation of calla lily (*Zantedeschia rehmannii*) *Folia Horticulturae*. 28. 10.1515/fhort-2016-0021.

Cultivation of Zantedeschia

Transplanted in Beds

Growth and Flowering in Open

Seedlings Planted in Pots

Flowering in Pots

Growing in Greenhouse

Flowering in Greenhouse

Black-eyed Beauty - Plants 30-40 cm high; leaves heavily white-spotted; spathe creamy black central mark in the throat.

Black Magic - Plants 70 cm high, leaves white-mottled; spathe yellow with black throat (Brickell, 1996).

Golden Africa - Plants 60 cm high; leaves unspotted; spathe bright yellow.

Majestic Red - Plants 65 cm high; leaves unspotted, pointed; spathe crimson.

***Rehmannii* Hybrids**

Dusty Pink - Plants 70 cm tall; spathe rounded, mauve-pink, throat darker (Brickell, 1996).

There are many garden forms, said to be hybrids of species : Angola – *Richardiaaurata* (*Z. albomaculata*× *Z. oculata*), *R. lathamiana* (*Z. elliottiana*× *Z. albomaculata*), *R. taylori* (*Z. elliottiana*× *R. aurata*) (Bailey, 1963).

4.2.0 Cultivars

Joung *et al.* (2015) found 'White Cutie' was selected specifically for use in cut flower production after investigation over seven years (2005 to 2011) of genetic and phenotypic characteristics, resistance against soft rot, and customer preference regarding the culture vigor and post-harvest quality. 'White Cutie' was early flowering (85.6 days to flowering) with white flowers (RHS W155C), although it had a mid-sized flower in which spathe height and width were 8.6 cm and 8.7 cm, respectively. It was multi-flowering (6.2 flowers per plant) and produced a very high number of cormels (13.4 per plant). Furthermore, it was resistant to soft rot disease.

Hoof (2003) developed a new and distinct cultivar named 'Red Desire', characterized by its upright and somewhat outwardly spreading plant habit, white-colored Spathes with dark pink-coloired throats and dark pink-coloured Spadices, and freely flowering habit.

Schram (2009) developed a cultivar named 'Snow White' which is characterized by dense dwarf habit, medium-green foliage, and large fragrant, white and yellow inflorescences.

Klaver (2010) developed a cultivar named 'Flamingo', characterized by its upright plant habit; glossy medium green-coloured leaves; large pink-colored spathes that are positioned above and beyond the foliage on erect scapes; and good inflorescence longevity.

Kapiteijn (2015) developed a cultivar named 'Captainsolo', characterized by yellow inflorescence, flowers occurring above the leaves and a short plant height. The new variety is commercially suitable for indoor plant production.

5.0 Genetics and Breeding

Selection for elite germplasm from natural or hybrid populations are major types of breeding programs in coloured calla lily. However, some Chinese organizations have initiated planned coloured calla lily improvement programs compared to New Zealand, the Netherlands and the USA. To date, about ten Chinese varieties have been bred through hybridization and released to the public since 1990. One reason is the ill-defined genetic background of the materials used in breeding practices.

Triploid hybrid plants were produced from crosses between diploid and tetraploid plants of the winter-dormant section in the genus *Zantedeschia*. In different crosses, the endosperm became degenerate and embryo development was restricted to the globular stage because of post-fertilization incompatibility. Using an embryo rescue technique, hybrid plants were produced from two crosses. These plants were confirmed to be triploids by counting chromosomes in root-tip cells. The triploid plants flowered in a greenhouse and are pollen free. Hybridizations among diploid plants in the winter-dormant section are generally compatible. However severe incompatibility was noticed in the diploid-tetraploid crosses due to imbalance in the number of chromosome (Yao *et al.*, 1996).

An integrated approach of *in vitro* pollination and fertilization, followed by embryo rescue has been applied in many crosses of *Zantedeschia*. Barriers occurring after successful embryo rescue are hybrid breakdown (*e.g.*, hybrid albinism) and F_1 sterility. Hybrid breakdown results in the loss of the hybrid before flowering and is a result of new unbalanced genome combinations. F_1 sterility of interspecific hybrids is very common and among others may be the consequence of reduced chromosome pairing during meiosis. In these cases fertility can be restored by polyploidization, enabling pairing of homologous chromosomes in the allopolyploid hybrid (Van-Tuyl and Halevy, 1997). Application of 2n-gametes (meiotic polyploidization) in interspecific breeding programmes may be of great importance for the introgression of characters from diploids to tetraploids.

Ngamau (2008) cultivated *in vitro* seeds of *Z. aethiopica* cv. Green Goddess under different salinity and temperature conditions and subsequently made a selection. He then acclimatized and cultivated them in a greenhouseand obtained plants tolerant to elevated NaCl levels. This is one of the few works that show the breeding of the calla lily with self-fertilization and a focus on tolerating adverse environments

Joung *et al.* (2015) artificially crossed two genotypes namely 'Childsiana' showing the multi-flowering characteristic and 'Wedding March' resistant to soft rot disease. Of the progenies obtained, 'White Cutie' was selected specifically for use in cut flower production after investigation over seven years of genetic and phenotypic characteristics, resistance against soft rot, and customer preference regarding the culture vigor and post-harvest quality. 'White Cutie' was early flowering (85.6 days to flowering) with white flowers (RHS W155C), although it had a mid-sized flower in which spathe height and width were 8.6 cm and 8.7 cm, respectively. It was multi-flowering (6.2 flowers per plant), produced a very high number of cormels (13.4 per plant) and was resistant to soft rot disease.

Snijder *et al.* (2004) screened six hybrids ('Back Magic', Galaxy', Pink Persuasion', Sensation', 'Treasure' and 'Florex Gold') using a non-destructive resistance test (leaf disk test) and scored different levels of susceptibility. Some hybrids were less susceptible than the very susceptible reference 'Florex Gold', while all hybrids were more susceptible than the partial resistant reference *Z. rehmannii*. The half-sib family of 'Pink Persuasion' was more resistant, but less vigorous than the half-sib family of 'Florex Gold'. They concluded that resistance and vigour may be negatively correlated in the diploid cultivars and that resistance should be bred from other sources than the used cultivars.

Plastome-genome incompatibility (PGI) is prevalent in several plants including the *Zantedeschia* species, a worldwide commercial flower crop native to South Africa. Generally, hybrids suffering from PGI appear less vigorous and more susceptible than normal plants. Previous reports revealed that the PGI level in interspecific hybrids is correlated with the relatedness of the parental species in the genus *Zantedeschia.* To provide a basis for utilizing and improving resources in breeding programs, a total of 117 accessions of colored calla lily (*Zantedeschia* hybrid), collected from New Zealand, the Netherlands and the United States, were genotyped using 31 transferable expressed sequence tags-simple sequence repeats (EST-SSR) markers from the white calla lily (*Zantedeschia aethiopica*) (Wei *et al.*, 2017a). A moderately high level of genetic diversity was observed, with 111 alleles in total, an observed/expected heterozygosity (Ho/He) of 0.453/0.478, and polymorphism information content (PIC) of 0.26. Genetic distance and STRUCTURE-based analysis further clustered all accessions into four subgroups (G-Ia, G-Ib, G-IIa and G-IIb), which mostly consisted of *Zantedeschia pentlandii*, *Zantedeschia elliotiana*, *Zantedeschia albomaculata* and *Zantedeschia rehmannii*, respectively. Significant genetic differentiation was observed between all inferred subgroup pairs, with the Fst ranging from 0.142 to 0.281. Finally, the accessions assigned into G-IIb (*Z. rehmannii*) were recommended as top priority parents in efficient *Zantedeschia* breeding program designs.

Wei *et al.* (2017b) also showed that the number of alleles and the polymorphism information content (PIC) in white calla lily ranged from two to four and from 0.110 to 0.613, respectively. Moreover, 24 (89 per cent) SSR markers were transferable to the cultivars of colored calla lily, in spite of the reproductive isolation between species of the two sections. These novel SSRs will be useful for further research on germplasm evaluation, comparative genomics, phylogenetic relationships and molecular breeding in the genus *Zantedeschia*.

5.1.0 Interspecific Hybridization

Interspecific hybridization between the two sections of the genus *Zantedeschia*, restricted embryo development to the globular stage and the endosperm degenerated. Embryo culture was used to rescue hybrid embryos. All the 178 hybrid plants produced were albino because of plastome-genome incompatibility (Yao *et al.*, 1995). Yao and Cohen (2000) also obtained albino, variegated and virescent hybrids through interspecific hybridization in the genus *Zantedeschia*. The inhibition of chloroplast development in these hybrids is due to plastome-genome incompatibility. In albino leaf sectors of variegated hybrids, the mRNA levels for five plastid-development-related genes were also severely reduced. In the green sectors of variegated hybrids, chlorophyll content and mRNA levels of the five genes were slightly reduced. F_1 hybrids between *Z. aethiopica* and *Z. odorata* were albino when *Z. odorata* plastids were present but virescent when *Z. aethiopica* plastids were present. When the F_1 hybrids with *Z. aethiopica* plastids were backcrossed to *Z. odorata*, the progeny were albino, virescent or green. *Z. odorata* plastids were inherited only from maternal parents in the F_1 progenies but inherited from either maternal or paternal parents in the backcross. The increased compatibility with, and inheritance of *Z. odorata* plastids in the backcross suggested that multiple genes play a major role in the plas-tome-genome incompatibility. An increase in the number of genes from the parent contributing the plastids improves chloroplast development in the hybrids of the genus *Zantedeschia.*

5.2.0 Mutation

Tetraploid plants were produced from shoot cultures of nine zantedeschia cultivars using *in vitro* colchicine treatment on a medium containing 0.05 per cent colchicine for 1, 2 or 4 days. Following the treatment, most shoots were killed but the surviving shoots were multiplied for several subcultures. These shoots were then rooted *in vitro* and transferred to a greenhouse. Plants were screened 2 months later by measuring stomatal length, and 110 out of 565 plants were selected as putative tetraploids with a stomatal length significantly greater than in diploid control plants. Chromosome counts carried out on root tips from 44 plants confirmed that 38 were tetraploids, 2 were chimaeras (predominantly tetraploid with a few octoploid cells) and 4 were diploid (Cohen *et al.,* 1996).

6.0 Growth and Flowering

6.1.0 Growth

Welsh *et al.* (1992) reported that *Zantedeschia* hybrid is a day neutral plant so it can be cycled out of season using protected cropping. Clark and Boldingh (1991) determined the growth of *Z. elliottiana* cycle in 3 distinct phases : the first phase (5-7 weeks from planting), roots and vegetative tissues developed concurrently, tuber dry matter (DM) declined to 50 per cent of their starting mass and flowering was completed. In the second phase (about 6-16 weeks after planting), there was rapid growth and maximum nutrient uptake, DM and number of leaves increased to their maximum, new tuberous tissue formed and root growth declined. The final phase was initiated with shoot senescence (16 weeks' growth). Tuber DM slowly declined in the absence of leaves and with the onset of dormancy. Singh *et al.* (1996) observed 5 phases of floral development in *Z. aethiopica* : pre-female, female, male, fruit development and fruit ripening. Plants are self-incompatible and separation of sexes is achieved by marked protogyny. *Z. aethiopica* is the only member of the genus in which the basal part of the spathe turns green during fruit maturation while the apical part withers away. The berries turn orange, soft and mucilaginous on ripening.

Subbarai *et al.* (2010) studied the efficacy of cytokinin alone to stimulate branching declined with the transition to dormancy (phase 1 = 3.8 ± 0.2 shoots; phase 3 = 1.0 ± 0.3 shoots). To sustain branching during this transition, a sequential application of gibberellin was necessary. Gibberellin alone failed to stimulate branching. The efficacy of gibberellin alone to stimulate flowering diminished with the transition to dormancy. Any flowering during this transition occurred only after the sequential application of cytokinin. Cytokinin alone failed to stimulate flowering.

Alleviating bud dormancy and enhancing flowering in *Zantedeschia*, achieved by the reciprocal cross-talk between cytokinin and gibberellin, contributes to the pool of evidence drawing common mechanisms between dormancy and flowering and may have commercial implications.

Fonseca *et al.* (2013) showed different behaviors between species in dry matter and nutrient uptake, where *Z. elliottiana* showed a higher biomass production and greater nutritional requirements in relation to *Z. rehmannii*. The percentage differences of nutrient uptake at the end of the cycle between them were evaluated in the following magnitude: 59 per cent N, 46 per cent S, 44 per cent K, 43 per cent Ca, 38 per cent Mg 38 per cent Zn, 33 per cent Mn 22 per cent P, 15 per cent B, 8 per cent Cu and 6 per cent of Fe. The amount of nutrients uptake in the plant in both species followed a decreasing order: K>N>Ca>P>S>Mg. As for the micronutrients:Fe>Zn>B>Mn>Cu.

Posada *et al.* (2012) showed that the flowers quality related to the length of peduncle was positively affected by shading, however the number of flowers, the flowering time and the dry mass allocation to flowers were not influenced by the applied treatments. Calla lily showed versatility to adapt to environmental conditions, which was manifested through physiological and morpho-anatomical changes in plants growing under manipulated spectra.

Janowska (2010) evaluated the effect of flurprimidol applied directly into soil on growth and flowering of six *Zantedeschia* cultivars with coloured spathes. Flurprimidol had an effect on flowering, quality of flowers and leaves of cultivars: 'Sunglow', 'Treasure', 'Pacific Pink', 'Black Eyed Beauty', 'Cameo' and 'Mango'. Independently of the cultivars the solution of retardant in concentrations of 112.5 and 150.0 mg/dm^3 reduced flowering and growth of peduncles, but had no effect on length of spathe. Flurprimidol had no effect on the number of leaves, but they had shorter petioles, more rounded leaf blades and higher leaf greenness index.

6.2.0 Flowering

Naor *et al.* (2004) found that the flowering pattern could be a result of a gradient in the distribution of endogenous factors involved in inflorescence differentiation (possibly GA_3) and in floral stem growth. This gradient along the tuber and shoot axes is probably controlled by apical dominance of the primary bud. They also studied the transition from the vegetative to the reproductive phase was characterized by a swollen, dome-shaped apex that transformed into a smooth elongated apex surrounded by the spathe primordium, at the tip of the elongating peduncle primordium. Floret primordia developed in inflorescences at a more advanced stage. The female florets located at the base of the primordial spadix, could be clearly distinguished from male florets located above them. BA did not show any effect on flower induction but, in the presence of GA, BA at concentrations up to 4.4 μM enhanced inflorescence differentiation.

Corte *et al.* (2017) indicated the increased vase life is achieved at alkaline pH by fluorescence addition of 20, 40 and 60 per cent, respectively. Likewise, the higher water consumption is also achieved at alkaline pH. It was concluded that fluorescence can affect the calla physiology for increasing vase life.

Lee *et al.* (2019) found the effectiveness of mulching materials which included wood chip (WC), rice hull charcoal (RHC) and perlite. The highest leaf number was recorded in RHC (12.2), followed by perlite treatment (11.5). Leaf length was greatest in RHC (142.9 cm), and floral stem length was highest in the perlite mulching (112.2 cm), which was longer than control by 7.3 cm. The fresh weight of cut flower was highest in perlite and WC treatments (210.4 and 201.9 g, respectively), while 176.9 g in the RHC treatment, compared to 144.4 g in the control. The yield of cut flowers, as expressed flower number per plant, was highest in RHC (5.1), which was about 38 per cent higher than the control (3.7). Bulb yield was significantly promoted by RHC and perlite mulches (270.5 and 247.0 kg 10 a^{-1}, respectively) compared to the control and WC (199.0 and 211.5 kg 10 a^{-1}, respectively). Therefore it is concluded that RHC mulching promoted the yield of both cut flowers and bulbs of calla lily, while soft rot incidence declined considerably.

6.3.0 Environmental Factors

6.3.1 Light

The effects of irradiance level, day length on growth and flowering of *Z. elliottiana* and *Z. rehmannii* were studied by Corr and Widmer (1990) in greenhouse. Plants grown with 45 per cent or 15 per cent of natural irradiance were taller than those grown under full natural irradiance but flowered at the same time and produced a similar number of flowers. Calla lily was subjected to full sun or 55 per cent or 67 per cent shade treatments for 2 or 3 years. A quadratic decline in yield in 3 of 4 cultivars of zantedeschia occurred as shade increased, but yield was similar for ambient light and 55 per cent shade. Scape length and spathe width increased as shade increased (Armitage, 1991). Plants grown with a night interruption (NI 22.00 h to 02.00 h) were taller than those under short days (of 8 h), but flowered at the same time and produced a similar number of flowers (Corr and Widmer, 1990). Ribeiro *et al.* (2009) found that the light type did not influence in vitro propagation of calla lily and there were a higher number of sprouts (2.0) in 45g/l of sucrose. There was higher number of leaves in white (3.8) and in blue light (2.9) with 15 and 60g/l of sucrose, respectively. Higher length of sprouts (3.7cm) was observed in 60g/l of sucrose and higher fresh weight of the aerial part (1.07g) in 60g/l of sucrose.

6.3.2 Temperature

Z. aethiopica cv. 'Childsiana', a dwarf cultivar of common white calla lily, is free branching and produced flowers year-round when grown below 18.5°C night temperatures (Tjia, 1989). Corr and Widmer (1990) studied the effects of air and medium temperature on growth and flowering. *Z. rehmannii* plants grown at media temperature of 20° or 25°C and an air temperature of 15° or 20°C flowered earlier than at an ambient growing medium temperature and an air temperature of 15°C, but plants of both species from all temperature regimes produced the same number of flowers over a 120-day cycle.

In *Zantedeschia rehmannii* Engl. tuber storage at 1-5°C for 30 days is suitable for breaking dormancy, and a temperature of 10°C is suitable for flower bud initiation.

Thus, effects of the combination of 1°C and 10°C during tuber storage on growth and flowering were investigated (Goto *et al.*, 2005). Tubers which were not given low temperature (<15°C) were stored for 50 days. Initial storage temperature was either 1°C or 10°C, and tubers were moved to alternate temperatures 10, 20, 30, and 40 days after storage. There were no significant differences among treatments in days to emergence, although days to flowering of plants stored at 10°C 20 days followed by 1°C 30 days were shorter than those of other treatments. Storage temperature and duration had no significant effect on anthocyanin content, cut flower weight, malformed flower percentage, cut flower length, and bract length. Percentage of flowering and number of flowers per plant decreased as storage duration at 10°C decreased. Flower production increased remarkably in plants stored at 1°C for 10 days followed by 10°C for 40 days and at 1°C for 20 days followed by 10°C for 30 days because secondary flowers were harvested. It is thought that dormancy is broken at any temperature for 50 days, however, storage at 10°C for >30 days is suitable for increase in flowering. The results suggest that tuber storage at 1°C then 10°C may be desirable when *Z. rehmannii* is forced in summer (Goto *et al.*, 2005).

Masayoshi *et al.* (2015) identified the main compounds responsible for low temperature-induced yellow pigmentation of the bracts of *Zantedeschia aethiopica* 'Wedding March'. On the basis of the area ratios estimated from absorbance at 400 nm in HPLC analysis, two flavonoids were identified namely, isoorientin and swertiajaponin. Two additional flavonoid compounds, isovitexin and swertisin were also identified, which does not contribute considerably to the yellow pigmentation. Flavonoids of *Zantedeschia* bracts seem to belong to the class of flavone C-glycosides.

6.3.3 Light and Temperature

Zantedeschia rehmannii cv. Pink and Flame (a hybrid of *Z. elliottiana*, *Z. maculata* and *Z. rehmannii*) were grown in 4 environmentally controlled growth chambers with 11 h day/13 h night. Three temperature treatments of 21°C (constant), 24/18°C (day/night) and 18/24°C, with fluorescent and incandescent lighting, were maintained. A fourth chamber with high intensity discharge (HID) lighting at 430 μmol/m^2/s was set at a constant 21°C for comparison. All plants received a preplant 100 ppm GA_3 dip for flower induction. Leaf and flower heights were found shortest at 18/24°C and tallest at 21°C constant (Reiser *et al.*, 1993).

Dry matter accumulation and partitioning were quantified by Funnell *et al.* (1998) under a range of temperature and regimes using plant growth analysis and obtained maximum values of relative growth rate (RGRM) with increasing temperature (from 13 to 28°C), with a base temperature of 2.1 + 2.7°C. The optimum temperature for growth was photosynthetic photon flux (PPF) dependent with maximum total plant dry mass occurring under high PPF (694 μmol/m^2/s) at 25°C. However, as the plant responded to PPF by altering leaf matter partitioning (LMP), final total plant dry mass was actually greater under the low PPF regime (348 μmol/m^2/s) at temperatures < 22°C.

6.3.4 Growth Regulators

Dipping of Zantedeschia tubers (cultivars Red Sunset and Orange Sunset) in 25 or 50 ppm GA_3 just prior to planting, almost doubled flower yield but quality was increased only at 25 ppm treatment (Dennis *et al.*, 1994). Treatment of cv. Galaxy tubers by immersing in GA_3 or promalin (GA_{4+7}) for 30 second was undertaken by Funnell *et al.* (1992). Treatment with GA_3 at 25 ppm resulted in production of 3.4 flowers/bulb, compared to 1.4 in control, while promalin at 100 ppm produced 5.9 flowers/bulb (Funnell *et al.*, 1992). Both GA_3 and promalin increased the number of primary flowers rather than the number of secondary ones. It was suggested that promalin and GA_3 induced floral shoot formation at an early stage was more effective in induction of more number of inactive buds to develop into shoots. In California, plants of *Z. elliottiana* and *Z. rehmannii* grown in greenhouse produced rhizomes and flowered only after soaking in 500 ppm GA_3 before planting. Whereas field produced rhizomes flowered without GA_3 treatment, but soaking them in GA_3 increased the number of flowers (spathe and spadix) (Corr and Widmer, 1991).

Nasr and Shalabi (1996) soaked the rhizomes of *Z. aethiopica* to GA_3 (0-1000 ppm) and paclobutrazol (PP_{333}) (0-1 per cent) for 0-30 minutes. Another 216 rhizomes were soaked in 500 ppm GA_3 for 30 minutes of which 108 were then dipped in PP_{333} at 0, 0.5 or 1 per cent. Fifteen days later the remaining 108 were treated with PP_{333} at the same concentrations, as a soil drench (240ml/pot). Increased time and concentration of soaking in GA_3 significantly delayed flowering time, lengthened flowering period, increased peduncle height, number and dry weight of flowers. These characteristics were inversely correlated with PP_{333} concentrations. As a flowering plant, best results were obtained with GA_3 at 500 ppm combined with 0.5 per cent PP_{333} as a soil drench. Reiser *et al.* (1993) in an experiment on *Zantedeschia rehmannii* cv. Pink, *Z. elliottiana* × *Z. maculate* cv. Yellow and *Z. albomaculata* cv. White treated with GA_3 as a preplant rhizome dip to induce flowering. The growth retardants paclobutrazol and uniconazole were applied at 2 and 4 mg *a.i.*/pot as soil drenches. Application of 100ppm GA_3 increased the flower number and reduced the ratio of flower deformities. Both the growth retardants at 2 mg *a.i.*/pot reduced foliage height more than flower height.

Yellow calla lilies (*Z. alliottiana*) were treated with paclobutrazol (bonzi) and uniconazole (sumagic) at 20 ppm a.i. as a 15 minutes and 6 to 10 seconds pre-planting dip respectively (White, 1990). Each retardant reduced stem height at anthesis by about 30 per cent compared with the untreated controls. Gonzalez *et al.* (1999) studied the effect of flurprimidol (0.56-6.75 mg/pot), paclobutrazol (0.20-7.50 mg/pot), daminozide (150-400 mg/pot) and ancymidol (5 or 10 mg/pot) on *Z. aethiopica*. Only some of the plants treated with ancymidol and daminozide flowered. All growth retardants reduced the size of the plant, causing deformations in leaves at high concentrations. Tjia (1989) suggested that when plants are grown under low light intensities, growth regulating chemicals, such as paclobutrazol, uniconazole or ancymidol, should be applied as soil drenches to reduce stretching of foliage and scapes.

Zantedeschia cv. Best Gold tubers were stored for 0, 3 or 6 months at 9°C and 70 per cent RH. Upon removal from storage, the tubers were immersed in promalin (GA_{4+7}) at 0-200 ppm before planting. Time to emergence and flowering decreased as storage duration increased. Increase in promalin concentrations increased the proportion of flowering buds, but higher concentrations were required with the increase in storage duration (Funnell *et al.*, 1993).

Treder (2005) obtained highest yield of 5-6 flowers/tuber in the variety 'Florex Gold' and 'Pink Persuasion'by the application of Gibberellic acid. The cultivars 'Cameo' and 'Black Magic' recorded 3-4 flowers/tuber. The GA_3 application did not affect the flower weight. There has been decrease in fresh weight of tubers lifted after flowering of cultivars with shorter flowers ('Florex Gold' and 'Pink Persuasion') whereas it increased in cultivars with long, big flowers ('Black Magic' and 'Cameo'). Arhips *et al.* (2015) concluded that the treatment with gibberellins (250 ppm, for 30 min.) on *Zantedeschia* 'Picasso' tubers reduced the duration of formation of floriferous stems and also determined the early flowering of the plants. Gibberellic acid negatively influenced floral stem length and average height of the plants. However, it favored the production of flowers/plant. Gibberellic acid did not significantly affect the number of leaves/plant.

7.0 Propagation

Calla lily is conventionally multiplied through rhizomes. It is also propagated through seeds, and by tissue culture (Welsh, 1990). Chen *et al.* (2000) suggested improved calla lily production *in vitro* by using larger plantlets as the source material and growing them in soil beds in a protected house.

7.1.0 Seed Propagation

Z. elliottiana can be easily propagated by seeds. The seeds are formed at the base and are sown in the month of November. Seeds germinate better at 21-27°C.

Hernandez *et al.* (2011) observed that treatment with GA_3 was mainly favorable when applied at a concentration of 100 mg/l. However, the highest values of germination (percentage) and rate of emergence (speed) was obtained by using as substrates organic matter with leaf mold. It was concluded that the growing substrate was important for seed germination of *Z. elliotiana* than $GA_{3.}$

7.2.0 Vegetative Propagation

Zantedeschia is propagated by division of rhizome or from suckers removed at potting time. Welsh *et al.* (1988) planted 3 sizes (1, 2 or 3 cm diameter) of freshly-lifted or dry rhizomes of *Z. aethiopica* cv. Childsiana in pots. Flower and foliage dimensions increased with rhizome size. Plants grown from dry rhizomes produced more foliage than plants from freshly-lifted green rhizomes.

7.3.0 Micropropagation

The objective of tissue culture is to obtain plants free from *Erwinia* spp. Bacteria in *Z. aethiopica*. Surface disinfected rhizome buds were cultured on MS medium supplemented with inositol (100 mg/l), thiamin, HCL (0.4 mg/l), sucrose (30 g/l), BA (3 mg/l) and agar (6 mg/l) at pH 5.8 and transferred to fresh medium every 3 weeks. With sodium hypochlorite (10 per cent) sterilization, 80 per cent of buds failed to grow because of bacterial contamination, whereas 80 per cent of those treated with ethanol (95 per cent) grew effectively in the medium (Ruiz-Sifre, 1996).

Bud explants of zantedeschia (cultivars Golden Calla, Pixie, Pink Persuasion and Black Magic) were cultured on MS medium supplemented with different concentration of zeatin. Shoot multiplication was induced at 1-3 mg/l of zeatin whereas rooting was induced at 0.05-0.2 mg/l (Xiao *et al.*, 1998).

In an investigation on micropropagation on growth and development of plantlets, Cohen (1982) reported regeneration of plantlets from rhizome explants containing apical bud (2-4 mm long) cultured in MS media supplemented with myoinositol, thiamine and BA. The plantlets were planted in soilless potting mix and small rhizome was produced in 6 months. Treatments of these rhizomes with GA_3 at 40 ppm stimulated growth and flowering. Tuber explants from four *Z. aethiopica* cultivars were cultured on MS medium supplemented with 1.0-2.0 mg/l BA and 0.1 mg/l NAA, promoted the induction of adventitious shoots and their growth. BA concentrations <0.2 mg/l improved the growth of adventitious shoots but reduced induction of calluses and adventitious shoots, resulting in low proliferation coefficients (Wu *et al.*, 1999). Pule *et al.* (2009) concluded that sterilization for 12 min was best for annual tuber bud, while 16 min was optimum sterilization time for biennial and three-year-old tuber bud. The best hormone combination was BA 2.0 mg/l + GA_3 0.3 mg/l for MS medium, which could induce a higher proliferation multiple of adventitious buds of 5.3.

Hlophe *et al.* (2015) found that MS medium supplemented with 4 mg/l BA and 1 mg/l IBA produced the highest mean number of axillary shoots (2.8 ± 0.5), and significantly higher shoot length (5.9 ± 0.8 cm) and fresh weight (0.7 ± 0.1 g) after four weeks. A 94 per cent survival rate was recorded when *in vitro* derived shoots were rooted and acclimatized *ex vitro* on a soil mixture containing 1 : 1 (v/v) vermiculite : sand. Rooting occurred spontaneously during acclimatization.

8.0 Soil and Climate

8.1.0 Soil

Zantedeschia prefer loam-based humus rich well-drained moist soil. The best soil mixture is with well-rotten FYM in equal proportions by volume. It prefers the pH range of 6.0-6.5 for optimum growth and yield. Good drainage is vital to avoid root rot and fungal diseases.

Chen *et al.* (1997) studied the effects of different growing media on the growth of zantedeschia cultivars Black Magic and Best Gold. In the first experiment, the rhizomes were grown in media treated with *Bacillus subtilis* strains 001, 004 or 114 alone or together with rice hulls. Shoot emergence after 4 weeks was greatest in Black Magic in the media treated with *B. subtilis* strain 114 and rice hulls. However, in Best Gold shoot emergence was more with *B. subtilis* strain 001 or *B. subtilis* strain 114 + rice hulls. In the second experiment, rhizomes were grown in a medium amended with various combinations of pig manure, dairy cattle manure, mushroom compost and poultry manure. The dairy cattle manure amendment gave markedly poorer results (in terms of rhizome emergence and size) than the other amendments.

8.2.0 Climate

Callas grows best in temperate to subtropical climate. The plants thrive best under good sunlight. Direct sunlight is essential for brighter blooms of coloured cultivars, however in warmer areas they require some shade to prevent early fading of colours. Most of the species need an average temperature of around 13-24°C for satisfactory growth and flowering. For the coloured hybrids, day temperature of 18-21°C and night temperature at 16°C is recommended. White calla lily (*Z. aethiopica*) needs the same day temperature as coloured lily and night temperature of around 13°C for their proper growth. The plants are severely affected by frost and need protection during winter.

9.0 Cultivation

9.1.0 Planting

Zantedeschia rhizomes are usually planted in spring. Rhizomes are lifted during the period from September to October. Nimura and Itoh (1997) suggested ridge planting to obtain quality bloom. The size and storage of rhizome at the time of planting markedly influence the growth and flowering of calla lily. Corr and Widmer (1991) reported that rhizomes with diameter 6.5 cm in *Z. rehmannii* produced more shoots, leaves and flowers than smaller sized rhizomes. The performance of rhizomes are better when rested during summer time. Some of the species having yellow or pink-coloured spathes perform best when grown without rest period. In beds, calla lilies are planted at 30-45 cm apart. In a 25 cm pot, two to three rhizomes can be easily accommodated without any adverse effect on growth and flowering.

Chen *et al.* (2000) reported that calla lily (*Z. elliottiana* cv. Super Gold) plants grown in soil beds, regardless of size, produced larger tubers than did those grown in pots. Tubers size >3 cm in diameter was developed on 25 per cent and 52 per cent of plants grown in pots and soil beds, respectively.

Four sterilants-bactericides (physan-20, fixed copper, phyton-27 [copper sulphate], and Virkon) were compared as pre-planting dips in rhizomes of *Z. elliottiana* (*cv.* Yellow, a susceptible cultivar) to reduce plant losses due to latent field-infected *Erwinia carotovora* soft rot during greenhouse forcing (Blom and Brown, 1999). All sterilant solutions were prepared in combination with promalin (GA) to enhance flowering. An additional group of rhizomes was inoculated with *E. carotovora* as a preplanting dip in combination with GA treatment but were not treated with a bactericide. More than 90 per cent of the inoculated rhizomes collapsed within 5 weeks after potting due to bacterial soft rot. With the uninoculated rhizomes, the copper-based compounds (fixed copper or phyton-27) provided better control of bacterial soft rot than either Physan-20 or Virkon only during the first 6 weeks of forcing. During the remainder of the forcing period, there were no differences in weekly losses of rhizomes with the 4 sterilants. Rhizomes treated with solutions of the copper-based compounds produced less flowers (0.5) per rhizome than either Physan-20 or Virkon. Pre-treatment of rhizome (*Z. aethipica*) with captan and mancozeb and antibiotics AEM1 and imiponem eleminated both external and internal fungi and bacteria (Kritzinger *et al.*, 1998).

9.2.0 Irrigation

The plant requires ample of water during the period of active growth. Experiments to improve the quality and yield of the cultivar, and to avoid root rot caused by *Phytophthora richardiae*, were conducted by Nimura and Itoh (1997) on *Z. aethiopica* cv. Weeding March and Childsiana by growing on high ridges and using subsurface irrigation (for 40 minutes in summer or 15 minutes in winter every morning). Cut flower yields of Wedding March and Childsiana were 1.2-1.5 and 3 times than those obtained with the conventional flow irrigation system, and mean cut flower lengths were 15 and 50 cm longer, respectively.

Salim *et al.* (2019) suggested that the electrical conductivity of the salt solution, which reflects the solution osmotic potential, plays a significant role in preventing the splitting phenomenon, rather than the specific effect of Ca^{2+} ions, which prevented fruit tissue cracking. Thus, in order to prevent the stem-base splitting in cut *Zantedeschia* flowers, it is recommended to pulse the stems for 24 h with any salt solution at a concentration that yields an electrical conductivity of at least 5600 µS cm^{-1}. The salt solutions should be prepared in tap water in the presence of chlorine (50 µg/l), to avoid pathogen development during pulsing. The results suggest a simple, very effective, and low cost treatment to prevent the stem-base splitting also in other geophyte cut flowers that suffer from this problem.

9.3.0 Manuring and Fertilization

Calla lily shows good response with little manuring and fertilization. Addition of well rotten cowdung at planting and liquid manuring during the vegetative growth is useful to the plants. Clemens *et al.* (1998) studied the effect of fertilizers applied to *Z. albomaculata* cv. Starlights plants on subsequent flowering of harvested, stored and replanted tubers during a second growth period. Shoot development was most advanced and time to flowering was shortened at low N. Most flowers and highest cut flower yield were obtained at low N. Plant survival during the initial growth period was highest at low N and P, and at intermediate levels of K. Advanced shoot development, early flowering and highest cut flower yield were obtained at low N. The fresh weight of harvested tubers was highest at intermediate levels of N.

Cho *et al.* (2013) observed higher concentration of calcium and silicate in the foliar application and basal application resulted in vigorous leaf growth, hardness of petiole and flower stalk and in fresh weight. The resistance to *Erwinia carotovora* subsp. *carotovora* was more effective in $CaCl_2$ foliar treatment than the basal fertilization treatment. Foliar treatment at 0.2 per cent $CaCl_2$ was the most effective against the resistance of *Erwinia carotovara* subsp. *carotovora*. Also, the level of resistant to *Erwinia carotovora* subsp. *carotovora* was higher in proportion to the silicate application rate both in foliar and basal application of fertilizer. Almeida *et al.* (2012) evaluated the use of different levels of cattle manure combined with mineral fertilizer on the growth and production of inflorescences of calla lily. The best result was obtained with 350 kg/1000m^2 of 10:10:10 NPK combined with 40 l/m^2 of cattle manure. This treatment provided greater number of inflorescences on plants of calla lily.

Carneiro *et al.* (2015) evaluated macronutrient accumulation in *Zantedeschia aethiopica*. Following acclimatization, micropropagated shoots were cultivated in pots with coconut fiber as a substrate and were fertilized with a modified version of Malavolta solution, as described. For each tested plot, evaluations were performed every 30 days for 12 months. They evaluated the dry mass of each plant part (leaves, rhizomes, roots and flower stalks) and then calculated the dry mass accumulation and the nutrient contents. The aerial parts of the plants showed macronutrient content and accumulation in the sequences K^+>N>Ca^{2+}>S>P>Mg and K>N>P>Ca>S>Mg, respectively, with the highest values being observed between 265 and 328 days after planting. The aerial parts showed greater accumulation of dry mass and macronutrients, whereas the flowers did not display significant macronutrient accumulation. The highest rates of macronutrient accumulation and growth occurred during the 210 days after transplanting, which corresponds to the pre-blooming period, indicating that fertilizers must be divided and applied at their highest levels prior to blooming.

9.4.0 Weed Control

Zantedeschia requires weeding at its early growth phase. Weed population suppressed the growth and flower yield.

Mackay *et al.* (1994) reported that weed competition reduced tuber yields of *Z. albo-maculata* cv. Chromatella by up to 50 per cent. The highest yields of tubers were obtained from post-emergence application of terbutryn (0.5 kg/ha), methazole (1.1 kg/ha), bentazone (1.0 kg/ha), methabenzthiazuron (1.6 kg/ha) and a phenmedipham + desmedipham (0.66 + 0.66 kg/ha) mixture gave high yields of tubers and also suppressed weed growth. Pre-emergence herbicide oryzalin (3.5 kg/ha), terbuthylazine (1.0 kg/ha), simazine (1.5 kg/ha) and propazine (1.0 kg/ha) were also useful.

In field trials, Ingle and Bussell (1991) applied 10 pre- and 14 post-emergence herbicides on calla lilies (*Z. elliottiana*). The lilies tolerated pre-emergence caragard (terbumeton + terbuthylazine) at 3 kg/ha, 2 kg simazine or diuron, 6 kg metribuzin, 1 kg atrazine and 1.5 kg oxadiazon. The lilies also tolerated post-emergence methazole and cyanazine at 1.6 kg, 2 kg methabenz thiazuron, 0.6 kg metribuzin, 0.3 kg (but not 0.6 kg) dicamba and 1.5 kg (but not 3 kg) oxadiazon.

9.5.0 Greenhouse Cultivation

Zantedeschia can be grown in greenhouse condition. A greenhouse temperature of 15-16°C is maintained after planting and soil temperature is kept at 15-18°C. Later during cultivation, once the leaves have unfurled, the greenhouse temperature is usually kept at 18 to 20°C. Depending on the amount of light, the temperature may be allowed to rise to a maximum of 25°C. More light benefits flower production and stem quality, so try to make as much light as possible available to the crop during the spring and autumn. Shading with chalk or shade cloths is usually done during the summer to keep both the greenhouse and soil temperatures sufficiently low. When the crop comes into flower during the spring and autumn, the greenhouse temperature can be lowered to 15 to 10°C at night. This slows crop development but intensifies flower colour. The best possible relative humidity during the day is 60 to 75 per cent. But it should not drop below 50-60 per cent. Neither the plants nor the tubers tolerate any frost.

10.0 Diseases and Pests

Species of *Zantedeschia* are susceptible to a number of diseases.

10.1.0 Diseases

Fungal Diseases

10.1.1 Leaf Spot

Leaf spot first cause small spots which later become large and irregularly decayed areas. Maxwell and Wykes (1997) isolated *Phoma* spp. from leaf lesions and *Fusarium, Pythium, Rhizoctonia* and *Alternaria*, and the commonest species *F. oxysporum* and *F. solani*. *P. coloratum* from roots and rhizomes of *Z. aethiopica*. The infected leaves should be removed and fungicidal sprays, though rarely done, are effective.

Basavand *et al.* (2019) for the first time reported bacterial leaf spot disease on calla lily caused by *Pseudomonas viridiflava*. During the early stages, spots were small, circular and brown, then water soaked spots appeared on the leaves, enlarged and coalesced to cover a large leaf.

10.1.2 Crown Rot

Symptoms of crown rot caused by *Pellicularia filamentosa* and *Sclerotium delphinii* are soft decay of the lower half of corm and feeding roots. Use of disease free corm for planting and soil sterilization has been suggested as preventive measures.

10.1.3 Powdery Mildew

Powdery mildew disease on *Z. elliotriana* was caused by *Leveillula taurica* (Chen *et al.*, 1996). This disease is noticed on all the aerial parts of the plant in late season. Spraying with wettable sulphur (0.2 per cent) or karathane (0.025 per cent) or bavistin (0.1 per cent) control the disease.

10.1.4 Root Rot

Root rot of *Z. elliotiana* is caused by *Pythium myriotylum* (Long, 1988). Infection by *Phytophthora richardiae* causes root rot disease characterized by yellowing of leaves, finally wilting, and turning brown. The flowers of the affected plants become malformed and turn brown.

Hirano *et al.* (1996) eradicated root rot (*Phytophthora richardiae*) of calla *Z. aethiopica* from infested soil by solarization. The soil was kept at 40°C to 5 cm and 15 cm depth for more than 300 hours and at 39°C to 30 cm depth for more than 200 hours during soil solarization.

Treating the dormant rhizome in hot water, formaldehyde or mercuric chloride is effective for control of root rot.

The efficacy of 11 fungicides for protection against fungal infections and/or controlling newly established infections of *Acremonium* sp. and *Alternaria* sp. on calla lily flowers was studied. Cyprodinil, EZ752, fluazinam, mancozeb, prochlorez and terbuconazole was found effective against *Alternaria* sp. infection. Chlorothalonil, difenoconazole, mancozeb, prochloraz and terbuconazole controlled new infections caused by *Alternaria* sp. (Braithwaite *et al.*, 1996).

Bacterial Diseases

10.1.5 Soft Rot

Bacterial soft rot caused by *Pectobacterium carotovorum* var. *aroideae* infects the base of the plant at soil level, corms are affected and finally the plant wilts. Bacterial soft-rot of calla is also caused by *Erwinia carotovora* subsp. *carotovora* (Cortes-Monllor, 1990; Horita, 1994; Wright, 1998). Maxwell and Wykes (1997) also identified five *Pseudomonas* and four *Erwinia* spp. for infected roots of *Z. aethiopica*. Mori *et al.* (1999) studied the relationship between growing temperature and the outbreak of bacterial soft rot, caused by *Erwinia carotovora* subsp. *carotovora*, in 8 cultivars of *Zantedeschia* in growth cabinet and reported that the plants grown at 25/20°C showed a lower rate of soft rot infection, although variation was observed among cultivars. The incidence and severity of soft rot, flower grades, and tuber yields of *Zantedeschia* plants were affected by the quantity of water received during the growing season, saw dust mulch, and enhance biocide applications to tubers before planting. Wright and Burge (2000) reported that irrigated plus saw dust mulched callas had 15 per cent less soft rot (*E. carotovora*) than the irrigation without mulch or the mulch without irrigation treatments. Dipping tubers in enhanced biocide before planting slightly reduce the severity of rotting in harvested tubers from non-irrigated plants. They also obtained 90 per cent more tuber yield from irrigated plants. The number of long stemmed flowers was increased 41 per cent by irrigation, to over 1.5 per tuber.

Wright *et al.* (2005) determined the effects of chemical drench treatments on the incidence and severity of soft rot. It was found that more calla plants died when grown in soil than in sawdust for each drench treatment. For both growing media, drenching plants at 28-day intervals resulted in the lowest numbers of dead plants during crop growth and rots in lifted tubers. They also observed that dipping tubers before planting in an aqueous solution of 0.08 per cent copper hydroxide, 0.12 per cent thiram, plus 0.03 per cent benomyl reduced plant and tuber infections from bacterial soft rot. Drenching plants during the growing season with an aqueous solution of 0.12 per cent thiram, 0.08 per cent copper hydroxide, plus 0.15 per cent quintozene gave further control against rots. Weekly removal of rotting foliage and tubers during the growing season also reduced levels of plant and tuber soft rots. This study has demonstrated that effective control of calla soft rot can be achieved using disease management that combines appropriate pathogen exclusion, elimination, and removal methods.

Wright *et al.* (2011) demonstrated that calla soft rot caused by *Erwinia carotovora* subsp. *carotovora* (Ecc) was favoured by wounding of plant tissue, and infection was greatly enhanced under conditions that impaired host resistance (anerobic

Diseases and Pests of Zantedeschia

Gray Mould

Soft Rot

Crown Rot

Aphids

Catterpillar

Mealy Bugs

Harvesting and Postharvest Management of Zantedeschia

Harvesting of Flowers

Harvesting Stage in Greenhouse

Display of Flowers in Outdoor

Display in Greenhouse

Harvesting and Postharvest Management of Zantedeschia

Harvesting of Bulbs

Flowers in International Markets

Cut Flowers in Markets

Packing in Polythene

Packing in Cardboard Boxes

Grading of Flowers

conditions caused by water-logging) and favoured bacterial development (28°C that is optimum for bacterial multiplication).

Viral Diseases

Spotted wilt virus (*Lethum australiense*) affects several species of *Zantedeschia*. Chlorotic to pale streaks and lesions between veins on the leaves are the symptoms of infection. Leaves may be wrinkled, distorted and the margins curled.

Bellardi *et al.* (1992) and Mertelik and Mokra (1999, 1998) reported the occurrence of tomato spotted wilt tospovirus in *Zantedeschia aethiopica* leaves showing chlorotic degeneration, malformation, mosaic ring spots and chlorotic lines. The same plants also had severe flower abnormalities. Mertelik *et al.* (1996) observed that tomato spotted wilt virus transmitted in *Zantedeschia* sp. by *Thrips tabaci*. Denkova *et al.* (1995a) reported tomato spotted wilt tospovirus on the basis of host range, serological tests by DAS-ELISA and ISEM experiments in *Z. aethiopica* and *Z. elliolata* in Bulgaria.

Chen *et al.* (2005) studied that calla lily chlorotic spot virus (CCSV) was weakly related to Watermelon silver mottle virus (WSMoV) in enzyme-linked immunosorbent assay (ELISA) and western blot tests. Liang *et al.* (1994) identified the dasheen mosaic potyvirus by immunolabelling in *Z. elliottiana* and *Z. aethiopica*. Two new potyviruses in zantedeschia were also reported by Denkova *et al.* (1995b).

10.2.0 Pests

Aphids, caterpilars, mealy bugs, bulb mites, and thrips are among the pests reported on calla lily.

Maxwell and Wykes (1997) identified the nematode *Meloidogyne javanica* as the causative agent of galled roots in *Z. aethiopica*.

Carvalho *et al.* (2018) reported that the bagging of calla lily inflorescences was efficient in the control of stingless bee (*Trigona spinipes*), regardless of packaging used, because under these conditions, no inflorescence presented damage. In control, 84 per cent of damaged inflorescences were observed. Differences in postharvest characteristics were observed and inflorescences remained for longer periods in the process of spathe opening, which is characterized by the measurement of their length and width, when packed.

11.0 Harvesting and Postharvest Management

11.1.0 Harvesting

Zantedeschia flowers are harvested with long stems, make them especially suitable for floral arrangement. Fully opened flowers, but before pollen shed, considered optimum stage for harvesting. The flowers remain fresh for few days, if the harvested blooms reach maturity.

Sales *et al.* (2015) showed that calla lily harvested at closed spathe and semi-closed spathe (1/3 open), showed spathe opening, although it did not achieve fully spathe expansion, had higher water uptake and hydration of flower stems, and increased water retention capacity by floral tissues until saturation, followed by a period of weight reduction caused by transpiration rates greater than absorption.

11.2.0 Postharvest Management

Plummer *et al.* (1990) studied the stages of flower development and postproduction longevity of potted *Z. aethiopica* cv. Childsiana. Flowers at macrobud stage opened in the postproduction environment. Plants with flowers at the macrobud stage (Stage I) and plants with spathes fully opened but with unshed pollen (Stage 5) had shelf lives of 26 and 11 days, respectively. Cut leaves of *Z. aethiopica* treated with aqueous solutions of GA_3 considerably extended their display life, whether applied as a 24 h pulse treatment, or as a brief postharvest dip (Skutnik *et al.*, 2001).

In contrast, a standard preservative solution (8-HQS + sucrose) was deleterious to zantedeschia foliage, reducing display life several fold from that of the water control. In an investigation on the postharvest life of cut inflorescence of *Z. Aethiopica* and *Z. elliotiana,*Tija and Funnel (1986) found that the inflorescence of the former species lasted for 6-7 days while that of the latter for 7 days. Spathes of *Z. elliotiana* turned green and those of *Z. aethiopica* turned brown on the edges. They also reported that holding scapes in 100 ppm 8-HQC or 100 ppm 8-HQC + 2 per cent sucrose did not improve spathe longevity.

Janowska *et al.* (2003) observed the long lasting decorative properties in *Zantedeschia Elliottiana* leaves conditioned in a solution of gibberellic acid at the concentration of 300 mg/l and then placed into water. The gibberellic acid limited the degradation of chlorophyll. Further, Janowska *et al.* (2004) found that gibberellic acid and 8-hydroxyquinoline citrate had different effects on the quality of cut flowers of *Z. elliottiana* cultivars: 'Black Magic' and 'Florex Gold'. Gibberellic acid at concentrations of 50 and 100 mg/l caused an elongation of peduncles of unconditioned inflorescences in 'Black Magic'. A similar response was observed in unconditioned flowers put in water. In cultivar 'Florex Gold', both concentrations of gibberellic acid caused a lengthening of flower peduncles irrespective of the duration of conditioning. The conditioning of 'Black Magic' flowers in 8-hydroxyquinoline citrate solutions was detrimental to their quality irrespective of the duration of the treatment.

Castro *et al.* (2014) found that the floral stems harvested with a closed spathe and semi-opened (1/3) continue the process of opening of the spathe, but do not reach the completed opened, however have greater durability, regardless of the storage condition.Sales *et al.* (2015) showed that Calla lily harvested at closed spathe and semi-closed spathe (1/3 open), showed spathe opening, although it did not achieve fully spathe expansion, had higher water uptake and hydration of flower stems, and increased water retention capacity by floral tissues until saturation, followed by a period of weight reduction caused by transpiration rates greater than absorption.

12.0 Biotechnology

12.1.0 Anther Culture

Anthers of *Z. aethiopica* were cultured on MS medium supplemented with various combinations of auxins and cytokinins. Plantlets regenerated through organogenesis were observed on medium containing 1.0 mg 2,4-D and 3.0 mg kinetin/l, and those regenerated through embryogenesis were observed on medium containing 0.5 mg NAA + 1.0 mg BA/l or 2.0 mg NAA + 2.0 mg BA/l. Low temperature pretreatment for 6 days at 5°C was more effective in inducing callus, embryoids and growth of regenerated plantlets than treatment at 10°C and the control. Root tip squashes of the regenerated plantlet showed the haploid chromosome number (Ko *et al.*, 1996).

Zhang *et al.* (2011) described the advances in the anther culture of *Z. aethiopica*. Frequency of anthers producing calli was around 4-5 per cent. About 87 per cent of the calli gave regenerants, of which 52 per cent were haploid, 36 per cent were diploid and the rest had other ploidy levels. In addition to chromosome counting, cytological examination of the microspore development and amplified fragment length polymorphism (AFLP) analysis of the regenerants showed that haploid as well as diploid plants originated from the microspores. Finally, 12 doubled haploid (DH) plants could be produced from each inflorescence.

12.1.1 Embryogenic Callus Culture

Han *et al.* (2019) conducted experiment involving the regeneration from embryogenic calluses through shoot formation. In this study, the combination of auxin and cytokinin showed positive effects on both the induction of embryogenic calluses, the formation of shoots as well as multiple shoots in calla. The regeneration system reported here can contribute to the development of breeding programs of calla in the future.

12.2.0 Molecular Markers

The use of molecular markers has become a very powerful tool for the management and utilization of crop genetic diversity in germplasm collections. For *Zantedeschia* spp., some marker types such as inter-simple sequence repeat (ISSR), amplified fragment length polymorphism (AFLP) and random amplified polymorphic DNA (RAPD) have been widely applied in establishing a DNA fingerprint for cultivar identification (Hamada and Hagimori, 1996; Lu *et al.*, 2012), quantifying levels of genetic variation and relatedness (Zhang *et al.*, 2009; Chen *et al.*, 2013; Zheng *et al.*, 2015), and describing gene flow and parentage in germplasm accessions (Li, 2011). Although these molecular markers have proven useful, one major weakness is that they are dominant markers that cannot distinguish heterozygous genotypes from homozygous genotypes. Moreover, these marker systems are labor-intensive, with unstable reproducibility. Simple sequence repeat (SSR) markers possess several advantages over the other molecular markers, including co-dominance, high polymorphism, and good reproducibility. The development of SSR markers can be based on short-length sequences from expressed sequence tags (EST) available at the National Center for Biotechnology Information (NCBI, https://www.ncbi.nlm.nih.gov/) and is feasible for species for which no prior sequence information exists, including understudied but economically important crops (Zalapa *et al.*, 2012). Fortunately, 209 SSRs loci in 182 assembled EST-unigenes of white calla lily have been identified, 166 of which had flanking sequences suitable for primer design (Wei *et al.*, 2012).

13.0 References

Almeida, E. F. A., Paiva, P. D. O. Frazão, J. E.M. Santos, F. H. S. Resende, F.A. and Campos, M.L. (2012) *Ornamental Horticulture.*, **18** : 129-134.

Arhip, C. I. and Draghia, L. (2015) *Bulletin UASVM Horticulture.*, **72**/1.

Armitage, A.M. (1991) *HortScience*, **26** : 1174-1176.

Bailey, L. H. (1963) *The Standard Cyclopedia of Horticulture*,Macmillan and Co., New York.

Basavand, E., Khodaygan, P. and Rahimian, H. (2019) *Journal of Plant Pathology.*, **101**: 393

Bellardi, M.G., Bertaccini, A. and Benni, A. (1992) *Informatore Fitopatologico*, **42** : 60-62.

Blom, T.J. and Brown, W. (1999) *HortTechnology*, **9** : 56-59.

Braithwaite, M., Knight, K.W.L., Saville, D.J., Alexander, B.J.R. and O'-Callaghan, M. (1996) *Proc. of the Forty Nine New Zealand Plant Protection Conference*,Quality Hotel Rutherford, Nelson, New Zealand, 13-15 August, 1996, pp. 157-160.

Brickell, C. (1996) *A-Z Encyclopedia of Garden Plants*, Dorling Kindersely, pp. 1067-1068.

Carneiro, Daniella N.M., Coelho, Livia L., Paiva, Patricia D.O., Almeida, Elka F.A. and Carneiro, Leandro F. (2015) *Australian Journal of Crop Science*, **9**: 761-766.

Carvalho, L. M., Ladeira, V. A., Almeida, E. F.A., Cecília, L.V.C.S., Brighenti, D.M. and Resende, E. (2018) *Ornamental Horticulture.*, **24**: 353-360.

Castro, M.L, Paiva, P.D. Landgraf, P.R. Pereirra, M. A. and Souza, R.R. (2014) *Ornamental Horticulture.*, **20** : 131-136.

Chen, C.C., Chen, T.C. Lin, Y. H. Yeh, S. D. and Hsu, H.T.(2005) *Plant Disease.*, **89** : 440-445.

Chen, C.W. and Lin, V.Y. (2000) *Plant Pathol. Bull.*, **9**: 107-114.

Chen, C.W., Tsai, Y.F. and Huang, S.C. (1997) *Bull. Taichung District Agricultural Improvement Station*,No. 56, pp. 41-50.

Chen, J.J., Liu, M.C. and Ho, Y.H. (2000) *HortScience*, **35** : 290-292.

Chen, Y.K., Chen, C.C. and Chen, L.C. (1996) *Ann. Phytopathologicval Soc. Japan*, **62** : 580-583.

Chen, Z., Xu, B.L., Pu, C.J., Xu, Q. and Tian, G. (2013) *J. Nucl. Agric. Sci.*, **27**: 552–556.

Cho, H.R., Joung, H.Y. Lim, K.B. and Kim, S.K. (2013) *Horticulture, Environment and Biotechnology.*, **54** : 364-371.

Clark, C.J. and Boldingh, H.L. (1991) *Scientia Hort.*, **47** : 125-135.

Clemens, J., Dennis, D.J., Butler, R.C., Thomas, M.B., Ingle, A. and Welsh, T.E. (1998) *J. Hort. Sci. Biotech.*, **73** : 755-762.

Clemens, J., Welsh, T.E. and Lawson, R.H. (1993) *Acta Horticulturae*, **337** : 161-166.

Cohen, D. (1982) *Comb. Proc. Intern. Pl. Prop. Soc.*, **31**: 312-6.

Cohen, D. and Yao, J.L. (1996) *Plant Cell Tissue Organ Cult.*, **47** : 43-49.

Corr, B.E. and Widmer, R.E. (1990) *HortScience*, **25** : 925-927.

Corr, B.E. and Widmer, R.E. (1991) *HortScience*, **26** : 133-135.

Corte, L. Manuel, J., Lopez, D. and Ernesto (2017) *Magallania.*, **66**, DOI : 10.15446/acag.v66n3.54539

Cortes-Monllor, A. (1990) *J. Agri. Univ. Puerto Rico*,**74**: 83-92.

Cruz-Castillo,J.G. and Cardenas-Alonso, M. (1998) *Proc. Interamerican Soc. Trop. Hort.*, **41** : 84-87.

Dabral, M., Punetha, P and Bohra, M. (2019) *International Journal of Chemical Studies.*, **7**(3): 269-274.

Denkova, S., Bakardzhivera, N. and Hristova, D. (1995a) *RastenievdniNauki*, **32** : 7-8.

Denkova, S., Derks, A.F.L.M. and Bakardzhieva, N. (1995b) *EastenievdniNauki*, **32** : 129-131.

Dennis, D.J., Doreen, J. and Ohteki, T. (1994) *Scientia Hort.*, **57**: 133-142.

Devecchi, M. (2003) *Acta Horticulturae.*, **614**, DOI : 10.17660/ActaHortic.2003.614.80

Fonseca, A. S. and Segeren, M. I. (2013) *Acta Horticulturae.*,10.17660/ActaHortic.2013.1000.52

Funnell, K.A., Go, A.R. and Lawson, R.H. (1993) *Acta Horticulturae*, **337** : 167-175.

Funnell, K.A., Hewett, E.W., Warrington,I.J. and Plummer, J.A. (1998) *J. Amer. Soc. Hort. Sci.*, **213** : 973-979.

Funnell, K.A., MacKay, B.R., Lawoko, C.R.O. and Chong, K.K. (1992) *Acta Horticulturae*, **292** : 173-179.

Gonzalez, A., Lozano, M., Casas, J.L., Banon, S., Fernandez, J.A., Franco, J.A. and Caballero, R. M. (1999) *Acta Horticulturae*, **486** : 333-337.

Goto, T., Kawajiri, K., Kageyama, Y. and Konishi, K (2003) *Acta Horticulturae*., DOI: 10.17660/ActaHortic.2005.673.33

Hamada, K. and Hagimori, M. (1996) *Scientia Hort.*, **65** : 215-218.

Hans., Song, I. and Kim, Jong, B. (2019) *Journal of Plant Biotechnology.*, **46** : 32-36.

Hernández, J. Jesús, M.D. Solís, M. Juan, Rodríguez, C. Arturo, Santos, G. and Alejandro (2011) *Revista Mexicana de Ciencias Agrícolas.*, **2** : 495-507.

Hirano, T., Nakagome, T., Takimoto, M., Ohsawa, U. and Kinbara, Y. (1996) *Res. Bull. Aichi ken Agri. Res. Centre*,No. 28, pp. 241-146.

Horita, H. (1994) *Annual Report Soc. Plant Protection*,North Japan, **45** : 100-103.

Ingle, A. and Bussell, W.T. (1991) *Proc. forty-fourth New Zealand Weed and Pest Control Conference*, pp. 138-142.

Lee, J.J, Jin, S.Y., Cheong, D.C., Kim, J.M. and Guak, S.H. (2019) *Acta Horticulturae.*, DOI: 10.17660/ActaHortic.2019.1255.19

Janowska, B. and Jerzy, M. (2004) *Electronic Journal of Polish Agricultural Universities.*, **7**.

Janowska, B. and Jerzy, M. (2003) *Journal of Fruit and Ornamental Plant Research.*, **XI**, 69-76.

Janowska, B. (2010) *Science Nature Technologies.*, **4**.

Joung, Young, H. Choi, MokPil, Han, Sook, K. Kim, Su, Goo, Hoe, D. Kang, YunIm, Choi, Jung, Y. Park, Kun, S. (2013) *Korean Journal of Horticultural Science and Technology.*, **31** : 366-370.

Joung, Young, H. Cho, Ryoung, H. Rhee, Hee, J. Shin, Ki, H. Park, and Kun, S. (2015) *Korean Society of Horticultural Science.*, **33** : 618-623.

Kapiteijn, A. S. N. (2015) United States Plant Patent., US 2015/0334909 P1

Klaver, T.C. M. (2010) United States Plant Patent., US PP21,134 P2.

Ko, J.A., Kim, Y.S. and Eun, J.S. (1996) *J. Korean Soc. Hort. Sci.*, **37** : 468-474.

Kritzinger, E.M., Vuuren, R.J. van, Woodward, B., Rong, I.H., Spreeth, M.H., Slabbert, M.M., Van-Vuuren, R.J. and Cassells, A.C. (1998) *Plant Cell Tissue Organ Cult.*, **52** : 61-65.

Kuehny, J.S., Holcomb, G.E., Chang, W.C. and Branch, P.C. (1998) *HortTechnology*, **8** : 353-356.

Lewis, D. H., Arathoon, H. S. Swinny, E. E. Huang, S. C. and Funnell, K. A. (2003) *Acta Horticulturae.*, 10.17660/ActaHortic.2003.624.19

Liang, Y.G., Yang, T.C. and Chen, C.T. (1994) *Report Taiwan Sugar Res. Institute*,No. 143, pp. 41-50.

Long, P.G. (1988) *New Zealand J. Exp. Agri.*,**16** : 165-166.

Lu, B., Zheng, Y.H., Peng, F., Shu, X.C. and Chen, X.X. (2012) *North. Hortic.*, **11**: 123–126.

Mackay, T.J., Harrington, K.C. and Popay, A.J. (1994) *Proc. forty seventh New Zealand plant protection confer.*,Waitangi, New Zealand, 9-11 August, 1994, pp. 163-167.

Manandhar, S. Woolley, D.J. and Funnell, K.A. (2015) *Acta Horticulturae.*, DOI : 10.17660/ActaHortic.2015.1104.25

Maroyi, A. (2019) *Asian Journal of Pharmaceutical and Clinical Research*, **12**(8): 30-36.

Masayoshi, N. Masaji, K. Tadao, K. and Kiyotoshi, I. (2015) *Natural Product Communications.*, 10. DOI : 10.1177/1934578X15010003.

Maxwell, A. and Wykes, B.J. (1997) *Proc. workishop on arum lily held at HMAS Stirling, Garden Island*, Western Australia, 7 August, 1997, pp. 40.

Mertelik, J. and Mokra, V. (1998) *Acta Virologica*, **42** : 347-351.

Mertelik, J., Gotzova, E., Mokra, V. and Baayen, A. van (1996) *Acta Horticulturae*,**432** : 368-375.

Moore, J.H. and Wykes, B.J. (1997)*Proc. workishop on arum lily held at HMAS Stirling, Garden Island,* Western Australia, 7 August, 1997, pp. 4-6.

Mori, G., Kubo, T. and Yamaguchi, T. (1999) *Environ. Control Biol.*,**37**: 93-96.

Naor, V. Kigel, J. and Ziv, M. (2004) *Plant Growth Regulation.*, **42** : 7-14.

Naor, V. Kigel, J. Ziv. M. and Flasihman, M. (2004) *Journal of Plant Growth Regulation.*, **23**: 269-279.

Nasr, M.N. and Shalabi, H.G. (1996) *Alexandria J. Agric. Res.*, **41** : 247-262.

Ngamau, K. (2008) *Acta Hortic.*, **766**: 155-162.

Joung, Young, H. Cho, Ryong, H. Rhee, Hee, J. Shin, Ki, H. Park. and Kun, S. (2015) *Korean Journal of Horticultural Science and Technology.*, **33** : 618-623.

Nimura, M. and Itoh, K. (1997) *Res. Bull. Aichi ken Agricultural Res. Centre*,No. 29, pp. 177-183.

Perry, P.L. (1989) *South African J. Bot.*, **55** : 447-451.

Petrus M. M. Hoof (2003) United States Plant Patent., US PP13,837 P2

Plumer, J.A. and Wykes, B.J. (1997) *Proc. workishop on arum lily held at HMAS Stirling, Garden Island*, Western Australia, 7 August, 1997, pp. 7-9.

Plummer, J.A., Welsh, T.E. and Armitage, A.M. (1990) *HortScience*, **25**: 675-676.

Posada, C. Fanor, Nieto, Paola, J. and Christiam, U. (2012) *Revista U.D.C.A Actualidad and DivulgaciónCientífica.*, **15** : 97-105.

Pule, H. Chongbo, S. and Ming, X. (2009) *Acta Agriculturae Zhejiangensis.*, **21** : 234-237.

Reiser, R.A., Langhans, R.W. and Lawson, R.H. (1993) *Acta Horticulturae*, **337** : 87-94.

Ribeiro., Oliveira, M. D. N. Pasqual, Moacir, Silva, Bortolotti, A. Rodrigues, and Antônio, V. (2009) *Ciência Rural.*, **39**: 2388-2393.

Ruiz-Sifre, G., Rosa-Marquez, E. and Flores-Ortega, C.E. (1996) *J. Agric. Univ. Puerto Rico*, **80**: 193-194.

Sales, T. S. De, P.D. Paiva, O. and Mariano, G (2015) *Ornamental Horticulture*, Vol.21 No.3, PP368-375.

Salim, S. Philosophhadas, S. and Meir, S. (2019) *Acta Horticulturae.*, DOI : 10.17660/ActaHortic.2019.1263.49

Schram, F. (2009) United States Plant Patent., US PP19,830 P2

Singh, Y., Baijnath, H. and Van Wyk, A.E. (1996) *South African J. Bot.*, **62** : 146-150.

Sivakumar, R. and Nair, A.G.R. (1992) *Fitoterapia*, **63** : 189.

Skutnik, E., Lukaszewska, A., Serek, M. and Rabiza, J. (2001) *Postharvest Biol. Technol.*, **21** : 241-246.

Snijder, R. C. and Tuyl, J. M. V. (2004) *Acta Horticulturae*, DOI : 10.17660/ActaHortic.2004.651.5

Subbarai, A. K. Funnell, K. A. and Woolley, D. J. (2010) *Journal of Plant growth Regulation.*, **29**: 487-449.

Tija, B. O. and Funnell, K. A. (1986) *Acta Horticulturae,* **181** : 451-458.

Tjia, B.O. (1989) *Acta Horticulturae*, **252**: 187-191.

Treder, J. (2005) *Acta Horticulturae*, DOI : 10.17660/ActaHortic.2005.673.94.

Van-Tuyl, J.M. and Halevy, A.H. (1997) *Acta Horticulturae*, **430** : 465-476.

Wei, Z.Z., Luo, L.B., Zhang, H.L., Xiong, M., Wang, X. and Zhou, D. (2012) *Amer. J. Bot.*, **99**: e493–e497.

Wei, Z. Z. Sun, Z. Z. Xiong, X. Wang, X. and Zhou, D. (2017b) *Acta Hortic.*, **1171**: 175-186.

Wei, Zunzheng, Zhang, Huali, Wang, Yi, Li, Youli, Xiong, Min, Wang, Xian and Zhou, Di (2017b) *Genes*, **8**: 168; doi: 10.3390/genes8060168.

Welsh, T.E. (1990) *Combined Proc. Inter. Plant Propagators Soc.*, **40** : 478-484.

Welsh, T.E., Clemens, J. and Goto, E. (1992) *Acta Hort.*, **319** : 335-340.

Welsh, T.E., Plummer, J.A. and Armitage, A.M. (1988) *Combined Proc. Inter. Plant Propagators Soc.*, **38** : 384-388.

White, J.W. (1990) *Bull. Pennsylvania Flower Growers*, No. 402, pp. 1-4.

Wright, P. J., Clark, G. E. and Koolaard, J. (2005) *Acta Horticulturae.*, DOI : 10.17660/ActaHortic.2005.673.109

Wright, P. J., Triggs, C.M. and Burge, G. K. (2005) *New Zealand Journal of Crop and Horticultural Science.*,**33**. DOI : 10.1080/01140671.2005.9514340

Wright, P.J. (1998) *New Zealand J. Crop Hort. Sci.*, **26**: 331-334.

Wright, P.J. and Burge, G.K. (2000) *New Zealand J. Crop Hort. Sci.*, **28** : 225-231.

Wright, J., Clark, G.E. and McLachlan, A.R.G. (2011) *Acta Horticulturae.*, DOI : 10.17660/ActaHortic.2011.886.56

Wu, L.F., Qu, Y.H. and Qu, S.P. (1999)*J. Southwest Agri. Univ.*, **21** : 423-426.

Xiao Yi-Hu, Tang Hao-Ru, Ge Cong, Mo Fan, Li Nan-Yu, Luo Ya (2019) *AIP Conference Proceedings.*, **2079**, DOI: 10.1063/1.5092385.

Hlophe, N. P. Moyo, M. Staden, J. V. and Finnie, J. F. (2015) *Propagation of Ornamental Plants.*, **15** : 73-78.

Xiao, T.J., Li, H., Long, C.L., Xia, L.F. and Li, D.Z. (1998) *Acta Botanica Yunnanica*,pp. 101-103.

Yao, J.L. and Cohen, D. (2000) *Theor. Appl. Genet.*, **101** : 400-406.

Yao, J.L., Cohen, D. and Rowland, R.E. (1995) *Plant Sci. Limerick*, **105**: 199-206.

Yao, J.L., Cohen, D. and Yao, J.L. (1996) *New Zealand J. Crop Hort. Sci.*, **24** : 297-201.

Yao, J.L., Rowland, R.E. and Cohen, D. (1994b) *South African J. Bot.*, **60** : 4-7.

Zalapa, J.E., Cuevas, H., Zhu, H., Steffan, S., Senalik, D., Zeldin, E., McCown, B., Harbut, R. and Simon, P. (2012) *Amer. J. Bot.*, **99**: 193–208.

Zhang, Y.C., Tang, G.G., Chu, Y.X. and Wu, A.Z. (2009) *Mol. Plant Breed.*, **7**: 827–832.

Zhang, X., Wu, Q. Li, X. Zheng, S. Wang, S. Guo,L. Zhang, L. and Custers, J.B.M. (2011) *Sci. Hortic.*, **129**: 335-342.

Zheng, Y.H., Zhang, Z., Lu, B., Han, F.G., Peng, F. and Hua, X.P. (2015) *Jiangsu Agric. Sci.*, **43**: 142–144.

8

Water Lily (*Nymphaea* spp.)

S.P. Thamaraiselvi, M. Ganga, N. Naveena, K. Rajamani and Debmala Mukherjee

1.0 Introduction

Aquatic gardens or water gardens are an element of beauty in any garden and the aquatic plants are important components in them. Among the aquatic plants, water lily (*Nymphaea* spp.) belonging to the family Nymphaeaceae is a hardy and attractive perennial plant. Because of their beautiful, showy, sweet-scented flowers, several members of Nymphaeaceae are of great ornamental value for aquatic gardening. The genus name '*Nymphaea*' is a Latin word meaning 'water lily' and has its references in Greek and Latin tradition. Nymphaea is widely distributed in many parts of the world and is extensively used as an aquatic ornamental plant owing to its diverse colour, shape, size and fragrance of the flowers. More than fifty species of water lily possess ornamental value and are categorized under the genus *Nymphaea* (Borsch *et al.*, 2007). Several species of *Nymphaea* have culinary importance as well as phytoremediation potential for mitigating heavy metal pollution (Tani *et al.*, 2006).

2.0 Importance and Uses

2.1.0 Aesthetic Uses

Lily-ponds are considered as an important component in a garden which paves beauty and lively look in landscaping. *N. nouchali* is regarded as the national flower of Bangladesh and Sri Lanka. Nymphaea is highly appreciated as an ornamental plant because of its spectacular flowers, and is widely used during traditional and cultural festivals in Sri Lanka. In various parts of India, the Nymphaea flower are widely used in religious ceremonies for worshipping purposes (Devi *et al.*, 2015). They are also popular as aquarium plants under the name 'dwarf lily' or 'dwarf red lily'.

2.2.0 Therapeutic and Other Uses

Nymphaea stellata is a well-known medicinal plant and is widely used in the Ayurveda and Siddha systems of medicines for the treatment of various ailments including diabetes, inflammation, liver disorders, urinary disorders, menorrhagia, blenorrhagia, menstruation problem, as an aphrodisiac, and as a bitter tonic (Raja *et al.*, 2010). The rural people in India have been using various parts of this plant species as food and medicine against different ailments, such as diarrhoea, piles and cough (Devi *et al.*, 2015).

Nymphaea plants as a whole are considered highly nutritious and have great medicinal value. The seeds and rhizomes of various species of water-lily have been used as food source by many primitive races in Australia, Madagascar, West Africa and Central America. Different parts of Nymphaea such as the rhizomes, young leaves and peduncle are used for food and vegetables. The flowers and stems are eaten raw, while the roots and seedpods are cooked either on an open fire or in a ground oven. The young leaves and unopened flower buds can be boiled and served as a vegetable. The seeds are high in starch, protein, and oil, suggested to be popped, parched, or ground into flour. Potato-like tubers can be collected from the species *N. tuberosa* (Peterson, 1977).

Uses of Nymphaea

Spectacular Display of Nymphaea in Aquatic Landscape

Nymphaea as Cut Flower

Water Lily in Pot

Water Lily grown in Pool

Cut Flowers of Nymphaea Used in Various Floral Arrangements

Water lilies also provide shade which helps to reduce the growth of algae in ponds and lakes. Like lotus, water lily is also considered as an important water purification plant. The roots of water lily can absorb the poisonous substances like mercury, lead, phenol, *etc.* and filter the microorganism in water. It plays an important role in decontaminating water and mitigating water pollution.

3.0 Origin and History

Waterlilies are hardy and tropical, aquatic perennials native to Australia, South Africa and the Northern Hemisphere. They are cosmopolitan in their distribution. The impressions of the aesthetic beauty of water lilies have been found in ancient Egypt, Rome, France and Asian traditions. The water lilies of Nile were called as lotuses by ancient Egyptians and the lotus designs were important in temple architectures. Egyptians have symbolised the opening of blue Egyptian water lily (*N. caerulea*) in morning and sinking during dusk and the *vice-versa* in white water lily (*N. lotus*) with that of death and life after death (Tresidder, 2004). Evidences of both flowers have been found in the tomb of Ramesses II. Romans believed in the drink of Nymphaea extract in vinegar. Water lilies were also used as ritual narcotics.

4.0 Taxonomy

Taxonomically, the plants under the order Nymphaeales are called as 'water lily' (Friedman, 2008). Water lilies are categorized into three families *viz.*, Hydatellaceae, Cabombaceae, and Nymphaeaceae (Borsch *et al.*, 2007; Biswal *et al.*, 2012). There is a tremendous variation in the floral organs among the families in the order Nymphaeales. The family Nymphaeaceae consists of six genera *viz.*, *Barclaya*, *Euryale*, *Nuphar*, *Nymphaea*, *Ondinea*, and *Victoria* (Les *et al.*, 1999; Biswal *et al.*, 2012). In the genus *Nymphaea*, the flowers are composed of 4 sepals, 50 to 70 petals, 30 to 40 carpels, and 120 to 250 stamens and these unique characters are considered as the most primitive angiosperm floral characteristics, as evident in various ancestral flowering plant fossils (Gandolfo *et al.*, 2004). Plants in the genus *Nymphaea* can be divided into two groups: Apocarpiae and Syncarpiae. The Apocarpiae group consists of three subgenera: *Anecphya*, *Confluentes* and *Brachyceras*, while the Syncarpiae group includes *Hydrocallis*, *Lotos* and *Nymphaea*. Based on habitat, there are two ecological groups in genus *Nymphaea* namely 'Tropical water lily' and 'Hardy water lily' (Huang *et al.*, 2009). The tropical water lily includes *Brachyceras*, *Anecphya*, *Lotos*, and *Hydrocallis* and hardy water lily includes *Nymphaea* spp.

In 2015, Devi and coworkers evaluated and analyzed a total of 40 morphological characters in 207 waterlily plants and distinguished *Nymphaea* into three components as described below.

Component	*Variation (per cent)*	*Rhizome Group*	*Species*
I	33	Triangular	*N. micrantha*, *N. caerulea*, *N. nouchali*, *N. manipurensis* and *N. manipurensis* var. *versicolor*
II	30	Tubular	*N. marliacea* and *N. mexicana*
III	12	Globular	*N. rubra* and *N. pubescens*

Kit Knots (2003) characterized the hardy and tropical water lilies based on their growth habit, flower and leaf characteristics.

Characteristics	*Hardy Water Lilies (Nymphaeas)*	*Tropical Water Lilies (Brachyceras)*
Flower colour	White, all shades of pink, red and yellow	White, all shades of pink, red, yellow, autumn, blue, purple and lighter green shades.
Flower shape	Many are cup-shaped and some star shaped and float on the water surface after bloom	Mostly star shaped and some cupped and always stand above the water surface
Leaf blade	Thick, leathery with smooth edges, upper surface plain green and slightly mottled when young.	Thin with toothed edges, upper surface plain green, mottled or whorled with maroon or bronze.
Growth habit	Horizontal growth from a fleshy rhizome and new suckers are produced from the eyes of rhizomes.	Vertical growth from a single crown and all the leaves and flowers spread out from this point.

Botanically water lilies are aquatic plants possessing rhizomes or stolons with angular and erect stem. The leaves emerge from the rhizomes and have long petioles, floating and round upto 30 cm in diameter. Leaves have typical 'V' shaped notches at the centre; the colour is green at the top and purple in the lower surface. Flowers emerge and float on the surface of water either during daytime or at night. Protogynous flowering is noticed in most of the species of *Nymphaea*. Male and female phases are separated by flower opening with female phase exhibited initially and male phase exhibited during the end of female phase (Povilus *et al.*, 2015). The shape of flower petals range from cupped and pointed to star shaped with a wide spectrum of colours *viz.*, pink, white, yellow, purple, red, blue and orange (Songpanich and Hongtrakul, 2010). Stamens are present in the centre of the flowers and the flowers are cross pollinated most often by beetles. The fruit is a berry and borne on curved or coiled peduncles. The plants reproduce by rhizomes and seeds.

5.0 Species and Cultivars

5.1.0 Species

Water lilies comprise of more than 50 species, distributed globally and among them five species have been reported to originate from China *viz.*, *N. candida* Presl., *N. lotus* L. var. *pubescens*, *N. alba* L., *N. stellata* Wild and *N. tetragona* Georgi (Beckett, 1984; Zhu *et al.*, 2012). The flower colour is diverse ranging from white, yellow to red and blue (Beckett, 1984). Zhu *et al.* (2012) has given an elaborative discussion on the ecological types of water lily as tropical water lily and hardy water lily. The tropical water lilies have greater ornamental and commercial values when compared to hardy water lilies. Mitra (1993) has reported five genera and 70 species in Nymphaeaceae family, commonly known as water lilies, they are distributed in almost all tropical and temperate regions of the world. Ten species of the genus *Nymphaea*, *viz.*, *N. alba L.*, *N. candida*, *N. nouchali*, *N. pubescens*, *N. rubra*, *N. tetragona*, *N. caerulea*, *N. marliacea*, *N. micrantha* and *N. alba var. rubra* are reported from India by Mitra (1990) and two more *viz.*, *N. malabarica* and *N. mexicana* were reported by

Species of *Nymphaea*

Nymphaea alba

Nymphaea mexicana

Nymphaea mexicana

Nymphaea candida

Nymphaea marliacea Chormatella

Nymphaea marliacea* var. *rosea

Nymphaea micrantha

Nymphaea nouchali

Nymphaea pubescens

***Source*: RHS Plant Selector.**

Cultivars of *Nymphaea* spp.

Perry's Double White

Colorado

Texas Dawn

Princess Elizabeth

Clyde Ikins

Fuchsia Pom Pom

Red Flare

Panama Pacific

Sultan

Plyfah

King of Blue

Myra

Source: https://www.internationalwaterlilycollection.com

Diversely Coloured Waterlilies

(a) ***Nymphaea* 'Hermine'**, (b) ***N.* 'Marliacea Chromatella'**, (c) ***N.* 'Wanvisa'**, (d) ***N.* 'giganteaHybrid 1'**, (e) ***N. colorata***, (f) ***N.* 'Muang Wiboonlak'**, (g) ***N.* 'Piyalarp'**, (h) ***N.* 'Agkee Sri Non'**, (i) Leaf of ornamental Victoria water lily.

Floral Organs of a Typical Tropical Waterlily

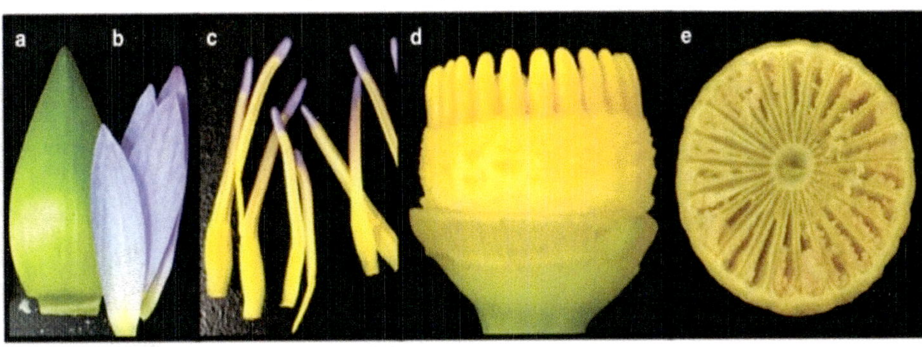

(a) Petal, (b) Sepal, (c) Stamen, (d) Carpels on the receptacle, (e) Numerous young seeds

Source: Chen *et al.* (2017) *Horticulture research,* 4 (1): 1-7.

Ansari *et al.* (2005) and Asharani and Biseshwori (2014) have reported *N. manipurensis* and *N. manipurensis* var. *versicolor* from Manipur.

Some of the commercially important species include, *N. alba, N. rubra, N. mexicana, N. nouchali, N. elegans, N. conardii, N. capensis, N. amazonum, N. candida,* N. *pubescens* and *N. lotus.*

Table 1: Description of Important Water Lily Species as given by Slocum (2005)

Common Name	*Botanical Name*	*Description*
Nilofar	*Nymphaea alba, N. venusta, N. splendens*	The flowers are white and with several small stamens inside.
Yellow Waterlily	*Nymphaea mexicana*	The floating lotus like flowers have yellow petals and pointed, star-like, greenish-yellow sepals
Blue Water Lily	*Nymphaea nouchali*	The flowers are usually violet blue in color with reddish edges. Some varieties have white, purple, fuchsia-colored flowers.
Pink Water Lily	*Nymphaea pubescens, N. purpurea*	Flowers are purple in colour and fragrant.
White Water Lily	*Nymphaea alba*	The flowers are white in colour with yellow stamens
Dwarf white water lily	*Nymphaea candida*	*Nymphaea candida* has a small white flower (10–20 cm across) with a yellow center
Red Water Lily	*Nymphaea rubra, N. rosea, Leuconymphaea rubra*	Flowers are intensely red or rose-coloured
West African lily	*Nymphaea micrantha*	Flowers usually white to light blue in colour

6.0 Genetics and Breeding

6.1.0 Genetics

Water lilies have been reported to occupy a critical space in evolutions of angiosperms. Pellicer *et al.* (2013) has reported the characteristics of four pantropical water lilies. Water lilies have been considered as potential model plants for basal angiosperms and have attracted many plant biologists to sequence its genome (Songpanich and Hongtrakul, 2010).

Table 2: Characteristics of Four Pantropical Water Lilies as Reported by Pellicer *et al.* (2013)

Species	*Genome Size*	*Chromosome*	*Classification*	*Distribution*	*Flowers in Diameter*
Nymphaea caerulea	567.24 Mb	2n= 28	*Nymphaea, Brachyceras*	East Africa	10–15 cm
Nymphaea colorata	489 Mb	2n= 28	*Nymphaea, Brachyceras*	Tropical East Africa	11–14 cm
Nymphaea minuta	449.88 Mb	2n= 28	*Nymphaea, Brachyceras*	Madagascar	2 cm
Nymphaea thermarum	498.78 Mb	2n= 28	*Nymphaea, Brachyceras*	Rwanda, Africa	10–15 cm

These four pantropical water lilies serve as the best model for basic information on the origin of angiosperms. The reports on water lily genome by Chen *et al.* (2017b) serve as valuable base for understanding the genetics of ornamental traits and will in turn accelerate the molecular breeding of water lilies. Chen and his team have sequenced the genome of waterlily and lotus with *Nymphaea* spp. The genome size of *Nymphaea* spp. has been reported to be 400 Mb – 2.7 Gb while, that of *Nelumbo nucifera* to be 929 Mb. Recently, the genome of *N. colorata* (2n = 28) has been revealed by Zhang *et al.* (2020), it has relatively small genome size (400 Mb approximately).

6.2.0 Breeding

Breeding of tropical water lilies is very much limited due to their inability to survive during winter seasons. Hence breeders have much interest in breeding cold stress tolerant tropical water lilies so that they can be grown in cooler areas. Most of the cultivated garden lilies are hybrids while others are natural mutants. To increase the success of improving plant resistance to cold stress, a lot of attempts have been made with plant molecular biotechnology (Gombos *et al.*, 1992; Vigh *et al.*, 1993; Worrall *et al.*, 1998). The hybrids or mutants with modified characters serve as breeding parents for the introduction of cold resistance into novel tropical water lilies with varied flower shape, colour and foliage (Chen *et al.*, 2017a). The two reproductive strategies *viz.*, sexual and asexual reproduction have eased the possibility of maintenance of introduced novel trait in the off springs (Bonilla-Barbosa *et al.*, 2000).

The breeding of water lilies is given attention in countries like Thailand, the United States, and China. The International Water Lily and Water Gardening Society is authorized to register new varieties of water lilies around the world (Kilbane, 2016). Almost all the registered novel varieties so far evolved have been either natural mutants or hybrids. The major constraints with traditional hybridization method are that it is laborious, expensive, time consuming, requires a rich germplasm collection, extensive screening and observations to be recorded for more than two years. Only after the phenotypic stability is achieved, registration can be attempted in International Water Lily Organization. Till recently hybridization has been the major technique for breeding novel water lilies.

6.2.1 Hybridization

Attempts on artificial crossbreeding in water lilies has been made as early as 1851 by a British breeder, Joseph Paxton, who successfully crossed two species from *Lotos* subgenus and obtained the first hybrid, Nymphaea 'Devonishire' (Perry and Robinson, 1996). The major breeding goal in water lily is to combine the characters of hardy water lilies and tropical water lilies so as to develop hybrids with characteristics of underwater part of the hardy water lily and above water portion retaining the characteristics of the tropical water lily. However, the cold stress tolerance in tropical types is the major challenge faced by breeders. Though, Thai breeders have evolved inter-subgeneric hybrids and named them as "Siam" series, these hybrids were mostly single coloured and lacked the fragrance of tropical water lilies (Rudall *et al.*, 2008).

Novel colour and flower shape were achieved in water lilies either through natural cross pollination or hybridization. In order to achieve the desired colour and shape, hand pollination has to be attempted as a means of controlled pollination. Songpanich and Hongtrakul (2010) have elaborated the hybridization techniques through their report on hardy water lilies. The flowers of waterlilies are protogynous in nature and the stigma become receptive on the first day of blooming. This enables the chances of cross-pollination from pollen of another flower/plant on the subsequent days after blooming. However, exceptions exist in this phenomenon though not in hardy water lilies. Before attempting to hand pollination, it is imperative to differentiate the flower that has opened on first day/ second day/third day.

The characteristics of the stigma and stamen and the avoidance of emasculation due to protogyny nature of waterlilies has also been described by Songpanich and Hongtrakul (2010). On the first day of flowering, the stamens are straight up with an easy view of the stigmatic disc and the presence of nectar. While on the second or third day of flowering, the stamens curve to the center of the stigma making it invisible. The inner rings of anthers are also open on the second day flower exposing the pollens when compared to the third day flower. The pollen of both second and third-day flowers can be used for hand cross-pollination. The step by step process in hand pollination techniques has also been described by Songpanich and Hongtrakul (2010) and by this method, development of 17 hybrids in hardy water lilies has also been reported. Songpanich (2009) has also reported on the viviparity of certain species of hardy water lilies.

Songpanich and Hongtrakul (2010) attempted hybridization in five water lily cultivars using the subgenus *Nymphaea* (hardy waterlily) as female parent and cultivars from subgenera, *Anecphya* and *Brachycera* as male parent. The cross between *Nymphaea* x *Anecphya* was unsuccessful, while *Nymphaea* x *Brachycera* yielded 20 inter-subgeneric hybrids. From these hybrids, a prominent blue-purple flowered hybrid was confirmed with PCR-RFLP marker for the important traits inherited from hardy waterlily (subgenus *Nymphaea*) and hence was named 'Siam Blue Hardy'.

In water lily (*Nymphaea*) hybrid breeding, breeders often encounter non-viable seeds, which make it difficult to transfer desired or targeted genes of different *Nymphaea* germplasm. Pre-fertilization barriers were the main factor in the failure of the hybridization of *Nymphaea* (Sun *et al.*, 2019). They studied the differences of stigma transcripts and proteomes at 0, 2, and 6 h after pollination (HAP). Moreover, some regulatory genes and functional proteins that may cause low pollen-pistil compatibility in *Nymphaea* were identified (Sun *et al.*, 2019).

7.0 Growth and Flowering

The Royal Horticultural Society has released the growing guide for water lilies and classified them into three types based on the spread as, dwarf species (30-60 cm); small/medium (60 cm) and vigorous (2.4 m). Bourgondien (2003) has provided the details of growth and flowering in water lilies. Tropical lilies bear blooms above the water surface and bloom only in conditions of >20°C while, that of hardy water lilies thrive well in frost conditions and bear blooms floating on the surface of water.

Flowering is profuse in tropical lilies when compared to hardy lilies. Waterlilies put forth foliage throughout the growing season. The leaves remain for upto 3-4 weeks during peak season. Hardy waterlilies generally flower during May till September. Flowers remain fresh for 3-5 days, open during day time and closes during night. After 3-5 days, the flowers sink into water and seed pod formation begins. Once the seeds are ripe, they are dispersed into the water and soil below. Flowering in water lilies is profuse during summer and the older flowers must be removed before they sink into water.

7.1.0 Effect of Temperature

Tropical water lilies thrive well and are preferred in tropical countries and hardy water lilies are cold hardy and hence preferred in temperate countries. Marie Lannotti (2020) has reported that the rhizomes of tropical water lilies can be stored in a container with moist media at 7-10°C. The optimum temperature reported in most of the studies for tropical water lilies is 20°C, while that of hardy water lilies can grow at lower temperature of 10°C. Water lilies can be best grown in direct sun and even in shade or partial shade condition. Four to six hours of bright sunshine is necessary for profuse flowering in water lilies. Else and Riemer (1984) has reported that *N. odorata* grows best in water between 15-18°C. It can even thrive well in water at 2-2.5 m depth and can tolerate a wide pH range (Wiersema, 1993).

8.0 Propagation

Water Lilies are propagated by two methods *viz.*, seeds and vegetative propagation by rhizomes.

8.1.0 Seed Propagation

In general, seeds of tropical lilies germinate easily when stored under refrigerated conditions at 0-4°C till sowing, while hardy water lilies are left into the water until the time of planting. Else and Riemer, (1984) observed that seed germination was most efficient in *N. odorata* at 13°C at a maximum depth of 1 meter. A successfully fertilized flower sets about 2000 seeds per flower within 15 days of pollination (Robert, 2019). These seeds are shed into the water source or carried away to different locations. Seed viability differs among the species. Viable seeds are larger and darker in colour while non-viable seeds are smaller and lighter in colour. Seeds of the temperate species, *N. alba* and *N. odorata* are reported to be desiccation sensitive (Else and Riemer, 1984; Estrelles, 2004), while those of tropical species *N. gigantea* are desiccation tolerant (Ewart, 1908). The seeds of tropical waterlily lack dormancy and hence readily germinate in water, while in case of hardy water lilies, physiological dormancy of seeds prevents the germination especially during winter (Baskin and Baskin, 2007).

The database of Kew Seed Information of Royal Botanic Gardens Kew (2018) on seed storage behaviour have categorized two species, *N. gigantea* and *N. maculata* as orthodox, while *N. alba* as recalcitrant and the two species *N. odorata* and *N. tuberosa* as 'uncertain'. In a recent report by (Dalziell, 2019) on seed storage physiology of eighteen native species of *Nymphaea*, it was established that the seeds of tropical species of *Nymphaea* have a complex germination behaviour due to morphophysiological or physiological dormancy. Moreover, it was apparent that the species, *N. immutabilis, N. lukei* and *N. macrosperma* did not germinate easily and also exhibited more difficult storage behaviour.

8.1.1 Seed Propagation *in vitro*

Khaenthong *et al.* (2019) studied the effects of seed coat scarification, seed priming, and cold stratification on germination of a tropical water lily hybrid, *N. rubra* 'Maeploi'. Germination of hydro primed seeds in filtered water was 68.3 per cent and a higher germination rate of 78.30 per cent was also observed in non-clipped seeds cultured for four weeks without cold stratification. This preconditioning treatment could serve as basic information for further *ex situ* conservation of tropical night-blooming waterlilies.

8.2.0 Vegetative Propagation by Rhizomes

Propagation by rhizomes is the commercial method of propagation in Nymphaea, as the plants are true to type and rhizome multiplication is also faster and easier when compared to seeds. The rhizome exhibits a horizontal or upright growth and four different types of rhizomes have been identified in hardy water lilies *viz.*, mexicana (upright growth), odorata (vigorous and fleshy), tuberosa (thin and vertical) and marilac (less vigorous and fleshy) (Bourgondien, 2003). All the rhizomes except 'mexicana' exhibit horizontal growth.

All tropical water lilies have rhizomes with upright growth. The rhizomes are divided from the mother plant and used for planting (Robert, 2019). Rhizome divisions are used for propagation and planted in submerged pots (Speichert and Speichert, 2004). The plants of *N. odorata* arise from dense fleshy rhizomes which is starchy storage organ that grow horizontally. The rhizome diameter range between 2 and 3 cm and are located beneath the soil surface (Northey, 2014). They branch out and each portion can produce a root and shoot system (Seago Jr *et al.*, 2000).

8.3.0 Micropropagation

Seed propagation in waterlilies has lower viability and seed storage method differ among different species and varieties. Micropropagation methods enable faster and uniform clonal planting materials. Lakshmanan (1994) attempted *in vitro* multiplication in *Nymphaea* hybrid 'James Brydon' using rhizome tips as explant. Sumlu *et al.* (2010) attempted to overcome the dormancy of seeds in *N. alba* under *in vitro* conditions. It was revealed that fresh seeds exhibited highest germination when cultured on MS medium supplemented with 1 mg/l BAP + 0.1 mg/l IAA. However, after five months of storage, the seeds could be germinated only in the presence of a dormancy breaking chemical *viz.*, 2mg/l of TDZ and germination frequency of 51.37 per cent. This study proved the significant role of TDZ in breaking the dormancy of *N. alba* seeds.

Selection of explants for micro-propagation is a major challenge in waterlilies. *In vitro* germinated seedlings have been used as explants in *Nymphaea alba* (Sumlu *et al.*, 2010), *Nymphaea tetragona* (Dang *et al.*, 2010) and *Nymphoides coreana* (Oh *et al.*, 2010). Nymphaea rhizome tip when used as explants pose the problem of surface sterilization and have poor response to growth regulator treatments. Seeds and rhizome divisions can be used as alternate explants in waterlilies. Seeds, excised embryos and ovary tissues can be effectively used as explants *in vitro* as reported by Nguyen (2016).

Multiplication of plants through seeds is difficult due to development of dormancy with the passage of time. The results of the study conducted by Sumlu *et al.*, 2014 that the fresh seeds of the species gave highest germination on MS medium containing 1 mg/l BAP + 0.1 mg/l IAA. However, the seeds that were stored for five months at 4°C failed to germinate on medium containing 1 mg/l BAP + 0.1 mg/l IAA, either used alone or combined with sucrose, IBA and GA_3 in different concentrations in the germination medium. After five months these seeds could be germinated only on germination medium that contained 0.05 to 4 mg/l TDZ, with highest germination on 2 mg/l TDZ with germination frequency of 51.37 per cent. No significant variation on germination was recorded in the light or dark; however, the seed germinated in the dark produced 2–3 times longer seedlings compared to those germinated under 16 h light photoperiod. This study signifies the role of TDZ to break the dormancy of *N. alba* seeds. As such the results indicate that TDZ could be used effectively to propagate *N. alba* from seeds which could help to conserve and multiply this plant species at its natural habitat.

9.0 Soil and Climate

9.1.0 Soil

Water lilies are best grown in clay soil as light weight organic soils or compost often float out of the containers. Garden loam mixture with high amount of clay and a small part of sand is mostly recommended for potting as well as repotting in *Nymphaea*. Clay soil also holds most of the nutrients and provides a slow release of it, while the sand allows for aeration of the medium. An equal quantity of peat moss or cocopeat can also be added instead of clay (Sacher 2006).

9.2.0 Climate

The hardy water lilies are best grown in temperate climates where the water freezes in winter. However, the tropical water lilies prefer tropical climate and cannot tolerate the cold winters. The optimum temperature reported in most of the studies for tropical water lilies is 20°C, while that of hardy water lilies can grow at lower temperature of 10°C. Water lilies can be best grown in direct sun and even in shade or partial shade condition.

10.0 Cultivation

Water lilies are cultivated in water bodies like lakes and ponds and commercially in submerged containers also. They are grown by planting in aquariums, clay containers or artificial ponds in the garden. Richards and Cao (2012) reported that seeds can be used to propagate *N. odorata* plants up to 90 cm in water and the seedlings grow submerged till the production of floating leaves.

10.1.0 Planting

Water lilies are grown in containers and submerged in the aquatic ponds or lily ponds. Growth of aquatic plants are better when planted in pots and placed in the bottom of the ponds rather than planting directly in the bottom (Al-Menaie and Al-Zalzaleh, 2000). Growing aquatic plants in containers restrict their growth and reduce the leaching of nutrients away from the soil medium (Bourgondien, 2003). Waterlily is planted at the edge of the container with the crown pointed at the center of the pot. The rhizome portions are buried in the soil with the crown of the plant at soil level. They are anchored with some heavy weight in order to produce newer roots sooner. Different types of clay containers which are familiar in lotus cultivation are widely used for planting water lilies also. The depth of planting can be maintained between 20 and 40 cm while planting initially and later they are grown between 30 and 60 cm depending on the cultivar. Rectangular pots of size 50 × 45 × 22 cm (h × w × d) were found to be the best for the growth and multiplication of *N. rosea* and *N. sunrise* (Al-Menaie *et al.*, 2012).

Table 3: Effects of Containers on Average Number of Leaves and Flowers Produced in *N. rosea* and *N. sunrise* (180 DAP)

Treatments	*Average Number of Leaves*		*Average Number of Flowers*	
	N. rosea	*N. sunrise*	*N. rosea*	*N. sunrise*
Circular pots	29.33	18.00	0.50	1.67
Rectangular pots	39.00	28.33	13.00	8.83
Baskets	26.88	28.20	1.38	2.16
S.Em ±	2.93	3.04	0.47	0.43

10.2.0 Manuring and Fertilization

Waterlily plants are fertilized at the time of potting and later on regular feeding is done at fortnightly or monthly intervals depending on the type of fertilizer used in order to sustain the vigorous growth of the plants. Addition of complex fertilizer tablets 14:14:14 or 17:17:17 at fortnightly intervals is helpful in initiation of flowering (Sacher 2006).

11.0 Diseases and Pests

11.1.0 Diseases

Water lily Crown rot and leaf spots are the most common diseases. The best method to control these diseases is to remove and destroy the infected plants.

11.2.0 Pests

Sucking pests *viz.*, waterlily aphid (*Rhopalosiphum nymphaeae*, waterlily beetle (*Galerucella nymphae*), Brown China mark moth (*Elophila nymphaeata*) and pond snails are the major pests of water lilies which affect the aesthetic appearance of the flowers. Waterlily pests are difficult to control and most of these pests are sporadic and do not usually occur every year on the same patch and hence the waterlilies

Symptoms of Nutrient Deficiency in Waterlilies

Phosphorus Deficiency | Potassium Deficiency | Magnesium Deficiency

Copper Deficiency | Zinc Deficiency | Boron Deficiency

Source: Staines and Sassoon, (2020) ***Water garden Journal***, 35(2): 8-10.

Propagation of Water Lily

Seeds | Rhizomes of Water Lily

Freshly Harvested Flowers of Nymphaea

Source: https://cutflowers.ces.ncsu.edu/wp-content/uploads/2019/03/CFQ-Winter-2019-Water-lilies-as-a-cut-flower.pdf?fwd=no

recover. Waterlily aphid and beetle can be wiped off the leaves and buds by a forcible spraying water over the foliage. This is feasible on small ponds where the lily leaves are within reach from the sides of the pond but then in larger ponds they are to be tolerated (Sacher 2006).

12.0 Harvesting and Postharvest Management

Water lilies are generally harvested during early morning hours. Harvesting of water lilies at proper stage is extremely important if the flowers have to be stored for longer durations. Generally, most water lilies remain open on the plant for 3 to 4 days, hence it is crucial to cut them at the right time to get maximum vase life. When buds are about to open, they start floating on the surface of the water in the morning, and open as the sun rises. Many hardy water lilies are day bloomers, so they open in the early morning and close in the evening. They are harvested either at unopened bud stage when buds have started showing colours between the sepals or when flowers are open on the first day and have fluid in the bottom of the receptacle cup. The flowers can be plucked from the rhizome by pulling, but it may lead to breakage of stems. Hence, it is recommended to harvest them by cutting the stems underwater. The vase life of water lily flowers varies from 4.0 to 5.0 days.

Nymphaea sp can be used as dry flowers. Hot-air oven drying is the best method for drying of water lilies. The suitable temperature for drying is 45-49°C for 120 hours. Among the various desiccants, fine-sand has proved to be best for drying waterlilies (Geetha *et al.*, 2002). In a study on the effect of growth regulators on post-harvest life of water lily, 50 mg/l GA_3 treatment recorded the best result for the preserving of *N. stellata* cut flower and could extend the vase life as compared to the control (distilled water) (Zhang *et al.*, 2004).

13.0 Biotechnology

Through traditional breeding approaches, the cold stress tolerance of tropical water lilies is difficult to improve. Hence, the cold stress-tolerant choline oxidase (CodA) gene was successfully transformed into the tropical water lily using pollen tube mediated transformation technique (Yu *et al.*, 2018). This transgene activated a series of physiological responses to combat cold and the transgenic tropical water lily exhibited a strong cold stress tolerance both under lab and field conditions. Flower pigments have been extensively studied in various species and cultivars of water lily and 34 flavonoids has been identified so-far by Zhu *et al.* (2012). Biotechnological improvement of water lilies is still in preliminary stage and the genome sequencing by Zhang *et al.* (2020) has paved way for transgenic research in waterlilies.

13.1.0 Genetic Transformation

Tropical water lilies cannot survive the winter in areas beyond 24.3°N latitude. The transgenic pipeline through the pollen-tube pathway was generated in water lily for the first time by Yu *et al.*, 2018. To improve cold stress tolerance of tropical water lilies, a gene encoding choline oxidase (*CodA*) driven by a cold stress-inducible promoter was transformed into a tropical water lily through the pollen-tube transformation. PCR and southern blot detection revealed that the

CodA gene had been integrated into the genome. Reverse transcription PCR showed that *CodA* gene was induced after cold stress treatment, and further quantitative real-time PCR revealed different expressions among six 4 lines and line 3 had the highest expression. Multiple physiological experiments showed that after cold stress treatment, both the conductivity and malondialdehyde (MDA) levels from transgenic plants were significantly lower than those of non-transgenic plants, whereas the content of betaine and the activity of superoxide dismutase, catalase, and peroxidase were higher than those from non-transgenic plants. These results suggest that expression of exogenous *CodA* gene significantly improved the cold stress tolerance of tropical water lilies through a wide range of physiological alterations.

14.0 References

Anonymous (2014) Wildflower Center, Lady Bird Johnson. "Nymphaea Odorata." Wildflower Center. NPIN: Native Plant Database - University of Texas at Austin, 2010. Web. 29 Nov. 2014.

Anonymous (2018) Database of Kew Seed Information of Royal Botanic Gardens Kew.

Al-Menaie, H.S., Al-Ragam, O., Al-Dossery, N., Zalzaleh, M., Mathew, M. and Suresh, N. (2012) *Am.-Eurasian J. of Ag. and Env. Sci.*, **12** (2): 148-153.

Al-Menaie, H. and M. Al-Zalzaleh (2000) *Kuwait Institute for Scientific Research, Report No. KISR* 6277.

Ansari, R., Jeeja, G. and Jeyalakshmi, S.K. (2005) *J. Palyno*, **41**: 139–152.

Asharani, D.S. and Biseshwori, T. (2014) *Phytotaxa*,**188**: 112–117.

Baskin, C.C. and Baskin, J.M. (2007) *Seed Sci. Res.*, **17** (4): 293-296.

Beckett, K.A. (1984) The concise encyclopedia of garden plants. *Orbis Publishing, London.*

Bonilla-Barbosa, J., Novelo, A. andOrozco, Y.H. (2000) *Aquat. Bot.*, **68** (3): 189-204.

Borsch, T., Hilu, K.W., Wiersema, J.H., Löhne, C., Barthlott., W. and Wilde, V. (2007) *Int. J. Plant Sci.*, **168** (5): 639-671.

Bourgondien, K.V. (2003) Introduction to growing water lilies and marginal plants.

Chen, F., Liu, X., Yu, C., Chen, Y., Tang, H. and Zhang, L. (2017a) *Hort. Res.*, **4** (1): 1-7.

Chen, Y.Y., Fan, X.R., Li, Z., Li, W. and Huang, W.M. (2017b) *AquaT. Bot.*, **140**: 55-61.

Dalziell, E.L. (2019) *Conservation physiology* (7).

Dang, J., Kumaria, S. and Tandon, P. (2010) *Nymphaea tetragona*, a rare and endangered plant of Meghalaya, India.

Devi, S.A., Biseshwori, T. and Pratap. J.H. (2015) *Braz. J. of Botany* : **38**(3).

Else, M.J. and Riemer, D.N. (1984) *J. Aquat. Plant Manage.*, **22**: 22-25.

Estrelles, E. (2004) *Proc. of 4th European Conference on the Conservation of Wild Plants.* Generalitat Valenciana and Universitat de Valencia, Valencia, Spain.

Ewart, A.J. (1908) *On the longevity of seeds.*

Friedman, W.E. (2008) *Nature*, **453**(7191): 94-97.

Gandolfo, M., Nixon, K. and Crepet, K. (2004) *PNAS*, **101** (21): 8056-8060.

Geetha, C., Shalini, M., Rajeevan, P. and Valsalakumari P.K. (2002) *Nymphaea stellata*: ISOH, pp.88.

Gombos, Z., Wada, H. and Murata, N. (1992) *PNAS*, **89** (20): 9959-9963.

Huang, G.Z., Deng, H.Q., Li, Z.X. and Li, G. (2009) Water lily. Beijing: China Forestry Publishing House. Pp.3.

Khaenthong, I., Phairin, U., Mattisinlapin, P., Subudom, S., Sripan, S., Boonthongto, A. and Malaithong, S. (2019) *Int. Sym. on Botanical Gardens and Landscapes*.1298.

Kilbane, T. (2016) *IWGS Water Gard. J.*, **31**: 14-17.

Knotts, Kit. (2003) How do you tell the difference between tropical and hardy waterlilies? www.victoriaadventure.org/waterlilies/differences.html

Lakshmanan, P. (1994) *PCTOC*, **36** (1): 145-148.

Lannotti, M. (2020) How to grow and care for Water lilies and lotus. https: / /www. thespruce.com/the-incomparable-waterlily.

Les, D. H., Schneider, E. L., Padgett, D. J., Soltis, P. S., Soltis, D. E. and Zanis, M. (1999) *Systematic Botany*,**24**: 28-46.

Mitra RL (1990) In: Fascicles of Flora of India (Eds. Nayar *et al.,)*, BSI, Kolkata, pp.11–25.

Mitra RL (1993) In: Flora of India, vol 1 (Eds. Sharma *et al.,*). BSI, Kolkata, pp. 430–431.

Nguyen, H. (2016) *In vitro* physiology of recalcitrant tissue cultured plants in the Nymphaeaceae, Alismataceae, and Orchidaceae. University of Florida.

Northey, A. (2014) *Nymphaea odorata*, fragrant water Lily, tuberous water lily, white water lily.

Oh, M.J., Na, H.R., Choi, H.K., Liu, J.R. and Kim, S.W. (2010) *Plant Biotechnol. Rep.*, **4** (2): 125-128.

Pellicer, J., Kelly, L., Magdalena, C. and Leitch, I.J. (2013) *Genome*, **56**(8): 437-449.

Perry, D. and Robinson, P. (1996) FP Water Gardening—Water Lilies and Lotuses. Timber Press, Portland, OR.

Peterson, L. (1977) *A field guide to edible wild plants of eastern and central North America*: Houghton Mifflin Harcourt.

Povilus, R.A., Losada, J.M. and Friedman, W.E. (2015) *Ann. Bot.*, **115**(2): 211-226.

Raja, M. M. M., Sethiya, N. K. and Mishra, S. H. (2010) *J. Adv. Pharm. Technol. Res.*, **1**(3), 311.

Richards, J.H. and Cao, C. (2012) *Aquat. Bot.*, **98** (1): 12-19.

Robert (2019) Water lily: The complete guide (Care, growing, propagation and more)

Rudall, P.J., Remizowa, M.V., Beer, A., Bradshaw, E., Stevenson, D.W., Macfarlane, T.D., Tuckett, R.E., Yadav, S.R.and Sokoloff, D.D. (2008) *Ann. Bot.,* **101**(7): 941-956.

Sacher, R. (2006) *Water garden journal online.*

Seago Jr, J.L., Peterson, C.A., Kinsley, L.J. and Broderick J. (2000) *Ann. Bot.,* **86**(3): 631-640.

Slocum, P.D. (2005) Timber press.

Songpanich, P. (2009) *Water garden journal online,* **4**(1).

Songpanich, P. and Hongtrakul, V. (2010) *Sci. Hortic.,* **124**(4): 475-481.

Speichert, G. and Speichert, S. (2004) *Encyclopedia of water garden*: Timber Press.

Sumlu, S., Atar, H. and Khawar, K. (2010) *Biotechnol. Equip.,* **24** (1): 1582-1586.

Sun, C.Q., Chen, F.D., Teng, N.J., Yao, Y.M., Shan, X. and Dai, Z.L. (2019) *BMC Plant Biol.,* **19**: 542.

Tani, M., Sawada, A., Oyabu, T. and Seiryo, K. (2006) *Sens Mater.,* **18**: 91-101.

Tresidder, J. (2004) *The complete dictionary of symbols: in myth, art and literature*: Duncan Baird.

Vigh, L., Los, D.A., Horvath, I. and Murata, N. (1993) *PNAS,* **90** (19): 9090-9094.

Wiersema, J. (1997) Nymphaea. In: *Flora of North America Editorial Committee* (*Vol.*3) *New York.*

Worrall, D., Elias, L., Ashford, D., Smallwood, M., Sidebottom, C., Lillford, P., Telford, J., Holt, C. and Bowles, D. (1998) *Science,* **282**(5386): 115-117.

Yu, C., Qiao, G., Qiu, W., Yu, D., Zhou, S., Yan, S., Yu, G., Jiang, J., Han, X., Liu, M., Zhang, L., Chen, F., Chen, Y. and Zhou, R. (2018) *Hortic Res.,* **5**: 73.

Yu, C., Qiao, G., Qiu, W. *et al.* (2018) *Hortic. Res.,* **5**(1): 1-11.

Zhang, L., Chen, F., Zhang, X. *et al.* (2020) *Nature,* **577** (7788): 79-84.

Zhang, Y., Luo, J. and Liu, J. (2004) *Acta Hortic. Sinica,* **31**(3): 392-394.

Zhu, M., Zheng, X., Shu, Q., Li, H., Zhong, P., Zhang, H., Xu, Y., Wang, L. and Wang L. (2012) *PLoS One,* **7**(4): e34335.

9

Crocus

Nomita Laishram, Arvinder Singh and K. Dutta

1.0 Introduction

Crocus is a group of ornamental bulbous plants belonging to the family *Iridaceae*. The name *Crocus* is derived from the Greek word 'krokos' meaning saffron yellow or turmeric yellow, which was meant for the species *C. sativus*. The genus contains more than 90 species. About 85 species were studied in detail by Mathew (1983). The genus is grouped by flower colour, flowering time and by species. They are native to North Africa and the Middle East, the Eastern Mediterranean and stretching to Central Asia and Western China. Some of the crocus species are well known in cultivation, others are poorly known. In India, it is being grown in Jammu and Kashmir and Himachal Pradesh.

2.0 Importance and Uses

2.1.0 Aesthetic Uses

Crocuses are very small plants which look quite insignificant in a large garden until and unless they are planted in mass. Crocuses are good outdoor plants. They are most suitable for planting in beds, borders, containers, mass plantings, ground cover, mixed groupings and in rock gardens. Also they do well in sink gardens and alpine houses. Spring flowering crocus can be used in mixed bed planting with annuals or other spring-blooming bulbs such as tulips, daffodils, muscari and hyacinths to extend the flowering period as crocus will bloom early than rest of the bulbs. Crocus can be used in the front of mixed borders as an addition to the lawn or in ground cover beds. They look best when planted in clumps or drifts, as they would be found growing in the wild.

3.0 Morphology

Saffron, a valuable spice is produced from *Crocus sativus*. Crocus grows up to a height of 8-15 cm depending on species or cultivar. The leaves are linear, dark green with a central white or silver mid-rib. The flowers have six tepals forming a narrow tube like structure at the base. The ovary is present below the ground surface. A crocus has three stamens which are used as spice in case of *Crocus sativus* and one style.

There exists a wide range of flower colours in different species of *Crocus*. White, yellow, orange, purple, red, blue, pink, mauve, violet and lilac can be commonly seen. However, bi-colored and tri-colored types also exist. In some of the species the petals are streaked and yet in some other, the petals are fringed. Seeds are borne in capsule below the ground and as the seed ripens, the peduncle elongates and the pods emerge to the surface of the soil.

Crocus survives by specialized underground storage organ known as the corm. The corm remains dormant during adverse climatic conditions and grows again on the onset of favourable conditions. It is a solid underground structure formed from thickened modified stem with a short, fleshy, vertical axis, covered with dried leaf bases. It has a basal plate from which new roots are formed. The corm will produce

Uses of Crocus

***Crocus ancyrensis* in Pots**

***Crocus sativus* in Garden**

In Flower Vase

***Crocus aureus* in Bed**

In Flower Arrangements

Species of Crocus

Crocus ancyrensis

Crocus aureus

Crocus balansae

daughter corms and multiply year by year. The crocus corm produces multiple flowers per corm. The blooms pucker up at night and also during cloudy weather in most of the species.

Its flowers start blooming in autumn before the leaves appear. The leaves of autumn crocus arise early in the spring when the flowering is over and persist till summer.

4.0 Species and Cultivars

Among the species, approximately 30 of them are cultivated. The varieties cultivated for garden decoration mainly represent five species *viz., C. vernus, C. chrysanthus, C. flavus, C. sieberi,* and *C. tommasinianus*. The spring flowering crocuses are more popular with gardeners.

4.1.0 Spring Blooming Crocus

Golden Bunch (*Crocus ancyrensis* (Herb.) Maw): It is also known as the Ankara crocus and is endemic to North and Central Turkey. It was named *ancyrensis* as it was first discovered in Ankara. The flowers are yellow in colour, 5-7 cm tall, rounded and bloom for a longer period. Medium- sized corm produces about 20 flowers per corm. The plant commonly flowers between the months of February to April. It commonly grows near rocks, bushes and pines. Its corm, rich in sugar and starch, is edible.

***Crocus aureus* Sibth. syn. *Crocus flavus* Weston, *Crocus luteus* Lam.:** It is known as yellow crocus or Dutch yellow crocus. It grows wild in Greece, Bulgaria, Romania and Turkey. The fragrant flowers are bright orange-yellow in colour attaining a height of about 5–6 cm. It naturalises well in gardens. The flower appears in March and forms a cup shaped structure.

Orange Monarch (*Crocus balansae* (J.Gay ex Baker) B. Mathew): The six-petaled flowers are vibrant golden-orange in colour with garnet striations on the reverse. It flowers best in full to partial sunlight. Flowering time is late March to early April. Flowers attain a height of about 10 cm. This species is also good for forcing indoors over the winter.

Silvery crocus (*Crocus biflorus* Mill.): It is native to south-eastern Europe and south-western Asia. It grows upto 10 cm height. The flowers are in shades of pale mauve or white, often with darker stripes on the outer tepals. The flowers appear early in spring.

***Crocus candidus* E.D. Clarke:** The flower appears in spring during April. Flower colour ranges from white, pale yellow to orange. Plant is 7-8 cm high.

Dutch Crocus (*C. vernus* (L.) Hill): It is perhaps one of the most popular species of garden crocus. It blooms early in the spring and is most hardy among all the available species. The flowers of Dutch Crocus are larger than any other species, hence it is commonly known giant crocus. Flower colour ranges from yellow, white and purple to striped or bronze often marked with contrasting streaks or blotches. The flowers bloom for about three weeks. The calyx-shaped flowers open only under full sun shine; they close up in rainy weather and at night. According

Species of Crocus

Crocus biflorus

Crocus candidus

Crocus vernus

Crocus tommasinianus

Crocus chrysanthus

Crocus corsica

Crocus sieberi

to Weryszko and Chwil (2011) *Crocus vernus* blooms from end of March until end of April. The flowers live for 3-4 days. Some of the varieties under this species are Joan of Arc, Little Dorrit, Queen of the Blues, Pickwick, Negro Boy, Purpureus Grandiflorus, Striped Beauty, Flower Record, Remembrance, Silver Coral, Grand Maitre, and Vanguard.

Early Crocus (*C. tommasinianus* Herb.): This species is also known as 'woodland crocus' or 'Tommy'. They are among the earliest to flower species. It blooms from late winter to early spring. The flowers are small, star-shaped with variable flower colour from bluish-lavender to red-purple.

The profuse flowering and spontaneous self-propagation makes this crocus a very good choice for naturalizing. Popular hybrids are Barr's Purple and White Well Purple.

Golden Crocus (*C. chrysanthus* Herb.): *C. chrysanthus* are also known as 'snow crocus'. It is a sweet-scented crocus species with orange-yellow flowers. Nowadays, hybrids are available in many colors, including pure white, pale blue, pale yellow, white with purple edges or blue with yellow centers. *Crocus chrysanthus* is the earliest blooming species among the crocus species. The flowers appear first before the foliage. This species produces smaller flowers in greater numbers. They are very much suitable for planting in rockeries and as ground covers. Some of the hybrids for garden cultivation are Advance, Blue Bird, Blue Pearl, E. A. Bowles, Goldilocks, Ladykiller, Princess Beatrix, Snow Bunting, Ard Schenk, Blue Peter, Cream Beauty, Prince Claus, Romance, Saturnus, Skyline, and Zwanenburg Bronze.

***Crocus corsicus* Vanucchi ex Maw:** The plant grows upto 8–10 cm tall. The fragrant slender flowers, one or two per plant, are bright lilac pink on the inside, pale pink with dark purple on the outside. The flowers appear in spring.

Cretan crocus (*Crocus sieberi* J. Gay): It is a small, early blooming crocus with brilliant orange colour flower which is mostly confined to the stamens and style, fading through the bottom third of the tepal. *Crocus sieberi* is a late winter-flowering crocus. It grows upto a height of 7–10 cm. Hybrids available under this species are Hubert Edelsten, Firefly, Ronald Ginns, Violet Queen, Tricolor and Bowles's White.

***Crocus versicolor*:** It is a vigorous March flowering species which flowers in various shades of mauves. One popular variety is 'Picturatus' also known as 'Cloth of Silver' or 'Zilverlakensi'. This is white or pale mauve with darker markings.

Tuscan Crocus (*Crocus etruscus* Parl.): It grows upto 8 cm tall. The flowers are lilac in colour with purple veins and prominent orange stigmas. The flower blooms in early spring during March- April.

4.2.0 Autumn Blooming Crocus

Some of the most common types of crocus for fall and early winter flowers include:

Iris-flowered crocus (*Crocus banaticus* Heuff. syn. *C. iridiflorus*, *Crocus byzantinus*): It is native to the Balkans. Flowers are purple to violet in colour and appear in autumn. The small inner tepals are surrounded by three larger outer tepals, reflexed when fully open, unlike the more symmetrical crocus species. The flowers are 7-10 cm tall and appear in October. The leaves appear after the flowering is over.

***Crocus cancellatus* Herb. syn. *Crocus nudiflorus* Smith., *Crocus schimperi*:** It is native to Turkey. The flower grows upto a height of 10 cm. Flowering starts from September to November before the foliage appear. The flowers are of blue violet in colour with deep colour veins. *Crocus cancellatus* is good for planting in large group throughout borders, gravel gardens and rockeries, or in container.

Bieberstein's crocus (*Crocus speciosus* M. Bieb.): This is the most popular autumn flowering garden species. Flower colour ranges from mauve and lavender to bluish-violet. Some improved varieties are Barr's Purple, Whitewell Purple, Lilac Beauty, Roseus, Ruby Giant, Albus and Artabir. This species prefers light shade or can be planted under deciduous shrubs.

Saffron crocus (*Crocus sativus* L.): *C. sativus*, also known as the 'Saffron crocus' or "Autumn crocus" flowers during autumn. The stigma of this crocus species is used as spice saffron. The saffron crocus produces purple flowers. The flower emerges before the leaves. The bright red stigma as soon as the flower open are harvested and dried. Approximately about 150 flowers are required to produce one gram of saffron. They need digging up and replanting into fresh soil every 4-5 years.

*Dadpour et al. (2012) studied the floral ontogeny of Crocus sativus and reported that a*fter production of leaves with helical arrangement, a determinate inflorescence forms which becomes completely transformed into a single terminal flower.

Cloth of Gold (*C. angustifolius* Weston): It is a popular early-winter flowering species that produces star-shaped, orange-gold flowers with a deep brown stripe running down the centre of each petal.

***Crocus pulchellus* Herb.:** This species produces pale lilac flowers, each with a yellow throat and contrasting veins of deep purple.

Kotschy's Crocus (*Crocus kotschyanus* K. Koch): It grows upto 10 cm tall. The flowers are pale lilac-blue in colour; blooms in autumn during September-October. Sword-like leaves appear when the flowering is over.

Smooth crocus (*Crocus laevigatus* Bory and Chaub.): These are strongly scented crocuses. It grows upto 10 cm high, with narrow sword-like leaves, and pale lilac or white flowers with a yellow centre. The flowers appear in autumn during October to December.

Long-flowered crocus (*Crocus longiflorus* Raf.): The plants grow upto 10 cm high. It produces pale lilac or purple blooms along with the sword-shaped leaves, in autumn during October-November. The flowers are most strongly scented.

***Crocus medius* Balb. syn. *Crocus ligusticus* M.G. Mariotti:** The grass-like leaves are fully grown in May. They are usually two or three and may reach a maximum height of about 10 cm. The solitary pale purple flowers bear three anthers with yellow-orange pollen and bright red and fringed stigmas. They are autumn flowering plant. The flowering period extends from September through October. The flowers are lightly scented.

***Crocus ochroleucus* Boiss. and Gaill.:** In this species four to six leaves appear along with flowers. Flowers are creamish white to yellow in colour and bearded at base. The flower blooms during October–December.

Hairy Crocus (*Crocus pulchellus* Herb.): The plant grows upto 10 cm high. The flowers are pale lilac-blue in colour and blooms in autumn (September to November), before the leaves appear.

***Crocus cartwrightianus* Herb.:** The small plant grows upto a height of 6 cm. The flowers are in shades of lilac or white with purple veins and prominent red stigmas, appear with the leaves in autumn and winter. *C. cartwrightianus* is the presumed wild progenitor of the domesticated triploid *Crocus sativus* – the saffron crocus.

5.0 Genetics and Breeding

5.1.0 Genetics

There are no wild forms and it exists solely in culture. Being triploid with chromosome number $2n = 3x = 24$, saffron never bears seed and it is propagated exclusively in a vegetative way by corms. The triploid genome of saffron causes the production of abnormal pollen triggering self-sterility. Cytological studies showed that among species collected from Iran, *C. michelsonii* was the closest to *C. sativa* L. A new cytotype *C. pallasii* subsp. *hausknechtii* with $2n = 2x = 12$ chromosomes from west of Iran was reported for the first time (Mohamad *et al.*, 2012).

5.2.0 Breeding

The literature on saffron breeding is very limited. There are known old works from Azerbaijan Republic more or less concerning with the breeding problems (Kapinos, 1965; Muzaferova, 1970; Agayev *et al.*, 1975). The urgency of saffron breeding problems and the necessity to solve them with the application of new extraordinary approaches was stated before (Agayev, 1994a,b). With respect to the clonal nature of saffron, it is believed that there is only one cultivar worldwide. Apparently, the lack of new cultivars of saffron at present may not be explained by the impossibilities of the improvement of this plant through clonal selection. To solve this problem, it is necessary for a researcher to be properly acquainted with the biology and genetics of saffron and work up subtle puzzled methods of the clonal selection especially for saffron, different from those of the other plants. Two main difficulties of saffron breeding through clonal selection explained by Agayev *et al.* (2007) are as follows.

a) Difficulty in the Recognition of the Clones

In any plantation saffron is represented by plants existing in highly different "ages" of individuals connected with different sizes of corms underground. Above the ground, these plants differ in the size and number of their flowers (at the stage of flowering), also in their number and size of leaves. If some plants are sharply different from the others in certain characters, interesting for the aim of breeding, a researcher cannot practically identify them. Naturally he does not see corms underground, and cannot elucidate the cause of the mentioned differences: whether these differences are due to the age (size) of corms, or because of the genotypes of plants. So, genetically different (if exist) and similar plants will continue to grow together and not be used as a subject for breeding.

b) Difficulty in the Multiplication of Clones and Bringing them to Cultivars

Suppose a farmer recognizes certain plant(s) which could be used as a clone with good economic characters. Multiplication of such clone(s) would be turned as insoluble problem. One saffron corm at planting with proper care produces on average 4 corms of middle sizes during vegetation (one year). At such intensity of propagation, it could be brought to about 1,000,000 corms only after 17 years. This amount could be enough for planting on the area of 2 ha. It is clear that a farmer will never accomplish such an exhausting work of many years. Therefore, farmers would not pursue the aim to make new cultivar of saffron even if they have been lucky to find some clones with very highly expressed economically valuable characters. Concerning to the researcher, in our opinion, he could be able to do it in the favourable conditions, namely in the case of worthy appreciation of his long term hard work on this way by existing law on breeding. That is why in all over the world saffron has been remained as one cultivar despite of existence in culture during hundreds and thousands of years.

Alternatively, genetically changed superior plants of saffron could be propagated rapidly via *in vitro* technique. Investigations in this direction are very promising. Unfortunately, the experiments pursuing rapid corm propagation of saffron have not been successful so far and a few *in vitro* developed corms had been produced. Matured corms of *in vitro* origin in mass production had not been produced.

With the object of breeding new cultivars of saffron with economically valuable traits, the following two principles were proposed by Agayev *et al.* (2007).

a) Searching, identification and separation of superior clones in existing plantations,
b) Creating new valuable forms experimentally.

Having an ancient history of cultivation, saffron apparently should contain a lot of genetically changed forms (clones) as the result of mutations in somatic cells. The task is to find and study them individually, to separate the economically valuable forms and to bring them to the new industrial cultivars.

In an attempt, corms of *C. sativus* were planted in the greenhouse, intensively growing apical meristems of plants were treated by modified injection method with the solutions of different concentrations of polyploidizing reagents (Agayev *et al.*, 2007). After treatments plants continued their vegetation during 1.0-1.5 months, having completely green leaves, and then their leaves became gradually yellow and dried. At the same way many plants of close diploid wild relatives of *C. sativus*, namely, *C. cartwrightianus* Herb., *C. hadriaticus* Herb., *C. pallasii* Goldb. and *C. asumaniae* Mathew and Baytop were treated. The increase of ploidy level of specified plants will allow them to accomplish wide range crossings and produce desirable hybrids for breeding purposes.

Lack of genetic variation restricts the use of traditional plant breeding based on selection (Shokrpour, 2019). Probable wild relatives could be an excellent source of genes to alter saffron traits by cross-pollination. In addition, an induced mutation approach with various mutagen agent treatments is an alternative to produce genetic variations. Recent advances in sequencing methods and next-generation sequencing (NGS), provide efficient approaches such as transcriptome sequencing along with proteome and metabolome information, which would help to exploit functional genomics toward genetic engineering of the economic traits of saffron.

6.0 Propagation

6.1.0 Vegetative Propagation

Crocus is commercially propagated by means of corms. Large and medium size corms are used for planting. Recommended corm size for planting is 8-9 cm. Larger corm gives multiple flower and they are not susceptible to flower blasting. The crocus corm that is planted in autumn gets completely used up during growth and flowering and produces new daughter corms. A corm of *Crocus sativus* survives for only one season, producing up to four to five daughter corms and ten 'cormlets' that eventually give rise to new plants (Ahmad *et al.*, 2014). Koocheki *et al.* (2016) reported maximum flower number and dried stigma yield in *Crocus sativus* with large-sized (>10 g) mother corms. Koocheki and Seyyedi (2019) reported maximum N, P, and K concentrations in corm mass when mother corms were originated from a 4-year-old farm and the minimum values were obtained in the mother corms originating from a 1-year-old farm. They further reported that mother corms planting, harvested from 4-year-old farms, at 20 cm soil depth can effectively alleviate the adverse effects of freezing injury in saffron.

According to Douglas *et al.* (2014), flower and fresh stigma production in *Crocus sativus* is highest in a 3 year management cycle, but if large progeny corms are desired, a 2 year management cycle would produce more large progeny corms. When the mother corm weight is greater than 33 g, it produce 4, 16 and 48 flowers in first, second and third year respectively. Mother corm weighing 11–15 g produces 1, 4 and 20 flowers in first, second and third year respectively. For the three respective harvest years, mother corm weighing greater than 33 g, produced 7, 30 and 69 progeny corms with a weight of 60, 406 and 427 g.

6.2.0 Micropropagation

Most of the crocus species are sterile triploid plants. They are propagated by corms as the flowers are sterile and fail to produce viable seeds. An *in vitro* multiplication technique like micropropagation can also be used for the propagation of crocus. The plants were able to form shoots or corms within 5 to 30 weeks from the start of culture (Gul Zaffar *et al.*, 2012). Various micropropagation studies have been conducted wherein the researchers have used various explants like style and perianth, leaf and corms *etc.*

Propagation of Crocus

Seedlings of Crocus

Planting of Crocus Corms

Growth of Seedlings in Container

Cultivation of Crocus

Cultivation of Crocus in Open

Planting of Crocus in Greenhouse

Harvesting and Diseases of Crocus

Harvesting of Crocus

Corm Rot of Crocus

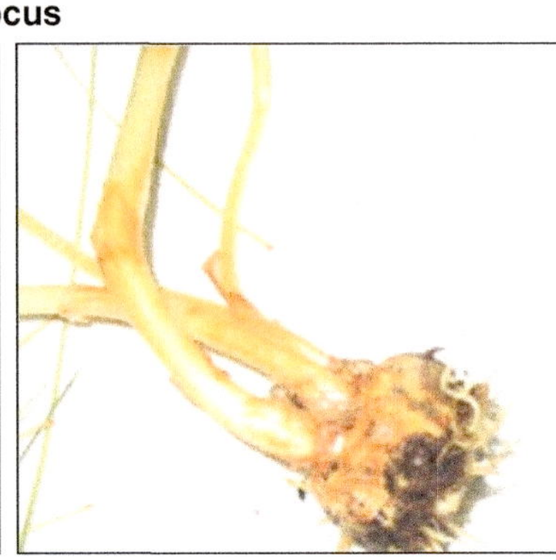
Fusarium Wilt

Ebrahimzadeh and Karamian (2000) reported that style and perianth explants produced stigma-like structures that proliferated forming up to 100 structures per explant. MS medium supplemented with 5 mg/l kinetin and 4 mg/l NAA was used as the basal medium. Almost all of the stigma-like structures formed directly from explants, instead of from callus.

Blazquez *et al.* (2004) used MS culture medium supplemented with 0.5 mg/l BAP and 0.1 mg/l 2,4-D for induction of somatic embryogenesis. Embryogenic calli were subcultured on MS medium containing 1 mg/l BAP and 0.05 mg/l NAA for multiplication in solid medium.

Karamian (2004) reported germination of maturated embryos on half-strength MS medium supplemented with 25 mg/l GA_3. Finally, complete plantlets were obtained by transferring germinated embryos into half- strength MS medium supplemented with 1 mg/l NAA and 1 mg/l BA at 20°C under 16/8 h (light/dark) cycle.

Leaf explants produced callus that regenerated somatic embryos and plantlets when cultured on 10 lM BA and 0.5 lM 2, 4-D. These were used for microcorm induction which was promoted by a half-strength MS medium plus 9 per cent sucrose (Raja *et al.*, 2007).

Mir *et al.* (2010) cultured various explants of *Crocus sativus* on different nutrient media and observed highest response (60 per cent) with half ovaries on media supplemented with 27 µM NAA and 44.4 µM BA. Maximum size (1.3 g) of microcorms were obtained from apical buds on the LS media supplemented with 21.6 µM NAA and 22.2 µM.

Ahouran *et al.* (2012) reported that during *in vitro* propagation of *Crocus cancellatus*, one to five cormlets emerged from every responding explant through direct organogenesis. Apical halves of corms were more highly responsive, produced a maximum multiplication rate with 3.45 ± 0.06 cormlets per explant in MS medium supplemented with 3 per cent sucrose and 2 mg/l NAA and 1 mg/l BAP. *In vitro* rooting of cormlets was induced on MS medium without plant hormones. Sivanesan *et al.* (2011) induced primary somatic embryos from the corm explant of *C. vernus* on MS medium supplemented with 1.0 mg/l 2,4-D and 0.5 mg/l TDZ with a mean number of 52 somatic embryos per explant. Sivanesan *et al.* (2012) further induced secondary embryo by culturing primary somatic embryos on SH medium fortified with 2.0 mg/l BA and 0.5 mg/l NAA.

Verma *et al.* (2016) reported highest frequencies of callus induction (100 per cent) and shoot regeneration (70 per cent with 7.2 shoots/callus) in *Crocus oliveri* ssp. *Oliveri*, using the MS medium containing 5 per cent (w/v) sucrose supplemented with 4 mg/l NAA + 4 mg/l TDZ and 2 mg/l IAA + 2 mg/l TDZ + 2 mg/l BAP. For micro propagation of *C. tommasinianus*, corms were used as explants. Callus initiation medium (10 mg/l NAA and 1 mg/l BA) induced callus formation with abundant somatic embryos. Embryos developed shoots when auxin content of medium was decreased for initiation of cultures (Freytag *et al.*, 2017).

7.0 Soil and Climate

7.1.0 Soil

Crocuses can be grown in a wide range of soils from light sandy to clay loam. However, deep, well drained, friable soils rich in organic matter and nutrients are most ideal for growing crocus. Water logging is detrimental for crocus as the corms may rot in water logged soils. When grown on heavy textured soil, addition of organic matter or making raised beds is recommended. Crocuses grow well in a neutral to near neutral pH and ideal pH should be in the range of 6-7.

Cardone *et al.* (2020) reported good flower and corm production in *Crocus sativus* when planted in media with loam and sandy loam texture, not very calcareous, with a sub-alkaline and neutral pH, low electrical conductivity, organic matter content between 5.46 and 8.67 g/kg, and active lime content between 21.25 and 26.25 g/kg.

Frederic and Weathers (2000) reported that on a dry weight basis, corm growth of saffron increased in aeroponics and hydroponics as compared with growth in soil suggesting the possibility that saffron can be grown aeroponically and hydroponically.

Crocus chrysanthus 'Ard Schenk' is highly adapted to intensive hydroponics for both interior decoration and public area decoration. In aquaponics, there is increased resistance to diseases and pests and human intervention is not needed for its control (Adriana and Maria, 2016).

7.2.0 Climate

Most of the crocus species are hardy and resistant to cold temperature. They can thrive in cold to moderate winter conditions and fail to grow in hot climates. A minimum of 15 weeks of chilling is required for breaking corm dormancy. Crocuses require good amount of sunshine for its growth and are drought tolerant. Some of the species can grow in partial shade too. However, there are some reports that suggest rainy autumn, mild winters and warm summer as the optimal climatic conditions for saffron crocus (Mollafilabi, 2003; Fernandez, 2004).

Mohamed *et al.* (2007) reported that in *Crocus vernus*, dry mass and cell size were higher at the lower temperature regime of 12/8°C (day/night), while leaf lasted longer than at the higher temperature regime of 18/14°C (day/night). At higher temperature, corm growth stopped before the first visual sign of leaf senescence, suggesting that corm growth controls leaf life duration in *C. vernus*.

In another experiment on *Crocus vernus*, prolonged growth at low temperatures results in the development of larger perennial organs and delayed foliar senescence. Corm growth stops before the first visual sign of leaf senescence indicating that leaf duration alone does not determine final corm size. A shift in carbon partitioning, not limiting carbon supply or leaf duration, which is responsible for the smaller biomass of the corm at higher temperatures (Lundmark *et al.*, 2009).

Saffron crocus grows well under temperate and dry environments, cold weather has an important role in its vegetative growth (Ghorbani and Koocheki, 2017). Maleki *et al.* (2017) suggested the main limiting factors in cultivation of *Crocus sativus* are high rainfall during reproductive stage, high relative humidity and low annual rainfall.

8.0 Cultivation

8.1.0 Planting Time

The best time for planting spring flowering crocus is early autumn (October-November) and those of autumn flowering crocus is late summer (August – September). Ideal planting time for spring flowering crocus is 6 to 8 weeks before the first expected frost. This allows the corm to develop a root system before going dormant. Flowering season in spring flowering crocus extends from mid December to early April whereas, autumn flowering blooms in September/October.

8.2.0 Plant Spacing and Planting Depth

The spacing will depend on species and varieties. A spacing of 3-6 cm between each corm is a good starting point. Greater spacing is required for large blooming varieties. For *C. spaciosus*, a spacing of 6-8 cm is ideal. For planting in pots, 5 to 6 corms should be planted in 10 cm pot, 7 to 9 corms in 12.5 cm pot and 10-12 corms in 15 cm pot. Since crocuses are small plants they look best when they are planted in groups. For dense plantings, corms should be spaced at 2.5 to 5 cm apart.

Yau and Nimah (2004) reported density of planting had no effect on numbers of corms produced per corm cluster, but it had a significant effect on corm weight, which had a large effect on flower production of *Crocus sativus*. They reported 20 cm × 20 cm spacing to be a optimum spacing for producing heavier corms and higher flower numbers per corm cluster. According to Temperini *et al.* (2009), number of flowers per square metre observed in the high density planting (179 corms/$m^{2)}$ was significantly higher by 10 per cent and 28 per cent in comparison with the number of flowers recorded in the medium (143 corms/$m^{2)}$ and low density (139 corms/m^2), respectively. Mohammad *et al.* (2011) reported a spacing of 10 × 20 cm as optimum for getting good yield and quality flower and corm production in *Crocus sativus*.

Planting depth varies with the size of the corm to be planted. The thumb rule is corms should be planted at a depth of three times their own size. Larger size corms should be planted around 12 cm deep or a little shallower in heavier soils. Smaller size corms can be planted at 6-8 cm depth. Corms should be planted with the basal plate in downward position and tip in upward position. After planting, light watering should be done. Galavi *et al.* (2009) observed that number of daughter corms significantly decreased by increasing planting depth.

8.3.0 Irrigation

The soil should have sufficient moisture at the time of planting of corms and there after light irrigation is required till sprouting. Later, irrigation is done once in 4-6 days depending on soil and weather conditions. Watering should be withheld during flowering period. Care should be taken not to waterlog the soil. Keep the soil evenly moist after flowering. Afterwards, irrigation should be withheld at least 4-6 weeks before lifting of corms. Seyyed *et al.* (2014) reported that flower number, fresh flower and dry stigma yields significantly decreased by irrigating at 50 per cent of saffron water requirement compared with 75 per cent or 100 per cent of saffron water requirement.

In *Crocus sativus* production, irrigation management plays a key role in nutrients uptake, daughter corm formation and stigma yield. The maximum flower number and dried stigma yields were observed when four irrigation rounds (irrigation in August, October, November and April) were performed. Although four irrigation rounds are highly recommended, applying at least two rounds (in August and before flowering in October) under limited water conditions is unavoidable (Koocheki *et al.*, 2016).

Dastranj and Sepaskhah (2019) observed a reduction in saffron yield due to irrigation water salinity or deficit irrigation and it can be compensated by the in-furrow planting method. They further reported that saffron yield is more sensitive to water stress than salinity stress. In general, they recommend the irrigation regime in basin planting to be 100 per cent saffron water requirement in salinity levels of 0.45-3 dS/m; however, in in-furrow planting, it is 75 per cent saffron water requirement in 0.45-1.0 dS/m salinity levels, and 100 per cent saffron water requirement in salinity level of 2-3 dS/m.

Shajari *et al.* (2020) reported that one week irrigation interval increased replacement corms yield by 20 per cent compared with two weeks irrigation intervals. Selection of moderate and light soil texture and shorter irrigation intervals (equal to 3000 m^3/ha) are important factors for improving flower and corm yield of saffron crocus.

8.4.0 Manuring and Fertilization

Fertilizer requirement varies with climatic conditions, irrigation methods and soil type. The soil should be amended with compost at planting. Application of bone meal or an organic phosphorous fertilizer encourages corm growth and flowering. A balanced fertilizer can be added in early autumn.

Mohammad (2008) recorded maximum flower fresh weight with longest stigmas with the application of urea @ 50 kg/ha + super phosphate @ 40 kg/ha + cow deep litter manure @ 20 t/ha in *Crocus sativus*. The highest corm multiplication rate (5.7 new corms/year) was recorded with the application of urea @ 50 kg/ha + cow deep litter manure @ 20 t/ha.

Omidi *et al.* (2009) reported that application of nitroxine can be used to reduce the application of nitrogen fertilizer in agro-ecosystem which will minimize environmental pollution. Foliar application of Fe (3 l/ha) also showed higher flower and corm yield than foliar application of K and Zn (Akbarian *et al.*, 2012).

Koocheki and Seyyed (2015) suggested composted cattle manure to be a better option than that of chemical fertilizer in terms of N and P use efficiency by 21 and 61 per cent, respectively in *Crocus sativus*.

Yarami and Sepaskhah (2015) reported that application of cow manure @ 60 Mg/ha increased saffron yield by about 23 per cent, due to improving soil fertility and supplying the nutrient requirements of plant.

Gresta *et al.* (2016) reported crop rotations of *Crocus sativus* with Faba bean as previous crop determined the highest performance in terms of flower number, stigma yield and replacement corms.

Shuwen *et al.* (2019) reported that presence of fungi such as *Ilyonectria macrodidyma, Mortierella humilis* and *Chaetomiaceae* in soil helps in good corm growth of *Crocus sativus*. They further reported that fungi affecting corm germination and blooming mainly included *Nectriaceae, Penicillium, Aspergillus* and *Saccharomycetales.*

8.5.0 Mulching

Application of organic mulch can be effective in increasing the flower and corm yields of crocus by improving the soil temperature, water availability and organic matter content of soil. Applying mulch with suitable materials may reduce unfavourable conditions and provide suitable soil microclimate for crop cultivation. Crop residues at the soil surface could lead to better root growth and higher yields. Mulching with a 4-5 cm thick layer of shredded bark mulch in autumn flowering crocus, or dried leaves in late summer crocus is beneficial. The mulch will conserve soil moisture, prevent weed growth and prevent the soil from splashing unto the delicate crocus flowers. The mulch should be removed after flowering to prevent the soil from staying too wet in winter. In cold areas, the mulch will protect the corm from cold.

Mulumba and Lal (2008) observed that soil moisture contents at field capacity and the plant available water capacity were positively affected by mulching rates. Moghaddam *et al.* (2012) found that fresh and dried flower yield and total corm yield of saffron crocus was increased by applying wheat straw mulch (8 t/ha). However, total corm numbers was decreased. Moghaddam *et al.* (2013) further reported that application of wheat straw as mulch in October increases flower number, fresh and dried flower yields. It also increased yield of replacement corms in Saffron Crocus in second year of planting.

Rohi *et al.* (2014) observed that using living mulch (watermelon) will accelerate the flowering process in *Crocus sativus*. Koucheki *et al.* (2016) reported that Persian clover and bitter vetch as cover crops improve saffron flowering and other agronomic traits, due to modified received radiation and soil physical and biological properties. Kader *et al.* (2017) illustrated that during the hot summer, soil temperature increases that causes soil water loss and negative impacts on crop yields.

Jafarimoghadam *et al.* (2018) reported that mean temperature during the summer season is decreased by mulching and cover crop management in saffron than the bare soil. In addition, the management practices succeed to improve soil physical properties, enhanced flower number and stigma dry yield.

8.6.0 Weed Management

Weeds compete with the crop for space, light, nutrients and water. Generally, weeding is done by mechanical ways *i.e.* hand weeding. Chemical weed control is essential for commercial operations and herbicides selection depends upon the nature of weeds infesting the crop and the sensitivity of the crop towards the herbicide.

Javad *et al.* (2013) reported lowest weed density (0. 7 and 0.8 species/$m^{2)}$ of narrow-leaf and broad-leaf species in field of *Crocus sativus* when vetch was used as cover crop. Maximum flower number was also observed with vetch cover cropping. Applying different crop residues and cover crops decreased density and dry weight of weeds. Cover crops increased growth and yield of saffron due to enhancement in nutrient and nitrogen availability as nitrogen fixation and improvement in soil characteristics.

Hossein *et al.* (2014) reported some of dominant weed species in cultivation of *Crocus sativus* as mouse barely (*Hordeum murinum*), wild barley (*Hordeum spontaneum*), hoary cress (*Cardaria draba*), and yarrow (*Achillea millefolium*). Chemical herbicides destroyed grasses and broadleaf weed, but it also destroyed saffron plant. Application of mulch was not be able to control weeds, however cover crops of barley caused least weeds dry matter weight similar to hand weeding.

8.7.0 Flower Forcing

Forcing is the technique for making the bulbs to flower out of season to match the market demand. The main objective of flower forcing is either promotion of flowering and or retardation of flowering. In general, flowering process in bulbous ornamentals involves five successive stages *viz.*, induction, initiation, organogenesis (differentiation of floral parts), maturation and growth of the floral parts, and anthesis.

Flowering geophytes require a sequence of temperature during their life cycle to flower. Spring flowering crocus require warm - cold - warm sequence of temperature to express active growth and complete their life cycle. In crocus, flower initiation occurs after harvesting of corms during storage. Crocus corms exhibit summer dormancy. Normally, spring flowering crocus blooms during March-April. By giving various temperature treatments, flowering can be obtained from November to May.

For forcing of corms, precooled bulbs should be stored at 9°C and non precooled bulbs should be stored at 17°C. For best presentability, maximum number of bulbs should be planted in a 15 cm pot. Crocus can also be forced hydroponically using special glasses.

Provide rooting room temperature for four to six weeks. A temperature sequence of 9°C - 5°C - 2°C should be given for forcing. The temperature of 9°C is given for the development of roots and it should be maintained until roots grow out of the bottom of the container. The 5°C temperature should be maintained until the shoots become about 2.5 cm long and a temperature of 2°C is used to retard the shoot growth. When the cumulative cold temperature treatment equals to 15-16 weeks, pots may be moved to greenhouse for forcing. Greenhouse temperature should be 13-15°C night and 15-17°C day. Crocus can be forced very fast. The time and duration of cold treatment should be in accordance to the scheduled date in which flowering is desired.

Molina *et al.* (2004) reported that the freshly formed replacement corms of *Crocus sativus* had no chilling requirements. Sprouting could be accelerated by a short curing at 30°C. A temperature of 23-25°C is also optimal for flower initiation. The optimal temperature for flower emergence was 17°C. Flowering can be forced from early September until the end of May. Molina *et al.* (2004) also demonstrated that when corms lifted 1 month before leaf withering, incubated for 55 days at 25°C, and then forced at 17°C, flowered by early September, 6 weeks earlier than in the open. Flowering could be further accelerated (up to 7 days) by curing the corms for 20 days at 30°C prior to 25°C storage. The flowering of corms lifted after leaf withering could be delayed until early December by extending the incubation at 25°C for 150 days.

Conditions for the cold-storage of saffron crocus (*Crocus sativus*) corms to delay flowering have also been characterized by Molina *et al.* (2005). Storage of corms at 2°C after flower initiation resulted in a time-dependent abortion of those flowers already initiated. The more advanced the stage of flower initiation at the beginning of cold-storage, the faster the rate of flower abortion. Corms stored in the cold before flower initiation, formed flowers when incubated after storage at 21–25°C. Corms lifted after leaf-withering, and stored at 2°C in 1 per cent oxygen for 70 days, could be forced to flower from early December until the end of January. Flowering could be further delayed until May by extending the duration of cold-storage.

Zahmati *et al.* (2018) reported that 23-25°C temperature in the greenhouse was suitable for vegetative growth whereas, farm temperature (10-18°C) led to further corm development in *Crocus sativus*. Munoz Gomez *et al.* (2002) stated that the storage of the corms at 30°C for 45 days increased the number of flowers as compared to corms forced to sprout directly at 17/10°C after leaf withering.

9.0 Harvesting and Postharvest Management

9.1.0 Corm Harvesting and Storage

The corms need to be harvested after flowering and store for next year. Corms should be lifted around four to six weeks after flowering and when foliage has died down completely. Spring flowering crocus dies down in summer after flowering is over while in autumn flowering crocus die down early in spring. After lifting, the corms should be placed in a ventilated tray for drying. The dried foliage should be removed from the corm and corm cleaned before drying. Corms should be placed in a warm room with good air flow at around 21°C for two to three weeks for curing. When fully cured (dry) the old corms can be twisted off to leave a clean basal scar, if the separation is not clean or the corms look diseased they should be discarded. As removal of husks encourages loss of moisture, only a dirty outer layer should be removed.

In permanent plantations, the corms should be dug out after every 3-4 years. Divide them, keeping only healthy corms and replant. Over time, large clusters of crocus may become overcrowded and produce fewer flowers. Dividing the corms into smaller clusters and replanting increase flower productivity.

If the crocus corms are planted in the lawn, mowing should be withheld in the area until the foliage begins to wither. This will help the corms to accumulate carbohydrates and store it for next year's flowers.

9.2.0 Breaking Dormancy of Corms

Corms dug in the warm parts of the years are dormant and require 3 to 4 months of cold storage (2-4°C) to break dormancy. Store corms preferably in well ventilated room with 75 per cent R.H. Room should be opened for 1 hour once or twice a week to prevent building up of toxic gases generated from respiration and rotting of sprouts. One to three weeks prior to planting, brought them to warmer conditions to break dormancy and to encourage root swelling. Root bud swelling indicated that corms are ready for planting.

9.3.0 Postharvest Management

Separation of the stigmas from the petals and stamens should be done as soon after picking as possible. When the flowers begin to get wilt, it is harder to pick out the stigmas from the other parts. If the flowers are not processed immediately after harvesting, these can be spread on a paper on a tray and stored in the refrigerator. Harvested saffron is inherently high quality, but the quality is degraded due to improper handling after harvesting and during packaging. If the separation of pistil, drying and packing is delayed the crop harvesting losses will be more due to microbial attact which may lower the market value of the product. The pistol should be separated from the flowers within 6-8 hours of removal from flowers, dried using a dryer and stored in a cool place (Munoz –Gomez *et al.,* 2005).

10.0 Diseases and Pests

10.1.0 Diseases

10.1.1 Fusarium Rot

Fusarium rot can attack the corms. Keep the soil well drained to prevent these diseases. Also, dividing and moving the corms after every three years of growing in one area, will help avoid disease building up.

Gupta and Vakhlu (2015) found that *Bacillus amyloliquefaciens* strain W2 was effective against corm rot caused by *F. oxysporum* R1. The disease incidence in pot assays was 93 and 40 per cent in absence and presence of *B. amyloliquefaciens* W2 respectively. *B. amyloliquefaciens* W2 can be used as a potent biofungicide for corm rot diseases in *C. sativus.*

10.1.2 Gladiolus Dry Rot (*Fusarium oxysporum* sp. *gladioli*)

Numerous small, black lesions develop on the corms which coalesce to large black areas. The corms then suffer from a dry rot and become shrivelled.

10.1.3 Gladiolus Scab (*Burkholderia gladioli* pv. *gladioli*)

Round sunken craters develop towards the base of the corms. Each crater has a prominent raised rim and is covered with shiny, varnish like coating.

10.1.4 Mould

Corms in storage are prone to rot and mould, if kept too moist. So proper curing before storage is recommended. Storage rot is caused due to blue mould. Buff coloured or pink resting bodies of the fungus develop on the corms. Under moist conditions, the blue mould develops on the lesions and the corm decay rapidly.

10.2.0 Pests

10.2.1 Rodents

Rodents like squirrels, mice and moles eat up the corms. The planting area should be protected with a wire barrier, such as chicken wire, to prevent digging and damaging in buried wire cages. Spreading blood meal around the planting bed also helps deter the pests.

10.2.2 Slugs and Snails

Slugs and snails feed on the foliage, especially in spring for the large flowering types. Remove the mulch away from the leaves in spring to reduce hiding places for the slugs. Place organic baits, such as iron phosphate around the area.

10.2.3 Nematodes

With the increase in area under cultivation, nematodes such as *Pratylenchus* spp., *Tylenchorhynchus* spp., *Bursaphelenchus fungivorus* and *Ditylenchus dipsaci* were reported to infect the crocus. Proper soil treatment, crop rotation and uprooting the infested plants and dispose it properly are suggested remeady.

11.0 Biotechnology

11.1.0 Molecular Marker

The quality of saffron mainly depends upon stigma processing and species origin (Sharaf-Eldin *et al.*, 2008). The phenological evaluation of *C. sativus* flowers obtained from corms from various regions of the world showed some variation in the color of the flowers as well as their fragrance and tepal lobes, but not in the pollen size (Caiola *et al.*, 2000). Improvements in DNA biology have provided new information for taxonomic analysis among accessions (Caiola *et al.*, 2004). A recent PCR-RAPD approach was used to analyze the extent of the polymorphisms between *C. sativus* genotypes grown in the Saudi Arabia (Sharaf-Eldin *et al.*, 2018). In this study, the DNA fingerprints of the stigmata of *C. sativus* genotypes [K1 and K2 = *C. sativus* var. *cashmerianus*, C1 = *C. sativus* (non-mutant), T1 = mutant (T0-2B), T2 = mutant (T1-2B), T3 = mutant (T4-2A)] were determined according to the floral parts, and a total of 10 decamer primers were used for PCR-RAPD analysis. Only three pairs of arbitrary primers showed polymorphisms (33.3 per cent–88.2 per cent) in the total genomic DNA extracted from these genotypes. Jaccard's similarity index (JSI) ranged from 0.88 to 1.0. An unweighted pair group method with arithmetic mean (UPGMA) similarity and dendrogram matrix showed that two genotypes (T1-2B and T4-2A) were closely related to each other and to the strain CM-cashmerianus, while the T0 of *C. sativus* genotype showed divergence. Thus, PCR-RAPD could be helpful in the identification of commercial saffron lines and could be a useful tool to supplement distinctness, uniformity, and stability (DUS) analyses for saffron genotypes to maintain their original identity.

11.2.0 Transcriptome Analysis

The characterization of saffron transcriptome is in fact a prerequisite to shed light on essential biological processes including the molecular basis of flavour and color biogenesis, genomic organization and flower development of *C. sativus* in particular and family Iridaceae in general. Although, a few studies have been carried out on the *C. sativus* plant to understand the flower development and apocarotenoid biosynthesis [Moraga *et al.*, 2004; Castillo *et al.*, 2005; D'Agostino *et al.*, 2007], molecular basis of these essential processes is still largely unknown. The study of Baba *et al.* (2015) presents first report on transcriptome sequencing and analysis of Crocus stigma and rest of the flower (Flower minus stigma) using high-throughput Illumina sequencing. The assembled transcripts were annotated to identify genes involved in various steps of the carotenoid/apocarotenoid pathway. The transcriptome aids in significant addition to the number of genes which are potentially involved in apocarotenoid biosynthesis. The transcriptome was further screened for the identification of the genes encoding transcription factors (TFs) so as to generate transcription factor database from Crocus which may help in unravelling the regulatory mechanism of apocarotenoid biosynthesis. Finally expression specificity of the assembled transcripts provided an accurate estimation of the biological processes involved in different tissues (Stigma or Flower). The Crocus transcriptome provides a platform for understanding the molecular basis of carotenoid/apocarotenoid pathway and various other biological processes pertaining to this crop.

12.0 References

Adriana, M.I. and Maria, B. (2016) *J. Hortic. Forest. Biotech.*, **20**: 7-10.

Agayev, Y.M. (1994a) In: *Abstracts of the 2nd Symposium on Saffron and Farming of Medicine Plants*, Gonabad, Iran, pp.12.

Agayev, Y.M. (1994b) In: *Abstracts of the 2nd Symposium on Saffron and Farming of Medicine Plants*, Gonabad, Iran. pp.13.

Agayev, Y.M., Muzaferova, R.S. and Savchenko, S.P. (1975) *Bulletin of Agricultural Science*, Moscow, **10**: 121-123.

Agayev, Y.M., Shakib, A.M., Soheilivand, S. and Fathi, M. (2007) *Acta Hortic.*, **739**: 203-207.

Ahmad, M., Zaffar, G., Habib, M., Arshid, A., Dar, N.A. and Dar, Z.A. (2014) *Sci. Res. Essays*, **9**: 13-18.

Ahouran, M., Hosseini, R. and Zarghami, R. (2012) *J. Crop Sci. Biotech.*, **15**: 47–51.

Akbarian, M.M., Sharifabad, H.H., Noormohammadi, G. and Kojouri, F.D. (2012) *Annals. Biol. Res.*, **3**: 5651-5658.

Blazquez, S., Olmos, E., Hernandez, J.A, Hellin, E., Fernandez, J.A. and Piqueras, A. (2004) *Acta Hortic.*, **650**: 261-267.

Caiola, M.G., Caputo, P. and Zanier, R. (2004) *Biologia Plantarum*, **48**: 375–380.

Caiola, M.G., Somma, D. D. and Lauretti, P. (2000) *Annals of Botany*, **58**: 73–82.

Cardone, L., Castronuovo, D., Perniola, M., Scrano, L., Cicco, N. and Candido, V. (2020) *Agronomy,* **10**: 1154.

Castillo, R., Fernandez, J. and Gomez-Gomez, L. (2005) *Plant Physiol.,* **139**: 674–689.

Dadpour, M.R., Naghiloo, S., Gohari, G. and Aliakbari, M. (2012) *Flora - Morphology, Distribution, Functional Ecology of Plants,* **207**: 257-263.

D'Agostino, N., Pizzichini, D., Chiusano, M.L. and Giuliano, G. (2007) *BMC Plant Biol.,* **7**: 53.

Dastranj, M. and Sepaskhah, A.R. (2019) *Sci. Hort.,* **251**: 215-224.

Douglas, M.H., Smallfield, B.M., Wallace, A.R. and McGimpsey, J.A. (2014) *Sci. Hort.,* **166**: 50-58.

Ebrahimzadeh, R.T. and Karamian, R. (2000) *Pakistan J. Bot.,* **32**: 141-150.

Fernandez, J.A. (2004) *Recent Research and Development in Plant Science,* **2**: 127–159.

Frederic, F.S. and Weathers, P.J. (2000) *J. Herbs, Spices, Medicinal Plants,* **7**: 25-35.

Freytag, C., Pabar, S.A., Demeter, Z., Simon, Á., Resetar, A., Molnar, A.V. Sramko, G. and Mathe, C. (2017) *Acta Biologica Cracoviensia Series Botanica,* **59**: 31–39

Galavi, M., Mousavi, S.R. and Ziyaie, M. (2009) *Asian J. Plant Sci.,* **8**: 375-379.

Ghorbani, R. and Koocheki, A. (2017) *Springer International Publishing,* Cham.

Gresta, F., Santonoeto, C. and Avola, G. (2016) *Sci. Hort.,* **211**: 34-39

Gul, Zaffar, Bhat, H.A., Iqbal, S. and Maying, B. (2012) *Souvenir and Abstracts 4th International Saffron Symposium,* October 22-25, pp. 22.

Gupta, R. and Vakhlu, J. (2015) *European Journal of Plant Pathology,* **143**: 123–131.

Hossein, Z.H., Reza, G., Hassan, R.M.M. and Hossein, R. (2014) *Saffron Agronomy and Technology,* **2**: 45-58

Jafarimoghadam, M., Ghasemi, E. and Khalesi, N. (2018) *ISTRO.*

Javad, S., Surur, K., Afsaneh, A.G. and Rahmatoahh, G. (2013) *Journal of Saffron Research,* **1**: 57-72.

Kader, M.A., Senge, M., Mojid, M.A. and Nakamura, K. (2017) *International Soil Water Conservation Research,* **5**: 302–308.

Kapinos, G.E. (1965) In: *Biology of development bulbous and cormous plants in Apsheron.* Azerbaijan Nat. Acad. of Sciences, Baku.

Karamian, R. (2004) *Acta Hortic.,* **650**: 253-259.

Koocheki, A. and Seyyedi, S.M. (2015) *Industrial Crops and Products,* **71**: 128-137.

Koocheki, A. and Seyyedi, S.M. (2019) *Industrial Crops and Products,* **138**: 6-8

Koocheki, A., Ebrahimian, E., Mohammad, S. and Seyyedi (2016) *Sci. Hort.,* **213**: 132-143.

Lundmark, M., Hurry, V. and Lapointe, L. (2009) *Journal of Experimental Botany,* **60**: 2203–2213.

Maleki, F., Kazemi, H., Siahmarguee, A. and Kamkar, B. (2017) *Ecological Engineering*, **106**: 140-153

Mathew, B. (1983) *Crocus*: A revision of the genus *Crocus* Timber Press.

Mir, J.I., Ahmed, N., Wani, S.H., Rashid, R., Mir, H. and Sheikh, M.A. (2010) *Physiology and Molecular Biology of Plants*, **16**: 369–373.

Moghaddam, P.R., Koocheki, A., Mollafilabi, A. and Seyyedi, S.M. (2012) *4th International Saffron Symposium Advanced in saffron Biology Technology and Trade*. Kashmir, India.

Moghaddam, R.P., Koocheki, A., Molafilabi, A. and Pirseyyedi, M. (2013) *Saffron Agronomy and Technology*, **1**: 53-67.

Mohamed, A., Peter, B., Minchin, E.H. and Lapointe, L. (2007) *Physiologia plantarum*, **130**: 67 – 76

Mohammad, A.A.A., Rezvani, M.P., Mohammad,A.A.A and Fallahi, J. (2011) *Agroecology*, **3**: 83-93.

Mohammad, E.A. (2008) *American-Eurasian Journal of Agriculture and Environmental Science*. **4**: 274-279.

Mohamad, Shakib Ali, Yousef Aghayev, Ali, Vatanpour-Azghandi, Mohsen, Mardi (2012) In: *Breeding of saffron (Crocus sativus L.) via clonal selection, using tissue culture techniques, interspecific hybridization and ploidy manipulation*, Agricultural Biotechnology Research Institute of Iran – ABRII, pp. 303.

Molina, R.V., Garcia-Luis, A., Coll, V., Ferrer, C., Valero, M., Navarro, Y. and Guardiola, J.L. (2004) *Acta Hort.*, **650**: 39-48.

Molina, R.V., Valero, M., Navarro, Y. Garcia-Luis, A. and Guardiola, J.L. (2004) *Sci. Hort.*, **103**: 79-91

Molina, R.V., Valero, M., Navarro, Y., García-Luis, A. and Guardiola, J.L. (2005) *The J. Hortic. Sci. Biotech.*, **80**: 319-326.

Molina, R.V., Valero, M., Navarro, Y., Guardiola, J.L. and García Luis, A. (2005) *Sci. Hort.*, **103**: 361-379.

Mollafilabi, A. (2003) *In: International Symposium on Saffron Biology and Biotechnology*, **650**: 195-200.

Moraga, A.R., Nohales, PF, Pérez, J.A. and Gómez-Gómez, L. (2004) *Planta*, **219**: 955–966.

Mulumba, L.N. and Lal, R. (2008) *Soil Tillage Research*, **98**: 106–111.

Munoz Gomez, R., de Juan Valero, A., Botella Miralles, O. and Moya Aparicio, A. (2002) *Crocus sativus*, pp. 200-212.

Muzaferova, R.S. (1970) In: *Proceedings of the Institute of Genetics and Selection of the Academy of Sciences of Azerbaijan SSR*, Baku, "Elm", pp.194-203.

Omidi, H., Badi, H.N., Golzad, A., Torabi, H. and Footoukian,M.H. (2009) *Journal of Medicinal Plants*, **8**: 98-109.

Munoz -Gomez RM, de Juan Valero A. Botella -Miralles, O.; Moya Aparicio, A. (2002) *ITEA Produccion Vegetal.*, **98**: 200 -212.

Raja, W., Zaffer, G. and Wani, S.A. (2007) *Acta Hortic.*, **739**: 291-296.

Rohi, M., Amirshekari, H., Kordenaeej, A. (2014) *Journal of Crop Production Research (Environmental Stresses in Plant Sciences)* **6** : 1-17.

Seyyed, A.K., Seyyedi, M., Eyni, M.J. (2014) *Sci. Hort.*, **180**: 147-155

Shajari, M.A., Moghaddam, P.R., Ghorbani, R. and Koocheki, A. (2020) *Sci. Hort.*, **271**: 109485.

Sharaf-Eldin, M.A., Alam, P., Elkholy, S.F. (2019) *Food Sci. Nutr.*, **7**: 247–255.

Sharaf-Eldin, M., Elkholy, S., Fernández, J. A., Junge, H., Cheetham, R., Guardiola, J. and Weathers, P. (2008) *Planta Medica*, **74**: 1316–1320.

Shokrpour M. (2019) In: *Advances in Plant Breeding Strategies: Industrial and Food Crops* (eds. J. Al-Khayri, S. Jain, D. Johnson), Springer, Cham, doi.org/10.1007/978-3-030-23265-8_17.

Shuwen, H., Liliqin, Y.X., Yumei, Y., Guifenb, Z., Dong, Y.C. and Xiaodonga, Y.Q. (2019) *Sci. Hort.*, **247**: 165-174.

Sivanesan, I., Lim, M.Y., Jo, E.H. and Jeong, B.R. (2011) *In: 2nd International Conference on Environmental Science and Development* IPCBEE Vol. 4 IACSIT Press, Singapore

Sivanesan, I., Son, M.S., Jana, S., and Jeong, B.R. (2012) *Propagation of Ornamental Plants*, **12**: 163-170.

Temperini, O., Rea, R., Temperini, A., Colla, G. and Rouphael, Y. (2009) *J. Food, Agric. Environ.*, **7**: 19-23.

Verma, S.K., Kumar, A., Cingoz, G.S., Uslu, E. and Gurel, E. (2016) *Biotechnology Reports*, **10**: 66-74

Weryszko,C.E. and Chwil, M. (2011) *Acta Agrobotanica*, **64**: 35-45.

Yarami, N. and Sepaskhah, R. (2015) *Agric. Water Manage.*, **150**: 57-66.

Yau, S.K. and Nimah, M. (2004) *Lebanese Science J.*, **5**: 112-120.

Zahmati, R., Amir, H., Mohammd, S. and Fotokian, H. (2018) *J. Biol. Sci. Biotech.*, **7**: 47-50.

10

Heliconia

S.L. Chawla, Sudha Patil, Roshni Agnihotri and R.D. Pawar

1.0 Introduction

Tropical ornamental plants have attracted the flower markets due to their beauty, exoticism and post-harvest inflorescences durability. Heliconias are tropical herbaceous plants of family Heliconiaceae which is a member of a large taxonomic category called the order Zingiberales. Heliconias have several common names including 'Lobster Claw', 'Crab's Claw', 'Parrot's Flower', 'Japanese Canna' or 'Parrot's Plantain', 'False Bird of Paradise' and 'False Plantain'. The genus '*Heliconia*' is derived from '*Helicon*', a mountain in southern Greece regarded by the ancient Greeks as a home of the *Muses* which shows the relationship between heliconias and bananas (Berry and Kress, 1991). Mainly, they are grown for their attractive foliage and brilliant flower spikes. Heliconias make great ornamental plants because the long hanging (or standing) "flower" which is really an inflorescence (a cluster of modified leaves/bracts containing the flowers). Owing to its unusual inflorescence, heliconia is categorized as 'Specialty Flower'.

2.0 Importance and Uses

2.1.0 Aesthetic Uses

Heliconias are used as an illustrative piece for tropical gardens because of its unique characteristics and uncanny resemblance to banana. Depending on species and varieties, heliconias have wide range of height from 75 cm to 6 meter. Due to its varied morphology in terms of plants as well as inflorescence, selection is based on aesthetic look of plant which includes both plant characteristics like colour, floral display, size, growing habit, foliage, *etc.* and site properties like exposure to sun, soil, wind, water, *etc.*

The inflorescence of heliconia is terminal and lasts for several days. India has an annual production of about 1 lakh heliconia stems which account for less than 1 per cent of the total flower production of country out of which 50 per cent of the production of heliconia comes from West Godavari district of Andhra Pradesh. Major heliconia producing nations include Barbados, Hawaii, Brazil and Venezuela.

Heliconia flowers are useful as cut flower due to wide variation in colour and other floral characteristics. *H. psittachorum* and *H. latispatha* × *H. psittachorum* are excellent for flower arrangement (Janakiram, 2007). Heliconias are also preferred by landscapers for its diverse uses like to define the space, to create boundary, screening for privacy, controlling erosion and hiding foundations. They are used in the floriculture industry either as cut flowers or pot plants. However, the use of heliconias in landscape design is not restricted to outdoor gardens; they are also used in window displays and indoor gardens with low light. Aside from their appearance, these plants can attract fauna such as bats, hummingbirds and even small primates. The smaller cultivars are easiest to keep as container plants.

2.2.0 Therapeutic and Other Uses

In Brazil, roots and seeds of some varieties are used for medicinal purposes. *Heliconia rostrata*, locally known as *Jhulonto Chingri* in Bangladesh, traditional medical

Uses of Heliconia

Heliconia in Garden

Heliconia in Border

Heliconia as Screening

In Front of Residential Area

Heliconia in Park

Heliconia in Pot

Home Decoration

In Flower Arrangement

healers are using several parts of this selected plant against different ailments for thousands of decades. A study by Shatil Shahriar *et al.* (2018) revealed *in-vitro* analgesic activity of the methanolic extract of *Heliconia rostrata*. Ethnobotanically, the traditional people of Malaysia used rhizomes of this plant to cure jaundice, intestinal pain and hypertension (Gunjan *et al.*, 2012). Interesting and innovative contributions were found in relation to the use and transformation of heliconias, *e.g.* for medicinal use.

3.0 Origin and History

Being a popular landscape plant and cut flower in tropical and sub-tropical areas of world, heliconias are native to Central and Southern America, the Caribbean Island and some of the Islands of south Pacific west to Indonesia. Heliconia (Heliconiaceae, order: Zingiberales) is among the showiest plants of the Neotropical rainforest and represent a spectacular co-evolutionary radiation with hummingbirds. According to Iles *et al.* (2017) extant diversity of Heliconia originated in the Late Eocene (39 Ma) with rapid diversification through the Early Miocene, making it the oldest known clade of hummingbird-pollinated plants.

4.0 Taxonomy

The combination of inverted flowers, a single staminode and drupe fruits is an exclusive characteristic of *Heliconia*. In the Zingiberales, a primarily tropical order of monocotyledons, most phylogenists currently recognize eight families: Musaceae, Strelitziaceae, Lowiaceae, Heliconiaceae, Zingiberaceae, Costaceae, Cannaceae and Marantaceae. Heliconia is the only genus in the monotypic family Heliconiaceae, but was formerly included in the family Musaceae, which includes the bananas (Judd, 2007). A new phylogenetic classification based on the cladogram is proposed that recognizes eight families, two superfamilies, and five suborders within the Zingiberales (Kress, 1990).

5.0 Species and Cultivars

5.1.0 Species

Heliconias are one of the showiest, distinctively tropical flowers with beautiful and brilliant colorful flowering bracts. The family Heliconiaceae contains a single genus, *Heliconia*, with approximately 180 species of Neotropical origin (Marouelli *et al.*, 2010). In this genus, 250-300 species are identified with great number of cultivars and 23 natural hybrids (Santos, 1978, Dahlgren *et al.*, 1985 and Castro *et al.*, 2007). However, there are taxonomic confusions and uncertainties about the number of species and the relationships among them. Marouelli *et al.* (2010) examined the genetic variability and the phylogenetic relationships of 124 accessions of the genus *Heliconia* based on RAPD markers. Molecular analysis using amplified fragment length polymorphism (AFLP) markers revealed two large, well-defined groups: the first grouped cultivars of the very closely related *H. caribaea* and *H. bihai* species with those of *H. orthotricha* and *H. psittacorum*, and the second group of *H. stricta* and *H. wagneriana* cultivars. The lowest percentage of polymorphism was found in *H. psittacorum* (17.65 per cent) and the highest was in *H. stricta* (55.88 per cent).

Using AFLP, phylogenetic analysis of the species studied, revealed the monophyletic origin of the Heliconiaceae family.

Heliconia genus is a group of highly variable and diverse plants. Species and cultivars are visually differentiated primarily on the basis of the colour and size of inflorescence bracts. Heliconia species like *H. stricta, H. rostrata, H. psittacorum, H. latispatha, H. angustifolia, H. bihai, H. chartacea, H. angusta and H. wagneriana* are particularly suitable as a cut flower because of their long-lasting characteristics (Criley *et al.*, 1993). Classification based on the plant stature includes small, medium and large types. Among them, *H. wagenariana, H. stricta* comes under large category. *H. psittacorum, H. rostrata* and *H. latispatha* fall in medium, whereas, *H. collinsiana* comes under the small category.

Following are some of the common species of *Heliconia:*

***H. stricta*:** The inflorescence of this species is beautiful which resembles lobster's claws. The crimson claw is rigid and retains shape and colour for weeks. The colour ranges from red, gold, orange, maroon to green, either in single or in combination. Smaller lobster claws are slimmer, shorter stemmed and more refined in shape than larger lobster claws. Flowers of this species are ideal for flower arrangement due to its size ranges from 12 to 30 cm long but light in weight.

***H. rostrata*:** Beautiful tropical herbs with banana like leathery green leaves. Traditional and most recognized heliconias with magnificent pendent inflorescences of alternating bracts each 6-10 cm long, scarlet red with cream to yellow tips. A deep red colour covers most of the bract with yellow green tips. It is one of the hardiest varieties, grows in height from 1-6 meter and blooms year-round.

***H. stricta* 'Dwarf Jamaican':** A real beauty, that is small (50-100 cm high) and does well in pots. Very beautiful about 12 cm dainty inflorescences that are rose coloured and evenly graded from pale to deep hues. Each bract is ridged with green on its upper edge, matching the tiny green and white striped sepals. Blooms year-round with a winter peak. Adapts beautifully to variable temperatures. Grows in full sun to 60 per cent shade.

***H. psittacorum*:** Originated from the Coast of Guyana, the *psittacorum* heliconias are small, dainty and exotically tropical and resemble the plant known commonly as 'bird of paradise' or 'parrot's beak'. They bloom abundantly round the year. Flower heads appear to be hand painted and glow with brilliant colours and greenish yellow flowers with black spots near apex. The *psittacorums* rarely exceed 1-2 meter in height. The floral bracts of *psittacorum* arise from the same point on the stem. *H. psittacorum* do not mind cool temperatures, but do best in warm, humid areas.

***H. humilis*:** Native of Trinidad and Brazil, having shiny green leaves and erect flower heads with salmon red boat shaped bracts changing to green towards tip and greenish yellow flowers.

***H. latispatha*:** Native of central and South America, showy plant with green broad oblong leaves and erect inflorescence with well separated boat shaped bracts with orange yellow colour at the base and red towards tip.

Species of Heliconia

Heliconia stricta

Heliconia rostrata

Heliconia psittacorum

Heliconia humilis

Heliconia latispatha

Heliconia angusta

Heliconia metallica

Heliconia collinsiana

Heliconia latispatha

Cultivars of Heliconia

H. chartacea
'Sexy Orange'

H. wagneria
'Red'

H. latispatha
'Orange'

H. wagneria
'Rainbow'

H. collinsiana
'Pendula'

H. latispatha
'Distance'

H. rostrata
'Parrot Beak'

H. psittachorum
'Golden Torch'

H. psittachorum
'Andromeda'

H. bihai **(Wild Plantain/Fire Bird):** A large perennial, oblong smooth textured pointed green leaves, greenish yellow flowers clustered in the axils of large stiff boat shaped crimson red flattened bracts with pointed tip and arranged in two ranks on effect inflorescence.

***H. distans*:** Native of tropical America having slender bracts, lemon yellow at base and crimson on top to apex.

***H. angustifolia*:** Striking perennial herb with erect inflorescence having boat shaped orange red bracts becoming darker as they mature with creamy white flowers. The seeds are first orange later turning blue.

H. collinsiana **(pendula):** Robust tropical perennial with lush growth and striking inflorescence pendent below the leaves with spreading crimson red bracts yellowish towards tip and covered with waxy powder with creamish flowers.

***Heliconia* × *nickeriensis*:** This hybrid is characterized by short heighted plant, with quick development, dispersed type of clump architecture and open growth habit. The dark green leaves contrast with the yellow-orange inflorescences which are easily visualized above its foliage.

5.2.0 Cultivars

The most important criteria for selecting cultivars are production throughout the year, market acceptance, long-term postharvest durability, production during the short period in the region, low maintenance requirement, best market price, adequacy of the packed inflorescence when considering the size and weight, and finally the novel appearance of the inflorescence (Criley, 2000 and Loges *et al.*, 2015). There are many varieties of heliconia, including smaller heliconia *psittacorum* types, and other 'clumping' types that grow in a cluster. Popular *psittacorum* types include, 'Golden Torch', 'Petra' and 'Halloween'. Claw II is the most common clumping heliconia.

6.0 Genetics and Breeding

6.1.0 Genetics

Cytological studies have been carried out on about 20 per cent of known heliconia species. In all New World species, the somatic chromosome number is mentioned as 2n = 24 (Andersson, 1984) which includes the main cut flower species of heliconia such as *H. bihai, H. psittachorum* and *H. caribae* (Cheeseman and Larter, 1935 and Mahanty, 1970). Darlington and Wylie (1955) mentioned that 2n= 16, 18, 22 and 26 have been reported in *Chromosome Atlas of Flowering Plants* however, Andersson (1984) dismissed these as miscounts after studying illustration in original reports and suggested the possible occurrence of anueploidy. The chromosome number 2n= 24 was again determined in 31 taxa of *Heliconia* representing nearly entire morphological spectrum of the genus. Evaluation of morphological and floral variability that exist in heliconia serves as a tool for identifying the varieties with potential for further breeding programmes and to exploit the ornamental attributes of the *Heliconia* for expanding tropical cut flower industry. The evaluation has been undertaken by several workers in India as well as in abroad. Kress (1983) studied self-incompatibility in populations of 19 species of Central American Heliconia. Pollen tube growth and in some cases fruit set were used to determine the extent of physiological self-incompatibility in each species.

6.2.0 Breeding

6.2.1 Selection

Heliconia is a cross pollinated plants and pollination by various agents (pollinators) resulted into development of some outstanding selections. For initiating any breeding programme, germplasm characterization is a must and some workers have achieved it through cataloguing of heliconia germplasm. Some varieties developed through selection by flower growers are depicted below:

Table 1: Varieties Evolved through Selection

Variety	*Parent*	*Characteristics*
Pink Blush	*H. psittacorum*	Unique large inflorescence form with thick long stem, excellent cut flower
Flamenco	*H. psittacorum*	Beautiful pink coloured large inflorescence
Pearl	*H. psittacorum*	A rare pale lemon to cream coloured inflorescence ideal for floral wedding arrangement, prolific producer of flowers even during the dry season
Fantasia	*H. psittacorum*	Attractive small inflorescence, red-orange bracts highlighted by a bright yellow band at the base
Rosie Q	*H. psittacorum*	Pink, apricot coloured
Red Devil	*H. psittacorum*	Reddest of all the nine new *psittacorum,* good producer of flowers
Rose Red	*H. psittacorum*	Rosy red coloured inflorescence
Exotic	*H. psittacorum*	Variegated leaves, height 1 to 1.5 m, early, red flower with good production
Cracker Jack	*H. psittacorum*	Fire-red to orange coloured inflorescence, small, good as filler in arrangements

6.2.2 Mutation Breeding

Study on mutation breeding was conducted by many scientists. Among the different mutagens, gamma radiation induced mutation can bring promising results in different *Heliconia* species. Kikuchi (2003) stated that *H. psittachorum* and *H. rostrata* were highly sensitive to gamma and e-beam radiations at 300 Gy. Radiations caused damage like browning which was followed by drying up of plant. Whereas, Sangwanangkul *et al.* (2007) reported that *Heliconia* 'Keanae' showed high sensitivity to 250 Gy in a study to find sensitivity of cut flowers and foliage towards electron beam irradiation.

6.2.3 Polyploidy

Raizer *et al.* (2019) characterized the 38 clones of *Heliconia chartacea* var. Sexy Pink plants that were obtained through induced polyploidization for various morphological and physiological characteristics. The characterization analysis was performed through 49 morphological descriptors and stomatal density using microscopy. Among the 38 clones, the genotype 35 exhibited the greatest morphological variations, with alterations in the position and coloring of the inflorescence, in addition to having the edges of the entire limbus. Overall, the induction of polyploidy allowed for clones to be obtained with a high variability for the characteristics of the leaf, pseudostem and inflorescence with various attributes that confer a more efficient post-harvest management to some genotypes in addition to favorable aspects for commercialized purposes as a cut flower.

6.2.4 Hybridization

In cross pollinated crops like heliconias, hybridization is an important breeding method for crop improvement. However, artificial hybridization is unsuccessful due to pre-fertilization barriers. Except some natural hybrids, no artificial hybrids have been reported in heliconia. Some interspecific hybrids were reported by Berry and Kress (1991). Most of the species of heliconia are self compatible but cross fertilization is unsuccessful (Berry and Kress, 1991; Kress, 1983 and Lee *et al.*, 1994). Pre-fertilization barriers at stigmatic, stylar or ovarian level prevent distant hybridization in heliconias. Due to the presence of strong crossability barriers between majority of heliconia species, pollen tubes growth is either inhibited at the stigmatic surface within the stylar tissues or within the ovary.

The self- compatible species of heliconias are *H. colgantea, H. collinsiana, H. curtispatha, H. danielsiana, H. imbricate, H. irrasa, etc.* while, the cross- compatible species include, *H. collinsiana* × *H. nutans, H. trichocarpa* × *H. collinsiana, H. imbricata* × *H. latispatha, H. latispatha* × (*H. latispatha* × *H. imbricata*)*, etc.* The species such as *H. trichocarpa, H. umbrophila, H. wagneriana and H. wilsonii* are partially compatible.

Some of the interspecific hybrids of heliconia are: Cinnamon Twist (*H. bihai* × *H. spathocircinata*), Alan Carle (*H. psittacorum* × *H. spathocircinata*), Bushmaster (*H. mariae* × *H. pogonantha holerythra*), Carib Flame (*H. bihai* × *H. caribaea*), Criswick (*H. bihai* × *H. caribaea*), Crocodile (*H. curtispatha* × *H. mariae*), Grand Etang (*H. bihai* × *H. caribaea*), Golden Torch (*H. psittacorum* × *H. spathocircinata*), Golden Torch Adrian (*H. psittacorum* × *H. spathocircinata*), Grenadier (*H. bihai x H. caribaea*), Guyana (*H. psittacorum* × *H. spathocircinata*), Harvey Bullis (*H. imbricata* × *H. spathocircinata*), Johnson Beharry V.C. (*H. psittacorum* × *H. spathocircinata*), Jose Abalo (*H. imbricata* × *H. latispatha*), Pedro Ortiz (*H. bourgeana* × *H. collinsiana*), Richmond Red (*H. bihai* × *H. caribaea*), Toucan (*H. clinophila* × *H. secunda*), Vermillion Lake (*H. bihai* × *H. caribaea*), Yellow Dolly (*H. bihai* × *H. caribaea*) and Yellow Parrot (*H. psittacorum* × *H. spathocircinata*).

7.0 Growth and Flowering

7.1.0 Effect of Growth Regulators

De Moraes *et al.* (2005) studied influence of benzyladenine on longevity of *Heliconia latispatha* and found positive correlation between benzyladenine concentration and increase in the vase life. Spraying the heliconia stalks with 300 mg/l benzyladenine increased the longevity by at least 1.85 fold compared to the untreated flowers. The growth retardant such as Paclobutrazol have been found effective in reduction of plant height and increasing the number of shoots in heliconia 'Red Opal' without affecting the inflorescence formation, but its effect is temporary (Castro *et al.*, 2016). Jadhav *et al.* (2015) concluded that MH @ 50 ppm proved to be the best treatment for obtaining higher yield of good quality flowers.

7.2.0 Effect of Environment

Geertsen (1989) investigated the influence of photoperiod (8, 12 or 16 h) and temperature (15, 18 or 21°C) on growth and flowering of *H. psittacorum* 'Tay' and revealed that by increasing the minimum air temperature from 15 to 21°C, the number of shoots emerging as well as the number of flowering stems produced per square meter were more than double along with increase in stem length and noticeable improvement in quality.

Flowering responses of *Heliconia psittacorum* × *H. spathocircinata* Aristeguieta 'Golden Torch' to temperature and photosynthetic photon flux (PPF) was studied by Catley and Brooking (1996) under controlled environmental conditions. It was revealed that temperature was more dominant than light in influencing production and quality of flowers, however temperature had no significant effect on new shoot production at both the ranges of 32°C day/20°C night and 24°C day/20°C night temperatures. An average of 9.3 shoots were recorded per plant over the 248 days of treatment, duration from shoot until inflorescence emergence was significantly less at 32°C day/20°C night than at 24°C day/20°C night (140 and 146 days, respectively). The number of leaves subtending the inflorescence increased at the higher temperature and flower stems were taller and thicker at 32°C day/20°C night and these dimensions increased further with increasing order of shoot appearance.

8.0 Propagation

Heliconia can be propagated through seeds, as well through division of rhizomes. For places where seeds are rarely produced, heliconias are easily propagated from a piece of rhizome (Lennox and Seddon, 1978). *In vitro* propagation of heliconia has been standardized.

8.1.0 Seed Propagation

Seeds if formed and germinated, take longer time to grow. In a study carried out by Simao and Scatena (2003), seeds of *Heliconia velloziana* took about four months to germinate, probably due to the undifferentiated embryo and hard endocarp. Seedlings had two first foliar structures as scale leaves and the primary root developed, but the adventitious roots were more conspicuous. Heliconia seeds are very attractive to birds and germinate best after passing through alimentary canal of birds. Dozens of bird species have been recorded eating heliconia seeds in Central America. In species of Heliconiaceae, there have been few detailed studies about seed and seedling morphology (Humphrey, 1896; Boyd, 1932; Tillich, 1995). The germination occurs about four months after the embedding. Some seeds may even germinate after six months.

8.1.1 Effect of Temperature on Seed Germination

Montgomery (1986) reported that the germination period of *Heliconia* may

Propagation of Heliconia

Fruits of Heliconia

Seeds

Germination of Seeds

Heliconia Rhizome

Division of Rhizomes

Planting of Rhizomes

Cultivation of Heliconia

Seedlings

Planting in Greenhouse

Cultivation in Open

range from three months to three years, depending on the degree of the embryo development at the time of the fruit maturity. Species from moist places usually take less time to germinate, while those from dry areas took longer, until the environmental conditions were favourable. Although, *H. velloziana* occurs in moist places, its seeds takes four to six months to germinate, suggesting that other environmental factors, *e.g.* temperature, or even other factors, such as the development of the embryo, could influence the beginning of the germination period. According to Criley (1988), the scarification did not hasten germination in species of *Heliconia*. Graven *et al.* (1996) also reported that the removal of the operculum did not increase or accelerate the germination in species of Musaceae.

8.2 Vegetative Propagation

Heliconias are usually grown by dividing the rhizomes, a modified underground stem, side shoots and suckers arising from the mother plant. Rhizomes produce both terminal and auxiliary bud. Each division should contain at least one terminal or auxiliary bud. Each clump produces around 5-20 side suckers in a year depending on the variety. In relation to the vegetative propagation, the rhizomes of the *Heliconia* species studied by Criley (1988; 1995) took about one month to develop young roots and four to six weeks to the activation of the buds. Hence, cultivation of heliconia for planting material is also profitable when selling as plant in nursery.

8.3 Micropropagation

Heliconia can be propagated through rhizome but it is time consuming process. Also because of the slow germination of its seeds which takes about three months to three years which makes a long term process. The most difficult aspect in the *in vitro* regeneration of heliconia was the elimination of *Pseudomonas solanacearum* from the explants of heliconia rhizomes, an omnipresent bacterial pathogen, which was directly brought in from the field. Regarding propagation, the studies were oriented to the use of growth regulators (BAP and NAA) in culture medium as this favors an increase in sprouts and it is possible to obtain satisfactory levels of propagation in heliconia cultivars (Raizer *et al.*, 2015; Rodrigues *et al.*, 2016). Ulisses *et al.* (2007) standardized a protocol for somatic embryogenesis in *Heliconia chartacea* cv. 'Sexy Pink' and cv. 'Sexy Scarlet' using sections of ovaries as explants.

9.0 Soil and Climate

9.1.0 Soil

Heliconias prefer sandy loam soil having slightly acidic to neutral pH and rich in organic material with good drainage. Chlorosis develops readily when plants are grown in soils with a high pH.

9.2.0 Climate

Heliconias thrive well in tropical climate and require a temperature of 25°C-35°C with 75-80 per cent relative humidity for its successful cultivation. Many varieties of heliconia come up well under partial shade making them ideal as intercrop in coconut gardens. It is well adapted to different agro climatic zones of the country and can be cultivated up to height of upto 1000 meter. Cool conditions slow down the rate of shoot emergence.

In tropical areas, heliconias grow well under 30 per cent green shade net. For topical conditions, the importance of the percentages of shade for the cultivation of heliconias is highlighted because it develops better in these conditions which influences the content of photosynthesized pigments and the growth (Gomes *et al.*, 2016; Souza *et al.*, 2016). Heliconias grown under the lowest photosynthetically active radiation (PAR) (30 per cent to 35 per cent of the open condition), show improved growth with increased biomass accumulation, plant height, suckering habit, appearance of flowering shoots, yield of inflorescence and light use efficiency (Nihad *et al.*, 2019).

10.0 Cultivation

10.1.0 Planting of Rhizome

The ideal time for planting the rhizomes is June to September and March. Rhizomes are planted in the pits of 30/40 cm^3 size at desired distance on raised beds of about 15-20 cm height and 80 cm width. Generally, 40 g is considered as the optimum size of rhizome. Rhizomes should be washed well with water and then a treatment of carbendazim fungicide (0.2 per cent) should be given. Plant spacing varies from species to species; dwarf heliconias as are planted at 1.0×0.8 m, medium at 1.25×1.0 m while taller at 1.5×1.5 m distance.

10.2.0 Manuaring and Fertilization

Heliconias need well decomposed FYM, compost and plenty of water supply during growing period. Well decomposed farmyard manure @ 5 kg/m^2 should be applied at the time of bed preparation and 20 g each of N, P and K/m^2/year at the time of planting the rhizomes. A top dressing with N @ 20 g/m^2 is done three months after planting. Flower yield could be optimized with appropriate fertilizer application. In second and third year, 60:40:20 g NPK/m^2 should be applied where nitrogen should be applied in 4 equal split doses. When plant fertilization and nutrition research is viewed internationally, it was found that the authors recommend the use of organomineral fertilization (combination of organic fertilizers with mineral fertilizer) since it better promotes the growth and development of heliconias (Albuquerque *et al.*, 2010; Beckmann Cavalcante *et al.*, 2011; Carvalho *et al.*, 2012; Farias *et al.*, 2013; Linares-Gabriel *et al.*, 2018). Although, heliconias need substantial amounts of water but poor drainage causes major root rot problem. Recommended water requirement for most of species is 1.5-2.0 liter per plant per day by drip irrigation system.

11.0 Diseases and Pests

Although heliconia is generally considered as a disease free low maintenance ornamental but sometimes pathogens such as *F. oxysporum* f. sp. *cubense*, *Pseudocercospora fijiensis*; nematodes of the genus *Meloidogyne*, *Helicotylenchus erythrinae*, *H. crenacauda*, *H. dihystera*, *Meloidogyne incognita*, *Pratylenchus zeae* and *Rotylenchulus reniformis*, defoliators such as *Cornops frenatum* and mites, particularly

Diseases and Pests of Heliconia

Heliconia Rust

Leaf Spots

Root Rot

Heliconia Bugs

Ants

Mites

Harvesting and Marketing of Heliconia

Full Bloom Stage

Harvesting of Flowers

Display of Cut Flowers

In Local Market

In International Market

specimens of *Tropicoseius venezuelensis* (Castro *et al.*, 2008; Lemos *et al.*, 2010; Sardinha *et al.*, 2012; Sobrinho *et al.*, 2012; Britto *et al.*, 2015; Vázquez-Euán *et al.*, 2019) affect Heliconia along with erratic attack of aphids, mealybugs and scales. Occasionally, snails chew the young leaves by scraping, making big irregular holes, adult and young devours the plant during nights.

Despite the affectations, some authors recommend that the use of synthetic insecticides should be done only in extreme cases (Lemos *et al.*, 2010). Spary Acephate 75 per cent WP @ 1g/lit or Triazophos 40 per cent EC @ 1 ml/lit for aphid and Profenophos 50 per cent EC @ 2 ml/lit for mealy bug. Hand picking during evening and bait of Metaldehyde 3 per cent (Snail kill) is effective to control snails while soil sterilization before planting and application of neem cake and carbofuran granules around root zone is best remedy for root knot nematodes.

Resistance breeding with host resistance that uses the ability of the host plant to prevent disease is the most economical and best method to control disease, but it usually takes a long time to develop. Today, biotechnological techniques that allow the transfer of genes from one plant species to another may hasten the development of resistance to serious diseases (Sewake and Uchida, 1980).

12.0 Harvesting and Postharvest Management

12.1.0 Harvesting

Heliconias take 8 to 15 month for flowering which varies from species to species. Harvesting is done when 3/4th bracts are open in case of pendent varieties and when 1/3rd lower bracts open in case of erect type varieties. Flower stalks should be cut near the ground leaving 7-10 cm in cool hours of the day to minimize desiccation after harvest. Inflorescence lasts longer when cut from irrigated plants than from plants under stress. Flowers should be cut at the stage of opening since further opening of the bracts does not occur after cutting or negligible.

12.2.0 Postharvest Management

Post harvest life of heliconia flowers varies according to variety. They are sensitive to temperature below 15°C and should not be placed in cold storage. Cold injury develops as blackening at the base of the bracts. As desiccation reduces the post harvest life hence, storage condition should be maintained at 90-95 per cent humidity. Flowers in general maintained their bract color and shape for 5-8 days in tap or deionized water at 23°C temperature.

According to the citations for pulse solutions, the studies focused on the use of sucrose, silver nitrate, calcium chloride, gibberellic acid, citric acid, and organic *Aloe* solutions (*Aloe vera and A. barbadensis*) and moringa (*Moringa oleifera*), as medium of preserving flower stems (Costa, 2015; Malakar *et al.*, 2019; Sardinha *et al.*, 2019; Shokalu *et al.*, 2019). The hardiness of this species has also been compared with rose stems in terms of resistance to moisture loss (Leyva-Ovalle *et al.*, 2011). A temperature of 10 to 19°C is considered ideal for storing heliconia flower stems for upto eight days on an average (Costa *et al.*, 2015; Darras, 2019; Silva *et al.*, 2019).

Other researchers recommend an in-depth study that uses techniques to confirm or reject the sensitivity of heliconia to ethylene (Bañuelos-Hernández *et al.*, 2016).

In general, the storage of heliconias has been little studied. The recommended temperature for the transport and storage of heliconias is around 10°C as they are sensitive to damage by cold. It is recommended to wax the stems to maintain the appearance of the flower stem (Alvarado *et al.*, 2019). The heliconia stems pulsed with 10 per cent sucrose for 60 minutes maintain the quality of the inflorescences. This treatment allows the stems to reach up to 20 days of longevity, showing the color and brightness appropriate to quality standards (Costa, 2015). Some authors recommend dry treatment in heliconia as an alternative, guaranteeing a reduction in the excessive use of water (Linares-Gabriel *et al.*, 2019).

Vase solution treatments of lipoic acid 100 mg/l + 8-HQC 250 mg/l + sucrose 3 per cent and Spermine 100 mg/l + 8-HQC 250 mg/l + sucrose 3 per cent effectively increased water uptake and retained fresh weight of heliconia inflorescence and maintained pigment (carotene) in the bracts. The same treatments also reduced the levels of catalase (CAT) and peroxidase (POD) enzymatic activities, decreased the lipid peroxidation (measured as TBARS) and improved *per cent* absolute integrity (PAI) in the bract. Maximum vase-life was recorded in heliconia inflorescence held in vase solution comprising of a-lipoic acid 100 mg/l + 8-HQC 250 mg/l + sucrose 3 per cent followed by spermine 100 mg/l and $CaCl_2$ 250 mg/l along with 8-HQC 250 mg/l+ sucrose 3 per cent.

Patel *et al.* (2017) reported the significant maximum water uptake and higher fresh weight retention on 7th,11th and 15th day, flower opening percentage, number of bracts open with maximum TSS and carotene content on 8th, 10th and 12th day, and useful vase life were recorded in flowers of Golden Torch variety of heliconia. With respect to vase solutions, they were found significant for all parameters on post harvest quality of heliconia during vase life. Flowers treated with 100 ppm GA_3 show significantly maximum water uptake, fresh weight retention, flower opening percentage, number of bracts open, TSS, carotene content and useful vase life (15.50 days).

Small and medium heliconias such as *H. psittacorum*, and *H. angusta* are bunched (5s or 10s) and packed in sleeves of plastic film or news paper or butter paper over the inflorescences to prevent bruising. Bunches of the smaller heliconias are packed on top of the heavier heliconias.

13.0 Yield

It gives economical yield upto 2-3 years and after that the clumps require shifting. Average yield of heliconia depends on species and varieties. Golden Torch yields highest *i.e.* 15-18 stems/clump/year while, *H. wagneriana* yields up to 5-6 stems/clump/year. Chawla *et al.* (2016) recorded the maximum number of spikes per plants (13.33) in Golden Torch followed by *H. psittacorum* Kenea Red (8.67).

14.0 Biotechnology

14.1.0 Markers

Different types of morphological markers have been utilized to differentiate among different species/genotypes. The DNA based markers are widely used for analysis of genetic diversity, phylogeny and molecular identification, they can be exploited for planning future breeding programs. AFLP marker was able to infer microevolutionary studies in *Heliconia bihai* (Melendez-Ackerman *et al.*, 2005) and among 67 genotypes by Isaza *et al.* (2012). Costa (2016) pointed out that identification of genotypes in Heliconia is mainly based on morphological characteristics, resulting in generation of synonyms. Therefore, in order to differentiate among genotypes and elucidate the origin of hybrids and triploids chromosome markers were used. The work resulted into cytogenetic characterization of 18 heliconia genotypes, including three hybrids, two triploids and their possible parent plants, by means of chromosome counting, staining with the fluorochromes CMA and DAPI, and FISH for 45S and 5S rDNA sites. The analyses revealed 16 diploid (2n=2x=24) and two triploid (2n=3x=36) genotypes, with new counts for eight genotypes, including the triploid 'Suriname Sassy'.

14.2.0 Somaclonal Variation

Rodrigues (2008) reported somaclonal variation in micropropagated *Heliconia bihai* cv. Lobster Claw I plantlets. The Stem apices were inoculated on MS culture media with the addition of 2.5 mg/l of benzylaminopurine (BAP) and 500 mg/l of sodium cefotaxime. After selecting the apex stem, it was sub-cultivated in MS media and supplemented with 4.0 mg/l of BAP to induce side buds. Variation of the chlorophyll in the leaf, low stature variant and pseudostem and petiole color variant were recorded, the later with ornamental potential.

Lia *et al.* (2019) analyzed the species diversity and phylogenetic relationship of *Heliconia* spp. collections of Purwodadi Botanic Garden (East Java, Indonesia) by rbcL gene sequences. The rbcL gene located in chloroplast genome is one of appropriate proposed marker for plant barcoding assessment and was found able to distinguish at higher level taxa between closely related genus and above.

15.0 References

Albuquerque, A.W., Rocha, E.S., Costa, J.P.V., Farias, A.P. and Bastos, A.L. (2010) *Revista Brasileira de Engenharia Agricola e Ambiental*, **14**(10): 1052-1058.

Alvarado, G.C., Arevalo-Galarza, M.L., Velasco-Velasco, J., Baltazar Bernal, O. and Salinas Ruíz, J. (2019) *Acta Hortic.*, **1237**: 221-226.

Andersson, L. (1984) *Nordic J. Bot.*, **4**(2): 191-193.

Andersson, L. (1985a) *Opera Bot.*, **82**: 1-123.

Andersson, L. (1985b) In: G. Harling and B. Sparre, (eds.) *Flora of Eucador.*, Swedish Res. Council Publ. House. Stockholm, Sweden. pp. 1-87

Bañuelos-Hernández, K.P., García-Nava, J.R., Leyva-Ovalle, O.R., Peña-Valdivia, C.B. and Ybarra-Moncada, M.C. (2016) *Postharvest Biol. Technol*, **112**: 159-169.

Barreiros, H. S. (1970) *Rev. Brasil. Biol.*, **30**, 571-573.

Beckmann-Cavalcante, M.Z., Amaral, G.C., Silva, A.A., Cavalcante, Í.H.L. and Lima, M.P.D. (2011) *Afr. J. Biotechnol.*, **10**(68): 15272-15277.

Berry, F. and Kress, W.J. (1991) Heliconia: An Identification Guide. Smithsonian Institution Press, Washington and London. pp. 1-334.

Boyd, L. (1932) *Trans. and Proc. Bot. Soc. Edinburgh*, **31**(1): 212-218.

Britto, E.P.J., Finotti, A.S. and De Moraes, G.J. (2015) *Exp. Appl. Acarol.*, **66**(2): 203-217.

Bruna, E.M., Kress, W.J., Marques, F. and da Silva, O.F. (2004) *Acta Amazonica*, **34**(3): 467- 471.

Cantor, M., Singureanu, V., Denisa, H. and Erszebet, B. (2014) *J. Hortic. Sci. Biotechnol.*, **1**: 22-27.

Carvalho, J.S.B., Martins, J.D.L., Ulisses, C. and Silva, W.L. (2012) *Hortic. Bras.*, **30**(4): 579-583.

Castro, A., Hawerroth, F., Taniguchi, C., *et al.* (2016) *Ornam. Hort.* **22**: 301-306.

Castro, C.E.F., Gonclaves, C. and May, A. (2007) *R. Bras. Hortic. Ornam. Campinas*, Brazil, **13**: 38- 62.

Castro, N.R., Coêlho, R.S.B., Laranjeira, D., *et al.* (2008) *Summa Phytopathologica*, **34**(2): 127-130.

Catley, J.L. and Brooking, I. R. (1996) *HortScience*, **31**(2): 213–217.

Chawla, S.L., Dhaduk, B.K. and Patel, R.B. (2016) *Indian J. Agric. Sci.*, **86**(6): 750-752.

Cheeseman, E.E. and Larter, L.N.H. (1935) *J. Genet.*, **30**: 31-52.

Costa, M.G.d., Leite, B.S.F., Loges, V., *et al.* (2016) *Euphytica*, **212**; 501–514.

Costa, R.R., Costa, L.C., Carneiro, G.G., *et al.* (2015) *Acta Hortic.*, **1060**: 201-206.

Criley, R. A. (1988) *Acta Hortic.*, **226**: 509-517.

Criley, R. A. (1995) *Rev. Bras. Hort. Orn.*, **1**: 14-21.

Criley, R.A. (2000) *Acta Hortic.*, **541**: 159-165.

Criley, R.A. and Broschat, T.K. (1992) *Hort. Rev.*, **14**: 1-51.

Criley, R.A. and Paull, R.E. (1993) *Acta Hortic.*, **337**: 201- 212.

Dahlgren, R.M.T., Clifford, H.T. and Yeo, P.F. (1985) *Springer-Verlag*, Berlin, pp. 350- 358.

Darlington, C.D. and Wylie, A.N. (1955) Chromosome atlas of flowering plants. London, UK: George Allen and Unwin Ltd.

Darras, A.I. (2019) *J. Hortic. Sci. Biotechnol.*, **95**(1): 1-7.

De Moraes, P.J., Finger, F.L., Barbosa, J.G., Cecon, P.R. and Cesa, L.P. (2005) *Acta Hortic.*, **683**: 369-374.

Dobkin, D.S. (1984) *Oecologia*, **64**: 245-254.

Estrada, S., Quintana, J.C., Jiménez, S.L.; *et al.* (2009) *Vitae*, **16**(2): 252-257.

Farias, A.P., De Albuquerque, A.W., Moura Filho, G. and Reis, L.S.(2013) *Revista Brasileira de Engenharia Agricola e Ambiental*, **17**(7): 713- 720.

Geertsen, V. (1989) *Acta Hortic.*, **252**: 117-128.

Gomes, R.J., Guiselini, C., Siqueira, G.M., Albuquerque Filho, J.C.C., Loges, V. and Pandorfi, H. (2016) *Ornam. Hortic.*, **22**(3): 318-325.

Graven, P., Koster, C.G., Boon, J.J. and Bouman, F. (1996) *Ann. Bot.*, **77**: 105-122.

Gunjan, M., Karna, L., Dayalan, K. and Sasigaran, P. (2012) *Int. J. Phytother. Res.* **21**: 26-41.

Humphrey, J.E. (1896) *Ann. Bot.*, **10**: 1-40. Iles, W.J.D., Sass, C., Lagomarsino, L., Martin, G. B., Driscoll, H. and Specht C. D. (2017) *Mol. Phylogenet. Evol.*, **117**: 150–167.

Isaza, L., Marulanda, M.L. and Lopez, A.M. (2012) *Genet Mol Res.*, **11**(4): 4552-4563.

Jadhav, S., Chawla, S.L., Agnihotri, R. and Gurjar, R.A. (2015) *The Bioscan*, **10**(4): 1509-1513.

Janakiram, T. (2007) *Heliconia. In*: Reddy, S.B.S., Janakiram, T., Kulkarni, B.S. and Mishra, R.L. (Eds.) *Hi-Tech Floriculture*. ISOH, Division of Floriculture and Landscaping, IARI, New Delhi-110012. pp.118-123.

Judd, W. (2007) Plant Systematics: A phylogenetic approach (3rd ed.). Sunderland: Sinauer Associates, Inc.

Kikuchi, O.K. (2003) *Radiat. Phys. Chem.*, **66**: 77-79.

Kress, W. J. (1983) *Evol.*, **37**(4): 735-744.

Kress, W. J. (1991) *Brittonia*, **43**(4): 253-260.

Kress, W.J. (1985) *Biotropica*. **17**(4): 302- 308.

Kress, W.J. (1990) *Ann. Missouri Bot. Gard.*, **77**(4): 698-721.

Kress, W.J. and Stone, D.E. (1982) *Grana*, **21**: 129- 148.

Kress, W.J., Stone, D.E. and Sellers, S.C. (1978) *Am. J. Bot.*, **65**: 1064- 1078.

Lee, Y.H., Ng, N.Y. and Goh, C.J. (1994) *Sci. Hortic.*, **60**: 167-172.

Lemos, W.D.P., Ribeiro, R.C., Lhano, M.G., Da Silva, J.P.S. and Zanuncio, J.C.(2010) *Entomotropica*, **25**(1): 43-47.

Lennox, G.W. and Seddon, S.A. (1978) Flowers of the Caribbean. Macmillan Education, Oxford England.

Leyva-Ovalle, O.R., Rodríguez-Goya, A.Y., Herrera-Corredor, J.A., Galindo-Tovar, M.E. and Murguía-González, J. (2011) *Trop. Subtrop. Agroecosystems*, **13**(3): 551-559.

Lia, H., Trimanto and Wahyudi, D. (2019) *Biodiversitas*, **20**(5): 1266-1283.

Linares-Gabriel, A., López-Collado, C.J., Hernández-Chontal, M.A.; Velasco-Velasco, J. and López-Romero, G. *et al.* (2018) *Ornam. Hortic.*, **24**(3): 7.

Loges, V., Texeira, M.C.F., Silva, S.S.L., Lago, P.G.P., Silva, S.A.C.G., Lima, T.L.D. and Castro, A.C.R.(2015) *Acta Hortic.*, **1104**: 455-462.

Mahanty, H. K. (1970) *Cytologia*, **35**: 13-49.

Malakar, M., Acharyya, P. and Biswas, S. (2019) *Acta Hortic.*, **1256**: 77-94.

Marouelli, L. P., Inglis, P. W., Ferreira, M. A. and Buso, G. (2010) *Genet. Mol. Res.*, **9**(3), 1377–1387.

Melendez-Ackerman, E.J., Speranza, P., Kress, W. J., Rohena, L., Toledo, E., Cortés, C., Treece, D., Gitzendanner, M., Soltis, P. and Soltis D. (2005) *Int. J. Plant Sci.*, **166**(5): 781-794.

Montgomery, R. (1986) *Heliconia Soc. Inter.*, **1**: 6-7. Nihad, K., Berwal, M.K., Hebbar, K.B., *et al.* (2019) *Hortic. Environ. Biotechnol.*, **60**(6): 799-808.

Patel, U., Patil, S. Sathyanarayana, E. *et al.* (2017) *Int. J. Chem. Stud.*, **5**(6): 400-402.

Pereañez, J.A., Jiménez, S.L.,Quintana, J.C., *et al.* (2008) *Vitae*, **15**(1): 157-164.

Quinaya, D.C.P. and d'Almeida, J.R.M. (2019) *J. Nat. Fibers*, **17**(10), 2020: 1488-1496.

Raizer, M.D.M., Quisen, R.C., Valente, M.S.F., Lopes, R. and Lopes, M.T.G.(2019) *Bioscience J.*, **35**(1): 222-235.

Rodrigues, P.H.V. (2008) *Sci. Agric.*, **65**(6): 681-684.

Rose, M.J. and Barthlott, W. (1995) *Plant Syst. Evol.*, **195**(1-2): 61-65.

Roy, S., Choudhury, P., Dhara, M. and Adhikari, L.(2019) *Research J. Pharm. and Tech.*, **12**(12): 6107-6110.

Sangwanangkul, P., Saradhuldhat, P. and Paull, R.E. (2007) *Post Harvest. Bio. Tech.*, **48**: 264- 271.

Santos, E. (1978) *Rodriguesia*, **30**: 99-201.

Sardinha, D.H.S., Rodrigues, A.A.C., Diniz, N.B., Campos Neto, J.R.M. and Reis, F.O. (2012) *Summa Phytopath.*, **38**: 159-162.

Sewake, K.T. and Uchida, Janice Y. (1980) Hawaii Institute of Tropical Agriculture and Human Resources. Research Extension Series 159: pp. 18.

Shatil Shahriar, S.M., Sammy, L.A., Rabbi, F., Bhuiyan, B., Rahman, M.A. and Nafiujjaman, Md. (2018) *Biol Eng Med*, **3**(6): 1-6.

Shokalu, A., Akintoye, H., Olatunji, M., Adebayo, A. and James, I. (2019) *Acta Hortic.*, **1263** : 497-502.

Silva, C.G.D., Krause, S., Botini, A.F., França, R.P.A.D. and Silva, C.A. (2019) *Ornam. Hort.*, **25**(1) : 80-86.

Simao, Daniela G. and Scatena, Vera, L. (2003) *Braz. Arch. Biol. Technol.*, **46**(1): 65-72

Smith, R.R. (1977) *Phytologia*, **30**: 251-261.

Sobrinho, C.C.D., Silveira, D.A., Cesar, F.B.D., de Oliveira, C.M.G. and Bittencourt, M.A.L. (2012) *Nematropica*, **42**(2): 351-355.

Souza, R.R., Cavalcante, M.Z.B., Silva, E.M., Amaral, G.C., Brito, L.P.S. and Avelino, R.C. (2016) *Comun. Sci.*, **7**(2) : 214-222.

Stiles, G. (1975) *Ecology.*, **56**(2): 285-301.

Stone, D.E., Sellers, S.C. and Kress, W.J. (1979) *Ann. Mo. Bot. Gard.*, **60**: 701-730.

Tillich, H. J. (1995) *In*: Rudall, P.J.; Cribb, P.J.; Cutler, D.F. and Humphries, C.J. (eds.). *Monocotyledons: systematics and evolution*. Royal Botanic Gardens, Kew. pp. 303-352.

Vazquez-Euan, R., Chi-Manzanero, B., Hernandez-Velazquez, I., Tzec-Simá, M., Islas-Flores, I., Martínez-Bolaños, L., Garrido-Ramírez, E.R. and Canto-Canché, B. (2019) *Agronomy*, **9**(10) : 666.

11

Agapanthus

M. Lakshmi Durga, Panigrahi Pratyusha Bhagavati and Anamika Gurung

1.0 Introduction

The genus *Agapanthus,* established by L'Heritier in 1788 belongs to the family Amaryllidaceae and order Asparagales (Duncan, 2002). Agapanthus is a herbaceous perennial and rhizomatous monocot, commonly known as blue lily, African lily, or lily of the Nile (Sharaibi and Afolayan, 2017). African lilies are common ornamentals in warm climates, grown for their large spherical flower clusters. The word 'Agapanthus' has been translated from the Greek word *'agape'* means love, and *'anthos'* meaning flower, connoting 'love flower' or 'lovely flower' (Notten, 2004; Snoeijer, 2004). It can also symbolize fertility, purity and beauty.

2.0 Importance and Uses

2.1.0 Aesthetic Uses

In ornamental gardens, Agapanthus is used in mass plantings, roadsides, specimen planting, pots, revegetation, coastal landscapes and erosion control in hilly areas. Blue and white agapanthus makes an excellent evergreen border to paths and driveways in areas exposed to the afternoon sun.

2.2.0 Therapeutic and Other Uses

It is one of the widely used and highly valued medicinal plants in South African folkloric medicine. It is used as a traditional medicine by South African women during pregnancy and to induce or augment labour and ensure delivery of healthy child (Veale *et al.,* 1992; Varga and Veale, 1997). It is also used by the traditional healers for the treatment of heart disease, paralysis, coughs, colds, and chest discomforts (Varga and Veale, 1997). The crude extract of *A. africanus* leaves contained one or more active principles with agonist activity on the uterine muscarinic receptors and promoted the synthesis of prostaglandins in the uterus. The root's decoction is given to sheep and goats as antidiarrhoea (Dold and Cocks, 2001). Agapanthus promotes an abundance of children and easy childbirth (Batten and Bokelmann, 1966). Pooley (2003) stated that bathing the newborn baby with the decoction of *A. praecox* makes them strong, keeps them free from bowel problems, and prevents crusts on the head. It was listed among the plant species traditionally used in Southern Africa for the treatment of various central nervous system-related ailments (Stafford *et al.,* 2005). Koduru *et al.* (2004) reported that the dried root powder contains anticancer property. Agapanthus rootstocks contain saponin and sapogenins that generally have anti-inflammatory (reduce swelling and inflammation), anti-oedema (oedema = swelling due to accumulation of fluid), antitussive (relieve or suppress coughing) and immunoregulatory (have influence on the immune system) properties.

3.0 Origin and Distribution

Agapanthus is endemic to South Africa, more specifically to the Cape Province of Good Hope (Leighton, 1965). Within its natural limits of distribution, agapanthus is

Uses of Agapanthus

In Garden

In Pathway

In Pot

As Border in a Garden

In Rock Garden

As Specimen in Garden

Along the Sides of Garden steps

In a Park

In Flower Vase

In Residential Area

In Flower Arrangement

hardy and durable to adapt various soil and environments. It is found at an altitude of 2000 m above mean sea level, with an annual rainfall higher than 500 mm.

The *Agapanthaceae* is a single-genus (monotypic) family containing the genus Aga*panthus*, and all the species are native to Southern Africa (Notten, 2004; Snoeijer, 2004).

Agapanthus was one of the first plants to be taken by the Dutch from the Cape to the Netherlands and flowered in the garden of van Beverningk in 1679 and was described by Breyne in 1680 as *Hyacinthus africanus* (Snoeijer, 2004; Jamieson, 2004). The species was listed next in 1687 by Schuyl in Hermann's *Horti Academici Lugduno-Batavi catalogus* [Catalogue of Plants in the Leiden Botanic Gardens].

It became a popular garden plant and spread throughout Europe (Jamieson, 2004), reaching England by 1692, where it was described growing in the garden of Hampton Court Palace (Snoeijer, 2004). One of the English ships rounding the Cape of Good Hope prior to the Dutch settlement may have brought a specimen to England. Now, agapanthus is widely distributed and prized as garden subjects and cut flowers. Some species are naturalised weeds in many parts of the world.

4.0 Morphology

Plants are perennial herbs, glabrous and solitary but often forming colonies. The plant has an underground stem, called a rhizome that is used as a storage organ. The roots, which grow out of the rhizome are white, thick and fleshy. The leaves grow directly from the rhizome, having sometimes a leek-like stem, deciduous or evergreen, linear, entire, upright to arching or even spreading, normal to dark green, dull or shiny, thin fleshy to leathery. The leaves are arranged in two opposite rows.

The peduncle is stout, up to 2 m high, always arising from between the leaves, round or flat, which is especially detectable at about ¼ of the peduncle below the inflorescence, in full light upright but in shady places bending over.

The inflorescence is enclosed by a single large bract. This bracts opens on one side to release the flowers. In some species and cultivars this bract also splits, giving the inflorescence the appearance of opening on 2 sides. The bract drops or remains attached to the inflorescence.

The inflorescence is a pseudoumbel (or pseudociadium) in which cymose partial inflorescences, namely helicoidcymes [bostryces], are arranged in a racemose order, hence it is a thyrse (thyrsus) and consists of a few to many bostryces. A thyrse (a racemose main axis bears cymose branches) is a frequent flowering behaviour among the angiosperm families.

The flower is funnel-shaped, salver-shaped, or tubular/cylindrical, 6 tepals (3 inner and 3 outer) grown together in a tube at the base with free segments, slightly fleshy, blue to violet-blue and with a darker coloured midrib or white. Stamens 6, inserted on the tube, ovary superior. Style canal present. Fruit a triangular capsule, lengthwise opening. Seed flat and winged, usually black. Pollination is entomophilous in nature. Fruit is a three seeded elongated capsule, the length of which is between 4.5-5 cm. initially its colour is green to gradually become grayish brown. When matured it ruptures along, releasing about 20-100 determined for all seeds, each of nearly 5-6 mm in length. These determined for all seeds are black, flattened have a winged back part for easier spreading with the wind.

5.0 Species and Cultivars

5.1.0 Species

The genus *Agapanthus* was established in 1788 by L' Heritier (Duncan, 2002). *Agapanthus* has been difficult to classify into distinct species (Notten, 2004) as there are few unique characteristics and are greatly variable depending on growing conditions and geographical location. Leighton (1965) revised the genus and recognized ten species: four evergreen species which are *A. africanus* (L.) Hoffmanns, *A. comptonii* F.M. Leight, *A. praecox* Willd., and *A. walshii* L. Bolus and six deciduous species, namely, *A. campanulatus* F.M. Leight, *A. caulescens* Spreng, *A. coddii* F.M. Leight, *A. dyeri* F.M. Leight, *A. inapertus* (L. Bolus) F.M. Leight, and *A. nutans* F.M. Leight. However, based on the nuclear DNA content and pollen morphology as well as variability, Zonneveld and Duncan (2003) recognized six species in total comprising of two evergreen species which are *A. africanus* and *A. praecox* and four deciduous species, *A. campanulatus*, *A. caulescens*, *A. coddii*, and *A. inapertus*. The genus has been difficult to classify due to high rate of hybridization within member species which has led to array of garden hybrids with similar macromorphological features (Notten, 2004). This accounts for the numerous misidentifications of the members of this genus. The recent study (Sharaibi and Afilayan, 2017) of micromorphological characterization of the leaf and rhizome of *Agapanthus praecox* subsp. *praecox* Willd was a positive step in resolving the taxonomic confusion that exists in the genus *Agapanthus*. The presence of anomocytic stomata, polygonal epidermal cells with wavy anticlinal walls, helical thickened vessels, and xylem tracheary elements with crystals as well as the absence of trichomes and secretory ducts in *A. praecox* subsp. *praecox* may be useful tools in the identification of this species from the closely related species of *Agapanthus*.

There are currently six accepted species, seven hybrids and numerous cultivars especially involving *A. praecox* and its subspecies (Snoeijer, 2004). The evergreen, perennial growth habit, showy flowers and long flowering season have made it popular for mass plantings in herbaceous borders, along drive ways and road side, banks and on traffic islands (Dawson and Ford, 2012).

Various species in the family are *A. africanus*, *A. campanulatus*, *A. caulescens*, *A. coddii*, *A. inapertus* and *A. praecox*. It is difficult for untrained people to differentiate between most species. The same common name is often applied to many different species. For example, 'ubani' is the Zulu name for *A. africanus*, *A. campanulatus* and *A. praecox*.

Agapanthus orientalis (syn. *Agapanthus praecox*) is the most common type of agapanthus. This evergreen plant produces wide, arching leaves and stems that reach heights of 1 to 1.5 m. Varieties include white flowering types such as 'Albus,' blue varieties like 'Blue Ice,' and double forms such as 'Flore Pleno.'

Species of Agapanthus

Agapanthus orientalis

Agapanthus campanulatus

Agapanthus africanus

Agapanthus inaperatus

Agapanthus praecox

Agapanthus caulescens

Cultivars of Agapanthus

Indigo Frost

Glen Avon

Lavender Haze

Snow Cloud

Silver Moon

White Heaven

A. campanulatus (Bell agapanthus) is a deciduous plant that produces strappy leaves and drooping flowers in shades of dark blue.

A. africanus is an evergreen variety that displays narrow leaves, deep blue flowers with distinctive bluish anthers, and stalks reach heights of no more than 50 cm. Cultivars include 'Double Diamond,' a dwarf variety with double white blooms; and 'Peter Pan,' a tall plant with big, sky blue blooms.

A. caulescens is a deciduous agapanthus and depending on the sub-species, colors range from light to deep blue.

A. inapertus ssp. *pendulus* 'Graskop,' also known as grassland agapanthus, produces violet-blue flowers that rise above tidy clumps of pale green leaves.

5.2.0 Cultivars

Agapanthus, a revision of the genus (Snoeijer, 2004), arguably the most comprehensive book on the genus, lists 625 cultivars worldwide. These are derived mainly from evergreen species and subspecies, and particularly from *A. praecox* subsp. *orientalis*. However, cultivars are derived from other taxa and through hybridisation between them. For example, *A. praecox* subsp. *minimus* and the deciduous species *A. inapertus* have been important parents. Exact details of the parentages of many cultivars are unknown, complex or only partially recorded. Flower colours of the cultivars are in numerous shades of violet, blue, lavender and purple, with several white-flowered selections. Most have flowers with six tepals, although there are a few semi-double selections with additional tepals. Snoeijer (2004) arranged cultivars in several groups including those with funnel-, trumpet-, salver-, and tubular-shaped flowers.

Queen Mum: It is a new giant variety with white flowers, striped with the palest of blue. It reaches two metres high and has mid-green leaves that grow to one metre.

Black Panther: It has dark purplish flower buds with the usual green foil of foliage.

Purple Cloud: Its color is amethyst purple, widely used at the back of the garden bed. Grey-green foliage grows to one metre high and the flowers reach to two metres and about 30cm in diameter.

Alba Roseus: It blooms in whitish pink and grows to a height of 70cm which is highly suitable to pots

Snowball: It is one of the lowest growing of all with evergreen clumps of short, upright dark-green leaves. During summer, quite large heads of white flowers are produced. It will make a fine edging plant with shy branching habit.

White Orb: It is a large, free-flowering white variety with huge flower heads. The evergreen foliage mounds will grow to 90cm.

Baby Pete: It is a long-flowering, drought-tolerant dwarf form with flower stems reaching up to 35 cm. It is cultivated from the Australian-bred 'Peter Pan'. Flowers are sterile with a non-invasive habit suitable to grow in pots, and as a low-growing garden border.

Cultivars of Agapanthus

Jacaranda

Brilliant Blue

African Skies

Bellarina

Twister

Flower of Love

Silver Baby

Moon Light Star

Purple Emperor

Source: www.gardensillustrated.com

Pavlova: It has creamy white flowers and blue-grey leaves. It arose as a single hybrid plant among a batch of *A. inapertus* seedlings and the pollen parent is probably *A. praecox*.

Sarah: It has distinctly arranged upward-pointing florets producing a candelabra-shaped umbel, bicoloured florets, soft blue flowers and dark green foliage.

Goldstrike: It arose as a chance variegated seedling from a batch of seed of a non-variegated plant. It is named after its striking golden yellow margin on the outside edges of the leaves, contrasting with grey and bluish toned longitudinal stripes. It has a compact growth habit and dark blue flowers on black stems.

Glen Avon: It is a hardy evergreen agapanthus cultivar bred in New Zealand with large flower heads consisting of hundreds of pale-blue flowers with darker stripes through the centres of the petals.

Some of the other cultivars include 'Lavender Haze', 'Snow Cloud' (pure white flowers with yellow anthers), 'Silver Moon' (silver edged leaves), 'White Heaven' (pure white flowers with upturned florets), 'Jacaranda', 'Brilliant Blue', 'African Skies', 'Bellarina', 'Twister' (bicolored agapanthus), 'Flower of Love', 'Megan's Mauve', 'Silver baby', 'Moon light star', 'Purple Emperor' (*Agapanthus inapertus* hybrid), 'Pretty Sandy', 'Bray Valley', 'Northern Star', 'Lilium's Lilac', 'Silver Lining' (*hybrid between Agapanthus praecox* x *Agapanthus inapertus*), 'Indigo Dreams'.

6.0 Genetics and Breeding

6.1.0 Genetics

The chromosome number and karyotype of four species of agapanthus namely *Agapanthus praecox* ssp. *orientalis* 'Big Blue', *Agapanthus comptonii* ssp. *longitubus*, *Agapanthus praecox* ssp. *Orientalis* and *Agapanthus praecox* ssp. *minimus* 'Storms River' studied by squash method showed that their chromosome number was 30 and their karyotype belonged to 2B (Gao *et al.*, 2011).

Through fluorescence *in situ* hybridization (FISH), one 45S rDNA signal was observed proximally to centromere of the chromosome 7, while for 5S rDNA sites, one signal proximally to centromere of chromosome 9 was observed. The 2C DNA content estimated for the species was 2C = 24.4 with 59 per cent of AT and 41 per cent of GC (Reil *et al.*, 2015).

6.2.0 Breeding

In *Agapanthus praecox* ssp. *orientalis*, flowering lasted 2 to 3 days and anther dehiscence lasted 3 to 5 hours. On the day of anthesis, the styles were shorter than the filaments. The seed set rate with different treatments showed that its breeding system could be determined as outcrossing with partly self-compatible and needed pollinators during the pollination (Suge, 1952).

A tall-growing *Agapanthus praecox* subsp. *orientalis* had the highest fertility (54–96 per cent seed set when sib crossed; >95 per cent pollen viability). When selfed, these agapanthus had low self-fertility (9.5 per cent seed set). A dwarf cultivar,

Agapanthus 'Sarah' was developed through controlled breeding programme by crossing unnamed *Agapanthus praecox* seedling, characterized by its strong stem, blue to lilac flower colour and leaf colour atypical for *Agapanthus praecox* as a female parent and a male parent, proprietary *Agapanthus praecox* seedling characterized by its profuse florets and leaf colour atypical for *Agapanthus praecox*. *Agapanthus* 'Sarah' has distinctly arranged upward-pointing florets producing a candelabra-shaped umbel, bicoloured florets, soft blue flowers and dark green foliage. This cultivar was found to be self-sterile and had the lowest female fertility when outcrossed. However, male (pollen) fertility was relatively high (Dawson and Ford, 2012).

6.2.1 Polyploidy

The successful induction of polyploids depends on choice of agent, its concentration, and the duration of treatment. Polyploid induction in agapanthus may produce sterile triploids in the F_1 generation that reduce the invasiveness of the plant. For each of the diploid species, a doubling of their genomes to produce tetraploids is required. The tetraploids are then to be back-crossed to the diploid progenitors, producing triploid plants. Colchicine treatment produced 50 per cent mortality in agapanthus seedlings, but oryzalin was more toxic, leaving only 12.5 per cent surviving. No entire tetraploids were created, but diploidtetraploid chimeras were found in 10 per cent (colchicine) and 5 per cent (oryzalin) of the surviving treated plants (Barrett, 2011).

The blue pigments in agapanthus have been well characterized as delphinidins (Bloor and Falshaw, 2000), thus this genus presents the best target for determining the genes responsible for expression. The structures of the two major anthocyanins in blue agapanthus flowers have been determined to be *p*-coumaroylated delphinidin diglycoside attached to a favonol triglycoside via a succinic acid diester link. These compounds represent unique examples of anthocyanin pigments where both types of co-pigment, an aromatic acyl group and a favonoid co-pigment, are attached covalently to the anthocyanin.

7.0 Growth and Flowering

According to Gillissen (1980), one bostryx consists of 3-7 flowers; but slicing down an *Agapanthus praecox* reveals bostryces of 8 and even 10 flowers in one inflorescence. These flowers open in a phased manner gives an impression of a long flowering period.

In agapanthus, the flower buds for next season develop just after the flowering period. The pedicel is thin, upright to horizontal or bent, at the base one thin bract. This bract can be easily seen throughout the complete inflorescence. Floral organs differentiation was comprised of 6 distinct stages including node differentiation, inflorescence bud differentiation, floret primordia differentiation, tepal primordia differentiation, stamen primordia differentiation, and pistil primordia differentiation (Zhang *et al.*, 2011).

8.0 Propagation

The three ways to propagate agapanthus are from seed, by division of offsets and multiplication *in vitro.*

8.1.0 Seed Propagation

The ripened seed usually germinates in 1-3 months at 18°C. They should be grown for an year in pots and then transplanted in permanent positions. Root bound pots can be divided every 4-5 years. Seedlings take 2-3 years to flower.

8.2.0 Vegetative Propagation

8.2.1 Division of Offsets

Agapanthus plants consist mostly of heterozygous seedlings due to the lack of an efficient vegetative propagation method of the cultivars (Supaibulwatana and Mii, 1997). Hence, an alternative mass propagation tool for this economic species would be very useful.

8.3.0 Micropropagation

Micropropagation techniques offer a viable alternative and innovative method of asexual propagation. Micropropagation is widely used for rare, endangered and medicinal plants for conservation and commercial propagation (Pattnaik and Chand, 1996). In order to apply micropropagation for clonal propagation, it is a prerequisite to develop a shoot-tip meristem-to-plant system. In the genus *Agapanthus*, however, very few studies on *in vitro* micropropagation have been reported. There are however, several reports describing plant regeneration through callus cultures and somatic embryogenesis from young flower buds of *A. africanus* (Supaibulwatana and Mii, 1997), and somatic embryogenesis from callus cultures of *A. praecox* spp. *orientalis* (Suzuki *et al.*, 2001a; Wang *et al.*, 2012), *Agrobacterium*-mediated transformation of *A. praecox* ssp. *orientalis* (Suzuki *et al.*, 2001b) and regeneration from callus-derived protoplasts of *A. praecox* ssp. *orientalis* (Nakano *et al.*, 2003) have also been achieved. Baskaran and Staden (2013) reported plant regeneration through an *in vitro* micropropagation system from shoot-tip cultures of *A. praecox*. Prolific shoot multiplication (47.3±1.96 shoots per explant) was achieved on Murashige and Skoog (MS) medium supplemented with 22.2 µM benzyladenine (BA), 2.9 µM indole-3-acetic acid (IAA), and 4.5 µM thidiazuron (TDZ). Shoots were rooted on half-strength MS basal medium supplemented with 5.7 µMIAA and 2.5 µM 2-isopentenyladenine (2iP) with 11.3±0.78 roots per shoot. The *in vitro* raised plants were established successfully in a 1:1 (v/v) vermiculite:sand mixture when maintained in a greenhouse with 100 per cent survival. The elongated shoots (more than 5 cm in length) were treated for rooting and acclimatization in a moistened (5.7 µM IAA and 2.5 µM 2iP) vermiculite:sand (1:1 v/v) mixture, first in the misthouse and then in the greenhouse.

9.0 Soil and Climate

9.1.0 Soil

Agapanthus grows best in the sub-tropics; in the tropics it grows at higher elevations. It grows best in fertile, moist and well-drained soil having pH range 5.5-7.5, in a sunny site that receives sun for most of the day. On heavy soils, mix in grit when planting.

Propagation of Agapanthus

Mature Pods

Seeds

Seedlings of Agapanthus

Propagated by Separation

Bulb

Sprouting of Bulbs

Source: https://davesgarden.com/community/forums/fp.php?pid=2144984#b

Cultivation of Agapanthus

Seedlings Ready for Planting

Cultivation in Open

Cultivation in Pots

Flowering in Greenhouse

Seedling in Greenhouse

Agapanthus can grow almost anywhere and survives well in poor soil (FloraBase, 2010). It grows on mountain coastal ranges (Rodriguez, 1997), rocky places on mountains (Burman and Bean, 1985), between rocks and in depressions on sheets of sandstone rock (Jamieson, 2004), subsurface wetlands (Zurita *et al.*, 2009), in bushland in remnant endangered ecological community, acidic sandy soil (Jamieson, 2004). *Agapanthus* spp. will grow in any kind of soil – clay, sand or gravel (Van der Spuy, 1971), permanently wet situations and dry, hot locations (Zonneveld and Duncan, 2006), from sea-level mountain tops and occur always in places where the rainfall is more than 500 mm a year in its native range.

9.2.0 Climate

It thrives in areas with cooler winter temperatures. It thrive in full sun and need 6-8 hours of sunlight each day. However, they do better in partial shade in hot climate areas. Agapanthus has fleshy roots and leaves and this can make them prone to frost damage. The deciduous varieties are hardiest, dying down in winter. Once they are established, they should withstand most conditions. Evergreen types are more tender and their leaves can be damaged by frosts. Therefore, a mulch of straw or fleece is advisable when young plants are establishing or extreme cold (below -5°C) is forecast. Established clumps of evergreen agapanthus can withstand -10°C to -15°C if the ground is well drained, but the number of flowers maybe reduced the following summer. Planting in beds against house walls can reduce the likeliness of frost damage.

10.0 Cultivation

It can be planted at 30 cm apart and with crowns 5cm below the ground. Agapanthus is well suited to being grown in pots, especially the evergreen varieties which can then be brought into a conservatory or greenhouse for the winter. Agapanthus doesn't like to be potted into pots that are too spacious as this will encourage leaf growth rather than flower production. Ideally, they perform best where root development is restricted but the plants are well watered and fed. Feed weekly or fortnightly with a balanced liquid feed during the growing season until flowers begin to show colour. August and September are the best time to boost with a fertilizer (high P and K) to enhance blossoming. A well-balanced N-P-K fertilizer should be applied as growth commences in spring.

Although plants flower well in pots if the roots are constricted, they should not be allowed to become completely root bound and should be split and replanted in fresh compost if the roots become too congested, otherwise flowering will suffer.

It may take two or three years for plants to establish before flowering really takes off, but after this they will grow in to long-flowering clumps. Agapanthus can be reluctant to flower if subjected to drought conditions after flowering. To ensure a good display the following year, keep plants moist until autumn after flowers start to feed, which will encourage the development of new flower buds. Cut down spent flower stems unless you are drying them to use for decoration. Mulch in autumn or cover the crown of the plant with straw or fleece to protect from cold. If clumps become too big, they can be lifted and split every four to five years. Deadheading of spent flowers is done to conserve the energy for next years' bloom. Weeding must be done periodically to keep the plots clean. Manual weeding is effective and it should be done at monthly interval. Atrazine @ 3 kg/ha in 1000 litres of water as a pre-emergent weedicide also keeps the crop weeds free.

11.0 Diseases and Pests

11.1.0 Diseases

11.1.1 Anthracnose

Murata *et al.* (2008) first reported the occurrence of anthracnose in agapanthus in Japan. The causal agent was identified as *Colletotrichum dematium*. Brown to grayish brown leaf blight lesions were observed on over-wintered leaves. Removal of the diseased plant and destruction of the plant debris and spraying of copper-based fungicide are suggested to control the disease.

11.1.2 Leaf and Stem Spot

Edwards *et al.* (2016) observed leaf and stem spot on *Agapanthus praecox* in northern Italy and leaf rot and spot on the same host in Melbourne, Australia. The disease is caused by a fungus *Fusarium agapanthi.* It is characterized by yellow, brown and even black spots that darken and pit agapanthus leaves. This disease can be controlled by removing heavily infected plant and spraying with fungicide like dithane M-45 or blitox @ 0.3 per cent at regular intervals. *Fusarium agapanthi* may also be more prevalent on other species of Agapanthus. It causes serious seedling damping-off in *Agapanthus africanus* (Guarnaccia, 2019).

11.1.3 Die-back

The agapanthus foliage may be attacked by the fungus *Macrophoma agapanthii* causing die-back of the leaves (Notten, 2006). Symptoms of the infection include brown leaf tips and lighter green foliage in the middle of the plant. The disease eventually kills the foliage. In severe cases, it can be combated with a fungicide like mancozeb or captan as a full cover spray.

11.1.4 Botrytis (*Botrytis cinerea*)

This fungus causes a grey mold on flowers, leaves, stems and buds. It thrives in cool wet weather conditions. Removal of affected plant parts, avoid over watering and opening of canopy for good air circulation controls the disease.

11.1.5 Powdery Mildew

This fungal disease occurs on the top of the leaves in humid weather conditions. The leaves appear to have a whitish or greyish surface and may curl. Usually occurs if agapanthus is grown in a shady area.

11.1.6 Rots

A number of pathogens cause root rots. Bulb rot and root rot are both agapanthus problems that start underground. The above ground shoot turns yellow, wilted and stunted. The roots and bulb get decayed and discolored. Crop rotation should be practiced as a control strategy.

11.1.7 Bacterial Disease

Xanthomonas axonopodis infecting agapanthus in Florida was first reported by Norman *et al.* (2002). During warm wet periods, a symptom of chlorotic, water-soaked, leaf-streaks was observed on agapanthus production in Florida.

11.1.8 Viral Diseases

The first report of natural infection of agapanthus by eggplant mottled dwarf virus (EMDV) was done by Zhai *et al.* (2014) from Latium region, Italy. The virus-infected African lily plants exhibit symptoms of bright mosaic consisting of white and yellow rectangular/oval to linear patches and stripes, often confluent, on leaves, flower stalks and buds. The virus is transmitted by the leafhopper vectors in a persistent, propagative manner and as well as by mechanical sap-inoculation.

Nerine virus X (NVX) infection in plants of *Agapanthus* spp. was reported from the Netherlands and Taiwan (Phillips and Brunt, 1980; Fuji *et al.*, 2006; Chiang *et al.*, 2010). The virus-infected African lily plants exhibit foliar chlorotic strip and mottle symptoms. The virus is transmitted by mechanical or sap transmission.

The occurrence of Tomato spotted wilt virus (TSWV) in agapanthus was first reported by Wilson *et al.* (2000). The infected leaves showed the symptoms of concentric ring and line patterns, irregular chlorotic blotches, and streaks. Symptomatic plants were severely stunted and failed to flower. Symptomatic leaves prematurely senesced, but young foliage subsequently produced was symptomless.

11.2.0 Pests

11.2.1 Agapanthus Borer

Agapanthus species are pest-hardy, neither being much attacked nor drastically affected by common garden pests. However, since the early 21st century *Agapanthus* in the far south of South Africa have fallen victim to a species of Noctuid moth, the *agapanthus* borer (*Neuranethes spodopterodes*). The larvae of the moth bore into the budding inflorescence and as they mature they tunnel down towards the roots, or emerge from the stem and drop down to feed on the leaves or rhizomes. A severe attack promotes rot and may stunt or even kill the plant; even plants that survive commonly lose most of their inflorescences and fail to produce the desired show of flowers.

11.2.2 Agapanthus Gall Midge

In 2016, a new species of gall midge, *Enigmadiplosis agapanthi*, was described damaging Agapanthus in the United Kingdom. The tiny (<5mm) gall midge lays eggs which develop into maggots inside the individual flower buds or inside the flower head sheath. The maggots can cause the bud to be deformed and discoloured and fail to open. The severity of this can range from a couple of buds failing to collapse of the entire flower head. Infestation can be confirmed by opening the buds or flower heads and looking for the presence of small maggots 1-3mm in length which are a

Diseases and Pests of Agapanthus

Anthracnose

Die-back

Powdery Mildew

Agapanthus Borer

Borer Adult Insect

Bud Damaged by Borer

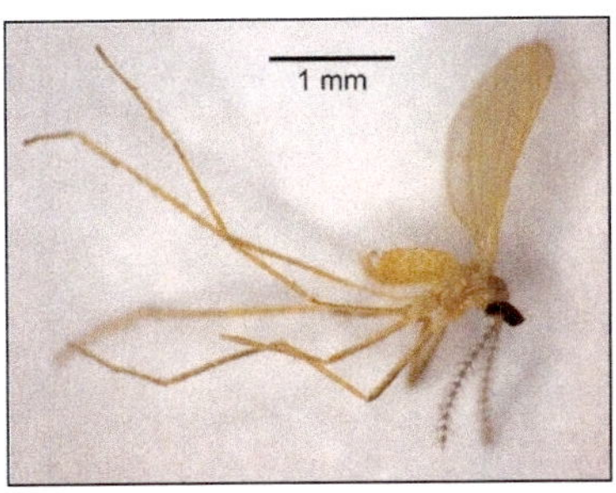

Agapanthus Gall Midge (*Enigmadiplosis agapanthi*)

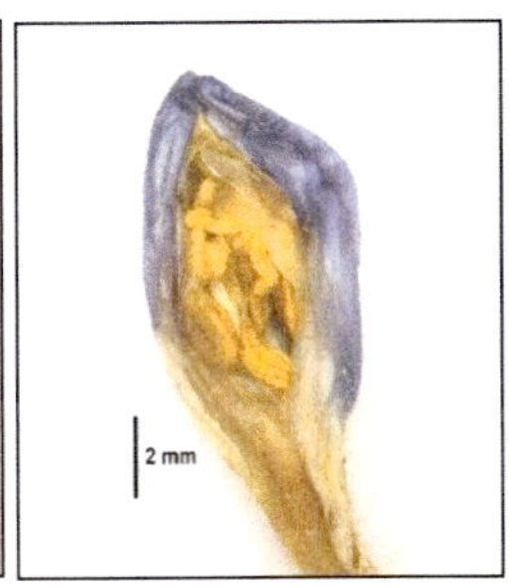

Petal Containing Gall Midge

Flower Attacked by Gall Midge

Harvesting and Marketing of Agapanthus

Harvesting Stage

Harvesting of Seeds

Harvesting of Bulbs

Flowers in Market

creamy yellow colour. The midge larvae leave the flower head to overwinter and pupate in the soil (Hayley *et al.*, 2015).

Currently, control measures attempt to interrupt the pest's life cycle:

- Remove and destroy infested flower heads
- Destroy badly infested plants
- Re-pot container grown plants, replacing growing media to remove pupating or overwintering larvae

Along with the above mentioned pests there some other minor pests which attack agapanthus such as mealy bugs, red spider mites, thrips and slugs.

12.0 Harvesting and Postharvest Management

Cut flowers are harvested when 2-5 florets open in a cluster. Cut the stems at a 45 degree angle with a sharp knife. Fill sterilised buckets with luke warm water and add flower food. Place the agapanthus in the buckets in a cool place. Leave over night to condition before using. Place away from direct sun light, radiators and drafts.

Agapanthus is very sensitive to ethylene, bacteria and cold. Ethylene causes major damages like flower and bud abortion, incomplete flowering and a short vase-life. During storage and rehydration, agapanthus faces water uptake problems caused by excessive bacterial growth due to insufficient hygiene in the water. During the post harvest treatment, the temperature in cold store and during transport should be above 10-12° C.

Basel treatment of partially opened flowers (2 to 5 florets open) with solutions containing 10-20 per cent sucrose and a bactericide improved bud opening. A 60 per cent to l20 per cent increase in vase life and a substantial reduction in bud abscission (to 20 per cent of the control) was obtained by treating the flowers with the anionic silver- thiosulphate complex (STS) and spraying the inflorescence with 30 ppm naphthalene acetic acid (NAA). These treatments also were effective in flowers subjected to short-term cold storage (4 days at l°C), was studied by Mor *et al.* (1984).

13.0 Biotechnology

13.1.0 Protoplast Culture

Protoplasts were isolated from leaf-derived embryogenic calluses of *Agapanthus praecox* ssp. *orientalis* 'Royal Purple Select' (2n=2x=32) with the yields of 0.8 to 1.5 $\times 10^6$ protoplasts per gram fresh weight of calluses. A plating efficiency of 0.8 per cent was obtained after 2 months of culture using a gellan gum-solidified medium containing 1 mg/l each of picloram (PIC) {4-amino3, 5, 6-trichloropicolinic acid} and benzyladenine (BA) under continuous illumination. Protoplast derived calluses produced somatic embryos at a frequency of 46.7 per cent on plant growth regulator (PGR) free medium, whereas 68.3 per cent of the 21 calluses regenerated adventitious shoots on a medium containing 1 mg/l BA (Nakano *et al.*, 2003).

13.1.2 Genetic Transformation

Suzuki *et al.* (2001) developed an efficient system for the production of transgenic plants of agapanthus via *Agrobacterium*-mediated genetic transformation. The system included co-cultivation of embryogenic calli with *A. tumefaciens* strain LBA4404/pTOK233 for 7 days in the presence of 20 mg l^{-1} AS, callus formation on a medium containing 1 mg/l PIC and 50 mg/l hygromycin, and somatic embryo formation on a medium containing 50 mg/l hygromycin. Similarly, Liu *et al.* (2011) also optimized system for *Agrobacterium*-mediated genetic transformation in *Agapanthus africanus* embryogenic calli which included 7 d pre-culturing, bacteria cell density (OD_{600}) 0.9, 5 min. infection incubation, 50 µmol/l acetosyringone (AS) in the pre-culture media and 100 µmol/l acetosyringone (AS) in the infection culture media. Under the optimized conditions, the genetic transformation of its embryogenic calli mediated by EHA105 strain increased.

Mori *et al.* (2007) carried out morphological characterization, and examination of the ploidy level, pollen fertility and stability of GUS gene expression in 12 transgenic plants derived from five independent lines at the flowering stage. Stable expression of a foreign gene (GUS reporter gene) in transgenic agapanthus plants were reported even after 5 yr of cultivation in pots.

An agapanthus gene, ApDFR1 encoding dihydroflavonol 4-reductase (DFR) gene contains four introns and five exons. ApDFR1 expression is specific to colored flower organs in *Agapanthus* sp. and flower color alteration was induced by heterologous expression of *ApDFR1* in petunia (Mori *et al.*, 2014).

The flower developmental mechanism of agapanthus follows the modified ABC model (Nakamura *et al.*, 2005). Hybridization signals of MADS-box gene in *Agapanthus praecox* ssp. *orientalis*, ApGLO and ApDEF, homologs of the *Antirrhinum majus genes* GLOBOSA and DEFICIENS, were observed in whorl 1 as well as in whorls 2 and 3. The transgenic plants that ectopically expressed ApGLO formed petal-like organs in whorl 1.There is possibility of molecular breeding for flower form alteration by genetic transformation with the class B MADS-box gene(s) of heterologous plant species (Nakano *et al.*, 2007).The class B genes, which belong to the MADS-box gene family, play important roles in regulating petal and stamen development in flowering plants. These genes exist in two different types termed DEFICIENS (DEF)- and GLOBOSA (GLO)-like genes, and the B-function is provided by heterodimers of DEF- and GLO-like gene products. These genes were ectopically expressed *Agapanthus praecox* ssp. *orientalis*, ApDEF or ApGLO. Gene ApDEF expressed in all the 4 whorls of non-transgenic plant, endogenous ApDEF products and transgenic ApGLO products may form heterodimers causing homeotic conversion in whorl 4 of transgenic plants (Okuzumi *et al.*, 2014).

14.0 References

Batten, A. and Bokelmann, H. (1966) *Books of Africa (Pty.) Ltd, Cape Town.*

Barrett, J. (2011) *Research Space at Auckland.*

Baskaran, P. and Staden, J.V. (2013) *South African Journal of Botany*, **86**: 46–50.

Bloor, S.J. and Falshaw, R. (2000) *Phytochem.*, **53**: 575-579.

Burman, L. and Bean, A. (1985) In: *Hottentot's Holland to Hermanus. South African Wild Flower Guide No. 5.* Botanical Society of South Africa.

Chiang, F.L., Chang, C.A., Cheng, Y.H. and Chen, C.C. (2010) *J. Taiwan Agric. Res.*, **60**: 89–100.

Dawson, M. and Ford, K. (2012) *New Zealand Garden Journal.*, **15**: 2-18.

Dold, A.P. and Cocks, M.L. (2001) *S. Afr. J. Sci.*, **97**: 375–379.

Duncan, G. (2002) *National Botanical Institute, Trident Press, Cape Town, South Africa.*

Edwards, J., Auer, D., de Alwis, S.K., Summerell, B., Aoki, T., Proctor, R.H., Busman, M. and O'Donnell, K.(2016) *Mycologia.*, **108**: 981-992.

FloraBase the Western Austalian Flora (2010) *South Afr. J. Bot.* Suppl., **4**: 21.

Fuji, S., Shinoda, K., Furuya, H., Naito, H. and Fukumoto, F. (2006) *Japan. Arch Virol.*, **151**: 205–208.

Gao, Y.H., Liu, F.Y. and Shang, A.Q. (2011) *Journal of Agricultural University of Hebei*, 4.

Gillissen, A.J.M. (1980) *Ph.D. Thesis, Department of Taxonomy of Cultivated Plants, Wageningen.*

Guarnaccia, V., Aiello, D., Polizzi, G., Crous, P.W. and Sandoval-Denis, M. (2019) *Phytopathol Mediterr.*, **58**: 127-137.

Hayley, J. and Andrew, S. (2015) *Royal Horticultural Society*, pp. 1-3.

Huxley, A., Mark, G. and Margot, L. (1992) *The Macmillan Press, Limited: London. The Stockton Press: New York.* 978-0-333-47494-5.

Jamieson, R. (2004) Available at http: //www.plantzafrica.com/plantab/agapanafric.htm.

Koduru, S., Grieson, D.S. and Afolayan, A.J. (2007) *Curr. Sci.*, **92**: 97–99.

Leighton, F.M. (1965) *S. Afr. J. Bot.* Supplementaery volume 1V.

Lui, H., Zhuo, L., Liu, Q. and Xu, Y. (2011) *Journal of Shanghai Jiaotong University-Agri.Sci.*, **29**: 62-67.

Mori, S., Oka, E., Umehara, H., Suzuki, S., Kobayashi, H., Hoshi, Y., Kondo, M., Koike, Y. and Nakano, M. (2007) *In Vitro Cell. Dev. Biol. Plant* **43**: 79-87.

Mori, S., Otani, M., Kobayashi, H. and Nakano, M. (2014) *Sci. Hort.*, **166**: 24-30.

Mor, Y. Halevy, A.H., Kofranek, M.A. and Michael, S.R. (1984) *J. Amer. Soe. Hort. Sci.*, 109: 494-497.

Murata, Y., Natsuaki, K.T. and Kobayashi, T. (2008) *J. Agric. Sci.-Tokyo University of Agriculture (Japan)*, **52**: 97-101.

Nakamura, T., Fukuda, T., Nakano, M., Hasebe, M., Kameya, T. and Kanno, A. (2005) *Plant Mol.Biol.*, **58**: 435-445.

Nakano, M., Tanaka, S., Oota, M., Ookawa, E., Suzuki, S. and Saito, H. (2003) *Plant Cell Tiss. Org. Cult.*, **72**: 63-69.

Nakano, M., Umehara, H., Hara, Y., Makino, M., Igarashi, M., Nakada, M., Nakamura, T., Hoshino, Y. and Kanno, A. (2007) *Mol. Breed.*, **20**: 425-429.

Notten, A. (2004) *South African National Biodiversity Institute.*

Notten, A. (2006) *PlantZAfrica – SANBI.*

Okuzumi, N., Otani, M., Otsubo, H., Meguro, S., Hara, Y., Umehara, H., Igarashi, M., Han, D.S., Nakano, M., Hoshino, Y. and Kanno, A. (2014) *Acta Hort.*, **1025**: 93-98.

Pattnaik, S. K. and Chand, P. K. (1996) *Plant Cell Rep.*, **15**: 846–850.

Phillips, S. and Brunt, A.A. (1980) *Acta Hort.*, **110**: 65-70.

Pooley, E. (2003) In: *Mountain flowers: A field guide to the flora of the Drakensberg and Lesotho*, The Flora Publications Trust, Durban, South Africa.

Reil, A.C., Franco, A.L., CAmpoS, V.R., Souz A, F.R., Zorzatto, C., Viccini, L.F. and Sousa, S.M. (2016) *Anais da Academia Brasileira de Ciências*, **88**: 1727-1734.

Rodríguez, P. (1997) *Soil Technol.*, **11**: 79-93.

Sharaibi O.J. and Afolayan, A.J. (2017) *Journal of Botany*, doi.org/10.1155/2017/3075638.

Snoeijer, W. (2004) In: *Agapanthus: A Revision of the Genus*, Timber Press, pp. 320.

Stafford G.I., Jager A.K., and Van Staden, J. (2005) *Journal of Ethnopharmacology*, **100**: 210–215.

Supaibulwatana, K.and Mii, M. (1997) *Plant Biotechnol. J.*, **14**: 23–28.

Suge, H. (1952) *Japanese J. Breed.*, **2**: 93-96.

Suzuki, S., Oota, M. and Nakano, M. (2001a) *Sci. Hort.*, **95**: 123–132.

Suzuki, S., Supaibulwatana, K., Mii, M. and Nakano, M. (2001b) *Plant Sci.*, **161**: 89-97.

Van der, S.U. (1971) *Hugh Keartland Publishers.*

Varga, C. A., Veale, D.J.H. (1997) *Social Science and Medicine*, **44**: 911–24.

Veale, D.J.H. Furman, K.I. Oliver, D.W. (1992) *J. Ethnopharmacol.*, **36**: 185–191.

Veale, D.J.H., Havlik, I., Oliver, D.W. and Dekker, T.G. (1999) *J. Ethnopharmacol.*, **66**: 257–262.

Wang, Y., Fan, X.,Wang, G., Zhang, D., Shen, X. and Wang, Y. (2012) *Propagation of Ornamental Plants* **12**: 148–154.

Zhang, D., Shen, X. and Zhuo, L. (2011) *Agricultural Sciences in China*, **10**: 1365-1373.

Zonneveld, B.J.M. and Duncan, G.D. (2003) *Plant Syst. Evol.*, **241**: 115-123.

Zurita, F., De Anda, J. and Belmont, M.A. (2009) *Ecol. Eng.*, **35**: 861-69.

12

Gloriosa

C. Aswath, V.A. Parthasarathy,
B. Biswas and K. Dutta

1.0 Introduction

Gloriosa, popularly known as glory lily, gloriosa lily or glory flower, is a fascinating and scientifically interesting genus. *Gloriosa superba* L. is a striking tuberous climbing plant with brilliant wavy-edged yellow and red flowers. The name *Gloriosa* comes from the word 'gloriosus' which means handsome and *superba* from the word 'superb' means splendid or majestic kind. It is a national flower of Zimbabwe and is also the state flower of Tamil Nadu. Plants are tall, weak-stemmed and tuberous rooted creeping plants, supporting themselves by means of leaf tendrils. Leaves are oblong, lanceolate or lance-ovate. The showy flowers are borne on long stalks and have reflexed petals with wavy edges and very conspicuous stamens. Ovary is 3-loculed, style long and bent upward near the base.

2.0 Importance and Uses

2.1.0 Aesthetic Uses

Gloriosa lilies have assumed an important place in flower trade and are widely grown for cut flower and tuber production both in the USA and Europe. Its place in the garden also looms ever larger. They are light climbers and are suitable for training on low walls or trellis. With proper training they can also be grown in pots.

2.2.0 Therapeutic and Other Uses

The tubers are rich in alkaloids (Subbaratnam, 1952; Thakur *et al.*, 1975). Determination of colchicine in *Gloriosa simplex*, by Ntahomvukiye *et al.* (1984), showed following percentages of colchicine in various plant parts : young leaves 2.36, flowers 1.18, adult leaves 8.7, tubers 0.54-0.92, seeds 0.37-0.46 and stems 0.33-0.41. Narain and Raina (1975) reported that 100, 75 and 50 ppm solutions of *Gloriosa* colchicine induced C-mitotic and C-tumour reaction. Mamatha *et al.* (1993) reported that glory lily (*G. superba*) is an important medicinal plant, and the alkaloids (mainly colchicine) in the seeds are used in the treatment of gout and rheumatism.

The seeds of this taxon are highly priced in the world market as they are the main source of colchicines and colchicoside (Akter *et al.*, 2014). *Gloriosa superba* is a medicinal and ornamental climber, its seeds and tubers contain valuable alkaloids *viz.*, colchicine and colchicoside as the major constituents, which are used to treat gout and rheumatism (Chitra and Rajamani, 2010).

3.0 Origin and History

Gloriosa superba L. (tribe Colchiceae, family Colchicaceae, Vinnersten and Manning, 2007) was first described by Linnaeus in 1737, based on material from southwest India (Malabar) and the name was validated in 1753 (cited by Maroyi *et al.*, 2013). It is native to South Africa, tropical Africa and Asia. *Gloriosa* originated in tropical Africa and Asia. The genus derives its name from the Latin word 'gloriosus', referring to the flowers. The first species known to horticulture was *G. superba*. It was known to pre-Linnaean writers under the name of *Methonica malabarorum*. Linnaeus

Uses of Gloriosa

In Garden

As Climber

In Pot

In Pergola

In Interior Decoration

In Flower Vase

In Flower Arrangement

gave its generic name *Gloriosa* and it remained till the mid-19th century despite the efforts of botanists to refer the genus back to *Methonica*. *Gloriosa carsoni* was first described by J. G. Baker in 1895. It was introduced into the USA in 1949 by R. S. Cammack. *Gloriosa virescens* was first described in 1825, based on a plant obtained from Mozambique in East Africa. Linnaeus published the species *G. simplex* in a rare work known as *Mantissa* in the 1760s. The first tubers of *G. plantii* are reported to have been sent to England by R. W. Plant from Natal in 1851 (Hayward, 1951).

G. superba is an inhabitant of tropical Africa and now found growing naturally in many countries of tropical Asia including Bangladesh, India, Sri Lanka, Malaysia and Myanmar. In India, it occurs commonly in tropical forests of Bengal and Karnataka (Sivakumar and Krishnamurthy, 2004). The plant thrives from Bundelkhand to humid Assam valley. It is known by different names in India such as 'Kalihari', 'Agnishikha', 'Languliata' and 'Nangulika'. There are several associated species of *Gloriosa* including *G. superba, G. simplex, G. grandiflora, G. lutea, G. plantii, G. lipolidii, G. longifolia, G. rothschildiana, G. virescens, G. sudanica, etc.* These species are distributed mainly in Africa (Ade and Rai, 2009).

4.0 Species and Cultivars

4.1.0 Species

4.1.1 Taxonomy

Species delimitation in *G. superba* complex is clearly controversial, which is also reflected in the high number of synonyms. It is the result of a poor understanding of the taxonomy and evolutionary relationships within the group. To solve this problem, numerical methods were used to re-evaluate morphological variation and species delimitation in the *G. superba* complex (Maroyi *et al.,* 2013). The morphological study has revealed the existence of four phenetic species in the *G. superba* complex. Recognition of these species is based on habit, inflorescence characters and distribution patterns. It is here proposed that four species should be recognized in the *G. superba* complex : *G. baudii* (Terracc.) Chiov., *G. carsonii* Baker, *G. superba* L. and *G. simplex* L. *Gloriosa superba* is the most widespread taxon, occurring in South Africa, tropical Africa and Asia. *Gloriosa simplex, G. carsonii* and *G. baudii* are confined to the African continent, with *G. simplex* wide spread in tropical Africa, *G. carsonii* common in south, central to east tropical Africa, and *G. baudii* having the most restricted range, confined to the arid regions of northern Kenya, Ethiopia and Somalia.

4.1.2 Species

The genus *Gloriosa* comprises of about 10 to 15 known species apart from *Gloriosa superba, viz. G. lulea, G. plantii, G. latifolia, G. longifolia, G. magnifica, G.rothschildiana, G.abyssinica, G. congijolia and G.simplex*. In all these the important species found in India are *G. superba* and *G. rothschidiana* (Hayward, 1951). Its distribution spreads to Nasik, Ratnagiri, Savanthwadi (Maharashtra), Uttar Karnataka, Hasson, Chikmangalur, Coorg, Mysore (Karnataka), Cannanore Palakkad, Trivandrum (Kerala), Tamil Nadu and Goa. In Tamil Nadu the largest area are under hilly region

Species of Gloriosa

Gloriosa superba

Gloriosa modesta

Gloriosa rothschildiana

Gloriosa aurea

Gloriosa simplex

Gloriosa lutea

Gloriosa lindenii

Gloriosa rigidiflolia

Gloriosa virescens

and is up to 6000 acres and spread over seven districts *viz.*, Karur, Tirupur, Dindigul, Salem, Ariyalur, Perembalur and Nagapattnam (Jasmine and Balakrishanan, 2018). According to Bailey (1963) also, there are five or more species. The species grown in the USA include *G. superba*, *G. rothschildiana*, *G. carsonii* and *G. plantii*. Important species commonly grown in Indian gardens are *G. superba* and *G. rothschildiana*.

There are ten accepted species of *Gloriosa*, ignoring hybrids, varieties and cultivars (Padmapriya *et al.*, 2016). They are:

Gloriosa aurea Chiov.

Gloriosa baudii (A.Terracc.) Chiov.

Gloriosa flavovirens (Dammer) J.C.Manning and Vinn.

Gloriosa lindenii (Baker) J.C.Manning and Vinn.

Gloriosa littonioides (Welw. ex Baker) J.C.Manning and Vinn.

Gloriosa modesta (Hook.) J.C.Manning and Vinn.

Gloriosa revoilii (Franch.) J.C.Manning and Vinn.

Gloriosa rigidifolia (Bredell) J.C.Manning and Vinn.

Gloriosa sessiliflora Nordal and Bingham

Gloriosa superba L.

Gloriosa carsonii

This species is taller and more freely branched than *G. superba*; it bears obtuse leaves and 20 or more flowers on a large plant (Hayward, 1951). Perianth segments, not more than 6.25 cm long, recurved, crisped, light yellow with bronzy tip changing to claret brown; style erect and stigmas three.

G. richmondensis

Leaves 15 cm long, 2 to 2.5 cm wide; flower yellow with a narrow red stripe down each petal, shape differs from *G. superba* being more squat with the petal tips incurving.

G. rothschildiana

Stems climbing, simple at first but afterward branched; leaves bright green, glabrous, oblong-acuminate, alternate or opposite, wider than *G. superba*; flowers solitary in the axils or peduncles 8-12 cm long, curved near the ovary; perianth segments oblong-lanceolate, lanceolate, recurved, over 8.70 cm long, crimson, dark purple mark near the base. More attractive than *G. superba*.

G. superba

Stems 1.5-3.0 m high; leaves ovate-lanceolate; flowers twisted, crisped with six recurved or reflexed; petals 5-7.5 cm long, less than 2.5 cm wide, opening yellow but changing to yellow-red and deep scarlet.

G. virescens

This species commonly known as Mozambique lily. Plants dwarf, about 1.5 m high, produce yellow flowers becoming deep yellow-red when exposed to the sun.

4.2.0 Types and Cultivars

Morphological characters of seventy garden cultivars belonging to species – *G. superba*, *G. lutea*, *G. plantii*, *G. carsonii*, *G. richmondensis*, *G. virescens*, *G. rothschildiana*, *G. magnifica*, *G. longifolia* and *G. latifolia* at the National Botanical Research Institute, Lucknow (India) were described by Narain (1977) and placed into the following types.

Type A : Flowers erect, perianth segments narrow, linear-lanceolate and highly crisped. The characters exhibited the influence of *G. superba* and *G. lutea*.

Type B : This type had all the floral characters of the type A, except that the perianth was reflexed.

Type C : Flower of this type was recurved as usual with narrow linear-lanceolate, erect and non-crisped perianth lobes. This group showed the influence of *G. virescens*.

Type D : In this type the flowers possessed all the characteristics described in type C except that their perianths are undulated or reflexed.

Type E : This included plants with oblanceolate perianth and erect or rarely undulated segments. This type showed the influence of *G. magnifica* and *G. latifolia*.

Type F : This was a very wide and diversified group of cultivars, and influence of *G. plantii*, *G. carsonii*, *G. richmondensis* and *G. rothschildiana* was apparent. Flowers were very showy with variable colours. Perianth is very wide, oblong and oblanceolate.

Many popular hybrid garden cultivars are Purple Prince, African Chief, Orange Gem, Orange Gift, Lavender Lady, and Orange Ball are grown for ample range of flowers and color combination of the perianth, which bloom for long during the autumn (Ade and Rai, 2009).

5.0 Genetics and Breeding

Traditional or conventional breeding has not been attempted so far as there are only local ecotypes of *G. superba* under cultivation and genetic wealth is limited. There are no named varieties in Gloriosa (Anandhi *et al.*, 2013).

5.1.0 Genetics

5.1.1 Cytology

It was observed that genus *Gloriosa* is monobasic and based on x = 11. Comparative morphological and cytological analyses of the elemental species of *Gloriosa* and the garden cultivars have been made by Narain (1977). Out of the 10 elemental species studied, *G. superba*, *G. lutea* and *G. plantii* are diploid (2n=22); *G. carsonii*, *G. virescens* and *G. richmondensis* are tetraploid (2n=44) and *G. rothschildiana*,

G. latifolia and *G. magnifica* are octoploid (2n=88). In general, octoploid species are comparatively short statured which constitute a medium group of plants.

Detailed karyomorphology of four morphotypes of *Gloriosa superba* is reported (Vijayavalli and Mathew, 1990). In gross karyomorphology they were similar, but with respect to total chromosome size and chromosome structural details of certain individual chromosomes, recognisable difference was noticed between and among the different morphotypes. This is pointed out to shed light on the role of intraspecific karyotype changes and its bearing on the plant morphological difference evident in the species.

Sultana *et al.* (2019) studied cytogenetically with orcein, CMA and DAPI-staining for authentic characterization. "Complex Chromocenter Type" of interphase nuclei with 3–4 bigger darkly stained heterochromatic regions. The prophase chromosomes were "Interstitial Type" with darkly stained region at different interstitial sites of chromosomes. *G. superb* has 2n = 22 chromosomes with heterogenous centromeric formulae 14 m + 8 sm. In addition to the regular chromosomes, 1–6 small chromosomes were found in several mitotic pro-metaphase, metaphase and anaphase stage which covered 43.57 per cent of the total cell count of this species suggesting the probable occurrence of B-chromosome. Presence of B-chromosome was probably the first report for this genus. After fluorescent banding, a pair of dot like CMA fluoresced bands were observed whereas no bright band was found in *G. superba* after DAPI-staining which suggested the coexistence of GC- and AT-base pairs in the genome.

5.2.0 Pollination

Ie Raux and Rabbertse (1997) made an extensive study on the reproductive biology of *Gloriosn superba.* They reported that ovule morphology, seed set after open- and hand-pollination, seed and ovule structure, as well as germination of *G. sllperba,* as possible factors affecting seed production. The protogynous flowers in the fran dose raceme are psychophilous and ornithophilous with a stigma receptivity period of at least four days. Development of the female gametophyte conforms to the Polygonum type. The inner integument acts as an endothelium due to the absence of the nucellus on the lateral sides of the embryo sac, where the nucellus has degenerated. A well-developed funicular obturator directs pollen tubes to a nucellar epistase where a partial self-incompatibility reaction appears to occur. A hypostase on the chalazal side contains transfer cells connecting the vascular tissue with the embryo sac. Seeds are tegmic, surrounded by a sarcotesta showing adaptations for endozoochory. Seed production was slightly better after cross-pollination, and also when 48 hour-old pollen was used, especially when pollination was done in the second flower phase. When the three factors were combined and applied to the proximal flowers in the inflorescence, seed production was significantly improved.

Pollen studies in *Gloriosa superba* Linn. revealed that pollen grains were oval in shape and pollen fertility in Acetocarmine stain was 96.5 per cent. Pollen grains remained viable for about 14 hrs after dehiscence. Germination was maximum in 10 per cent sucrose solution (98.61 per cent). Pollen tubes grew to a length of 14.24 μm (micrometre) per hr in distilled water and fruit set was 90 per cent under artificial cross pollination, 100 per cent under artificial self pollination but only 40 per cent under open pollination indicating that inadequate pollination is the main reason for poor fruit set in this corp (Mamatha *et al.*, 1993). The comparison of the different methods of pollination indicated good efficiency of cross-pollination. Pollen can be stored even several months in low temperature (Krause, 1988).

Complex modifications of angiosperm flowers often function for precise pollen placement on pollinators and to promote cross-pollination. Daniels *et al.* (2020) explored the functional significance of the unusually elaborate morphology of *Gloriosa superba* flowers, which are divided into one hermaphrodite meranthium and five male meranthia (functional pollination units of a single flower). They used controlled pollination experiments, floral measurements, pollen load analyses and visitor observations in four populations of *G. superba* in South Africa to determine the breeding system, mechanism of pollination and role of flower in the promotion of cross-pollination.

Raina and Gupta (1999) studied the effects of seven pollination treatments : natural pollination, bagging, autogamy, idiogamy, open crossing, allogamy and honey bee (*Apis mellifera*) pollination on fruit set, fruit size, number of seeds/fruit and/plant, and seed yield/fruit and/plant. No genetic self- or cross-incompatibility was observed. Although flower colour and shape favour cross pollination, self-pollination gave better results. Controlled selfing between flowers on the same plant (idiogamy) gave a significantly higher seed yield (9.2 g/plant; 681.73 kg/ha) than natural pollination (4.31 g/plant; 319.26 kg/ha).

From the studies, Daniels *et al.* (2020) were able to establish that *G. superba* is self-compatible, but reliant on pollinators for seed production. Butterflies, in particular the pierid *Eronia cleodora*, were the primary pollinators (>90 per cent of visitors). Butterflies brush against the anthers and stigma during nectar feeding and pollen is carried on their ventral wing surfaces. Butterfly scales were positively correlated with the number of pollen grains on stigmas thus demonstrating that the styles were orientated towards clearings in the vegetation and we confirm that the highest proportion of initial visits was to hermaphrodite meranthia pointing towards clearings.

5.3.0 Genetic Resources

5.3.1 Genetic Variability

Genotypic and environmental effects on growth and colchicine content of *G. superba* were studied by Jatisatienr *et al.* (1996). Genotype environment interaction was not significant for any of the traits investigated. Some traits were affected by heritability more than by environmental effects. Morphological differences occurred between plants from Thailand and India. The colchicine content in the seeds showed low genetical effects; however, it is suggested that it might be possible to select for high tuber colchicine content, especially in Thai *G. superba* plants in which tuber colchicine content is nearly 80 per cent that of the seed.

Farooqi *et al.* (1999) estimated genetic variability and association of some quantitative characters with yield in glory lily (*Gloriosa superba*). High genetic variability was observed for all the characters studied. PCV values were higher than GCV values and heritability values were high for all the characters. However, high genetic advance was recorded only for number of leaves per plant. Correlation and path analysis studies showed significant positive correlations between dry weight of pods, number of branches per plant and seed yield per plant, which in turn and high correlation with colchicine percentage. Therefore, plant with more branches and higher dry weight of pods per plant should be given emphasis in selecting plants for higher yield of seeds and colchicine content in glory lily.

Eighteen accessions of *Gloriosa superba* collected from various geographical regions of the important *Gloriosa* growing states *viz.*, Tamil Naduand Andhra Pradesh were evaluated for total sugar, starch, protein, total phenol, peroxidase, catalase and poly phenol oxidase to assess its suitability for seed yield as well as tolerance to leaf blight. The accessions exhibited remarkable variations for all the traits studied. The maximum peroxidase, catalase and minimum poly phenol oxidase were observed in GS 15 while GS 12, recorded the highest total sugar content. The accession GS 15 excelled in the accumulation of starch, soluble protein and total phenol in tuber and thus can be utilized for extraction of colchicine and colchicoside (Chitra and Rajamani, 2010).

5.4.0 Interspecific and Intergeneric Hybrids

Gloriosa spp., *Littonia modesta* and *Sandersonia aurantiaca*, all belonging to the family Colchicaceae, are tuberous plants native to South Africa. *Gloriosa* spp. have red or yellow flowers, in which tepals are reflexed turning upwards and backwards. *L. modesta* has orange, bell-shaped flowers. *S. aurantiaca*, a monotypic species, has orange, lantern-shaped flowers. Although these plants have recently become popular as cut-flowers, they have few variations in plant form, flower color and shape. Interspecific and intergeneric hybridization may widen variations in horticultural traits of these plants. However, according to Kuwayama *et al.* (2005), there have been only 2 reports on wide hybridization in the family Colchicaceae: *S. aurantiaca* x *L. modesta* and *L. modesta* x *S. aurantiaca* (Morgan *et al.*, 2001), and *S. aurantiaca* x *G. rothschildiana* (Nakamura *et al.*, 2005). Kuwayama *et al.* (2005) examined the production of interspecific and intergeneric hybrid plants using 6 genotypes of *Gloriosa* spp., 1 genotype of *L. modesta* and 2 genotypes of *S. aurantiaca*. In some of the plantlets derived from intergeneric cross-pollination, early confirmation of the hybridity was accomplished by flow cytometry analysis of relative DNA content of nuclei (Kuwayama *et al.*, 2005). All the hybrid plants, *L. modesta* x *S. aurantiaca*, *L. modesta* x *G. superba 'Lutea'*, and *S. aurantiaca* x *G. superba* 'Lutea', showed a climbing habit like those of *L. modesta* and *G. superba* 'Lutea' (Amano *et al.*, 2008). Plants of *L. modesta* x *S. aurantiaca* and *L. modesta* x *G. superba* 'Lutea' were taller and shorter than their respective parents, whereas, plant height of *S. aurantiaca x G. superba* 'Lutea' was nearly intermediate between the parents. Leaves of all the hybrids had a tendril at the tip like those of *L. modesta* and *G. superba* 'Lutea'. Flower morphologies of all the hybrids were nearly intermediate between the parents. Flower colors of all the hybrids were similar to the seed or pollen parent. Although hybrids of *L. modesta* x *G. superba* 'Lutea' showed low pollen fertility as assessed with acetocarmine staining, the other two kinds of hybrids had nondehiscent anthers or no fertile pollen grains. Chromosome observation in root tip cells revealed that all the hybrids had a diploid number of chromosomes: *L. modesta* x *S. aurantiaca* (2n=2x=23), *L. modesta*, *G. superba* 'Lutea' (2n=2x=22), and *S. aurantiaca* x *G. superba* 'Lutea' (2n=2x=23). Novel morphological characteristics of the hybrids may be valuable for future breeding of colchicaceous ornamentals (Amano *et al.*, 2008).

Cross breeding attempts involving *G. superba*, *G. lutea*, *G. plantii* (diploids), *G. carsonii*, *G. virescens*, *G. richmondensis* (tetraploids) have been carried out in several combinations. The only successful hybrids were *G. lutea* x *G. plantii* and its reciprocal, *G. carsonii* and its reciprocal, *G. virescens* × *G. richmondensis* and *G. carsonii* × *G. richmondensis*. Morphological analysis and cytogenetical behaviour of all the F_1 hybrids had been worked out. It was also observed that the tetraploid *Gloriosa* species had higher crossability compared with diploid species. Further, *G. superba* had always failed to produce hybrids, although it had been involved in numerous hybridization combinations (Narain, 1979). Mamatha *et al.* (1993) recorded 90 per cent fruit set when flowers were artificially cross-pollinated, 100 per cent with artificial self-pollination but only 40 per cent with open pollination indicating that lack of pollen transfer may be the cause of poor fruit set.

No large variations in horticultural traits are found in Gloriosa. Amano *et al.* (2009) examined intergeneric hybridization using 6 genotypes of *Gloriosa* spp., 1 genotype of *L. modesta* and 2 genotypes of *S. aurantiaca* to obtain wider variability and to develop novel cultivars in those groups. Following intergeneric cross-pollination, putative hybrid plantlets were obtained via ovule culture in various combinations. Early confirmation of the hybridity of ovule culture-derived plantlets was accomplished by flow cytometry and random amplified polymorphic DNA analyses. Several intergeneric hybrids have so far been produced flowers and subjected to morphological characterization. All the hybrids examined, *i.e. L. modesta* x *S. aurantiaca*, *L. modesta S. aurantiaca* 'Phoenix', *L. modesta* x *G.* superba 'Lutea', *L. modesta* x *G.* 'Marron Gold', *S. aurantiaca* x *G.* superba 'Lutea', *S. aurantiaca* x *G.* 'Marron Gold' and *S. aurantiaca* 'Phoenix' x *G.* 'Marron Gold', had novel morphological characteristics compared with their parents, some of which were horticulturally attractive. The results obtained in the series of studies indicate the validity of intergeneric hybridization in the improvement programs of colchicaceous ornamentals. Amano *et al.* (2009) suggested examining to develop a rapid and efficient micropropagation system and to restore fertility by artificial chromosome doubling of intergeneric hybrids that had been produced in series of experiments.

Intergeneric hybrid plants between Colchicaceous ornamental plants, *Sandersonia aurantiaca* and *Gloriosa rothschildiana*, have successfully been produced via ovule culture (Nakamura *et al.*, 2005). After 5 days of reciprocal cross-pollination, a few pollen tubes were observed in the ovary. Although seeds were obtained in both reciprocal cross-combinations, they did not germinate under ex vitro conditions. Ovules with placental tissues isolated 14 days after cross-pollination of *S. aurantiaca* × *G. rothschildiana* were cultured on a medium containing 0.01 mg/l each of α-naphthaleneacetic acid (NAA) and 6-benzyladenine (BA), on which 41.5 per cent of ovules swollen and produced callus-like structures within 10 weeks. When such swollen ovules were transferred to a medium containing 0.1 mg/l each of NAA and BA, 7.5 per cent of the initially cultured ovules produced rhizome-like structures within 6 weeks. Among the rhizome-like structures, those derived from two independent ovules (3.7 per cent of the initially cultured ovules) produced multiple shoots following transfer to a medium containing 0.25 mg/l NAA and 2.5 mg/l BA. Multiple shoot-derived plantlets were established on a plant growth regulator-free medium, and they were successfully transplanted to pots. Early verification of their hybridity was accomplished by flow cytometry (FCM) analysis, chromosome observation and rDNA analysis (Nakamura *et al.*, 2005).

GISH analysis is a powerful tool for identifying the genome constitution of intergeneric hybrids in colchicaceous ornamentals (Nakazawa *et al.*, 2011; Kishimoto *et al.*, 2014).

Some intergeneric hybrids produced by the crosses among several colchicaceous ornamentals, *Gloriosa superba* 'Lutea' ($2n = 2x = 22$), *G.* 'Marron Gold' ($2n = 4x = 44$), *G.* 'Verschild' ($2n = 7x = 77$), *Littonia modesta* Hook. ($2n = 2x = 22$), and *Sandersonia aurantiaca* Hook. ($2n = 2x = 24$), were subjected to genomic *in situ* hybridization (GISH) analysis in order to clarify their genome constitutions (Nakazawa *et al.*, 2011). Chromosome preparation was made from root tip cells of *L. modesta* × *G. superba* 'Lutea', *L. modesta* × *S. aurantiaca*, *S. aurantiaca* × *G. superb* 'Lutea', *L. modesta* × *G.* 'Marron Gold', *S. aurantiaca* × *G.* 'Marron Gold' and *G.* 'Verschild' × *S. aurantiaca*. Total DNA of one parent was labeled with digoxigenin or biotin and used as probe, and chromosomes were counterstained with 42 -6-diamidono-2-phenylindole (DAPI). For all the nine intergeneric hybrids, chromosomes from each parent could be clearly distinguished by GISH analysis. The results obtained by GISH analysis study may be important for further progress in breeding of colchicaceous ornamentals (Nakazawa *et al.*, 2011).

Diploid and triploid intergeneric hybrids obtained by crosses among *Gloriosa superba* 'Lutea' ($2n = 2x = 22$), *G.* 'Marron Gold' ($2n = 4x = 44$), *Littonia modesta* ($2n = 2x = 22$), and *Sandersonia aurantiaca* ($2n = 2x = 24$) were analyzed for their meiotic chromosome pairing in pollen mother cells by genomic in situ hybridization (GISH) with digoxigenin-labeled total DNA of one parent as probe. Chromosomes from each parent could be clearly distinguished in pollen mother cells of all the five intergeneric hybrids by GISH. For three diploid hybrids, *L. modesta* × *G. superba* 'Lutea' ($2n = 2x = 22$), *L. modesta* × *S. aurantiaca* ($2n = 2x = 23$) and *S. aurantiaca* × *G. superba* 'Lutea' ($2n = 2x = 23$), 0.04–0.27 autosyndetic bivalents (intragenomic pairing of non-homologous chromosomes) and 0.13–0.36 allosyndetic bivalents (intergenomic chromosome pairing) were observed per pollen mother cell, indicating that there are some homologous chromosomal regions within each genome and among the genomes of *Gloriosa*, *Littonia* and *Sandersonia*. Differences in the average number of allosyndetic bivalents per pollen mother cell among different genome combinations may reflect the evolutionary distances among the three genera, and *Gloriosa* and *Littonia* may be closely related to each other, while *Sandersonia* may have relatively distant relationships with *Gloriosa* and *Littonia*. For two triploid hybrids, L. *modesta* × *G.* 'Marron Gold' ($2n = 3x = 33$) and S. *aurantiaca* × G. 'Marron Gold' ($2n = 3x = 34$), no allosyndetic bivalents were observed. Based on the results obtained in the present study, possible utilization of the diploid and triploid intergeneric hybrids for further breeding of colchicaceous ornamentals is possible (Kishimoto *et al.*, 2014).

5.5.0 Mutation Breeding

Anandhi *et al.* (2013) reported the mutation breeding work being carried out at Tamil Nadu Agricultural University. Correlation and path coefficients for 17 morphological characters were studied on morphologically distinct mutants of *Gloriosa superba* in the second generation of vegetative mutant (VM2) along with their mother genotypes collected form Mulanur. The highest and positive correlation for dry seed yield/plant was observed with number of seeds/pod, number of leaves/plant, dry pod weight, fresh seed weight/pod, fresh pod weight, plant height and number of secondary branches/plant. The path analysis of component characters on dry seed yield/plant of *G. superba* in the VM2 generation exerted a direct positive effect through these characters: number of leaves/plant, dry pod weight, number of seeds/pod, fresh seed weight/pod.

6.0 Growth and Flowering

The growth and development of *Gloriosa* is somewhat similar to that of gladiolus in its periodicity. In nature, after flowering, with the advent of cold weather, the aerial part of the plant gradually withers and the tubers become dormant which start sprou-ting with the rise in temperature in summer. Carow (1976a) studied the development of *G. rothschildiana* and found that shoot development was related to the decrease in dry weight of the mother tuber which lost 60 per cent of its dry weight and 75-90 per cent of its water content within 34 days of planting, the actual amount varying with planting date. Shoot growth was faster during the summer months when the light intensity was high. Under these conditions more shoots of greater weight developed and daughter tubers continued to grow so long as the plant had leaves. Length of primary lateral shoot is important with regard to cut flower production. Although tuber weight was significantly correlated with length of main stem and mean length of lateral shoots of the 2nd order, 102 days after planting in summer, there was no correlation between tuber weight and mean length of lateral shoots of the 1st order. Subsequent studies revealed that the number of sinks developed on lateral shoots of the 1st order between 69 and 102 days was considerably increased in plants with 60 g tubers as compared with plants from smaller tubers. Inversely, shoot growth in this time was promoted in plants with 10 g tubers compared with those grown from 60 g tubers. Carow (1977c) considered these changes as sink affected correlation. In another experiment, Carow (1976b) observed that the number of single flowers was closely correlated with tuber weight. Yield, however, was modified by planting date. Weight of daughter tubers could be increased by removal of aerial sinks because tubers, flower buds and lateral shoots compete for assimilation products (Table 1).

Carow (1977b) also reported that flower formation takes place at the growing shoot if tuber weight exceeds a critical value. First flower buds can be started in the period of strongest dry matter decrease in the mother tubers. Flower differentiation is possible in complete darkness as well as under continuous light, and number of flowers increases with the increase in tuber weight. Carow (1980b) also found that pre-sprouting storage of tubers at 10°C and 50-70 per cent relative humidity for 60 days resulted in 100 per cent sprouting in 19-23 days. Tubers could be kept for 3-4 weeks at 18-20°C under black plastic without any inhibitory effect on their later sprouting. Henrard (1982) observed that tubers which rooted most vigorously before stem growth or the growth of replacement tubers also produced the best growth of aerial parts and flowers. According to Zimmer (1975), storage of tubers at 10°C for 3 months after harvest increased the shooting percentage of tubers subsequently held at 20°C and 25°C, but not in those held at 15°C and 30°C. The average time to shooting was reduced for all holding temperatures.

Table 1: Number of Flowers/Plant Dependent on Planting Date and Tuber Weight (Calculated on regression equations)*

Fresh Weight of Mother Tuber (g)	*Planting Date*						
	1 Feb	*15 Mar*	*23 Mar*	*1 May*	*10 Jul*	*18 Oct*	*3 Dec*
10	0	4	6	7	6	2	0
20	3	6	8	10	9	4	0
30	5	9	10	14	13	5	1
40	7	11	13	18	16	6	2
50	9	13	15	22	20	8	3
60	11	15	17	26	23	9	4
80	0	19	0	0	30	12	6
100	0	24	0	0	37	15	8
Correlation	0.93	0.93	0.86	0.95	0.85	0.93	0.91
Coefficient							
Regression Coefficient	0.22	0.22	0.22	0.38	0.34	0.14	0.10

* Carow (1976b)

Pod and seed development of *G. superba* was studied by Rajagopalan *et al.* (1995) in field-grown plants at weekly intervals starting from 7 days after pollination (DAP). Pods attained their maximum length, girth, fresh and dry weights at 63 DAP. Seed dry weight and 100-seed dry weight were also highest at this time. The relative growth rates (RGR) of pods increased gradually to reach a peak at 28 DAP and then decreased. Similarly, seed RGR was highest at 42 DAP. The colchicine content (per cent) and colchicine yield of seeds increased to reach a peak at 70 DAP and then decreased. Harvesting pods at 70 DAP is recommended for giving maximum seed and colchicine yield.

Talia *et al.* (1996) reported that tuber growth depended on initial tuber size with greater weight increases in large tubers than in small ones.

Application of alar at 4000 ppm gave the maximum (11.29 per cent) increase in leaf area. Maximum fresh (5.17 g) and dry weights of stems (3.06 g) were recorded in the control treatment. Length of tubers was maximum (6.50 cm) with 1000 ppm cycocel, and girth with 2000 ppm alar (2.25 cm) treatments, whereas the fresh and

Cultivars of Gloriosa

Purple Prince

Orange Gem

Lavender Lady

Simplex

Citrina

Plantii

Propagation of Gloriosa

Gloriosa Pods and Seeds

Emergence of Seedlings

Propagation of Gloriosa

Tubers

Planting of Tubers in Rows

Planting of Tubers in Open

Cultivation of Gloriosa

Cultivation in Open

Greenhouse Cultivation of Gloriosa

Staking of Gloriosa

Growth and Flowering in Open

dry weights of tubers were highest with 3000 ppm cycocel treatment (Suparna *et al.*, 1993).

7.0 Propagation

Propagation of Gloriosa is done by seed and division of tubers.

7.1.0 Seed Propagation

All the species except *G. rothschildiana* produce seeds readily in nature, and seeds after the proper ripening period can be grown into little tubers in another season. In *G. rothschildiana*, seeds are produced more readily by inter-clonal pollination (Hayward, 1951).This species is propagated commercially from seeds, by cross-pollination in June-July, harvested in September-October and sown at the end of March. In an investigation, germination was found the best (86 per cent) and most rapid (8 days) when seeds were kept in light at 30°C after 60 days of darkness in moist boxes (Carow, 1979). Krause (1986) also reported higher seed yield from cross-pollination than from self-pollination. The medium used for raising seedlings influences the germination and growth of seedlings. In peat substrate, seeds germinated 10 days earlier than in the leaf mould and the seedlings were of better quality.

G. superba seeds (already soaked in water for 24 h) were subjected to pretreatments consisting of running water leaching, distilled water soaking, and growth regulators – gibberellin (GA_3), BA [benzyladenine], Ethrel [ethephon] and thiourea for 36 h. Percentage germination was best (67.5 per cent) in 4000 ppm thiourea whereas germination rate was fastest in 3000 ppm thiourea (1.64 days). GA_3 gave the earliest germination (20.25 days), while 2000 ppm thiourea and running water treatments were best at inducing the early appearance of the first pair of leaves (8.5 days). Distance between the first leaf and the base was maximum (2.38 cm) in 300 ppm GA_3 (Supari *et al.*, 1993).

The coefficient of vegetative propagation of gloriosa is very low so a great number of plants can be obtained in a short time from seeds only. The yield of seeds is high if the flowering period falls in summer (Krause, 1988). There are about 60 seeds in one capsule but that number depends on the length of fruit. Seeds ought to be sown not earlier than in the next spring. Peat substrate is the best medium for sowing and cultivation of seedlings. Tubers can be harvested 12–14 weeks after pricking. The weight of two-armed tubers is 5–6 g only but in the next season of growing daughter tubers will reach even 20 g in the weight (Krause, 1988).

Only 10-15 per cent of seeds germinated after three weeks following cold treatment (Mahajan *et al.*, 2016). Soaking seeds in hot water for one hour results in maximum germination of 32.75 per cent and vigor index (565.92) (Anandhi and Rajamani, 2012).

7.2.0 Vegetative Propagation

Gloriosa produces tubers during growing season which are detached from the mother tubers. Tubers should be lifted carefully as they are very brittle and liable to break easily. Zimmer (1975) noted that when freshly harvested tubers of

G. rothschildiana were placed in damp peat at 30°C, 96.9 per cent produced shoots in about 6 weeks.

Carow (1975) observed that the meristem of this species normally differentiates into a single stem, roots and only 2 daughter tubers. Propagation of clonal material from tubers is thus very slow.

It is commercially propagated by tubers which are 'V' or 'L' shaped sourced from the wild especially from the forest areas and hillocks (Anandhi and Rajamani, 2012). It has been reported that more than 500 tonnes of wild tubers are collected every year and used for planting in Tamil Nadu alone. About 800 Kg of tubers are required to be planted in one acre. The sprouting of tubers is irregular and in a period of 30 days they sprout only to an extent of 60 per cent (Anandhi *et al.*, 2016).

Shimasaki *et al.* (2009) examined efficient methods for tuber formation of *Gloriosa superba* using stem branches under cultivation in green house. They studied the effect of longitudinal division of stem for explants derived from branches under cultivation. The number of tuber formed increased with higher concentrations of sucrose and showed maximum at sucrose 80 g/l + TDZ (thidiazuron) 3 µM. The number of tuber formed from decapitated stem explants (DS) after decapitation showed more than the double (8.9/explants) of the apical stem explants (AS). In trehalose treatment, the number of tuber formed in the AS was approximately similar to that of sucrose treatment. In longituidinal section (LS) explants, the number of tuber formation showed maximum (9.0) by treatment with trehalose 40 g/l + TDZ 3 µM and decreased with treatment of trehalose 80 g/l + TDZ 3 µM. It becomes clear that LS have higher ability for tuber formation. Combining TDZ 3 µM with 80 g/l sucrose or 40 g/l trehalose are highly effective for multiple tuber formation in stem cultures of *G. superb*.

Ethrel at 500 ppm recorded maximum sprouting percentage (100 per cent), earlier sprouting of tubers (6.33 days), maximum plant height (99.32 cm) and maximum number of leaves per plant (34.04) and plant girth (1.81 cm) (Anandhi *et al.*, 2016). The tubers weighing more than 25 g when treated with GA_3 @ 250 ppm was found to be superior in improving the sprouting percentage of tubers and other growth characters like plant height, stem girth, number of branches, number of leaves, days to first flowering and number of flowers per plant in glory lily (Sivasankar *et al.*, 2016).

8.0 Soil and Climate

Glory lily prefers sandy loam soils on the acidic side having proper drainage. It requires humus and slight fertilization with 5-7-5 mixture of NPK before planting. In greenhouse, the tubers are often grown in peat and sand with fertilizer added via soluble solution. It is adapted to different soil texture and climatic variation. The plant grows in sandy-loam soil in the mixed deciduous forest in sunny positions. It is very tolerant of nutrient poor soils. It occurs in thickets, forest edges and boundaries of cultivated areas in warm countries upto height of 2530 m (Ade and Rai, 2009).

Talia *et al.* (1996) observed that substrate had no significant effect on weight increase by large tubers but the soil : compost mixture gave better results than soil alone in tubers initially weighing up to 5 g. Red, loamy soils with good drainage and a pH range of 6.5-7.5 are suitable for cultivation of this species. The ideal temperature for cultivation of this crop is between 25 and 32°C during day and 15-20°C during night. Regions with high humidity (>70 per cent) is vulnerable to *Curvularia* blight (leaf blight) causing severe mortality of the vines. (Padmapriya *et al.,* 2016). According to Scholz (1965) temperatures favourable for growth and flowering are 15-20°C during the day and 10-15°C at night. The relative humidity should be high. Bright sun induces more brilliant colour of the flowers, stronger and sturdy vines but reduces stems and height of the plant.

9.0 Cultivation

9.1.0 Planting

Gloriosa tubers are usually planted during the month of April-May in subtropical or mild temperate climates, while in March-April in tropical conditions. Carow (1976a) reported that tubers planted in summer grew faster than winter planted tubers. He (1977b) also noted that tubers planted in May-July produced the best flowers rather than February-March planting. Tubers may be planted every week in green-house or subtropical outdoor culture and a continuous sequence of bloom obtained. According to Hayward (1951), best results are obtained by planting of sprouted tubers. A thoroughly dormant tuber will grow in time, but may take several months unless sprouted before planting. Hetman and Pudelska (1997) compared flower yield of three *G. rothschildiana* clones in an unheated plastic tunnel over 3 years. Cut flower yields and quality were comparable to those obtained under greenhouse conditions, averaging 19-21 flowers/plant for clones 2 A and B-7. Clone F-12 gave significantly lower yields than this. Daughter tuber production was also comparable to that under greenhouse conditions.

The size and weight of tubers at the time of planting greatly influence the growth and development of gloriosa. Stem diameter, shoot weight, length of main stem, weight of daughter tubers, number and weight of roots, and number of flowers per plant were found to be closely correlated to mother tuber weight (Carow, 1977a). Plants raised from large tubers gave higher and more uniform yield per plant than those from small tubers (Krause, 1981a). Apeles Dias and Caballero Ruano (1985) studied the growth and flowering in *G. rothschildiana* raised from tubers ranging in weight from 20 to 150 g and reported that sprouting percentage, total number of flower stems, total number of flowers, precocity, flower stem length and lateral stem percentage gene-rally increased with increasing tuber weight, but there was little difference amongst tubers weighing over 50 g. Total flower production ranged from 440 to 679 flowers/100 tubers. Tubers weighing < 50 g were not suitable for cut flower production. In an experiment on 4 tuber sizes, 27 cm (195g), 21 cm (110 g), 16 cm (75 g) and 13 cm (50 g) in length and plant population of 6, 8, 10 and 12/m^2, respectively, Escher *et al.* (1973) concluded that increasing tuber size was associated with higher flower yield/plant, earlier flowering, longer flower stem and more new tubers. However, larger number of small tubers produced more flowers per unit area, and tuber growth relative to the original size was greater than that of larger tubers.

In Tamil Nadu, field is prepared by formation of ridges measuring 45 cm width, 45 cm height and of convenient length depending upon the local soil types for planting of tubers. For conventional irrigation system, farmers adopt furrow planting. Planting of tubers is done during July-August. The tuber rate varies from place to place and in practice, farmers use 1-2 ton of healthy tubers to plant one hectare of land. The tubers are planted at a spacing of 45-60 cm. As a prophylactic measure against soil borne pathogens, application of *Trichoderma viride* and *Pseudomonas fluorescens* (each at 5 kg/ha.) is recommended (Padmapriya *et al.*, 2016).

9.2.0 Irrigation

The crop is capable of withstanding drought. In Tamil Nadu, majority of glory lily farmers adopt drip irrigation. The first three months are critical and irrigation must be ensured two or three times in a week. Later during peak flowering (between October and December), irrigation can be given twice in a week. As flowering, pollination, pod set and pod maturity occurs concurrently and continuously, irrigation must be ensured uniformly covering these stages of growth. In the early stage of crop growth, frequent weeding is required to control the weeds which will otherwise compete with plants for moisture and nutrients and will restrict the growth of the plant. (Padmapriya *et al.*, 2016).

Adequate watering is necessary during the vegetative growth, however, the frequency of watering should gradually be decreased with the maturity of the plants. It should be withheld after the flowering is over, otherwise tubers may rot. Overwatering is also harmful to the plants and causes yellow- or brown-coloured patches on the leaves which often fall prematurely. Removal of mother tuber on the 30th day and thereafter markedly increased the shoot and tuber growth and flower production (Carow, 1976a).

9.3.0 Manuring and Fertilization

Though gloriosa makes satisfactory growth with little manuring and fertilization, addition of well-decayed manure, bone meal and fertilizers to the soil ensures a vigorous plant, strong tuber and better flowering. Liquid manuring once at the active phase of vegetative growth and another just before flowering is also very useful to the plants. Carow (1976b) reported that application of 0.4-0.8 g of a complete fertilizer per litre substrate increased the number of blooms per plants, long flower stem and large size petals. He (1980a) further noted that application of 400 mg nitrogen along with 150 mg P, 600 mg K, 150 mg Ca and 75 mg Mg plus trace elements per litre gave the best result, both regarding the number of blooms and the formation of new tubers. The maximum benefit was obtained when N was supplied in the first 6 to 8 weeks after planting. In a 2-year greenhouse trial, *G. rothschildiana* seedlings pricked out into 10 cm pots in May-June received NPK fertilizer at 1-3 g/l subsrate and the plants were top dressed 4 times before mid-September or early October. The best weight increment of newly formed tubers was obtained by applying 1-2 g fertilizer/l substrate plus top dressing with florovit (3.2 per cent N) at 0.1 per cent applied twice to the foliage at 14-day intervals and then to the substrate at 0.2 per cent. The fertilizer rate at 3 g/1 reduced tuber weight (Krause, 1981b). In another trial, Krause (1983) recorded the maximum flower yield

(28 flowers/plant) with NPK at 250:150:550 mg/l in one year and at 550:150:250 mg/l in another year (23.6 flowers/plant). An experiment to study the influence of nitrogen on the duration of the crop and seed yield in *G. superba*, showed that nitrogen application had a favourable effect in enhancing vegetative growth and prolonging crop duration. Seed yield and its components increased with increase in the dose of nitrogen from zero to 160 kg per hectare (Kumaraswamy *et al.*, 1994).

Gloriosa superba is an exhaustive crop and prefers more nutrients from the soil for growth and development of plants, flowering, seed set, pod maturity besides, to meet the nutritional requirement for tuber growth and multiplication of daughter tubers for the next generation. The tubers have shallow root system comprising mainly fibrous root which can absorb water and nutrients available in the localized zone (Padmapriya *et al.*, 2016). Small fibrous roots can absorb the nutrients doses for which nutrients have to be applied in several split doses. The fertilizer dose applied should suffice the nutrient requirement during various critical stages of crop growth. The maximum seed yield was obtained with 150:50:100 kg of NPK per ha, along with vermicompost @ 5t/ha (as basal) combined with basal dose of $ZnSO_4$ @ 25kg/ha, $FeSO_4$ @ 50 kg/ha, borax @ 10 kg/ha, Sodium molybdate @ 0.5 kg/ha at the time of planting along with foliar application of $FeSO_4$ (1 per cent), $ZnSO_4$, (0.5 per cent) boric acid (0.2 per cent) and giberellic acid as foliar spray twice @ 200 mg/kg at critical stages of crop growth (Padmapriya *et al.*, 2016).

The growth parameters *viz.*, plant height, number of primary and secondary branches, number of leaves and leaf area were the highest when manured with a combination of FYM @ 25 t/ha, consortium biofertilizer (2 kg/ha) along with recommended dose of inorganic fertilizers @ 112.5:37.5:56.25 kg NPK per ha when compared to other treatments (Gayathiri and Anburani, 2011).

9.4.0 Staking

When the plants are about 30-40 cm tall, they should be staked or tied to wires or allowed to climb on some frame or bush. For cut flower the vines may be trained on fence wire, and the flowers will be produced on wiry stems in the axils of the leaves as the plant grows taller.

10.0 Diseases and Pests

10.1.0 Diseases

Gloriosa has been observed to have few diseases and pests. Koenig and Leselmann (1974) isolated a virus from *G. rothschildiana* which was named as gloriosa stripe mosaic virus.

Stored tubers are affected by a dry rot which may be controlled by treating the tubers with fungicides before storage.

10.2.0 Pests

Black caterpillars attack the plants. They lay mass of eggs on the undersurface of a tender leaf or in the crown of the growing shoot. Within a very short time, the newly hatched caterpillars with voracious appetites damage the leaves and growing

Diseases and Pests of Gloriosa

Leaf Blight

Root Rot and Tuber Rot

Caterpillar

Harvesting of Gloriosa

Harvesting of Gloriosa

Harvesting of Gloriosa Bulbs

Harvesting Stage

Gloriosa Flowers in Market

points. Yadav (1972) also recorded *Polytela gloriosae* larvae infesting gloriosa plants. These larvae carry cytoplasmic polyhedrosis virus (CPV). Spraying with 0.2 per cent metacid or malathion effectively controls these pests. Fourteen insect pests were observed by Swamy and Raja Gopal (1995) as pests of *G. superba* during the course of studies at the campus of University of Agricultural Sciences, Bangalore. Of these *Polytela gloriosae* was considered as a major defoliator, causing severe losses during cultivation.

11.0 Harvesting and Postharvest Management

11.1.0 Harvesting

Gloriosa flowers are harvested when perianths start to loose apart and enlarge buds curl upwards gradually. Flowers are harvested with stalks. If harvested at this stage, they last longer in vases or during transportation.

11.2.0 Yield of Flowers and Seeds

Cut flower yields of *Gloriosa* vary with species, cultivars and size of tubers. Krause (1983) obtained 28 flowers/plant in *G. superba*. Hetman and Pudelska (1997) obtained 19-21 flowers/plant of *G. rothschildiana* clones 2A and B-7. Raina and Gupta (1999) obtained 681.70 kg/ha gloriosa seeds from controlled selfing between flowers on the same plant. Plants treated with grown regulating substances produced 533.3 kg/ha dry seeds (Mamatha *et al.*, 1993).

11.3.0 Postharvest Management

The effect of cold storage on the vase life and quality of freshly harvested cut flower stems of *Gloriosa superba* was investigated (Hettiarachchi and Balas, 2005). A significant difference resulted between "dry" (inflorescences stored in polythene bags) and "wet" storage in a preservative solution 'Standard Vase Solution' ($NaHCO_3$ 125 mg/l, $CaCl_2.2H_2O$ 99 mg/l, $CuSO_4.5H_2O$ 1.2 mg/l introduced by van Meeteren *et al.* (1999) for flower quality features during vase life period. 'Wet' storage (4°C for 7 d) of fully open *Gloriosa* flowers markedly improved flower keeping quality. Compared to other treatments, flowers kept at 4°C ("wet" storage) for 7 days maintained a higher photosynthetic yield (0.76) during vase period. There is a significant decrease of chlorophyll fluorescence yield with increased storage temperature. Although visual quality could be maintained for up to 10 days in cold storage at 4°C, flower quality decreased notably after 7 days. The results indicated that 'wet' cold storage at 4°C for 7 days has the potential to be used for delaying inflorescence senescence, prolonging vase life and postharvest quality of *Gloriosa* cut flowers (Hettiarachchi and Balas, 2005).

12.0 Biotechnology

12.1.0 Tissue Culture

12.1.1 Somatic Embryogenesis

An efficient protocol for somatic embryogenesis and plant regeneration from rhizome explants of *Gloriosa superba* L. was developed using various plant growth regulators (PGRs), *Ulva lactuca* extracts (ULE), and phycomolecule-coated *U. lactuca* silver nanoparticles (ULAgNPs) (Mahendran *et al.*, 2018). Callus was initiated from rhizome explants on the Murashige and Skoog (MS) medium supplemented with various concentrations of 2,4-dichlorophenoxyacetic acid (1.0–5.0 mg/l), α-naphthaleneacetic acid (NAA) (0.5 mg/l), ULE (10–50 per cent), and ULAgNPs (0.1–0.5 mg/l). Callus was cultured on MS medium fortified with various concentrations and combinations of 6-benzylaminopurine (BAP) (0.5–2.5 mg/l), kinetin (KIN) (0.5–2.5 mg/l), ULAgNPs (0.1–0.5 mg/l), and 20 per cent ULE along with 0.5 mg/l NAA for the development of somatic embryos (SEs). The maximum percentage of embryo maturation (100 per cent) was observed on the medium containing 0.5 mg/l ULAgNPs, 2.0 mg/l BAP, 0.5 mg/l abscisic acid (ABA), 0.5 mg/l silver nitrate ($AgNO_3$), and 20 per cent ULE. The highest percentage of embryo germination (86.1 per cent) was noticed on the MS medium containing 0.3 mg/l ULAgNPs, 5.0 mg/l gibberellic acid (GA_3), 2.0 mg/l BAP, 0.5 mg/l adenine sulfate (AdS), and 20 per cent ULE. Well-rooted plantlets were successfully acclimatized in the greenhouse with 70 per cent survival rate. Results suggest that *U. lactuca* extract-derived silver nanoparticles could be used as biostimulants for the enhancement of somatic embryogenesis and plant regeneration rate in *G. superba* (Mahendran *et al.*, 2018).

Organogenesis was obtained from embryogenic callus in MS medium supplemented with BAP, Kn and IBA. Further subculturing of embryogenic callus in MS medium containing 2,4-D, BAP and Kinetin resulted in globular structures while addition of only BAP resulted in heart shaped somatic embryos on the callus (Mahajan *et al.*, 2016).

12.1.2 Micropropagation

In vitro micropropagation system of *G. superba* was reported by many researchers (Samarajeewa *et al.*, 1993; Custers and Bergervoet, 1994; Sivakumar and Krishnamurthy, 2004; Hassan and Roy, 2005; Akter *et al.*, 2014; Gopinath *et al.*, 2014; Anandhi *et al.*, 2016).

When stem tips of *G. rothschildiana* were cultured on MS medium at 25°C with 16 h light (1000 lx), adventitious shoots were produced. Plants were regenerated which formed tubers upon subculture. The basal plate of regenerated plants exhibited a strong differentiation activity (Goi and Guo, 1998). Okumura *et al.* (2000) reported that multiplication rate was high and growth was good when *G. rothschildiana* cv. Rosequeen meristems cultured on MS liquid medium supplemented with 0.2 mg NAA/l and 1.0 mg BA/l as the primary culture. As the secondary culture, the plantlets from meristems were transplanted on MS solid medium supplemented with 0.5 mg NAA/l and 2.5 mg BA/l. The multiplication rate of plantlets was relatively high and complete plantlets were obtained. Shimasaki *et al.* (2000) reported thidiazuron (TDZ) was effective on shoot and tuber formation in both apical and stem segment cultures *in vitro*. The optimal shoot proliferation was obtained at 3 and 10 µM TDZ (average number : 7.8/segment, formation rate : 96 per cent). Jadhav and Hegde (2001) reported initiation of friable callus from shoot apices of *G. superba* on basal MS medium supplemented with 2,4-D + kinetin + casein hydrolysate (CH)

+ coconut water (CW) (20 per cent). The subculture of callus on the same medium after 4-5 weeks showed an induction of a large number of somatic embryos, and the development of embryoids in plantlets took place when the embryogenic callus was transferred to basal MS medium supplemented with BA, CH/CW (20 per cent).

Various explants have been used for micropropagation of *G. superba*. Gopinath *et al.* (2014) used root explants for micropropagation using MS basal medium supplemented with different concentration and combination of phytohormones *viz.* 2,4-dichlorophenoxy acetic acid (2,4-D), Indole- 3-acetic acid (IAA), 6- Benzyl aminopurine (BAP), Gibberellic acid (GA), Indole-3-propionic acid (IPA), α-napthalene acetic acid (NAA) and Indole- 3-butyric acid (IBA). The maximum percentage of yellowish callus (94.40 per cent) were obtained from the root explants on MS medium supplemented with concentration of (2.0 mg/l), (1.0 mg/l), (0.75 mg/l) for 2, 4-D, IAA and NAA, respectively. The development of roots from the callus were highly regenerated (93.80 per cent) on MS medium with combination of GA (8.0 mg/l), IAA (4.0 mg/l) and BAP (2.0 mg/l). Rooted callus were transferred into half strength of MS medium containing BAP (3.0 mg/l), IBA (1.0 mg/l) and IPA (0.75 mg/l) to produce dark greenish shoots (90.60 per cent). The rooted shoots were transferred into small polythene bags which contain a sterilized cow dung powder, sand and red soil in the ratio of 1:2:3 and kept in a mist house. After acclimatization in the mist house, the regenerated plantlets were hardened in the greenhouse and transferred into soil, which showed 80 per cent survival rate (Gopinath *et al.*, 2014). Best shoot multiplication in sprouted tubers (88.67 per cent) was observed in MS supplemented with 2.0 mg/l BAP, 0.5mg/l Kn and 1.0 mg/l GA_3 with 8.7±0.18 average number of shoots per explant. Long and healthy roots were observed in MS medium supplemented with 2.0 mg/l IBA resulting in 84.50 per cent of root growth and average of 7.47±0.29 roots per explants (Mahajan *et al.*, 2016).

Factors affecting *in vitro* propagation and microtuberization were reported (Yadav *et al.*, 2012). Surface sterilization of tuber explants with 0.1 per cent mercuric chloride ($HgCl_2$) for 5 min eliminated the contamination effectively with highest survival rate. Among the various combinations used, Murashige and Skoog (MS) medium with 2.0 mg/l 6-benzylaminopurine (BAP) + 0.5 mg/l α-naphthalene acetic acid (NAA) containing 3 per cent sucrose with 16-h photoperiod exhibited the greatest *in vitro* tuberization (3.2) with the highest shoot regeneration frequency (90 per cent). The longest tuber regeneration occurred on MS media containing 4 per cent sucrose. Transfer of *in vitro*-regenerated shoots to half-strength MS medium with 1.0 mg/l indole-3-butyric acid (IBA) + 0.5 mg/l NAA showed maximum root induction (66.6 per cent). The *in vitro*-grown plantlets were successfully acclimatized and transplanted to sterilized soil and sand mixture (3:1) in the glasshouse with 70 per cent survival. The colchicine content was determined in the tubers of *ex vitro* plants by HPLC using the same retention time (1.5 min) as that of the standard colchicine. This revealed that the micropropagation protocol for rapid mass production could be used as raw material for colchicine extraction and provides a basis for germplasm conservation and genetic improvement of *G. superb*.

12.1.2.1 Genetic Fidelity of Micropropagated Plants

With ever increasing demand, there is a pressing need to conserve *Gloriosa* through biotechnological approaches. Large number of plants can be produced in vitro under aseptic conditions, but there is always a danger of producing somaclonal variants by tissue culture technology (Yadav *et al.*, 2013). Thus, the genetic stability of micropropagated clones was evaluated using random amplified polymorphic DNA (RAPD) and inter simple sequence repeat (ISSR) analysis. A total of 80 (50 RAPD and 30 ISSR) primers were screened, out of which 10 RAPD and 7 ISSR primers produced a total of 98 (49 RAPD and 49 ISSR) clear, distinct and reproducible amplicons. The amplification products of the regenerated plants showed similar banding patterns to that of the mother plant thus demonstrating the homogeneity of the micropropagated plants. This is the first report that evaluates the use of genetic markers to establish genetic fidelity of micropropagated *G. superba* using RAPD and ISSR, which can be successfully applied for the mass multiplication, germplasm conservation and further genetic transformation assays for colchicine production to meet the ever increasing demand of this medicinally potent plant for industrial and pharmaceutical uses.

12.1.3 *In vitro* Tuber Induction

Propagation of *G. superba* by the continuous production of *in vitro* tubers is a good option for large scale production of planting materials as reported for several tubers (Ghosh *et al.*, 2007). *In vitro* tubers can be planted directly in the soil without acclimatization. Storage and transport of tubers are also easier, facilitating germplasm exchange across national borders (Ghosh *et al.*, 2007).

An efficient system for *in vitro* tuber production of *Gloriosa superba* has been established from leaf and nodal explants. Root tuber induction was more successfully induced from nodal explants than leaf explants (Madhavan and Joseph, 2010). Root tuber induction from nodal explants was most successful on Murashige and Skoog (MS) medium supplemented with 3 mg/l 1 α-naphthalene acetic acid (NAA) and 1 mg/l 1-phenyl-3-(1.2,3-thidiazol-5-yl-urea) (TDZ). Root tubers on high cytokinin to auxin ratio induced shoot initials. Shoot initiation and multiplication was optimal in 2 mg/l Kn+1 mg/1 NAA containing MS media and rooting of separated shoots was achieved on 1/2 MS medium with NAA and Indole-3-butyric acid (IBA) as plant growth regulators. Root tubers could be maintained in culture for 2 years without losing their regenerative potential (Madhavan and Joseph, 2010). Another protocol was reported by Ghosh *et al.* (2007) using non-dormant tubers of *G. superba* L. on MS medium without addition of plant growth regulators. Among the different basal media tested MS medium was found to be suitable for induction and development of secondary tubers; one in vitro tuber per explant was obtained after 6 weeks and 3 tubers per explant after 12 weeks of culture. These tubers produced healthy green shoots that rooted on basal medium. Best rooting was noted in half strength (half organics and inorganics) MS medium with one-fifth nitrates (Ghosh *et al.*, 2007).

Surface sterilized seeds of *Gloriosa superba* were soaked overnight in 1 per cent GA_3 on the next day seeds were planted on germinating media containing MS basal salts with 0.5 mg/l GA_3 and 1.0 mg/l BA, 1 per cent sucrose and 0.8 per cent agar. 72.5 per cent of seed germination was observed (Kolar and Basha, 2014). The germinated seeds were transplanted on MS basal medium supplemented with 1.0 mg/l BAP, 0.05 mg/l GA_3, 9.5 mg/l NAA and 6 per cent sucrose which led to 90 per cent tuber induction within 6 weeks of culture (Kolar and Basha, 2014).

12.1.4 Genetic Transformation

Very little work has been reported on genetic transformation in Gloriosa, Glory Bai and Agastian (2013) investigated for callusing, direct regeneration and induction of hairy root using soil bacterium *Agrobacterium rhizogenes*. Friable callus was initiated from corms on basal MS medium amended with 2,4-D (1 mg/l) and kinetin (1 mg/l). Callus generated from such effective combinations were co-cultivated with the bacterium *in vitro*. Infection was facilitated by acetosyringone (AS) (20 mg/l) supplemented during co-cultivation. Extensive hairy roots were induced from the callus within 15-18 d. These roots were then established on MS basal broth medium. Alkaloids like colchicine and colchicoside were extracted from hairy roots and compared with authentic standard colchicines in HPLC. The findings indicated that nitrogen, inoculum and ferric ions influenced colchicine production (22.7 mM) and hairy root growth (Bai and Agastian, 2013).

12.2.0 Markers

Large number of studies on use of Markers including genomic studies has been reported in the last two decades.

12.2.1 Phylogeny and Diversity Analysis

Diversity within and among the population of *Gloriosa superba* collected from five different location of Tamil Nadu State, India, were explored by using Inter Simple Sequence Repeat (ISSR) DNA sequencing method (Jasmine and Balakrishnan, 2018). A total of 86.27 per cent of polymorphic bands were seen in 94 reproducible bands generated from 12 number of ISSR primer ranging between 200bp to1000bp of band width. ISSR primer produces the bands with an average polymorphism of more than ninety nine per cent among all the ecotypes. The dendrograms drawn for the analysis of genetic similarity in all the ecotypes. The studies reveals that, genetic variations among different accessions of *Gloriosa superb* was very well identified with the help of ISSR finger printings technique (Jasmine and Balakrishnan, 2018).

Thirty six ISSR primers were used to analyse the diversity (Sahana *et al.*, 2019) and among them 16 primers showed 83.56 per cent polymorphism and produced 213 amplicons. This indicates that there is a high level of variation at the genetic level among these accessions. The primer UBC-807 showed the highest PIC value (0.958), which represented the high efficiency of the individual primer. *Gloriosa* accessions GSU-6 and GSU-7 are highly similar with 68 per cent and GSU-7 and GSU-16 are distinctly similar with 39 per cent. Cluster analysis was done based on similarity coefficient value obtained from banding pattern of ISSR marker. The 16 Gloriosa accessions were grouped into two major clusters, of which one cluster contained all 15 accessions whereas as another cluster contained GSU-16 accession alone. This represents that the Gloriosa accession GSU-16 is genetically distinct from other accessions based on banding pattern (Sahana *et al.*, 2019).

12.2.2 Identification of Hybrids and Mutants

Plantlets derived from ovule culture following intra- or intergeneric crosses among Colchicaceous ornamentals, *Gloriosa* spp., *Littonia modesta* Hook. and *Sandersonia aurantiaca* Hook., were subjected to flow cytometry (FCM) and random amplified polymorphic DNA (RAPD) analyses in order to verify their hybridity (Amano *et al.*, 2007). For crosses between plants with apparently distinct relative fluorescence intensity (RFI) of nuclei, *i.e.*, intrageneric crosses between *Gloriosa* genotypes with different ploidy levels, and most intergeneric crosses, detection of hybrid plantlets could be readily accomplished by FCM analysis. The results of hybrid identification by RAPD analysis supported those of FCM analysis. In addition, RAPD analysis allowed the verification of the hybridity of intra- or intergeneric cross-derived plantlets, which could not be identified as hybrids by FCM analysis due to the similarity of RFI of the parents or appearance of the RFI peak in an unexpected position. Totally, 110 independent hybrid plantlets (60 intrageneric and 50 intergeneric hybrids) have so far been identified by FCM and/or RAPD analyses. Thus, FCM in combination with RAPD analyses offer simple and rapid means for the early detection of intrageneric and intergeneric hybrids in Colchicaceous ornamentals (Amano *et al.*, 2007).

12.2.3 Genomics

Colchicum autumnale L. and *Gloriosa superba* L. are well known sources of colchicines. The complete chloroplast genome of two colchicine medicinal plants is reported for the first time by Nguyen *et al.* (2015). Deletion of *ycf15* gene occurred only in *Colchicum* but not in *Gloriosa* and suggests this as a potential marker for delineating the two species (Nguyen *et al.*, 2015).

According to Nguyen *et al.* (2015), in *C. autumnale*, the circular double-stranded cpDNA sequence of 156,462 bp consists of two inverted repeat (IR) regions of 27,741 bp each, a large single-copy region (LSC) of 84,246 bp, and a small single-copy region (SSC) of 16,734 bp. The cpDNA sequence of *G. superba* is longer than that of *C. autumnale* (157,924 bp), which consists of two IRs (28,063 bp), an SSC (16,786 bp), and an LSC (85,012 bp). Significant structural differences between them were observed in the ycf15 gene. *ycf15* gene was absent from *C. autumnale* cpDNA and affected the length of the chloroplast genome between the species. Furthermore, this gene loss event was specific to the expanded genus of *Colchicum sensu* Vinnersten and Manning. Therefore, this gene may be an effective and powerful molecular marker for identifying the *Colchicum* genus within the family.

13.0 References

Akter, S., Roy, P.K., Mamun, A.N.K., Islam, M.R., Kabir, M.H. and Jahan, M.T. (2014) *Nuclear Science and Applications*, **23** : 55-59.

Amano, J., Kuwayama, S., Mizuta, Y., Oomiya, T., Nakamura, T. and Nakano, M. (2007) *J. Japan. Soc. Hort. Sci.*, **76**(1) : 73–78.

Amano, J., Nakazawa, D., Kuwayama, S., Mizuta, Y., Okuno, H., Watanabe, Y., Godo, T.,Han, D. S. and Nakano, M.(2009) *Plant Biotechnol.*, **6** : 535–541.

Amano, J., Kuwayama, S., Mizuta, Y., Godo, T., Okuno, H. and Nakano, M.(2008) *HortSci.*, **43** : 115-118.

Anandhi, S., Rajamani, K. and Jawaharlal, M. (2016) *African J. Agrl. Res.*, **11** : 217-220.

Anandhi, S. and Rajamani, K. (2012) *Global J. Res. Med. Plants and Indigen. Med.*, **1**(10) : 524-528.

Anandhi, S., Rajamani, K., Jawaharlal, M., Maheshwaran, M. and Gnanam, R. (2013) *Int. J. Agric. Sci. Res.*, **3** : 85-92.

Anandhi, S. and Rajamani, K.(2017) *European J. Med. Plants*, **21** : 1-13.

Apels'Diaz, M. and Caballero Ruano, M. (1985) *Jornadas Tecnicas*. Almeria, Spain, pp. 4-7.

Bailey, L.H. (1963) *The Standard Cyclopedia of Horticulture*, Macmillan Co., New York.

Carow, B. (1975) *Gartenwelt*, **75** : 151-154.

Carow, B. (1976a) *Angewandte Botanik*, **51** : 157-166.

Carow, B. (1976b) *Gartenbauwiss.*, **41** : 56-64.

Carow, B. (1977a) *Angewandte Botanik*, **51** : 157-166.

Carow, B. (1977b) *Acta Hort.*, **68** : 59-63.

Carow, B. (1977c) *Gartenbauwiss.*, **42** : 175-177.

Carow, B. (1979) *Deutscher Gartenbau*, **33** : 845-846.

Carow, B. (1980a) *Deutscher Gartenbau*, **34** : 458-459.

Carow, B. (1980b) *Deutscher Gartenbau*, **35** : 2200-2202.

Chitra, R. and K. Rajamani (2013) *J. Med. Plants Res.*, **7** : 2122-212.

Chitra, R. and K. Rajamani. (2010) *Research J. Med. Plant*, **4**(1) : 21-25.

Custers, J.B.M. and Bergervoet, J.H.W. (1994) *Scientia Hortic.*, **57** : 323-334.

Daniels, R.J., Johnson, S.D. and Peter, C.I. (2020) *Ann. Bot.*, **125** : 1137–1149.

Escher, F., Strech, H. and Labebusch, H. (1973) *Gartenwelt*, **73** : 460-462.

Farooqi, A.A., Gowda, H.R. and Hedge, L. (1999) *J. Medicinal and Aromatic Plant Science*, **21** : 316-319.

Gayathiri, M. and Anburani, A. (2011) *Asian J. Hort.*, **6** : 358-360.

Ghosh, S., Ghosh, B. and Jha, S. (2007) *Sci. Hort.*, **114** : 220–223.

Glory Bai, A.L. and Agastian, P. (2013) *J. Academia and Industrial Res.*,**2**(1) : 68-73.

Goi, M. and Guo, Z.G. (1998) *Acta Hortic.*, **25** : 170-183.

Gopinath, K., Karthika, V., Gowri, S. and Arumugam, A. (2014) *The Scitech J.*, **1** (8) : 27-30.

Hassan, S.A. and Roy, S.K. (2005) *Plant Tissue Cult.*, **15**(1) : 67-74.

Hayward, W. (1951) *Plant Life*, **7** : 145-152.

Henrard, G.E.J. (1982) *Bull. des Recherches Agronomiques de Gembloux*, **17** : 59-73.

Hetman, J. and Pudelska, K. (1997) *Zeszyty Problemowe Postepow Nauk Rolniczch,* **449** : 75-83.

Hettiarachchi, M.P. and Balas, J. (2005) *Acta Hortic.,* **683** : 455-462.

http: //dx.doi.org/10.21746/aps.2016.12.005

Ie Raux, L.G. and Rabbertse, P.J. (1997) *S. Afr. J. Bot.,* **63** : 191-197.

Jadhava, S.Y. and Hegde, B.A. (2001) *Indian J. Exp. Biol.,* **39** : 943-946.

Jasmine, J.A.P. and Balakrishnan, V. (2018) *Res. Plant Biol.,* **8** : 17-21.

Jatisatienr, C., Jatisatienr, A., Kamolrat, B. and Pank, F. (1996) *Proc. Beitrage zur Zuchtungsforschung Bundesanstalt fur Zuchtungsforschung an Kulturpflanzen,* **2** : 131-135.

Kishimoto, T., Yamakawa, M., Nakazawa, D., Amano, J., Kuwayama, S. and Nakano, M.(2014) *Euphytica,* **200** : 251–257.

Koenig, R. and Lesselmann, D. (1974) *Phytopathologische Zeitschrift,* **80** : 136-142.

Kolar, A.B. and Basha, M. G. (2014) *Int. J. Pharm. Bio. Sci.,* **5**(2) : 300-310.

Kozak, D. (2000) *Annales Universitatis Mariae Curie Sklodowska Sectio EEE, Horticultura,* **8** : 35-41.

Krause, J. (1981a) *Prace Institutu Sadownictwa i Kwiaciarstwa w Skierniewicach,* B, **6** : 21-26.

Krause, J. (1981b) *Prace Institutu Sadownictwa i Kwiaciarstwa w Skierniewicach,* B, **6** : 27-32.

Krause, J. (1983) *Prace Institutu Sadownictwa i Kwiaciarstwa w Skierniewicach,* B, **8** : 101-110.

Krause, J. (1986) *Acta Hortic.,* **177** : 353-360.

Krause, J. (1988) *Acta Hortic.,* **226** : 555-558.

Kumar, C.N., S.K. Jadhav, K.L. Tiwari and Q. Afaque. (2015) *Biotechnol.,* **14**(3) : 142-147.

Kumaraswamy, B.K., Bhojappa, K.M. and Faroogi, A.A. (1994) *Indian J. Forestry,* **17** : 249-251.

Kuwayama, S., Mizuta, Y., Nakano, M. and Nakamura, T. (2005) *Acta Hort.,* **673** : 421-427.

Madhavan, M. and Joseph, J.P. (2010) *Plant Arch.,* **10** : 611-615.

Mahajan, R., Kapoor, N. and Billowria, P. (2016) *Annals of Plant Sciences,* **5**(12) : 1466-1471.

Mahendran, D., Kavi Kishor, P.B., Geetha, N. And Perumal, V.(2018) *J. Appl. Phycol.,* **30** : 1425–1436.

Mamatha, H., Farooqi, A.A., Joshi, S.S. and Prasad, T.G. (1993) *Acta Hortic.,* **331** : 371-376.

Mamatha, M., Farooqu, A.A. and Prasad, T.G. (1993) *Acta Hortic.,* **331** : 365-370.

Mamatha, M., Farooqu, A.A., Joshi, S.S. and Prasad, T.G. (1993) *Acta Hortic.*, **331** : 371-376.

Maroyi, A. (2012) *African J. Biotechnol.*, **11** : 14732-14737.

Morgan, E.R., Burge, G.K., Seelye, J.F., Hopping, M.E., Grant, J.E., Warren, A.G.F., and Brundell. D. (2001) *Euphytica*, **121** : 343–348.

Nakamura, T., Kuwayama, S., Tanaka, S., *et al.* (2005) *Euphytica*, **142** : 283–289.

Nakazawa, D., Kishimoto, T., Sato, T. *et al.* (2011) *Euphytica*, **181** : 197–202.

Narain, P and Raina, S.N. (1975) *Cytologia*, **40** : 75-77.

Narain, P. (1977) *New Botanist*, **4** : 75-84.

Narain, P. (1979) *New Botanist*, **6** : 103-110.

Nguyen, P.A.T., Kim, J.S. and Kim, J.H. (2015) *Planta*, **242** : 223–237.

Ntahomvukiye, D., Hakizimana, A., Nkiliza, J. and Puyvelde, L. Van. (1984) *Plantes Medicinales et Phytotherapie*, **18** : 2427.

Okumura, Y., Ohishi, K. and Morioka, K. (2000) *Res. Bull. Aichi Ken Agricultural Res. Centre*, **32** : 169-174.

Padmapriya, S., Rajamani, K. and Sathiyamurthy, V.A. (2016) *Int. J. Curr. Pharm. Rev.Res.*, **7**(1) : 43-49.

Raina, R. and Gupta, L.M. (1999) *Acta Hortic.*, **502** : 175-179.

Rajagopalan, A., Khader, J.B.M.M.A. and Abdul Khader, J.B.M.M. (1995) *Plant Physiol. and Biochem.* New Delhi, **22** : 30-32.

Sahana, K.S., Gnanam, R., Rajesh, S. and Rajamani, K. (2019) *Int. J. Curr. Microbiol. App. Sci.*, **8**(6) : 2125-2134.

Samarajeewa, P.K., Dassanayake, M.D. and Jayawardena, S.D.G. (1993) *Indian Exp. Biol.*, **31** : 719-720.

Samarajeewa, P.K., Dassanayake, M.D. and Jayawardena, S.D.G. (1993) *Indian J. Exp. Biol.*, **31** : 719-720.

Scholz, H. (1965) *Deutsche Gartnerborse*, **65** : 179-182.

Shimasaki, K., Sakuma, K. and Nishimura, Y. (2009) *Acta Hortic.*, **812** : 245-250.

Shimasaki, K., Suzuki, S. and Fukumoto, Y. (2000) *Environ. Control Biol.*, **38** : 207-210.

Sivakumar, G. and Krishnamurthy, K.V. (2004) *Russian J. Plant Physiol.*, **51** : 790-798.

Sivasankar, S., Manivannan, K. and Muruganandam, C. (2016) *Int. J.Curr. Res. Life Sci.*, **7**(12) : 594-596.

Subbaratnam, A.V. (1952) *J. Sci. Industr. Res., India*, **11B** : 446-447.

Sultana, S.S., Dash, C.K., Alam, S.S. and Hassan, M.A.(2019) *Nucleus*, **62** : 31–38.

Supari, M.R., Farooqi, A.A. and Prasad, T.G (1993) *Indian J. Forestry* **16** : 123-126.

Suparna, M.R., Farooqi, A.A. and Subbaiah, K.T. (1993) *Indian J. Forestry*, **16** : 54-57.

Swamy, B.C.H. and Rajagopal, D. (1995) *Indian J. Forestry*, **18** : 158-160.

Talia, M.A.C., Cristiano, G. and De Lucia, B. (1996) *Colture Protette*, **25** : 99-102.

Thakur, R. S., Potesilova, H. and Santay, F. (1975) *Planta Medica*, **28** : 201-209.

Vijayavalli, B. and Mathew, P. M. (1990) *Cytologia,* **55** : 531-533.

Yadav, K., Aggarwal, A. and Singh, N. (2012) *J. Crop Sci. Biotechnol.,* **15** : 297–303.

Yadav, K., Groach, R., Aggarwal, A. and Singh, N. (2015) *AsPac J. Mol. Biol. Biotechnol.,* **23** : 243-252.

Yadav, K., Aggarwal, A. and Singh, N. (2013) *Fitoterpia,* **89** : 265-270.

Yadav, R.L. (1972) *Plant Sci.,* **4** : 125-126.

Zimmer, K. (1975) *Deutscher Gartenbau,* **29** : 1226-1227.

13

Crinum

B. Hemla Naik, Basavaraj Dalawai and S.Y. Chandrashekhar

1.0 Introduction

Crinum belongs to the bulbous family Amaryllidaceae. Crinums are large, showy plants with umbels of lily-like flowers. It is commonly called as Crinum Lily, Giant Crinum, Giant Crinum lily, Cape Lily, Poison Lily, Seashore Lily, Grand Crinum Lily, Spider Lily, Wide Leaved Crinum, *etc.* In Hindi, it is known as Chindar and Pindar; Shukhdarsan and Sudarshan in Kannada and Madhuparnika, Vriskani, Nagadamani and Sudarsana in Sanskrit. The word 'Crinum' has originated from the Greek word 'Krinon' which means white lily. As most species have white or whitish flowers the name seems especially appropriate. The *Crinum* species are widely grown as popular ornamental bulbous plants for landscaping. Members of the genus *Crinum* are highly cherished for their ornamental, economical and medicinal values.

2.0 Importance and Uses

2.1.0 Aesthetic Uses

The *Crinum* spp. are well known for their ornamental values and are often used in home indoors as well as in landscape gardens with various forms and styles. It is suitable for planting in pots, boarders, to make edge for specimen plant, water pools, canals, garden drives, along wall and window sills. Crinums are often planted as bedding plants and their fragrant blossoms are used as cut flowers, along the walls and windowsills. Depending on the species and variety, the flower colors in crinum may vary from white, pink, deep rose, or even striped. The highly attractive white and variegated flowers are often used as cut flowers for vases and bouquets and are even good for mass effect in the garden. Its spectacular foliage is also markedly used for ornamental purpose. The crinums are ideal for planting in or near water features. The dwarf species of *crinum* can be successfully cultivated even in a small garden.

2.2.0 Therapeutic and Other Uses

Crinum is also well known for its therapeutic properties and has long been utilized to treat several diseases around the globe. The immense medicinal value of the Indian Crinum is a signal of their remarkable potential globally. (Chahal *et al.*, 2020). Crinum is an internationally acclaimed traditional herb and has been registered for its ethno- medicinal properties. It has several medicinal properties and is often used to treat various disorders. The leaf extract is being used for treating vomiting, ear-aches, skin infection and as expectorant; anthracia and swelling toxicity, adenolymphitis, laryngopharyngitis, headache, arthralgia spasm and numbness, falls and bruises, fractures and venomous snake bites. The bulbs are crushed and applied onto piles and boils to cause suppuration. The roasted bulbs may be used as a rubefacient in rheumatism (Jayaweera, 1981). *Crinum asiaticum* possesses antimicrobial activities (Win, 2011). Some of the medicinal properties of Crinum includes; antioxidant, antimicrobial (antibacterial, antifungal), antiviral, antitumor, antiemetic, anthelmintic, analgesic and anthelmintic activities *etc.* The

Uses of Crinum

Crinum Planted on the Circular Driveways

Planted along the Paths

Crinum Planted as Border Plant

Pot Plant

Crinum in Sunken Gardens

Crinum for Bedding

Crinum Planted along the Road

***Crinum americanum* Planted Near Pond**

***Source*: https://pondinformer.com/swamp-lily-crinum-americanum/;http://pza.sanbi.org/crinum-macowanii**

seeds are used as purgative and diuretic. It also has antiparasitic and insecticidal activity. Squamous bulbs are used as antidotes and pain killer especially in joint pain. They are also helpful in treatment of superficial infections, swelling sores, sarcoptidosis, mammary abscess, laryngalgia, toothache and pain of rheumatic joints, injuries caused by falls, fractures and venomous snake bites (Patel, 2017).

3.0 Origin and History

The genus *Crinum* belongs to the Amaryllidaceae family and comprises approximately 160 species distributed throughout the tropics and warm temperate regions of the world in Asia, Australia, Africa and America. The genus *Crinum* L. was formerly included in the family Liliaceae. However, the Angiosperm Phylogeny Group (APG) has re-evaluated the taxonomic position of this genus and placed in the family Amaryllidaceae (APG III, 2009). Linnaeus established the genus *Crinum* in 1737 by recognising four species, *viz*. *Crinum latifolium*, *C. asiaticum*, *C. americanum* and *C. africanum* (Nordal, 1977). The pantropical genus *Crinum* L. consists of about 112 species distributed in tropical Africa, America, Asia and Australia (Govaerts *et al.*, 2012). The genus is most diverse in Africa, particularly sub-Saharan Africa. Biogeographical analyses place the origin of *Crinum* in southern Africa (Meerow *et al.*, 2003; Kwembeya *et al.*, 2007). The crinums are found growing in a wide range of habitats from aquatic to desert. Each species is adapted to its specific habitat and this should be taken into account by the gardener. *Crinum* species are found in tropical, subtropical and warm temperate regions in Asia, Australia, Africa and the America. Most of them are summer growers and have large, showy, fragrant flowers.

4.0 Taxonomy

The systematic study of the genus *Crinum* has been carried out from time to time by several scientists to resolve the confusion in nomenclature. *Crinum* species were earlier described under other genera especially under *Amaryllis* and many species of other genera were described under *Crinum* (Hannibal and Williams, 1998). Between 1800 and 1830, a number of *Crinum* species were collected from India, West Africa, the Caribbean, mostly from the British colonies, which initiated the competition in naming these newly collected species. This led to many misinterpretations such as the broad petalled *Crinum* species like *C. zeylanicum* being included in the genus *Amaryllis* (*Amaryllis zeylanica*). Even during these times, forms with narrow petals like *C. asiaticum* were readily recognized as *Crinum* (Hannibal and Williams, 1998).

Crinums are evergreen, bulbous, perennials with large, tunicated bulbs which produce a neck or a pseudostem made up of the sheathing bases of the old leaves. The leaves are linear to sword-shaped, arranged in a rosette or rarely in two opposite rows, very attractive long, green, red, yellow shiny, around 1.0-1.5 m in length. The inflorescences stalk arise laterally on a long, solid peduncle and are umbellate *i.e.*, flower stalks radiate from one central point with two spathe valves *i.e.*, bracts and one to many flowers. The flowers have short or long stalks with a long perianth tube and linear to broadly lanceolate segments that are spreading or held together in a trumpet shape. Flowers are generally white in colour, some time with diversity in colour and are originating in clusters on thick and succulent stalk. Flower stalk is about 1-1.2 meter long which bears beautiful flowers in clusters. Around 20-30 flowers are arranged in umbel. Flowers are hermaphrodite with perianth segments consisting of 6 tepals, slightly united at the base. 6 stamens, 3 carpels, 1 stigma present, united inferior ovary, placentation axile. Fruit is globose nearly 5 cm across filled with large seeds with 2-5 cm wide in *Crinum asiaticum*. Sometimes fruits are sub-globose, beaked with the persistent remains of the tube bulging with large seeds and eventually bursting irregularly to release the seeds. The seeds few, large, green, rounded or irregularly compressed with a more or less impervious, smooth or distinctly hairy seed coat (Patel, 2017 and Sumona Afroz *et al.*, 2020).

5.0 Species and Cultivars

5.1.0 Species

Crinum is a large genus belonging to the Amaryllidaceae family, containing about 180 species of perennial plants that exists in the Globe (2014). However, the World Checklist of selected plant families listed 105 species in the genus *Crinum*. Among these, some important species which are under cultivation across the globe includes; *Crinum americanum* L.; *C. asiaticum* L; *C. bulbispermum*; *C. latifolium* L.; *C. macowanii* Baker; *C. moorei* Hook. f.; *C. pedunculatum* (Syn. *C. asiaticum* var. *pedunculatum*); *C. thaianum* Schulze; and *C. viviparum*.

Crinum americanum: The Southern Swamp Lily, is native to swamp and wetlands of the south eastern United States. The species grows in small groups in still-water habitats, prefers a water depth of 1-6 inches, but will grow up to a 12-inch depth. It is ideal for growing in water gardens, along streams, ponds or bogs. It spread through its large rhizomes and can even be grown in container if kept moist to wet.

C. asiaticum: Plants are perennial herb with large tunicated bulbs. It is widely distributed in western India throughout south-eastern Asia to southern Japan, the Indo-Pacific Islands to northern Australia. Widely planted in the gardens for its beautiful flowers and highly valued for its medicinal properties.

C. bulbispermum: It is also known as Orange river lily. It is a highly attractive garden plant suitable for growing in soggy soils in swamp or water gardens. The leaves are greyish green or pale bluish green colour. It is a large bulbous plant that can grow up to a height of 1 m. meter. The plant produces attractive grey green gracefully arching leaves. 5-16 lily-type flowers are produced in umbel which is white with a pink to red stripe in each petal.

C. macowanii: Commonly known as 'Cape coast lily', ideal for landscaping or garden planting. It is one of the easiest species to cultivate and produce large number of seeds. The plants can flower within three years under optimum conditions. The flowers are large, bell-shaped, strongly sweet-scented, white with dark pink stripes, produced in umbels of 5 to 25 flowers on the tip of a long stalk (Archer and Condy 1999b).

C. moorei: Commonly known as 'The Natal lily', has large, white to pinkish red, open tubular flowers. Plants are tall with almost evergreen leaves emerging from a

Species of Crinum

Crinum acaule

Crinum bulbispermum

Crinum campanulatum

Crinum lugardiae

Crinum graminicola

Crinum moorei

Crinum americanum

Crinum paludosum

Crinum macowanii

Crinum album

Crinum asiaticum

Crinum amoenum

Source: Pacific Bulb Society (https://www.pacificbulbsociety.org/pbswiki/index.php/Crinum) Photography by: Alani Davis and Jay Yourch

thin pseudostem, is most suitable for any garden. This is the only one of the South African species which grows and matures new leaves each season, therefore the leaf tips are neat and not truncated as in the other species. Five to ten large flowers are born on a long erect peduncle. The perianth segments are white or pale pink and form a very wide open funnel, adding to its attraction. In addition, the species prefers shade for its growth and development (Batten, 1986).

C. latifolium : The plants are smaller and less spreading compared to the other *Crinum* spp. The flower scapes are glaucous, slightly pigmented and grow up to 60 cm in height. Flowers open in the late afternoon and are quite fragrant while open at night. The flowers are short lived, they only last a night and wilt by the dawn of the next morning. The flower buds have a pale pink stripe outside; the interiors may have a pink stripe as well or may be pure white. This species sets seeds easily, when selfed or cross pollinated. It is possible that this form, whatever its true origins, is involved in the mysterious origins of some classic hybrids such as Peachblow, Alamo Village and Louis Bosanquet.

C. ligulatum: Is native to Madagascar. It has narrow, spreading, arching foliage that is medium flat green and white salverform flowers with burgundy stamens tipped with deep red-orange pollen covered anthers.

C. variabile: Grows on rocky or sandy, seasonally stream beds from the northwest Cape to the western Karoo. Plants grow from 40 to 60 cm high and form clumps. Flowers are white to pale pink with deeper pink or green keels and are heavily scented. In the wild it blooms between January and May. It also performs better in summer with least water in cultivation and tends to bloom better when grown in large pots or in the ground.

5.2.0 Cultivars and Hybrids

A large number of hybrid cultivars are available in Crinum. The Pacific Bulb Society has classified the crinum hybrids on the basis of their colors and stripe:

White Flowered Hybrids: Carolina Beauty, Burbank Hybrid, Bundrant's Light Pink Powellii, Catherine, Hannibal hybrid, John's Bid, Joyful, Long Shot, Ollene, Louis Bosanquet, Maiden's Blush, Princess Elizabeth, Seven Sisters, White Mogul, White Queen, *etc.*

Pink Flowered Hybrids: Alamo Village, Appleblossom, Bradley Giant, Cape Dawn, Claude Davis, Emma Jones, Fiesta, Fragrant Lady, Glory, Patricia Hardy, Peachblow, Peculiar Pink, Pink Flamingo, Pink Mystery, Pink Trumpet, Summer Glow, Spring Joy, *etc.*

Rose Red Flowered: Antares, Birthday Party, Bradley, Burgundy, Carnival, Eagle Rock, Elizabeth Traub, Ellen Bosanquet, Infusion, Lolita, Lorraine Clark, Rose Parade, Sunbonnet, Super Elle, *etc.*

Striped Flowered Hybrid: Alice Elizabeth, Crinum × amabile, Amelia Garza, Asian Rose, Bambino, Candy Ruffles, Cortes, Crinum × digweedii, Old City Cemetery, Exotica, Regina's Disco Lounge, Royal White, School Bells, Schreck, Stars and Stripes, Stryped Spider Lily, Thaddeus Howard, Zan Gypsy, *etc.* The description of some of the promising crinum hybrids has been given below:

Cultivars/Hybrids of Crinum

Exotica

Alice Elizabeth

Antares

Birthday Party

Amelia Garza

Bambino

Big Baby Boy

Elegance

Flamingo

Hildebrandtii

Merry Me

Maurine Spinks

Source: Tikorangi Jury Garden

Cultivars/Hybrids of Crinum

Bradley

Asian Rose

Star Bright

Bayou Traveler

Carolina Beauty

Fragrant Lady

Improved Peachblow

C. mauritianum x *C. lineare*

Patricia Hardy

Pink Flamingo

Thaddeus Howard

White Prince

Source: Tikorangi Jury Garden

☆ **Amelia Garza:** *C. americanum* × *C. scabrum*, it has the largest flower among the crinum hybrids, about 25 cm in diameter. The huge, white, star-shaped flowers have a light pink stripe running down each tepal/sepal.

☆ **Antares:** Ellen Bosanquet × J. C. Harvey, produce sweet-scented, huge, rich-red flowers.

☆ **Asian Rose:** *C. asiaticum* × *C. scabrum*, attractive foliage and spidery, striped flowers.

☆ **Carolina Beauty:** *C. bulbispermum* × *C. americanum*, the scapes are tall, thin not sturdy and require staking. The flowers are fragrant, the bulb tolerates cold winter temperatures and it is a repeat bloomer.

☆ **Fragrant Lady:** It is a hybrid of unknown ancestry with pale pink, fragrant flowers.

☆ **Improved Peachblow**: *C. erubescens* × *C. album*. It has large, fragrant flowers often born on tall, sturdy scapes and an attractive rosette of foliage.

☆ **Patricia Hardy:** *C. rautanenianum* × *C. moorei* var. *schmidtii*. It has tidy, narrow, arched foliage and light pink flowers on tall scapes.

☆ **Pink Flamingo:** *C. bulbispermum* × C. 'Pink Perfume', it has compact foliage and nicely shaped light pink flowers.

☆ **Eagle Rock:** It is a hybrid of unknown ancestry but based on its appearance and bloom season it seems related to the C. × *powellii* group. It produces dark pink flowers on very tall scapes up to 180 cm that may topple if not staked.

☆ **Thaddeus Howard:** It is a rare red-flowered form of C. × *herbertii* (*C. bulbispermum* × *C. scabrum*). The variable flowers often have random white streaks.

☆ **Maurine Spinks:** C. × *baconi* (*C. americanum* × *C. zeylanicum*). Because of the *C. zeylanicum* parent the stripes on this hybrid are usually darkest at the tips and fade towards the centre.

☆ **Birthday Party:** C. 'Bradley Giant' × *C. scabrum*, the flowers are large, wide open, fragrant and rose-red in colour.

☆ **Bradley:** *C. scabrum* × *C. flaccidum*, flowers are open-faced, fragrant, pink in colour often born on tall scapes.

☆ **Ellen Bosanquet**: *C. scabrum* × C. 'J.C. Harvey'. It produces large bell-shaped flowers of dark rose-red on stems up to 90 cm tall.

☆ **Burgundy**: *C. scabrum* × *C. moorei*, flowers are large, bell-shaped and dark rose-red in colour.

☆ **Exotica:** *C. scabrum* × *C. asiaticum, striped flowers*

☆ **Infusion:** Ellen Bosanquet × C. 'J.C. Harvey'. It offsets readily and is more compact than the parents. The hot pink flowers are large and well formed.

6.0 Genetics and Breeding

6.1.0 Genetics

The somatic chromosome number for most of the *Crinum* species has been mentioned as 2n= 22. The chromosome number of some of the species under genus *Crinum* have been reported as 2n =18, 22+2f, 24, 33 and 72 (Traub, 1962). Its chromosomes are of moderate size and amenable to analysis but, unfortunately, like other genera in this family, meiosis takes place within the bulb and is relatively inaccessible for examination. The common basic number in crinum is x = 11, however few accessions with x = 12 have been derive from the Accra Plains of Ghana. Comparative chromosome morphology shows that a uniform pattern prevails throughout the x = 11 species but that of the x = 12 plants is quite distinct and includes three pairs of telocentric. There is every indication that unique karyotype is closely correlated with the distribution of the species in an area of particular ecological interest (Jones and Smith, 1967).

The cytological analysis of *C. augustum* has revealed 2n=22 chromosomes, while, in *C. rattrayii* a triploid chromosomal number (2n=33) has been found. The finding of 22 somatic chromosomes for *C. amoenium* and *C. asiaticum* by previous studies has been confirmed. Two apparent polyploid series-one starting with 6 and the other with 11 as the basic number have been suggested for this genus (Bose, 1965). The need for extensive cytological studies and hybridization work in order to know the nature of polyploidy and speciation in Crinum has been stressed.

6.2.0 Breeding

6.2.1 Hybridization

William Herbert, an English botanist and naturalist is regarded as the godfather of crinums. He attempted the first crossing between *C. scabrum* and *C. bulbispermum* in 1819 and developed one of the greatest *Crinum* hybrids, *C.* × *herbertii*. These hybrids were described by Herbert in 1837. The another important relatively cold hardy hybrid, *C.* × *powellii* was developed by C. B. Powell in 1858 by crossing *C. moorei* and *C. bulbispermum*. *Crinum americanum*, a semi-aquatic bulb poorly suited for the general garden, has also contributed to the development of large number of attractive interspecific hybrids in *Crinum*. The interspecific hybrids in crinum were first described by Herbert (1937). He has listed three hybrids namely, *C.* × *brownii* (*C. americanum* × *C. bracteatum*); *C.* × *digweedi* (*C. scabrum* × *C. americanum*), and *C.* × *parkeri* (*C. americanum* × *C. erubescens*). Varieties of *C.*× *digweedii* became popular cultivars. Lehmiller (1992) successfully developed 14 interspecific hybrids of *C. americanum* . Only 2.3 per cent of cross pollinations yielded viable hybrid seed when *C. americanum* was the seed parent. Crinums are closely related to *Amaryllis belladonna* as well as to *Ammocharis, Brunsvigia* and *Nerine*. The bigeneric hybrid between *Crinum* and *Amaryllis* has been successfully developed and is called X-*Amarcrinum*.

7.0 Growth and Flowering

7.1.0 Environment

The plant flowers during the rainy season in the plains and summer in the hills. The bulbs are planted in February to March both in the hills as well as plains. The plant will grow and flower for years, if watered in the summer months and top dressed with fresh loamy soil around new vegetative growth. Bulbs may be planted in April to get flower during the rainy season. The plants can grow in semi shaded or sunny condition. It requires warm and moist condition for the best growth with attractive foliage. They can survive four to five years in pots and for long time in garden without re-potting/replanting. When more dense in plant space then side shoots are to be separated and re-potting and plant multiplication can be done.

Mori *et al.* (1992) studied the effects of temperature on growth and flowering of *Crinum* x *Powell*. They observed that when the plants with inflorescences at carpel formation stage were exposed to chilling at 15°/12°C (day/night) for 30 days or at 6°C for 45 days and then shifted to a glasshouse kept above 20°C, they flowered normally. However, the plants did not flower when they were maintained above 20°C under outdoor condition till November and then moved to a glasshouse. This suggests that the plants have a chilling requirement for flowering.

7.2.0 Growth Regulators

Hassanein *et al.* (2009) studied the effect of foliar sprays GA_3 (0, 50, 100 and 150 ppm) and calcium (0, 50, 100 and 150 ppm) as well as their interaction on growth, flowering, chemical composition and bulb production of *Crinum asiaticum* plants. All combinations of GA_3 and calcium had a positive effect on growth, flowering and bulb production. All concentrations of GA_3 or calcium decreased the number of days to full blooming and extending the duration of flower on plant and flower vase life. Further it was revealed that the $GA_3 \times$ calcium combination at 150 ppm showed the most beneficial effect on growth and flowering.

8.0 Propagation

The Crinum can be propagated through sexual and asexual means. Overcrowded side shoots can be separated carefully without damaging the bulbs and then re-potted or planted directly in field. However, they can be propagated in three ways; by using bulbils from the stem, bulblets around the stem base and through scales.

8.1.0 Seed Propagation

Crinum seeds should be sown soon after their ripening. The seeds have to be sown in moist soil just on surface or pushed a little into the soil. Keep the soil moist and warm. Seeds take around 30 days to germinate in the presence of favourable environmental conditions. The Crinum raised through seed take several years to produce bulbs and take 3-5 years to flower. The ideal temperature for seed germination is 15-21°C.

8.2.0 Vegetative Propagation

Crinum is propagated by using its mature adventitious bulbs followed by proper water supply and other needed supply made for its successful growth. The mature mother bulbs after 3-4 years of planting produce several new small adventitious bulbs (bulblets) which can be used to propagate it by vegetative modes. The small daughter bulbs developed nearby main large bulb are separated by digging carefully near of large main bulb. The bud development occurs on the upper surface of the daughter bulb from which the new plant grows in the following year. The bases of old leaves are thin and dry, which cover the bulb. These papery leaves are called husks. The husks overlap each other and meet to form a point at the top. While, the new daughter bulb is formed on the top of old one, small new bulbs called bulblets are produced from the base. These bulblets are axillary buds on the bulb which is a compressed thickened stem and as the resting perpetuating organ (Kumar *et al.*, 2007).

The bulbs of crinum are dug up in the spring and immediately planted in the soil to prevent them from drying out. Treat the bulbs with fungicide and plant one third of the cut bulb with the basal plate downward then they should be planted at a depth of 20 cm at the end of spring. After the bulb transplanting regular irrigation is essential for re-establishment of the bulbs planting.

8.3.0 Micropropagation

Micro propagation is another method, a successful option for large-scale propagation of many species of *Crinum*. *C. variabile* was successfully propagated *in vitro* using twin-scale explants. Shoots developed in the axis of twin-scales when placed on a Murashige and Skoog (MS) medium. Plant growth regulators are not required for the induction of shoots but benzyl adenine (BA), in the absence of naphthalene acetic acid (NAA), promoted their outgrowth. The inclusion of 5g/l activated charcoal improved development by increasing the size and frequency with which shoots formed bulblets. These bulblets were used to initiate further cultures by splitting them in half. Such a system provided for the continuous multiplication of *C. variabile* bulblets which readily transferred to the soil (Fennell *et al.*, 2000). Tissue culture techniques have been successfully applied for the regeneration and growth of bulblets from bulb scale segments of *Crinum macowanii* Bak. *in vitro*. Shoots were successfully induced on twin scales taken from the basal plate region of flowering size bulbs on MS medium containing 0–20 mg/l NAA and BA and a modified MS medium containing 1.25 mg/l ancymidol, 0.1 mg/l NAA and 0.1 mg/l kinetin. A total of 700–1000 bulblets can be obtained from each initial bulb within 12 months (Slabbert *et al.*, 1993). Callus cultures were initiated from bulb-scales and flower buds of *Crinum asiaticum* on MS medium. Bulb- scales calli did not show organogenetic responses. Somatic embryogenesis was achieved from compact flower bud calli. Beside 2, 4- D, NAA and BAP fortified media were equally found effective in inducing regular somatic embryo (Mujib *et al.*, 1996).

Propagation of Crinum

Bulbs

Separation of Side Shoots and Potting

Micropropagation

Different Stages of Development of Shoots from the Scales in *Crinum malabaricum*

Source: Priyadharshini *et al.* (2020). Plant Cell, Tissue and Organ Culture (PCTOC) 141: 593–604.

Cultivation of Crinum

Cultivation in Open

Cultivation in Protected Structure

Diseases and Pests of Crinum

Anthracnose

Botrytis Blight

Larva

Adult

Larva Boring the Leaf

Larva Feeding on Stem

Black-and-Yellow-Striped Amaryllis Caterpillar (*Brithyspancratii/Brithyscrini*)

9.0 Soil and Climate

9.1.0 Soil

Matured bulbs should be grown in well prepared sites; it may be pots or the field by following proper supply of water and other requirements to develop new plantlets. It can be grown in almost all type of well-drained soil. It is well adapted to grow in porous soil with moderate water availability. The soil should be rich in organic matter and fertile. Heavy soils with drainage problems should be avoided for crinum. The *Crinum* seeds can be placed in a well-drained, sandy medium with plenty of compost and never saturated. The pot mixture consisting of Sand: Garden Soil: Vermicompost: Coco Peat mixed in a proportion of 1:1:1:1 is used for pot culture.

9.2.0 Climate

Semi-shaded and sunny conditions are required for *Crinum* depending upon the species. Plants grow best in full sun but will tolerate light shade and will thrive in a wet part of the garden. Too much sun can cause leaf scorching in Crinum. *Crinum moorei* does well under trees and is a most useful species for the 'dry shade'. Some species (*Crinum calamistratum*) are adapted to aquatic condition and some for rocky or sandy (*Crinum variabile*). *C. campanulatum* needs to be placed in a pot containing water. The bulbs of *C. asiaticum* can survive freezing temperatures, but freezing will kill off the foliage. The plants will grow back when spring arrives.

10.0 Cultivation

10.1.0 Planting

Bulbs or side shoots can be used for planting. Crinums usually have large bulbs, sometimes as much as 60 to 80 cm long with numerous fleshy roots. If planted in beds, the soil should be dug to a depth of 60-80 cm, mixed well with sufficient quantity of well decomposed farm yard manure or compost. It can also be grown in large pots containing soil rich in organic matter. Planting may be done to a depth below the ground level, twice the size of the bulbs. The bulbs are planted 30-60 cm apart. The bulbs of small species can be spaced between 45 cm, while a spacing of 80 cm is maintained for large species. A hole of about 25 cm deep is dug for each bulb and the bulb is buried upto the place where the stem emerges from the neck of the bulb. The stem should be above the ground and the neck should be under the soil. The pot should also be strong enough to resist the pressure of the roots as they expand and press against the sides of the container. The young bulb should not be pushed out of the soil. Light irrigation has to be given immediately after planting the bulbs.

10.2.0 Irrigation

Crinum requires light watering frequently. However, water logging condition near the plant will adversely affect its growth and development. Depending upon the soil and climatic condition, it needs alternate days of irrigation during the establishment phase. During the first summer season, watering is needed to keep the soil around the bulbs lightly moist. Once crinums are well established they are relatively drought tolerant, though the plants grow best with consistent moisture. Potted crinums need to be watered frequently to keep the soil moist. Though it is considered drought resistance, proper moist soil is required unless dormant condition.

10.3.0 Manuring and Fertilization

Crinum as a garden plant should be supplied with slow-releasing fertilizer such as bone meal with regular watering. In the garden, most species will grow well in any soil to which plenty of compost has been added. The plants will benefit greatly from occasional feeding with liquid fertilizer and compost added to the growing medium. Crinums flower best when they are fed twice a year with a balanced fertilizer. Apply a high-potassium liquid fertilizer every two weeks from planting until six weeks after flowering.

10.4.0 Pruning

The faded flowers of Crinum should be removed and this will encourage the plant to continue to produce more flowers. Cut the Crinum flower stalk near the base for cut flower and keep them in a vase of water to enjoy its beauty for longer period. Prune all the damaged foliage infested with insects or disease.

10.5.0 Mulching

Crinum plants needs to be mulched with any farm waste or black plastic sheets to maintain the soil moisture during the summer months. Mulching is also helpful in maintaining the microclimate of the plant to get the quality blooms and foliage.

11.0 Diseases and Pests

11.1.0 Diseases

11.1.1 Anthracnose

The infected leaves develop small, tan to dark coloured spots that decay with age. Severely infected plants may also experience dieback of flowers and defoliation. The disease can also be controlled by spraying copper-based fungicidal treatments and by maintaining proper sanitation.

11.1.2 Powdery Mildew

The incidence of this disease is severe during the wet and moist weather. The infected *Crinum* lilies develop small, white to grey-coloured spots. As the disease progresses, white covering of fungal growth can be seen on the entire surface of the infected foliage. Severely infected foliage will begin to show foliage decay, as well as stunted growth, leaf curl and distortion and yellowing of foliage. Spray of Karathan (2 ml/l) is effective against this disease.

11.1.3 Botrytis Blight

This disease targets most parts of the crinum lily, including the foliage, stems and flowers. Initially, necrotic areas of brown-colour with dying tissue develop on the infected plant parts. Plants will eventually show wilting and defoliation and necrosis.

11.1.4 Virus

The infested plants, especially from large or retail nurseries can be a problem and hence, it should be checked carefully for any viral symptoms. Unfortunately, it is not easy to distinguish between viruses and some fungal diseases. The suspected plants should be destroyed to stop the virus from spreading by root disturbance or insects.

11.2.0 Pests

Crinum lilies are rarely attacked by insects. The black and yellow striped amaryllis caterpillar (*Brithys pancratia*) is only the serious pest which is specifically adapted to Crinum because of its poisonous alkaloids. Occasionally snails and slugs may render especially the thin leaved species an unsightly mess in a short time. However, the bulbs of large plants are seldom entirely destroyed by the caterpillars and new leaves grow out quickly. Some of the other common pests of Crinum includes nematodes, mealybugs and grasshopper. Use any broad-spectrum insecticide containing modern pyrethrums can be used. In a garden, the amaryllis caterpillar often necessitates regular spraying in summer.

12.0 Harvesting and Postharvest Management

The first flowering can be expected after three (*C. macowanii*) to eight years (*C. graminicola*) if it is propagated by seeds. The plants perform best in a permanent position like any members of Amaryllidaceae; do not react well to any disturbance of the roots. The vegetatively propagated plants will produce flowers every year during rainy season. On an average 4-5 stalks can be expected per clump in a year and subsequently it increases depending upon sucker production. Further, each stalk consists of 5-10 florets depending upon the cultivars. The flowers/florets in the stalk should be removed as soon as they fade. Blooms, left in the stalk will set the seed, which diverts energy from flower production and plant growth. Hence, the flowers should be cut-off or pinched off. Alternatively, cut the stalks when the blooms first open and use them in floral arrangements as cut flower.

13.0 References

Afroz, S., Rahman, M.O. and Hassan, M.A. (2018) *Bangladesh J. Plant Taxon.*, **25**(2): 257–271.

APG III. (2009) *Bot. J. Linn. Soc.* **161**: 105–121.

Archer, R.H. and Archer, C. (1996) *Botihalia*, **26**: 153.

Archer, R.H. (1997) *Herbertia*, **52**: 90-94.

Archer, R.H. and Condy, C. (1999a) *Flowering Plants of Africa*, **56**: 36-40.

Archer, R.H. and Condy, C. (1999b) *Flowering Plants of Africa*, **56**: 30-35.

Batten, A. (1986) *Flowers of southern Africa*. Frandsen, Sandton.

Bose S. (1965) *Cytol.*, **30**: 349-353.

Bryan, J. (2002) *Bulbs*. Timber Press, Portland.

Chahal, S., Lekhak, MM., Kaur H., Shekhawat, M.S., Dey, A., Jha, P., Pandey, D. K. and Kumar, V. (2020) *S. Afr. J. Bot.*,**136**: 7-15.

Craib, C. (1997) *Herbertia*, **52**: 101-108.

Dassanayake, M. D. and W. D. Clayton, (2000) A Revised Handbook to the Flora of Ceylon. Vol. XIV., Oxford and IBH Publishing Co. Pvt. Ltd. (New Delhi and Calcutta): 18 -20.

Du Plessis, N. and Duncan, G. (1989) *Bulbous plants of southern Africa*. Tafelberg, Cape Town.

Fennell, C.W. (2001) *S. Afr. J. Bot.*, **67**: 74-77.

Fennell, C.W. and van Staden, J. (2001) *J. Ethnopharmacol*, **78**: 15–26.

Fennell, C.W., Crouch, N.R and Staden, J.V. (2000) *African J. Bot.*, **67**: 74 – 77.

Govaerts, R., Snijman, D.A., Marcucci, R., Silverstone-Sopkin, P. A. and Brullo, S. (2012) *World checklist of Amaryllidaceae*. Facilitated by the Royal Botanic Gardens, Kew.

Hannibal, L.S. and Williams, A. (1998) The Genus Crinum (Amaryllidaceae). <http: //www.crinum.org>.

Hassanein, M. M. (2009) *J. Agric. Sci. Mansoura Univ.*, **34**: 6555 – 6565.

Hesse, G. (2001) Veld and Flora, **87**: 136.

http: //www.indiaagronet.com/horticulture/contents/bulbs_tubers.html

http: //www.sahyadrica.com/2013/06/the-crinum-chance.html June 28, 2013.

https: //en.wikipedia.org/wiki/Crinum#cite_ref-WCSP_Crinum_3-0

Jager, A.K., Adsersen, A. and Fennell, C.W. (2004) *S. Afr. J. Bot.*, **70**: 323-325.

Jayaweera, D.M.A. (1981) *Medicinal Plants used in Ceylon*. Part I. National Science Council of Sri Lanka, Colombo, pp. 1–61.

Jones, K. and Smith, J.B. (1967) *Caryol.*, **20** (2) 163-179.

Kumar, S., Awasthi, V. and Kanwar, J.K. (2007) *Hort. Sci. (Prague)*, **34** (2): 77-83.

Kwembeya, *E.G.*, Bjora, C.S., Stedje, B. and Nordal, I. (2007) *Taxon*, **56**: 801–810.

Lehmiller, D.J. (1992) *Acta Hortic.*, **325**: 591-596.

Lehmiller, D.J. (1997) *Herbertia*, **52**: 44-65.

Linnaeus, C. (1781) *Herbertia*, **4**: 91-100.

Meerow, A.W. and Snijman, D.A. (1998) In: K. Kubitzki (ed.), *Families and genera of vascular plants*, **3**: 83-110.

Meerow, A.W., Lehmiller, D.J. and Clayton, J.R. (2003) *Bot. J. Linn. Soc.*, **141**: 349-363.

Mori, G., Imanishi, H. and Sakanishi, Y. (1992) *J. Japan. Soc. Hort. Sci*, **61**(3): 651-657.

Mujib, A., Bandopadhyay, S., Jana, B.K and Ghosh, P. D.(1996) *Indian J. Plant Physiol.*, **1**(2): 84-87.

Nordal, I. (1977) *Norwegian J. Bot.*, **24**: 179–194.

Patel, D. K. (2017) *Int. J. Environ. Sci. Nat. Res.*, **6**(1): 555-678.

Priyadharshini, S., Kannan, N., Manokari, M. and Shekhawat, M.S. (2020) *Plant Cell, Tissue Org. Cult.*, **141**: 593–604

Slabbert, M.M., de Bruyn, M.H., Ferreira, D. I. and Pretorius, J. (1993) *Plant Cell Tissue Org. Cult.*, **33**: 133–141.

Traub, H.P. (1962) *Pl. Life*, **18**: 50-54.

Win, Z.Z. (2011) *Univ. Res. J.*, **4**(1): 123–137.

www.pacificbulbsociety.org

14

Hymenocallis

M.R. Dhiman

1.0 Introduction

Hymenocallis, commonly known as spider lily, basket lily, and Peruvian daffodils, is a bulbous perennial, herbaceous plants with elegant, long-tubed, and snowy white flowers. The sweet scented 8-10 cm long flowers are borne on 60-80 cm leafless flower stalks during mid-summer. The flowers are pancratioid with the stamens united into a distinctive corona and the long and narrow tepals which give them their characteristic 'spidery' look. The bulbs are roundish or oblong, covered by a tunic. They have variable number of distichous leaves range from 5-7 mm to 15 cm wide, which are sessile, linear, glaucous or lanceolate and bright green (Preuss, 2002; Grossi, 2007). In India, it is commercially cultivated in western parts, especially in Gujarat and Maharashtra and occupies a premium position in the flower market of Mumbai. Spider Lily is gaining popularity among the farmers of Gujarat due to its hardy nature, easy maintenance, consistent yield and regular high returns per unit area.

2.0 Importance and Uses

2.1.0 Aesthetic Uses

Hymenocallis is an indispensable choice for garden, cultivated for its white, fragrant spider shaped flowers for varied uses such as loose flower for making venis, decoration purposes, bouquets, worshipping, garland making *etc.* It is also attractive as a cut flower, but the flowers do not last for long time. The white delicate flowers of spider lily have spiritual significance and disseminate the message of 'Alchemy' (Mother, 1995). These plants are very well suited for border plantings in the gardens, alongside the boundary walls and water channels, in herbaceous border, alongside the lawn and also in beds in the gardens. Spider lily is fast growing, hence a very affordable solution for any garden. It creates a brilliant, tall ground cover in short span of time. It is well suited for planting as a specimen in small gardens. Flowers and foliage both attract attention. Spider Lily makes a spectacular addition to any landscape. Besides being potential commercial loose flower crop and garden plant, the genus also has medicinal, ecological and nutritive value.

2.2.0 Therapeutic and Other Uses

H. littoralis bulbs possess pancratistatin (20-24 mg/kg fresh weight; Pettit *et al.*, 1995) which is used for preparing anticancer drug (Backhaus *et al.*, 1992) together with lycorine, narciclasine, 7-deoxynarciclasine littoraline, lignin and 13 other known alkaloids (Lin *et al.*, 1995). The bulbs and leaves of *H. expansa* contain (+)-tazettine, hippeastine and haemanthidine (Antoun *et al.*, 1993).

3.0 Origin and History

This genus has its origin in South America and Africa. The word 'Hymenocallis' is derived from the Greek word *'hymen'* meaning 'membrane' and *'kalos'* meaning 'beauty', referring to the membrane that joins the stamens together and forms the cup of the flower (corona). *Hymenocallis* is native to warmer regions of the new world

Uses of Hymenocallis

In Garden

Along the Pathways

In Pots

As Bedding Plant

Along the Side of Water Bodies

In Garden for Mass Display

In Flower Arrangement

In Rock Garden

i.e. southern part of North America (Flory, 1976; Smith and Flory, 1990), southern United States (Florida, Indiana, Kentucky, Virginia, Georgia, Texas, *etc.*). It has also spread down to the Northern regions of south America including Venezuela, Guyana, Brazil, Peru, Guatemala and the warmer reaches of the Andes (Flory, 1976; Smith and Flory, 1990).

There is evidence that these wide spread bulbs were once found in the genus *Ismene* before they were move to Hymenocallis in 1812. The first species moved to *Hymenocallis* was *H. narcissiflora*. Introduced in 1796, *H. narcissiflora* was a very popular greenhouse plant with the Victorians and therefore has been gifted with numerous English common. Sealy (1954) traced the history of the species now included in *Hymenocallis* both before and after Salisbury (1812), separated the American spider lilies from *Pancratium* and established the genus Hymenocallis on the basis of the "beautiful membrane" that form the stamina cups. Small (1933) provided the first comprehensive review of the south-eastern United States species in the 20th century.

It is believed that most of the cultivated hymenocallis are old heirloom flowers spread widely by sailors in the 16th century; they were originally brought from the tropical of the Antilles and the Spanish Main. These beautiful flowers are sold in nurseries as *H. littoralis, H. caribbaea, H. rotata,* or *H. tenuiflora* (Ogden, 2007). Hymenocallis is a minor component in the bulb trade, however, the beauty of the spider lily flowers, their fragrance and vigorous growth hint at the potential value of the genus (Hanks and Jones, 1987). The interest in the species greatly increased and it is now grown in large number, particularly in many of the gardens of United States and India.

4.0 Taxonomy

The monocotyledonous genus *Hymenocallis littoralis* belongs to the family Amaryllidaceae. The plants are 30-70 cm tall, bushy, with broad strap-shaped leaves. The flowers are fragrant white umbels, each flower with slender recurved petals and elongated stamens emerging from a central cup. The perianth tube is long and there are six segments which are united at the base by a thin membranous cup of the corona. The flowers are borne in a bunch of 10-12 which open one by one in 4-5 days. Common flower colours are white and yellow. The tepals are adnate attached to the staminal cup. Each flower's tube is 14 to 17 cm long or longer. Anthers are introse, globular ovary with less than four ovules per locule. The leaves arise in 'V' shape from the bulbs. The bulbs are roundish or oblong from which thick whitish roots emerge, covered by a tunic.

5.0 Species and Cultivars

5.1.0 Species

The genus *Hymenocallis* consists of 70 species (Chase *et al.*, 2009; Tapia-Campos *et al.*, 2012), which may be evergreen or deciduous. The evergreen species are widely grown in the tropics and subtropics, whilst the deciduous species are more uncommon, being mainly grown in subtropics and warm temperate areas. Many of these deciduous species were briefly included in the genus *Ismene*. Some important species are described below.

Hymenocallis calathina: It is white flowered species commonly called as Peruvian daffodil. It is native to Peru and Bolivia. The leaves are narrow, 30 cm long, flowers white and very fragrant, and many on the stalk, opening in succession in spring. The fringed cup has a green throat and the umbel lasts for two weeks.

H. crassifolia: Commonly called the coastal Carolina spider lily. This species is distinguished from other species of the genus by its stiff, coriaceous, liguliform leaves. Flowers are white, borne in clusters of four on a 60 cm long stalk. The flowering time is summer. It may be evergreen in the south, in a line from Picayune to Lucedale, but elsewhere it will go dormant.

H. eucharidifolia: This species is native to tropical America. The leaves are broad, green hosta-like foliage. Snowy foliage held in up-facing groups of three or more. Flowers are white, funnel shaped, with green tube, 4 or 5 in a cluster on a tall stem, flowering in spring.

H. macrostephana: This is a South American species having strap-shaped, 90-95 cm long leaves. Spathe is long which bears white flowers in cluster with green tube.

H. littoralis: This species is a potent loose flower crop in western India, flowering throughout the year up to 5-6 years from single planting. In hilly areas, in greenhouse it flowers from July to October and bulbs remain dormant between November and March (Biswas, 2006). This species possesses some anti-viral activity (Berghe *et al.*, 1978).

H. narcissifolia: This deciduous lily in native of Andes of South America. Flower petals are similar to spider lily but also possesses large frilly cup with orange incurved stamens. Bear two to nine fragrant flowers on 60 cm tall stalk.

H. latifolia: This species has the broadest leaves, 4 to 9 cm wide, and the largest number of flowers per scape, nine to 15. It's having orange instead of yellow pollen. This species is found growing in sandy coastal strands of South Florida, including the Keys. It is widely available in the nursery trade and is grown as a low-maintenance perennial (Nelson, 2003).

H. occidentalis: This species has no rhizome and relatively thin-textured, glaucous leaves. The bracts at the top of the scape are 4 to 7 cm in length with long, narrow acuminate tips. This species blooms in July and August, with the leaves withering immediately afterward.

H. rotata: This species is robust, with two to four flowers per scape, each with a staminal cup 6 cm or more in diameter that spreads flat when the flower is fully open. It has deep green, arching to nearly erect, strap-shaped leaves that are 40 to 100 cm long. This lily bloom from April to June.

H. palmeri: This species has one flower per scape and tepals are shorter than the floral tube. The leaves are smaller, rarely more than 40 cm long and only 0.4 to 1 cm wide. This species blooms in summer, from May to September.

H. caribeae: It is an evergreen species native of West Indies. The flowers are fragrant, 9-10 cm in diameter, intricately designed. White flowers are borne on top of 40-60 cm tall stalks in mid-summer.

H. pimana: The bulbs of this species contain toxic and bitter tasting alkaloids, used by mountain Pima of Mexico as a famine food after detoxification by boiling in lye.

H. ovata: This species is native of the West Indies. The leaves are strap shaped and 25 cm long, flowering in September-October; flowers white, fragrant with greenish tube on a 30 cm long stalk.

H. speciosa: This species grow widely in the West Indies. It contains toxic substances even though cultivated widely. The leaves are up to 60 cm long and 10 cm wide. Flowers are white and highly fragrant.

5.2.0 Cultivars

There are very few cultivars registered in genus *Hymenocallis*. Some attractive and popular cultivars are as follows:

***Hymenocallis* × 'Tropical Giant'**: This exotic cultivar produces spectacular fragrant flowers. Flowers resemble spidery daffodils with 6 narrow petals and a large cup; a unique selection for the garden or container, blooming during July in most areas. It grows at a fast rate, and under ideal conditions can live for approximately 8 years.

***Hymenocallis* × 'Bellum'**: A hybrid with beautiful cream coloured flowers having golden centres.

***Hymenocallis* × 'Mystery'**: The leaves are huge, about 5 cm wide and 60 cm long, similar to Crinum leaves. Bulbs are as large as oranges. The fragrant blooms are excellent as cut flowers.

***Hymenocallis* × 'Zeylanicum'**: Produces very beautiful flowers, similar to *Hymenocallis x 'Zwanenburg'*, with narrow linear foliage about 30 cm long.

***Hymenocallis* × 'Sulphur Queen'**: The flowers have a citrus fragrance, soft yellow in colour with green striped throats. Flowering in the month of June and lasting over a week as cut flowers.

***Hymenocallis* × 'Zwanenburg'**: It is one of the most beautiful of all *Hymenocallis*; the gracefully postured petals are lasting as cut flowers or as pot plant. The leaves are identical to *Hymenocallis calathina*.

***Hymenocallis* × 'Pax'**: It is an improved, long lasting, yellow hybrid.

***Hymenocallis* × 'Sofforthiae'**: This is a popular fragrant hybrid of *Hymenocallis*, producing bright golden yellow blooms which are very beautiful. It blooms in June or July, and the flowers are an excellent choice as cut flowers because of their wonderful aroma and lasting quality.

***Hymenocallis* × 'carribea'**: A cut flower hybrid with flowers lasting up to one week.

Species of Hymenocallis

Hymenocallis calathina

Hymenocallis crassifolia

Hymenocallis coronaria

Hymenocallis latifolia

Hymenocallis littoralis

Hymenocallis occidentalis

Hymenocallis rotata

Hymenocallis palmeri

Hymenocallis narcissifolia

Hymenocallis caribaea

Hymenocallis speciosa

Hymenocallis macrostephana

Cultivars of Hymenocallis

***Hymenocallis* x 'Tropical Giant'**

***Hymenocallis* x 'Sulphur Queen'**

***Hymenocallis* x 'Variegata'**

***Hymenocallis* x 'Zwanenburg'**

***Hymenocallis* x 'Zeylanicum'**

Hymenocallis x festalis

Propagation of Hymenocallis

Bulbs of Hymenocallis

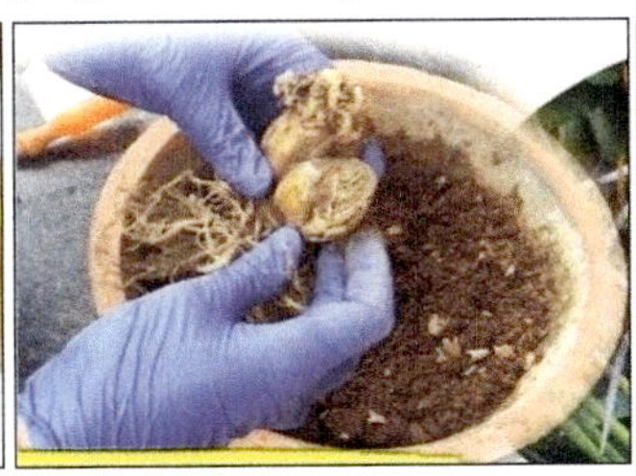

Propagation by Division of Bulbs

Sprouted Bulbs of Hymenocallis

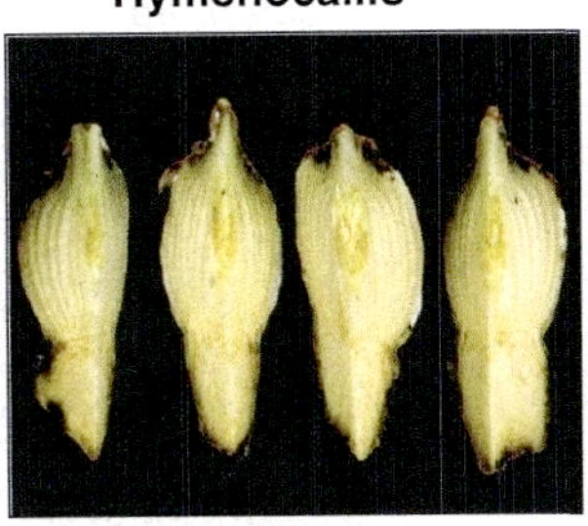

Propagation by Chipping

Propagation by Scoring

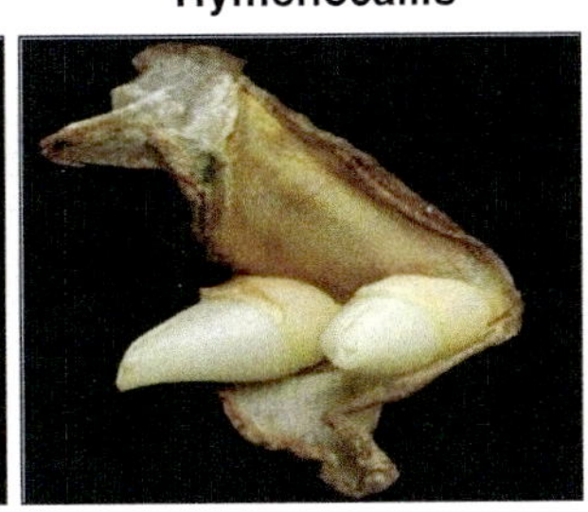

Propagation by Twin Scaling

Source: Peter J.M. Knippels (2012). Advanced *in Vivo* Propagation Techniques for Specialty Bulbs, *Floriculture and Ornamental Biotechnology,* Global Science Books.

Ismene × festalis (syn ***Hymenocallis × festalis***): It is a Mexican hybrid between *Hymenocallis x narcissifolia* and *Hymenocallis x longipetala.*

6.0 Genetics and Breeding

6.1.0 Genetics

Cytological investigations conducted by Smith and Flory (1990) recorded the chromosome numbers of *Hymenocallis henryae* as 2n=2x=38. Jee and Vijayavalli (1999) reported the karyotypes of eight taxa of *Hymenocallis* viz., *H. harrisiana* (2n=74), *H. littoralis* (2n=46), *H. speciosa* (2n=54), *H. rotata* (2n=46), *H. occidentalis* (2n=46), *H. daphe* (2n=42, 44 and 46) and suggested that polyploidy and aneuploidy have contributed significantly in the evolution of chromosomes in the genus and n=23, representing the basic chromosome number for this genus. Sharma and Bal (1956) considered both 22 and 23 as possible basic numbers because of the presence of 2n=46 and 44 numbers in the studied hymenocallis species. Sato (1938) believed that 11, the basic number for the genus and x=23 is derived from the duplication of one chromosome in primary diploid (2n=22+1) accompanied by doubling (2n=46) and secondary balance. Raina and Khoshoo (1971) observed telocentric chromosomes on several species of hymenocallis their numbers ranging from 2-82. *Hymenocallis littoralis* and *Hymenocallis tenuiflora* are telocentric species of the genus. It is apparent that hymenocallis is an evolutionally dynamic genus in which speciation is rapid. This is because of remarkable variation in chromosome numbers (ranging from 24-110) and strong tendency of aneusomaty in both somatic and gametic tissues.

6.2.0 Breeding

The present day *Hymenocallis* are the result of multiple hybrids between species and cultivars. Desirable flower size, form and fragrance have been obtained, and the colour range has been increased from white, yellow, bright golden yellow and cream except blue. It has been suggested that physical and chemical mutagen treatments may be utilized in treating bulb scales of diploid spider lilies in attempts to obtain more variants. The important breeding objective in *hymenocallis* includes different flower architecture, the intensity and consistency of scent, arrangement of stamens, orientation of flowers and longer life of all parts of the flower. Breeding for extending the blooming season is very important. Spencer (1955) exposed the bulbs of *Hymenocallis calathina* to 5200 r of X-rays at 170 r per min which resulted in 1.2 days earlier flowering. The number of flowers per scape is important breeding goal for cut flowers, pot or bed planting. *Ismene* × 'Sulphur Queen' is a hybrid between the white-flowering vigorously-growing *Ismene narcissiflora* and the yellow-flowering *Ismene amancaes* – both are Peruvian species. The cross between these species was first recorded in 1837 by the English bulb specialist Rev. William Herbert; the same year he published Amaryllidaceae, an attempt to arrange the Monocotyledonous Orders". Using *Ismene amancaes* as the seed parent and pollen from *Ismene nacissiflora* (which was called as *Ismene calathina*) Herbert named the resultant hybrid *Ismene* x 'spofforthiae' and it was also later referred to as *Ismene amancaes* var. *sulfurea*. The Dutch bulb breeding company, Van Turbergen later experimented with the same two parent species and produced several hybrids; the most distinctly yellow of these resultant clones they called 'Sulphur Queen'. Development of plants with excellent growth habit and cold hardiness is also very important. Both closed bud and the open flower stage are of interest for their aesthetic value in *Hymenocallis*.

7.0 Growth and Flowering

7.1.0 Effect of Bulb Size

The growth and flowering of *Hymenocallis* is significantly influenced by bulb grade, storage temperature, duration and growing conditions. In *Hymenocallis* × *festalis*, bulb grade is reported to exert a marked effect on the number of florets produced, which increased from 2 in 9/10 cm-circumference bulbs, to 7 in 18/19 cm-grade (Hanks and Jones, 1987). They obtained only 25 per cent flowering with 9/10 cm bulbs, while 90-100 per cent flowering with bulbs of more than 13/14 cm in circumference. In an attempt to break dormancy in *Hymenocallis harrasiana* and *Hymenocallis jaliscensis*, bulb stored at 18-22°C for 5-7 months and grown at a temperature of 10-35°C, decreased the time to leaf initiation and number of scapes (Borys *et al.*, 2009).

7.2.0 Effect of Temperature

The growth of the plants and flowering is optimum around 20-30°C day temperatures and 15-20°C night temperatures. However, flowering in *Hymenocallis littoralis* is not much affected even up to 45°C temperature. The earliest flowering was obtained when plants were transferred to 15-20°C and long day length of 16 hours was provided (Mori *et al.*, 1991).

7.3.0 Effect of Relative Humidity

The relative humidity of 50-60 per cent enhances the growth and flower production. Higher humidity may, however, cause disease infestation. Atmospheric CO_2 enrichment by 400 and 700 ppm of *Hymenocallis littoralis* resulted in 48 per cent increase in above-ground plant biomass and 56 per cent increase in underground biomass (Idso *et al.*, 2000). Further, it also led to an increase in the concentrations of five bulb constituents that possess anticancer and antiviral properties.

7.4.0 Growth Regulators

Parmar *et al.* (2009) and Chaithra *et al.* (2020) concluded that treatments of GA_3 (200 ppm) and NAA (100 ppm) were effective in increasing growth, yield and flowering of spider lily. Hanks and Jones (1987) while working with *Hymenocallis x festalis*, reported that ethephon effectively delayed flowering up to 3-6 days and reduced the scape length to its half.

8.0 Propagation

8.1.0 Seed and Vegetative Propagation

Hymenocallis is traditionally propagated by seeds and bulb offsets (Borys *et al.*, 2005). It can also be propagated through bulbs, division of clumps, twin scaling and cuttings. Two groups of Hymenocallis can be identified based on their ecology;

xerophytes and meso-hydrophytes (Borys *et al.*, 2005 and 2008). In both groups the seeds lack dormancy; the xeric group (caduceus leaves) is polyembryogenic, their seeds show a sequential emergence of embryogenic roots followed by rapid formation of small rooted, leafless bulbils, which are forced by contractile roots into deep soil layers and enter a deep dormancy period. During the following rainy season, the bulbs form feeder roots and primary leaves. The meso-hydrophyte group is evergreen and their seeds produce only one embryogenic root, terminating in a leafy and rooted bulbil of continuous growth (Borys *et al.*, 2005).

Leszczynska *et al.* (2005) utilized scales and rooted basal plates for the propagation of some Mexican spider lilies. Both scales and rooted basal plates were able to form bulbils in a variable number and size per scale or bulb. The bulbs were produced in different positions, the scales formed bulbils at the end of veins, while in the rooted basal plates, the bulbils were produced either between scales or on the upper parts of the remaining scales. The central part had to be removed to avoid the re-appearance of the dominant meristem and the bulbils were separated when the basal plate decomposed.

8.2.0 Micropropagation

Propagation of spider lily through tissue culture is a recent method. Backhaus *et al.* (1992) developed a protocol for the large scale *in vitro* multiplication of *Hymenocallis littoralis* using bulb scales as explant. After, surface sterilization, the scales with a small portion of the basal plates attached were cultured in dark on basal medium (BM) composed of Murashige and Skoog (1962; MS) salts supplied with 3 per cent sucrose, 100mg/l *myo*-inositol, 0.4 mg/l thiamine-HCl and 7 g/l agar; 1-9 bulblets were formed after 5 weeks and were sub-cultured in a proliferation medium containing the BM medium to which 2 mg/l 6-benzylaminopurine (BAP) and 0.02mg/l IBA were added. On an average, 8 new shoots were produced. After proliferation, the bulbs were placed in a rooting medium composed of the BM medium plus 6g/l agar. The rooted bulbs were successfully transferred to an intermittent mist in greenhouse. Explants of *Hymenocallis speciosa* are cultured on solid White's medium, supplemented with NAA and BA at 25°C in the dark. The regeneration of bulblets by scales is markedly increased by NAA at 0.01mg/l and BA at 5 mg/l. Regenerative capacity is higher on the abaxial side (Yanagawa and Ito, 1988).

9.0 Soil and Climate

Spider lily prefers sunny locations. The plant can tolerate the broadest possible range of growing conditions. It can grow in wet boggy areas, as well as in dry areas. Mountain Pima of Mexico was once adversely affected by a severe drought which reduced the crop (Lafferriere, 1992). The spider lily prefers a well drained sandy loam soil with rich organic matter and abundant irrigation. The pH of the soil should be around 6 to 7. Acidic soil can be neutralized by addition of lime and it should be done before planting. A moist environment is required for growing of this plant. In order to minimize infestation of soil-borne pathogens, disinfestations of soil should be considered for commercial production of cut flowers.

10.0 Cultivation

10.1.0 Planting Density

The soil should be prepared well in advance before planting. The field should be dug 45-60 cm deep and mixed well rotten compost in the soil. Hill grown hymenocallis prefers greenhouse cultivation. The bulbs are planted at a depth of 7 cm. Furrows should be made for bulb planting at 90 × 20 cm spacing *i.e.* 90 cm distance between two rows and 20 cm spacing between plants in row (Krushikar, 1988). Thus, planting density is around 45,000 bulbs per hectare of land. Kullur *et al.* (2018) concluded that the spacing of 90 × 60 cm, superior over all the treatments in terms of all flowering and quality parameters in *Hymenocallis speciosa*. Planting is done in the month of October-November (Krushikar, 1988).

10.2.0 Fertilization

Hymenocallis needs adequate organic matter and nutrition for proper growth and flowering. Higher levels of nitrogen and phosphorus were found to be very effective in increasing the growth of the plants and flower production. Kejkar and Polara (2017) reported that optimum growth and flower yield in *Hymenocallis littoralis* was obtained with the application of 400 Kg N/ha and 200 Kg P_2O_5/ha. De Hertogh and Le Nard (1993) gave 2-3 times application of 14 Kg of 12:10:18/ha of NPK every season. Deficiencies of these nutrients, on the other hand, resulted in marked reduction of plant height and number and size of flowers. The effect of potassium was less pronounced.

10.3.0 Irrigation

Spider lily also needs liberal watering for growth and flowering. However, the frequency of watering depends on the type of soil and stage of growth of the plant. Normally around 25 irrigations per year are required though the frequency largely depends upon prevailing weather conditions (Patel, 2000). Lack of moisture in the soil delays sprouting of bulbs. The plants are irrigated properly, especially during the blooming period.

10.4.0 Mulching

Mulching operation can be performed either by covering the ground with a polythene sheet or by dry grass, which will preserve moisture and keep the field weed-free, although plastic or polythene mulching increases soil temperatures. Staking of plants is practised to keep the plants straight. Weeding is a common intercultural operation in the field. Oxyflurofen @ of 0.5Kg/ha is the most effective herbicide for controlling weeds.

11.0 Diseases and Pests

Hymenocallis species are free from most of the insect-pests and diseases but a few attacking these are discussed below along with their suitable control measures.

11.1.0 Diseases

11.1.1 Phyllosticta Leaf Spot (*Phyllosticta hostae*)

Phyllosticta leaf spot occurs all through year with more severity during late September to early October. Small, circular or oval, brown-reddish to black spots occurred on the leaves in the early period. Spots sometime form to be concave, oval spots, occasionally surrounded by a water-soaked, discoloured halo. The spots gradually enlarge to form big patches. Subsequently, the diseased leaves became yellow with blighting at the tip (Yi *et al.*, 2015). Repeated applications of fungicides such as Dithane M-45, Captan, Mancozeb or thiophanate-methyl based treatments will help to control infection levels and can prevent new infections in healthy plants.

11.1.2 Cercospora Leaf Spot (*Ceroscospora amarillidis*)

It is a common disease throughout the world. It appears as large reddish brown to dark brown spots on leaf surface and progress by formation of yellow haloes and thin papery centre. The spots further increase in size, infecting entire leaf and ultimately result into death of leaves. This indirectly affects the crop production by causing a decrease in total photosynthesis in the plant. This disease can be controlled by spraying Carbendazim at the rate of 2 ml/l of water.

11.1.3 Anthracnose (*Colletotrichum gloeosporioides* syn. *Glomerella cingulata*)

The infection predominantly develops on leaves in the form of small grey-brown spots which spreads and finally results in premature drying. The disease can be managed effectively by Bordeaux mixture (0.8 and 1.0 per cent) or copper oxychloride (0.2 and 0.3 per cent).

11.1.4 Leaf Tip Blight (*Curvularia eragrostidis*)

Prajapati and Joshi (2012) reported this disease first time from India in *Hymenocallis littoralis*. The chief symptoms of the disease include the appearance of small straw, yellow coloured leaf spot which later coalesced to form necrotic blight, with dark yellow halo interfacing diseased and healthy leaf area. The disease then spreads downwards covering more than half the length of the leaves and all the foliage of the plants display severe necrotic blightening. Remove all affected leaves and burn them or place them in the garbage. Mulch around the base of the plant with straw or other natural mulch to prevent fungal spores in the soil from splashing on the plant. Thoroughly spray the plants with copper based fungicide.

11.1.5 Fusarium Leaf Spot (*Fusarium oxysporum*)

Fusarium leaf spot on *Hymenocallis littoralis* was reported for first time in Malaysia by Baker *et al.* (2019). Brownish to blackish spots surrounded by red haloes were observed on the infected leaves. At severe stage, the disease resulted in defoliation and affected the aesthetic value of the plant. This disease can be effectively managed by foliar spray of Chlorothalonil 500 SC (Kavach, Deconil).

Diseases and Pests of Hymenocallis

Phyllosticta Leaf Spot

Leaf Blight

Anthracnose

Red and White Lily Caterpillar

Adult and Egg mass

Source: Van Zandt *et al.* (2013) Southern Lepidopterists' News 35: 5-13.

Harvesting of Hymenocallis

Commercial Cultivation

Harvesting Stage of Flowers

Bundles of Spiderlily Flower

11.1.6 *Phoma narcissi*

This disease is observed in *Hymenocallis narcissifolia* causing bulb rot. It can effectively be controlled by bulb dressing with a mixture of thiophanate methyl 70 WP (Topsin M, Rocko), Chlorothalonil 500 SC (Kavach, Deconil) and Carbendazim with Prochlozol 380 EC (Saniewska, 1997).

11.2.0 Pests

11.2.1 Red and White Lily Caterpillar (*Polytera gloriosae*)

It is also called as spotted ballworm and a dangerous insect of *Hymenocallis*. The caterpillar feeds vigorously on the foliage as well as the flower buds causing a heavy damage to the crop. Patel and Patel (1997) studied the biology of this pest in spider lily and suggested Spark 36EC (0.045 per cent) spray followed by Chlorphyrophos 20 EC (0.03 per cent) or Alphamethrin 10 EC (0.005 per cent) and Quinalphos 20 A.E. (0.03 per cent) sprays. Some species of aphids have been observed to damage the spider lily. A major insect infecting this crop is large bulb fly (*Lampetia equestris*) which can be controlled through 2h hot water treatment at 43.5°C. Stem nematode (*Ditylenchus dipsaci*) infests on this crop, which can be brought under check by using healthy bulbs, soil sterilization and field sanitation.

12.0 Harvesting and Postharvest Management

12.1.0 Harvesting of Flowers

Harvesting of spider lily flowers at the proper stage is important. The optimum stage of harvesting is a mature bud when buds have started turning white just to open the next day. Harvesting is done in the early morning hours between 6 and 9 am. The flower stalk length, number and size and colour of the florets are some of the vital characteristics for maintaining quality. The flower stalk length should be straight and stout and just after flower harvest, the stalk can be placed in water in a cool place. The buds of spider lily can be stored in cold storage for duration of one week with pre-storage pulsing with 250 ppm 8-HQC and 2 per cent sucrose for 45 minutes and tissue paper packaging at 2°C temperature (Bhatt, 2007). Patel *et al.* (2008) reported that during cold storage the flower of spider lily can be stored with acceptable quality for 10 days with news paper wrapping and for 15 days with tissue paper wrapping in storage at 20°C.

12.2.0 Harvesting of Bulbs and Storage

Hymenocallis littorallis plants can thrive well continuously for 5-6 years at same site and there is no need of bulb harvesting every year. After 5-6 years, a group or cluster of 7-8 bulbs will develop, hence the old crop should be uprooted. After harvesting, the adhering soils is removed, the bulbs are kept in a well ventilated shady place for a week for drying. De Hertogh and Le Nard (1993) advocated lifting of the bulbs before the first frost and storing them immediately at a well ventilated place at 25°C temperature for drying and then storing at 20°C after grading. During storage of bulbs, pre-planting sprouting and decay of bulbs are the serious problems.

Pre harvest spray of MH 3000 ppm was found to be effective in reducing sprouting and spoilage of spider lily bulb (Bhosale and Barad, 2015).

13.0 Biotechnology

13.1.0 Molecular Markers

Catherine *et al.* (2010) studied the phylogeny of the South-eastern United States *Hymenocallis* based on ISSR fingerprinting and morphological data. The Neighbor Joining (NJ) analysis showed several supported clades with morphological synapomorphies. *Hymenocallis choctawensis* was sister to *H. franklinensis* (83 per cent bootstrap (bt) support); these taxa share the morphological characters of oblanceolate leaves, few ovules and the presence of rhizomes. *Hymenocallis coronaria* was sister to *H. liriosme* (63 per cent bt) and these taxa share the morphological feature of a prominent yellow "eye" in the middle of their staminal cups and both reproduce asexually by offsets rather than by rhizomes. *Hymenocallis godfreyi, H. rotate, H. tridentata* formed a clade (99 per cent bt) with the morphological synapomorphy of a 2n = 48 chromosome number. *Hymenocallis henryae* was sister to *H. palmeri* (82 per cent bt) with several morphological synapomorphies including subcircular scapes, hyaline margins on their leaves, scape bracts that do not enclose the flowers in bud, and tepals that are colored green. *H. puntagordensis* and *H. latifolia* were sister taxa (84 per cent bt) with the synapomorphy of persistent (versus deciduous) leaves and free filaments that are green distally. Rhizomes appear to have evolved twice within the southeastern United States species of hymenocallis: once in the *H. franklinensis, H. choctawensis, H. gholsonii, H. pygmaea, H. crassifolia, H. duvalensis* clade and once in the clade containing *H. godfreyi, H. rotata* and *H. tridentata*.

Microsatellite loci were isolated from the allotetraploid aquatic plant *Hymenocallis coronaria*. A repeat-enriched genomic library was constructed and primer pairs designed, resulting in six polymorphic loci. A total of 230 individuals were genotyped, and allelic richness per locus ranged from three to 11, while observed heterozygosity ranged from 0.017 to 0.570 (Markwith and Scanlon, 2006). Analysis of 24 populations using seven universal cpDNA markers and six nDNA microsatellite markers indicate that the genetic structure for both genomes correlates with the location of a major gap in the species' range, showing deep divergences in haplotype and population dendrograms between eastern and western regions (Markwith and Parker, 2007).

14.0 References

Antoun, M.D., Mendoza, N.T., Rios,Y.B., Proctor, H. H., Wickramaratne, D. B. M., Pezzuto, J. M. and Kinghorn, A. D. (1993) *J. Nat. Prod.*, **56**(8): 1423-1425.

Backhaus, R.A., Pettit, G.R., Huang, D.S., Pettit, G.R., Groszek, G., Odgers, J.C., Ho, J. and Meerow, A. (1992) *Acta Horti.*, **306**: 364-366.

Bakar, M.H.A., Azni, N.N.M., Kee, Y.J. and Mohd, M.H. (2019) *J. Plant Path.*, **101**: 779.

Berghe, D.A., Leven, M., Mertens, F. and Vlietinck, A.J.(1978) *J. Nat. Prod.*, **41**(5): 463-471.

Bhosale, N. and Barad, A.V. (2015) *HortFlora Res. Spectrum*, **4**(1): 48-51.

Biswas, B. (2006) *Bulbous Ornamentals and Aquatic Plants (ed. Bhattacharjee, S.K.), Pointer Publisher, Jaipur*, pp. 211-217.

Borys, M.W., Leszczynska-Borys, H. and Galvan, J.L. (2005) *Acta Horti.*, **673**: 191-198.

Borys, M.W., Leszczynska-Borys, H. and Galvan, J.L. (2008) *Acta Horti.*, **766**: 391-398.

Borys, M.W., Leszczynska-Borys, H. and Galvan, J.L. (2009) *Acta Horti.*, **813**: 631-636.

Catherine, M. B., Dollie, R. and Gerald, L. S. (2010) *Castanea*, **75**(3), 368-380.

Chaitra, P.V., Hemla, N.B., Chandrashekar, S.Y., Ganapathi, M. and Kantharaj, Y. (2020) *Int. J. Chem. Stud.*, **8**(5): 1748-1751.

Chase, M.W., Reveal, J.L. and Fay, M.F. (2009) *Bot. J. Linn. Soc.*, **161**: 132-136.

De Hertogh, A.A. and Le Nard, M. (1993) *The physiology of flower bulbs, Elsevier, Amserdam, Holland.*

Flory, W.S. (1976) *Nucleus*, **19**: 204-227.

Grossi, A. (2007) *The Plantsman*, **6**: 222-226.

Hanks, G.R. and Jones, S.K. (1987) *Sci. Hortic.*, **30**: 301-313.

Idso, S.B., Kimball, B.A., Pettit, G.R., Garner, L. C., Pettit, G. R. and Backhaus, R. A.(2000) *Amer. J. Bot.*, **87**(6): 769-773.

Jee, G. and Vijayavalli, B. (1999) *Caryologia*, **2**(1-2): 59-64.

Krushikar, I. (1988) *Krushi Jivan*, **20** (12): 4-7.

Kullur, L.R., Prasad, V.M., Singh, D. And Rajwade, V.B. (2018) *The Pharma. Innov. J.*, **7**(10): 558-559.

Lafferriere, J.E. (1992) *Ecol. Food Nutr.*, **28**(1-2): 1-9, 44.

Lafferriere, J.E. (1998) *Econ. Bot.*, **52**(2): 207-208.

Leszczynska-Borys, H., Borys, M.W. and Galvan, J.L. (2005) *Acta Horti.*, **673**: 691-697.

Lin, L.Z., Hu, S.F., Chai, H.B., Pengsuparp, T., Pezzuto, J.M., Cordell, G.A. and Ruangrungsi, N. (1995) *Phytochemistry*, **40**(4): 1295-1298.

Markwith, S.H. and Scanlon, M. (2006) *Mol. Ecol. Notes*, **6**(1): 72 – 74.

Markwith, S.H. and Parker, K.C. (2007) *Conserv. Genet.*, **8**, 949–963.

Mori, G., Imanishi, H. and Sakanishi, Y. (1991) *J. Jap. Soc. Hortic, Sci.*, **60**(2): 382-393.

Mother. (1995) *Flowers and their message, Shri Aurobindo Ashram Press, Pondichery*, India.

Murashige, T. and Skoog, F. (1962) *Physiol. Plant.*, **15**: 473-497.

Nelson, G. (2003) *Florida's best native landscape plants, University, Press of Florida, Gainesville*, Florida, pp. 411.

Ogden, S. (2007) *Garden Bulbs for the South*, Oregon, USA, pp. 396.

Parmar, A.B., Patel, H.C., Chavda, J.C. and Parmar, J. R. (2009) *The Asian J. Hortic.*, **4**(1): 170-172.

Patel, G.D., Patel, R.B., Patel, B.N., Singh, A. and Dhaduk, B.K. (2008) *J. Ornam. Hortic.*, **11**(3): 224-227.

Patel, K.S. and Patel, M.B. (1997) *J. Appl. Horti., Navsari*, **3**(1/2): 115-124.

Patel, R.L. (2020) *Ph.D. Thesis*, GAU, Gujarat.

Pettit, G.R., Pettit, G.R.III., Backhaus, R.A., Boyd, M.R. and Meerow, A.W. (1995) *J. Nat. Prod.*, **58**(1): 37-43.

Prajapati, V.P. and Joshi, D.M. (2012) *J. Myco. Plant Patho.*, **42**(2): 276-277.

Preuss, K.D. (2002) *Bulbs*, **4**: 27-31.

Raino, S.N. and Hkoshoo, T.N. (1971) *La Cellule*, **68**(3): 239-259.

Salisbury, R.A. (1812) *Trans. Horti. Soc. London*, **1**: 338-341.

Saniewska, A. (1997) *Plant Prot.*, **37**(2): 387-389.

Sato, D. (1938) *Cytologia*, **9**: 203-242.

Sealy, J.R. (1954) *Kew Bull.*, **2**: 201-240.

Sharma, A.K. and Bal, A.K. (1956) *Cytologia*, **21**: 329-352.

Small, J.K. (1933) *Manual of the Southeastern Flora*, New York, **Pp**.321-324.

Smith, G.L. and Flory, W.S. (1990) *Brittonia*, **42**(3): 212-220.

Spencer, J.L. (1955) *Amer. J. Bot.*, **42**(10): 917-920.

Tapia-Camos, E., Rodriguez-Dominguez, J.M., Revuelta-Arreola, M., Tuyl, J. M. V. and Barba-Gonzalez, R. (2012) *Floricul. Ornam. Biotech.*, **6**: 129-139.

Yanagama, T. and Ito, I. (1988) *J. Jap. Soc. Horti., Sci.*, **57**(3): 454-461.

Yi, R.H., Gan, L.J., Chen, J. and Xu, X.L. (2015) *J. Plant Prot. Res.*, **55**(4): 438-445.

15

Importance and Uses of Annual Ornamental Plants

K. Dutta and T.K. Bose

1.0 Importance and Uses of Annuals

1.1.0 Importance as Ornamental Plants

Annuals and biennials, treated as annual are important group of ornamental plants, extensively grown in all parts of the world, outdoors and indoors, as house plants and in greenhouses, depending on the environment and purpose of cultivation. They contribute colour and beauty in gardens and suitable for growing in beds, herbaceous border, shrub borders, edges, rock gardens and other locations in private gardens, parks and other public gardens, botanic gardens, in gardens around residential and commercial complexes, around monuments and public buildings. Annuals are widely grown in pots and hanging baskets and very effective in beautifying gardens, open spaces, on stands, used for beautifying residential buildings, apartments, garden rooms, balcony, roof gardens, patios, living rooms, window boxes. Pot plants of different annuals are the major exhibits in flower shows in all parts of the world. Annual climbers can be trained on arches, trellis and fences. Cut flowers from large number of species and cultivars of annuals are very popular as bouquets, in vases and various floral arrangements in residential houses, offices, institutions, temples, for decorating dias and surroundings, areas in public functions and meetings, in marriages and other social ceremonial occasions. Cut flowers of annuals and perennial ornamentals are the major items in local and global markets. Several annuals like *Amaranthus tricolor, Kochia trichophylla, Senecio cruentus, etc.* are grown for their showy foliage. Apart from cut flowers, loose flowers of many annuals are also extensively used in making floral ornaments for temples, decorating pandals, idols, widely used in public and private ceremonies, creating a charming atmosphere befitting the occasion. Loose flowers are universally used for worshipping God. Aroma extracted from scented flowers and natural dye extracted from petals of annuals and perennials are now widely used commercially. In addition to the use of fresh flowers in bouquets and vases, dry flowers of annuals are also important component of pot-pourri, an item in great demand in flower-trade. Intensive cultivation of annuals for cut-flower and loose flower production outdoors, in greenhouses, and also in pots is a highly remunerative practice in most parts of the world. In the developing countries thousands of farming families are dependent on intensive flower cultivation, particularly annuals for their livelihood which also provides employment to women, for flower harvesting, grading and making floral ornaments for marketing. Annual flowers constitute major items in flower markets in rural and urban areas including cities and towns, providing income to the growers, agents and retailers, the year round. Annual flowers are therefore of immense importance not only for the beautification of gardens and living rooms but also of great commercial value, providing employment and livelihood to thousands of families in India. In seed industry of horticulture crops, a major turnover is from production and sale of seeds of annuals, particularly in developed countries.

1.2.0 Therapeautic and Other Uses

Pharmacological studies have confirmed that *Calendula officinalis, Althaea rosea* exhibit a broad range of biological effects, some of which are very interesting for possible future development. It has been demonstrated to possess multiple activities such as anti-HIV, cytotoxic, anti-inflammatory, hepatoprotective, spasmolytic and spasmogenic, amongst others. *Zinnia elegans* contains numerous secondary metabolites of different classes including sterols, flavonoids, sesquiterpenes, diterpenes and hydrocarbons (Gomaa *et al.,* 2016). *Antirrhinum majus* contained amino acids, pigments, oils, anthocyanidins, flavanols, flavones, aurones, flavanones, cinnamic acids and many other compounds. *Matthiola incana* is utilized in the traditional medicine of several countries (Iran, India, Ecuador, Bolivia, Italy) for the treatment of various ailments, including inflammations and cancer, particularly breast and testicular cancer.

Omokhua (2018) reported the extracts of the *Tithonia diversifolia* to possess antiviral, antidiabetic, anthelmintic, anti-inflammatory, antispasmodic, antiproliferation and antimalarial activities as well as analgesic properties. Moldovan (2017) reported that *Rudbeckia triloba* extracts indicate significant antioxidant activity, especially of macerate of petals (resulted by applying Folin-Ciocalteu, total flavonoids, and DPPHÅ" assays). Also, leaves, inflorescences, and the inflorescences' essential oil of *Rudbeckia triloba* were found to be rich in monoterpene hydrocarbons (especially, α-pinene in leaves and inflorescences and β-phellandrene and germacrene D, in inflorescence essential oil). The chemical constituents of *Callistephus chinensis* flower (CCF) and their anti-diabetic activity and aim to find new natural source for the development of α-glucosidase inhibitory agent for dealing with diabetic complications was reported by Zhang *et al.* (2013). *Celosia argentea,* with its active compounds, possesses wide-reaching biological activities such as hepatoprotection, tumour treatment, anti-diarrhoea, anti-diabetes, anti-oxidant, anti-hypertension, and for treatment of a number of eye diseases (Tang *et al.,* 2016). Many other species of annual ornamentals are used as medicinal plants in different parts of the world.

1.3.0 Use of Annuals as Natural Dyes

Natural or botanical dyes are extracted from a number of colourful annuals which are eco-friendly, biodegradable and non-toxic. Cellulose fibres extracted from jute, bamboo, ramie, cotton *etc.* can easily absorb natural dyes extracted from annual flowers.

These natural dyes re-emerged as potential viable sources as an alternative to synthetic dyes (Mrijalini *et al.,* 2011). Natural flora/fauna is full of exquisite colours, fascinating and attracting human beings towards a vast portfolio of possibilities. A large number of plant sources have been identified for extraction of colour and their diversified use in textile dyeing and other disciplines (Dawson, 2009; Shahid, 2013). The natural colour extracted from annuals is calendula (*Calendula officinalis*); lemon yellow; coreopsis (*Coreopsis* spp.) gold, yellow, and orange; hollyhock (*Alcea rosea*) orange-yellow; dark red or purple flowers will yield purple and maroon shades;

Importance and Uses of Annuals

Annuals in Outdoor

Home Decoration

In Border of a Garden

In a Residential Area

As Edges in a Park

Around a Specimen Tree in Garden

In a Private Garden

As Herbaceous Border

Importance and Uses of Annuals

Annuals in Pots

Hanging Baskets

In Balcony

In Window Boxes

Annuals in Rock Garden

Seedling Production in Polyhouse

Kochia in Landscaping

***Amaranthus tricolor* in Garden**

Importance and Uses of Annuals

Annuals in Greenhouse

Annuals in Market

Importance and Uses of Annuals

Annual in Pot

Annuals in Flower Arrangements

marigold (*Tagetes* spp.) bright yellow; yellow cosmos (*Cosmos sulphureus*);yellow and gold *etc.* (Chris McLaughlin, 2018).

1.4.0 Annuals for Rock Garden

A rock garden in a sunny location may not necessarily be entirely composed of cacti, succulents and hardy shrubs. In a rock garden hardy, dwarf annuals which thrive and flower in porous soil are usually selected. Ageratum, alyssum, arctotis, brachycome, french marigold, melampodium, nasturtium, phlox, portulaca, verbena, dwarf zinnia, *etc.* are some annuals suitable for this purpose.

1.5.0 Early Blooming Annuals

There are occasions when display of colour in a garden becomes necessary before the time of normal flowering in most of the annuals. Ageratum, alyssum, balsam, browalia, candytuft, celosia, dianthus, gomphrena, marigold, portulaca, salvia, sunflower, zinnia *etc.* are early blooming annuals in winter months in tropical and sub-tropical gardens.

1.6.0 Annuals as Cut Blooms and Loose Flowers

A group of annuals bearing flowers on long stalk, having comparatively longer vase-life are used as cut bloom for floral decoration. Spikes of antirrhinum, statice, stocks, larkspur, clarkia, panicles of gypsophila flowers of annual chrysanthemum, calendula, cornflower, dianthus, gaillardia, gomphrena, small and miniature sunflower, celosia, coreopsis, sweet sultan, zinnia, *etc.* are popular cut-flowers. In tropical climate, carnation is grown in gardens as winter annual from seed and produces excellent cut-flower in various colours. Marigold (both African and French) and China aster are commercially grown for making garlands and floral ornaments with those flowers, loose flowers of annual chrysanthemums are also used for similar purposes.

1.7.0 Annuals in Pot

Although all types of annuals can be grown in pots, some are widely grown in containers. The potted annuals when in flower, are very popular for the purpose of decoration and extensively grown for exhibiting in flower shows. The types and cultivars, medium to small in height, bushy in appearance and flower profusely for a long period, are usually grown in pots. Ageratum, antirrhinum, China aster, balsam, brachycome, bracteantha, carnation, calendula, celosia, cineraria, clarkia, clianthus, dianthus, gazania, impatiens, lupin, marigold, nasturtium, petunia, pansy, phlox, portulaca, sunflower, salvia, verbena, zinnia *etc.* are popular annuals in pots. The low-growing spreading annuals, like mesembryanthemum, pansy, phlox, petunia, portulaca, verbena, are grown in shallow pans or hanging baskets; container grown annuals are popular in window boxes, balcony gardens and used as house plants in different locations. *Amaranthus tricolor*, coleus and kochia are attractive pot plants for their showy foliage.

2.0 Climate and Soil for Cultivation of Annuals

As annuals are the major group of plants for a colourful display in the gardens of tropical areas in winter months, range of temperature prevailing during growth and flowering largely influences the beauty of the gardens with blooming annuals. It is also relevant to mention that many winter annuals which require exposure to comparatively lower temperature for flowering also grow and flower well in the tropics where winter is mild and continues for a longer period. Several summer flowering annuals like *Amaranthus* spp. balsam, coreopsis, gaillardia, gomphrena, kochia, sunflower, yellow cosmos, zinnia, *etc.* are also grown in summer months and early rains in well-drained soil. In mild subtropics, annuals are grown outdoors mainly in spring and summer, depending on the range of temperature. In mild climate, prevailing in any parts of India, annuals and biennials grow and flower all the year round except in severe winter; if protected from excess moisture in beds in the rainy season, produce flowers of better quality than those grown in winter in the tropics. Prolonged mild climate and bright sun also promote seed formation in annuals. In temperate climate, annuals are grown outdoors mainly in summer, in frost-free, warm temperate areas, frosty-hardy plants, however, are grown in different seasons. *Calceolaria, Delphinium, Digitalis, Lathyrus, Lupinus, Matthiola, Penstemon, Ranunculus, Euostoma etc.* are popular annuals in temperate climate. Apart from outdoor cultivation of ornamentals in favourable environment, protected cultivation of flowering and foliage plants, particularly those of commercial importance including annuals, is now a highly developed technology for production of quality flowers and plants. Many annuals with beautiful flowers of bushy growth, medium or dwarf in height, long lasting flowers are extensively grown in pots, in greenhouses for commercial purpose. Ornamental plants, in general, and annuals in particular, grow and flower well in well-drained, loamy soil rich in organic matter. Wet or soggy soil, even for a few days will adversely affecst growth, may often cause rotting of roots and collar region, and promotes fungal and bacterial infections, resulting in decay of the plants.

3.0 Site and Layout of Annual Bed

Most of the annuals require full sun; shade for a few hours in the morning or mid-day suppresses growth and flowering; western sun, however, is less important. The annuals are grown in mixed border or only one type or cultivar in a bed. A strip of green grass in front of a mixed border or in between two beds improves the display of colour.

The space for a mixed border should be large and of informal design. Plants varying in height and flower colour are planted in groups, tall plants at the back and shorter ones gradually towards the front. Tall salvia, hollyhock, seedling dahlia, lady's lace, annual chrysanthemums should go at the back; plants of intermediate height, *e.g.* antirrhinum, cornflower, salvia, larkspur are grown in the next group; the third group may consist of dianthus, aster, wall flower, anchusa, calendula; alyssum, phlox, pansy, brachycome and candytuft should be selected for the front patches. In a small bed, usually one type of annual of attractive colour of *Salvia, Antirrhinum, Celosia plumose, Dianthus, Tagetes, Petunia etc.* medium or short height, is grown depending on the size and position of the bed. In a formal garden, due to the symmetry in garden design, beds of geometrical shape, types of annuals and their colours are selected and grown for maintaining the symmetry.

4.0 Cultivation

A sketch of a mixed annual border on a sheet of paper will be of assistance to amateurs who are diffident about risking their own ideas directly on the field.

4.1.0 Raising of Seedlings

Seeds of annuals are sown in beds, boxes, shallow seed pans or plug trays. The seed compost should be light and porous consisting of one part each of garden soil and finely screened dry leaf mould and coco peat, where peat is not available. The ingredient and composition of seed or pot compost vary depending on availability of materials. In most countries, compost for different purposes are readily available for use. The media for sowing seeds should be sterilized to prevent infection of soil-borne diseases which often cause severe damage to the seedlings and plants after transplanting. The compost should be slightly moist before sowing, otherwise the water will not soak readily and the seeds will float on water. Seeds may be sown thinly with a cover of fine layer of screened leaf mould or coir dust. Protection of seed pan or seed bed from rains after the seeds have been sown is absolutely necessary. After emergence of seedlings, exposure to sunlight should be gradually increased, so that before transplanting they become sufficiently hardy to bear the full-day sun. The seedlings are severely damaged by fungal diseases in shady location. The damage from excess moisture, high humidity and cloudy weather can be effectively minimized by fungicide application. Most of the seedlings become ready for transplanting at four-leaved stage. Time of sowing seeds will depend on the season of flowering in a particular climate and the number of days required for the annuals to flower. Double and large flowering petunia, cineraria, carnation and stock, which require about 4 months to flower, should be sown earlier. For summer flowering in the subtropics and mild temperate climate, seeds of annuals are sown in February- April, depending on the environment. In tropical climate, sowing of seeds for summer flowering annuals is done in January-February.

4.2.0 Preparation of Beds and Transplanting of Seedlings

The beds should be thoroughly dug, at least 30-40 cm deep and the soil brought to fine and uniform tilth. Decomposed organic manure should be mixed thoroughly at the time of preparation of the soil. At the time of transplanting, small holes are made and the seedlings planted in those holes with the root just covered with soil. The soil should be gently pressed to keep the seedling erect and watered through a fine rose can. While taking out the seedlings from the seed bed or pan, care should be taken to minimize the damage of roots. Roots system of seedlings raised in plug trays is less damaged when removed for transplanting, light watering is done whenever necessary; over-watering or drying of soil for the first few days after transplanting causes death of the seedlings.

4.3.0 Irrigation and Fertilization

The common mistake of daily watering, whether the plant requires moisture or not, causes more damage than anything else. After the seedlings have started new growth, they are irrigated thoroughly. Weeds are removed from the bed when it is just moist. The soil is allowed to come to a semi-dry stage before it is again irrigated. Loosening of soil helps in aeration and absorption of water much below the root zone. This simple practice is the secret for successful cultivation of annuals. Chemical fertilizers including micronutrients should be used in low concentration. Mixtures of chemical fertilizers in the form of solution, if applied in low concentration during the vegetative phase of plant growth, improve the plant growth and also the quality and quantity of flowers. Fertilizers can also be used on the soil as dry powder, followed by thorough watering.

4.4.0 Diseases and Pests in Annuals

Most of the annual flowering plants, being soft and succulent, are often infected by diseases and attacked by pests for which both preventive and curative measures need to be taken for their control. Preventive measures for minimizing infection of diseases include use of disease-free fresh seeds, or hardy and healthy planting materials, optimum soil moisture, bright location and use of well decomposed organic matter. Application of effective fungicides would control most of the diseases, if used at the early stage of infection. Among the fungal diseases, damping off, caused by *Rhizoctonia* spp., *Pythium* spp., at the seedling stage, leaf spot and blight, caused by a number of pathogens (*Altenaria* spp., *Cerospora* sp., *Septoria* sp.), inflorescence blight, flower bud rot, powdery mildew (*Oidium* sp., *Aeveillucla* sp.), collar rot (*Rhizoctonia* sp. *Phytophthora* sp., *Pythium* sp., *Sclerotium* sp.), *Uromyces* sp., and rust disease caused by *Puccinia* are the major diseases of annuals. Grey mould, a common disease in many annuals, also causes damping-off of seedlings and rotting of flowers, wilt (*Fusarium* sp. *Verticillium* sp.) and downy mildew (*Peronospora* sp.) severely damage many annuals. Bacterial soft rot, caused by *Pectobacterium* sp. and *Bacillus* sp, is responsible for severe losses in annuals. Several species of annuals are severely affected by viruses. Aster yellow virus (*Chlorogenus callistephii*) causes yellowing of leaves, stunted growth, and change in flower colour. Several annuals including marigold are affected by cucumber mosaic virus (*Marmor cucumeris*), causing stunted growth and deformation of leaves. Viral diseases are often transmitted by insects like aphids, thrips, spider mites and looper larvae, and the incidence of viral diseases can be minimized by controlling the insect vectors.

Common pests on annuals including aphids, red spider mites, looper larvae., leaf miners and nematodes can be controlled effectively by using proper insecticides. Root exudates of marigold plants repel nematodes.

5.0 References

Bose, T.K., Singh, L.J., Sadhu, M.K., Maity, T.K. (2015) *Ornamental Plants and Garden design in Tropics and Subtropics*, Vol 2: 1-952.

Chris McLaughlin (2018) *Natural Dyes From Your Flower Garden*, (*https: //www.fix.com/blog/natural-plant-dyes/*).

Dawson, T.L. (2009) *Color Technol.*, **125**, pp. 61.

Gomaa, A.A.R., Samy, M.N., Desoukey, S.Y. and Kamel, M. (2019) *Journal of advanced Biomedical and Pharmaceutical Sciences*, **2**: 29-37.

Moldovan, Z., Buleandrã, M., Oprea, E. and Mînea, Z. (2017) *Journal of analytical methods in chemistry*, pp. 27.

Mrijalini, K. Nazarpoor, L. and Karimi (2011) *J. Clean. Prod.* **19**, pp.1045.

Omokhua, A.G., Abdalla, M.A., Van Staden, J. and McGaw, L.J. (2018) *BMC complementary and alternative medicine*, **18**: 1-15.

Shahid, M., Ul-Islem, S. and Mohammad, F. (2013) *J. Clean. Prod*, **53**, pp. 310.

Tang, Y., Xin, H.L. and Guo, M.L. (2016) *Revista brasileira de farmacognosia*, **26**: 787-796.

Zhang, X., Liu, Z., Bi, X., Liu, J., Li, W. and Zhao, Y. (2013) *EXCLI journal*, **12** : 956.

16

Gomphrena

C. Aswath, P. Pal and M. Mitra

1.0 Introduction

Gomphrena species is an edible, ornamental and medicinal plant commonly known as 'Globe Amaranth' or 'Bachelor Button', belongs to the family Amaranthaceae. *Gomphrena globosa*, commonly known as 'Globe Amaranth', '*makhmali*', and '*vadamalli*', is an edible plant. The round-shaped flower inflorescences are a visually dominant feature and cultivars have been propagated to exhibit shades of magenta, purple, red, orange, white, pink, and lilac. Within the flowerheads, the true flowers are small and inconspicuous (Jiang *et al.*, 2011).

It is an annual herb native to Central America now cultivated worldwide as an ornamental, traditional medicine and as a source of betacyanins for use in the food and cosmetic industry. It has escaped cultivation in gardens and is now found in waste grounds and disturbed sites in tropical and subtropical countries. It is listed as an invasive species by CeNBIO, and classified as an invasive plant in Cuba, Hawaii and Costa Rica (Febles, 2017).

As a tropical annual plant, *G. globosa* blooms throughout summer and early fall. It is very heat tolerant and fairly drought resistant, but grows best in full sun and regular moisture (Silva *et al.*, 2012). The plant fixes carbon through the C4 pathway (Herold *et al.*, 1976). At maturity, the flowerheads are approximately 4 cm long and the plant grows up to 24 inches in height. *Gomphrena globosa* is an outcrossing species that is pollinated by butterflies, bees, and other insects. Floral volatiles likely play a significant role in the reproductive success of the plant by promoting the attraction of pollinators (Jiang *et al.*, 2011; Anon., 2020).

2.0 Importance and Uses

2.1.0 Aesthetic Uses

Gomphrena is grown in beds, borders, rockeries and pots. Mass planting is very effective and instead of using only one colour, a combination of pink and white or purple and white in contrasting colours can be planted. Pot-grown plants are useful for indoor decoration. It is also grown for cut flowers which last long. The dried flowers are used in flower arrangements and indoor decoration.

2.2 Therapeutic and Other Uses

The *Amaranthaceae* family comprises of many species, which are used in nutrition and in traditional folk medicine for the treatment of several diseases such as high blood pressure and used for oliguria, heat and empacho besides, infections, inflammation and fever. *Gomphrena* species found all over the world. It has been recognized in different traditional system of medicines for the treatment of various diseases of human being (Tarnam *et al.*, 2014). *Gomphrena* species is an edible, commercial ornamental and medicinal plant and used as food colourant. The phytoconstituents present in it are flavonoids, phytosterols, phenolics and terpenoids. The different parts of this plant are traditionally claimed to be used for the treatment of broad spectrum of ailment including baby gripe, oligouria, body

Uses of Gomphrena

Gomphrena in Pot

In Border of a Garden

In Rock Garden

Gomphrena in Bed

In a Mix Annual Bed

In Flower Arrangements

In Interior Decoration

In Flower Vase

sore, malaria, bacterial infections, jaundice, urinary problems, high cholesterol, cough, fever, diarrhea, liver disorders, kidney disorders and cooling. Due to changes in the lifestyle and associated decline of the use of plants, the ethnobotanical knowledge is continuing to decline. Hence, further work to isolate and characterise the active compounds responsible for this activities in the plant is recommended (Tarnam *et al.*, 2014).

In northwest Argentina, plant extracts of both *Gomphrena maritiana* and *G. boliviana* are used to treat stomachache, gastrointestinal disorders, traumatic injuries, inhibition of tumour growth and several bacterial diseases (Pomilio *et al.*, 1991; 1992). *G. globosa* is used as an indicator plant for *potato* × *potexvirus* (Codero, 1988).

Phytochemicals like betacyanins have been reported for significant antioxidant activity and the different phytochemicals recognised are responsible for various activities such as anticancer, analgesic, anti-inflammatory and cytotoxic (Chandana and Pradeep, 2018).

2.2.1 Beta Cyanins

Even though betacyanins are associated mostly with *Beta vulgaris* L., *G. globosa* contain betacyanins making this plant an excellent source for obtaining succompounds (Roriz *et al.*, 2017a). In one of the earliest studies, Minale *et al.* (1967) reported the characterization of seven betacyanins from *Gomphrena globosa* inflorescences. Two of these pigments, gomphrenin-I and gomphrenin-II, have been shown to be 6-O-β-d-glucopyranosides of betanidin and isobetanidin, respectively. The other five pigments were proved to be hydroxycinnamoyl (trans feruloyl, trans p-coumaroyl or cis p-coumaroyl) derivatives of gomphrenin-I or gomphrenin-II. For gomphrenins-III, -V and -VI, the number and position of the acyl group in the molecule have also been established (Minale *et al.*, 1967). The betacyanins gomphrenin I (betanidin 6-O-β-glucoside) and its acylated forms gomphrenin II (betanidin 6-O-[62 -O-(E-4-coumaroyl)-β-glucoside]) and gomphrenin III (betanidin 6-O-[62 -O-E-feruloyl-β-glucoside]) were isolated from flowers of *Gomphrena globosa* and their structures elucidated by NMR spectroscopy and ion spray mass spectrometry by comparison with betanin (betanidin 5-O-β-glucoside) and its feruloyl conjugate lampranthin II (betanidin 5-O-[62 -O-E-feruloyl-β-glucoside]), which were isolated from *Beta vulgaris* roots and *B. vulgaris* cell suspension cultures, respectively (Heuer *et al.*, 1992). Isorhamnetin 3-O-robinobioside, choline and betaine were isolated from whole plants of *Gomphrena martiana* and identified by spectral data. Structural assignment of the disaccharide as robinobiose was mainly based upon 13C-nmr and mass spectra of its trimethylsilyl derivative (Buschi and Pomilio, 1982).

Moreover, these compounds are in use as colorants in the food industry, offer the possibility to replace the chemical synthetic counterparts, such as azo-dyes. They are excellent in medicinal properties. In order to replace conventional extraction techniques, microwave (MAE) and ultrasound assisted extraction (UAE) processes were optimized by the response surface methodology (RSM) to maximize the recovery of betacyanins from *G. globosa*. The optimum processing conditions for

MAE (t =8 min; T= 60 °C; Et=0 per cent; and S/L =5 g/l) conducted to an extraction yield of 39.6 ± 1.8 mg/g dry weight, while for UAE (t=22 min; P=500 per cent; Et=0 per cent; and S/L =5 g/L) a value of 46.9 ± 4.8 mg/g was achieved, validating the UAE more suitable to obtain the target compounds. This technique can provide extracts rich in betacyanins and with high potential to be used as natural colorants (Roriz *et al.*, 2017a). They (Roriz *et al.*, 2017b) subsequently described a novel mechanical process for the pigmented parts of *Gomphrena globosa* L. The effects of the variables of the maceration extraction of betacyanins have not been properly described. They optimized the conditions that maximize betacyanins extraction from *G. globosa* as an alternative source. Assisted by response surface methodology, an experimental design was developed for testing the extraction variables (time, temperature, ethanol-water proportion and solid-liquid ratio). The responses used were betacyanins quantification (by HPLC-PDA-MS/ESI and spectrophotometric analysis), the extraction-yield and the colour intensity of the produced powder. The betacyanins identified were gomphrenin and isogomphrenin II and III. The highest betacyanins content (45 mg/g) was obtained by 165 min, 25°C, 0 per cent of ethanol and 5 g/L of solid-liquid ratio. The betacyanins content from the floral parts of *G. globosa* is higher than those normally found in other sources highlighting its industrial application.

Currently, there are some examples of natural colourants with commercial use. However, these colourants are usually under-exploited, besides being obtained from a reduced number of plant or algal species (Roriz *et al.*, 2018). Accordingly betalains obtained from an alternative plant species, *Gomphrena globosa*, which have a powerful colouring activity besides being strong antioxidants, as a novel ice-cream colourant was studied (Roriz *et al.*, 2018). For comparison purposes, other ice-cream formulations were prepared, namely without colourants, added with commercial betalain or with *Beta vulgaris* extract. Besides evaluating the Hunter colour parameters L*, a* and b*, the nutritional parameters, individual sugars and fatty acids profiles were also studied. These parameters were evaluated throughout time, up to a maximum of 60 days of freeze (–22 °C) storage. Betacyanin quantification of each formulation was also performed to determine its maintenance along storage. In general, ice-creams prepared with *G. globosa* were similar (considering nutritional, colour, individual sugars and fatty acids profiles) to those including *B. vulgaris* extract, thereby validating the suitability of this alternative plant as a source of food colourants, particularly as ice-cream colourants. Furthermore, the positive effects induced by the addition of this natural colourant were maintained throughout storage time, as indicated by the markers distribution in the linear discriminant analysis (Roriz *et al.*, 2018).

2.2.2 Antimicrobial Compounds

The antimicrobial activity of extracts and constituents of *Gomphrena martiana* and *G. boliviana* were determined in order to identify the compounds responsible for the folk-medicinal use of these plants (Pomilio *et al.*, 1992). Each extract was evaluated against 20 microorganisms, including Gram-positive and Gram-negative bacteria, spore-forming Gram-positive bacteria, an acid-fast bacterium, a fungus and two yeasts. Fractionation of each petroleum ether (PE) extract yielded five

5,6,7-trisubstituted flavones that were separately tested showing high activity against *M. phlei* (minimum inhibitory concentration (MIC) 15, 20 and 75 µg/ml) approaching that of commercial bactericides. Other natural and synthetic flavonoids with diverse structures were also tested to define structure-activity relationships. Each EtOH extract was subsequently fractionated and monitored by bioassays leading to isorhamnetin 3-O-β-robinobioside (MIC 50 µg/ml) in both instances (Pomilio *et al.*, 1992).

3.0 Origin, Species and Cultivars

3.1.0 Origin

It is an annual herb native to Central America now cultivated worldwide as an ornamental, as well as having uses in traditional medicine and as a source of betacyanins for use in the food and cosmetic industry.

3.2.0 Species

The genus belongs to the family Amaranthaceae and consists of about 90 species (Brickell, 1996) of annual, biennial and herbaceous perennial in Australia and tropical Central and South America. Only one species, *Gomphrena globosa* with its cultivars is in general cultivation. The plants are 30 to 40 cm high, bushy with erect stems and hairy, light green, oblong-ovate leaves tapering towards the base. Round button-like flower heads are borne on long, thin but strong stems above the foliage. They have tiny, conspicuous, colourful bracts. The colour of flowers is red, yellow, purple, pink or white.

According to Stevens (2019), the genus comprises approximately 120 species found in the America, Antartica and Indo-Malaysia. *Gomphrena* is a genus in the Amaranthaceae, a family comprising 180 genera and 2000 to 2500 species of herbs and shrubs (Stevens, 2019). Although native to Central America, *G. globosa* was first described from India (Flora of Panama, 2020). *Gomphrena meyeniana* is an extremely variable species from the Andean highlands, which has attracted the attention of many botanists because it is the world's highest-elevation C4 eudicot and because of its wide morphological variability (Bena *et al.*, 2020). It has the typical high-Andean plant morphology, with small leaves tightly clustered on a thick rootstock. The large range of morphological variation within this species coupled with the varying opinions on the existence of several species or infra-specific taxa and the lack of molecular information has made the clarification of the *G. meyeniana* complex a challenge. Bena *et al.* (2020) performed broad molecular sampling to identify the phylogenetic position of *G. meyeniana* within Gomphrena and to perform a multivariate analysis to objectively differentiate taxa based on morphological characters. The ITS and trnL-F regions were analyzed individually and in combination using Bayesian inference and maximum parsimony methods. To analyze the morphological characters they performed a clustering method (partitioning around medoids with the Gower's dissimilarity algorithm). The molecular analyses supported the monophyly of the *G. meyeniana* complex, but did not support segregation into varieties. The morphological analyses supported recognition of three varieties that can be easily distinguished by three simple

Species of Gomphrena

Gomphrena globosa ***Gomphrena meyeniana***

Seed Propagation of Gomphrena

Gomphrena Seeds **Seedling in Plug Tray**

Propagation of Seedlings **Flowering from Seedlings**

characters: the presence of leaves on the flowering shoot, the habit of the flowering shoot, and the pilosity of the sepals. The varieties of *G. meyeniana* accepted are *G. meyeniana* var. *meyeniana*, *G. meyeniana* var. *caulescens*, and *G. meyeniana* var. *flaccida*. A dichotomous key to identify the intra-specific taxa was presented and illustrated. *Gomphrena meyeniana* var. *tucumanensis* and *G. cmeyeniana* var. *conwayi* are synonymised with *G. meyeniana var. caulescens* and *G. meyeniana var. meyeniana*, respectively (Bena *et al.*, 2020).

It is an outcrossing species that is pollinated by butterflies, bees and other insects. Floral volatiles likely play a significant role in the reproductive success of the plant by promoting the attraction of pollinators (Ashwini *et al.*, 2019).

3.3.0 Cultivars

There are many different cultivars of globe amaranth (Mahr, 2019). The variety 'Bicolor Rose' has pink and white blooms, 'Buddy Purple' is short (eight to 12 inches tall) and has intensely purple flowers, 'Gnome Series' has flowers in pink, purple and white, and grows only six inches tall, 'QIS Series' has flowers in white, purple, lilac, carmine, rose and red that are especially good as cut flowers, 'Strawberry Fields' (a *G. haageana* type) produces large, brilliant red flowers on 20inch tall plants.

There are many patents on Gomphrena varieties. The following are some of the patented varieties of Gomphrena.

SAKGOMO05 (Mori *et al.*, 2015a)

A Gomphrena variety distinguished by having lavender flowers with white tips and a spreading plant growth habit. SAKGOMO05 originated from a controlled cross in July 1994 in Misato, Japan between the proprietary female Gomphera breeding line 94A-417 having a white flower color and upright growth habit and the proprietary male line 94T-414A having a smoky pink flower color and a creeping growth habit, producing seventeen F seeds. In April 1995, the seventeen F seeds were sowed and seven plants were cultivated. In July 1995, three plants were then chosen having a purple flower color and upright habit. Between July 1996 and February 2009, seeds from the threechosen plants were cross-pollinated and seeds were obtained. In February 2009, 2000 seeds were sowed and 1000 plantswere cultivatedbased on whether plants had pink flowers, lavender flowers, and lavender flowers with white tips, andwhether plants had an upright plant growth habit or a spreading plant growth habit. In April 2009, a single plant wasselected from this group having lavender flowers with whitetips and spreading plant growth habit.

SAKGOMO03 (Morin and Tanaka, 2015b)

A Gomphrena variety distinguished by having white flowers and a spreading plant growth habit. The present invention comprises of a new and distinct variety of Gomphrena, botanically known as *Gomphrena globosa*. SAKGOMO03 originated from a controlled cross during 1998 in Misato, Japan between the proprietary female Gomphera breeding line 95G-11A-6E-1 (unpatented) having white flowers and a semi-creeping plant growth habit and the proprietary male Gomphera breeding line 95G-1 1A-2A 3 (unpatented) having magenta flowers and a semi-creeping plant growth habit.Thirty F_1 seeds were obtained and evaluated from the original cross.

From these hybrid seedlings selection was made for white flowers and spreading plant type.

SAKGOMOO4 (Mori and Tanaka, 2015c)

A Gomphrena plant particularly distinguished by having apurple flowers and a spreading plant growth habit. SAKGOMOO4 originated from a controlled cross in July1998 in Misato, Japan between the proprietary female Gomphera breeding line 95G-11A-6E-1 (unpatented) having white flowers and a semi-creeping plant growth habit and the proprietary male Gomphera breeding line 95G-1 1A-2A 3 (unpatented) having magenta flowers and a semi-creepingplant growth habit.From this cross, a single plant having the spreading habit with purple flowers was selected.

GOMO70101 (Kitahara and Sakazaki, 2014)

A new and distinct cultivar of Gomphrena plant named GOMO70 101, characterized by its mounding plant habit;vigorous growth habit; freely branching habit; large pubes cent leaves; and large long-lasting inflorescences with numerous red purple-colored flowers. The new Gomphrena plant was selected as a single flowering plant from within the progeny of a cross-pollination of an unidentified selection of *Gomphrena macrocephala*, not patented, as the female, or seed, parent with an unidentified selection of *Gomphrena hybrida* as the male, or pollen, parent, in a controlled environment in Higashiomi, Shiga, Japan on Aug. 12, 2006.

"BALBOA" (*Gomphrena leontopodioides*) (Johnston *et al.*, 2011)

A new and distinct cultivar of Gomphrena plant named Balboa, characterized by its tall and upright plant habit; vigorous growth habit; freely branching habit; large grey green colored leaves; and purple-colored flowers with a white lanulose colored reverse giving a bi-color appearance to the flower.

4.0 Growth and Flowering

The growth and flowering of gomphrena is influenced by environmental factors. Hong and Jeong (1988) reported that high summer temperature affects flowering of gomphrena hybrid. Certain growth substances have also been used to regulate these processes. Soil application of PP333 (2 per cent granular paclobutrazol) or single spray of 10-80 ppm resulted in dwarf plants (Shanks, 1981).

In an experiment with *G. globosa* under natural conditions, exhibited enhanced stem length (160 per cent) by coaddition of the *Pseudomonas fluorescens* S_{272} culture broth and H 12 polysaccharide, suggesting enhanced adhesion of *P. Fluorescens* S_{272} cells by polysaccharide, useful for promoting plant growth by increasing antibiotic effect (Nakata *et al.*, 2000). Gomphrena usually starts flowering after about 75 to 80 days.

5.0 Propagation

Gomphrena is propagated by seeds which may be sown in raised nursery beds, or pans during January-March for flowering during summer and rainy season (April to September). Stefanisand Gortzig (1979) reported that seeds of *G. globosa* cv. Buddy were sown directly in pot and attractive plants were produced

Cultivation of Gomphrena

Cultivation of Gomphrena in Open

Gomphrena in Pot

Gomphrena in Garden

Mulching in Gomphrena

Protected Cultivation of Gomphrena

Gomphrena in a Public Garden

the year-round. Sowing of seeds at monthly interval from February-July, recorded long blooming period with late sowing (Hong and Jeong, 1988). According to Mahr (2009), Gomphrena seeds are to be sown indoors six to eight weeks before the expected last frost in the spring. Seeds are to be soaked for one or two days before planting to speed germination. Globe amaranth can also sometimes self-seeded. Thus, once globe amaranth is grown in the garden, they may continue to appear each spring. In view of the heterozygous nature, the flower color of these seedlings may be different than that of the parent plants (Mahr, 2019).

6.0 Soil and Climate

6.1.0 Soil

Gomphrena is a hardy annual and easy to grow. It grows well in any type of light and well-drained garden soil with moderate fertility status, cannot stand waterlogging and prefers open sunny situation. Gomphrena tolerate alkaline soil (Brickell, 1992). Salinity level of soil influences the physiology and mineral composition of gomphrena (Sharma *et al.*, 1989).

6.1.1 Salinity

Salinity in soil as well in irrigation water affects plants. With depletion of ground water level, salinity has become universal problem. Salicylic Acid (SA) is considered as one of the useful compounds for plants that plays an important role in plant resistance to environmental stresses such as salinity stress. For this purpose and to evaluate the effect of salicylic acid (SA) and salinity on morpho-physiological and qualitative characteristics of Gomphrena) *Gomphrena globosa*), an experiment with three levels of SA (0, 200, 400 ppm) and three concentrations of sodium chloride (0,150, 300 mM) was conducted (Kamali *et al.*, 2012). The interaction effect of salinity and SA showed that by increasing salinity stress up to 300 mM, stomotal conductance, flower number, shoot and root dry weight, chlorophyll index and RWC decreased 75, 96, 48, 48, 82 and 38 percentage respectively, and EL increased 99 per cent compared whit control. While spraying with SA reduced the rate of electrolyte leakage and improved other measured traits. Foliar spraying of salicylic acid leads to maintain of turgidity and leaf dimension that it is due to increasing the relative humidity content. Also by increasing the photosynthetic pigments and protection of them during stress conditions like salinity stress, it leads to improving the physiological traits of plants and increase the tolerance of plant during stress condition. The best concentration of salicylic acid for foliar spraying is 200 ppm that can be effective for improving the tolerance of plants to salinity.

6.2.0 Climate

As a tropical annual plant, blooms continuously throughout summer and early fall. It is heat tolerant, fairly drought resistant and tolerates poor soil, but grows best in full sun and regular moisture and once a favorite in British gardens. Though the flower yield and quality are primarily varietal characters, they are also greatly influenced by climatic factors. The climatic factors like photoperiod, temperature, relative humidity and also soil moisture influence both vegetative and reproductive phases of the plant, ultimately leading to variation in the performance of genotypes (Ashwini *et al.*, 2019).

The effect of light and CO_2 on both the endogenous and I-aminocyclopropane-l-carboxylic acid (ACC)- dependent ethylene evolution from metabolically active detached leaves and leaf discs of *Gonphrena globosa* L. was reported (Grodzinski *et al.*,1983).Treatment with varying concentrations of ACC did not appear to inhibit photosynthesis, respiration, or stomatal behavior. In all treatments, more ethylene was released into a closed flask from ACC-treated tissue, but the pattern of ethylene release with respect to light/dark/CO_2 treatments was the same. Leaf tissue in the light with a source of CO_2 sufficient to maintain photosynthesis always generates 3 to 4 times more ethylene than tissue in the dark. Conversely, the lowest rate of ethylene release occurs when leaf tissue is illuminated and photosynthetic activity depletes the CO_2 to the compensation point. Ethylene release in the dark is also stimulated by CO, either added to the flask as bicarbonate or generated by dark respiration. Ethylene release Increases dramatically and in parallel with photosynthesis at increasing light intensities in this C_4 plant. Ethylene release appears dependent on CO_2 both in the lit and in the dark. Therefore, it is suggested that the important factor regulating the evolution of ethylene gas from leaves of Gomphrena may be CO_2 metabolism rather than light per se (Grodzinski *et al.*, 1983).

7.0 Cultivation

7.1.0 Transplanting and Spacing

When seedlings are 3 to 4 cm in height, they are planted in bed at a distance of 20 to 30 cm. To obtain compact beds, dwarf cultivars are transplanted 18 to 20 cm apart, each way. It does best in full sun and prefers moist soil, but tolerates dry conditions once established (Mahr, 2009). If plants are started indoors from seed, seedlings should be hardened off prior to transplanting. They are planted at a spacing of approximately 30 cm to 45 cm in the garden. Because the hairy foliage of globe amaranth easily collects soil, mulching around the plants help for a cleaner appearance (Mahr, 2009). Hemalatha *et al.* (2014) recommended a spacing of 30x30 cm under Karnataka conditions.

7.2.0 Irrigation

Seedlings need watering immediately after transplanting. Water should be applied to ensure continuous supply of moisture. Normally, plants are irrigated by flooding once a week, during dry spell. Foliar application of hydrogen peroxide on plants can improve shoot and root dry weight and alleviate adverse effects of water deficiency (Goldani and Kamali, 2011). With the delay in irrigation, stomatal conductance, chlorophyll index and root volume decreased significantly; the lowest of this characteristics was in the irrigation interval after 10 days. In control treatment (interval 4 day irrigation) with increasing hydrogen peroxide to 2.5 mM, shoot dry weight and total root length increased by 20 and 91 per cent, respectively and with increasing hydrogen peroxide to 5 mM, chlorophyll index was increased 30.8 per cent compared to 0 mM Hydrogen peroxide. Foliar application of hydrogen peroxide decreased drought stress because of water deficiency (Goldani and kamali, 2011).

7.3.0 Manuring and Fertilization

It is not a heavy feeder and responds well to a small amount of fertilizer. Well-rotten farmyard manure at the rate of 10 kg/m^2 is applied at the time of bed preparation. Smith and Treaster (1991) recorded good growth and flowering with the application of municipal sludge compost to a depth of 1.5 - 2 inch deep at the time of planting. NPK fertilizer @ 15-20 g/m^2 should be applied for improvement in growth and flowering.

A study was conducted in Romania to evaluate the effect of the application of vinasse (byproduct obtained after the evaporation of residual water from beer yeast factory) on the growth and morpho-anatomy of *Gomphrena globosa* as well as on the biotic and enzymatic potential of the soil (Chelariu *et al.*, 2009). Vinasse was applied at concentrations of 3, 5 and 10 t/ha. Vinasse application stimulated the growth and flowering of *G. globosa*. Vinasse also increased soil reaction to moderate alkalinity.

Sulfur(S) regarded as the fourth key element is mainly taken by the plant roots. However, some plants can also absorb atmospheric sulfides, which may be of great importance for ameliorating the environment and for farming as a green organic S fertilizer used to balance insufficient soil S content for intensive cultivation. H_2S and mainly SO_2 are emitted to air as a result of the rapid industrialized and economic development (Wang *et al.*, 2009). Globe amaranth (*Gomphrena globosa* L.) might be one of the plants that can use atmospheric sulfides for its growth. Therefore, the effects of sulphate deprivation from root on its growth, S status and other elements concentration under hydroponic culture were explored firstly. Based on measurements of plant growth, biomass, nitrogen (N), phosphorus (P), potassium (K), calcium (Ca), magnesium (Mg), S, iron (Fe), manganese (Mn), copper (Cu), zinc (Zn), boron (B), and molybdenum (Mo) concentration, the results showed that S concentration in flower, shoot and root of plant without root sulfate supplied was increased with plant growth and development, symptoms of S deficiency disappeared and other elements concentration in plant tended to be nearly the same as the root sulfate-supplied plants. The interesting results might imply that globe amaranth may be able to live on the atmospheric sulfides as sulfur source (Wang *et al.*, 2009).

Barnes *et al.* (2014) reported that 'Las Vegas Purple' globe amaranth plants were grown in silica-sand culture to induce and photograph symptoms of nutritional disorders. Plants were grown with a complete modified Hoagland's all-nitrate solution (macronutrient concentrations in mM) 15 NO_3-N, 1.0 PO_4-P, 6.0 K, 5.0 Ca, 2.0 Mg, and 2.0 SO4-S plus (micronutrient concentrations in μM) 72 Fe, 18 Mn, 3 Cu, 3 Zn, 45 B, and 0.1 Mo. Nutrient-deficiency treatments were induced with a complete nutrient formula minus one of the nutrients. Reagent grade chemicals and deionized water of 18-megohms purity were used to formulate treatment solutions. Boron toxicity was also induced by increasing the element 10× higher than the complete nutrient formula. Plants were monitored daily to document and photograph sequential series of symptoms as they developed. Typical symptomology of nutrient disorders and critical tissue concentrations are presented. Out of thirteen treatments, eleven exhibited symptomology. Symptoms of N, P, and S deficiencies and B toxicity manifested early; therefore, these disorders may be more likely problems encountered by growers. Based on this research, growers will be able to recognize and avoid nutritional disorders of globe amaranth plants grown in peat-based substrates.

7.4.0 Pinching

Among the various horticultural techniques pinching and balanced nutrients are the main factors which govern the quality and yield of Gomphrena. Pinching the terminal portion of shoots is removed early, emergence of side branches starts and more number of flowers are produced. Pinching at 15 days after transplanting along with Vermiwash @ 1:5 dilution foliar spray was significantly increased the growth, flower yield parameters *viz.*, plant height, number of primary branches, number of productive branches, stem girth, inter nodal length, number of leaves, chlorophyll content index and dry matter content, days taken for first flowering, number of flowers per plant, flower yield plant, flower yield per plotand flower yield/ha. In general, pinching at 15 days after transplanting along with Vermiwash @ 1:5 dilution foliar spray could adjudged as the best treatment in performance of Gomphrena (Sendhil Nathan *et al.*, 2019).

Pinching has been found effective in regulating and enhancing flower production and also to encourage branching. Rajasekaran *et al.* (1983) reported that the plants pinched 15 days after transplanting produced large number of flowers (602.5/plant) and of bigger size (2.85 cm) while the non-pinched plants produced flowers (518.5/plant) of 1.5 cm diameter. The stalk length was, however, reduced in pinched plants.

7.5.0 Weed Control

Weeds become a problem at the early stage of cultivation of gomphrena. If weeds are not removed in time growth and productivity will be markedly reduced. Manual weeding is practised. Gomphrena being a shallow-rooted crop, deep interculture should be avoided.

8.0 Diseases and Pests

8.1.0 Diseases

Fungal Diseases

Only few diseases occur on this host. Leaf-spot is caused by *Cercospora gomphrenae*. Rane and Wick (1990) reported *Alternaria* (*A. gomphrenae*) to be a serious problem in commercial cultivation of gomphrena. Benomyl, iprodione and mancozeb significantly control the disease.

Leaf spot on *G. globosa* cv. Buddy and Globosa caused by *Nimbya gomphrenae* was first reported from Taiwan. A temperature range of 16-32°C was relevant for infection of pathogen and temperature as high as 32°C was favourable for the development of the disease. Spraying with 10 ppm pyrifenox was effective in controlling the malady (Chou and Wu, 2000).

Diseases and Pests of Gomphrena

Alternaria Leaf Spot

Ring Spot

Beetle Attack

Aphids Infestation

Harvesting and Packaging

Freshly Harvested Flowers

Packaging of Flowers

Viral Diseases

Gomphrena is also susceptible to several viral diseases. The affected plants should be removed as soon as they are detected. Subbarayudu and Wilcoxson (1967) reported that mannose can inhibit the infection of red clover vein mosaic virus. The plants are very susceptible to tomato bushy stunt virus infection when grown at high temperature and long photoperiod as reported by Redolfi *et al.* (1977). Naqvi and Mahmood (1975) observed that gomphrena is a symptomless carrier of bottle gourd mosaic virus. Srivastava *et al.* (1977) also reported that *G. globosa* is susceptible to yellowing of veins disease, causing stunted growth and deterioration in bloom quality. It is easily transmitted by white fly (*Bemisia tabaci*).

8.2.0 Pests

Cutworms, aphids and beetles attack the plants at any stage of its growth. Spraying with suitable insecticide controls these pests. Frommhold (1976) observed that spraying with silicone oil NMI-500 or silicone oil NMI-15 reduced the attack of aphids.

9.0 Harvesting and Postharvest Management

9.1.0 Harvesting

Flower heads of gomphrena are harvested when they attain full size. It is harvested with stems for table decoration or dry flower industry. Few leaves on the flowers stalk should be retained for longer life of the flowers.

Rajasekaran *et al.* (1983) obtained 602.5 flowers/plant from pinched plants while non-pinched plants produced 518.5 flowers. *G. globosa,* under frameless plastic covers, produced a yield 1.5 or 1.7 times higher than in the open (Levai and Farkas, 2000).

9.2.0 Postharvest Management

Good grading and bunching can enhance the quality of flowers and fetch better price than flowers of equal quality but of poor grade. Stalk length of flower diameter and faded colours are to be considered during grading. The stem ends, after bunching, should be trimmed and placed in a preservative solution. Though flower heads remain fresh for long time, use of floral preservatives prolongs vase life.

10.0 Biotechnology

10.1.0 Micropropagation

Work on tissue culture and micropropagation have been poorly reported in Gomphrena. Ruiz and Valadez (1985) obtained callus from hypocotyl tissue of gomphrena on MS medium with 3, 5 or 7 per cent sucrose and various concentrations of NAA, BA and 2,4-D. Media with 10^{-5} M NAA + 10^{-8} M BA induced shoot growth but few or no roots. Sucrose at 5 or 7 per cent was required for shoot and root growth.

Leaf and stem segments of *G. officinalis* from aseptically grown seedlings were cul-tured on gelled MS medium, supplemented with BA alone or combined with

NAA (Mercier *et al.*, 1992). Callus production was obtained after 10 to 15 days of culture and transferred to fresh medium every 30 days. The combinations of BA (5.0 or 10.0 mg/l) and NAA (0.1 mg/l) were the best for shoot regeneration. The medium containing 10 mg/l IBA was optimum for root formation in shoots.

10.2 Molecular Markers

Very few reports on DNA markers is available in *Gomphrena globosa.* RAPD analyses were carried out with 20 decamer primers from Kit A of Operon Technology (Fatinah *et al.*, 2012). DNA was amplified using master cycler gradient Eppendorf with 35 cycles. RAPD products were separated on 1,5 per cent agarose gels and detected by staining with ethidium bromide. There were 374 bands generated in 18 random primers. The number of monomorphic bands, polymorphic bands, and the percentage of polymorphism were 21 bands, 353 bands, and 94,38 per cent respectively. The high number and percentage of polymorphic bands revealed genomic DNA variation. This variation is in accordance with phenotypic variation detected in this experiment. Therefore, it can be concluded that, based on DNA polymorphism detected by RAPD, Amaranth family can be classified into two sub families namely Amaranthoideae and Gomphrenoideae (Fatinah *et al.*, 2012).

Polymorphism was detected by using RAPD-PCR method in a study on the genetic relationships among 6 species belonging to 4 genera of Amaranthaceae family collected from Baghdad province, namely, *Amaranthusretroflexus* L., *A. Spinosus* L., *A. Albus* L., *Alternanthera sessilis* (L.) R.Br. ex DC. *Gomphrena globosa* L. and *Iresineherbstii* Hook.exLindl (Jawad, 2017). Two main groups, the first one which separated at the similarity value of (0.54) included *Amaranthus retroflexus* and *A. spinosus*, while the second group included the other *Amaranthus* species and contained two subgroups, the first subgroup included *A. albus*, and the second subgroup that appeared related correlation at similarity range of (0.41) contained 2 species; *Alternathera sessile* and *Gomphrena globosa* (Jawad, 2017).

11.0 References

Anonymous. (2020) https: //en.wikipedia.org/wiki/Gomphrena_globosa.

Ashwini, A.R.K., Patil S. and Chethan Kumar, S. (2019) *J. Pharmacogn. Phytochem.*, **8(4)** : 541-545.

Barnes, J., Whipker, B., McCall, I. and Frantz, J. (2014) AGRIS, **982** : 69-74.

Bena, M.J., Ortuño, T., and Acosta, J.M. (2020) *Syst. Bot.*, **45**(2) : 294-305.

Brickell, C. (1992) *Encyclopedia of Gardening*. Dorling Kindersley, London, pp. 180.

Brickell, C. (1996) *Encylopedia of Garding*, Dorling Kindersley, London, pp. 478-479.

Buschi, C.A. Pomilio, A.B. (1982) *J. Nat. Prod.*, **45(**5) : 557–559.

Chelariu, E.L., Draghia L., G. Bireescu, L. Bireescu and M. Brânzã. (2009) : *Lucrãri'tiinifice, Universitatea de 'tiine Agricole 'iMedicinãVeterinarã "Ion Ionescu de la Brad" Ia'i, SeriaHorticulturã*, **52** : 615-620.

Chandana, E. and Kumar, P. (2018) *WorldJ. Pharm. Pharm. Sci.*, **7**(8), 386-394.

Chou, H.H. and Wu, W.S. (2000) *Plant Pathol. Bull.*, **9** : 23-28.

Codero, M. (1988) *Ciencia y Tecnicaen la Agricultura Hortalizas*, **7** : 7-16.

Fatinah, A.A., Arumingtyas, E.L.andMastuti, R. (2012) *J. Trop. Life Sci.*,**2** : 81-86.

Febles, E.V. (2017) https: //www.cabi.org/isc/datasheet/25617#tosummary OfInvasiveness.

Frommhold, I. (1976) *Archiv fur Gartenbau*, **24** : 319-326.

Goldani M. and Kamali, M. (2011) *Plant Prod. Technol. (Agric. Res.)*, **10** : 65-81.

Grodzinski, B., Boesel, I. and Horton, R.F. (1983) *Plant Physiol.*, **71** : 588-593.

Hemalatha, R., Gopinath, G. and Srikanth, L.G. (2014) *Environ. Ecol.*, **32** : 881-84.

Herold, A., Lewis, D.H. and Walker, D.A. (1976) *New Phytologist.*, **76**(3): 397–407. doi: 10.1111/j.1469-8137.1976.tb01475.x. ISSN 1469-8137.

Heuer, S., Wray, V., Metzger, J.W. and Strack, D. (1992) Phytochem., **31** : 1801–1807.

Hong, Y.P. and Jeong, J.H. (1988) *Res. Report Rural Development Administration, Horticulture*, Korea Republic, **30** : 104-110.

Jawad, M.M. (2017) *Proc. 8th International Conference on Agricultural, Environment, Biology and Medical Sciences*, Dec. 21-22, 2017 Dubai (UAE). pp.27-33.

Jiang, Y., Zhao, N., Wang, F. and Chen F. and Feng (2011) *J. Am. Soc. Hortic. Sci.*, **136**(1) : 16–22. doi: 10.21273/JASHS.136.1.16. ISSN 0003-1062.

Johnston, M., Harrison, D. and Joyce, D. (2011) US Patent No. US PP22.263 P2 dated Nov. 22, 2011.

Kamali, M., Kharrazi S.M., Selahvarzi Y. and Tehranifar A. (2012) *J. Hortic. Sci. (Agric. Sci. Technol.)*., **26** : 104-112.

Kitahara, A. And Sakazaki, U. (2014) US patent No. US 2014/0130218 P1 dated May, 8, 2014.

Levai, P. and Farkas, Z. (2000) *Inter. J. Hort. Sci.*, **6** : 148-150.

Mahr, S. (2009) https: //pddc.wisc.edu/2015/07/23/globe-amaranth/

Mercier, H., Vieira, C.C.J. and Figuciredo-Ribeiro, R.C.L. (1992) *Plant Cell Tissue Organ Cult.*, **28** : 249-254.

Minale, L., Piattelli, M., De Stefano, S, (1967) *Phytochem.*, **6** : 703–709.

Mori, K. and Tanaka, M. (2015a) US Patent No. US 2015/0208567 P1. dated July 23, 2015.

Mori, K. and Tanaka, M. (2015b) US Patent No. US PP25, 903 P3. dated September 22, 2015.

Mori, K. and Tanaka, M. (2015c) US Patent No. US PP25, 927 P3. dated September. 22, 2015.

Nakata, K., Harada, N., Sumitoma, K. and Yoneda, K. (2000) *Biosci. Biotechnol. Biochem.*, **64** : 459-465.

Naqvi, S.Q.A. and Mahmood, K. (1975) *Curr. Sci.*, **44** : 104.

Pomilio, A.B., Buschi, C.A, Tomes, C.N. and Vialeb, A.A. (1992) *J. Ethnopharmacol.*, **36** : 155-161.

Pomilio, A.B., Buschi, C.A., Sola, G.A.R. and Rumi, L.S. (1991) *Fitoterapi*, **62** : 511-514.

Rajasekaran, L.P., Shanmugavelu, K.G. and Nambisan, K.M.P. (1983) *South Indian Hort.*, **31** : 248-249.

Rane, K.K. and Wick, R.L. (1990) *Canadian J. Plant Pathol.*, **12** : 442-444.

Redolfi, P., Pennazio, S., Appiano, A. and Martin, C. (1977) *Phytopath. Z.*, **90** : 43-50.

Roriz, C.L., Barreira, J.C.M., Morales, P. (2018) *LWT - Food Sci. Technol.*, **92** : 101-107.

Roriz, C.L., Barros, L., Prieto, M.A. (2017a) *Food Chem.*, **229** : 223-234.

Roriz, C.L., Barros, L., Prieto, M.A. (2017b) *Ind. Crops Prod.*, **105** : 29-40.

Ruiz, A.Q. and Valadez, J.M. (1985) *J. Nat. Prod.*, **48** : 976-980.

Sendhil Nathan, R., Vijay R.B., Sureshkumar R. and Rajkumar, M. (2019) *Plant Arch.*, **19** (Suppl.): 1002 -1005.

Shanks, J.B. (1981) *Proc. Seventh Annual Meeting, Plant Growth Regulator Working Group*, Longmost, Colorado, USA, Great Western Sugar Co. pp. 46-51.

Sharma, S.S., Yamdagni, R. and Arora, R.K. (1989) *South Indian Hort.*, **37** : 337-340.

Silva, R.L., Valentão, P., Joana, F. (2012) *Food Chem.*, **135**(2): 756– 763. doi: 10.1016/j.foodchem.2012.05.015. PMID 22868155.

Smith, E.M. and Treaster, S.A. (1991) *OARDC Special Circular - Ohio Agri. res. Development Centre*, No. 137.

Srivastava, K.M., Singh, B.P., DwadashShreni, V.C. and Shrivastava, B.N. (1977) *Plant Dis. Reptr.*, **61** : 550-554.

Stefanis, J.P. and Gortzig, C.F. (1979) *New York State Flower Industries Bulletin*, **102** : 7-9.

Subbarayudu, S. and Wilcoxson, R.D. (1967) *Phytopath.*, **57** : 1292-1295.

Tarnam, Y.A., Ilyas, M.H.M. and Begum, T.N. (2014) *Int. J. Pharma Res. Rev.*, **3(1)** : 58-66.

Wang, M.Y., Wu, L.H. and Zhang, J. (2009) *Plant Soil Environ.*, **55**(11) : 484–493.

17

Cosmos

Sapna Panwar, Reeta Bhatia Dey, B.K. Tejukumar and C. Vaishali

1.0 Introduction

Cosmos is an annual flowering ornamental with colorful daisy-like flowers that are produced on long slender stems. The genus '*Cosmos*' has derived its name from the Greek word '*kosmos*', which means 'beauty' (Smith, 1972), 'orderly', 'beautiful', and 'ornamental'. The cosmos flowers often symbolize order and harmony, mainly due to their harmoniously flower petals. It is grown for diverse purposes like annual, border plant, pot plant, cut flower *etc.*

2.0 Importance and Uses

2.1.0 Aesthetic Uses

It is annual flowering plant mainly grown for landscaping purposes like bedding and container gardening. Cosmos is a wonderful accent and border plant preferred for mixed flower gardens (http://www.public.asu.edu/). It is one of the choicest plant for herbaceous borders. Cosmos flowers are capable of adding beauty to any place and can greatly enhances the festive mood with its vibrant colors. *Cosmos bipinnatus* possesses long stems that make it good for cut flower production. There are also reports of using cosmos foliage in flower arrangements. Cosmos possesses pleasant aroma which attracts butterflies and beneficial insects which makes it excellent preference for pollinator gardens.

2.2.0 Therapeutic and Other Uses

Besides ornamental uses, it also possesses nematocidal properties. The flower extract of *Cosmos* has antioxidative, and antigenotoxic activities (Jang *et al.*, 2008). The cosmos flowers are used as an eco-friendly dye in woolen industry (Kale *et al.*, 2006). The flower essence of *Cosmos bipinnatus* is known for its healing properties. It is also being used in traditional medicines, *C. bipinnatus* flowers are used in treatment of jaundice, intermittent fever and splenomegaly (Jang *et al.*, 2008). South African Tribe 'Basotho' traditionally uses *C. bipinnatus* for headaches and stomach disorders, whereas, another Afrikaans ethnic group utilizes it to manage head lice and bed bugs (Ashafa and Olajuyigbe, 2014). Leaf extracts of *C. bipinnatus* exhibit larvicidal and pupicidal activities against aquatic *Culex quinquefasciatus* larvae and pupa, (Modise and Ashafa, 2016).

3.0 Origin and History

The genus *Cosmos* belongs to the family Asteraceae and is native of Mexico. It is believed that Mexican priests used to grow this flower crop and given the name cosmos because of the evenly placed and orderly arrangement of petals. It is considered as an invasive plant in many parts of the world. It has naturalized across North America, South America, the West Indies, Italy, Australia, and Asia. Cosmos is the flower for the birth month of October and symbolizes balance and harmony (Hansen and Drost, 2020). *C. sulphureus* was first described by Cavanilles (1791) from plants cultivated in the Real Jardín Botánico de Madrid.

Uses of Cosmos

Herbaceous Borders
(https://sallynex.com)

Pot Plant
(https://www.healthbenefitstimes.com)

Cut Flower
(https://www.richardjacksonsgarden.co.uk/)

Flower Essence
(https://www.powerfloweressences.com)

Nectar Plant in Butterfly Gardens
(https://www.joyfulbutterfly.com)

Bedding Plant
(https://www.lucasgreenhouses.com/)

Cosmos Planted in Window Boxes
(https://in.pinterest.com/pin/131308145362763444/)

Cosmos Planted Along the Fence
(https://in.pinterest.com/pin/517069600950170841/)

Cosmos Planted along the Walkways
(https://i.pinimg.com/nals/1f/62/43/1f6243120a439eae3046dd567ee77940.jpg)

4.0 Taxonomy

The genus *Cosmos* consists of 36 species, of which 28 species are endemic to Mexico. The important species under genus *Cosmos* includes *Cosmos sulphureus, Cosmos bipinnatus,* and *Cosmos diversifolius*. *Cosmos atrosanguineus* (Chocolate cosmos), another important species, is often classified ambiguously as *Bidens atrosanguinea*. Oku *et al.* (2008) established phylogenetic relationship of chocolate cosmos among the genera *Cosmos, Bidens* and *Dahlia* based on plastid subtype identity (PSID) sequences. It was revealed that chocolate cosmos has a closer relationship with *Cosmos* than with *Bidens* or *Dahlia*.

5.0 Species and Cultivars

5.1.0 Species

The most common ornamental species which are commercialized as garden plants in different countries of Europe and Asia includes *Cosmos sulphureus, Cosmos bipinnatus,* and *Cosmos diversifolius* (Rodríguez-Bernal *et al.,* 2013). Although there are many important species, some of these are *Cosmos pacificus, Cosmos purpureus, Cosmos crithmifolius, Cosmos parviflorus etc.*, the description of important species is as follows.

Cosmos bipinnatus

Commonly known as 'Mexican aster', 'garden cosmos' or 'sensation cosmos'. It is an annual herb with plant height ranging from 1-2 meter. Leaves are simple, finely cut and thread-like and are pinnately cut into deep lobes. Stems are thin and weak in nature. It produces showy flowers in capitula with small disc in spectacular colour range including red, lavender, pink, white, violet and rose (Gilman and Howe, 1999).

Cosmos sulphureus

Commonly known as 'yellow cosmos' or 'sulphur cosmos'. Plant height ranges from 1.5-2 m. It has foliage that is not as finely divided as that of *Cosmos bipinnatus*. Species name *sulphureus* is in reference to the orange-yellow colours of the plant's flower (Smith, 1972). It produces very attractive heads of golden yellow to red-orange.

Cosmos atrosanguineus

Commonly known as 'chocolate cosmos'. A unique species that usually produces dark red or maroon-coloured flowers with a strong scent of chocolate. The leaves of the species resembles to that of miniature dahlias (Sherff, 1932).

5.2.0 Cultivars

Some of the important cosmos cultivars belonging to three important species are as follows

Cosmos bipinnatus: Candy Stripe Mix, Daydream, Gazebo Mix, Seashells Mix, Sonata Mix, Sonata White, Sonata Pink Blush, Sonata Carmine, Sonata Carmine

Important Species of Cosmos

Cosmos bipinnatus

Cosmos sulphureus

Cosmos atrosanguineus

***Source*: https://pfaf.org/; https://www.healthbenefitstimes.com; https://www.plant-world-seeds.com/**

Important Cultivars of Cosmos

Crest Red Cosmos

Day Dream Cosmos

Gloria Cosmos

Candy Stripe Cosmos

Seashell Cosmos

Psycho White Cosmos

Important Cultivars of Cosmos

Tetra Versailles Red Cosmos

Yellow Cosmos Klondyke Mix

Picotte Cosmos

Source: https://aggie-horticulture.tamu.edu/

Black Magic

Dark Secret

Spellbound

Source: https://www.rhs.org.uk/

Propagation of Cosmos

Seeds

Germinationn of Seeds

New Emerging Seedlings

Source: https://silverfallsseed.com/product/cosmos-bipinnatus

Pinching of Cosmos

***Source*: https://forum.gardenersworld.com/discussion/1012966/cosmos-pinching/p2**

Premium, Sonata White Premium, Dazzler, Radiance, Sensation, Double Click Mix, Pied Piper Red, Picotee, *etc.*

Cosmos sulphureus: The common varieties are Lady Bird Dwarf, Klondike, Cosmic Yellow, Cosmic Orange, Cosmic Red, Limara Lemon, Red Crest, Lemon Crest, Sunset, Zluty, Carpet Mix, *etc.*

Cosmos atrosanguineus: Black Beauty, Black Magic, Dark Secret, Eclipse, Mystique, New Choco, Pinot Noir, Spellbound, *etc.*

6.0 Genetics and Breeding

6.1.0 Genetics

The basic chromosome number for the genus *Cosmos* is recognized as x=12. *Cosmos bipinnatus* and *C. sulphureus* are diploid (2n=2x=24), while *Cosmos atrosanguineus* is tetraploid (2n=4x=48) in nature (Oku *et al.*, 2008). Castro-Castro *et al.* (2017) reported that *Cosmos nitidus* and *C. ramirezianus* are diploid species (2n=2x=24) while, *C. pseudoperfoliatus* is tetraploid (2n=4x=48).

6.2.0 Breeding

6.2.1 Hybridization

Inter-specific hybridization between *Cosmos atrosanguineus* (female parent) with *C. sulphureus* (male parent) was carried out by Oku *et al.* (2008). The interspecific hybrids were successfully developed following *in vitro* embryo rescue. The resultant hybrid was found to be intermediate between *C. atrosanguineus* and *C. sulphureus* and it produced tuberous root similar to *C. atrosanguineus*.

The popular varieties developed though interspecific hybridization includes: 'Chocamocha' (interspecific hybrid between *C. atrosanguineus* and an unknown species) and 'Strawberry Choco Sanse 41' (interspecific hybrid between *C. atrosanguineus* and *C. sulphureus*. Similarly, F_1 hybrids with reddish orange flower colour were developed as a result of interspecific hybridization involving *C. sulphureus* and *C. caudatus*. In F_5 generation, bright pink flowered individuals were obtained and their offspring bred true to type for the colour (Samata *et al.*, 1977).

6.2.2 Mutation Breeding

The irradiation of dry seeds of *C. bipinnatus* with acute dose of X-ray (15 kR) resulted in mutants with varied flower forms (Rana *et al.*, 1963). It also revealed that dominant mutation is responsible for doubleness. Somatic mutations were induced in *C. bipinnatus* through physical mutagens like X-ray and gamma rays (Gupta and Samata, 1967). Bose *et al.* (1969) irradiated the dry seeds of *C. bipinnatus* with 10 and 20 kR of X- rays. The plants with bifurcation and fasciation were observed in M_1 and M_2 generations. Moreover, it also showed dwarfness, varied flower form, size and colours. Kaicker *et al.* (1969) utilized gamma rays and EMS for induction of mutations in *C. sulfureus*. Mutants with additional number of ray florets, additional petaloid segment from the base of the ray florets and convertion of flat shape ray florets into tubular shape ray florets were obtained. Lakshmi *et al.* (1982) observed one spontaneous mutation in cosmos. Mutant having flower head but without ray florets was identified, showed drastic reduction in capitulum diameter, size of the floret of the outermost whorl, the disc florets of mutants were similar to those of the normal plants.

6.2.3 Polyploidy

Chhonkar *et al.* (1957) carried out experiment on induction of polyploidy using colchicine (0.01 to 0.5 per cent) from 2 to 10 hours. The induced tetraploids showed higher number of petals (16) in some flowers as compared to control (8), also petals were broader along with fringed margins. However, the colchicine treatment resulted in reduced stem elongation. Verma *et al.* (2017) successfully induced autotetraploids in *C. sulphureus* following colchicine treatment of the seedlings. Young seedlings were treated with different concentrations of colchicine (0.15, 0.2 per cent) for different durations using the cotton-swab method.

7.0 Growth and Flowering

7.1.0 Environmental Factors

7.1.1 Temperature

Cosmos is a heat-loving plant and perform best above 16°C. The growth and flowering are inhibited when temperature falls below 13°C. *C. sulphureus* is highly sensitive to temperature than *C. bipinnatus* and flower at temperature range of 21- 29°C (Post, 1949). High temperature regimes lead to increase in plant height and reduction in flower size. Under long day conditions, flowering do not occur regardless of temperature (20 or 13°C). However, it has been observed that flowering occurred rapidly under short days at 20°C, but no flowering was observed at 13°C. When plants were initially exposed to short days conditions at 13°C for 30 days and then 20°C, flowering was delayed but it resulted in more flower number and longer stems. Whereas, when short days were given at 20°C for 30 days, then 13°C, early flowering was observed than the 13/20°C regime, but with less flower number and short stems (Biebel, 1936).

7.1.2 Light

Cosmos grows well in full sunlight. According to photoperiod requirement, cosmos is considered as a facultative short-day plant (SDP) (Dole and Wilkins, 1999) which means it will flower best when day length is 14 hours or less. Under long day conditions (over 14 hours of light per day) slow blooming and decreased flower production were reported. It has been observed that photoperiod effect is more on young plants as compared to mature plants which flowers regardless of photoperiod. Based on the photoperiod response, *C. bipinnatus* is classified as a facultative SDP, whereas, *C. sulphureus* seems to be an obligate SDP.

7.1.3 Growth Regulators

Growth regulators play an important role in overall growth and development by modifying or regulating physiological processes in plants. Application of B-Nine (daminozide) aids in reduction of internodal length. Moreover, it is reported that if B-Nine applied at the time of flower peduncle elongation, it prevents inflorescence drooping (Samata *et al.*, 1974). No height control measures are required for cut flower production in the field (Dole and Wilkins, 1999).

8.0 Propagation

8.1.0 Seed Propagation

The propagation of cosmos is primarily by seeds. The seeds can be sown directly or seedlings can be raised in nursery. It is seeded at 46.5 g/100 m^2 (Armitage, 1993). Seedlings should be provided with long days until it develops six pairs of leaves. Deep sowing of seeds should be avoided as it can drastically reduce the germination percentage. The temperature requirement for seed germination is presented in Table 1.

Table 1: Temperature Requirement for Seed Germination of Cosmos

Temperature (°C)	*Days Required for Germination*	*Reference*
18°C	8-14	Armitage, (1993); Post, (1949)
21-22 °C	7-10	Armitage, (1993); Post, (1949)

Source: Dole and Wilkins (1999).

8.2.0 Micropropagation

The Chocolate cosmos is propagated through division of tuberous roots. Propagation of Chocolate cosmos by division of tuberous roots is a slow process as only four to six new tubers can be obtained per year from the mother tuber. Hosoki *et al.* (2003) successfully optimized the protocol for effective micropropagation of Chocolate cosmos via repeated nodal-culture and division of axillary shoots. A multiplication rate of 3.7 per 17- day culture period was achieved on MS medium containing 0.88 µM benzyladenine (BA).

9.0 Soil and Climate

9.1.0 Soil

It requires well-drained soils with pH range from neutral to slightly alkaline. Sandy loam soil is good for its cultivation although the crop can also be raised under low fertility soil regimes. Field with good drainage facilities is required so as to avoid the risk of root and stem rot. The field should be prepared well in advance for direct seeding. It is recommended to analyze the soil nutrients status before seeding. In case of greenhouse cultivation of cosmos, peatlite medium would be suitable.

9.2.0 Climate

Cosmos is a sun loving plant. In general, it grows well at a temperature above 16°C. However, temperature below 13°C inhibits its growth and flowering.

10.0 Cultivation

10.1.0 Planting Distance

Planting distance is an important factor affecting the growth of the plants. Optimum spacing provides proper aeration which encourages healthy plant growth. Close spacing will lead to development of diseases, especially stem cankers. The recommended planting distance is presented in Table 2.

Table 2: Planting Distance in Cosmos

Sl.No.	*Planting Distance*	*Reference*
1.	30 x 76 cm	Post, 1949
2.	23 x 30 cm	Armitage, 1993

Source: Dole and Wilkins (1999).

10.2.0 Support

Cosmos varieties which are tall in nature requires proper support so as to prevent bending of stem. Therefore, two layers of support nets are necessary for the field grown plants. However, no support is required for the plants grown in the greenhouses (Dole and Wilkins, 1999).

10.3.0 Irrigation

Young seedlings should be thoroughly irrigated once a week. Avoid over watering of the well-established plants. It is drought tolerant and can also tolerate drier conditions/summer heat. It is recommended to maintain the moist medium and allow proper drying cycle in between watering for good overall development of root system. Over watering results in less flowering and damaged flowers.

10.4.0 Fertilization

It is suggested to lightly fertilize the plants. Excess of fertilizers are not recommended. Application of high doses of nitrogenous fertilizers may result in

Cultivation of Cosmos

Cultivation in Open — Cultivation in Greenhouse — Planting in Pot

Diseases and Pests of Cosmos

Powdery Mildew — Bacterial Leaf Blight — Phyllody

Source: Kitazawa *et al.* (2014). *J. Gen. Plant Pathol.,* 80:499–503.

Harvesting and Marketing of Cosmos

Harvesting of Seeds — Harvesting of Flowers — Flowers in Market

Source: https://www.istockphoto.com/photo/cosmos-plant-for-sale-with-price-tag-gm487980740-73915867.

tall and spindly growth. Sometimes, it also results in scarce and few blooms. Hence, avoid planting in soils which are heavily amended with fertilizers and compost. The new transplants should be fertilized with low dose of nitrogen (50 to 75 ppm) and thereafter, a dose of 150 to 200 ppm nitrogen is recommended for well-established plants (Armitage, 1993).

11.0 Diseases and Pests

11.1.0 Diseases

11.1.1 Stem Blight or Canker

The causal organism is *Diaporthe stewarti*. This disease affects the mature plants. It is characterized by formation of brown spots on stem and branches, which finally girdle them. Further, the portions beyond these cankers wilt and die. Moreover, affected stems become weak and break down. The disease is not of common occurrence. For disease management, the affected plants should be removed and burnt.

11.1.2 Leaf Spot

It is caused by *Cercospora* and *Septoria*. The disease is characterized by formation of dark brown irregular spots on leaves which then coalesce giving a blighted appearance. The disease is of common occurrence during rainy weather. Spray Bavistin (0.1 per cent) or Dithane Z-78 (0.2 per cent) for its control.

11.1.3 Powdery Mildew

It is caused by *Erysiphe cichoracearum*. It appears as powdery growth on leaves and young branches. The disease is of common occurrence. Spray with Dinocap (0.02 per cent) or Carbendazim (0.05 per cent) is recommended.

11.1.4 Dieback

The causal organism is *Phoma herbarum*. The symptoms of forming cankers appeared on twigs and stems, which result in dieback symptoms and finally death of plants. Spray of Bavistin (0.2 per cent) is recommended.

11.1.4 Bacterial Wilt

The causal organism is *Pseudomonas solanacearum*. It causes wilting resulting in the death of plants. The infected plants should be rouged out and 3-4 years rotation may be followed.

11.1.6 Phyllody

Causal organism, *Phytoplasma aurantifolia*, is observed in *Cosmos bipinnatus* (Nikooei *et al.*, 2017)

11.2.0 Pests

11.2.1 Aphids

Aphids are very small insects which sucks leaf sap. The first damage is on leaves which ultimately leads to plant stunting. They also secrete honeydew which attracts ants and lead to the growth of sooty mold. Use of insecticidal soap is one of the control measure.

11.2.2 Japanese Beetles

These are large in size, brown, black, and green in colour. They are voracious eater. Control with row cover until plants approach flowering, is suggested.

11.2.3 Leafhoppers

They suck sap on the underside of leaves which causes stippling and browning. At flowering stage, use of insecticides is recommended.

12.0 Harvesting and Postharvest Management

The proper stage of harvesting of fresh flowers is when the petals on the first flower are just begin to open, but has not yet flattened. For prolonging the shelf life of flowers, it is recommended to harvest the flowers when the bud is showing colour. For cut flower purpose, cut the flowers along with stems. Vase life of cosmos flower is relatively short *i.e.* 4–6 days in water however, it can be extended with the addition of preservatives. Generally, storage is not recommended, but if required, a storage temperature of 2–4°C is recommended.

13.0 Biotechnology

Rodríguez-Bernal *et al.* (2013) studied the genetic diversity among 7 cosmos species based on random amplified polymorphic (RAPD) and inter-simple sequences repeats (ISSR) markers. RAPD markers revealed 91.7 per cent polymorphism and genetic diversity was 0.33, whereas ISSR markers revealed 65.6 per cent polymorphism and genetic diversity was 0.22. Dendrograms was generated by using unweighted pair group method with arithmetic mean which revealed 2 major clusters. The first cluster comprised of 3 species *viz Cosmos sulphureus, Cosmos pacificus*, and *Cosmos diversifolius*, while the second cluster comprised of 4 species *Cosmos purpureus, Cosmos crithmifolius, Cosmos bipinnatus*, and *Cosmos parviflorus.*

Li *et al.* (2019) compared the *LEAFY* (*LFY*) and *UNUSUAL FLORAL ORGANS* (*UFO*) like genes (*CbFLY* and *CbUFO*) which were isolated from wild-type and *gh* cosmos. It revealed that *CbUFO* may play significant role in inflorescence differentiation and flower initiation. However, the reduced expression of *CbUFO* in the *gh* cosmos could be associated with the flower-defective phenotypes. Expression analysis revealed that *CbUFO* may promote the conversion of inflorescence meristem into floral meristem in early stage of ray flower formation; however it does not act in later stages of growth.

14.0 References

Armitage, A.M. (1993) *In*: *Specialty Cut Flowers*. Varsity Press/Timber Press, Portland, Oregon. pp. 77–81.

Ashafa, A.O.T. and Olajuyigbe, F.O. (2014) *Iran. J. Pharm. Res.*, **13**: 1417-1423.

Biebel, J. (1936) *Proc. Am. Soc. Hortic. Sci.*, **34**: 635-643.

Bose, S., Mukherjee, D. and Bosu, S. (1969) *Indian J. Hort.*, **26**: 76-79.

Castro-Castro, A., Vargas-Amado, G. and Castañeda-Nava, J.J. (2017) *Acta. Bot. Mex.*, **118**: 41-52.

Cavanilles, A.J. (1791) *Icones et descriptions plantarum.*

Chhonkar, V. S., Khanna, A. N. and Sinha, R. B. (1957) *Indian J. Hort.*, **14**: 64.

Dole, J.M. and Wilkins, H.F. (1999) *Floriculture: Principles and Species*, Prentice- Hall, Inc., New Jersey, pp.613.

Gilman, E. F. and Howe, T. (1999) *Cosmos bipinnatus*. The University of Florida Extension Service, IFAS.

Gupta, M. N. and Samata, Y. (1967) *Rad. Bot.*, **7**: 225-240.

Hansen, S. and Drost, D. (2020) *Cosmos in the Garden*, Utah State University.

Hosoki, T., Kobayakawa, H., and Ohta, K. (2003) *Acta hortic.*, **625**: 261-264.

http: //www.public.asu.edu

Jang, I.C., Park, J.H., Park, E. (2008) *Plant Foods Hum. Nutr.*, **63**: 205-210.

Kaicker, U. S., Joshi, P.K. and Singh, H.P. (1969) *Delhi Garden Magazine*, **8**: 33-35.

Kale, S., Naik, S. and Deodhar, S. (2006) *Nat. Prod. Radiance*, **5**(1): 19-24.

Lakshmi, N., Vardhan, Z. V. and Rao, N.B. (1982) *Curr. Sci.*, **51**: 1071-1072.

Li, F., Lan, W., Zhou, Q., Liu, B., Chen, F., Zhang, S., Bao, M. and Liu, G. (2019) *Int. J. Mol. Sci.*, **20**(10): 2503.

Modise, S. A. and Ashafa, A.O. T. (2016) *Trop. J. Pharm. Res.*, **15** (5): 965-972.

Nikooei, M., Hemmati, C. and Bagheri, A. (2017) *J. Plant Prot. Res.*, **57**(3): 314–317.

Oku,T., Takahashi, H., Yagi, F., Nakamura, I. and Mii, M.(2008) *J. Hortic. Sci. Biotech.*, **83**(3): 323–327.

Post, K. (1949) *Florist Crop Production and Marketing*. Orange Judd Publishing, New York, pp. 428-430.

Rana, R. S., Jain, H.K. and Nerwal, S.K. (1963) *Indian J. Genet.*, **23**: 325-330.

Rodriguez-Bernal, A., Pina-Escutia, J.L and Vazquez-Garcia, L.M. (2013) *Genet. Mol. Res.*, **12** (4): 6257-6267.

Samata, Y., Inazu, K. and Takahashi, K. (1977) *Japan. J. Breed.*, **27**(3): 223-236.

Samata, Y., Kosei, I. and Hideo, E. (1974) *Bull. Faculty of Agri. Univ. of Tamagawa*, **14**: 51-58.

Sherff, E. E. (1932) *Field Museum Publication Botany*, **8**: 401–447.

Smith, A. W. (1972) *Gardener's dictionary of plant names.*

Verma, R. C., Dass, P., Shaikh, N and Khah, M.A. (2017) *Chromosome Botany*,**12**(3): 41-45.

18

Lisianthus

Reeta Bhatia, Madhubala and Kanwar Pal Singh

1.0 Introduction

Lisianthus [*Eustoma grandiflorum* (Raf). Shinn] is a promising new flower crop, that has been introduced to world market. It is known by several common names including 'Texas Bluebell'; 'Prairie Gentian'; 'Prairie Rose' and 'Tulip Gentian'. Lisianthus is an herbaceous annual or biennial plant, with slightly bluish green waxy leaves. The plants bear showy, bell shaped flowers, either solitary or paniculate on long straight stem. The multi-branched stems of lisianthus bear several flower buds that open successively. Each flower lasts more than a week after opening and, since there are several flower buds on a stem, others continue to open for several more days. Though lisianthus is a relatively new flower crop when compared with other commercial flowers, yet it had made significant impact in the global flower trade (Wazir, 2014).

2.0 Importance and Uses

2.1.0 Aesthetic Uses

Lisianthus production has been regarded as one of the remunerative business in the floriculture trade. They are highly valued as specialty cut flowers and are often grown as bedding and flowering pot plant. In the past few decades, lisianthus has gained tremendous popularity owing to its long vase life (approx. 2-3 weeks), diverse range of colors (including true blue), unique flower forms, and its suitability to grow under varying agro climatic conditions. Presently, it has emerged among the top ten cut flowers of the world, it ranked fifth in Japan, eleventh in the USA, and tenth in the global cut flower trades (Harbaugh *et al.*, 2000; Kawakatsu and Fukuta, 2012; USDA-NASS, 2012). In European and Asian markets, it has already been listed among the top selling cut flowers. Lisianthus flowers are in great demand mainly due to its rose-like blooms, excellent post-harvest life and availability of true-blue flowers.

It is also widely used as a flowering potted and bedding plant. Lisianthus have tremendous potential for use in landscaping and garden display owing to its vibrant colour display and eloquent flowers. The display of lisianthus in the gardens of Disney World had attracted tremendous attentions of the visitors (Parke, 1986). Its flowers are used for preparation of bouquets, buttonholes, corsages and floral crowns. In western weddings, lisianthus is one of the most sort flowers, especially for preparation of wedding bouquets and floral crowns, and decoration of the wedding tables, stages and wedding arches. This crop has created a niche in the floriculture industry and is regarded as one among the few flower crops that could be used as a cut flower, potted plant, and bedding plant.

3.0 Origin and History

The lisianthus has travelled a long way from its origin as a North American wildflower to its current use as a flowering potted plant, bedding plant and cut flower. This crop was first introduced for commercial cultivation in United States by the Japanese Seed company 'Sakata Seeds' in the early 1980's (Sakata, 1982). It

Uses of Lisianthus

In Pot

In Bouquet

Indoor Decoration

Decorating Wedding Arch

In Buttonhole

In Garden Display

Bridal Bouquet

Lisianthus Floral Crown

In Flower Vase

was initially registered by the name *Lisianthus russellianus* in the seed catalogues of United States during early 1980s. Later its scientific name was recognised as *Eustoma grandiflora* (Raf.) Shinn, with *Eustoma andrewsii; E. russellianum* and *Lisianthus russellianus* as its synonymous names (Bailey and Baily, 1976; Everett, 1981). Although, lisianthus originated in United States itself, it was re-introduced as 'new flower' from Japan (Rickett, 1966). It belongs to the family Gentianaceae and is native to the warm regions of the southern United States and northern Mexico (Popa *et al.*, 2004).

Being a moderately cold-tolerant annual or biennial plant, it is often found growing in dry climate near water, mainly in the prairies regions. The natural habitats of this crop includes, the northern part of Mexico, Texas, Oklahoma, Kansas, Nebraska, Colorado, Wyoming, and South Dakota (Shinner, 1957; Wood and Weaver, 1982). The wild types of lisianthus produce only blue or purple shaded flowers, however, the intense breeding efforts in the Japan has led to the development of varieties bearing diverse range of flower colours including, white, mauve, pink, yellow and even patterned flowers. 'Maurine Blue' (Harbaugh and Scott, 1996) and 'Florida Blue' (Harbaugh *et al.*, 1996) were the first heat-tolerant lisianthus varieties that can be grown at 28- 31°C without rosetting. The Park Seed Company has mentioned lisianthus in its catalogues as early as 1887 (Parke, 1986). However, it was only because of the systematic breeding efforts of the Sakata Seed Company' that this crop was re-introduced into the cut flower trades of USA in early 1980s (Halevy and Kofranek 1984). The Japanese breeders have put tremendous efforts in selection and breeding of lisianthus for over 50 year before its actual commercialisation.

In Europe, lisianthus was introduced as a garden plant from the USA (Nebraska and Louisiana) in 1835. However, the commercial cultivation of lisianthus as cut flower started only after 1980s when Japanese hybrids were introduced to Europe (Halevy and Kofranek, 1984). Lisianthus was introduced into Japan from the USA in the 1930s for ornamental use (Fukuta, 2016; Ohkawa, 2003), and is being cultivated since 1960. It is among the most prominent cut flowers of Japan (Ohkawa and Sasaki, 1999). The popularity of this flower increased gradually due to its beautiful and eloquent flowers, long vase life, and presence of blue flowers, which were rare in the cut flower trade. Presently, lisianthus is widely cultivated in countries like USA, Netherlands, Europe, Israel and South Africa.

In India, this is relatively a new introduction as a specialty flower. It has been recommended for growing under the moderate climate of mid Himalayan region (Wazir, 2014) and Nilgiris area (Anithaa *et al.*, 2012). The agro-techniques have been standardized for its successful cultivation under mid Himalayan region and Nilgiris area.

4.0 Taxonomy

Lisianthus is botanically known as *Eustoma grandiflorum* and is a member of family Gentianaceae. The name 'Eustoma' has been derived from two Greek words 'Eu '(beautiful, good, well), and 'Stoma' (mouth). The plants under the genus *Eustoma* Salisb are annual to short-lived perennial herbs, growing up to 80 cm in height. Leaves are often simple, opposite, sessile, often clasping, glabrous, more or less glaucous and ovate to linear-lanceolate in shape. The flowers are showy, either solitary or paniculate, often borne terminally on long pedicels. Calyx have deep cleft, the lobes are attenuate, erect and glabrous. Corolla is deeply campanulate, the lobes are large, erect, glabrous, purple to lavender, pink to white. The stamens are 5 or 6 in number, inserted on the corolla throat; anthers are versatile, linear, yellow. Styles are erect, slender, glabrous; stigmas are broadly 2-lobed, their upper surfaces densely and minutely glandular-pubescent. The fruits (capsules) are ellipsoid, 2-valved, glabrous. Seeds are minute, numerous, reticulate, more or less globoid.

The plants under this genus have been described as both an annual and biennial (Halevy and Kofranek, 1984; Wilkins and Grueber, 1983). Under the native environment, lisianthus behave as biennial. The seeds germinate during summer, and seedlings are often exposed to high temperature conditions that lead to rosetting, such plants will survive the winter in the rosette form and initiate flowering in the following year. However, if seeds germinate in the spring and seedlings developed under cool spring conditions, the plants would start flowering in the same year and thus classified as an annual.

5.0 Species and Cultivars

5.1.0 Species

The taxonomical analysis conducted by Turner (2014) have revealed that the genus *Eustoma* contain only two partially sympatric species, *E. exaltatum* (a widespread, small-flowered type) and *E. russellianum* (a more restricted, large-flowered type). The plants of *E. russellianum* are biennial and first produces a rosette which later bolts into a single flowering stem after cold treatment, then dies after branching. In contrast, *E. exaltatum* grows more like a perennial and produces additional shoots each season. In both the species flowers are mostly purple in colour, funnel-shaped to campanulate. The corolla lobes of *E. exaltatum* range up to 2.5 cm in length, whereas those of *E. russellianum* are 5 to 6 cm in length. Both the species are believed to be inter-fertile and often produce fertile progeny, and natural hybrids

Species of Lisianthus

Eustoma exaltatum

***Eustoma russellianum* (Syn *E. grandiflorum*)**

Promising Hybrids of Lisianthus

Rosita 3 Red Improved

Voyage 2 Apricot

Rosita 1 Pink Picotee

Rosita 2 Blue Picotee

Borealis Blue

Voyage 2 Green

Rosanne 1 Green

Rosita 3 Pink Improved

Excalibur Rose Pink

Echo Lavender

Mariachi Grande White

Doubline Rose Pink

Source: https://sakataornamentals.com

among these two species have been rarely detected. Another, Mexican species, *E. barkleyi* Shinners, has been identified under the genus *Eustoma*, but later it has been treated as a form of *E. exaltatum* [f. *barkleyi* (Shinners) B.L. Turner].

E. russellianum: They are the large flower types, corolla lobes mostly 2.5-5.0 cm long, 20-30 mm wide; calyx lobes mostly 1.5-2.5 cm long. The plants have bluish green, waxy leaves and showy bell-shaped flowers in shades of blue pink or white, borne singly or in panicles on an upright plant. Depending on where it grows, it may behave as an annual, biennial or perennial plant. It is native to the southern United States south to Mexico, where it is often found growing near streams or creek-beds and prefers moist, sandy soils.

E. exaltatum: It is commonly known by the name 'Seaside gentian'. The plants of this species are herbaceous and wide spread in its distribution. It produces large, purple, bell-shaped flowers throughout the growing season, peaking in spring through summer. Leaves are mostly ovate to ovate-lanceolate in shape, 6-15 mm wide, 2-5 times as long as wide; corolla lobes mostly 1.5-2.5 (3.0) cm long, 5-15 mm wide; calyx lobes mostly 1.0-1.5 cm long. This species is native to southern United States, Mexico, Central America, and the West Indies. In its natural habitat, it is often found growing in wet places, such as alkaline marshes, stream beds, and saline coastal areas. It is moderately tolerant of inundation by salt water, and highly tolerant of salt winds.

E. exaltatum f. barkleyi: This species is found distributed in northeastern Mexico. Leaves mostly lanceolate, 2-6 mm wide, 5-12 times as long as wide; corolla lobes mostly 1.5-2.5 (3.0) cm long, 5-15 mm wide; calyx lobes mostly 1.0-1.5 cm long.

5.2.0 Cultivars

A number of F_1 hybrids have been bred by different multinational companies like Sakata Seeds (Japan), Miyoshi Seeds (Japan), Pan American Seed Company (USA) and Gold Smith Seeds (USA). A large number of F_1 hybrids with diverse floral morphology (*e.g.*, double or single), corolla shape (fringed or non-fringed), flower color, wider adaptability, tolerance to heat and reduced sensitivity to rosetting are available in the market. The cultivar selection in lisianthus is generally made on the basis of color, floret number (single or double) and market demand.

Sakata Seeds have classified the lisianthus cultivars in to 4 different groups (0, 1, 2 and 3) based on number of nodes at the time of flowering and their suitability to flower under different seasons of northern hemisphere.

Group 0 (9-10 nodes): The cultivars of this group flowers during winter months (November-March) *e.g.*, Borelis.

Group 1(12 nodes): The cultivars of this group are ideal for spring flowering (March-June). The F_1 hybrids of Doublini, Echo, Rosanne 1 and Rosita 1 series are classified under Group 1.

Group 2 (15 nodes): These are mid-summer (June-August) F_1 hybrids, and cultivars of 'Excalibur', 'Mariachi', 'Rosanne 2', 'Rosita 2' and 'Voyage 2' series are classified under this group.

Group 3 (18 nodes): They includes late summer/fall flowering F_1 hybrids which flowers during August-October. Mariachi Grande White, Rosita 3 and Voyage 3 series belong to this group.

Some of the promising hybrids of lisianthus have been described as hereunder.

Borealis: It belongs to Group 0/Standard Double Flower, ideal for winter harvest with high yield and less incidences of blindness as compared to other types. It is available in five colours namely 'Apricot', 'Blue' 'Green' 'White' and 'Yellow'.

Echo Series: Echo series is the world's first all-double lisianthus released for cut flower production. This series belongs to Group I/Standard double flower and ideal for late spring/summer harvesting. In this series 7 vibrant colour range with strong flower stalks are available.

Doublini: They belong to group I/Spray double flower types, petite rose-shaped, long lasting and produce several branches per stem. This series often resembles mini rose without the thorns and are eye-catching element in any flower bouquets. Three colours *i.e.*, white, blue and deep rose are available in this series.

Rosita1: It belongs to group 1/Spray double flowers and ideal for spring flowering. Flowers are medium sized, rose shaped, with thick petals and strong stems, Rosita 2 and Rosita 3 are similar to Rosita 1, however, Rosita 2 is suitable for spring and early summer flowering, and Rosita 3 is suitable for late summer and fall flowering.

Mariachi Series: This series belongs to Group II/Standard quadruple flower, have extra-double, flowers that resemble camellia flowers. This series is suitable for winter sowing and early spring to summer harvest, and are available in 10 different attractive shades.

Voyage 2: This series has beautiful, large, double flowers with lightly-fringed petals. Voyage 2 is available in 8 different attractive colors ranging from white, green, yellow, pink, deep rose, mauve and blue.

Excalibur 2: Belongs to Group II/Standard double flower, suitable for growing under low latitudes areas, where temperatures are a bit higher for growing any other lisianthus. This series is easy to grow, less sensitivity to rosette and available in pure white, green, yellow, deep pink and blue picotee shades.

Rosanne 1: Rosanne belongs to Group 1 and have standard double flowers. It is a unique series with exceptional flower colors and patterns and have excellent vase life with more number of useable buds. This series is available in brown, black and green colours. Rosanne 2 is ideal for spring and early summer flowering, while Rosanne 3 is suitable for late summer and fall flowering.

F_1 Soiree Series: Soiree is the earliest flowering double flower series, that flowers even under the low temperature condition. The flower size is medium to large and available in six shade including white, pink, pink blush, pink picotee, blue blush and blue picotee.

F_1 Bolero Series: F_1 Bolero series is early flowering and less susceptible to rosetting under warm growing conditions. This series is adaptable to different cropping schedules from winter to summer and available in White, Green, Blue Blush and Blue Picotee colors.

F_1 Largo Series: The series belongs to the mid-late flowering group and give excellent performance through summer to autumn. The most remarkable characteristic of Largo series is its flower which are large with unique rim colors. Largo series is available in three colors; Blue Picotee, Lilac Pink Picotee and Pink Picotee.

F_1 Purple Flamingo: It is mid-late variety with large sized flowers, thick petals and long vase life. Color development in this series starts when flowers are still at bud stage. F_1 Vintage Blue Picotee and F_1 Purple Flamingo are the two shades available in this series.

F_1 Papillon Series: The series produce numerous small-medium sized rose-like flower buds on the top of the stems. It belongs to the mid-late flowering group, and has excellent performance through summer to autumn with very strong stems and perfect vase life. F_1 Papillon Rose Pink, F_1 Papillon Pink Blush, F_1 Papillon Blue Blush and F_1 Papillon White, Shallot Green and Luna Rose are some of the important hybrids of this series

F_1 Nightingale Series: Nightingale series is suitable for summer production and known for its intense rosy red or dark purple petals. This series need good light condition for good color appearance. F_1 Nightingale, F_1 Nightingale Midi, F_1 Nightingale Dark Purple are the three promising hybrids of this series.

F_1 Chaconne Series: It belongs to mid to mid-late flowering group. It produces large flowers with thick petals that have excellent vase life. This series is available in white, Blue Picotee, Purple, Apricot and Purple (50/50) and Blush Blue and Pink (50/50) and yellow shades.

F_1 Minuet Series: They produce small double blooms with very attractive rose-like flower buds. They are classified under the intermediate flowering group. F_1 Minuet Dark Purple and Minuet Apricot Rose, F_1 Minuet White Imp. and F_1 Romance Green Imp. are the popular hybrids of this group.

ABC Series: These are double-flowered lisianthus F_1 hybrids with high petal count and are ideal for year-round cut flower production. The flowers are long lasting and are available in vibrant range of colours including white, green, yellow, lavender, misty blue, purple, blue, blue rim, deep rose and misty pink.

Flare Series: This is regarded as the best spray-type lisianthus series and produces a spray of flowers on the top of each stem that opens within a short window, giving a bouquet effect to every stem.

Sapphire Series F_1: This is the most compact and extremely branched series ideal for pot use.

Florida Series F_1: This is the most heat-tolerant series for garden use; resists rosetting in both plug stage and after transplanting; ideal for growing as bedding and potted plant.

The newly introduced lisianthus germplasm was evaluated under the mid hills of Himachal Pradesh (HP) at the scientific research facility as well as on farmer's field by Wazir (2014). The cultivars 'Art Marine', 'Bolero Blue Picottee', 'Gavotte Yellow', 'Purple Flamingo' and 'Bolero White' have been recommended for commercial cultivation under mid-hills of HP. The performance of lisianthus as an emerging new cut flower has been evaluated under Nilgiri areas by Anithaa *et al.* (2012). Three cultivars namely, 'Echo Pink', 'Blue Picotee' and 'Bolero White' were found suitable for commercial cultivation under Nilgiri condition. Bhatia *and* Sindhu, (2019) have recommended 'Echo Double Blue, 'Echo Double Pink' and 'Echo Double Champagne' for commercial cultivation under mid Himalayan region.

6.0 Genetics and Breeding

6.1.0 Genetics

Lisianthus is diploid in nature, and the recent cytological analysis have estimated its diploid somatic chromosome as 2n=2x=36. As most of the breeding work in this crop has been under taken by private companies, very limited information is available on inheritance of various traits. The systematic research on inheritance of flower color in lisianthus have revealed that blue petal color is dominant over white or pink, and pink is dominant over white (Harbaugh, 2006). Two types of possible gene actions, simple recessive or epistasis, have been reported to control the development of white colour in lisianthus flowers. The recessive gene action is involved in the development of cream color, while pure white flowers are as a result of epistasis. The crossing between a pure homozygous, white and a blue/purple flower often leads to the development of light blue flower colour. The inheritance of bicolored flowers is complex and may involves more than one gene pair. The rim color is highly unstable trait in lisianthus and varies with environmental conditions. Sometimes, flowers with different hues ranging from pure white with a tinge of blue tip to completely blue, may develop on the same stem. It has been observed that plants exposed to cold temperatures may have blue flowers, while the rim flowers often develop under warmer temperatures. Generally, eight to nine generations of selfing are required to develop a completely homozygous bicolor lines that have stable color patterns over a range of environmental conditions.

6.2.0 Breeding

The breeding work in lisianthus has been undertaken mostly by the multinational seed companies like 'Sakata Seeds-Japan', 'Miyoshi Seeds-Japan', 'Pan American Seed Company- USA' and 'Gold Smith Seeds- USA'. The lisianthus cultivation in Japan has stated as early as 1933 (Ohkawa, 1994). In the beginning, the flower growers of Japan initiated the improvement work in this crop. In 1963, the first purebred lisianthus cultivar, 'Shihai' was bred by Fukukaen Nursery and Bulb Co., Ltd., Japan (Fukuta, 2016). The systematic breeding work in this crop for the development of F_1 hybrids was started by the Sakata Seed Company way back in 1977 and the first F_1 cultivar of lisianthus was developed in 1981. Initially, the main focus of the lisianthus breeders was to develop varieties for pot or bedding plant use only. In USA, the first cut flower F_1 hybrid 'Colorado Blue Bell' of this crop was released by Charles Weddle in 1984. In the early 1990's, three dwarf lisianthus cultivars namely, 'Blue Lisa-Pan American Seed Co.', 'Little Belle Blue-US Dept. of Agriculture' and 'Mermaid Blue-Sakata Seed' were released.

Development of F_1 hybrids with uniform and early flowering, reduced sensitivity to rosetting, tolerance to heat, novel flower color and forms, flower size, double flowers, enhanced basal branching and resistance to diseases are some of the important breeding goals in lisianthus.

6.2.1 Hybridization

Lisianthus are easy to breed compared to many other commercial flower crops. The crossing is quite easy in this crop and hybrids are developed using hand emasculation and pollination techniques. The pollen production is abundant, the flowers can be easily emasculated and pollinated, and large number seeds (1000 to several thousands) are developed in a single capsule. The only disadvantage in the breeding of lisianthus is the long seed to seed generation time (often 9 to 12 months). It is cross-pollinated crop, the flowers are self-compatible but show severe inbreeding depression on selfing. In most cases, the stigma is positioned well above the anthers. The flowers are protandrous, the pollen dehiscence starts with the opening of flowers and continue till one or two days after anthesis. The stigmas generally become receptive only 3-5 days after anthesis. Since, anther and stigma mature at different times, emasculation is easily accomplished in this crop without accidental selfing (Harbaugh, 2006).

Inbreeding depression is a serious concern in this crop, few generations of selfing may result in severe loss in vigour of the plant. Hence, sibling crosses are recommended for developing a strong and viable homozygous inbred line for hybrid breeding. For hybridization, the flowers are emasculated one or two days prior to anthesis before the anthers start dehiscing. At receptive stage, the stigmas appear to be plumpy and slightly sticky, hence the pollen readily get adhered to the stigma. For pollination, the pollen may be collected from freshly dehisced anthers and applied to the receptive stigma using a pollination brush. Alternatively, the pollen can be stored at room temperature (22-26°C) in packets and inside the sealed jars. Seeds usually mature in 8 to 10 weeks after pollination during the summer, but may require longer during the winter. Each seed pod contains sufficient number of seeds, 1000 seeds or more (several thousand). The seed pods begin to turn yellow with maturity and often split longitudinally when fully ripe. The seed pods should be harvested well before they started to split, else seeds may be lost. In humid climate, the splitting of seed pods may invite fungal infection and deteriorate the quality of seeds.

Lisianthus pollen often stain clothes and even their own petals on accidental dispersal during transport and thus reduce the decorative value of the flowers. Recently, Sakata Seeds have developed world first pollenless lisianthus, these pollen-free type flowers do not have any stamen. These pollenless varieties also have the advantage of long-lasting flowering. They have developed three pollenless varieties namely 'Single-flowered White', 'Single-flowered Blue Picotee', and 'Single-flowered Pink Picotee'. In past few decades, breeders have developed several cultivars in lisianthus with desirable traits such as uniform flowering, lack of rosetting, heat tolerance, flower color, and size and form, including double flowers.

6.2.2 Interspecific Hybridization

The lisianthus plantlets have a tendency to rosette (heat-induced resting stage) when they are exposed to high temperatures (more than 25°C). Interspecific hybridization can be valuable tool for development of heat tolerant cultivars in this crop. The wild species of lisianthus *i.e., E. exaltatum* is well adapted to high temperature conditions under its natural habitat, and can grow without rosetting even above 35°C. Interspecific hybridization has been attempted between *E. grandiflorum* and *E. exaltatum* in order to develop heat-tolerant cultivars, and to reduce the rosette problem. The resultant hybrids showed wide variation for heat tolerance and colour, and some of the hybrids were more heat tolerant than their parents (Barba-Gonzalez, 2017).

6.2.3 Polyploidy Breeding

Tetraploids (2n=4x=64) have been successfully induced in lisianthus using colchicine. The induced tetraploid plants of lisianthus had stronger stems and reduced plant stature Griesbach and Bhat (1990).

7.0 Growth and Flowering

The flowering in lisianthus is influenced by several factors including, genotypes, light (duration and quality), temperature and the age of the plugs. Lisianthus plants are extremely sensitive to temperature and light during flower initiation, especially when they have three pairs of leaves. A number of growth regulators have been utilized in this crop for regulating the growth and flowering. Lisianthus flowers can be made available all year round by using selected cultivars grown in appropriate climatic conditions.

7.1.0 Temperature

Temperature plays a crucial role in the growth and flowering of lisianthus. For better growth and quality flower production, the optimum day and night temperatures should be maintained between 20-24°C and 16-18°C, respectively. During flower initiation, the night temperature must be maintained above 16°C. Some cultivars of lisianthus are extremely susceptible to high temperatures, and may respond to heat stress by rosetting or premature flowering. Higher temperatures (above 28°C) during the first four weeks of transplanting can induce rosetting in susceptible varieties. Most of the commercial lisianthus cultivars have a tendency to rosette when seedlings are exposed to temperatures above 25°C (Harbaugh *et al.*, 1992; Ohkawa *et al.*, 1991). However, cultivars differ significantly in their sensitivity to high temperatures (Harbaugh *et al.*, 1992). If temperature is higher, it is beneficial to maintain the night temperatures approximately 4°C cooler than day temperatures. Although, the cooler growing temperature can add to stem strength and caliper, but it inhibits plant development and greatly add to crop time.

7.2.0 Light

The optimal light levels for lisianthus flower are 4000 to 6000 ft candles. Plants may produce more florets under high light intensities, but excessive light may result in less intense flower colours. The light quality also influences growth and flowering in lisianthus. It has been observed, that selective reduction of far red (FR) wavelengths of sunlight using the YXE-10 FR-absorbing film significantly reduced the height and internode length of lisianthus plant. However, these photo films had no significant impact on the development of flowers. Lisianthus is a quantitative long day plant and hence, light is an important component for its successful production

Due to the complex response of lisianthus cultivars to photoperiod, it has been classified differently, both as day-neutral plant and as a quantitative/facultative long-day plant. Halevy and Kofranek (1984) considered lisianthus as a day-neutral plant as they successfully obtained flowering in pink, blue, and white color variants of lisianthus even under an 8 hr photoperiod. While, other researchers considered lisianthus a facultative long-day plant (Grueber *et al.*, 1984; 1984; Roh *et al.*, 1989). Harbaugh *et al.* (1992) studied the effect of cultivars, temperature, and day length during the seedling stage on per cent rosetting in lisianthus. They observed that more plants rosetted under high temperature *i.e.*, 28°C, when seedlings were grown under short days compared to long days (Table 1). However, once the seedlings bolted, flowering was influenced by photoperiod × temperature interactions, and the response was cultivar dependent (Harbaugh, 1995). The flowering in lisianthus can be greatly accelerated by providing long photoperiod (>14 hours). During late spring and early fall season, it become difficult to produce the quality flowers with adequate stem length. Hence, in order to delay pre-mature flower bud initiation, it is recommended to treat the lisianthus plugs with short days (< 9 hours) or maintain the night temperature below 16°C during the seedling stage (Harbaugh, 2005). Further, the results of Islam *et al.* (2005) also indicated that lisianthus is a quantitative long-day plant (LD), and more specifically high light, enhanced flower bud initiation, development and subsequent growth. An initial SD period is preferred to increase the number of branches, number of flowering buds and flowers, stem diameter and shoot dry weight. The Aged plugs often exhibit delayed flowering and produce flowers with shorter stem lengths. Hence, in order to maximize the stem length only actively growing plugs should be planted.

Table 1: The Effect of Lisianthus Cultivar, Temperature, and Day Length during the Seedling Stage (14-43 days after sowing) on the Percentage (per cent) Rosetted Plants

Daylength	Temperature	Yodel White	Lisa Pink	Maurine Blue
SD (12 hr)	28°C	100	42	0
SD (18hr)	28C	42	0	0
SD (12 hr)	12°C	0	0	0
SD (18 hr)	12°C	0	0	0

Source: Harbugh, 2006.

7.3.0 Growth Regulators

Rosetting is a major challenge during the commercial production of lisianthus. Gibberellic acid sprays (15-25 ppm) were found useful for rapid development stem, when applied after rosetted plants began to bolt, however failed to hasten flower initiation in rosetted lisianthus (Hisamatsu *et al.*, 1999). Roh *et al.* (1989) also observed improvement in stem elongation in lisianthus cut flowers with GA_3 sprayings (250 ppm). Several plant growth regulators including, Ancymidol, Daminozide Paclobutrazol and Uniconazol have been successfully used in lisianthus for effective regulation of height in pot plants and bedding plants. Ancymidol drenches at 0.25 to 0.5 mg a.i. per pot was effective in controlling height in lisianthus (Adriansen 1989). Daminozide foliar sprays applied once at 5,000 ppm or twice at 2,500 ppm (Tjia and Sheehan, 1986); Paclobutrazol drench with 2-4 ppm (Adriansen, 1989); Uniconazol foliar sprays at 5 or 10 ppm (Starman, 1991; Whipker *et al.*, 1994) have been recommended for regulation of growth in lisianthus. Several factors such as temperature, day length, and cultivar govern the overall response of lisianthus plant to the applied growth regulators.

8.0 Propagation

8.1.0 Seed Propagation

Lisianthus is mainly propagated through seeds. In hilly areas, the seeds can be sown from mid-December to February under controlled environmental conditions. The Lisianthus seeds are very small (19,000 seeds per gram), and seedling growth is very slow. For these reasons, it is advisable to purchase plug seedlings from commercial nurseries. In markets, most of the seed companies are supplying pelleted seeds of this flower crop. Initially, seeds can be sown in pots or 392 or 406-cell plug trays, and then seedlings can be transplanted into 48-cell trays to continue growth before planting. Care must be taken not to cover the seeds after sowing. Seed germination can be hampered upon covering the seeds with the soil/germination media. Throughout seedling phase, air temperature should be maintained at 18 to 21°C (night) and 21 to 24°C (day). Seedling emergence is delayed if temperature is not maintained carefully during the germination period. A large fluctuation between day and night temperatures also affects seedling growth and development, resulting in rosetting and delayed flowering.

Lisianthus seeds require light to germinate (up to 2,500 ft candles). A minimum of 100–300 foot candles light should be provided when seeds are sown in germination chamber. After germination, the plugs are shifted to a greenhouse with high light (2,500-3,000-foot candles). Photoperiod is also an important factor for regulating germination and seedling growth because of its effect on cotyledon expansion, leaf and hypocotyl development, and photosynthesis (Harbaugh, 2005).

In lisianthus, the initial development of seedling is very slow, thus proper fertilization is extremely important. The seedlings should be fertilized lightly with 100-150 ppm Nitrogen using Calcium Nitrate based fertilizers. Capillary watering is preferred over normal watering during germination phase as it provides uniform

Propagation of Lisianthus

Newly Emerged Seedlings | Seedlings in Plug Trays | Transplanted in Field

Different Stages of Micropropagation of Lisianthus

Cultivation of Lisianthus

Lisianthus in Greenhouse | Full Bloom under Greenhouse | Full Bloom at ICAR-IARI, RS Katrain, Kullu, HP

Harvesting and Grading of Lisianthus

Optimum Stage of Harvesting | Freshly Harvested Cut Stems | Grading of Lisianthus Flowers

moisture and promote germination. During seedling phase, exposure of plugs to any kind of stress should be avoided as it may induce rosetting.

Although, lisianthus is commercially grown from seeds, propagation through seeds is a highly challenging task. There is a huge time gap of 3-5 months between seed sowing and transplanting of seedlings when seeds are sown under open field conditions. Further, lisianthus being a outcrossing in nature, the seedling population may exhibit lots of variability with respect to flowering time, stem length and flower qualities (Ordogh *et al.*, 2006).

8.2.0 Vegetative Propagation

Vegetative propagation through shoot tip cuttings can be a viable alternative to seed propagation in lisianthus. It will ensure true to the type plants with great uniformity. The genetic characteristics of a mother plant are maintained by this technique. In hilly areas, the terminal cuttings (5 cm in length) with 2-3 pairs of leaves can be obtained from healthy mother plants during months of February. The cuttings can be treated with 250 mg/l IBA for 5 minutes to promote rooting (Bhatia and Sindhu, 2019). The rooted cuttings get ready for transplanting within two months' time period, whereas, a time gap of 3-4 months is generally observed from seed sowing to transplanting. However, this technique has not been commercialised yet, and further research is required to establish its economic feasibility.

8.3.0 Micropropagation

Micropropagation protocols has been successfully established in this crop by several workers (Ordogh *et al.*, 2006; Esizad *et al.*, 2012). Large scale multiplication of selected plants with superior characters is possible using this technique. Further, it allows for the production of uniform and disease-free planting material. In lisianthus, *in vitro* shoots can be easily regenerated from stem, leaf and petals. The shoot regeneration has been obtained from petals (Ruffoni *et al.*, 1996), while, adventitious shoot formation from leaf explants has been successfully achieved by Dong *et al.* (2003). Popa *et al.* (2004) compared the *in vitro* regeneration response of three different explants including, shoot tips, sections of stems and leaf sections of lisianthus on different media. The best response was observed in shoot tips and the highest number of shoots were induced on MS medium supplemented with 3 mg/l BAP. Lots of research has been conducted to standardize protocols for synthetic seeds in this crop. Somatic embryos have been successfully induced in this crop and artificial seeds has been obtained after encapsulation of these somatic embryos with alginate layer (Ruffoni and Bassolino, 2016).

9.0 Soil and Climate

9.1.0 Soil

Lisianthus prefer well-drained soils, which are rich in organic matter for its successful cultivation. The soil pH must be maintained at 6.5 to 7.2. The low pH (<6.2) often leads to micro-element toxicities, especially zinc, and may result in undesirable leaf chlorosis and necrosis and poor stem growth. A pH above 7.5 often reduce the intensity of flower colour in lisianthus. The soil EC should be maintained at about

1.5 mmhos/cm. It also requires high calcium and adequate phosphorous levels for growth. Adjust the pH and add calcium and phosphorous prior to planting. High salt levels can delay flowering and can promote rosetting in susceptible varieties.

9.2.0 Climate

The Lisianthus can be produced in open fields, naturally ventilated polyhouses or in low cost polytunnels that simply protect the plants from wind and rain or in climate-controlled greenhouses. It prefers moderate climate, and can be grown at an altitude range of 1000-1800 meters above mean sea level. Lisianthus can be easily grown in areas where a minimum air temperature is above 15°C. Air temperatures below this level can inhibit plant development and greatly add to crop time. Crop quality is greatest if the daytime high temperature is less than 25°C, although plants will tolerate much higher temperatures. Some cultivars of Lisianthus are extremely susceptible to high temperatures, and may respond by rosetting or premature flowering. Though lisianthus can be grown under open-fields, most growers in moderate climates prefer to grow it in greenhouses equipped with shade cloth systems. These greenhouses provide protection from extremes in the environment that can devastate a crop. Further, it has been noticed that production in full sun can result in 60 per cent shorter stems than the same varieties grown under greenhouses or in outdoor shade nets.

10.0 Cultivation

10.1.0 Preparation of Soil

Soil preparation prior to planting is the most important part of lisianthus production. It should be grown in disease-free soils. Lisianthus is highly susceptible to soil-borne diseases. Soil sterilization with formaldehyde is recommended. Formaldehyde can be applied as spray or soil drench, cover the bed immediately with polythene sheets for 7 days. After removing the polythene cover, allow the soil to aerate for another 6-8 days. The planting beds should be thoroughly irrigated one week prior to planting to remove any traces of the chemical. Soil solarization, Methyl Bromide (25-30 g/m^2) or Basamid/Dazomet (30-40g/m^2) can also be used for soil disinfection.

10.2.0 Planting

Lisianthus seedling takes 8-12 weeks (65-90 days) from seed sowing to final transplanting in the field. The seedlings with three to five pairs of true leaves are generally transplanted. Seedlings must be transplant before they become root-bound. Any growth restriction in the seedling stage can adversely affect the stem quality of the mature plant. In India, under mid-Himalayan conditions, lisianthus can be planted from April to June for outdoor production. Under climatic controlled glass houses, all-year-round production is feasible if supplementary lighting is used for the crop planted between July and December.

10.3.0 Planting Density

It is necessary to provide adequate distance for proper development. The planting density in lisianthus depends upon the light conditions. A planting density of 64 to 96 plants/m^2 is generally recommended for this crop. Planting density of 84-96 plants/m^2 is used in areas of high light intensity, while in areas of low light, planting density of 64 plants/m^2 is maintained. When growing in areas with fluctuating levels of light, growers can utilize higher plantings density in the summer, and lower in the winter.

10.4.0 Crop Timing

Crop cycle from transplant to harvest is related to variety and environment. In general, lisianthus production from transplant requires up to 15 to 18 weeks under low light and temperatures, but can be as fast as 11 to 12 weeks when transplanted under optimal light and temperatures conditions. Though, the best quality cut stems in lisianthus are produced when it is grown at lower temperatures, however, it extends the crop time and may not be economically feasible.

10.5.0 Fertilization

Lisianthus is a heavy feeder and grows best with high fertility levels in the soil. High nitrogen and potassium fertilization are required for good growth. Fertilization using a 15:0:15 N:P:K fertilizer is recommended. Nitrogen should be predominantly in the nitrate form; potassium should be equal to nitrogen. Feeding with liquid fertilizer at 200 ppm N and K at every irrigation or every other irrigation is recommended. When growing outdoors a three-month slow-release fertilizer has been recommended. Calcium deficiency can cause tip burn of young foliage, bud abortion and poor stem strength. Use supplemental calcium during production unless the soil has high calcium content; calcium nitrate may be used as one component of the fertilizer solution to provide calcium. Fertilizer applications should be discontinued as buds become visible.

10.6.0 Irrigation

Lisianthus irrigation requires careful attention. Excessive irrigation increases plant susceptibility to soil-borne fungal pathogens. On the other hand, drought stress can cause premature flower initiation resulting in short, weak stems. Generally, the lower the light and temperature, the less water the plants require. Most growers establish the crop with overhead irrigation, and then switch to drip irrigation after roots are established in the beds. This crop also responds positively to overhead irrigation during periods of high heat and light. Increasing humidity in the greenhouse with an early afternoon irrigation can increase stem length in regions with high heat and light. Avoid high humidity and overhead irrigation after buds have formed however, since botrytis can develop.

10.7.0 Supplementary Light and Shading

Light intensity and day length significantly affect crop growth, crop time and flower quality in this crop. The optimal light levels for Lisianthus flower are 4000 to 6000 ft candles. In regions with low light intensity, supplemental high-intensity lighting may be necessary to produce a quality crop. The high light intensity improves the bud count and flower development, however, excessive light (over 7000 ft candles) reduces stem length in lisianthus. Lisianthus crop planted in areas with excessive light require shading to increase stem length. The light levels should be reduced 1,500 to 2,000 ft candles as flower color begins to show to prevent the fading of flower color.

During winters, when day length is shorter than 12 hours, supplemental light (using incandescent or HID) can accelerate production cycle in lisianthus. Long day (greater than 14 hours) treatment or night interruption from 10 p.m. to 2 a.m. can accelerate flowering and reduce the harvest time.

10.8.0 Pinching and Staking

Some grower performs single pinching after 20-25 days of transplanting. Generally, pinching causes shorter stems and delays flowering. To encourage a good second flush, plants should be cut back after the first picking. One to two layers of support netting (4 x 6 in./cm) are recommended. A minimum of two layers of netting (15 x 20 cm) is recommended to support the 3-4 foot (90-120 cm.) long stems.

11.0 Diseases and Pests

11.1.0 Diseases

A number of fungal, bacterial and viral disease including; botrytis blight, downy mildew, pythium root rot, fusarium crown and stem rot, fusarium wilt and powdery mildew, bacterial crown rot and wilt, and phytoplasmas have been reported in lisianthus by McGovern (2016).

11.1.1 Botrytis Blight

This disease is caused by B. *cinerea* and is commonly known as gray mold rot. It is one of the most devastating diseases of lisianthus during greenhouse production. The symptoms of gray mold on lisianthus include crown rot, damping off, stem cankers, and stem, leaf, and flower blights. It can also cause postharvest deterioration of flowers and often renders the flowers unmarketable. On petals, it begins as small, bleached spots which rapidly enlarge to blight entire flowers. The affected flowers often senesce and abscise. It causes brown lesions on stem which may remain localized or enlarge and kill the entire plant. Localized lesions which may be somewhat concentric often girdle the stem and lead to wilting, and death of tissue above the infection site. Excess moisture, cool temperatures and lack of ventilation invite the infection of botrytis. The incidence of Botrytis disease can be minimized by reducing the relative humidity and avoiding overhead irrigation. Simultaneously, heating and venting greenhouses during the evening can also be practiced to reduce humidity.

11.1.2 Fusarium Crown and Stem Rot

It is caused by *Fusarium avenaceum* (*Gibberella avenacea*) that primarily attacks the crown and stems of lisianthus, but may also infect the taproot and large feeder roots near the soil line. The symptoms of this disease begin as water-soaked areas in the stem at the soil line which become sunken and brown. The above-ground symptom of this disease appears as a gradual loss of green coloration in leaves, which is followed by tan leaf flecks, browning of leaf veins and a tan discoloration of entire leaves. As the disease progresses, the rot extends into the root system and the plant eventually wilts and dies. The plants are more susceptible at the start (transplant) and at the end when they are beginning to flower. The pathogen can attack seed, seedlings, and mature plants. To avoid the incidence of this disease, it is advisable to start the crop with healthy and disease-free plugs, and maintain good sanitation at the production site. Prompt removal of infected plants, steam sterilization and crop rotation can also minimize the incidence of this disease. fludioxonil and triflumizole are effective in reducing both the incidence of and mortality due to Fusarium.

11.1.3 Fusarium Wilt

It is a widespread disease of lisianthus caused by *F. oxysporum* f. sp. *Eustomae*. The symptoms of this disease begin as a dark discoloration of roots and the stem base. The pathogen moves upward through the xylem causing yellowing of leaves and wilting; vascular discoloration is visible when infected stems are cut longitudinally. As the disease progresses white masses of spores (sporodochia) are produced on stems, the leaves become tan, and plants wilt entirely and die. The pathogen may also cause pre- and postemergence damping-off.

11.1.4 Downy Mildew

It is caused by *Peronospora chlorae,* that affect both seedlings and mature plants. The disease mainly infects the developing tissues of shoots and buds, and often results in stunting and distortion. The symptoms begin as small yellow spots which gradually enlarge and turn into tan-brown blotches. Grey masses of sporangia are often produced within the mature blotches on the lower surface of leaves. Lack of ventilation, excess air humidity and excess fertilizer; especially high nitrogen may promote the incidence of downey mildew disease. The spray of Maneb (Manzate ditahne M-22) fungicide is recommended for effective control of this disease.

11.1.5 Powdery Mildew

Powdery mildew of lisianthus has been reported to be caused by *Leveillula Taurica* (*Oidiopsis sp.*). The symptom of this disease includes, appearance of yellow blotches on both the upper and lower leaf surfaces and stems which eventually become covered with profuse white fungal growth (mycelium, conidiophores, and conidia). Floral infection and leaf distortion and premature abscission may also occur.

11.1.6 Pythium Root Rot

It is associated with *Pythium species.* In this disease, the root systems become darkly discoloured and rotted, and infected plants often show rapid wilting. The outer rotted and discolored layer of *Pythium*-infected roots is often easily pulled off leaving behind the central, light-coloured stringy root fibre. This disease may also cause pre- and postemergence damping-off. Application *Trichoderma lignorum* can significantly reduce the incidence of damping-off caused by *Pythium* sp.

11.1.7 Rhizoctonia Crown and Stem Rot

It is caused by *Rhizoctonia solani*, the pathogen often infects the roots, stems, and foliage of lisianthus. The symptoms appear as dark discoloration and rot in stems at the soil level, gradually leading to wilting and rapid collapse of the entire plant. Under humid conditions, mycelial growth of *R. solani* may rapidly envelop and blighted leaves and shoots, and often referred as Rhizoctonia aerial blight. It may also cause pre- and postemergence damping-off. Soil disinfection with chloropicrin is reported to reduce the incidence *R. solani* in lisianthus crop.

11.1.8 Bacterial Crown Rot

It is caused by *Burkholderia gladioli*, infect plants of different age groups ranging from young seedlings to flowering plants. It causes tan veinal necrosis in the leaves, rot in the tap root, crown, and stem, which often leads to overall wilting and death of the plant. This disease can be evaded by using pathogen-free seeds, and avoiding unnecessarily wounding to the plants.

11.1.9 Burkholderia Wilt

This wilt disease is associated with Burkholderia *caryophylli*. The pathogen often causes destruction of the water-conducting tissues and often results in root rot and yellowing, wilting, and browning of the foliage. The lower leaves are infected first. The severely infected stems have a typical tan to yellow-brown discoloration in the cortical tissue. The infected plants must be rapidly removed to prevent the further spread of the disease.

11.2.0 Phytoplasma Diseases

The phytoplasmas causes yellowing of foliage, abnormal development of floral parts into leafy structures (phyllody), stunting and proliferation of shoots (witch's brooms), and green coloration in flowers (virescence). The phytoplasma disease can be restricted only by controlling the insect vector populations through chemical or biological means. Alternatively, the vector may be excluded by physical barriers such as nylon mesh screen, vector disorientation and repellence in field-grown crops through the use of reflective mulches, and elimination of, and avoidance of growing near, alternate hosts especially established, infected crops.

11.3.0 Viral Diseases

Viruses cause the usual symptoms - leaf yellowing, mottling, chlorosis and distortion and poor flowering. Tomato spotted wilt and Impatiens necrotic spot viruses are spread by thrips, bean yellow mosaic virus by aphids, tomato yellow leaf curl virus by whiteflies, and tobacco mosaic virus by humans. Initiating a spray program as soon as the problem occurs can eliminate the pest quickly, and thus prevent the spread of viral disease.

11.3.1 Impatiens Necrotic Spot Virus (INSV)

It is one of the most common viruses that infect lisianthus. The symptoms include down turned tips, stunted growth, clustered terminal leaves. In advanced infections small brown spots appear on the terminal leaves. Once infected, there

Diseases and Pests of Lisianthus

Downey Mildew

Botrytis Blight

Fusarium Wilt

Fusarium Crown and Root Rot

Bacterial Crown Rot

Aphids

Fusarium Wilt

Botrytis Blight

Thrips

Source: McGovern R.J. (2016).

are no chemical controls, the infected plants should be immediately removed to limit damage. The control fungus gnats can also prevent the spread of this virus.

11.4.0 Pests

Aphids, thrips, whiteflies, red spider mite and leaf miners cause significant damage to this crop. They also act as vectors to several viruses. Initiating a spray program as soon as the problem occurs can eliminate the pest quickly and thus prevent the spread of viral disease.

11.5.0 Physiological Disorders

11.5.1 Rosetting

Rosetted plants have a cluster of leaves with very short internodes on the stem. The most common cause of rosetting is high temperatures, especially during young plant production. Under ideal temperatures, the lisianthus seedlings generally bolt after they develop three to four pairs of true leaves. However, lisianthus is very sensitive to high temperatures from the time seeds absorb water during germination through the development of three to four leaf pairs. The general rule is that the earlier and longer seedlings are exposed to heat stress, and the higher the temperature to which seedlings are exposed, the greater the percentage of plants that will rosette (Harbaugh, 2006).

Susceptibility to rosetting varies with variety and can be affected by conditions during seed production. Although, the rosetted plants will usually begin to grow eventually, but the stem quality is often reduced. Rosetted plants also take unacceptably long time to flower and the results are unacceptable for commercial production. Plants that exhibit rosetting can be induced to grow with one or two spray applications of gibberellic acids (10-100ppm). Through continuous breeding efforts cultivars with good resistance to rosetting has been developed. New series such as Avila, Balboa and Catalina have demonstrated good resistance to rosetting.

11.5.2 Tipburn

The occurrence of tipburn is a serious problem during the production phase of lisianthus and may deteriorate the flower quality, and cause huge economic losses. It is often seen under stress conditions, especially in spring when high light follows a dark period with sudden temperature and humidity fluctuations. It is also related to poor root development from delayed transplanting, overly wet or dry soil, high salt levels or diseases.

Tipburn is often associated with calcium (Ca) deficiency. Tipburn in lisianthus is mainly due to the inability of the plant to translocate adequate amounts of Ca to the tips of the upper leaves, which is associated with an increase in the distribution of Ca in the roots. To avoid tip burn, top dressing with 700-1,400 ppm organic calcium at 2 ml/plant (7-8 times 2-3 days apart) is recommended. The foliar sprays of 325-350 ppm calcium chloride are also found effective for effective control of tip burn.

12.0 Harvesting and Postharvest Management

The lisianthus cut flowers are harvested when one or more flowers are open. Growers often remove the first and sometimes the second flowers that open before harvesting. This increases the number of flowers open at one time per stem. Flowers should be harvested in the mornings, when flower and plant tissues are cool. After harvesting, cut stems should be immediately placed into a bucket containing tepid water (20°C) and shifted to a cool shady place for rehydration. The cut stems of lisianthus can be stored at 4°C and treated with floral preservatives for extending the shelf life.

The use of floral preservatives with sucrose can increase the vase life by 40-50 per cent and promotes a deeper flower color for unopened flowers. Pulsing with 10 per cent sucrose for 24 hours after harvest has been reported to increases vase life of this flower. The vase life of cut lisianthus ranges from 12-15 days depending upon cultivars. Sharifzadeh *et al.* (2014) tested the effect of 8-hydroxyquinoline citrate (8-HQC), sucrose and peroxidase inhibitors namely catechol (CH) and p-phenylenediamine (PD) on delaying senescence and extending vase life of cut lisianthus flowers. The results indicated that the peroxidase inhibitors in combination with 8-HQC and sucrose increases vase life of lisianthus by improving water uptake and delaying fresh weight loss. They found 10 mM CH along with 200 ppm 8-HQC mg/dm^3 and 3 per cent Sucrose as the most effective treatment for vase life extension (13.3 days), increasing water uptake, and delaying fresh weight loss in lisianthus. The effectiveness of ascorbic acid in prolonging the vase life of cut lisianthus flowers has been revealed by Azizi *et al.* (2015). They observed the longest vase life (15.50 days) of cut lisianthus flowers with 200 mg/l of ascorbic acid.

Lisianthus cut stems are susceptible to geotropism (bending away from gravity). Flower buds often show upward bending when stored horizontally under warm temperature conditions. This is mainly due to uneven distribution of auxins. The incidence of negative geotropism can be greatly reduced by maintaining the flowers at low temperatures during transport and storage. Where it is not feasible to maintain low temperature, it is advisable to handle flowers upright. Treatment with low concentrations of the auxin transport inhibitor, naphthyl phthalamic acid (NPA), is very effective in preventing geotropic bending, but this material is neither commercially available nor registered for this use.

13.0 Biotechnology

The doubled haploidy is a valuable tool to fix homozygosity and reduce the breeding cycle in any crop. This technique has been utilized to develop completely homozygous lines for hybrid breeding in lisianthus. Zhou *et al.* (2014) successfully induced doubled haploids in lisianthus cultivars using anther culture technique. The shoot regeneration from anther derived calli was obtained on MS medium containing 8 mg/l 6-benzyladenine and 0.1 mg/l naphthaleneacetic acids. Novel somaclonal variants have been isolated in tissue-cultured population of lisianthus (Semeniuk and Griesbach, 1987). These variants were found to be dwarf or miniature, and were more branched compared to the parent.

The novel flower colour patterns have been generated in lisianthus by altering flavonoid biosynthesis using genetic engineering techniques. The level and pattern of pigmentation in purple-flowered lisianthus cultivars was altered by the introduction of a lisianthus CHS cDNA in an antisense orientation, and genetically modified lisianthus lines with altered flavonoid biosynthesis were successfully generated (Deroles *et al.*, 1998).

There are very limited reports on basic molecular research on lisianthus, especially the research related to the molecular mechanism of response to drought stress. Recently Anxia *et al.* (2020) established the transcriptome profiling of lisianthus under drought stress using transcriptome second-generation sequencing technology. A total of 1,02,014 non-redundant genes were obtained, including 2,929 coding genes which were found to belong to 79 transcription factor families after analyzing. This study has provided insight into the candidate molecular resources for subsequent studies on lisianthus response to drought stress in the future.

14. References

Adriansen, E. (1989) *Greenhouse Grower*, **7**(4): 50-54.

Anithaa, K., Selvaraj, N., Jegadeeswari, V. and M. Sharathkumar (2012) *Acta Hortic.*, **1241**: 293-297.

Anxia, A.X., Qiang, Z., Xuping, L., Lufeng, L., Jie, C., Tingting, L., Wenlue, L. Xiahong, L. and Guanlin, Z. (2020) *Mol. Plant Breed*, 11.

Azizi, S., Onsinejad, R. and Kaviani, B. (2015) *ARPN J. Agric. and Biol. Sci.*, **10**(11): 417-420.

Bailey, L. H. and Baily, E. Z. (1976) Hortus Third, MacMillan, New York.

Barba-Gonzalez, Tapia-Campos, R., E., Lara-Bañuelos, T. Y., Cepeda-Cornejo, V., Dupre, P. and Arratia-Ramirez, G. (2015) *Acta Hortic.*, **1097**: 93–100.

Bhatia, R. and Sindhu, S. S. (2019) *Ind. J. Hort.* **76** (4): 714–720.

Deroles, S. C., Bradley, J. M., Schwinn, K. E., Markham, K. R., Bloor, S. J., Manson, D. G. and Davies, K. M. (1998) *Mol. Breed.*, **4**: 59-66.

Dong, D. K., Song, Z., Yun, Z., Xiang, Z., Guo, Z. and Ping, W.L. (2003) *Plant Physiology Communications.* **39**(6): 494-498.

Esizad, S.G., Kaviani, B., Tarang, A., and Zanjan, S.B. (2012) *POJ*, **5**(3): 314-319.

Everett, T. H. (1981) The New York Botanical Garden Illustrated Encyclopedia of Horticulture, **6**. Garland Pub. Co., New York, London.

Fukuta, N. (2016) *In*: M. Shibata (ed.). Japanese history on flower breeding. Yushokan, Tokyo. pp./ 289–312.

Griesbach, R. J. and Bhat, R. N. (1990) *Hort. Sci.*, **25**(10): 1284-1286.

Grueber, K.L., Corr, B.E. andWilkins, H.F. (1984) *Eustoma grandifloum* (*Lisianthus russellianus*). Minn. State Flor. Bull. **33** (6), 10-14.

Halevy, A.H. and Kofranek, A.M. (1984) *Hort. Sci.* **19**: 845–847.

Harbaugh, B. (2006) In: Anderson. N.O. (Ed.), *Flower breeding and genetics; Issues, Challenges and Opportunities for the 21st Century*, Springers, Netherlands, *pp.* 644–663.

Harbaugh, B. K. and J. W. Scott. (1996) *Hort. Sci.*, **31**: 1055-1056.

Harbaugh, B. K., Bell, M. L. and Liang, R. (2000) Hort. Technol. **10**(4): 812-815.

Harbaugh, B. K., Roh M. S., Lawson, R. H. and Pemberton, B. (1992) *Hoff Sci.*, 27(8): 885-887.

Harbaugh, B. K., Scott, J. W., and Rubino, D. B. (1996) *Hort Sci.*, **31**: 1057-1058

Harbaugh, B.K. (1995) Flowering of *Eustoma grandiflorum* (Raf.) Shinn. cultivars influenced by photoperiod and temperature. *Hort Sci.* **30**: 1375–1377.

Hismatsu, T., Koshika, M., Oyama, N. and Mander, L. N. (1999) *J. Jap. Soc. Hortic. Sci.*, **68**: 527-533.

Islam, N., Patil, G. G., Hans Ragnar Gislerod, H.R. (2005) *Sci Horti.*,**103**: 441-451.

Kawakatsu, K. and Fukuta, N. (2012) JARQ, **46**(3): 269–275.

McGovern, R.J. (2016) In: McGovern R., Elmer W. (eds.) Handbook of Florists' Crops Diseases, Handbook of Plant Disease Management. Springer, Netherlands, https: //doi.org/10.1007/978-3-319-32374-9_20-1

Ohkawa, K. (2003) In: Ohkawa, K. (Ed.). Practical technique of floral horticulture-Lisianthus. Seibundo Sinkosha, Tokyo, Japan, p./ 14–19.

Ohkawa, K. and Sasaki, E.(1999) *Acta Hortic.* **482**: 423-426.

Ohkawa, K., Yoshizumi, T., Korenaga, M. and Kanematsu, K. (1994) *Hort. Sci.* **29**: 165-166.

Ordogh, M., Jambor-Benczur, E. andTilly-Mandy, A. (2006) *Acta Horti.*, **725**: 457-460.

Parke, M. (1986) *Horticulture*, 32-34.

Popa, G., Balan, V. and Plocon, C. (2004) *Buletinul Universitatii de Stiinte Agricole si Medicina Veterinara Cluj Napoca Seria Zootehnie si Biotehnologii*, **60**: 319-322.

Rickett, H. W. (1966) Wild flowers of the United States. Volume 3. Texas. McGraw-Hill Book Comp., New York.

Roh, M. S., Halevy, A. H. and Wilkins, H. F. (1989) In: Halevy, A. H. (ed.). Handbook of flowering. Vol. VI. CRC Press, Boca Raton, Fla. p 322-327.

Roni, M.Z.K., Islam, M.S. and Shimasaki, K. (2017) A timeline for *Eustoma grandiflorum* seedling production based on an *in vitro* germination protocol. POJ. **10**: 232-236.

Ruffoni, B., Giovannini, A. and Allavena, A. (1996) *Plant Phys and Biochem Special issue FESSP Congress*: 56.

Ruffoni, B. and Bassolino, L. (2016) In: Germana, M. A. and Lambardi, M. (eds.) *In Vitro Embryogenesis in Higher Plants, Methods in Molecular Biology*, vol. 1359, Springer, Netherlands, DOI 10.1007/978-1-4939-3061-6_17.

Sakata, T. and Co. (1982) Sakata's Reliable Seed Catalog 1982/83. Yokahama, Japan

Semeniuk, P and Griesbach, R. J. (1987) *Plant Cell Tiss Org Cult.*, **8**: 249-253.

Sharifzadeh, K., Asil, M.H., Roein, Z. and M. Sharifzadeh (2014) *J. Hort. Res.* **22**(1): 41-47.

Shinner, L. H. (1957) *The Southwestern Naturalist*, **2**: 38-43.

Starman, T. W. (1991) *HortSci.* **26**(2): 150-152.

Tjia, B. and Sheehan, T. J. (1986) *HortSci.*, **21**(1): 147-148.

Turner, B.L. (2014) *Phytologia* **96**(1): 7-11.

USDA-NASS (2012) Floriculture: U.S. Summary. 8 pp http: //www.nass.usda.gov/ Statistics_by_State/Ohio/Publications/Reports_by_Title/flor2012.pdf.

Wazir, J.S. (2014) Evaluation of eustoma/lisianthus cultivars for assessing their suitability as prominent new cut flower crop under mid hill conditions of H.P. *Int. J. Agric. Sci. and Vet. Med.* **2**(1): 105–110.

Whipker, B. E., Eddy, R. T. and Hammer, P. A. (1994) *Hort Sci.*, **29**(11): 1368.

Wilkins, H. F. and Grueber, K. L. (1983) *Minnesota State Flor. Bull.* **32**(6): 10-11.

Wood, D. E. and Weaver, R. E. (1982) *J. Arnold Arbor.*, **63**: 441-487.

Zhou, X., Mo, X., Qu, S., Tian, M., Wu, X., Wu, M., Li, J., Luo, Y., Wang, J. and Gui, M. (2014) *Hortic. Environ. Biotechnol.* **55**: 125–128.

19

Campanula

Tarak Nath Saha, P. Naveen Kumar, Swati Saha, Ganesh B. Kadam and K.V. Prasad

1.0 Introduction

Campanula is an important cool season perennial flowering plant of Campanulaceae family, grown mainly in cooler region of the globe. It is also known as Bellflower as the fully opened flower resembles the shape of a bell, although some are known as 'Harebell', 'Canterbury bell', 'Carpathian bell' or 'Blue bell'. The word 'Campanula' is derived from Latin means 'Little bell'. The floral lobes are shallow to deeply cut, thus enhancing the cupped, bell, or starry appearance. Flowers of these plants/species are available in the shades of blue, purple, white and pink. Some members of the genus *Campanula* are short day, some are long day and some are day neutral plants.

Campanula is relatively a new commercial ornamental plant to India. However, they are commercially cultivated as cut flower as well as potted ornamental plants in USA and Europe. In Europe and America, it is widely used to add splash of colour in the garden during spring and its flowering continues up to early summer. The plants are easy to grow and can sustain winters. Campanulas are low maintenance plants which make them suitable for beds, borders, pots, rock garden, and hanging baskets.

2.0 Importance and Uses

2.1.0 Aesthetic Uses

The plants, due to their adaptability and growth habit are widely used in gardens. There are garden-worth varieties for each purpose like beds, borders, meadows, and rock gardens. Tall bell flowers such as *Campanula lactiflora* and 'Kent Belle' are ideal for planting in the middle and back of the border or paired with other perennials. *Campanula carpatica* and other small-sized bell flowers are great choices for the front of the bed, in rockeries, or as annualized plants in containers. Low growing creepers such as *C. poscharskyana* and *C. portenschlagiana* make good edging plants along walks or beds, and are especially striking on crannied rock walls. The spreading habits of *C. glomerata*, *C. punctata*, and *C. takesimana* make them appropriate as ground covers or for naturalizing but are not well-suited to small or formal gardens. *Campanula lactiflora*, *C. latifolia*, and *C. takesimana* thrive well in light shade, making them perfect choices for transition zones between sun and shade (Hawke, 2008). During the late 1980s, some Danish growers adapted *C. carpatica* and *C. poscharskyana* to pot culture and production schedule for growing them in and out of greenhouses (Jensen *et al.*, 2005).

2.2.0 Therapeutic and Other Uses

A few species, *Campanula glomerata*, *C. persicifolia*, *C. rotundifolia*, *C. bononiensis*, *C. sibirica* and *C. patula* have been used locally for preparation of traditional drugs in Russian folk medicine (Barnaulov *et al.*, 1983, 1984). In Italy, similar use has been made of *C. medium*, *C. cervicaria*, *C. rotundifolia*, *C. latifolia* and *C. trachelium* (Gastaldo, 1978). Drugs are being prepared from roots and the above-ground parts of this flowering plants, that are dried and prepared as broths or infusions. Preparations

Uses of Campanula

Cut Flower
(https://i.pinimg.com)

Vase Decoration
(https://wellgrowhorti.com)

Pots
(https://www.waitrose garden.com)

Garden
(https://veddw.com)

Cottage Garden
(http://holicoffee.com)

Hanging Basket
(https://www.primrose.co.uk)

Hedge
(https://www.jparkers.co.uk)

Rock Garden
(https://www.nargs.org)

Window Box
(https://in.pinterest.com)

Interiorscaping
(https://www.flower council.co.uk)

Vertical Walls
(https://in.pinterest.com)

Window Screening
(https://in.pinterest.com)

are being used to treat epilepsy, nervous diseases, coughs, headache, rheumatism, and inflammation. The roots and leaves of *C. rapunculoides* can be used as a vegetable (Becker-Dillingen, 1950).

3.0 Origin and History

The genus *Campanula* includes about 400-600 species that are naturally distributed in temperate zone of northern Hemisphere including Mediterranean region, Caucasus and European Alps (Kuss *et al.*, 2007). A few of these beautiful plants have been cultivated as garden and rockery plants for a long time. During the late 1980s, some Danish growers adapted *C. carpatica* and *C. poscharskyana* for pot culture in a combination production schedule in and out of greenhouses. Some selection and breeding took place yielding better and more flowers; also the cultivation practices were improved to give better post-production shelf life. These species still comprise a large part of the about 10 million *Campanula* plants produced in Denmark each year (Jensen *et al.*, 2005).

4.0 Taxonomy

Campanula is (mostly) perennial flowering shrub although annual and biennial species/varieties are also common. It belongs to the family *Campanulaceae*, Order *Asterales*, Superorder *Asteranae* and Class *Magnoliopsida*. The annuals were included in sect. 'Annuae' by Boissier (1875), but later were separated by Damboldt (1976) into four subgenera: *Megalocalyx*, *Sicyocodon*, *Roucela* (Dumort.) Damboldt and *Brachycodonia* (Fedorov) Damboldt. Phylogenetic analyses based on nuclear ITS and chloroplast *trn*L–*trn*F data revealed that *C. cremnophila* is distinct from other species of this group, forming a subclade with *C. garganica* and *C. poscharskyana* (Bogdanovic *et al.*, 2019). However, comprehensive information about different campanula species, to discover useful characters to define them taxonomically, and to define their geographical distribution is still lacking (Saez and Aldasoro, 2003).

Different species behave as annuals to herbaceous perennials with plant height ranging from 5 cm to 2 m. Foliage colour ranges from light to dark green and even yellow. The leaves are estipulate, alternate (rarely opposite or whorled), simple, entire or variously toothed or dissected, petiolate or sessile. The bell flowers have characteristic basal and stem leaves. Basal leaves may be oval, round, lanceolate, or heart shaped, and are usually larger, long petioled, and often toothed. Stem leaves are usually simple, smaller in size, and short stalked or sessile (Hawke, 2008). Roots are often long tap roots sometimes with fleshy starch rich parts. All parts of the plants contain milky white latex. Most species are hardy to very low temperatures, especially when kept on the dry side (Jensen *et al.*, 2005).

Generally, the flower colour is blue to violet, but white and pink are also found in many species with seldom existence of yellow colour variants. Flowers are borne singly or clustered inflorescences, and may be upward-facing or nodding in habit and are held above the foliage. The flowers are five lobed and contain five stamens and a five- or three-parted stigma. Flowers tetracyclic, perfect, with a specialized method of protandrous secondary pollen presentation, pedicellate or rarely sessile; bracts leaf like or reduced, rarely absent; pedicels often bracteolate.

Calyx synsepalous, adnate to ovary and forming a hypanthium (free in *Cyananthus*); lobes (3-10), valvate, sometimes with a reflexed appendage in each sinus. Corolla sympetalous, actinomorphic or zygomorphic, most often some shade of blue or violet, or white; lobes (4-10), valvate. Stamens equaling number of corolla lobes, anti-sepalous, distinct, coherent, or connate; anthers tetra-sporangiate, dithecal, introrsely dehiscent by longitudinal slits, and basifixed. Gynoecium syncarpous, 2-10 locular with axile placentation, ovary at least partly inferior (superior in Cyananthus), sometimes crowned by an annular or tubular nectary; style solitary, pubescent with pollen-collecting hairs below apex; stigma typically with as many lobes as ovary locules (Hong *et al.*, 2011).

Fruit is a capsule, commonly apically loculicidal or laterally poricidal, or a berry. Seeds are usually small, numerous; embryos are small, straight; endosperm copious, oily or rarely starchy. Many species contain white latex in the leaves and stems (Hong *et al.*, 2011 and Pizzetti and Cocker, 1968).

5.0 Species and Cultivars

5.1.0 Species

Although there exists more than 400 species in Campanula. It is not possible to describe all the important species which are prevalent in certain areas of the world and are used in cultivation and breeding. Therefore, some of them described by various workers are presented in tabular form for easy understanding (Bailey, 1942; Pizzetti and Cocker, 1968; Hay and Beckett, 1971; Beckett, 1985, 1987; Brickell, 1994, Jensen *et al.*, 2005).

Sl.No.	*Species*	*Native/Spread*	*Characteristics*
1.	*Campanula americana*	North America	It has flowering spikes that may reach 2 m high with saucer-shaped flowers bearing long curved styles.
2.	*Campanula carpatica* (Tussock Bellflower)	Eastern Europe	Hardy herbaceous perennial with compact habit grows to a height of 30 cm during bloom. Flowers are blue, terminal, solitary, largest among the small campanulas and bell-shaped with wide mouth.
3.	*Campanula formanekiana* (Greek bells)	Greece and Macedonia/ Mediterranean	It is monocarpic species often propagated by seeds, first dicovered in 1931. The plant with tall main raceme (up to 75 cm) is supported by spreading basal, prostrate, and lateral racemes.
4.	*Campanula garganica* (Adriatic Bellflower)	Italy	Herbaceous mat-forming perennial grows up to 10 cm and bears pale-blue starry bell-shaped flowers.
5.	*Campanula glomerata* (Dane's Blood)	Europe	Hardy (moderately) herbaceous perennial suitable for border and grows up to 45 cm tall. It bears funnel-shaped violet-blue flowers in terminal.
6.	*Campanula lactiflora* (Milky Bellflower)	Caucasus Mountains	Hardy herbaceous perennial growing up to 90-180 cm. It is free-flowering and flowers are arranged in loose panicles. Plant bears white flowers which are slightly tinted (pale-blue) and propagated by seeds.

Sl.No.	Species	Native/Spread	Characteristics
7.	*Campanula lasiocarpa* (Mountain Harebell)	North America	The plants are found growing on sandy gravel slopes at high altitudes (1200 m). The plant bears large bell-shaped blue flower.
8.	*Campanula latifolia* (Giant bellflower)	Europe	It is a common, tall (1 to 1.5 m) perennial species growing in woodlands. Flowers are purple or dark blue (around 6 cm long), hairy and found in loose racemes.
9.	*Campanula medium* (Canterbury bells)	Italy and Switzerland/ Alps	Sun loving hardy biennial growing upto 1m used in garden borders. Flowers are large, upright blue-violet colour. White form also exists.
10.	*Campanula portenschlagiana* (*C. muralis*)	Dalmatia and South Europe	Hardy herbaceous perennial grows upto a height of 20 cm with spread of 45-60 cm and is long-lived invasive species. Flowers are bell shaped, and deep blue-purple in colour. Propagated through division and seeds.
11.	*Campanula pyramidalis* (Chimney bells)	Austria and Dalmatian region/Alps	Very tall, biennial, growing upto 2 m when flowering. During flowering it looks like pyramidal raceme with numerous light blue to white, star-shaped flowers. Its compact cultivar are used as a potted plant.
12.	*Campanula rotundifolia* (Bluebell-of-Scotland,roundleaf harebel)	Europe, Siberia and North America	Dwarf hardy herbaceous perennial with quick spreading habit. Grows upto a height of 30 cm. Plant bears stalkless and pendulous sky blue flowers.
13.	*Campanula scabrella* (Rough Bellflower)	North America	Generally found growing in dry rocks exposed to full sun. The plant height is up to 15 cm and usually terminates in a single erect flower (violet-purple).
14.	*Campanula scouleri* (Pale Bellflower)	North America	Generally found growing in soft, stony soil in woodlands. The plants are perennial 10–30 cm tall. Flowers are panicled, outward angled, pale blue colour.

5.2.0 Cultivars

Campanula medium has been grown for many years, but older cultivars were biennials, requiring a long production cycles and cold or short days followed by long days for flowering. The long production time made the crop unacceptable for large-scale greenhouse cut flower production. However, the cultivars of 'Champion series' (Champion Blue and Champion Pink) offers a production time of 18-23 weeks in the greenhouse with no cold requirement for flower development (Dole *et al.*, 2001). The other popular varieties are 'Pink Octopus', 'Blue Waterfall', 'Dalmatian Bellflower', 'Champion', 'Champion Pro', 'Champion II' and 'Fairies' Thimbles'. Below is the list of promising varieties/types that thrives in most gardens with moderate climates (https://www.gardenloversclub.com).

Korean Bellflower: It is also known as "Elizabeth bellflower" and bloom from early summer till early fall. These bellflowers have blooms that are a stunning shade of deep, rich, royal purple. They have long, bell-shaped flowers that hang upside down in clusters from their stems.

Species of Campanula

Korean Bellflower

Campanula garganica

Campanula rotundifolia

Campanula persicifolia

Campanula poscharskyana

Campanula glomerata

Campanula portenschlagiana

Campanula medium

Campanula Lactiflora

Campanula carpatica

Campanula cochleariifolia

Campanula Latiloba

Source: www.en.wikipedia.org, www.temperate.theferns.info, www.seedcorner.com, www.gardenloversclub.com

Species of Campanula

Campanula lasiocarpa

Campanula Sarastro

Campanula Trachelium

Campanula americana

Campanula pyramidalis

Campanula latifolia

Campanula scabrella

Campanula formanekiana

Campanula cashmeriana

Campanula carpatica

Campanula scouleri

Campanula alliariifolia

Source: www.plantopedia.com, www.calscape.org, www.dorsetperennials.co.uk, www.interistroy.ru, www.science.halleyhosting.com, www.etsy.com, www.in.pinterest.com, www.plant-world-seeds.com, www.dorsetperennials.co.uk, www.kelways.co.uk, www.commons.wikimedia.org

Cultivars/Varieties of Campanula

Champion

Champion Pro

Champion Pink

Champion Blue

Champion Pro White (F_1)

Champion Pro Pink (F_1)

Pink Octopus

Blue Waterfall

Dwarf Double

Source: www.bluestoneperennials.com, www.thompson-morgan.com, www.in.pinterest.com, www.pinterest.ca, www.harrisseeds.com, www.fleuroselect.com, www.sakataornamentals.com

Adriatic Bellflower: Also called "Dickinson's Gold," these bellflowers are known for their deep green foliage that produces tiny clusters of flowers that vary in color from blue to light lavender and have teeth-like petals.

Bluebell: These spunky little flowers contain attractive, trumpet-shaped blooms that hang loosely from central foliage that grows around 2-3 feet in height. Their flowers are exceptionally delicate looking due to their slender design. These summer-blooming flowers are well prized in gardens.

Peach-Leaved Bells: One of the most popular varieties of bellflowers, these flowers are not peach in color at all but range from white to a light purple. They are prized by gardeners because they bloom nearly the entire year. They attract bumblebees and honeybees that help them pollinate and spread to other areas of the garden. There are two common varieties: 'Chettle Charm' and 'Blue-Eyed Blonde'.

Serbian Bellflower: This flower is often called 'the blue waterfall' because it produces a show-stopping amount of small, lavender blooms. Each flower is perfectly star-shaped and they bloom from spring to early fall. It is exceptionally hardy, requires little to no maintenance and resistant to garden pests.

Dalmatian Bellflower: The flower is named for the South-eastern European country of Croatia. The flower stay green all through the colder season and bloom best with continual trimming once the leaves begin to turn brown.

Carpathian Bellflower: These flowers come in shades of blue and purple. They take long time to bloom and look stunning in a rock garden, border edge or a raised pot.

Clustered Bellflower: As the name implies, these flowers grow in small clusters of flowers that have bell-shaped blooms. However, they have short blooming period.

6.0 Genetics and Breeding

6.1.0 Cytology and Pollination

The morphological characters are the main criteria for the identification of a genus, but cytological characters like basic number of chromosomes, size of the chromosomes and shape of the chromosomes are also very important. In *Campanula,* there is a correlation between a certain combination of morphological characters and the number of chromosomes of the species concerned. The genus Campanula is highly heterogeneous in terms of morphological and cytological aspects. Therefore, the genus seems to be a collective genus. Although a splitting up of the genus is pointless, the delimitation should be considered in connection with other genera of the family Campanulaceae (Love, 1963). Only 40-50 per cent of the species of the genus have been investigated morphologically as well as cytologically. Gadella (1964) concluded that that, beside some rare chromosome numbers (2n = 24, 26, 28, 36, 56, 58), also some cytological series exist, each of which has its own basic number: x= 8, 10, 15, 17. Within each series the species usually show a great morphological resemblance.

On the basis of self-pollination study Gadella (1964) concluded that self-fertilization is a rarely occurring phenomenon in the genus Campanula. Most of the species are self-sterile and insect pollination is the rule, self-pollination is the exception. The flowers of most individuals in *C. rapunculoides* are self-incompatible when they first open but become more self fertile as the flowers age. Moreover, when both cross and self pollen are deposited onto the stigmas of older flowers, the cross pollen tubes grow faster and sire a disproportionate number of the seeds. Therefore, self-fertilization occurs only after most opportunities for outcrossing have exhausted.

6.2.0 Interspecific Hybridization

Campanula species are produced for both indoor and outdoor environments, but only a few species *e.g. C. portenschlagiana, C. poscharskyana* and *C. carpatica* are used for large scale production. Inbreeding depression was reported for *Campanula americana*. Ovule culture technique has been utilised to overcome post-fertilisation barriers in interspecific hybrids of Campanula species. In the selected inter-specific crosses of *C. portenschlagiana* x *C. poscharskyana* and *C. medium* x *C. formanekiana*, abnormally coloured seedlings were observed and interspecific hybrids developed poorly after germination on soil. Histological analysis of embryo and endosperm development revealed that majority of ovules did not contain an embryo. Parent-of-origin effects on germination were observed for interspecific hybrids of *C. medium* and *C. formanekiana*. (Roper *et al.*, 2015)

7.0 Growth and Flowering

The blooming period of campanula is very long which extend from June-July to early fall in the northern hemisphere. The production time varies from 18-23 weeks and depends upon management practices followed. It is advisable to follow prompt transplanting and exposure to long days. It can successfully be grown under protected conditions where the growing environment encourages long and thick stem. Further, its greenhouse cultivation also reduces the growing cycle/ time. Campanula species are usually cut back after the cold treatment and prior to forcing (Dole *et al.*, 2001).

7.1.0 Light

According to Harold Wilkins, the original species *C. medium* (Canterbury Bells) has a complex but remarkable environmental sequence. Flowers are initiated only after vernalization under short days (SD) followed by long days (LD). Only older seedlings perceive the cold and SD signals. Stem elongation occurs without flowering when plants are treated with gibberellic acid (GA_3) under SD. If these elongated plants are cold-treated, they flower under LD at the same time as the vernalized plants grown under SD and subsequently placed under LD (Dole *et al.*, 2001).

As blue bell is an obligate Long Day Plant (LDP) it is possible to programme the plants. Whitman *et al.* (1995) elaborated the following schedule for large scale production. Under photoperiods of 12 hours or less, plants remain vegetative and form a compact rosette of leaves. Flower induction occurs when the photoperiod exceeds 14 to 16 hours. After flower initiation has occurred and buds are visible, the buds will continue to develop even if the plants are subsequently placed under short days. The elongation associated with long-day bolting will decrease, which results in a shorter plant. Limited Inductive Photoperiod (LIP), in which plants are first exposed to long days for a limited period, then transferred back to short days for height control. Although this technique can be adapted for height control of Carpathian harebell, it should be noted that it usually delays bloom time and reduces the number of flower buds formed.

Long days (LD) can be provided either by extending the day length to 16 hours, or by night-break lighting for 4 hours from 10:00 p.m. to 2:00 a.m. Incandescent, high pressure sodium light, cool white fluorescent and metal halide lamps are all effective. Minimum intensity should be 10 foot candles. When using incandescent lamps, about 1.5 watts of lighting per square foot of growing space is required. The time to flower after beginning of LD depends on the forcing temperature. This is calculated about 10-11 weeks at 15°C, 8-9 weeks at 18°C or 7- 8 weeks at 21°C. Dole *et al.* (2001) determined the photoperiod and light intensity requirements for campanula (Champion Blue and Champion Pink) and concluded that for greenhouse production highest profits were obtained from campanula grown in the 8-hour initial photoperiod and transferred at 8-9 true leaves into the 16-hour final photoperiod. The type of supplemental lighting used to provide LD will influence plant height. Incandescent lights cause more elongation than fluorescent due to the higher proportion of far-red light emitted by incandescent lamps (Whitman *et al.*, 1995).

7.2.0 Temperature

Bluebells are naturally quite compact, but a variety of cultural techniques can be used to control plant height. When grown at cooler temperatures, the plants tend to be more compact and additional height control may not be necessary. DIF (temperature differential) plays an important role in its development. DIF is the difference between day and night average temperatures and is expressed positive (+) or negative (-). Plants grown under positive DIF conditions will be taller than those grown under zero DIF or negative DIF.

7.3.0 Use of Growth Regulators

Growth regulators can also be used to control plant height in campanula. However, it delays flowering in most of the species/varieties. The plant responds well to chlormequat (Cycocel) at a concentration of 1000-2000 ppm. Repeated applications at 10-12 day intervals may be necessary, depending on the vigour of the plants. The treatment is usually given about one week after the start of LD (Whitman *et al.*, 1995). Daminozide or ancymidol are recommended during visible bud stage (Adams, 1991; Madson and Madson, 1986). B-Nine application physiologically reduces the plant height by reducing the internodes length in *C. isophylla* and *C. poscharskyana*. Uniconazole (Sumagic) at 5ppm and paclobutrazol (Bonzi) at 30ppm were found to have dwarfing effect on *C. glomerata* (Pilon, 2007).

8.0 Propagation

Campanula is commonly propagated by seeds, cuttings and rhizomes. However, propagation by cutting is common for the perennial species while seed propagation is mainly used for the annual and biennial species.

8.1.0 Seed Propagation

Generally seeds are sown during autumn or spring. The seed germinates well in plug tray and light is required for germination. Seedlings will emerge in 14-20 days. A temperature range of 20-22°C is ideal for good germination. After germination, the photoperiod should be maintained at less than 14 hours to maintain vegetative growth (Whitman *et al.*, 1995). Through seed propagation, though some seedling variation occurs, but the crops are quite uniform.

For greenhouse production, campanulas are propagated exclusively by seeds. The seeds are generally raised in plug trays and the seed germinate well at 18-20°C within 5-10 days. Immediately after germination, nitrogen @ 100 ppm (from a complete fertilizer) is applied to invigorate the seedlings. The seedlings are provided temperature range from 20-22°C. When seedlings start covering the tray, a media EC of 0.7-1.0 dS/m (2:1 dilution) or 1.0-1.5 dS/m should be maintained using a complete fertilizer. When the seedlings in the plug tray attained 2-3 true leaves, it is ready for transplanting. It is advisable not to allow the plants in the plugs getting root-bound and overgrown (Dole *et at.*, 2001).

8.2.0 Cuttings and Division

Campanulas can also be propagated by cuttings and takes few weeks for rooting. The basal cuttings comprising of soft green herbaceous part roots quickly. It is recommended to grow them in a shady place. Moe (1977) stated that cuttings from stock plant of *Campanula isophylla* grown under favourable condition with light (10 Klx) and CO_2 concentration (900 vpm/ppm) enhanced root formation, growth and shoot formation.

Some of the species like *C. glomerata* and *C. punctata* can be propagated through division. This is usually done in spring once new growth appears. The clumps are lifted by carefully removing the surrounding soil all around the plant. It is usually separated into two or more sections and replanted in desired location. After planting adequate watering is needed (https://www.gardeningknowhow.com).

8.3.0 Micropropagation

Micropropagation provides an alternative to classical propagation methods, thus enabling large-scale cultivation and preservation of wild populations and endangered species. Stamenkovi *et al.* (2012) standardized *in-vitro* multiplication protocol from shoots of aseptically germinated seeds. The study revealed that the highest multiplication rate (7.7 shoots per explants) was achieved in the 3rd subculture on half-strength Murashige and Skoog (MS) medium containing 1.0 µM benzylaminopurine (BA). Very good rooting (>90 per cent) of excised shoots was achieved on media containing either indole-3-butyric acid (IBA) or indole-3-acetic acid (IAA). Paunescu (2010) standardized an efficient micropropagation system for regeneration and *in-vitro* conservation of *C. Polymorpha* using single-node aerial stem segments from mature flowering plants. The best rate of shoot production (average 14.8 shoot/explant) was achieved after 5 weeks of culture on media supplemented with 0.1 mg/l NAA (1-naphthylacetic acid) and 1.0 mg/l BAP. The established

Cultivation and Growing of Campanula

Seeds

Seedlings

Plug Seedling

Propagation by Cuttings

Transplanting

In vitro Propagation

Bed Preparation

Transplanting

Single Stem Production

Harvesting Stage

In Full Bloom

Greenhouse Production

Source: www.sakataornamentals.com, www.cutflowers.ces.ncsu.edu, www.holex.com, www.dreamstime.com, www.gardenersworld.com, www.kayabaparts.ru, www.minnesotawildflowers.info, www.tomsgardens.wordpress.com, Paunescu, A. (2010).

protocols offer a valuable micropropagation method for mass propagation of other Campanula species of horticultural importance.

9.0 Soil and Climate

9.1.0 Soil

Campanula grows best in well drained soil with pH range of 6.0-7.5. It does not prefer compact soil, therefore addition of organic matter in the growing media is beneficial for growth, flowering and longevity. It does not tolerate soggy soils, although some species such as *Campanula lactiflora*, *C. latifolia*, and *C. takesimana* prefer moist soils. Extremely wet soils, especially in winter months, will cause root rot. Fertile soils may encourage too much growth for fast-spreading species such as *C. glomerata*, *C. punctata*, and *C. rapunculoides* (Hawke *et al.*, 2008). The species *C. poscharskyana* can be cultivated in dry soils (Madson and Madson, 1986).

9.2.0 Climate

The plants grow well in sunny location and little shade in the morning hours can affect flowering severely. As it is a cool season flowering plant, it prefers temperature range of 17-18°C (day) and 14-16°C (night). It can tolerate mild frost to some extent. For seed germination, temperature and light plays a vital role in campanula. Forcing schedule is recommended at lower temperature (16-20°C) to avoid small size flowers. Improved campanula varieties are particularly suited to unheated greenhouse production, which encourages long, thick stems and provides protection from weather (Dole *et al.*, 2001).

Campanula is basically long day plant. Light is essential for germination, therefore the small seeds should not be covered. The flowering behaviour is proportional to the direct exposure to sunlight. Supplementary lighting of 400-500 foot candle in UK/USA is highly beneficial for improving flower quality during winter months (Whitman *et al.*, 1995). For commercial production, plants should be exposed under High Intensity Discharge (HID) grow lights. Supplementary HID lights are used in Northern Europe (Dole *et al.*, 2001).

10.0 Cultivation

10.1.0 Raising of Seedlings

Seedlings can be prepared in a raised nursery beds. The dimension can be kept as per the requirements. The site should have full sunlight away from any shade. The soil should be well drained and rich in organic matter. The top soil should be free from weeds and well rotten manures/leaf mold should be mixed and levelled. Seeds should be sown evenly and thinly in lines and cover with fine soil for germination. Keep the soil moist and seeds will germinate in three weeks.

10.2.0 Spacing

The plants should be spaced at 10-15 cm for single stem production and 25-30 cm for pinched production. Under protected cultivation system, longest stems are produced with single stem production. Pinched production works best for field or unheated greenhouse production and 6-10 stems can be harvested per plant (Dole *et al.*, 2001). For commercial field cultivation planting density of 32/m^2 should be followed (Madson and Madson, 1986), whereas for cut flower spacing of 15cm × 15cm or 25cm × 25cm is advisable (Jain *et al.*, 2017).

10.3.0 Mulching

A slight moisture stress may severely affect the growth and flowering in campanulas. Mulching using organic mulches (straw, bark or shredded) help in retaining soil moisture and maintain even soil temperatures. In a garden the mulch also gives a natural look to the bed and improves the soil texture with time. To prevent possible rot it is advisable to keep mulches away from main plant.

10.4.0 Fertilization

For successful cultivation it is advisable to go for moderate fertilizer application. For commercial cultivation NPK@ 20:10:20 is useful for plant growth. Study also suggested that use of calcium nitrate fertilizers is also beneficial. In the standing crop iron deficiency leads to tip burn and boron deficiency may cause distortion and tip abortion (Dole *et al.*, 2001). For garden culture application of NPK @ 5:10:10 or 10:10:10 once in the spring and once during the mid-summer is advisable.

10.5.0 Irrigation

Campanulas are very precise in water requirement. As with most of the campanula species, plants should not be overwatered and should be grown on the dry condition. They are also susceptible to wilt and require frequent irrigation (Dole *et al.*, 2001). Watering should be done when the top third of the soil has dried out by bottom-up method to reduce the chance of diseases. Excessive watering causes rotting of lower leaves, yellowing of leaves, withering of buds and flowers, and root or crown rot.

10.6.0 Staking

Since campanulas have generally weak stems at the base, netting should be provided particularly during periods of low light intensity for supporting the plants. Netting is highly recommended for field-grown campanulas as their bell-shaped flowers catch hold of water and fall over (Dole *et al.*, 2001).

11.7.0 Pot Culture

Campanulas are successfully grown in pots of various sizes. The varieties/ species with short stature and form mound like shape are suitable for pot culture. Plants can be propagated by seeds/seedlings/cuttings or rhizomes. Multi-mix potting compost (cocopeat, vermicompost, leaf mold, peat, perlite, vermiculite, *etc.*) with moist well drained media are suitable for pot culture. To avoid transplanting shock it is advisable to plant in the afternoon. The media should be watered on a regular basis without becoming soggy. Liquid fertilizer mix can be given every week to keep the plants healthy. For better growth and flowering in potted campanulas it is advisable to provide artificial light with at least 13 hours illumination per day. In summer, pots should not dry out in hot sun. Whiteman *et al.* (1995) reported that the plants can be placed pot-to-pot after transplanting, until leaves begin to touch. Final spacing should be about 3 to 4 plants per square foot, depending on plant size at the start of flower induction. Regular deadheading of withered flower stalk promotes blooming. However, heavy pruning is recommended when all the flowers have dried. It will help the plant to re-grow and flower again. Repotting is recommended every three years.

11.0 Diseases and Pests

11.1.0 Diseases

Bluebells are highly susceptible to various fungal diseases like damping-off or root rot caused by *Pythium* or *Rhizoctonia*. Use of a well drained medium will help to reduce these problems. Flower spotting (*Botrytis*) can be especially prevalent on field-grown campanula. Root and stem rot (*Fusarium* species, *Rhizoctonia solanii* and *Sclerotinia sclerotiorum*) can be a major problem during production. Overgrown seedlings are susceptible to root and crown rot (Dole *et al.*, 2001). Leaves may also become infected by *Botrytis cinerea*, so it is necessary to keep the foliage dry (Whitman *et al.*, 1995). Some workers have also reported the incidence of botrytis, stem rot, rust, powdery mildew, aster yellows, tomato spotted wilt virus (TSWV) *etc.*

11.2.0 Pests

Few insects are attracted to Carpathian harebell, but spider mites, aphids, white flies and slugs occasionally causes problem. Spray of acaricide (Omite) and systemic insecticide (imidacloprid) control the infestation. Thrips are especially prevalent and can be particularly damaging to dark purple flowers. Fungus gnats are typically a problem during propagation (Dole *et al.*, 2001).

13.0 Harvesting and Postharvest Management

Campanula stems should be harvested when 2-3 lower flowers have opened. For maximum shelf-life, a reduction in fertilization near the end of the crop is recommended. Two or three weeks before harvest, it is advisable to irrigate with clear water and reduce fertilizer dose especially nitrogen. Research on Carpathian harebell suggests that the production temperature regime associated with best keeping quality is zero DIF. The use of either positive or negative DIF decreased postharvest shelf-life, but more research is needed in this area (Whitman *et al.*, 1995).

The flower responds well when placed in warm water (37°C) immediately after harvest followed by pulsing (5 per cent sucrose) for 24-hour. Fresh cut, un-stored flowers will have a vase life of 10-14 days. Cut stems can be stored at 1-2°C for one week in water. Longer storage of up to three weeks in water or up to two weeks dry storage will result in a shorter vase life of 6-9 days. Generally, lower storage temperatures are recommended as flowers continue to open even during cold storage. Dry storage may be more effective at temperatures close to 0-1°C. In the arrangements or bouquet, use of 1-2 per cent sucrose solution is advisable (Dole *et al.*, 2001).

Disease and Insect Pests

Rust

Slug

Harebell aphid

Harvesting and Postharvest Management

Harvesting Stage of Flowers

Harvesting in Greenhouse

Flowers in Market

Pod Setting

Dried Pods

Seeds of Campanula

Source: www.houzz.com, www.wildflowerfinder.org.uk, www.influentialpoints.com, www.alamy.com, www.bladmineerders.nl, www.rocketfarms.com

The campanula flowers are sensitive to ethylene, which causes browning and shriveling of open petals; buds may also discolor and wilt. Application of silver-thio-sulphate (6 ppm) shortly before harvest maximizes postharvest life (Whitman *et al.*, 1995). Other chemicals which can be used are Ethylbloc (1-MCP) (Dole *et al.*, 2001), and 8-HQC (Bosma and Dole, 2002).

14.0 Biotechnology

Campanula portenschlagiana germplasm was evaluated horticulturally and using a molecular marker. Flowering of *C. portenschlagiana* collected in the United States (CPC) was accelerated by 16-hour long day (LD) photoperiod at 7.5°C. However, flowering of *C. portenschlagiana* collected from Japan (CPJ) was not affected by LD at 7.5°C. CPC and CPJ were grouped closely when polymorphic RAPD bands were analyzed. Seedlings of *C. portenschlagiana* from Germany (CPG) were grouped distantly from CPJ and CPC and were significantly different in growth habit. The results of greenhouse forcing and randomly amplified polymorphic DNA (RAPD) analysis, species and cultivars should be characterized and identified morphologically and using molecular markers (Joung *et al.*, 2001).

Investigation on genetic relationships among seven species in *Campanula* L. by inter-simple sequence repeat molecular markers (ISSR) revealed that 28 of 40 primers can be amplified to clear bands. All the 28 ISSR primers amplified 164 bands, and the proportion of polymorphic fragment was 98.8 per cent. The value of genetic similarity (GS) indexes of seven species based on the ISSR data varied from 0.421 to 0.945, and the highest genetic similarity was 0.945 between *C. aristata* and *C. modesta*, which indicated close relative relationship. Cluster analysis revealed that all the species could be divided into four groups: *C. colorata*, *C. cana* and *C. calcicola* in the first group, *C. calcicola* in the second group, *C. crenulata* in the third group, and *C. aristata* and *C. modesta* in the last group. It provided reference for the molecular marker assisted identification, resource conservation, flower variety development and breeding of Campanula (Bao *et al.*, 2018).

Postharvest biotechnological study conducted on three Campanula species revealed that *C. portenschlagiana* (Cp) is ethylene sensitive, *C. formanekiana* (Cf) with intermediate sensitivity and *C. medium* (Cm) as ethylene insensitive. Jensen *et al.* (2016) identified key elements in ethylene signal transduction, specifically in Ethylene Response Sensor 2 (ERS2), Constitutive Triple Response 1 (CTR1) and Ethylene Insensitive 3- Like 1 and 2 (EIL1 and EIL2) homologous. A natural mutation was identified in Cmeil2 causing a frameshift which resulted in difference in expression levels of EIL2, with more than 100-fold change between Cf and Cm in young flowers. This study showed that the naturally occurring 7 bp frameshift discovered in Cmeil2 is the key gene in the ethylene signaling pathway, correlates with ethylene insensitivity in flowers. The transfer of the EIL2 mutation to other plant species shall provide a novel tool to engineer ethylene insensitive flowers.

15.0 References

Adams, R. (1991) *Grower Talks*. Geo J. Ball Publishing, Geneva, Illinois, USA.

Bailey, L.H. (1942) *The Standard Cyclopedia of Horticulture* **1**, The Macmillan Co., New York, USA.

Bao, Z.Y., Qu, J.P., Yang, Z.J., Lu, L., Liao, M.L., Zhou, M.X., Peng, Z.S (2018) *Guangxi Zhiwu/Guihaia,* **38**(6): 818-824.

Barnaulov, O.D., Limarenko, A. Yu and Teslov, L.S. (1983) *Rastit Resur,* **19**: 20–27.

Barnaulov, O.D., Manicheva, O.A., Teslov, L.S. (1984) *Khim Farm Zh.,* **18**: 853–857.

Becker-Dillingen, J. (1950) *Handbuch des gesamten Gemiisebaues*. Parey, Berlin.

Beckett, K.A. (1985) *The Concise Encyclopedia of Garden Plants,* Orbis Publishing Ltd., London.

Beckett, K.A. (1987) *The RHS Encyclopaedia of House Plants including Greenhouse Plants,* Salem House Publishers, Massachusetts, USA, pp. 130-131.

Bogdanovic, S., Resetnik, I., Jericevic, M. *et al.* (2019) *Plant Systematics and Evolution.,* **305**: 687–703.

Boissier, P.E. (1875) *Flora Orientalis* **3**. Genevae and Basilae.

Bosma, T. and Dole, J. M. (2002) *Hort Sci.,* **37**(6): 954-958.

Brickell, C. (1994) *The RHS Gardener's Encyclopedia of Plants and Flowers,* Dorling Kindersley Ltd., London, pp. 453-454.

Damboldt, J. (1976) *Notes from the Royal Botanical Garden*, Edinburgh, **35** : 39–52.

Dole, J., Cavins T. and Bosma, T. (2001) *Greenhouse Product News*. pp. 28-35.

Gadella, W. J. (1964) *Botanical Museum and Herbarium, Utrecht*, pp.1-115.

Gastaldo, P. (1978) *Compendium of Italian Officinal Flora, Part 23. Fitoterapia,* **49**: 27–38

Hawke, R. G. (2008) *Chicago Botanic Garden*, **31**: 1-6.

Hay, R. and Beckett, K.A. (1971) *The reader's Digest Association Ltd.,* London, pp. 109-113.

Hong, D., Song, G., Lammers, T. G. and Klein, L. L. (2011) *Flora China.,* **19**: 504-564.

Jain, R., Hiremath, V. M. and Misra, S. (2017) *In*: Misra, R.L. and Misra, S. (eds.), *Commercial Ornamental Crops Cut Flowers,* Kruger Brentt Publisher, U.K., pp. 85-92

Jensen, G., Nielsen, R. H. and Andersen, A. S. (2005) *Combined Proceedings International Plant Propagators' Society*, **55**: 310-318.

Jensen, L., Hegelund, J. N., Olsen, A., Lütken, H. and Müller, R. (2016) *BMC Plant Biol.,* **16**: 117.

Joung, Y.H., Roh, M.S., Kim, T.I. and Song, J.S. (2001) *Acta Hortic.,* **546**, 421-425.

Kuss, P., Ægisdo, H.H., Stöcklin, J. (2007) *Perspectives in Plant Ecology, Evolution and Systematics,* **9**: 37–51.

Lovasen-Eberhardtz (2000) *Index Florae Croaticae Pars 3. Nat Croat* **9** (1): 65.

Love, A. (1963) *Regnum Vegetabile,* **27**: 45-51.

Madson, P. and Madson, K. (1986) *Bedding Plants International News*. **16**: 7.

Moe, R. (1977) *Scientia Hortic.,* **6**: 129-141.

Paunescu, A (2010) *Analele Universităþii din Oradea - Fascicula Biologie,* **XVII** (2): 235-238.

Pilon, P. (2007) *Greenhouse product news,* **17**(13): 1.

Pizzetti, I. and Cocker, H. (1968) *Flowers-A Guide for Your Garden.* Harry N. Abrams, Inc. Publishers, New York. pp. 174-182.

Roper, A.C., Lutken, H., Christensen, B., Boutilier, K., Petersen K. K. and Muller R. (2015) *Euphy,* **203**: 643–657.

Saez, L. and Aldasoro, J. J. (2003) *Bot. J. Linn. Soc.,* **141**(2): 215–241

Stamenkovi, V., Jelencic, B. and Tkalec, M. (2012) *Period. Biol.,* **114** (1): 77–82.

Whiteman, C., Heins, R., Cameron, A. and Carlson, W. (1995) *Professional Plant Growers Association News,* **26**(4): 2-3.

www.gardenloversclub.com

www.gardeningknowhow.com

20

Astilbe

K. Ravindra Kumar and M. Lakshmi Durga

1.0 Introduction

Astilbe Buch.-Ham.ex D. Don is a rhizomatous, hardy, herbaceous clump forming and shade loving perennial of the family Saxifragaceae (Armitage, 1996). It is valued by the horticultural industry for their distinctive fern-like foliage, attractive flowers, and low maintenance. It is commonly called as 'False Spirea', 'False Goat's Beard' and 'Feather Flower'. The coloured panicle blooms as a tall and fluffy plume above frilly foliage that is highly suitable for ornamental gardening. It is one of the top ten landscape perennials. Astilbe is named by a Scotland botanist, Lord Hamilton, 'A' means 'without', 'stilbe' means 'brilliance', which referred to matte, dull astilbe leaves. Plants are resistant to deer and rodents (Skelly, 2014).

2.0 Importance and Uses

2.1.0 Aesthetic Uses

Astilbe can be used as a cut flower, in fresh and dry floral arrangements and as a hedge in landscape gardens. Being a shade tolerant plant with attractive flowers, they make great companions with other shade loving plants such as hosta and hellebores, with contrasting foliage and coordinating blooms. They are often planted to lighten up the shaded portion or soften a sunny spot in a garden. The astilbe flower beds are highly suitable to grow in the garden during the summer months of temperate regions. Wide ranges of colours from white to dark purple are available in this for ornamental landscaping. In addition, different varieties bloom at different times and are available in varying heights. These plants can be used as border, container, massing, rock garden and woodland garden for ornamental purpose. The low growing cultivar 'Pumila' is used as ground cover.

2.2.0 Therapeutic and Other Uses

The rhizome of *Astilbe chinensis* possesses anti-elastase and anti-tyrosinase activity which protects the human skin from wrinkling and hyper pigmentation (Lee, 2009). Astilbic acid from *A. chinensis* is beneficial in regulating various inflammatory processes (Moon *et al.*, 2005). The dried rhizomes of *Astilbe thunbergii* have been used for the treatments of sword cut, animal bitten wounds, frost-bite, burn, suppurative dermatitis or skin inflammatory diseases (Kimura *et al.*, 2007). The ethno-medicinal use of *Astilbe rivularis* was reported by Mandal *et al.* (2009) and proclaimed that the species have free radical scavenging potentiality and anti-lipid peroxidation.

3.0 Origin and History

Recognized species of the genus astilbe are endemic to Eastern Asia, with the exception of sub-dioecious *A. biternata*, which is native to North America (Trader *et al.*, 2006). *Astilbe* was brought to Europe from Japan in the early 19th century by Carl Peter Thunberg and Philipp Franz von Siebold, and since then it is under cultivation in shady gardens. 'Peach Blossom', the first cultivar introduced in 1903, is still leading the market. *A. simplicifolia* has been recognized as the parent species from which all the members of the genus were derived (Olson and Antonovics, 2000).

Uses of Astilbe

Shrubbery Border

Divisional Line

As Specimen

As Hedge

In Pots

Bankside Plant in Water Garden

Species of Astilbe

Astilbe chinensis

Astilbe japonica

Astilbe simplicifolia

Astilbe thunbergii

Astilbe koreana

Astilbe rivularis

4.0 Taxonomy

Astilbe is a genus of herbaceous perennials, belonging to the family Saxifragaceae. It comprises of 18 to 40 species. The foliage are fern like, glossy, dentate, compoundly pinnate and produces upright stems with a height ranging from 30 cm to 120 cm. Inflorescence is fluffy, feathery plumes, extending above the foliage and are available in variety of colors. The individual flowers are mostly monoecious, numerous, small to inconspicuous on a paniculate, rhombic or pyramidal inflorescence (Trader, 2006). Flower consists of 5 small lobed calyx, 4 petals, 8-10 stamens, 2-3 carpels with numerous seeds in the dehiscent follicle (Bailey, 1935). The fruit of Astilbe is a capsule. Aerial growth arises from a crown of several buds and subtended by a mass of fleshy storage roots. There are more than 25 species scattered around the world with hundreds of hybrids available in the market.

5.0 Species and Cultivars

Extensive breeding of astilbe was done to develop large flower types, flower color, and unique foliage. The selections and cultivars currently utilized in horticulture were derived from the breeding efforts of a famous German hybridizer, George Arends (1862-1952). Most of the selections and varieties were derived mainly from four species for having dense flowers with longer petals: *A. chinensis, A. japonica, A. simplicifolia,* and *A. thunbergii* (Randhava, 2005). The most popular hybrid groups are Arendsii hybrids (*Arendsii hybrida*), Japanese hybrids (*Astilbe japonica hybrida*), Chinensis astilbe (*Astilbe chinensis*) and its types, and *Astilbe simplicifolia* (Namyatova, 2019).

A new species of *Astilbe* named *A. tsushimensis* Kadota was discovered in Tsushima Island of Japan (Kadota, 2012). It is distinguished from other species by wider and white spathulate petals, coarsely dentate, large leaflets and carpels connate in the upper part.

Japanese Astilbe (*A. japonica*)

It is an earliest flowering type. Plant is low (≤ 80 cm height), compact, with shiny ornamental foliage. Inflorescence is panicle with pink or white flowers. Modern varieties of Japanese astilbe have cold tolerance. Varieties include 'Deutschland' (*Astilbe japonica* Deutschland) with white flowers, 'Rheinland' varieties (*Astilbe japonica* Rheinland) with pink flowers, 'Europe varieties' (*Astilbe japonica* Europe) with pale lilac, and, 'Montgomery astilbe' (*Astilbe japonica* Montgomery) with fluffy panicles of bright red or burgundy hue.

Hybrid Arends

The species is named after the famous German breeder G. Arends. Plants are vigorous, spreading, medium tall (1 m height) and foliage is dark green with burgundy piping. Inflorescence is spherical or conical, with cream pink yellow flowers. It blooms longer than other species, for 30-40 days from July to August. The most popular varieties include 'Amethyst' (Lavender pink), 'Bridal Veil'(White), 'Cattleya' (Rose purple), 'Boogie Woogie' (pink), 'Gloria' (pink) 'Midget' (pink), 'Diamond', 'Weisse Gloria' (White), 'Rubin' and 'Diamond Glut'.

Cultivars of Astilbe

Younique Carmine

Astilbe Chinensis 'Pumila'

Astilbe × arendsii 'Amethyst'

Younique Ruby Red

Astilbe × arendsii 'Fanal'

Ostrich Plume

Astilbe x arendsii 'Chocolate'

Astilbe japonica 'Peach blossom'

Younique Silvery Pink

Astilbe japonica 'Deutschland'

Astilbe japonica 'Rheinland'

Astilbe chinensis var. *taquetii* 'Purple Lance' syn. 'Purpurlanz'

Cultivars of Astilbe

Astilbe chinensis
'Maggie Daley'

Astilbe chinensis
'Vision'

Astilbe japonica
'Europa'

Astilbe Younique White

Astilbe japonica
'Bonn'

Astilbe x arendsii
'Bridal Veil'

'Younique Salmon'

***Astilbe x arendsii* 'Chocolate'**

Bressingham Beauty

'New Wave'

Japanese Hybrids

Hybrid Arends

Chinese Astilbe (*A. chinensis*)

The Chinese Astilbe (*A. chinensis*) is considered to be more heat and drought tolerant and can be used for mass effect, ground cover or as a perennial border. It grows to a height of 90-110 cm. The basal leaves are large, long-petioled and the stem leaves are smaller, with short petioles. Inflorescence is dense, 30-35 cm in length. Flowers are small, usually lilac, with scarce occurrence of pink and white flowers. Two variants are available *viz.*, dwarf ranging from 15 to 25 cm height (*var.* Pumila *hort.*) and varieties with cone-shaped inflorescences (*var.* Taquetii). This species of *Astilbe* perfectly grows in full sun. Popular varieties in this species are *Astilbe chinensis taquetii* 'Purpurlanze' (lilac), 'Vision in Pink' (pink) and *Astilbe pumila hybrid* 'Vision in Red' (dark purple).

Astilbe simplicifolia and *Thunberg hybrids*

Hybrids of *Astilbe simplicifolia* poorly tolerate dry air and heat. Plants are low (1.5-50 cm) in height and inflorescence is drooping and feathery. The varieties include 'Praecox Alba' with white loose candles of inflorescence, 'Pinkish Bronze Elegans', named after the bronze shade of leaves, 'Coral Straussenfeder' and 'white Professor van der Wielen' (Thunberg hybrid).

Astilbe koreana

This species is distinguished from other species by having one rhizomatous leaf and dense multicellular short glandular hairs on the young shoot, stem, and inflorescence (Han and Oh, 2019).

6.0 Growth and Flowering

According to the flowering time, astilbe was classified as early (late June-early July), medium (July), and late (August) groups. It takes 27-36 days from visible inflorescence to first flowering irrespective of the photoperiod (Runkle *et al.*, 1998). Flowering time varies between the species *i.e.*, *A. chinensis*, a late flowering species, flowers during July to August. *A. japonica* and other early hybrids flowers in May-June. *A. thunbergii* flowers in May. With these variations in flowering time, the whole sequence of garden color is possible with this genus. Grantzau and Fischer (1990) reported that the cultivars 'Brautschleier', 'Fanal' and 'Washington' flowered from mid to late June with most of the flowers bloomed from early July to early August. Else Schluck, Glut, Montgomery, Lilly Goos and Superba cultivars flowered from mid-July.

Plants may be forced to flower in greenhouse as potted plants after completion of cold treatment. Vernalization is required at a temperature of 15.5°C for 10-18 weeks from last week of December to May (Halevy, 1987). A constant water supply with no shade and fertilizer is required during forcing. However, a light intensity of less than 32 klx is required during the commencement of flowering. Runkle (1998) reported that under long day conditions, flowering percentage of *Astilbe chinensis pumila* progressively increased from 0 to 100 as cold treatment increased from 0 to 15 weeks. However, flowering percentage under short days was low regardless of cold treatment. Taller inflorescence was noticed under long day conditions.

6.1.0 Plant Growth Regulators

Foliar application of 100 mg/dm^{3}benzyl adenine at third leaf stage and seventh leaf stage accelerated the flowering in the field by 2 weeks and at the concentration of 200 mg/dm^3, increased the flowering abundance in *Astilbe* × *arendsii* (Pogroszewska and Sadkowska, 2007).

7.0 Propagation

Astilbe is commercially propagated by division even though seed propagation and rhizome separation is possible (Beattie, 1994).

7.1.0 Seed Propagation

One of the Astilbe species, *A. chinensis pumila* yields homozygous plants when propagated by seed. Otherwise, seed propagation is only recommended for the development of new varieties as the results are variable and take many seasons to produce flowers. Stratification is must for germination of seedlings. For germination, seeds are to be sown in a container (15 cm wide) with a mixture of peat and sand (1: 1) and covered with 0.4 inch snow. This is followed by covering the container with a transparent bag and kept in refrigerator for twenty days for better seed germination. Once the shoot is emerged, it is transferred to a light warm (17-22°C) place and given light irrigation. Seedlings are transplanted at 2-3 leaf stage during May-June. Cho *et al.* (2016) studied the effects of different environment and chemicals on seed germination in *Astilbe koreana*. They observed that seed germination was good at 15 °C under light conditions; however no germination took place in dark conditions. Treating the seeds with 20 mM KNO_3 increased the per cent seed germination within shortest period of time. Generally, the weight of 1000 seeds was 40 to 50 mg.

7.2 Vegetative Propagation

7.2.1 Propagation by Division

Division is one of the simplest and easiest methods of astilbe propagation. The best time for division is in early spring (March-April), before the bud break followed by August. Old crowns, less than one year of age are selected as stock plants for division. Larger and more vigorous growing crowns of the *Astilbe* × *arendsii* and *Astilbe japonica* types are best divided with the help of a knife. *Astilbe simpficifolia* types tend to have more numerous, but smaller buds, and can often be separated by pulling them apart by hand. Replanting the eye at the correct depth is critical for the success of Astilbe propagation. The eyes should be planted at or just below soil (medium) level. With most cultivars, planting a single eye in the spring should yield a salable sized plant with as many as 10 eyes by the end of the growing season. Leclerc *et al.* (2005) found that there is no influence of inflorescence removal on propagule enhancement in *Astilbe*× *arendsii.*

7.2.2 Propagation by Rhizomes

Separation of rhizomes is also an easy method of propagation of astilbe. The mature plants perform better if divided after every 3 to 4 years. Rhizomes are spaced at 40 to 75 cm and soil packed firmly around the new plants.

Propagation of Astilbe

Seeds

Seedlings

Growth of Seedlings

Astilbe Plants

Separation of Plants

Astilbe Rhizomes

Planting of Rhizomes

Growth of Plants from Rhizome

Cultivation of Astilbe

Cultivation in Open

Cultivation in Greenhouse

7.3.0 Micropropagation

Shoot tip culture is being practiced for pink and white flowers but not for red flowers as their tendency to sport naturally. Tip cuttings are not possible as the shoots are compressed in to buds or eyes. Gu *et al.* (2003) reported callus induction and shoot regeneration in *Astilbe tabularis* using leaf as explant. The tender leaves were cultured on MS medium supplemented with 1.0 mg/l 6-BA + 0.08 mg/l IBA for callus induction and the callus was transferred to 2.80 mg/l 6-BA + 0.02 mg/l IBA + 0.90 mg/l GA_3 for shoot bud differentiation. The well-developed micro-shoots were cultured on $1/6^{th}$ MS of basal media supplemented with 0.02 mg/l IAA and 0.04 mg/l IBA for rooting. For germplasm conservation in *in vitro* for 27 months, the cultures were maintained on $1/4^{th}$ MS basal media supplemented with 0.1 mg/l 6-BA and 2.50 mg/l phloridzin.

8.0 Soil and Climate

Astilbe prefers to grow in shade at cool climatic areas. The plants are tough, hardy and can survive winter, even in harsh climates. The plants will bloom in full shade, but they require some sunlight to achieve their full size. They thrive best in fertile, moist, humus-rich sandy soil with an optimal acidity of 5.5-6.5. Browning of plant tissues can be observed in dry soils (Blanchette and Beattie, 2001). After harvesting, the cut stems can be incorporated in the soil as organic manure. As a perennial, astilbe can be grown at the same place for 20 years.

9.0 Cultivation

9.1.0 Fertilizer

In spring (March-April) plants should be fed with nitrogenous fertilizers and in mid-June with potassium based nutrients. The plants need phosphorus to bloom, hence phosphoric acid application is required to be done before flowering. Alternatively, N:P:K formulation with 5:10:5 or 10:10:10 can also be applied. After fertilization, the soil is loosened and mulched to conserve the moisture and fertilizer use efficiency. The supplemental application of fertilizers needs to be fixed based on the native soil nutrient status. Astilbe is a moisture loving plant suitable for waterside plantings, shady woodland gardens and boggy areas.

9.2.0 Intercultural Operations

Earthing up is necessary as daughter rhizomes are developed at the top of soil and the mother rhizomes gradually die-back. Mulching will protect not only the root system from overheating, but also retain the soil moisture and get rid of weeds. Deadheading is not practiced in astilbe as it can't put forth recurrent blooms after removal.

10.0 Diseases and Pests

Among the diseases, cercospora leaf blight, powdery mildew, tobacco ring spot virus affects the crop adversely. Aphids, asiatic garden beetles, slugs, tarnished plant bug, vine weevil, meadow spittlebug, strawberry crimp nematode and root-knot nematode are the major pests affecting this crop.

10.1.0 Diseases

10.1.1 Cercospora Leaf Spot

Small flecks with yellowish halo appear on the leaves and turn brown and coalesce. They cause the leaves to wither and die. Crop rotation and removal of affected plant parts is recommended.

10.1.2 Powdery Mildew

The disease occurs on the top of the leaves in humid weather conditions. The leaves appear to have a whitish or greyish surface and may curl. Provide good air circulation for wider spacing and keep the weeds under control.

10.1.3 Tobacco Ring Spot Virus

Plants may have a yellowish tone and become stunted, infected plants should be removed.

10.1.4 Fusarium Wilt

This pathogen infects the vascular or water-conducting tissues of plants, causes wilt symptoms by impairing water flow. The symptomatic plants initially wilts on one side of the plant, leaves twisted, turns yellow to brown and hang down prior to drying up. Vascular tissue discolours and plant exhibits drought symptoms despite adequate soil moisture. Wilt control involves removal of infected plants and associated roots and soil.

In 2013, Southern blight (*Sclerotium rolfsii*) was first reported in astilbe in Japan (Sasai *et al.*, 2015). Recently, Pavlyuk *et al.*, 2020 identified *Puccinio stelemandschurica* as a disease causing agent of rust infection in astilbe cultivars. The symptoms of this disease includes, decrease in plant height and foliage density, deformation of inflorescence, appearance of pustules, galls, yellow spots on the leaves and flower stalks.

10.2.0 Pests

10.2.1 Aphids

Greenish, red, black or peach coloured sucking insects can spread disease as they feed on the undersides of leaves. They leave a sticky residue on foliage that attracts ants.

10.2.2 Asiatic Garden Beetle (*Maladera castanea*)

Velvety, cinnamon-brown beetle grubs eat the roots and the adults damage the foliage and blossoms. Cleaning up fall debris to prevent the inhabitation of grubs in the soil. Feeding on the foliage can be prevented by a spray of acephate, azadirachtin or malathion, applied early in the season as a soil drench, will provide season-long systemic control.

10.2.3 Slugs

These pests leave large holes in the foliage or eat leaves entirely. They leave a slime trail, feed at night and are mostly a problem in damp weather. Hand pick at night, use of attractants like cornmeal and placing barriers like diatomaceous earth helps to get rid of slugs.

10.2.4 Tarnished Plant Bug

These insects cause distorted leaves and flower buds. The adults are about 1/4 inch long, oval shaped and flat. They are greenish brown with reddish brown markings on their wings. There is a small but distinct yellow tipped triangle in the center of the back behind the head.

10.2.5 Vine Weevil

This insect cuts irregular notches in leaf margins and grubs feed on plant roots, sometimes causing the death of the plant. Handpick adults at night, shake the plants over newspaper to dislodge them.

10.2.6 Meadow Spittle Bug

It settles in the axils of the leaves and forms foamy salivary secretions. Leaves become wrinkled and covered with yellow spots. As a result, the plant withers partially or completely. A spray of malathion controls this pest.

10.2.7 Strawberry Crimp Nematode and Root-knot Nematode

Strawberry crimp nematode is a leaf pest, it parasitizes on the buds, leaves and flowers, making them deformed and covered with yellow or brown necrotic spots. The growth of plants is slowed down due to nematode infestation. The root knot nematode affects the roots, forming small root-knot galls that contain nematodes. Infected plants develop and bloom poorly, sometimes they even die.

11.0 Harvesting and Postharvest Management

11.1.0 Harvesting

The ideal stage of flower harvesting is half to two-third florets open. Flowers are moderately sensitive to ethylene (Dole *et al.*, 2017). Inflorescence blooms for 2-3 weeks. Flower opening is acropetal and even half-dry flowers add beauty to the garden. After harvest, stems should be cut off at the ground level, and the area covered up for the rhizome division. After frost, clumps are harvested, graded and stored in moist moss peat, sphagnum or shredded wood at 0-2°C and 85-90 per cent RH for long time periods until next planting. In early autumn, the rhizomes are divided so that each section has a growth bud for the next season planting.

11.2.0 Postharvest Management

Astilbe inflorescence has many small florets and a short vase life. The negative effects of the dry period from market to the consumer can be prevented by sleeving the flowers to restrict transpiration and desiccation. Xylem blockage in astilbe is a plant induced obstruction that prevents water uptake from vase solution (Loubaud, 2003). Vandoorn *et al.* (1993) reported that improved longevity of cut stems (6.6 ± 2.1 days) was noticed when flowers are cut and stored in water at 20°C. Astilbe responds well to a 10 min 'pulse' with a 0.02 per cent detergent solution. Yamazaki, (2020)

Diseases and Pests of Astilbe

Powdery Mildew

Tobacco Ring Spot Virus

Cercospora Leaf Spots

Aphids

Tarnished Plant Bug

Astilbe Weevil

Harvesting and Marketing of Astilbe

Harvesting Stage

Harvesting in Open

Astilbe in Market

Astilbe Cut Flowers

Astilbe Pots Ready for Market

investigated the effects of trehalose and sucrose on gene expression of senescence related genes in *Astilbe ×arendsii* 'Gloria purpurea'. Vase life was extended from 3.7 days to 12.4 days when flowers are treated with 4 per cent sucrose + 2 per cent trehalose.

12.0 Biotechnology

Zhu *et al.* (2013) conducted phylogenetic analysis in an Asian North American disjunct genus using sequences of nuclear ribosomal internal transcribed spacer (ITS) and plastid matK, trnL-trnF and psbA-trnH regions. The monophyly of the genus was well described by both ITS and plastid sequences.

13.0 References

Armitage, A.M. (1996) *Stipes Pub.Champaign.*, 167-176.

Bailey, L. H. (1935) *Standard Cyclopedia of Horticulture*, Macmillan, New York, 422-423.

Beattie, D.J. (1994) *Georgia Commercial Rower Growers Association Newsletter.*, **1**: 18.

Blanchette, L. and Beattie, D. (2001) *Combined Proceedings-International Plant Propagators Society.*, **51**: 449-456.

Cho, J.S., Kwon, H.J. and Lee, C.H. (2016) *Korean J Plant Res.*, **29**(3): 339-346.

Dole, J., Stamps, R., Carlson, A., Ahmad, I., Greer, L. and Laushman, J. (2017) *Assn. Specialty Cut Flower Growers, Oberlin, OH.*

Grantzau, E. and Fischer, C. (1990) *Deutscher Gartenbau*, **44** (4): 226-228.

Gu, D., Lu, S., Zhu, J. and Chen, X. (2011) *J. Zhejiang Univ. Sci. B (Science Edition).*, **38**: 689-694.

Halevy, A.H. (1927) *Astilbe* In: *Handbook of flowering*, Vol. I : 5521-522

Han, S.H. and Oh, B.U. (2019) *J Asia Pac Biodivers.*, **12**: 302-310.

Kadota, Y. (2012) *J. of Japan. Bot.*, **87**: 307-313.

Kalkman, E.C. (1986) *Acta Hort.*,**181**: 389-392.

Kimura, Y., Sumiyoshi, M. and Sakanaka, M. (2007) *J. Ethno pharmacol.*, **109**: 72-77.

Leclerc, M., Caldwell, C.D., Lada, R.R., and Norrie, J. (2005) *Hort.Sci.*, **3**: 756-759.

Lee, S., Sandesh, S., Shruti, S. and Seo, S. (2009) *Amer. J. Pharmacol Toxicol.*, **4**: 127-129.

Loubaud, M. and Van Doorn, W.G. (2004) *Postharvest Biol. and Technol*, **3**: 281-288

Moon, Chul, T., Lin C.X., Lee, J.S., Kim, D.S., Bae, K.H., Son, K.H., Kim, H.P., Kang, S.S., Son, J.K. and Chang, H.W. (2005) *Biol. Pharm. Bull.*, **28**: 24-26.

Mandal, P., Mishra, T.K. and Basu, P.K. (2009) *Can. J. Pure Appl. Sci*, **3**: 649-654.

Namyatova, E. (2019) https: //floristics.info/en/garden/4288-astilbe-growing-planting-and-care.html

Olson, M.S. and Antonovics, J. (2000) *Amer. J. Bot.*, **87**: 837-844.

Pavlyuk, N.A., Shelekhova, O.M. and Nesterova, S.V. (2020) *Ìèêîëîãèÿ è ôèòîïàòîëîãèÿ*, **5**: 378-383.

Pogroszewska, E. and Sadkowska, P. (2007) *Acta Sci. Pol.*, **6** (1): 29-33.

Randhava, L. (2005) In: *Chicago Botanical Garden Online Index*. www.chicagobotanic.org/plantinfo/pp/PPAstilbe.html.

Runkle, E.S., Heins, R.D., Cameron, A.C. and Carlson, W.H. (1998) *Hort Tech.*, **8**: 207-209.

Sasai, Y., Tachi, S., Ozawa, A., Yoshizawa, Y., Ono, F., Kagiwada, S., Ishikawa, S. and Horie, H. (2015) Annual Report of the Kanto-Tosan Plant Prot. Soc. **62**: 97-100.

Skelly (2014) *University of Nevada Cooperative Extension Fact Sheet.*, **14**: 06.

Trader, B.W. (2006) *Ph.D Thesis*, Virginia Polytechnic Institute and State University.

Vandoorn, W.G., Perik, R.J. and Belde, P.J. (1993) *Postharvest Biol. Technol.*, **3**: 69-76.

Zhu, W.D., Nie, Z.L., Wen, J. and Hang Sun. (2013) *Bot. J. Linn. Soc.*, **171**(2): 377-394.

Yamazaki, K., Suzuki, T., Iigo, M., Aiso-Sanada, H., Kurokura, T. and Yamane, K. (2020) *J. Hort.*, 182.

21

Bellis perennis

R.K. Dubey and Simrat Singh

1.0 Introduction

Bellis perennis L. (common daisy) is a herbaceous perennial herb belonging to the family 'Asteraceae'. The word *Bellis* has been derived from the Latin *'bellus'* meaning 'pretty'. It is often grown as an annual or biennial and is found in all types of mowed or grazed grassland, along the roadside verges, riverbanks, dune-slacks, and lake margins (Frederic *et al.*, 2010; Panda, 2004). It is also commonly known as *'bruisewort'* and occasionally *'woundwort'*. The bellis flowers also exhibit the phenomenon of heliotropism where the flowers face themselves with the position of the sun in the sky.

2.0 Importance and Uses

2.1.0 Aesthetic Uses

The daisy is an important ornamental plant, widely grown as edging alongside the borders of lawn, pavements or flower beds, for carpet bedding, in window boxes and also cultivated in pots as a winter annual and also for cut flowers.

2.2.0 Therapeutic and Other Uses

Bellis perennis has potential wound healing properties and has been used as a herb for the treatment of sore throat (Uysal *et al.*, 2010), headache (Uzun *et al.*, 2004), common cold (Cakýlcýoglu *et al.*, 2010), bleeding, rheumatism, muscular pain, purulent skin diseases, as expectorant, diuretic, and anti-inflammatory (Siatka and Kasparova, 2010; Avato and Tava, 1995) in the traditional medicine. In addition, being astringent, it has been used for the treatment of common cold, stomach aches, eye diseases, eczema, skin boils, gastritis, diarrhoea, bleeding, rheumatism, and for curing the infections of the upper respiratory tract in traditionalmedicine (Karakas *et al.*, 2014). It also possess anti-microbial activity of poly-acetylenes with many other synthetic derivatives that were isolated (Avato *et al.*, 1997). In folk medicine, an infusion made from the flowers is used in curing pleurisy and upper-respiratory ailments and a poultice, prepared from the boiled leaves is used for the treatment of abscesses, bruises, sores, boils, *etc.*

3.0 Origin and History

The daisy is believed to be a native of western, central, and northern Europe, however, it is commonly observed as an invasive plant in North America (Tutin *et al.*, 1976). Daisy is well known as a traditional wound healing herb (Al-Douri and Al-Essa, 2010) and has been mentioned in the ancient literature for the treatment of bruises, broken bones, and wounds by crusaders in the Middle Ages (Mitich, 1997). In the Medieval times, *Bellis perennis* or the English daisy was commonly known as 'Mary's rose', and is also considered to be a flower of children and is a symbolic of innocence. Often the name 'Daisy' was used as a girl's nick name.

4.0 Taxonomy

The genus *'Bellis'* comprises 15 species of perennials from Europe and

Uses of Bellis

In Garden

In Rock Garden

In Pot

In Home Garden

In Border

As Cut Flowers

Species of *Bellis*

Bellis perennis

Bellis annua

Bellis longifolia

Bellis rotundifolia

Bellis sylvestris

Bellis cordifolia

Mediterranean region. The *Bellis* group has been included in subtribe 'Asterinae' (tribe Astereae) along with another 117 genera representing more than 3,000 species (Bremer, 1994). The flower heads are either yellow, violet or creamish white and are composite, in the form of a pseudanthium, consisting of many sessile flowers about 2–3 cm in diameter, with white ray florets (often tipped red) and yellow disc florets. Each inflorescence is borne on single leafless stems, 2–10 cm tall. The capitulum, or disc of florets, is surrounded by two rows of green bracts known as "phyllaries". The achenes are without pappus. The leaves are ovate, simple with dentate margins and appear alternate in arrangement.

5.0 Species and Cultivars

Bellis species are classified as annuals or perennials. The daisy varieties may be kept in three following groups:

(i) The Pompons (pomponette): These have smaller flowering forms with an average height and spread of 10-15cm.

(ii) The Monstrosa (carpet series or goliath): These are exceptionally large-flowered types having height and spread ranging 15-20 cm.

(iii) The Cockscomb types: These are Monstrosa forms but consist of several smaller inflorescences fused together in the form of a crest.

White carpet daisies have the same growing habit as the carpet series, but the plants are compact, large and white-flowered with the height and spread across 15 cm (Pizzetti and Cocker, 1968; Brickell, 1989).

5.1.0 Species

Some of the *Bellis* species having good ornamental values have been described as follows:

Bellis annua

Commonly known as annual daisy, belongs to the group of annual and biennial plants. This species is found in Morocco, Algeria and Tunisia. It is a deciduous plant with simple spoon-shaped leaves with entire margins. The plants grow to a height of approximately 12 cm. The flowers have numerous petals that bloom from February to June. The flowers are arranged solitary. This species prefers a sunny to partial shade site and grows best in pebbly, loamy soil with adequate moisture holding capacity.

B. longifolia

It is a perennial species, height ranging from 5 to 20 cm. The roots are fleshy with leafless stalk. Leaves are bright green, oblong-lanceolate and narrowing at the base gradually. Flowering period extends from April to July. The species grows on rocky soils at an altitudes from 1,200 to 2,000 meters.

B. perennis

This species is a native of the meadows and pasture lands spread across Europe into western Asia. The species name '*perennis*' in Latin means 'everlasting'. This is a slow-growing and clump-forming evergreen species having spatula shaped 3-8 cm long leaves. The flower receptacle is conical possessing one seriate ray flowers, yellowish disc flowers, white or pink ligules and obovate achenes (Davis *et al.*, 1988). It is a hardy herbaceous perennial and is a parent of many of the fine flowers types exhibiting different flower colours and inflorescence forms such as white, pink or red flowers in singles, semi-doubles and doubles, and sometimes with spots, streaks, marbling or bi-colouration.

In the temperate regions of India, it flowers perpetually from February to November under sheltered positions whereas, in the Indian tropics and sub-tropics, bellis blooms from December to March, the peak being during spring season. Many of its forms and varieties are sterile. Ligules (outer petals) are generally white with mostly pinkish reverse while, the central yellow disc is composed of ray florets. The flowers close by the dusk and reopen at the dawn. It grows either prostrate or straight but almost stem-less and spreads rapidly. This species is propagated through seeds or division. *Bellis perennis* varieties inheriting long and winter-flowering habit, are most suitable for pot cultivation.

B. rotundifolia

This species is a native of southwest Spain, Morocco and Algeria. It is perennial frost sensitive species, grows 10-15 cm in height and grown usually as a non-hardy annual. The leaves are simple, arranged in basal rosettes with ovate margins, glabrous, serrated, 1.8 cm to 3.7 cm in length and with 5 cm long leaf stalks. The flower stalks which appear during March to August, rise well above the leaves and bear many-petalled single flowers on slender stems, followed by achenes. Ligules (outer petals) are white and lilac and the central disc florets yellow. This species prefers a sunny to half shady site and can withstand temperatures down to -6.6 °C. It grows best in pebbly and loamy soil that is moderately moist.

B. sylvestris

A native of southern Europe, similar to *B. perennis* but grows considerably tall up to 45 cm in height, bearing narrowly obovate to oblong-spathulate leaves with winged stalks, floral heads 2-4 cm wide, ray florets white but tinged red-purple or entirely of that colour. Flowers appear during spring to summer.

5.2.0 Cultivars

Several attractive cultivars have been developed that are suitable for commercial bedding plant production and also for the beautification of diverse landscapes. Some of the promising cultivars developed by breeders from ornamental plant breeding companies have been described here as follows:

Habanera series: This series represents the largest flowered Bellis available with fully double flowers on short sturdy stems. It has been found perfect for spring production with pansies.

Mars series: The flowers of this series bloom without vernalization treatment and thus, are excellent for both fall and early spring markets. Its plants are compact, therefore more plants can be cultivated per unit area. It possess semi double flowers

Series of *Bellis perennis*

Habanera Series

Mars Series

Speedstar Series

Tasso Series

Bellissima Series (Pink)

Bellissima Series (Red)

Cultivars of *Bellis*

Alice

Monstroa

Etna

Giant Double

Rob Roy

China Pink

with a yellow center, flower size measuring approx. 3 cm across with an average plant height of 12 cm.

Speedstar series: This series can also be produced without vernalization for both fall and spring sales, has a compact plant habit and leaf size that makes "Speedstar" a wonderful companion to pansies for fall and spring production. The series has semi double flowers with a yellow center and an average height of 12 to 14 cm.

Tasso series: These are large flowered pomponette series with double flowers measuring 4 cm, have a uniform plant growth habit and flowering of the mutual colours with an average height of 12 cm.

Bellissima series: This series has vigorous cultivars that have been named 'Fleuroselect', comprises 'Bellissima Red', 'Bellissima Rose', 'Bellissima Rose Bicolor' 'Bellissima White' and 'Bellissima Mix'. These cultivars available in different colours form compact, low-growing rosettes reaching 10-15 cm in height and 18-20 cm in diameter. It produces masses of large; fully double flowers in the early spring. Commercially, Bellissima cultivars bloom for long durations, up to 18-20 weeks, when produced under optimum conditions.

Hen-and-Chickens-daisy: An unusual form of daisy, flower large white or pale-pink flowers, with extended flowering period, is also very popular.

Some of the varieties with attractive flowers are *Bellis perennis* var. *alba* (pure white), 'Alice' (pale-pink), 'Brilliant' (large double red), 'China Pink' (smaller bright pink), 'Dresden China' (miniature, double, pink), 'Etna' (brilliant red), 'Giant Double' (mixed colours), 'Liliput' (miniature, double, crimson), 'Longfellow' (deep pink), var. *lutea* (cream), 'Pomponette' (miniature, double, white to crimson with small quilled petals), 'Quilled Mixed' (larger than miniature, double, mixed colours), 'Red Buttons' (miniature, double, carmine-red), 'Rob Roy' (miniature, double, red), *etc.*

6.0 Genetics and Breeding

6.1.0 Genetics

Bozdað *et al.* (2011) published data on the chromosome number and morphology of *Bellis* species of Turkey. The chromosome number was determined in *B. sylvestris*, *B. perennis* and *B. annua* as 2n = 36, 18 and 18, respectively. The plants from the Asteraceae family are largely pollinated by insects (rarely by birds), and studies (Andersson, 2008; Elomaa *et al.*, 2018) have indicated that the unique architecture of the capitulum inflorescence has a selective advantage to attract pollinators and induce a greater plant vigor. Moreover, the diversity in inflorescence form and varied colors generated by breeders is also a key attraction to consumers contributing significant worth in the floriculture industry

6.2.0 Hybridization

The hybrid varieties in daisy are evolved with the following objectives:

1. To increase the flower size and doubleness
2. More number of flowers with uniform flowering

3. Better resistance to drought, insect-pests and diseases
4. Wider adaptability to changing environmental conditions

Top cross or inbred x variety cross (A x Variety) is employed in *Bellis perennis* that is a cross between an open pollinated variety and inbred line. Speciality of F_1 hybrids in *Bellis perennis* is production of more number of flowers per season. Self-incompatibility is applied for production of F_1 seeds in this crop.

7.0 Growth and Flowering

7.1.0 Effect of Temperature

Daisy thrives well in areas that are fairly wet and soggy for most part of the year. Seedlings (plug plants) can be grown in the greenhouse at 18°C to 21°C during stage 2 (starts from fully developed cotyledons and ends with the fully developed true leaf or true leaf pair), and during Stage 3 (starts from the fully developed true leaf or true leaf pair and ends with 80 per cent of the young plants being marketable) and 4 (all young plants are ready for sale and in the process of being hardened off, lasts about 7 days), the temperatures are kept between 15°C to 18°C. There are three common practices followed to produce flowers on bellis. The first method is to plant plugs (ready to transplant seedlings) into the final container during early to mid fall, allow them to bulk up slightly followed by vernalization treatment during the winter and force them to bloom in the early spring using low production temperatures of 10-15.5°C for 4-6 weeks. The second practice follows the transplanting of vernalized seedlings within the plugs in late winter and forcing them at 12.7-15.5°C for 5-7 weeks. However, the drawback of this practice is the plants are often seen in full bloom before the pots are completely filled out. The third and most common practice entails transplanting an non-vernalized plug into the final container during the late winter and forcing it at 15.5-18.3°C for 5-7 weeks. It should be kept in mind that, bellis are cool-season plants and prefer to be grown at cool temperatures. They can tolerate temperatures in the 4.4-7.2°C range.

7.2.0 Effect of Growth Regulators

With their compact growth habit, daisy usually does not need any height management. However, in the winter months, during periods of low light levels, when grown at higher plant densities or when grown with luxuriant nutrient levels, excessive plant growth might occur, requiring some type of height management strategy. The growth of bellis can often be controlled by providing adequate spacing between the plants. However, it may be necessary, although not common, to use plant growth regulators (PGRs) to control the growth of English daisies. Application of B-Nine (2,000 ppm) and Sumagic (3 ppm) sprayed twice at weekly intervals provide adequate height control.

8.0 Propagation

Bellis perennis can be propagated either through seed or by division. These can be divided in early spring or after they have finished flowering.

8.1 Seed Propagation

The seeds of daisy possess low seed viability and germination rate of seeds is low (5 per cent). In mild-winter areas, the seeds are sown in the garden in late summer. Daisy seeds are very small, 4,900-6,600 seeds per gram and hence 3-5 seeds are sown per plug. Seeds in the plugs (288- or 220-cell plug trays) or in the media (with germination mix or medium-grade vermiculite) start with the radicle breaking through the testa and the roots touching the medium. It ends with fully developed cotyledons. In colder regions, seeds should be sown in September or early spring. Use a well-drained, disease-free plug media with a pH of 5.5 to 6.3 and EC of 0.50 to 0.75 mmhos/cm.

The covering is optional, but it helps maintain a suitable environment around the seed during this phase and can improve the uniformity of germination. Seed flats should be moistened and moved to a warm environment where the temperatures can be maintained at 18.3-22.2°C for germination. Starting bellis inside a germination chamber will increase both the germination rate and per cent germination while decreasing the time necessary for all of the seeds to sprout. It is beneficial to maintain high humidity (95 percent) until the cotyledons have emerged. The seeds should be germinated in 3-7 days. Once germinated, seedlings can be grown with temperatures from 60 to 68° F (15.5-20°C). Following germination, the moisture levels are reduced, allowing the growing medium to dry out slightly before watering to promote rooting. Fertilizers are usually applied once the true leaves are present. Apply 75-100 ppm nitrogen every third irrigation or 50 ppm with each irrigation using a balanced, water-soluble source. At these temperatures, Bellissima will finish the plug stage in 5-7 weeks.

8.1.0 Effect of Temperature on Germination of Seed

Daisy seeds can also be sown indoors by maintaining 21.1-24°C temperature and 95-97 per cent relative humidity within the growing medium until germination, which takes 10-15 days. Seeds are sown in boxes or in the nursery beds in the open in late spring and when the seedlings are ready in 4-5 weeks for transplanting, these are planted during autumn for flowering the following spring. After germination reduce the temperature to 15 to 20°C and keep adequate ventilation. Light is not required for germination.

8.2.0 Micropropagation

Karakas and Turker, (2013) developed an efficient *in-vitro* plant regeneration protocol for *B. perenis* via adventitious shoot development from leaf, pedicel and petiole explants. Two different explant sources (field-grown plants and *in vitro* -grown plantlets) were compared in terms of their regeneration capacity. It was observed that field-grown plants were the best as explant source.

9.0 Soil and Climate

9.1.0 Soil

Daises require a well-drained, soil or a growing substrate having a pH range of 5.5 to 6.2, with 0.5 to 0.75 ds/m EC for its proper growth and development.

Propagation of *Bellis*

Seeds New Emerging Seedling Growth and Flowering

In vitro Regeneration Process of *B. perennis* by Adventitious Organogenesis.

(A) Shoot formation from pedicel explants on MSMO medium containing 0.5 mg/ITDZ + 0.5 mg/l IAA after 3–4 weeks in culture. (B) Shoot elongation of regenerated shoots on MSMO medium containing 0.5 mg/IGA$_3$. (C) Rooting of the regeneratedshoots on medium containing 1.0 mg/IIAA after 3 weeks in culture. (D) Regenerated plants transferred to foam cups containing sterile potting soil under high humidityconditions. (E) Regenerated plants transferred to under growth room conditions. (F) Regenerated plants transferred to big plastic pots containing sterile potting soil undergrowth room conditions. (G) Flowering of a regenerated plant after transferring to room conditions.

Source: Karakas and Turker, 2013.

9.2.0 Climate

Bellis like cool environment but freezing temperatures during their growth and development period should be avoided. Daises are responsive to temperature fluctuations as exposure to 12 °C and above will result in bigger foliage growth, thin flowering stems with a poor flowering initiation. Temperatures below 6°C delay the buds development. However, freezing temperatures should be avoided. Full light can promote plant growth with perfectly green leaves and numbers of flowers also increase. Bellis can tolerate temperatures in the range of 5°C to 7°C. It is advisable to keep the plants at 10°C to 12°C for 4 weeks before sale.

Bellis plants do not have a vernalization requirement, but exposure to cold temperature is beneficial to reduce the time to bloom slightly and increase the number of blooms produced. English daisies will flower under any photoperiod and can be forced into bloom under natural day lengths. Bellis can be grown at 10-12°C outdoors for 5-7 weeks during mid-December. If these are planted in the greenhouse, the vents are opened to escape out heat and promote circulation of outside fresh air.

10.0 Cultivation

10.1.0 Planting

Seedlings of daisy are generally transplanted as a bedding or pot plant during mid-October toearly November under subtropical conditions and blooms from early to mid-summer. *B. perennis* generally blooms in the temperate regions from early to mid summer, although when grown under ideal conditions, they have a very long flowering season and will even produce a few flowers in the middle of mild winters. The optimum planting density of Bellis is generally kept at 20 plants per square meter that will produce healthy plants with an average height of 15 to 25 cm and average spread of 12 to 20 cm.

10.2.0 Fertilizer and Manures

Daisies have a moderate fertilizer requirement. When the cotyledons are fully developed, water soluble fertilizers are applied starting with a low concentration of 50 ppm N given twice a week. As the true leaves develop, the fertilizer rates are increased to 100 ppm N. Weekly application of 150-200 ppm N along with the complete balanced fertilizers give encouraging results during rapid growth and development of plants. It has been reported that uptake of the minerals is higher when the plants are exposed or grown at 5°C. High N levels should be avoided, and also ammoniacal forms of N should not be applied. Fertilizer application should be restricted or stopped completely after fall period (October), however, during spring 150-200 ppm N along with K_2O at 1:1.5 (N: K_2O) should be applied. Roots of daisy are sensitive to high salt levels in the substrate. Providing moderate to high fertility levels causes them to appear lush and leafy. Maintain the media EC at 1.50 to 2.00 mmhos/cm and pH at 5.8 to 6.5. Growers commonly deliver nutrients using either a constant liquid fertilization program, feeding at rates of 50-75 ppm nitrates, or a controlled release fertilizer (CRFs) incorporated at a rate equivalent to three-fourths of N per cubic feet yard of growing medium.

Cultivation of *Bellis*

Cultivation in Open

Planting in Rows

In Greenhouse

Diseases and Pests of *Bellis*

Botrytis

Rust

Aphid Infestation

Harvesting and Marketing of *Bellis*

Harvesting Stage

Cut Flowers of Bellis Perennis

Marketed as Pot Plants

10.3.0 Weed Control

Weeds are a problem in daisy cultivation and manual weeding is cumbersome and very expensive. Lenacil at 2.2-4.4. kg/ha applied one week after planting was found safe and effectively controlled autumn germinating annual broad-leaved weeds. Bellis bedding plants are irrigated thoroughly and the growing media is then allowed to dry slightly between each successive irrigations.

11.0 Diseases and Pests

11.1.0 Diseases

Botrytis is likely to occur late in the crop cycle once the leaves are fully expanded and canopy closes shortly before the plants began to bloom. In most cases, Botrytis can be prevented or reduced by providing adequate spacing and good air circulation at all times, maintaining a relative humidity below 70 percent. It is advisable to sell the plants when the flower buds just begin to open to avoid overcrowding and disease infestation.

11.2.0 Pests

No serious insect-pests damage has been recorded on daisies. However, sometimes seedling damping-off, gray mould, leaf spots, anthracnose, downy mildews have been found infecting the crop but are of insignificant importance. Aphids are the most troublesome insect pests of bellis. A preventative drench application of Marathon 60WP (OHP) can be done to ensure the suppression in the population of aphids during the production cycle.Whiteflies and leafhoppers may also be observed feeding on bellis.

12.0 References

Al-Douri, N.A. and Al-Essa, L.Y. (2010) *J. Pharm. Sci.,***3**: 100–108.

Andersson, S. (2008) *Am. J. Bot*. **95**: 1072–1078.

Avato, P. and Tava, A. (1995) *Phytochem.*, **40**(1): 141-147.

Avato, P., Vitali, C., Mongelli, P. and Tava, A. (1997) *Planta Med.*, **63**: 503–507.

Bozdag, B., Ozdemir, C. and Sepet, H. (2011) *Caryologia* **64**: 251–255.

Bremer, K. (1994) *Asteraceae, Cladistics and Classification*. Timber Press, Portland.

Brickell, C. (1989) *The RHS Gardeners Encyclopedia of Plants and Flowers*, Dorling Kindersley Limitedf, London, UK. pp. 444-445.

Cakýlcýoglu, U., Sengun, M.T., Turkoglu, I. (2010) *J. Med. Plants Res.*, **4**: 567–572.

Davis P.H., Mill, R.R. and Tan, K. (1988) *Flora of Turkey and the East Aegean Islands* (Suppl.).Edinburgh Univ. Press, Edinburgh, **10**: 201-210.

Elomaa P., Zhao Y. and Zhang T. (2018) *Hort. Res.***5**: 36.

Frederic P. M, Agnes, F. V. and John, M. B. (2010) *Bellis perennis*. Alpha script Publishing. Pp. 1-64.

Karakas, F.P. and Turker, A.U. (2013) *Ind Crops Prod.*, **48**: 162-170.

Karakas, F.P., Sohretoglu, D., Liptaj, T., Stujber, M., Turker, A.U., Marak, J., Calýs, I. and Yalcýn, F.N. (2014) *Pharm. Biol.*, **57**: 951–955.

Mitich, L.W. (1997) *Weed Technol.*,**11**: 626–628.

Panda, H. (2004) *Handbook on Medicinal Herbs with Uses*. Asia Pacific Business Press, India, pp. 188–189.

Pizzetti, I. and Cocker, H. (1968) *Flowers - A Guide for Your Garden*, Harry N. Abrams, Inc., Publishers, New York, USA. pp. 112-117

Siatka, T. and Kasparova, M. (2010) *Molecules*, **15**: 9450–9461.

Tutin, T.G., Heywood, V.H., Burges, N.A., Moore, D.M., Valentine, D.H., Walters, S.M. and Webb, D.A. (1976) *Flora Europaea*. vol. 4. Cambridge University Press, Cambridge, Great Britain, pp. 111

Uysal, I., Onar, S., Karabacak, E. and Celik, S. (2010) *BioDiCon*, **3**: 15–22.

Uzun, E., Sariyar, G., Adsersen, A., Karakoc, B., Otük, G., Oktayoglu, E., Pirildar, S. (2004) *J. Ethno pharmacol.*, **95**: 287–296.

Index

C

G

H

I

Q

R

S

T